Aquaculture
Farming Aquatic Animals and Plants

Edited by

John S. Lucas
Centre for Marine Studies
University of Queensland
Australia

Paul C. Southgate
School of Marine Biology and Aquaculture
James Cook University
Australia

Fishing News Books
An imprint of Blackwell Publishing

Editorial offices:
Blackwell Publishing Ltd, 9600 Garsington Road, Oxford OX4 2DQ, UK
Tel: +44 (0)1865 776868
Iowa State Press, a Blackwell Publishing Company, 2121 State Avenue, Ames, Iowa 50014-8300, USA
Tel: +1 515 292 0140
Blackwell Publishing Asia Pty Ltd, 550 Swanston Street, Carlton, Victoria 3053, Australia
Tel: +61 (0)3 8359 1011

First published 2003 by Blackwell Publishing Ltd

Library of Congress Cataloging-in-Publication Data is available

ISBN 0-85238-222-7

A catalogue record for this title is available from the British Library

Set in Times
by Prepress Projects Ltd
Printed and bound in Great Britain using acid-free paper
by Ashford Colour Press, Gosport

For further information on Blackwell Publishing, visit our website: www.blackwellpublishing.com

Contents

List of contributors

Dr Trevor Anderson
GFB Fisheries Ltd
Australia

Dr Peter Appleford
Victorian Department of Natural Resources and Environment
Australia

Assoc. Professor Michael Borowitzka
Biological Science and Biotechnology
Murdoch University
Australia

Dr Allan Bremner
Allan Bremner and Associates
Australia

Dr Tomás Cabrera
Escuela de Ciencias Aplicadas del Mar
Universidad de Oriente
Venezuela

Dr Laura Castell
School of Marine Biology and Aquaculture
James Cook University
Australia

Dr Simon Cripps
WWF International
Switzerland

Professor Sena De Silva
School of Ecology & Environment
Deakin University
Australia

Dr Don Fielder
School of Life Sciences
The University of Queensland
Australia

Emeritus Professor Nigel Forteath
Omlas Pty Ltd
Australia

Dr Gideon Hulata
Department of Aquaculture
The Volcani Center
Israel

Dr Clive Jones
Freshwater Fisheries & Aquaculture Centre
AFFS Fisheries and Aquaculture
Australia

Dr Darryl Jory
BioCepts International Inc.
USA

Dr Sagiv Kolkovski
Mariculture Research and Advisory Group
Fisheries Western Australia
Australia

Avi Koren
AAA Advanced Aquaculture
Israel

Dr Martin Kumar
Aquatic Sciences Centre
South Australian Research and Development Institute
Australia

David 'Dos' O'Sullivan
Dosaqua Pty Ltd
Australia

Dr Leigh Owens
School of Biomedical Sciences
James Cook University
Australia

Dr Michael Poxton
Aquaculture Associates International
UK

Dr John Purser
School of Aquaculture
University of Tasmania
Australia

Dr Mike Rimmer
Department of Primary Industries
Northern Fisheries Centre
Australia

Dr Yitzhak Simon
Ministry of Agriculture and Rural Development
Israel

Dr Victor Suresh
Aquaculture Feed Industry
Indiaa

Dr Douglas Tave
Urania Unlimited
USA

Professor Clem Tisdell
School of Economics
The University of Queensland
Australia

Professor CK Tseng
Chinese Academy of Sciences
People's Republic of China

Dr Craig Tucker
National Warmwater Aquaculture Center
Mississippi State University
USA

Preface

This textbook seeks to convey to its readers the contributors' enthusiasm for aquaculture and their accumulated knowledge. The contributors are recognised internationally in their fields. While it is not possible to comprehensively cover the ranges of aquaculture theory, practices and cultured organisms in one textbook, it is our earnest hope that this text will give readers a broad understanding of these topics. The first part of the text introduces aquaculture with a series of 'theory and practice' topics, ranging from traditional topics such as ponds and pumps to contemporary environmental issues, nutrition physiology and genetic engineering. The second part of the text consists of chapters dealing with specific organisms, or groups of organisms, which illustrate the variety of culture methods used in aquaculture. It also provides examples of biological and other factors that make these organisms suitable for culture. The aquatic animals and plants treated in the text are but a small proportion of the hundreds of commercially cultured species; however, they constitute the most significant commercial components of world aquaculture production. They include the four major groups of cultured organisms – fish, crustaceans, bivalve molluscs and seaweeds; the three broad categories of aquatic environments – fresh, brackish and seawater; and the broad latitudinal zones – temperate, subtropical and tropical regions.

We express our sincere gratitude to the authors for their commitment in contributing chapters and, in some cases, for their understanding. Mr Michael New, President, European Aquaculture Society, Past-President, World Aquaculture Society, kindly assisted by reviewing Chapters 1 and 23. We also wish to express our gratitude to our wives, Helen and Dawn, for their substantial contributions.

John Lucas
Paul Southgate
April 2003

1
Introduction

John Lucas

1.1 What is aquaculture?

> Give a person a fish and you feed them for a day; teach them how to grow fish and you feed them for a lifetime.
>
> (from a Chinese proverb)

Aquaculture is at an exciting stage of development. World aquaculture production is increasing at a very rapid rate. It is increasing much more rapidly than animal husbandry and capture fisheries, the other two sources of animal protein for the world's population. There is widespread recognition that seafood production from fisheries is at or near its peak, and that aquaculture will become increasingly important as a source of seafood production, and ultimately the main source. There is widespread public interest in aquaculture. This is the context in which this textbook is written and we trust that it will convey some of the excitement of the rapidly developing discipline of aquaculture.

The term 'seafood' is used inclusively in this textbook, i.e. for all animal and plant products from aquatic environments, including freshwater, brackish and marine environments. The term 'shellfish', according to common usage, is used to describe aquatic invertebrates with a 'shell'. In this way, bivalve and gastropod molluscs, decapod crustaceans and sea urchins are combined, while recognising the great diversity of morphology and biology within this grouping. The two groups that overwhelmingly constitute shellfish are the bivalves (oysters, mussels, clams, etc.) and decapod crustaceans (shrimp, crayfish, crabs, etc.). The other major group of aquatic animal that is cultured is fish, also known as finfish. 'Fish farming' is used in the sense of aquaculture of fish, crustaceans, molluscs, etc., but not plants.

There are many different forms of aquaculture and, at the outset of this book, it is important to establish what aquaculture is and what distinguishes it from capture fisheries.

> The definition of aquaculture is understood to mean the farming of aquatic organisms, including fish, molluscs, crustaceans and aquatic plants. Farming implies some form of *intervention* in the rearing process to enhance production, such as regular stocking, feeding, protection from predators, etc. Farming also implies individual or corporate *ownership* of stock being cultivated.
>
> (FAO, 2001a)

> For statistical purposes, aquatic organisms which are harvested by an individual or corporate body which has owned them throughout their rearing period contribute to *aquaculture* while aquatic organisms which are exploitable by the public as a common property resource, with or without appropriate licences, are the harvest of *fisheries*.
>
> (FAO, 2001a)

The two essential factors that together distinguish aquaculture from capture fisheries are:

- intervention to enhance the stock
- ownership of the stock.

Thus, a structure to which fish are attracted and caught [e.g. a fish-aggregating device (FAD) floating in the open ocean] may be owned, but this does not confer ownership of the stock of attracted fish. Furthermore, the FAD facilitates capture but does not enhance the fish stock that is being captured. This is fisheries production. Hatchery production of juvenile salmon is aquaculture: they are owned by the hatchery and sold as fingerling fish. Their ultimate capture, after being released into rivers to which they eventually return to breed, is a fishery. The released fingerlings enhance the stock, but they become a common property resource.

Hydroponics, the cultivation of terrestrial plants with their roots in dilute nutrient solutions instead of soil, is not aquaculture. Hydroponics is an alternative method for growing terrestrial plants.

Activities constituting aquaculture production, according to FAO (2001a), are:

- hatchery rearing of fry, spat, postlarvae, etc.
- stocking of ponds, cages, tanks, raceways and temporary barrages with wild-caught or hatchery-reared juveniles
- culture in private tidal ponds (e.g. Indonesian 'tambaks')
- rearing molluscs to market size from hatchery-produced spat, transferred natural spatfall or transferred part-grown animals
- stocked fish culture in paddy fields
- harvesting planted or suspended seaweed
- valliculture (culture in coastal lagoons).

1.2 Origins of aquaculture and agriculture

Agriculture first developed about 10 000 years ago in the Middle East, when human populations changed from hunting–gathering to cultivating wheat and barley. Farming wheat and barley then rapidly spread to adjacent lands. Subsequently, there were independent origins of farming cereal crops on other major landmasses. Rice cultivation began in Asia about 7000 years ago. Sorghum and millet cultivation and maize cultivation developed somewhat later in Africa and America respectively. These changes from hunting–gathering to farming cereal crops caused profound changes in lifestyle, from a nomadic to a settled existence. They resulted in greatly increased productivity from the land for human consumption and increased human populations per unit land area as a consequence. Whether quality of life improved in the early farming communities is debatable: diet became less varied and conditions became more favourable for disease.

The origins of aquaculture are much later. Culture of common carp (*Cyprinus carpio*) was developed some hundreds of years BC in China, where the carp is a native species. The first aquaculture text is attributed to a Chinese politician, Fan Lei, and is dated about 500 BC (Ling, 1977). Fan Lei attributed the source of his wealth to his fish ponds: so his fish culture was more than a hobby. However, in Africa, America and Australia, aquaculture was not practised until it was introduced in recent centuries.

The origins of aquaculture, even with carp, are thousands of years later than the origins of agriculture. The late origin of aquaculture compared with agriculture is partly because humans are terrestrial inhabitants and cannot readily appreciate the parameters of aquatic environments. There are several environmental factors that may profoundly affect aquatic organisms, such as:

- very low solubility of O_2
- high solubility of CO_2
- pH
- salinity
- buffering capacity
- dissolved nutrients
- toxic nitrogenous waste molecules
- turbidity
- heavy metals and other toxic molecules in solution
- phyto- and zooplankton concentrations
- current velocity.

These can be rigorously measured only with modern instrumentation. Many of the diseases that afflict aquatic organisms are quite unfamiliar to us. Furthermore, virtually all the animals used in aquaculture are ectothermic (their body temperature is variable and strongly influenced by environmental temperature) (see section 3.3.1). Their metabolic rates, and all functions depending on metabolic rate, are profoundly influenced by temperature in ways that we do not experience.

The difficulties of appreciating the influences of these environmental factors still apply today, causing aquaculture programmes to have a relatively longer development period than other forms of food production. 'Even when tested technologies are adopted, the construction of physical facilities (particularly pond farms), solution of site specific problems, the building up of the productivity of the system and, above all, attainment of skills by workers take considerable time' (Pillay, 1990). In agriculture we are much more readily able to appreciate the parameters influencing the success or otherwise of the output, and we have a very long history of attaining the skills needed.

A further major consequence of the late origin of aquaculture is that there has been relatively little genetic selection compared with the highly selected plants and animals of agriculture. Modern agriculture is based on organisms that are vastly different from their wild ancestors, and in many cases their wild ancestors no longer exist. This selection for desirable traits took place steadily and without any scientific basis over thousands of years of domestication. It was more intense last century with scientific breeding programmes. Modern agriculture would be totally uneconomic and the current world population would starve without these domesticated and genetically selected agricultural plants and domesticated animals. The majority of aquaculture, by contrast, is based on plants and animals that are still 'wild'. There are a few fish species that can be considered as domesticated:

- common carp
- Atlantic salmon
- rainbow trout
- tilapia species
- channel catfish.

Their breeding is based on broodstock that have been subject to intense genetic selection. Many other aquaculture species are based on broodstock obtained from natural populations. In some cases the life cycle has not yet been 'closed', i.e. the species has not been reared to sexual maturity and then spawned on a regular basis under culture conditions. Until the life cycle is closed, there is minimal potential for selective breeding.

1.3 Aquaculture and capture fisheries production

Fishing activities, whether they are spearing individual fish, collecting shellfish from a rocky shore or coral reef, using a cast net, or capturing schools of fish with huge nets from factory trawlers that ply the world's oceans, are all hunting–gathering regardless of the degree of technology. As fisheries production currently exceeds aquaculture production, hunting–gathering activities remain the principal source of seafood. These fisheries suffer problems that are fundamental to hunting–gathering:

- variable recruitment and consequent unpredictability of stock size
- difficulties in assessing stock size and its capacity for exploitation
- difficulty in regulating exploitation to match the stock size
- relatively low productivity.

The natural productivity of the world's water masses, fresh, brackish and marine, is huge, but finite; and a finite amount of plant and animal products can be harvested by fishing. For instance, the mean harvest from oceans that can be obtained for human consumption or processed for use in fish meal is ~2.5 kg per hectare of ocean surface per year. Furthermore, this huge but finite amount of harvest is within our current fishing capacity. Many of the world's major fisheries range from being heavily exploited to heavily overexploited, and production from fisheries has reached a plateau of ~90 million mt/year, around which it now fluctuates annually (Fig. 1.1). Total fisheries production of aquatic animals (fish, crustaceans, molluscs and miscellaneous animals) increased from 85.5 million mt in 1990 to a peak of 93.6 million mt in 1997, then slumped to 86.3 million mt in 1998 and rose again in 1999 (FAO, 2001b). There was a mean increase of < 1% per year in fisheries production over the decade to 1999 (Table 1.1).

There are two further factors in fisheries

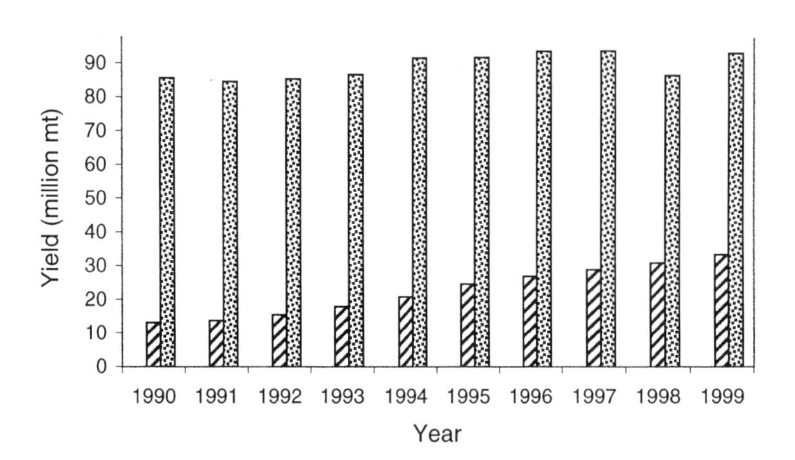

Fig. 1.1 World production of aquatic animals by capture fisheries and aquaculture per year over the decade 1990–99 (data from FAO, 2001a,b). Stippled bars, capture fisheries production; hatched bars, aquaculture production.

production. About one-third of fisheries production is used to make fish meal, i.e. dried fish products, based on sardines, anchovy, etc. Fish meal is used as a source of animal protein and lipids in feeds, primarily feeds used in animal husbandry, but also feeds for aquaculture. Thus, the effective annual production from fisheries for direct human consumption is in the order of 60 million mt/year. The other 30 million mt subsequently finds its way into the human diet by indirect processes. A further factor that does not appear in fisheries production statistics is the substantial proportion of the initial catch that is non-target catch and is discarded.

In contrast to fisheries, aquaculture production of animals, and animals and plants, grew at a mean rate of 9.9% over the same period (Fig. 1.1 and Table 1.1). In 1999, aquaculture production of animals and plants was 43 million mt compared with 94 million mt from fisheries. The increase in aquatic animal and plant production from 104 to 137 million mt over the decade to 1999 came largely from aquaculture (Table 1.1). Over this period, aquaculture increased from 16.2% to 31.1% of total seafood production, and it will continue to increase in relative importance. It is clear that further increases in production from aquatic environments will come largely from aquaculture. Unlike fisheries, aquaculture is not limited by the natural productivity of the world's water masses. It is therefore not surprising that aquaculture production has been increasing much more rapidly than fisheries.

A factor that is also evident in Table 1.1 is that aquatic plants (very predominantly seaweeds) contribute substantially to aquaculture and fisheries production, particularly to the former. Aquatic plant production from aquaculture and fisheries was ~9.5 million mt and ~1 million mt, respectively, in 1999. The relative proportions by weight and value of fish, molluscs, crustaceans and plants from aquaculture in 1999 are shown in Figs 1.2 and 1.3. Fish constitute about half the weight and value of aquaculture production. Plants and shellfish each constitute about one-quarter of the weight. There are, however, major changes between relative weights and values of plants and shellfish. Molluscs decline in relative value. Plants decline even more in relative value. Crustaceans (mainly shrimp) show a very large increase in value. This has significant implications for countries such as China (People's Democratic Republic of China) (see later consideration).

To put aquaculture and fisheries production in perspective of providing animal protein for the world's current population, slaughtered meat production from livestock is in the order of 200 million mt/year compared with about 60 million mt/year from fisheries (for direct human consumption) and 30 million mt from aquaculture in 1999. This 90 million mt from aquaculture and fisheries is pre-slaughtered weight, and slaughtered weight (after removal of viscera, heads and shells) is probably around 50–60%. This value is not easy to estimate as it varies markedly with the kind of animal and country of consumption. Consequently, seafood makes up around 20% of all animal protein production/year. Aquaculture is a modest 7% of all animal protein production/year based on these calculations. With fisheries production/year about static and unlikely to increase much, and although

Table 1.1. Aquaculture and capture fisheries production in 1990 and 1999, aquaculture as a percentage of total aquatic animal and plant production in these years, and mean per cent increases per year of aquaculture and fisheries over this decade (FAO, 2001a,b)

	1990 (thousand mt)	1999 (thousand mt)	
Aquaculture animal production	13 074	33 310	mean increase/year = 9.9%
Aquaculture production of animals and plants	16 826	42 770	mean increase/year = 9.9%
Fisheries production of animals	85 511	92 867	mean increase/year = 0.91%
Fisheries production of animals and plants	86 741	94 069	mean increase/year = 0.89%
Total animal production (aquaculture + fisheries)	98 585	126 177	
Aquaculture as percentage of total	13.3%	26.4%	
Total animal and plant production (aquaculture + fisheries)	103 566	136 839	
Aquaculture as percentage of total	16.2%	31.3%	

mt, metric ton = tonne.

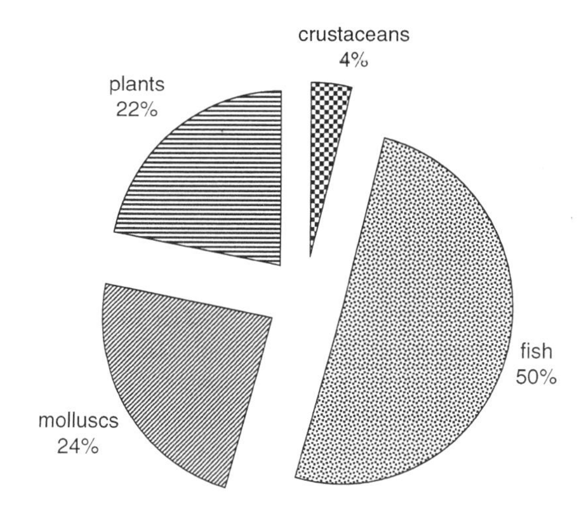

Fig. 1.2 Relative proportions of aquaculture production of fish, molluscs, crustaceans and aquatic plants by weight in 1999 (data from FAO, 2001a).

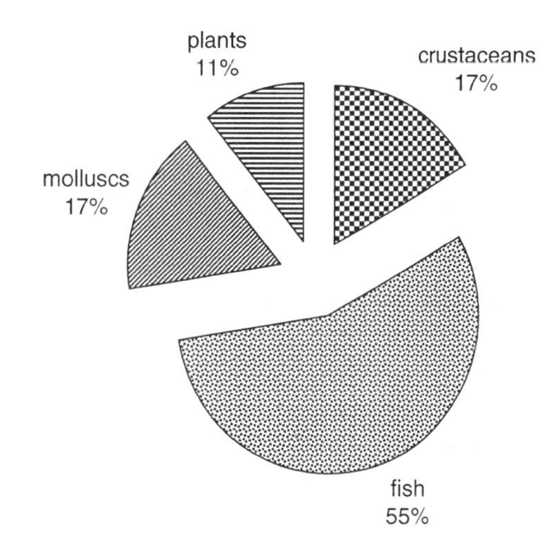

Fig. 1.3 Relative proportions of aquaculture production of fish, molluscs, crustaceans and aquatic plants by value in 1999 (data from FAO, 2001a).

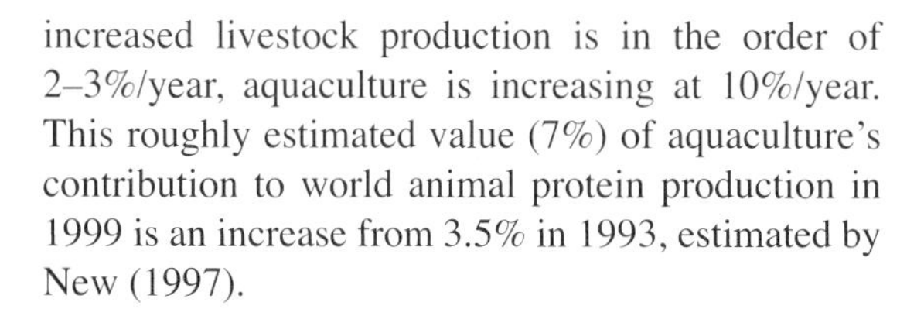

increased livestock production is in the order of 2–3%/year, aquaculture is increasing at 10%/year. This roughly estimated value (7%) of aquaculture's contribution to world animal protein production in 1999 is an increase from 3.5% in 1993, estimated by New (1997).

1.4 The 'Blue Revolution'

The rapid increase in aquaculture production has led to suggestions of aquaculture undergoing a 'Blue Revolution', which will transform the productivity of the ocean and other aquatic environments with new technology (e.g. Holmes 1996; Entis 1997) (Fig. 1.4). This envisages a revolution in productivity similar to the 'Green Revolution' in agriculture. The Green Revolution occurred 'where concentrated research developed the basis for the agricultural practices in use today (e.g. mechanisation, heavy fertilisation, heavy pesticide use, irrigation, genetically improved stocks, advanced feed formulations)' (Hopkins, 1996).

The great increase in animal and plant aquaculture production in the decade 1990–99 came mainly from increases in freshwater and marine aquaculture. There was a greater absolute increase in the marine

Fig. 1.4 Maricultura Tropical Shrimp Farm, a modern shrimp farm, Brazil (ABCC – Brazilian Shrimp Farmers Association).

environment (Table 1.2). In terms of specifically animal production, however, and hence animal protein production, the increase was substantially greater from freshwater aquaculture: 4 million mt greater than in the marine environment. The large increase in seaweed production in the marine environment is important as a source of income, but it is of low value as an immediate source of protein.

There is a further very important statistic about the growth in aquaculture production over the decade to 1999; this relates to the most needy countries in the world today. Dividing countries between low-income, food-deficit countries (LIFDCs) (FAO classification) and the developed countries, 24 out of the 26 million mt increase in aquaculture production for 1990–99 came from the LIFDCs. Aquaculture production in the LIFDCs increased at the remarkable rate of 13.2%/year on average over the decade (Table 1.3). Compared with this, the 4.2%/year increase in the more affluent countries was modest. The 'revolution' in aquaculture is occurring in the developing countries, often those with less access to technology, not in the higher technology countries. At this stage, the Blue Revolution may more appropriately be named the 'LIFDC expansion'. China alone showed an increase of almost 22 million mt over the 1990–99 decade at an outstanding rate of 16.1%/year on average. It was the driving force of the LIFDC aquaculture expansion. China accounted for 70% of the total weight of aquaculture production in 1999 (Table 1.3).

The 10 major countries in terms of the amount of aquaculture production in 1999 are shown in Table 1.4. All are Asian countries and, according to FAO classification, half are LIFDCs. The huge gap between China and the following countries in notable. If this ranking were purely for animal production, the USA and Norway would be ranked 8 and 9, respectively, at the ~1.5% level.

Although the developing Asian countries have aquaculture industries of high-value products, such as shrimp and scallops, for lucrative export markets, a high proportion of aquaculture in these countries continues to be from traditional pond culture of freshwater fish, especially carps and other cyprinids (Chapter 14). These accounted for an increase of 9 million mt during 1990–99. Carps and many other freshwater fish are low-value fish, and these are a major component of China's production. These low-value products are part of the reason for China's percentage of world aquaculture output being 49% by value compared with 70% by weight in 1999 (FAO, 2001a). Further factors are that China provides three-quarters of the world's seaweed production and a very high proportion of the molluscs (mainly bivalves). Both are relatively low value (Figs 1.2 and 1.3).

Freshwater fish are often cultured together as complementary species (polyculture; section 2.3.5). These low-value fish species are herbivores, omnivores and detritivores, feeding low in the food chain and requiring little supplementary input of feeds. This aquaculture often involves simple ponds, basic technology and low stocking densities. The production is often enhanced by using inexpensive organic fertilisers such as farm animal faeces and crop residues to feed the cultured fish, and to promote primary production in the ponds (integrated culture; section 2.3.6). In rural communities where animal protein is scarce and is prohibitively expensive from other sources, cultured fish form the major if not the exclusive source of animal protein.

A number of factors are responsible for the outstanding increase in aquaculture production in China (Cen & Zhang, 1998). However, the major factor has been a great increase in the surface area of ponds, other freshwater environments and shallow

Table 1.2 Aquaculture production in 1990 and 1999 by environment

	1990 (thousand mt)	1999 (thousand mt)	1990 → 1999 increase (thousand mt)
Animal and plant production			
Freshwater	7626	19 390	11 764
Brackish water	1312	1986	674
Marine	7888	21 394	13 506
World total	16 826	42 770	
Animal production			
Freshwater	7626	19 390	11 764
Brackish water	1297	1968	671
Marine	4152	11 952	7800
World total	13 075	33 310	

FAO (2001a).

Table 1.3 Total aquaculture production in the world in 1990 and 1999, with China indicated individually, and LIFDCs and developed countries shown separately

	Population (millions)	1990 (thousand mt)	1999 (thousand mt)	Mean % increase/year
World	5978	16 826	42 770	9.9
PDR China	1243	7 953	30 044	16.1
% of world total		47.3%	70.2%	
LIFDCs	3664	11 602	35 270	13.2
% of world total		69.0%	82.5%	
Developed countries	2 314	5 223	7 501	4.2
% of world total		31.0%	17.5%	

LIFDCs, low-income, food-deficit countries.
FAO (2001a).

Table 1.4 The top 10 countries in terms of weight of total aquaculture production in 1999

Sequence	Country	Percentage of world production (weight)	Asian country	LIFDC
1	China	70.2	Y	Y
2	India	4.8	Y	Y
3	Japan	3.0	Y	
4	Philippines	2.2	Y	Y
5	Korean Republic	1.8	Y	
7	Indonesia	1.5	Y	Y
8	Bangladesh	1.4	Y	Y
9	Thailand	1.4	Y	
10	Vietnam	1.4	Y	

LIFDC, low-income, food-deficit country; Y, yes.
FAO (2001a).

coastal environments committed to aquaculture. Listed below is a series of important factors.

- The Chinese government deliberately embarked on a large-scale programme to develop aquaculture.
- China has identified huge areas of potential sites for aquaculture and is prepared to use them to a large extent: 2.6 million ha of suitable coastal sites and 17.5 million ha of inland freshwater sites.
- Mariculture utilisation of shallow sea and mudflat areas was 25.1% in 1995 compared with 3.9% in 1978, and offshore areas used for mariculture have been extended offshore from 10 m to 50 m depth.
- Utilisation for aquaculture of inland waters was 25.5% in 1995 compared with 4.3% in 1978 (this 4.3% already consisted of 100 500 ha).
- Productivity/unit area, although still generally low, has been increased through research, extension and better technology.
- The number of species in culture has greatly increased, including high-value species for international markets such as shrimp, crabs, scallops, abalone, eels, rainbow trout, soft-shell turtles and bullfrogs.
- Sixty exotic species were introduced from abroad, 20 new species have been domesticated from wild populations and about 30 species have been improved through hybridisation and stock selection programmes.
- There are huge numbers of fingerlings and juveniles produced from many large hatcheries.

The result of this deliberate programme to promote aquaculture is that China leads the world in being the only country in which aquaculture production substantially exceeds fisheries production. China has already achieved the anticipated time when the balance of production between aquaculture and fisheries shifts to the former.

It may not seem surprising that China, as the country with the largest population, is the greatest aquaculture producer. However, aquaculture production is not just related to population size. Some production versus population size data are presented in Table 1.3. China's production is 21.9 kg/person/year (= 60 g/person/day). Production for all LIFDCs is 8.8 kg/person/year and for western countries it is 3.1 kg/person/year.

There is a complete contrast between aquaculture in the developing Asian countries and in developed countries. Over 60% of fish production in developed countries is based on high market value species of carnivorous fish (Fig. 2.6). These fish are reared in monoculture at high stocking densities and need inputs of high-protein feeds. High-protein feeds are expensive and usually require fish meal as sources of animal protein and lipids for the carnivorous diet. Thus, this form of aquaculture uses low market value products from fisheries as feed to increase their value in the final products of culture. This is inefficient (section 2.3.1).

1.5 Diversity of aquaculture

Two groups, freshwater cyprinid fish and seaweeds, dominate world aquaculture production. However, as well as these two large groups and the modest number of aquaculture species treated in this textbook, there is a huge diversity of species that are cultured. FAO (2001a) provides quantity and commercial value data on aquaculture production of some hundreds of species of fish, shellfish and algae that are cultured for human consumption. Even these three very broad categories of organisms do not encompass the whole range of aquacultured species, e.g. there are also turtles, crocodiles, bullfrogs, sea-squirts and sea urchins (although crocodile meat does not get a mention in the FAO statistics). In addition to aquaculture for human consumption, there are international and local industries producing live feeds for hatcheries, e.g. dried brine shrimp cysts, live 'blood worms'. There is a huge worldwide aquaculture industry for ornamental fish, especially tropical freshwater and marine fish. Other aquarium-related industries produce aquarium plants and freshwater and marine invertebrates, including coralline algae/coral-encrusted substrates for tropical marine aquaria. Pearls and sponges are cultured and sold for their traditional uses.

In particular, the Asian countries with large and traditional aquaculture industries are involved in culturing a wide diversity of species. Cen & Zhang (1998) indicate that at least 110 species (including introduced species) are cultured in China. These include freshwater and marine fish, shrimps, prawns, crabs, bivalves, gastropods, a

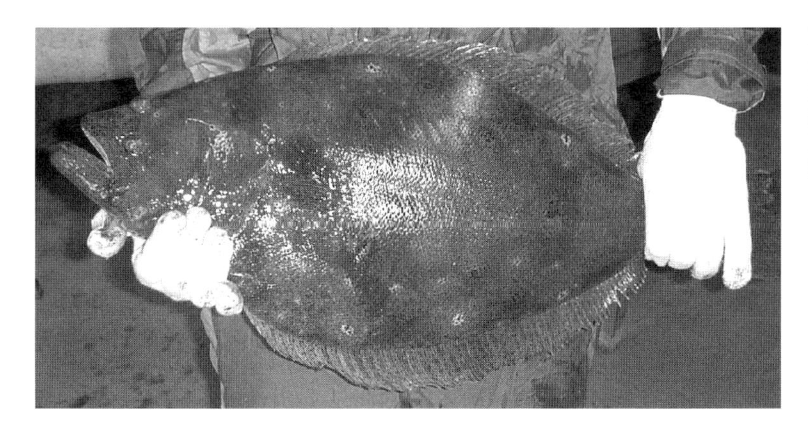

Fig. 1.5 Japanese flounder (*Paralichthyes olivaceus*) broodstock used to produce seed for stock enhancement (photograph by Dr Kotaro Kikuchi, Abiko Research Laboratory).

variety of seaweeds, and special products such as sea cucumbers, sea urchins, bullfrogs and soft-shell turtles. Aquaculturists in Taiwan have developed techniques for larval culture of more than 90 species of freshwater and marine fish (Liao *et al.*, 2001). In Japan in 1997, 284 hatcheries together produced fingerlings of 88 fish species (Fushimi, 2001). Fingerlings of 73 species, amounting to 168 million fingerlings, were released into the environment in stock enhancement programmes.

This raises another aspect of the diversity of aquaculture: links between aquaculture and fisheries. Stock enhancement of restricted freshwater environments for subsequent fishing (either recreational or commercial) is a standard procedure (e.g. with carp). Stock enhancement of marine populations in an attempt to improve fishery catches is an important aspect of Japanese aquaculture (Fig. 1.5). Stock enhancement of marine populations is also practised in Taiwan, and to a limited extent in countries such as Canada, France, Norway, the UK and the USA. Stock enhancement programmes

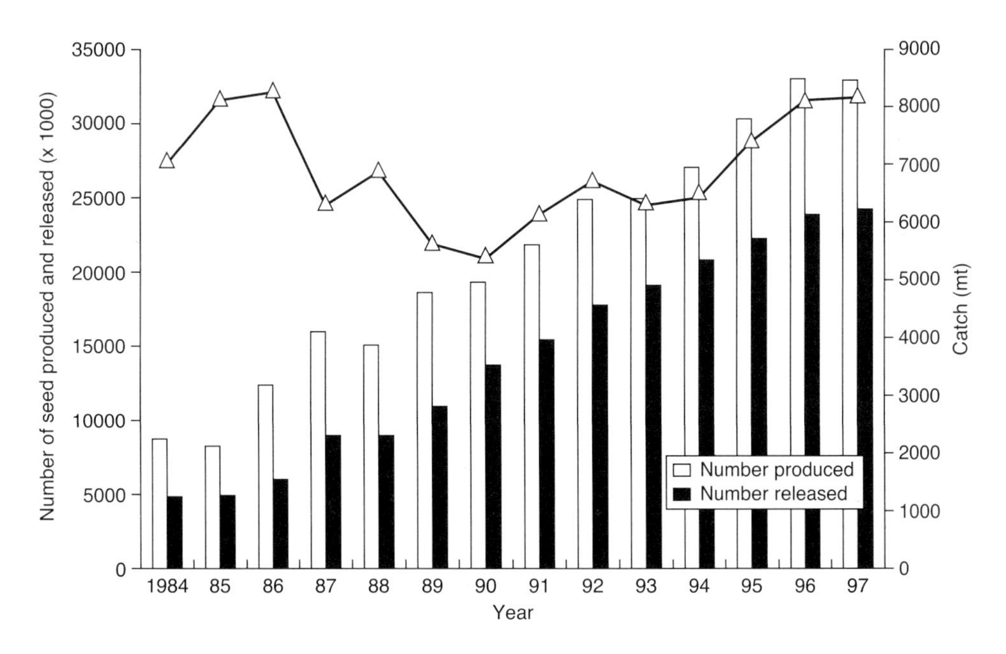

Fig. 1.6 Annual fluctuations in the catch of Japanese flounder (*Paralichthyes olivaceus*) versus 'seed' production and release in stock enhancement. Note upturn in declining fishery from 1990. Reprinted from Fushimi (2001) with permission from Elsevier Science.

with 'seed' from aquaculture involve predominantly fish, but they are also used for shellfish such as scallops and marine lobsters (Hilborn, 1998). These restocking programmes are not unequivocally successful. The best that can be shown for many is increasing fisheries that are approximately correlated with increasing input to the population of cultured 'seed' (Fig. 1.6). Of nine stock enhancement programmes in marine environments assessed by Hilborn (1998), only one appeared to be clearly successful in economic terms, i.e. the additional economic value of catches from the enhanced stock exceeded the cost of the stocking programme.

References

Cen, F. & Zhang, D. (1998). Development and status of the aquaculture industry in the People's Republic of China. *World Aquaculture*, **29**(2), 52–6.

Entis, E. (1997). Aquabiotech: a blue revolution? *World Aquaculture*, **28**(1), 12–15.

FAO (2000). *1998 Fisheries Statistics: Capture Production*, Vol. 86/1. Food and Agriculture Organization of the United Nations, Rome.

FAO (2001a). *1999 Fisheries Statistics: Aquaculture Production*, Vol. 88/2. Food and Agriculture Organization of the United Nations, Rome.

FAO (2001b). *1999 Fisheries Statistics: Aquaculture Production*, Vol. 88/2. Food and Agriculture Organization of the United Nations, Rome.

Fushimi, H. (2001). Production of juvenile marine finfish for stock enhancement in Japan. *Aquaculture*, **200**, 33–53.

Hilborn, R. (1998). The economic performance of marine stock enhancement projects. *Bulletin of Marine Science*, **62**, 661–74.

Holmes, R. (1996). Blue revolutionaries. *New Scientist*, **152** (2059), 32–7.

Hopkins, J. S. (1996). Aquaculture sustainability: avoiding the pitfalls of the Green Revolution. *World Aquaculture*, **27**(2), 13–15.

Liao, I. C., Su, H. M. & Chang, E. Y. (2001). Techniques of finfish larviculture in Taiwan. *Aquaculture,* **200**, 1–31.

Ling, S. W. (1977). *Aquaculture in Southeast Asia: A Historical Review*. University of Washington Press, Seattle.

New, M. B. (1997). Aquaculture and the capture fisheries: balancing the scales. *World Aquaculture*, **28**(2), 11–30.

Pillay, T. V. R. (1990). *Aquaculture Principles and Practices*. Fishing News Books, Oxford.

2

General Principles

Peter Appleford, John Lucas and Paul Southgate

2.1 Introduction

2.1.1 General aspects

Him, steeped in the odor of ponds.
Stanley J. Kunitz (1985)

A great deal of aquaculture involves growing species in ponds (under highly desirable conditions of minimal odour!), but obviously this is only part of the wide range of practices. The large number of aquaculture practices are not treated systematically in any other chapter of this book and hence they are considered here. There are differences in the structures used, the intensities of culture, the degree of water exchange and the factors to be considered in selecting suitable species and farm sites for aquaculture. To a considerable extent they are inter-related (Fig. 2.1). Choosing a site for an aquaculture venture will be strongly influenced by, among other things, the intensity of culture, the amount of water exchange required and the biological characteristics of the selected species (e.g. life cycle stages, diet, growth rate). The method of culture will depend on the availability and characteristics of farm sites. Culture structures, site and species selected for aquaculture will depend on the economics of the venture. The aquaculture venture will fail, despite having a species that is very suitable for culture and good technology on a good farm site, if there is insufficient demand for the product and if it cannot be sold profitably (Chapter 12).

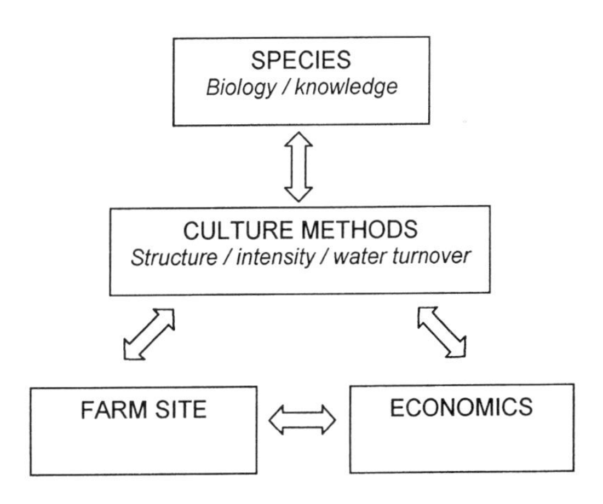

Fig. 2.1 The interrelationships between cultured species, culture methods, farm site and economics in an aquaculture venture.

Many aquaculture ventures have failed because at least one of these inter-related components has not been properly considered before undertaking the venture.

2.1.2 Aquaculture systems

Owing to the great diversity of aquaculture operations, the description of types of aquaculture systems may be complex and sometimes confusing to the novice. Usually culture systems are classified according to three criteria.

(1) Type of culture structure. Culture structure describes what encloses or supports the aquaculture organisms. Broadly, aquaculture structures include ponds, tanks, raceways, cages, pens, racks, long-lines and floats.
(2) Water exchange. Water exchange describes the amount of water exchanged or the control over water flow to the system. Broadly, the levels of water exchange are static, open, semi-closed and recirculating (closed).
(3) Intensity of culture. Intensity of culture reflects the number of aquaculture organisms per unit area or water volume and also the ability of the natural productivity to support the crop. Broadly, the intensity of culture is described as intensive, semi-intensive or extensive.

These criteria will be described in more detail in sections 2.2, 2.3 and 2.4.

The type of system used for aquaculture production is a combination of the above criteria. For example, there may be a pond system that is

- extensive and static: to grow major carp in ponds in China and India
- semi-intensive and semi-closed: to grow silver perch in ponds in Australia
- semi-closed and intensive: to grow shrimp in ponds in Asia and the Americas
- open and intensive: to grow Atlantic salmon in seacages in Norway and Canada.

2.2 Structures used for aquaculture

Although a particular kind of structure will be considered under each heading, one structure is rarely used for the whole lifespan of a cultured species, except in some extensive culture. For instance, the intensive culture of a particular fish species from gametes to harvest may successively involve the following structures:

- a small fertilisation tank
- an embryonic development tank with gentle flow-through of water
- a larval tank with gentle circulation
- a fingerling tank with gentle flow-through of water
- a pond for growth of fingerlings to large juveniles
- a cage for grow-out to harvest.

2.2.1 Ponds

Ponds are broadly defined as earthen impoundments for holding aquatic species. Ponds are the oldest aquaculture structure because of the simplicity of basic pond culture in freshwater. Pond culture can be undertaken with nothing more than convenient natural ponds. Purely harvesting from natural ponds is not aquaculture – it lacks the component of enhancing production; but enhancing production may involve nothing more than adding crude organic fertilisers or removing predators or competitors of the cultured species.

A pond may be a simple hole in the ground (sunken pond) or an enclosed waterway in a valley or stream bed where only one or two walls are constructed (barrage ponds), or it may be above ground (embankment pond). Embankments may also be used to divide a large freshwater or brackishwater mass into adjacent subunits, e.g. ponds of varying shapes according to the natural topography. Embankment ponds may be any size or shape, but rectangular ponds are most common as they limit wasted space between ponds. Ideally, water input and discharge from ponds are facilitated by gravity to minimise construction and operating costs. Ponds are most commonly used for culture of fish and crustaceans.

Cheap simple ponds are the most widely used freshwater and brackishwater aquaculture systems. The main requirements for ponds are:

- a reliable supply of good-quality water (preferably gravity fed)
- relatively impermeable soils for construction
- well-structured soils with good organic matter content to support pond ecosystems
- gravity drainage.

In general, ponds are cheaper to construct per unit area than tanks and cages, and may be inexpensive to run, depending on pumping costs. Ponds tend to have the lowest stocking densities of the culture structures; however, density varies according to whether the system is extensive, semi-intensive or intensive.

Ponds are generally maintained with a light algal bloom to minimise water exchange. More detail on pond culture techniques is given in section 19.4.

2.2.2 Pond site requirements

Assessment of the soil is a very important consideration in aquaculture pond site selection, development and management. It is the soils at the site that will determine the natural productivity of the pond, water-holding capacity, fertilisation requirements, water quality and construction.

Poor site selection will result in ponds that cannot be managed to suit the requirements of the cultured animals or that require an unacceptable level of inputs and maintenance to do so. The main soil properties important for aquaculture use are:

(1) physical: texture, strength, stability, water-holding ability
(2) physicochemical: ion exchange capacity, acid–alkaline reaction, leaching effects and absorptive/binding capacity
(3) biological: organic matter and nutrient transforming biomass.

More information on the properties of soil is available from texts on aquaculture pond construction (e.g. Boyd, 1991; Yoo & Boyd, 1994).

In some circumstances, particularly with small ponds and in areas where the soil is porous, the ponds may be lined with impermeable sheeting, e.g. thick plastic sheeting. Soil substrates may then be added to these lined ponds to provide the functions of soil.

2.2.3 Pond layout

In large pond farms, there is a carefully planned layout of ponds and embankments to include channels between ponds for inflow of water and outflow of effluent water. There may be tens to hundreds of ponds on a large farm, with the surface area of each pond ranging from ~0.5 to 10 ha (Fig. 1.4).

Ponds are generally constructed in one of four main configurations or layouts, each with its advantages and disadvantages.

(1) Series. Ponds are constructed consecutively so that water flows through each pond prior to its discharge. This layout allows easy movement of stock from one pond to another and maximal use of water. A disadvantage of constructing ponds in series is the decrease in water quality through consecutive ponds and the inability to isolate individual ponds, preventing the isolation of disease.
(2) Parallel. In a parallel pond layout, water is distributed from a common supply channel into each pond individually. Effluent water is collected into a common drainage channel. Stock movement is more difficult and water usage increases; however, this is offset by superior control over the water exchange for individual ponds and simpler water quality management.
(3) Radial. Constructing ponds in a radial design is uncommon and is only employed when there is a specific production planning purpose. In a radial layout, ponds are constructed so that each inner pond feeds into multiple outer ponds or larger outer ponds in a concentric pattern. This pond layout may be employed when there is a continual supply of smaller animals (e.g. fingerlings) and, as the animals grow and require more space, they are moved to the outer ponds until they are eventually harvested.
(4) Inset. In an inset pond layout, one pond is placed inside another larger pond. This method is used when a nursery pond(s) is set inside a grow-out pond.

2.2.4 Pond design

There are many pond designs, reflecting their range of uses. The most complicated pond design is one in which water must be delivered to the pond by pumping. This is usually because the pond is above the level of the water source. The following sections describe various components of ponds with pumping.

The description is particularly applicable to shrimp ponds, but there are general features that are applicable to all such ponds (Yoo & Boyd, 1994).

Size and shape

Ponds are usually rectangular to allow for the maximal utilisation of land. Typically, the length to width ratio is 2–3:1. The factors governing the size of the pond are the biological requirements of the animals and access to the pond for husbandry. For example, if a pond is to be harvested by seine netting then a recommended maximal width is 20 m. Similarly 20 m is about the maximum reach of many feed dispensers. If, however, the cultured animals come to the edge to feed and can be harvested using a collection sump, the pond may be much larger. A disadvantage of larger ponds with greater biomass and higher feed inputs is that they may be more difficult to manage.

Probably the greatest disadvantage of a rectangular pond is poor water circulation. Circular ponds have been trialled for high-value intensive crops with good results; however, the trade-off is less pondage per unit farm area (Wyban & Sweeney, 1989).

Walls

Wall design requires consideration of both wall height and wall slope. Wall height is governed by the depth of the pond, wave height and freeboard. Freeboard is the additional height above normal water and wave height to allow for extreme storm events resulting in raised water levels or greater wave height. Depth of ponds is generally in the range of 0.8 to 1.8 m. These depths allow adequate light penetration for primary productivity and enough depth to reduce temperature fluctuations while decreasing the chance of both oxygen and temperature stratification.

The photosynthetically productive depths of ponds are limited by the penetration of light through the water and, consequently, ponds are generally shallow compared with their surface area. This is because, unlike smaller culture systems, in which aeration can be used to promote gas exchange, and, unlike systems with continuous water flow, the pond water mass depends on gas exchange at the surface for both O_2 uptake and CO_2 removal at night. Gas exchange and diffusion are slow processes and it is possible to have strong stratification of dissolved oxygen (DO) through the pond water column. This stratification increases with water depth and is therefore a limiting factor on water depth, despite some aeration systems for ponds (section 3.5.2).

Settlement of the wall must also be considered when constructing the pond. Settlement generally varies from 5% to 10%, depending on the soil type used for construction. It is recommended that 10% additional height be allowed to take account of settlement (Yoo & Boyd, 1994).

Wall slope depends on the stability of the soil at the site, the erosion protection and what is deemed as acceptable maintenance. Working with stable soils to achieve minimal erosion, an internal wall slope of between 1:2 (vertical–horizontal) and 1:3 is recommended, and the external wall slope should be between 1:2.5 and 1:3.5. In practice, both slopes are often decreased to ~1:1 (internal) and 1:2 (external) to reduce the capital cost of construction. This also increases maintenance costs. If erosion protection is included, e.g. vegetation above the water line and stone below the water line, slopes can be reduced further. Wall reinforcing can allow the use of vertical walls, although this is more expensive.

Walls must also have a top width to allow appropriate access to the pond. Top widths depend on the height of the wall and the activity to be conducted around the pond. Recommended top widths (Yoo & Boyd, 1994) are related to wall height, but generally do not consider the activity around the pond. Farm machinery is often used around the pond during normal operation, and these vehicles will require a top width greater than that recommended.

Floors

The main considerations in pond floor design are pond drainage and harvesting. If a pond is to be drained to collect culture stock, the bottom must be smooth to prevent water collecting in hollows. A bottom gradient of 0.3% to 0.6% is used to effectively drain water to the end of the pond or to a central drain (Wang & Fast, 1991). Additional channels (30–50 cm wide and 10–20 cm deep) may be cut to facilitate drainage. At the lower end of the drainage system there may be a harvest sump, usually about 10–20 cm deep and occupying 0.5–1.0% of the total pond area.

Water supply (inlets) and drainage system (outlets)

Inlets and outlets are important components of ponds and need to be secure, easy to operate and adequately sized. The exchange of water through a pond is controlled via the inlet. Inlets may be pipes, channels or sluices (an artificial channel for conducting water). Flow through the inlets can be regulated using valves (pipes) or boards (channels and sluices) or by the use of pumps.

In many ponds, water must be filtered to prevent the inclusion of organisms that may be predators, competitors or vectors for disease. To do this, the water is screened using either a simple nylon sock or bag over a pipe or a screen/filter box in the channel or sluice. Socks and screens must be sized appropriately to remove anticipated predators (section 19.4.5, Fig. 19.6).

Outlets serve a number of functions and are used to:

- adjust and hold the water level
- drain or partially drain the pond
- drain water from specific levels of the pond
- provide an overflow in a continually flowing pond
- screen and collect culture stock.

Outlets are usually pipes or weir gates, which are also known as 'monks'. Pipes are cheap, simple and convenient for small ponds; however, they are not adequate for larger ponds. Weir gates are vertical control boxes attached to waste pipes (Fig. 2.2). They are a combination of screens to regulate animal movements, boards to regulate water level and flow, and pipes to move the water from the pond. Landau (1992) and Yoo and Boyd (1994) provide detailed descriptions of pond outlets.

Fig. 2.2 A weir gate or 'monk'. The two grooves at the front of the gate contain screens to regulate animal movements. The two grooves at the back contain a dual set of boards that regulate water level and flow. Water passing through the screens and over the boards is drained away from the pond using pipes at the back of the construction.

2.2.5 Tanks and raceways

Tanks are second to ponds as the most commonly used structures for aquaculture. Tanks are generally situated above ground on a solid base and may be used indoors or outdoors. Tanks have the advantage of allowing the use of land normally unsuitable for aquaculture, as the water is contained within the structures with no contact to the surrounding soils. There is a wide range of dimensions and sizes of tanks, corresponding to their wide range of uses in culturing microalgae, macroalgae, and various life cycle stages of fish and invertebrates. They range in size from tens of litres to hundreds of cubic metres.

Raceways are basically elongated tanks in which water enters at one end and exits at the other. They generally consist of elongate, narrow and shallow systems with continuous water flow. The limited cross-section of the raceway together with the strong flow rate are designed to keep continuous unidirectional flow along the raceway. It is a particularly suitable system for some fish, such as salmonids, that live in shallow streams and swim against the current (Fig. 2.3). Raceways generally require large amounts of water per unit volume of the system. Freshwater raceways, the most common form, need to be sited near a freshwater spring or a substantial stream that does not dry up. As with other aquaculture, they need an unpolluted source of water. A potential problem with raceways is the deterioration of water quality along their length.

Tanks are most commonly used for culturing the early developmental stages of fish, bivalves and

Fig. 2.3 A concrete raceway system in New Zealand used to culture Chinook salmon smolt for seacage farming and ranching operations (photograph by Dr John Purser).

crustaceans, and for culturing high-value fish species. Raceways are also used for culturing high-value fish. As with ponds, there is a variety of tank systems ranging from flow-through systems, in which the tank system is simply a confinement for the animals, to recirculating tank systems, in which the water is used, treated and re-used while maintaining a high density of animals. The majority of raceways are flow-through.

Structure

Tanks and raceways may be constructed from a variety of materials but are most commonly constructed from concrete or synthetics, such as fibreglass. The major considerations in material selection are:

- strength
- abrasiveness to the animals
- cost
- possible chemical residues
- shape.

The shape must allow good water flow to maintain uniform water quality and prevent build-up of wastes in 'dead' places. The shape of tanks varies greatly from hemispherical to flat-bottom and square. Tanks that are circular in the horizontal plane with conical bottoms are generally considered to provide the best circulation and self-cleaning properties. The flow of water in raceways, being linear from inlet to outlet, facilitates cleaning and more uniform water quality.

Inlets and outlets

As discussed for ponds, the major function of the inlet for tanks and raceways is to regulate water exchange. Generally, water flow into tanks is regulated using a valve, whereas flow to raceways may be regulated by either valves or sluices.

Inlets in tanks or raceways, however, must also provide water movement. The delivery of water into the tank or raceway must be such that it facilitates an even movement of water throughout the structure, maintaining uniform water quality. In raceways this is often achieved by delivering water at multiple points along the input end of the raceway either through perforated pipes or over spillways. Maintaining uniform water quality may also be assisted by having multiple inlets along the length of the raceway.

Water in tanks is generally supplied through pipes and the number of input points, combined with the direction inlet water, will govern water movement and assist with aeration in some systems.

The functions of outlets from tanks and raceways include:

- maintaining water level
- retaining cultured animals
- allowing drainage of the structure
- removal of wastes.

Outlets for tanks usually consist of a vertical or horizontal pipe, a screen to retain stock (or a collection trap for stock) and pipes to direct water either back to a sump or to waste. However, given the diversity of tank systems and the flow design for these systems, there are many outlet designs.

2.2.6 Cages

Originally, the cages used for aquaculture consisted of poles or stakes driven into the sediment of shallow lakes or bays with netting stretched around them. These are still in use and are referred to as net pens or hapas (section 16.6.4). Modern cages are floating structures with a net suspended below. They may be square or rectangular (Fig. 2.4) or round (Fig. 15.7). Floating cages may be small and of limited strength (Fig. 18.9) or they may be many thousands of cubic metres in volume and designed for use in the open ocean (Fig. 23.2). Cages are used for fish culture in their grow-out phase, that is the months or years up to their market size.

Cages are intermediate to tanks and ponds for capital costs, and are relatively cheap to operate, requiring maintenance but no pumping. Cages allow no control over water quality and therefore require a good site with adequate exchange of high-quality water, allowing intensive farming at stocking densities up to 15–40 kg/m^3. Being in the ambient water, cages have problems with:

- fouling of the meshes by seaweeds, bivalves, sponges, etc., which reduce the flow of water through the cage
- predators, e.g. seals, birds
- parasites
- diseases, which are difficult to manage, as cultured stock are difficult to observe and diseases are not easily treated except with medicated feeds
- algal blooms.

Cages are typically sited in protected inshore ocean locations, such as bays, lochs and fjords. Cages may, however, also be used in large freshwater lakes and ponds (Fig. 14.5). Seacages are generally used in groups, either individually or linked together, and anchored to the substrate. In these groups, they can be serviced and the fish fed and managed from a base facility on the adjacent shore or from a proximate floating facility.

Fig. 2.4 Square cages used for salmonid culture on the west coast of Scotland. Note the raft system in the background used for mussel culture.

The design of cages varies depending upon their use and location.

(1) Ponds. These cages are small, (approximately 4 × 6 m and 1.5 m deep) and are secured to stakes driven into the bottom of shallow ponds. The cages are located within a very 'friendly' environment and are accessed easily from the pond wall or a floating walkway.
(2) Lakes, estuaries or protected bays. These cages tend to be much larger (10–40 m diameter) and are secured to the bottom of the waterway by anchoring lines. The forces associated with these environments are greater than those found in ponds as a result of tidal currents, wind-generated waves (< 1 m) and swells. Again, these cages are easily accessed from shore or by a short trip in a small powerboat.
(3) Open seas. These cages must undergo the rigours of oceanic swells and wind-driven waves in excess of 4 m. The cages are very large, and may be 50 or more metres in diameter. Often located far offshore, these cages require attached working areas.

Irrespective of the use of the cage, there are structures that are common features. The aims of these structures are to hold and protect culture stock, and to maintain the position, size and shape of the cage.

(1) Float. The float keeps the cage at the surface of the water and helps maintain the shape of the cage in the vertical and horizontal planes. Floats may be large-diameter rubber hose or high-density polyethylene tubes/pipes.
(2) Collar. The collar maintains the shape of the cage in the horizontal plane. It may simply be a ring of metal placed at the bottom of the cage to weigh the cage down in the required shape or it can be a complicated design involving flotation and weights.
(3) Nets. There are several types of nets for cages.
    - Main net. This net holds the cultured stock. It must resist ripping by objects and predators, and be able to hold the biomass when nets are lifted for harvest. The mesh must be designed to prevent cultured fish being caught in the netting. Besides holding the fish, this

net must also permit adequate water flow through the cages to maintain water quality.
- Predator net. This net keeps sea-borne predators away from the cultured stock to reduce predation. It is placed outside the main net.
- Covering net. This net is placed over the top of the cages to prevent birds landing, fouling, scavenging and preying on cultured stock.
- Jump net. This net projects vertically out of the water around the main net, preventing fish escaping.

(4) Moorings. These are used to secure the cage at the selected site. Moorings are often specific to the type of cage and for larger cages are generally specified by the manufacturers.

Beveridge (1996) provides further details on cages and cage culture.

2.2.7 *Pens*

Pens, enclosures or hapas are used in shallow water, typically in ponds, to create a restricted environment for culture of fish and some crustaceans. They are not usually large, being in the order of tens of square metres or less. The walls of the enclosures may be closely spaced stakes, such as bamboo stems or mangrove branches, or wire and other mesh. This system of culture is practised mainly in developing countries. One interesting exception to the shallow water pen is the use of mesh fences or walls to enclose bottom-dwelling scallops. These pens are of sufficient height to prevent the scallops from swimming over the wall. They may use floats to allow the mesh to rise and fall with the tide (Quayle & Newkirk, 1989). Similar enclosures may also be used for culture of gastropod molluscs (section 22.3.3).

2.2.8 *Substrates, racks and suspended culture*

Surfaces are used, as would be expected, for culture of bottom-dwelling and attached species. This category is taken to apply to species that are grown in the field for most of their culture, rather than in tanks with artificial surfaces, e.g. abalone culture that occurs totally in tanks.

The techniques used in this category are quite varied, reflecting the diverse range of organisms that are cultured and their varying environmental requirements.

Some bivalves permanently attach to hard substrates. Their larvae settle and metamorphose on attractive substrates, which are put out in the field at appropriate places and times to obtain juveniles. The juveniles may then be grown on the original substrate or are removed to another culture system. The secondary culture systems may be:

- mesh trays on horizontal racks above the substrate in the intertidal zone
- suspension in various kinds of nets, ropes, trays and baskets, hanging down at intervals from a horizontal long-line on the surface
- suspension on vertical ropes below rafts (Fig. 2.5).

More on these techniques and their application to bivalve mollusc culture can be found in Chapter 21.

Other benthic organisms, e.g. seaweeds and sponges, may be grown tied to lines stretched above the substrate.

2.3 Intensity of aquaculture

Intensity of aquaculture describes the various densities of organisms per unit volume or per unit area. It is meaningful in comparisons between the levels of culture of a species or related species. It is, however, meaningless in terms of comparisons of densities of organisms from different groups. For instance, culture of tilapia at 100 kg/m^3 of water in a recirculating system is considered to be intensive culture; culture of shrimp at 50 individuals/m^2 (1–2 kg/m^3) in ponds is considered to be intensive culture. There are, however, characteristics of different intensities of culture that are not specific to groups, and the consideration of culture intensity will be based on these.

Intensity of culture will broadly consider the inputs into the system to maintain adequate growth of the cultured organisms. It comes under the term

Fig. 2.5 Raft culture of blue mussels (*Mytilus edulis*) on Loch Etive, Scotland.

'intensity' because the greater the intensity (or density) of cultured organisms the greater the requirement for inputs into the system.

2.3.1 Natural aquatic ecosystems

All ecosystems, natural and artificial, need some form of energy input to sustain them. Even if there is perfect recycling of organic matter, there will be loss of energy through metabolism and that energy must be replaced.

Natural aquatic ecosystems typically consist of primary producers, various levels of consumers (primary, secondary, etc.) and decomposers. They are self-supporting with recycling of nutrients and input of energy from the sun. They are characterised by long and complex food webs. There are very large energy losses in this food chain. Energy transfer from one level to another in the food chain is in the order of 10% as a 'rule of thumb'. The following is a hypothetical example of six levels in a food chain:

(1) sun's energy + inorganic nutrients →
(2) phytoplankton (e.g. flagellates and diatoms) →
(3) zooplankton and filter feeders (e.g. copepods, invertebrate larvae, bivalves) →
(4) larger zooplankton (e.g. fish larvae, planktivorous fish) →
(5) juvenile fish →
(6) large carnivorous fish.

Thus, in the example consisting of six levels in a food chain: for levels (1) to (6).

(1) 100% → (2) 10% → (3) 1% → (4) 0.1% → (5) 0.01% → (6) 0.001%

This is a very crude approximation, but it is clear that the final stage of this food chain, for example the large carnivorous fish, has magnitudes less energy (and food) available to it than lower levels in the food chain. The energy lost at each step (approximately 90%) is lost through:

- metabolism (the metabolic energy required to keep the organism functioning, e.g. locomotion, feeding, digestion of food, circulation of body fluids, etc.)
- metabolic wastes excreted and faeces released by the organism
- energy expenditure on reproduction
- energy released into the environment as heat, because metabolic processes are not 100% efficient.

These substantial declines in energy with each step in the food chain have an obvious implication for aquaculture; it is more efficient to culture primary producers or animals that feed at a low stage in the food chain. This is an important factor in developing nations, where aquaculture products are valued as a major source of animal protein. With the exception of aquaculture of species such as shrimp, which are directed at international markets, most aquaculture in developing countries is derived from extensive or semi-intensive culture based on species that feed low in the food chain (Fig. 2.6a). Not only is it inefficient to produce carnivorous species through a multilevel food chain, but productivity will be very poor per unit area or volume from such extensive or semi-intensive systems.

In many developed nations, however, the demand is for aquaculture products that are high in the food chain. In fact, the largest demand is for top or near-top carnivorous fish and shrimp, as these are regarded as the 'quality' products (Fig. 2.6b). These carnivores must be fed high-protein feed, typically with fish meal as a major source of protein. Thus, one kind of less valuable seafood is used to produce a more valuable seafood, with the consequent loss of energy.

2.3.2 *Intensive aquaculture systems*

In intensive aquaculture systems all the nutrition for the culture stock comes from introduced feeds, with no utilisation of natural diets. Intensive systems may be in:

- ponds (e.g. for shrimp in tropical/subtropical regions)
- cages (e.g. for marine fish culture in the Mediterranean region) (Fig. 2.7)
- raceways (e.g. for trout species in temperate regions)
- tanks (e.g. for eels in Japan).

The peak stocking density achieved in each case depends upon being able to maintain the water quality conditions required by the cultured organism. Generally, stocking densities are lowest in ponds, followed by cages and with greatest densities achieved for raceways and tanks.

Intensive aquaculture systems are a complete contrast to natural systems. They are characterised by:

- very simple food chains: feed → cultured organisms
- low energy losses from feed input, with high food conversion ratios from specialised artificial feeds

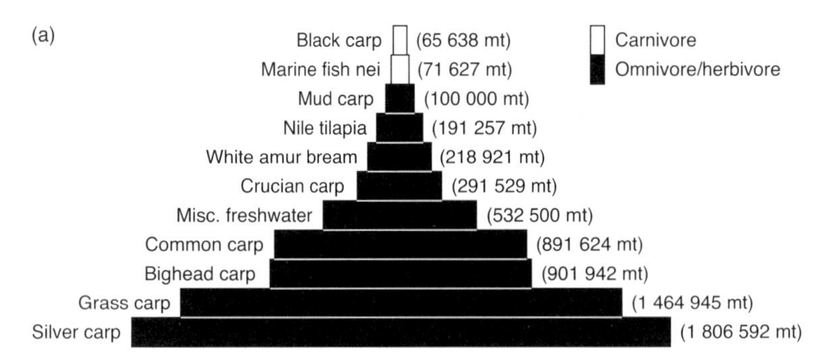

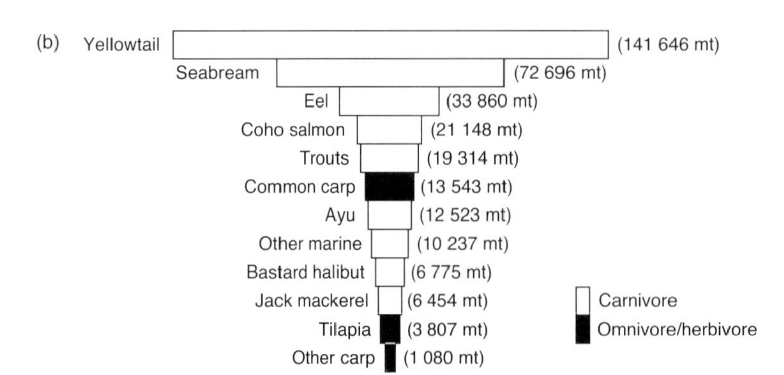

Fig. 2.6 Production pyramid of major farmed fish according to feeding type. (a) China in 1993; (b) Japan in 1993.The widths of the bars are relative, not absolute, production. Based on data from FAO (1995), modified from Tacon (1996) with permission of the World Aquaculture Society.

Fig. 2.7 Underwater view of a seacage used for culturing Atlantic salmon (*Salmo salar*) in southern Australia (photograph with permission of the Tasmanian Aquaculture and Fisheries Institute).

- no recycling of energy and totally non-self-supporting
- the requirement for high inputs of energy (e.g. feed, nutrients, aeration, filtration, pumping)
- high yields per unit area or volume.

Water quality is usually maintained by high water exchange rates and, in some cases, by mechanical means. In intensive culture in indoor tanks, particulate waste removal, gas exchange and oxygen production are all undertaken by mechanical means. In outdoor intensive systems with a soil substrate and phytoplankton there is settlement of particulate wastes, decomposition by bacteria and gas exchange enhanced by mechanical aeration. Stocking density (mass of culture stock per volume or area of water, expressed as kg/m^3 or kg/ha) in intensive systems varies greatly with the type of system and the cultured organism, but is always relatively high.

2.3.3 Extensive aquaculture systems

Intensive and extensive aquaculture systems are contrasted in Table 2.1. Extensive aquaculture differs markedly, largely being part of a natural ecosystem and depending upon it for maintenance of water quality and most of the animals' food and other requirements. An extensive aquaculture system, therefore, has limited inputs to maintain animal growth and survival, i.e. it may have some basic organic fertilisers, such as animal and plant wastes, but no aeration, etc. These systems usually have a low stocking density, generally < 500 kg/ha, and the natural productivity of feed (plants and animals) within the system and natural gas exchange is sufficient to support the cultured organisms. Extensive aquaculture is used for pond culture and for organisms grown on or in various substrates, e.g. bivalve and seaweed culture. Seaweed culture differs, of course, from animal culture in terms of food requirements, but the seaweeds are dependent on the culture environment for inorganic nutrients.

Apart from seaweeds and bivalves, a considerable amount of extensive aquaculture is of low-value fish, such as carps and tilapia. This is possible because of the low costs of production associated with extensive culture. Furthermore, the production statistics per country and species show that this is the major source of aquaculture production in the world (FAO, 2001).

Extensive versus intensive aquaculture is rather like the difference between

- feed-lot cattle, which are totally dependent on the feed-lot environment and the feed provided (intensive culture), and
- free-range cattle roaming through natural vegetation on a large cattle station or ranch, feeding on natural pasture and with minimal husbandry (extensive culture).

Table 2.1 Comparison of the general characteristics of extensive and intensive aquaculture systems

	Extensive aquaculture	Intensive aquaculture
Establishment costs	Low	High
Culture systems used	Natural water bodies and simple containment structures	Fabricated culture systems, including tanks, pond systems, raceways, seacages, etc.
Technology level	Low	High
Degree of control over the environment, nutrition, predators, competition and disease agents	Very low	High
Typical source of seed stock (plant or animal)	From nature	From domesticated broodstock (possibly genetically selected)
Operating costs, stocking rates and production levels	Low	High
Dependence on local climate and water quality	High, may be some crude adjustment of pH with lime and related substances	Low to very low
Monitoring of water quality	Nil	Undertaken regularly
Food source for cultured animals	Natural food organisms; often some input of animal and plant wastes	Pelletised, fabricated feeds, which must be nutritionally complete
Production (kg/ha/year)	Low (per unit area or volume) (20–500)	High: maximising output of product in minimum surface area, water volume and time 5000–100 000+*
Production costs	Low to very low	High to very high

*The highest values are extrapolations from square metres.

In each case there is a harvest of beef, but the harvest per hectare is magnitudes higher from the feed-lot system. None the less, both systems are profitable. Profitability depends on the difference between total cost of production and total value of product.

2.3.4 *Semi-intensive aquaculture systems*

There is no abrupt cut-off point between extensive and intensive aquaculture (Table 2.2). Semi-intensive aquaculture is used as an approximation to describe the middle ground. Semi-intensive aquaculture systems rely to an extent on natural productivity; however, there is more supplementation of the natural system. Supplementation may take many forms, including:

- addition of aeration to maintain dissolved oxygen levels
- addition of inorganic or organic fertilisers to improve natural productivity
- addition of prepared feeds (supplemental feeding).

Semi-intensive culture is almost exclusive to ponds and allows for an increase in the stocking density within the pond.

Further comparisons of the three levels of culture intensity are given in Tables 16.2, 19.2 and 20.3.

2.3.5 *Polyculture*

In some semi-intensive and extensive culture systems, such as ponds, the target species will share the environment with other related species, e.g. a target fish with other fish or a target shrimp with crabs. Depending on the degree of niche overlap, these other species may compete with the target species for food, habitat, etc. On the other hand, they may facilitate its growth and health through supporting activities, e.g. a detritivore removing benthic waste from the environment before it creates anoxic conditions in the lower water column.

Table 2.2 Examples of extensive, semi-intensive and intensive grow-out aquaculture showing the continuity between levels

Harvest rate (kg/ha)	Examples of cultured animal and structure	Culture methods	Level of aquaculture
90	Freshwater crayfish, large-mouth bass Dam	Simple stocking	Extensive
> 10 000	Marine mussels Raft	No supplementary feeding, suspended in water currents to filter feed	Extensive
900	Carps, catfish, tilapia Pond	Some supplementary feeding and pond fertilisation, possible emergency aeration	Semi-intensive
6000	Shrimp Pond	Almost complete diet, regular water exchange, aeration	Intensive
15 000	Sea bream, sea bass Atlantic salmon Cage	Complete diet	Intensive
> 60 000*	Eels, barramundi Battery tank systems in warehouse	Complete diet, recirculation and rigorous water quality control	Intensive

*Extrapolated.

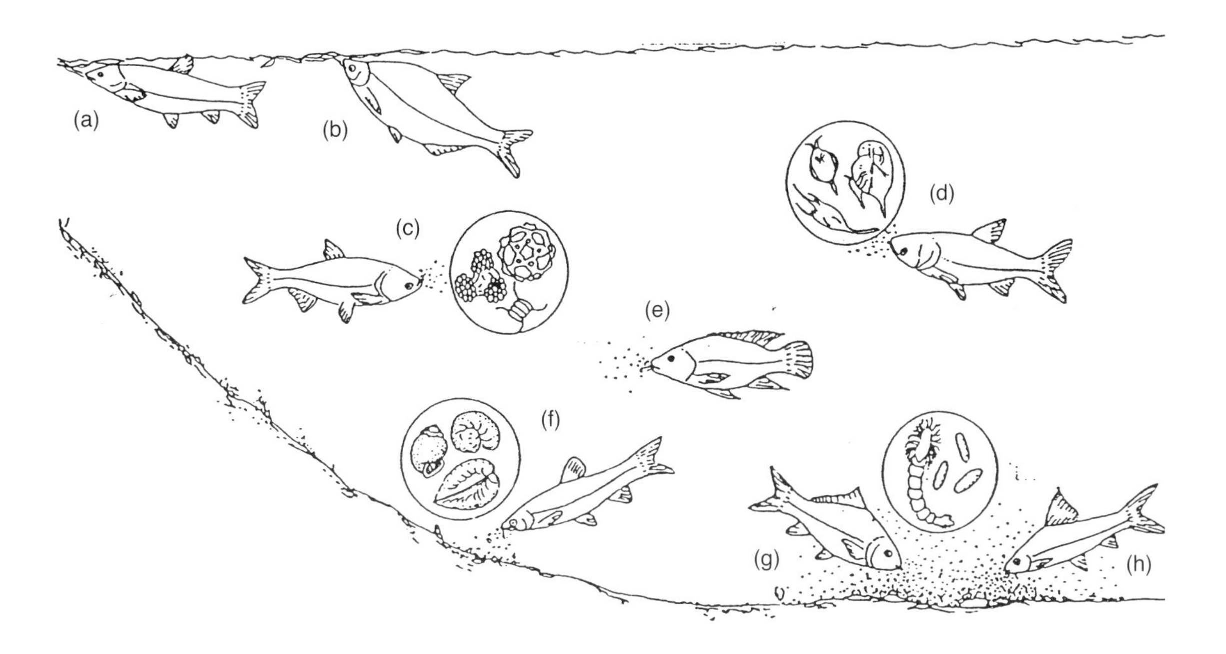

Fig. 2.8 Natural food resources utilised by major fish species cultivated in Chinese carp polyculture ponds. Grass carp (a) and wuchang fish (b) feed upon terrestrial vegetation and aquatic macrophytes; silver carp (c) graze upon phytoplankton; bighead carp (d) consume zooplankton; tilapias (e) feed upon both kinds of plankton, green fodders and benthic organic matter; black carp (f) feed on molluscs; and common carp (g) and mud carp (h) consume benthic invertebrates and bottom detritus. Reproduced from Zweig (1985) with permission from the Royal Swedish Academy of Sciences.

Polyculture describes the deliberate culture together of complementary species, i.e. species occupying complementary niches, especially in regard to their food and feeding (Fig. 2.8). Polyculture increases production per unit area or volume of the pond by maximising the utilisation of all nutritional niches within the pond. A greater proportion of the primary production is utilised as the target species are linked to more of the food webs than a single species, resulting in increased productivity. They also have the effect of reducing the complexity of the food webs because they 'cover' a high proportion of primary production at its sources with their feeding. In these ways, polyculture systems may average fish production of > 3 mt/ha/year in ponds in which there is heavy addition of manure fertiliser (Billard, 1999).With supplemental feeding there may be yields up to 16 000 kg/ha/year. Carp polyculture is described in greater detail in section 14.5.2.

A new form of polyculture that is being researched is the use of effluent from, for example, shrimp ponds for the culture of complementary species. The effluent, rich in nutrients and sediment and potentially an environmental hazard, may be passed through further ponds containing animals or plants that extract the nutrients and sediment from the effluent. There may be bivalves, which filter the water of sediment, microalgae and some bacteria and macroalgae, which extract nutrients from the water, or even mangroves for the production of wood. The objective is not only to restore the quality of the effluent water, to the extent that it can be recycled into the aquaculture system, but also to produce a commercial crop from the effluent-treating organisms, e.g. using a bivalve of commercial value (section 13.2.4, Fig. 13.6).

This developing form of polyculture differs from traditional polyculture in that the complementary species are isolated in different parts of the system.

2.3.6 *Integrated agri-aquaculture systems*

Aquaculture was traditionally developed on an integrated basis, and much aquaculture in Asian countries still operates in this way. Integration involves:

- growing a variety of aquatic species in a single body of water
- water re-use for successive aquaculture species or other crops
- integration of aquaculture with other farm production or by-products.

Traditional integrated aquaculture usually occurs on a family farm in which there are plant crops, animal rearing and aquaculture. Aquaculture is extensive and the cultured animals are low in the food web. Ponds are fertilised with faeces from pigs, ducks, etc., whatever is reared or used on the farm (Fig. 2.9). Wastes from crops, e.g. stubble from cereal crops, may be added to the ponds as food for benthic feeders and detritivores. Ponds are not drained but re-used for successive aquaculture crops. In some cases, the aquaculture animals (fish and freshwater crustaceans) are added to fields where the crop, typically rice, is grown partially submerged in water. The cultured animals feed on the plant waste materials and grow, and are harvested after the crop, when the field is drained. A double crop, plant and animal, is obtained from the field.

Modern aquaculture in developed countries has disintegrated aquaculture. The move has been towards developing monoculture systems for high-value species with single use of water. However, the limited availability of water and realisation of the global importance of water resources has led to a move towards aquaculture integration in countries where monoculture industries have developed.

Integrated agri-aquaculture systems (IAASs) have been variously defined (see Cohen, 1997; Edwards, 1998). The basic premise of IAASs is the multiple use of water for both traditional terrestrial farming and aquaculture in a profitable and ecologically sustainable manner. As a result, integrated agri-aquaculture embraces a diversity of practices, systems and operations.

Water is one of the world's most precious and poorly utilised natural resources. Integration of farming practices to enhance productivity and water-use efficiency will contribute to the ecologically sustainable development of natural resources. IAASs allow for irrigated farming systems and the multiple use of the same water, typically for fish production first and then for irrigation.

Presently, IAASs most commonly occur in countries with very limited water resources. In many

Fig. 2.9 Fish (tilapia) culture integrated with pig production in Thailand. The pigs are in the shed on the left and their faeces are washed into the tilapia pond through pipes (photograph by Dr Victor Suresh).

developing countries of the world and in Israel, IAASs are highly developed and make optimal use of the available water. In Israel, fresh surface water and brackish groundwater have been utilised in integrated farming of a variety of fish species with a variety of land crops (Cohen, 1997; Chapter 5). In Asian countries, fish, rice crops and ducks have been integrated to better utilise available water, land and nutrients (Huazhu *et al.*, 1994).

In developed countries, such as the USA, Australia and those in Europe, IAAS technology has generally been limited to small-scale systems linked to irrigation farming. Opportunities commonly include use of traditional farm water storage facilities (dams), irrigation channels, ground and surface water and inland saline groundwater.

2.4 Static, open, semi-closed and recirculating (closed) systems

2.4.1 Static systems

As outlined in Chapter 1, much global aquaculture production uses traditional pond culture methods. These ponds are static, with no exchange of water during the culture period. There may be some topping up to offset evaporation.

Static pond culture is usually extensive because of major problems in maintaining water quality under conditions of a large biomass of cultured animals per unit volume of static water (Table 2.3). Increasing biomass requires increasing inputs of fertilisers and supplementary feeds to maintain productivity. This, in turn, requires management for such water quality problems as unacceptable levels of toxic N compounds and low DO levels at night. With supplementary aeration it may be possible to maintain DO with a higher biomass and achieve greater productivity (section 2.4.4). Aerators are, however, often not available or feasible in rural regions where static pond culture is employed.

2.4.2 Open systems

In these systems the environment is the aquaculture farm, i.e. the culture organisms are confined or protected within the farm in a vast amount of water (e.g. a lake or an ocean) so that water quality is maintained by natural flows and processing. There is no artificial circulation of water through or within the system. The following are two examples of open systems.

(1) Cage systems are classified as open systems when they are placed within a large body of water such as an ocean or an estuary. In these cage systems the fish are generally at high density and artificial feed is supplied. Water quality is, however, maintained by natural currents and tides. Therefore, these are intensive open systems (Table 2.3).
(2) Bivalve culture on racks or long-lines placed in the open water is an open system. Natural

Table 2.3 Example of aquaculture in combinations of different levels of culture intensity together with open, static, semi-closed and recirculating (closed) systems

	Culture intensity		
System	Extensive	Semi-intensive	Intensive
Open	Long-line culture of scallops Table oysters in mesh baskets	Enclosures within freshwater lakes for culturing fish, such as tilapia	Seacage culture of fish, such as sea bream, sea bass, Atlantic salmon
Static (ponds)	Carp polyculture in ponds Milkfish culture in lakes	Freshwater fish polyculture with aeration	Saltwater crocodiles
Semi-closed	Mudcrab culture in tidal ponds in mangroves	Pond culture of shrimp at 20 animals/m^2	Pond culture of shrimp at > 50 animals/m^2
Closed	X	X	Indoor culture of high-value fish

X, very uncommon.

currents and tides maintain water quality. The bivalves gain their food by filtering phytoplankton from the water flowing past. These are extensive open systems (Table 2.3).

Open systems tend to have low operating costs, as there is no requirement for pumping. Capital costs vary greatly depending on the type of culture, with bivalve culture systems generally of low cost and intensive fish cage culture systems of high cost. Generally, sites for open water culture are not available for freehold purchase and must be leased from the appropriate government agency (e.g. oyster leases).

Open systems are prone to problems that either do not apply to or are more difficult to mitigate than in other culture systems. A major problem associated with site selection for open systems is lack of control over water quality. The quality of water depends on local factors and cannot be moderated. It is therefore essential that the farmer is aware of all extremes of water quality occurring at the site, i.e. water temperature, salinity, algae blooms, etc., prior to developing an aquaculture facility. Seasonal variations in environmental factors may result in large variations in growth rates, and local differences in the environment may also cause major differences in growth and survival. Open systems are also more prone to predation and disease. Predation can be controlled through the addition of some protective devices; however, methods of predator control in most countries must be non-destructive and are generally expensive to operate and maintain. Predation may also extend beyond wild animals to human interference or poaching. Although interfering with aquaculture equipment and stock is an offence in most countries, enforcement is difficult, and the responsibility to protect equipment and stock often falls to the system operator.

2.4.3 Semi-closed systems

Semi-closed systems involve ponds, tanks and raceways where culture water is confined in discrete units. These systems fall distinctly between static and open systems in terms of water exchange with adjacent water sources. There is a degree of water exchange in semi-closed systems that is substantially greater than in static systems and much less than in open systems. Another major difference between semi-closed and open aquaculture systems is that the culture water is continuously or frequently brought to the farm. The water source may be freshwater, brackish or marine. Characteristics of freshwater and brackishwater sources are outlined in Tables 2.4 and 2.5 and aspects of marine water supply are outlined in section 2.7.

Water is drawn from a reliable source and flows to and through the farm, driven by gravity, tidal exchange or pumping. In these systems, water is exchanged to maintain water quality. As the farm is not located within a natural aquatic environment, there is a degree of control over water quality, but only to the extent that water flow can be increased, decreased or stopped. If the source of water becomes contaminated or of unacceptable quality, the farmer can only stop the water flow, leaving the stock in water with deteriorating quality. The benefits of semi-closed systems vary from

- enhancing production from ponds by exchanging some water while maintaining some reliance on natural processes of the pond ecosystem, to
- complete reliance of water quality on water exchange (e.g. raceways) resulting in a large increase in production.

In the latter, water use per unit of production is extremely high. In these systems water may make a single or multiple passes through the culture structures.

If water is exchanged by pumping, the costs of pumping may be high, depending on the height water has to be pumped and the volume of water exchanged (section 2.5.1).

In large semi-closed systems with semi-intensive to intensive culture (Table 2.3), water flow is generally high to very high. It is generally recommended that water exchange is in the range 5–10% per day for semi-intensive ponds (Boyd, 1991) and up to 30–40% per day for intensive pond (Yoo & Boyd, 1994). An exchange rate of 1% per day is equal to 100 m^3/ha per metre of average pond depth (Yoo & Boyd, 1994). That is, an exchange rate of 5% per day in a 1-ha pond of average depth 2 m requires a total of 1000 m^3 of water per day. In addition to this, water loss from evaporation and seepage must be

Table 2.4 Freshwater sources for semi-closed culture

Water source	Advantages	Potential problems
Lakes and reservoirs	Large volumes available	Susceptible to climactic changes and pollution Predators, competitors and pathogens may be present
Streams or shallow springs	High oxygen content	Highly variable supply and water quality as a result of climatic changes Susceptible to pollution Predators, competitors and pathogens may be present
Deep springs	Quite constant supply Sediment free	Low oxygen Supersaturation with N_2 gas Sulphides
Wells	Quite constant supply Sediment free	Yields difficult to predict High pumping costs May deplete groundwater Supersaturation with N_2 gas

Table 2.5 Brackishwater source for semi-closed culture

Water source	Advantages	Potential problems
Estuary	Reliability May rely on tides for water exchange in small ponds	Seasonal variation in salinity Tidal affect on pumping period (if pump is sited above low-tide level) Variable suction lift with tidal range Susceptible to pollution Predators and competitors will be present Pathogens may be present High sediment load

considered. Boyd (1991) provides further details on determining exchange rates.

2.4.4 Recirculating (closed) systems

Recirculating systems are usually characterised by minimal connection with the ambient environment and the original water source (Table 2.6). These systems have minimal exchange of water during a production cycle, hence the description as 'closed' systems. Water is added to offset the effects of evaporation or incidental losses or, more frequently, to maintain water quality. Some water is discharged and replaced each day in most recirculating tank systems with intensive culture (Losordo, 1998a,b). This arises from aspects of the regular maintenance system, such as removing accumulated solids from filters. Water quality in completely closed tank systems with intensive culture is much more difficult to maintain than in systems in which there is a regular 5% or more replacement per day (Losordo, 1998a,b). Even with some limited water exchange each day, water quality within a recirculating tank system will only be maintained by artificial manipulation. Losordo (1998a,b) and Losordo *et al.* (2001) reviewed these systems and their critical considerations, status and future.

The cost of construction and production in intensive recirculating tank systems has limited the commercial development of these systems for grow-out production. However, the possibility of high yields with year-round production close to markets drives their development. Some of the advantages and potential disadvantages of recirculating systems are outlined in Table 2.6.

The artificial means of waste processing and some typical components used in recirculating systems are shown in Fig. 2.10. Feed input, animal metabolism, wasted feed and faeces production all impact upon water quality. Parameters that require regulating in an intensive recirculating tank system are:

- particulate matter (settleable, suspended and fine waste solids) in the system resulting from feed and faeces
- nitrogenous wastes (unionised ammonia, ionised ammonia, nitrite and nitrate, which are often expressed as NH_3-N, NH_4-N, NO_2-N and NO_3-N respectively)
- dissolved gases (O_2, CO_2 and N_2)
- pathogens
- pH
- alkalinity.

For more detail on each of these aspects of water quality refer to Chapter 3.

At high stocking densities without recirculation technology, a water exchange in excess of 100%/h would be required to maintain water quality during maximal production. This is an unsustainable amount.

In recirculating tank systems, water quality is

Table 2.6 General characteristics, advantage and potential problems of recirculating aquaculture systems in indoor conditions

Characteristics	Advantages	Potential problems and disadvantages
Limited water exchange outside system	Rigorous control is possible	High capital costs
Rigorous monitoring and control of crucial water quality parameters	Unaffected by weather	High running costs (pumping, filtration, maintenance, etc.)
High densities of culture animals	Predators and competitors are easily excluded	Requires very careful management (no buffering capacity)
High yield per unit volume and per unit surface area of water	Pathogens are less likely to infect	A pathogen outbreak will be severe
	Efficient feeding	
	Minimal environmental impacts	
	Requires minimal water	

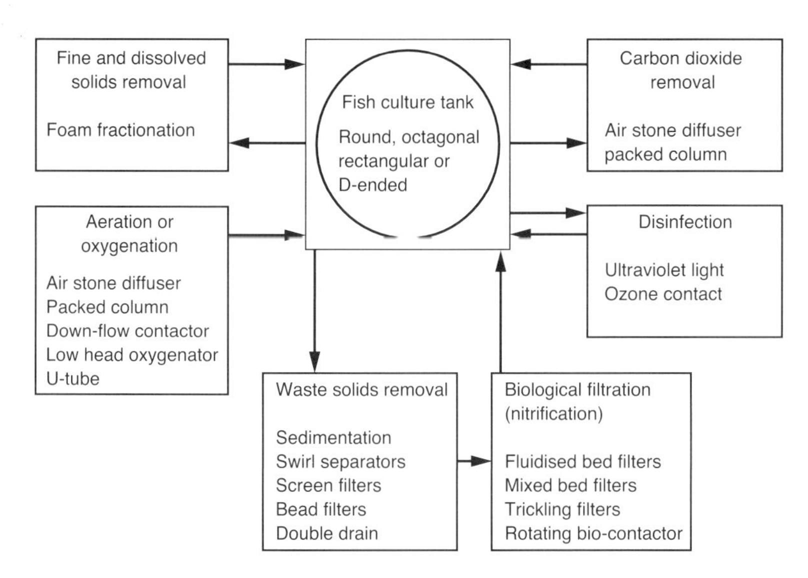

Fig. 2.10 The water treatment processes and typical components of a major recirculating system for intensive culture. Reproduced from Losordo *et al.* (2001) with permission of the World Aquaculture Society.

maintained by pumping the culture water through specialised filtration and aeration equipment. The components of intensive recirculating tank systems are described below. Some further descriptions of the various components of these systems are provided by Losordo *et al.* (2001).

Waste solids filtration

The method used for removal of solid wastes depends on the type of waste. Settleable solids sink and may be removed by gravity and flow on a continuous basis or by siphoning on a regular basis. Alternatively, settleable solids can be kept in suspension with continuous agitation from aeration, water flow and stock movement, and removed using a mesh screen or allowed to flow into settlement basins or tanks. Settleable solids can also be removed by centrifugal force using a swirl separator or hydrocyclone.

Suspended solids that will not settle out by gravity are generally removed from recirculating tank systems using mechanical filtration.

(1) Screen filters are made of a fine mesh through which water flows and the suspended solids are retained on the screen. The screen filters are then periodically or continuously cleaned. Types of self-cleaning screen filters include rotating screen filters and rotating drum filters (Fig. 3.9).
(2) Particulate filters, include sand filters and expandable granular media filters. Particulate filters also require cleaning, a process known as 'backwashing'.

Fine and dissolved solids cannot be removed using gravity or mechanical processes. Foam fractionation, or protein skimming, is often employed to remove the solids which adhere to the water–air interface created by vigorous bubbling of air through the water.

Ammonia-N and nitrite-N removal

There are numerous technologies available for removing ammonia-N from water, including air stripping, ion exchange, biological filtration and removal by algae (Fig. 3.8). Air stripping requires the water pH to be adjusted to 10 prior to stripping and then readjusted to culture levels near 7 prior to water re-entering the rearing tanks. Ion exchange technology is costly. For these reasons, biological filtration is still the most widely used method. Biological filters are available in a variety of forms including trickle filters, submerged filters, rotating biological contactors, packed tower filters and fluidised bed filters (section 3.2.7, Figs 3.2 and 3.3).

Dissolved oxygen

Aeration is the dissolving of oxygen from the atmosphere into water. It is an important aspect of recirculating tank systems, as it is necessary to maintain dissolved oxygen (DO) at the required level for the cultured animals. It is also important to maintain dissolved carbon dioxide below 20 mg/L to maintain favourable pH, reduce stress and maximise growth (section 3.2.4). There are several types of aeration devices commonly used in recirculating tank systems, including diffuser aerators, mechanical aerators, vertical pump aerators and packed column aerators.

Oxygenation is the process of pure oxygen transfer to water and is used when oxygen consumption is greater than the capacity to transfer oxygen through aeration using air (which is only 21% oxygen). Oxygen can be supplied as compressed oxygen gas and liquid oxygen from commercial sources or it can be produced on-site by oxygen generators. The choice is generally one of cost and reliability of electricity supply to produce oxygen on site. As the transfer of gas to water using diffusers is poor (less than 40% for oxygen), specialised components to deliver oxygen to recirculating tank systems have been developed. These components can effectively transfer greater than 90% of oxygen to the water. A variety of oxygenation devices are available.

Pathogens

Continuous disinfection of water in a recirculating tank system can help to limit the introduction and spread of disease, and the build-up of pathogenic bacteria, e.g. *Vibrio* species, to levels at which they cause mortality in the stock. Ultraviolet (UV) irradiation and ozonation are two methods of continuous disinfection used in recirculating tank systems (section 3.5.2).

This section has dealt primarily with recirculating tank systems used for intensive culture, but larger-scale recirculation systems are now being developed for pond culture. In ponds with intensive culture, e.g. shrimp culture, it is difficult to operate without regular water exchange. However, pond systems in which there is very limited or no water exchange throughout a growing season are now being developed (section 19.10.5).

2.5 Plumbing and pumps

This section falls within the scope of aquaculture system engineering. Further detailed information on aspects of these topics may be obtained from an appropriate text, e.g. Wheaton (1977), Huguenin & Colt (1989) and Lawson (1994). Wheaton (1977) is still regarded as the classic text in this field.

2.5.1 Pipes

General

Pipes, channels and pumps are basic components of most aquaculture systems, except those that are open and static. Water can be conducted around an aquaculture facility using either pipes or open channels. The advantages of pipes include the ability to pump water at pressure and up a pressure gradient. The disadvantages of pipes compared with open channels include larger construction costs per unit water flow, larger frictional losses causing higher pumping costs and lower achievable rates of water flow. However, channels cannot deliver water at pressure or up a pressure gradient. As such, pipes are more commonly used for the delivery of water, except when very large flow rates are required, and channels are employed largely as drains.

The aim of pipe selection is to obtain the optimum size of pipe (usually expressed as internal diameter in mm), the correct pipe material, and the best pressure ratings to meet the pumping pressure and the flow capacities required for the facility.

When selecting pipes for a given flow there is a choice between

- smaller pipes with higher flow velocity and pressure and
- larger pipes with lower flow velocity and pressure.

The advantage of using smaller pipes is that the capital cost of construction is much reduced, with the cost of pipes increasing rapidly with pipe diameter. The disadvantages of using smaller pipes are that they require greater water velocities for a given water volume and produce higher frictional forces and pumping costs. Higher frictional forces also result in larger pressure heads to be pumped against, increasing the strength (or class) of pipe required and, to an extent, increasing the capital cost of construction. Selection of smaller-diameter pipes during initial construction also greatly reduces the ability to expand the capabilities of the system in the future.

In addition, the higher the flow rate the greater the possibility of water hammer occurring (p. 32). As a rule, it is better to have the water velocity as low as possible within the constraints of the capital costs of construction.

If the flow rate and flow/water velocity required are known, then the required pipe can be determined. Manufacturers of thermoplastic piping will provide charts, known as nomograms, to facilitate the correct choice of pipe by cross-referencing flow rate and water velocity. Pipes, however, are manufactured in discrete sizes and pressure-class ratings (i.e. the maximum pressure a pipe can work under), not all of which are readily available or provide the same selection of fittings. In general, the velocity criteria will not provide a clear decision on the best pipe to use and will result in a choice of two sizes. The wise decision is to be conservative and select the larger pipe.

Another conservative measure in system design is to duplicate pumps, major plumbing and other components of the system. Even in the best-maintained systems there may be abrupt equipment failures. Duplication allows the problem component to be bypassed without having to shut the system down while repairs or replacements are effected. Furthermore, there are a number of routine maintenance activities that can be conducted without otherwise having to shut the whole system down. Some of these activities are:

- servicing pumps and other components in the system
- back-flushing filters to remove sediment or otherwise renewing filter surfaces
- clearing intake pipes of bio-fouling
- flushing general plumbing to clear out accumulated sediment.

Frictional losses

Frictional losses are due to factors such as flow type, flow rate or velocity, pipe length and the roughness of the internal surface of the pipe. In aquaculture systems, flow is turbulent, creating internal eddies that collide with the side of the pipe, causing increased internal friction and, as a result, increased energy losses. Flow rate depends on the water requirements of the facility, while flow velocity results from the required flow and the pipe diameter. Pipe length is a result of facility layout and design. Roughness of the internal surface of the pipe is due to sand scour factors, scale formation, sediment deposit and bio-fouling. Frictional losses will also vary with salinity and temperature, but these changes are only slight.

Frictional losses are generally measured in head (elevation units, m) or pressure difference. Frictional losses for pipes are available in nomograms and flow charts from the manufacturer for standard conditions. Wheaton (1977) and Yoo & Boyd (1989) provide further information on how each of the above factors affects friction and, therefore, how decisions on pipe size and flow rates may affect the operating of the system.

Fittings

Plumbing fittings are the pieces that

- join pipes or in-line components to the line
- redirect and split the flow
- change the pipe diameter.

All these fittings provide resistance to the flow and must be allowed for in the calculation of friction within the system.

Losses due to fittings depend on:

- type of fittings or transition (e.g. bends, couplings)
- abruptness of flow change
- materials and condition of inner surface
- velocity of water movement.

In addition to fittings, there may be dramatic head losses from in-line components (e.g. UV sterilisers and sand filters). These effects on flow rate are given by the manufacturers and are described by Wheaton (1977) and Yoo & Boyd (1994).

Choice of plumbing materials

For the majority of applications, and in particular for delivery of water over extended distances with pipe diameter of less than 200 mm, the choice of plumbing materials comes down to two types of general-purpose, cost-effective thermoplastics: rigid PVC (polyvinylchloride) and high-density polypropylene. For special applications, e.g. in outdoor areas exposed to high UV levels, high temperature (> 50°C) or high abrasion due to water-borne sand, and in high-impact areas, especially adjacent to large pumps, acrylonitrile butadiene styrene (ABS) pipe can be used. When the internal diameter exceeds 200 mm, and especially in high-impact areas, concrete and steel tend to be used as these are more durable and cost-effective.

Two common problems in plumbing design

Two of the more common problems in plumbing design are water hammer and bio-fouling.

Water hammer is a transient pressure phenomenon caused by rapid stopping or changing in direction of flow (Wheaton, 1977; Yoo & Boyd, 1994). The change in flow creates a wave of backpressure, resulting in a loud bang, hence water 'hammer'. It is the noise that may be heard in domestic plumbing when taps are turned on. The pressure wave is added to the existing pressure in the system and has the ability to rupture plumping or permanently damage piping, pumps and in-line equipment. The magnitude of the pressure wave is related to the velocity of water in the pipe, pipe length and rate of change of flow.

The risk of water hammer increases with:

- water velocity
- the length of uninterrupted pipe
- the abruptness of flow change.

As a result, it is recommended that water velocity in pipes be kept below 1.2 m/s. It is also important to be aware of high-risk areas in plumbing design. These include design parameters creating flow change at the end of a long length of uninterrupted pipe.

In aquaculture systems, water hammer often originates from rapid opening of a line resulting in a high-velocity flowing down the pipe. This water eventually encounters a 90° bend or, perhaps, a mostly closed valve, resulting in a sudden change in flow and water hammer. Rapid opening of a line can result from either turning on a valve too quickly or having a pump switch on. In these circumstances it is important to turn valves on slowly. Similarly, the use of soft-start pumps reduces the risk of water hammer from pumps starting. The risk of water hammer can also be mitigated by ensuring that the receiving pipe is full of water, so reducing the resultant flow from the pump or ensuring that there are no sharp bends or closed valves in the path of the flow. As with opening a valve too quickly, closing a valve in a flowing line too quickly will enhance the risk of water hammer. This is easily corrected by identifying and marking critical valves or using valves that require several turns to shut them, e.g. gate valves, thus extending the time of closing.

The adverse impact of bio-fouling organisms, algae, sponges, barnacles and other shelled animals, has been indicated for cages. Bio-fouling also occurs on the inner surfaces of aquaculture plumbing. It is a major factor influencing performance in seawater systems, but has a lesser effect on freshwater systems. Bio-fouling roughens surfaces and thereby increases the friction on the inner surfaces of pipes and fittings. It also blocks protective grills and pipe cavities. Considerable effort is required to reduce bio-fouling in seawater systems, and it can only be controlled by regular maintenance. It is important during maintenance to remove both the fouling organisms and their calcareous shells, if present. Many methods, e.g. flushing with freshwater and hot water, and anoxia (by having stagnant water in the plumbing for a considerable period), simply kill the organism but do not remove the shells. The most common and effective method of controlling bio-fouling is to remove the fouling organisms by forcing a rigid structure through the pipe under pressure. These devices are known as 'pigs' or 'snakes', and access for pigs or snakes into and out of the system must be taken into account during system design, e.g. by having T-junctions instead of 90° elbows. Under pressure, the pigs or snakes scrape the internal surface of the pipe, removing both organisms and exoskeletons.

2.5.2 Open channel design

When high water flows are required either in to or out of a system, and there is no need to have a pressurised flow, as water can flow by gravity, it is cost effective to replace pipes with open channels. The advantages of open channels over closed pipes have already been considered, but there is a further advantage of the high flow capability allowing them to handle high transient flows (i.e. burst pipes and emergency draining of tanks or ponds). There are also the advantages of ease of access for routine cleaning and disinfection, and potential to be used as a harvest sump.

Drains are generally rectangular, semicircular or U-shaped. They are generally earthen or concrete. Concrete is often used in earthen channels to protect areas vulnerable to erosion, such as the portion of the channel immediately after the delivery pump. Concrete channels are also used for tank systems, both indoors and outdoors. The volume of water that can be moved by a channel is related to the size, shape, construction material and slope of the channel. Further information on how each of the above factors affects the maximum flow for a channel design may be obtained from Yoo & Boyd (1994).

2.5.3 Pumps

An important aspect of aquaculture systems is the correct choice of pump. Although this task is best performed by qualified personnel, it is important to know the types of pumps available and understand factors affecting pump selection. Pumps and pumping are covered in detail by Wheaton (1977) and Yoo & Boyd (1994).

The variety of pumps used in aquaculture can generally be classified into two groups:

- mechanically driven (positive displacement and rotodynamic pumps)
- air driven (airlift and air/water ballast pumps).

The majority of pumps used for water delivery in aquaculture systems are rotodynamic pumps. These pumps work by continuously imparting energy on the water using a rotating impeller, propeller or rotor. There are three types of rotodynamic pumps.

Centrifugal (radial flow) pumps

Water enters the impeller axially (along the axis of rotation) and exits radially in centrifugal pumps. The pressure is due to the centrifugal force of the impeller on the water and the water being contained by the casing of the pump. Centrifugal pumps have a high-lift/low-volume characteristic, being able to pump against heads > 15 m. They also have the best suction lift characteristics; however, they perform best if located as near to the water source as possible. Although centrifugal pumps will draw water, they will do so only if they are primed, that is if the line on the suction side and the pump is full of water.

Axial flow pumps

In axial flow pumps, water enters and leaves along the shaft axis (axially). The pressure is produced by the action of propeller blades (vanes) directly on the water. Axial flow pumps are used for low pressure head, generally < 7 m, and high volume rate, > 1 m^3/s. A limitation of these pumps is that they will not draw water, so the propeller must always be submerged in the water.

Mixed flow pumps

In mixed flow pumps, water enters axially and discharges axially and radially. The pressure developed is due to a combination of centrifugal force and lift due to propellers. These pumps are suitable for medium pressure head, 5–15 m, and flow rate applications.

Pump components

Pumps are an indispensable part of an aquaculture system, and understanding their structure and maintenance is important for aquaculturists. There are two main elements to a rotodynamic pump. These are the rotatory element (impeller or propeller on the shaft) and the stationary elements (casing, stuffing box and shaft bearings).

Centrifugal pumps (Fig. 2.11) have the following components:

(1) Casing. The casing converts water velocity leaving the impeller into pressure head. There are two types of casings used in centrifugal pumps. Volute casings spiral out from the centre, increasing casing volume with the spiral, resulting in a decrease in the velocity of water flow and the creation of pressure. Diffuser casings direct water through a set of diffuser vanes. The diffuser allows the conversion into pressure of a greater amount of energy imparted on the water, resulting in higher efficiency.
(2) Shaft and bearings. The shaft transmits torque of the motor to the impeller. The shaft is supported by bearings to centre the shaft and impeller both radially and axially and to allow for the rotation of the shaft. Bearings may be either sealed or water lubricated.
(3) Mechanical seals (or stuffing box). The mechanical seals prevent water from leaking around the shaft, which can result in deterioration of the bearings, shaft and motor. Seals are usually rings cut in soft plastic and fitted tightly around the shaft and shaft sleeve. Seals are cheap and readily replaced. Spare seals for crucial pumps are essential to have in stock.
(4) Impellers. There are three types of impellers and the impeller best suited to the pump is generally governed by the requirements of water to be pumped. Open impellers have vanes that are attached to a central hub and supported by ribs. A small area of faceplate results in large clearances, which allow water-borne solids to be pumped. However, open impellers are not very efficient because they allow slippage of the impeller through the water. Semi-closed impellers have a complete faceplate to which the vanes are attached. This means that the efficiency of the pump is increased, but they do not handle solids as well as open impellers do. Enclosed impellers have a complete faceplate attached to each side of the vanes. Enclosing the water reduces slippage and thus increases the efficiency of the impeller. However, the small pathways for water movement result in a poor ability of these pumps to handle solids within the water.

Axial flow pumps are simpler in construction than centrifugal pumps, basically being propellers inside a pipe with a motor-driven shaft. As described above, these pumps are best for low head, although multistage pumps can create head of up to 40 m, and they are used for high-volume pumping. The low head pressure of these pumps means that they are not

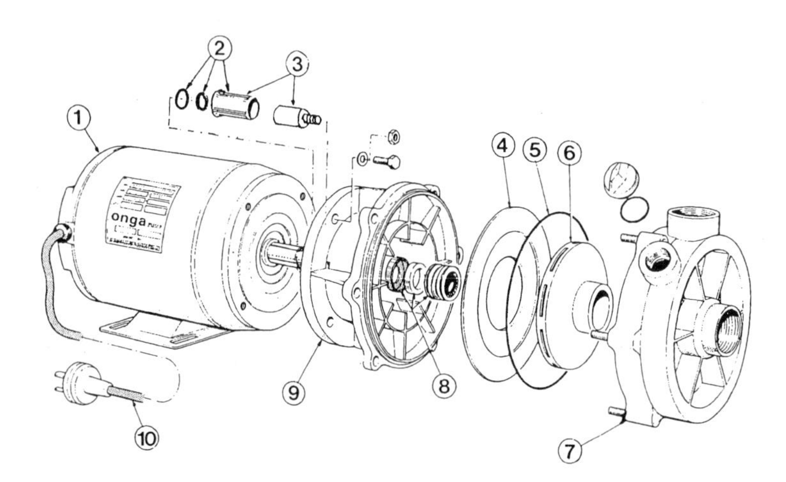

Fig. 2.11 A diagram showing the assembly and major components of a centrifugal pump. The major components include: (1) motor; (2/3) shaft sleeve assembly; (4) baffle; (5) 'O' ring; (6) impeller; (7) casing; (8) seal; (9) yoke; (10) power lead. Reproduced with permission of Onga Pty Ltd.

used for pipe distribution, their main use being for discharge into open channels or directly into ponds.

Pump selection

Apart from, or as well as, using the services of an expert, aquaculturists are required to select the type, make and model of pump, and alternative impeller size and speed, that will suit their system. This decision is aided by performance data for pumps supplied by the manufacturers. For a given pump body, data are presented for a number of different impeller sizes and pump speeds (Wheaton, 1977; Yoo & Boyd 1994).

Manufacturers will generally provide pump performance data in the form of head/flow capacity (*H–Q*) curves. *H–Q* curves are the graphical representation of the flow output of the pump against the total head (friction and elevation forces) against which the pump is acting. The shape of the *H–Q* capacity curve is critical. The curve must decrease uniformly from the high-head/low-flow to the low-head/high-flow conditions. If the curve has a flat spot or decreases abruptly at low flow the pump may not work smoothly in this range. In addition to the *H–Q* curve, the manufacturer's information will also often include efficiency data, allowing decisions about whether the selected pump will operate near its maximal efficiency in the system. The 'brake power' of the pump may also be represented. Brake power represents the power required to operate the pump at a certain flow and head. From this it is possible to determine the operating costs of different pumps by multiplying the operating time (hours) by the brake power for the pump and then multiplying this value by the kilowatt per hour charge of the electricity supplier.

Net positive suction head

The net positive suction head (NPSH) and total dynamic head (TDH) are required to determine the size of pump required (Wheaton, 1977; Yoo & Boyd, 1994). The latter will be treated in the next section.

In order for rotodynamic pumps to work, the water must be drawn to the pump under pressure. NPSH is the pressure on the suction side of the pump. The NPSH present within a system is the available NPSH (NPSH-A). The minimum NPSH required for a pump to operate effectively is the required NPSH (NPSH-R). NPSH-A can be calculated using the following formula:

$$\text{NPSH-A} = H_a + H_s - H_{fs} - H_{vap}$$

where H_a is atmospheric head (pressure), the pressure of the atmosphere on the surface of the water pushing the water to the pump. H_s is static suction head, the elevation difference between the pump and the water supply. It is negative if the water source is below the pump and positive if the water is above the pump. H_{fs} is the sum of the frictional losses from the pipe and plumbing on the suction side of the pump. H_{vap} is the vapour pressure of the fluid.

However, the values required for calculating NPSH-A are not all fixed, i.e.

- H_{fs} varies with flow rate and bio-fouling.
- H_s varies with height of tide, river, reservoir.
- H_a varies with weather conditions.

For this reason NPSH-A must be calculated for a range of conditions and be greater than NPSH-R for all of them, with a safety margin of at least 1 m. Failure to do so is the single greatest source of pumping problems. If NPSH-R is greater than NPSH-A, the pump will produce little or no flow or will cavitate because of vaporisation.

Total dynamic head

TDH is also referred to as system head, differential head, generated head and total head. Total dynamic head is the height difference between the water source and the outlet plus the frictional losses, i.e.

$$\text{TDH} = H_{st} + H_{ft} + H_v$$

where H_{st} is the total static head, the difference between elevation of discharge and water supply; H_{ft} is the total frictional head of pipes, fittings and in-line components on the suction and discharge sides of the system; and H_v is the velocity head.

When calculating TDH, it is important that the

range of flows used in the system is included, as frictional losses in pipes and fittings and the velocity head are proportional to water flow and velocity. Static head will also vary with the height of water in the water source, e.g. with tide height. Determining TDH over the range of flows and static head will enable the 'working range' of the pump to be determined.

A pump may be selected after establishing the flow rate, TDH and NPSH-A requirements.

2.6 Site selection and development

The primary objective of aquaculture system design is to ensure that a facility is productive in a reliable and cost-effective manner. As a result, the design of a system must allow for:

- maximal growth and survival of the cultured species
- optimal market price for the product
- minimal operating/maintenance costs.

Prior to designing a system, it is important to recognise for whom and what the system is to be designed. There are three major users of an aquaculture system: the target species, the operators of the system and the owners. To accommodate the target species, the design must allow the animal to grow and survive optimally. In order to facilitate this, the operators require a system allowing minimal operating and maintenance requirements. It must also be reliable. The owners require that the system be cost-effective to construct and operate. An appropriate design needs to:

- meet the species' requirements
- allow flexibility
- provide for low maintenance
- allow for the use of inexpensive materials without compromising quality
- incorporate back-up systems where possible
- allow for future expansion.

As seen earlier in this chapter, aquaculture systems vary immensely (outdoor ponds, indoor tanks, seacages, long-lines, etc.), but similar forms of aquaculture production have facilities in common. For instance, a semi-closed tank or pond system will include:

- water supply
- culture structure
- aeration
- power
- buildings
- layout.

Within the above constraints, the final system choice and design, as always, is a balance between species and user requirements, available capital for construction and site constraints. When designing an aquaculture facility it is important to consider the layout and design of the farm to minimise constructions and operating costs, and to ensure reliable operation and production. As a result, the following issues need to be considered at the earliest stages of site selection and planning.

The general site selection considerations and designs of systems for culturing particular species or groups are covered in other chapters, e.g. Chapters 13–22. In this chapter only farm layout will be discussed in detail.

2.6.1 Movement of stocks around the site

Handling stock causes stress, reducing growth and increasing susceptibility to disease. It should therefore be minimised. Movement handling can be reduced by effective farm design. Movement can be facilitated by linking culture sites. Ponds can be linked via canals or by building ponds within ponds. Seacages have been designed in which cages are sewn together for stock transfer. Tanks systems can also be connected through the overflows to allow stock to swim from tank to tank as required. Alternatively, stock handling can be reduced by growing animals in a single structure. It must be noted, however, that if the animals require handling for other husbandry reasons, e.g. grading, then using a single structure is of limited value.

2.6.2 Production patterns

When designing a facility, it is important to know the stages of production and the changes in the load (amount of product) within the various components of the system over the crop cycle and throughout the year. For example, if a system is stocked once a year, the load on a system at stocking is vastly different from the load at final production. As a result, the water and system requirements at the start of the crop will differ from those at the end of the crop.

It is also important to know whether the quantity and quality of water available throughout the year will vary. Consideration must be given to how these will be managed in relation to the crop cycle. It may be necessary to construct a dam and reservoir to enable excess water to be stored for drier months or when the required flow rates are higher.

Therefore, it is necessary to try to define the pattern of production as accurately as possible at the outset, allowing the sizing of components and the determination of project costs and site layout. Similarly, it is important to be able to vary pumping capacity to reduce excess capacity and therefore operational costs at times of low water requirement.

2.6.3 Water flows

Apart from ensuring the quantity and quality of water available to the system, the distribution of water around the site must also be considered. It is generally appropriate to utilise the shortest routes for water delivery and removal to save capital costs and, if piping is used, to reduce pressure losses due to friction and therefore pumping costs. In addition, it is preferable that the water supply system be controlled from a centralised point. In a pond system this may be as simple as being able to turn the water supply to a pond on or off without travelling to the pump house. Where possible, water flows to tanks, raceways and ponds should be able to be adjusted individually.

2.6.4 Use of existing topography: water head/pump capacity

All requirements of the farm are important when considering the topography. Topography can be utilised to reduce pumping costs, collect and store water from natural watersheds and provide sites for buildings.

'Water head' is the ability or the potential for water to fall, i.e. water will drop from a higher point to a lower point through gravity with no requirement for pumping. The provision of pumping is expensive in both capital construction and running costs. It is important to use water head efficiently in farm design. If using ponds or raceways, the design must utilise the slope of the land so that water may be supplied and drained or drained by gravity, i.e. let the water 'fall' down rather than pumping it up. The greater the height that water must be pumped and the greater the distance of piping, the greater the operating costs without improving productivity.

2.6.5 Utilisation of land and/or water

As land or water is generally expensive to purchase or lease, it is important to minimise non-productive space wherever possible. Minimisation of wasted space is as important for recirculation systems as it is for pond farms or marine leases. In general, it is considered prudent to minimise the distances between culture structures (ponds, tanks, raceways, long-lines) and the size of car parks or working areas and buildings.

However, although it is important to minimise unproductive space, there must be sufficient working space to allow:

- access to culture structures and pipework for routine activities
- vehicle access to work areas
- pumps and treatments systems for maintenance and adjustment for routine work (e.g. repairing equipment and cleaning/drying nets) and non-routine work or emergencies to be conducted
- animal husbandry (e.g. grading) and product processing (e.g. harvesting, slaughtering, packing)
- a buffer zone around the aquaculture facility, including an 'environmental protection zone'.

It is important to locate harvesting and processing

facilities as close as possible to the harvest site. This will minimise transport requirements, but more importantly it will aid in maintaining product quality, especially in hot climates. Processing facilities must be kept away from hatchery facilities to prevent contamination of the hatchery area by diseases from the on-grown stock (section 2.7.2).

Apart from the areas directly related to growing the product, areas for the delivery, storage and handling of goods such as feeds, fertilisers and chemicals must be considered. Thus, road design must meet all access requirements while minimising construction costs. Consideration must be given to the loads entering and exiting the property as well as the types of vehicles used. Although small vehicles may be used on the farm, transport companies, e.g. feed companies, may deliver using much larger vehicles. When deliveries are purchased in bulk, sufficient space must be available to store and handle the goods. In conjunction with access it is important to ensure that the working yard and turning area for the delivery of bulk supplies and the transport of harvested product is sufficient.

Expansion is often the target for aquaculture ventures, and the ability to expand facilities and operations beyond the initial development must be considered during the site selection and layout design. Where possible, adjacent land or water or land and water must be available to lease or buy in the future. Alternatively, if land or water is cheap enough, it can be purchased to allow for future expansion. The site must be engineered so it can be easily modified for increased capacity. For example, consider using larger pipes initially to avoid replacement costs later.

Although it may be thought of as 'unproductive space', a 'buffer zone' may be required as part of the licensing requirements from environmental agencies for some forms of aquaculture. A buffer zone around a pond farm may be required as protection against escapees, predators and vermin. Spare ground is also needed for disposal of the pond substrate wastes between crops. In seacage and high-biomass bivalve aquaculture, e.g. mussel rafts, there is the potential for heavy fouling of the benthos beneath the facilities (section 4.3.1; section 21.6). Pollution can be minimised if the cages and rafts can be moved to new areas within the lease.

2.6.6 *Minimisation of construction costs*

When constructing buildings of similar grades, it is generally cheaper to build larger buildings and then divide them into smaller sections and rooms. However, one needs to consider which activities can be conducted within close proximity to each other. Similarly, larger ponds are cheaper to build per unit area than smaller ponds. The manageable size of the pond, however, and the optimum size for growth performance of the cultured species must be considered.

The specification of buildings (i.e. the strength and sizing of materials) and pond walls (i.e. wall height, width and internal and external slopes) is also important. Costs of construction increase rapidly if overspecification occurs. As such, it is important to consider the appropriate specifications for the required tasks.

2.7 Hatchery systems

Hatcheries are those aquaculture facilities associated with reproduction, larval rearing and supply of juveniles to farms. A single hatchery often supplies a large number of farms, although many aquaculture industries (e.g. the oyster industry) still rely heavily on natural recruitment as a source of culture stock; natural recruitment can be unpredictable and unreliable. As well as providing independence from natural recruitment, other benefits of hatchery production include the potential for genetic improvement of culture stock (Chapter 7) and the reduced potential for conflict with capture fisheries.

Hatcheries have similar site requirements to nursery and grow-out sites (section 2.6). Of particular importance is the quality of intake water, which must be free from industrial, urban and agricultural contaminants. The water intake pipe is usually screened to remove foreign items such as seaweeds from intake water. Intake water should also have low levels of suspended solids, to minimise filtration requirements and bacterial contamination, and a temperature as close as possible to that to be used for larval rearing to minimise water heating/cooling costs. A hatchery should also be able to pump water regardless of the state of the tide and should be located close to the farms it supplies.

Marine aquaculture hatcheries may be positioned to use either an oceanic or estuarine water supply. Oceanic water is generally considered to be of higher quality, but such sites may be subject to high wave action and erosion that may lead to problems with water supply. Use of such sites may also result in public resistance on aesthetic grounds. Estuarine water sources are subject to rapid changes in temperature and salinity, which can result from heavy rainfall. Rainfall may also increase levels of suspended solids and bacteria in intake water. Both factors affect water quality and will impact on larval well-being. Freshwater hatcheries generally use either groundwater (e.g. from an aquifer using a well or bore) or surface water (e.g. streams, rivers and lakes) as water sources. It is important that marine hatcheries also have a reliable source of freshwater for cleaning and washing.

2.7.1 Water treatment in hatcheries

Water used for larval rearing is usually treated to reduce particulate matter and bacteria, and heated/cooled to optimise water temperature. Water treatment in hatcheries usually includes the following steps:

(1) Water storage. Water is usually stored in large tanks (e.g. 20–100 000 L) before being pumped into the hatchery. This also allows suspended solids to settle from the water column.
(2) Coarse filtration. Water is passed though filtration equipment (e.g. sand filter), which removes larger particles and filters water to around 20 μm.
(3) Heating. Water may be heated before use. This may be done by heating a common water source or 'header tank', from which larval culture tanks are supplied. However, water may also be heated using heat exchange systems (section 3.5.1) or larval culture tanks may be heated individually using immersion heaters.
(4) Fine filtration. Before entering larval culture tanks, water is usually passed through fine filtration equipment (e.g. cartridge filters), which filters the water to ~1 μm. This is generally achieved using a series of filters of different sizes in sequence (e.g. 10 μm, 5 μm and 1 μm). Water used for live food culture (e.g. microalgae; section 9.3.1) is filtered to finer levels of 0.2–0.45 μm.
(5) Ultraviolet (UV) treatment. Water to be used for larval culture and live food culture may also be passed through a UV steriliser to reduce levels of bacteria. Water must always be filtered to remove suspended particles before being passed through a UV steriliser because particulate matter is the major substrate for bacteria in culture water.

General considerations with regard to water quality and water treatment in freshwater hatcheries for salmonids are given in section 15.3.2.

2.7.2 General layout of hatcheries

Although hatcheries differ considerably in layout according to the type of animal propagated, they have a number of common design requirements. These include:

- an area for holding broodstock
- a dedicated spawning area
- dedicated food production area(s)
- a dedicated larval rearing area
- an area for early nursery culture.

Hygiene is a major concern in aquaculture hatcheries. Disease can be very costly to a hatchery and may result in the loss of live feed cultures as well as larvae, juveniles and broodstock. Hatcheries must be designed to minimise the possibility of disease transfer from one area of the hatchery to another and to facilitate quarantine of a particular area if required. To achieve this, each part of a hatchery should be separated from the others. For example, live foods are usually cultured well away from areas housing culture stock (broodstock or larvae). Larval culture and broodstock holding areas are similarly separated. As well as disease considerations, the isolation of broodstock holding areas minimises noise and other disturbances that may stress the broodstock and impact on reproductive output. Appropriate hatchery design can help minimise the risk of disease and facilitate hatchery efficiency and productivity (Fig. 2.12). Some important design considerations include:

Fig. 2.12 A large well designed fish hatchery/nursery system (Good Fortune Bay Fisheries, Queensland, Australia).

- separate access to all hatchery areas
- antiseptic foot baths between areas to minimise risk of disease transfer
- larval tanks raised off the ground to allow access for washing or sterilising floor areas
- well-drained floors to minimise standing water
- air lines, water supply and power located overhead to allow easy access.

2.7.3 *Hatchery management*

Perhaps the two most important management factors in hatcheries are water quality and feeding regimen. Both influence the rate of growth and the health of the larvae.

Water entering larval culture tanks should be free from competing organisms and predators and have low levels of suspended sediment and bacteria. Both are achieved by filtration, as outlined in section 2.7.1. Water should also be of an appropriate salinity and temperature. The quality of water in larval culture tanks declines over time as a result of larval attrition, metabolic waste products and uneaten food. Maintenance of water quality parameters within an acceptable range is achieved by regular water changes. Water changes generally occur every 2–3 days in static culture systems, such as those used for bivalve larvae (section 21.4.2), or more frequently in flow-through larval culture tanks used with fish and crustaceans (e.g. section 18.3.2). Items used in maintaining larval culture tanks include nets, sieves, hoses and brushes for cleaning, etc. It is good practice to have one set of equipment for each tank to minimise the potential for disease transfer between tanks. If this is not possible, equipment must be sterilised (usually in a chlorine solution, then rinsed free of chlorine and dried) between use.

2.8 Selecting a new species for culture

Which species will I grow? What type of system will I use to grow the selected species?

Criteria for the selection of new species for commercial aquaculture have been outlined by Avault (1996, 2001).

2.8.1 *Selecting an appropriate species*

The choice of aquaculture species is often a balance between biological knowledge and economics (Fig. 2.1). The biological knowledge required to allow successful culture of the species is diverse and will be outlined in the following section.

Prior to selecting a species for culture, it is important to consider the economics of the species. There is a wide range of uses of aquaculture products and, therefore, there is a myriad of potential aquaculture species. Aquaculture is much more than production of protein for human consumption, although this is the most frequent objective of new culture ventures. Other production objectives of aquaculture include:

- industrial products, e.g. agar/alginate
- pharmaceutical products, e.g. UV-resistant compounds
- augmenting wild stocks for conservation, wild fisheries or recreational fisheries
- ornamental species for the aquarium industry
- crocodiles hides
- food/feed components for culture organisms, e.g. phytoplankton and zooplankton.

Thus, the characteristics of the selected species will vary according to the production objectives. The use of the product may determine the market price, and this will in turn influence the economic viability of the venture.

2.8.2 Requirements of a suitable culture species

Aquaculture projects frequently fail as a result of an inadequate understanding of all facets of the biology and economics of the target species, or by selection of species that do not have the appropriate characteristics. Avault (1996) listed the following issues to be considered when selecting an aquaculture species.

Water temperature and water quality

Each species has specific requirements for various water quality parameters. These parameters include temperature, dissolved oxygen, salinity, pH and ammonia/nitrite/nitrate nitrogen (section 3.3). It is important to understand not only the tolerance ranges for these parameters, but also the optimum levels for growth, survival and reproduction.

Growth rate

Species showing rapid growth to reach market size in a short period of time are preferred for aquaculture. Historically it has been considered inappropriate to culture an animal that requires more than 2 years to reach market size. However, growth rate must be considered in relation to risk and economics. Risk is associated with the chance of losing the product to disease or system failure during the grow-out period. If, however, the risk of product losses is low or the product is of substantially higher value than the product from similar species, or both these desirable conditions pertain, then species with slower growth may be cultured effectively.

Feeding habits

A species' feeding habits can greatly influence the profitability of culture. Feeding habits can generally be divided according to production phase, i.e. hatchery/nursery, juvenile and grow-out phases. In general, feeding habits change markedly after the hatchery/nursery phase, with the juvenile and grow-out animals having similar feeding habits. Bivalves, however, have similar feeding habits throughout their life cycle. Feeding habits include the foods and the position in the water column in which the animal feeds. Feed requirements affect feeding costs, with animals feeding higher in the food chain generally requiring more expensive diets (section 2.3). The ability to accept an artificial diet is also of importance, as live feeds are expensive to produce in terms of space, labour and consumables (section 9.5). At the other end of the spectrum, extensively cultured grazing gastropods and filter-feeding bivalves feed from their environment without supplementation. The position in the water column at which the animal feeds can affect the type of aquaculture system that can be used to culture the animal. For example, if a fish is a benthic feeder, it is unlikely that it can be effectively cultured in a floating cage.

Reproductive biology

A reliable source of juveniles or 'seed' is fundamental to all aquaculture ventures. Seed must be available in the required quality and quantity. Although some aquaculture industries have been developed using wild-caught seed, e.g. table oysters and milkfish, it is highly desirable that the target species can be bred in captivity (section 2.7). In addition to being able to reproduce in captivity, it is also desirable that a culture species has high fecundity, either a large number of eggs per spawning or multiple spawnings per season. High fecundity helps offset the high costs of maintaining and spawning broodstock.

Hardiness

To achieve an acceptable level of production, cultured species will experience conditions that differ considerably from their natural environment. In culture, animals will generally experience social crowding, poorer water quality and handling. All these parameters will create stress. The cultured species should be able to adapt to these stresses and maintain high survival and optimal growth. It is important to consider the effects of these stresses on all stages of the life cycle. Combined with the ability to adapt to the stresses of culture is an ability to resist disease. The likelihood of disease occurrence and proliferation is higher in culture than in the wild, and a species susceptible to suffering mass mortalities from low levels of disease is not desirable.

Marketing

Prior to undertaking an aquaculture venture, it is important to thoroughly investigate the markets for the target species. As previously described, there is a range of uses for aquaculture products, all of which may be considered when selecting the target species. Some species may also have several marketing opportunities. For example, certain fish species may be sold as fingerlings for restocking into the wild, or they may be grown to plate size or grown to a larger size for filleting. A table oyster may be sold to oyster farms as seed or sold after grow-out as adults in the shell or shelled and smoked. It is important to identify the best market(s) for the product and access the market value of the product. A decision on the economics of culture can then be made.

Economics

At all stages of selecting an aquaculture species, the underlying aim of aquaculture is profit (Chapter 12). Thus, all the costs of production, including stock, feed, electricity, interest on money borrowed, labour, etc., and the returns from sale of the product must be estimated with a costs and return analysis. Economists have produced a variety of software that allows prospective aquaculture farmers to undertake all the appropriate economic calculations. The economic feasibility of the culture can be determined by combining market analysis and biological feasibility.

There may be alternative production systems for rearing a species. It may be possible to match the type of production system, and its associated costs, to the species and its market price to achieve profitability. For example, growing fish in an intensive, recirculating tank system may cost US$4–8 per kilogram to grow to market size. The market price for this species would need to be significantly higher than the production cost to ensure economic feasibility. If not, then the farmer must investigate an alternative cheaper method of production, e.g. semi-intensive pond culture, which may reduce production costs per kilogram of market size fish.

It is not just a matter of having a positive balance of profit to costs. If the rate of return on the investment in an aquaculture venture is not greater than the prevailing standard rate of return from investments, then the venture is effectively unprofitable. The committed funds would be better used for a lower-risk investment (section 12.2).

2.8.3 Compromise

The ideal culture species possesses all the above characteristics; however, few if any species are ideal, and generally there is some compromise in terms of these characteristics.

Often the choice of a species for aquaculture production will lead to two situations.

(1) A species has an existing market and/or can be marketed quite easily. In this situation the farmer may enter an established market and *the decision to culture is based upon the economic attractiveness*. If there is biological information to allow complete control over the life cycle, then, given economic feasibility, successful culture is nearly assured. If the understanding of the life cycle is not complete, problems in culture may be experienced. This may lead to failure, even though a ready market exists.

(2) Biological feasibility of culture is high, but there is no ready market or market price is low. In this situation, *the decision is based upon the fact that the control of the life cycle is complete rather than on economic attractiveness.* In this case there is a need to establish a market, which may involve considerable cost. If a market is established, given that culture is economically feasible, the venture must be successful.

If a product does not suit a market, culturing the product will be unsuccessful; however, an alternative to changing the species may be a matter of changing the market niche or undertaking an innovative marketing campaign. This approach has been exemplified for a number of fish species, e.g. tilapia in Asia (Chapter 16) and channel catfish in the USA (Chapter 17).

Fig. 2.13 While a ready market exists for sea urchins in South-East Asia, where they are prized for their gonads, further research is required to develop reliable and economic culture methods.

2.9 Developing a new cultured species

If the aquaculture species of choice is cultured widely in the region and culture techniques are well developed, then development of the culture species is relatively simple and the major concerns regarding success will be economic. In this situation, establishing culture is a matter of applying existing technologies to the chosen farm site. Some trial and error with husbandry may be required; however, provided that the training of employees is satisfactory, success is likely. If the species is cultured widely throughout other regions, then development may be more difficult. Difficulties may arise from population differences, with the species requiring more technological development, or from regional climate differences impacting upon traditional or developed culture technologies. Development of culture for such a species should begin at the pilot trial of the development protocol detailed below.

If the target species is a new species to the aquaculture industry, or if you propose to culture the species under very different conditions (e.g. intensive vs. extensive), then it will be necessary to develop culture techniques (Fig. 2.13). It is recommended that the following development protocol be used to determine the suitability of the species.

2.9.1 Stages of development

Many species fail at various points during the following development protocol. The key is not to make large investments in species destined for failure.

Screening

The aim of this stage is to make an informed decision about the culture potential of the species. During this stage, as much information as possible should be collected: life history, general ecology and husbandry. This information should then be considered in respect of the proposed site.

Provided the species looks promising, any legal constraints to culture should be investigated, e.g. permits, translocation issues. This will inform prospective farmers of the legal possibility of culturing the animal. Given that culture initially looks to be feasible from biological, site and legal perspectives, potential markets should be assessed. In assessing potential markets, the farmer should determine whether the species is consumed regionally or globally, and the present demand for the product. If a ready market does not exist, the farmer may wish to investigate if a potential market niche exists.

Finally, after determining a potential market for the species, the economics of culture should be assessed. It is wise at this stage to produce some preliminary figures on the potential returns on investment. These figures should be realistic and not work on the lowest possible production costs and highest possible production rates and market price. By adopting mid-range values, the reality of culturing this animal is more likely to be reflected. Potential investors should err on the conservative side. Determining the return on the investment should incorporate all economic aspects, including capital investment on land purchase and infrastructure, depreciation of assets and interest rates on borrowed money, as well as annual production costs. In addition, the potential return on the invested capital if it was used in an alternative investment strategy should be calculated (Chapter 12). Software programs are available commercially to produce the appropriate economic information.

Based upon the desirable characteristics for a culture species, the biological potential of culture, the availability of a market and the economics of culture, the target species may be accepted or rejected at this stage. If the species is accepted, development should proceed by moving to the research stage.

Research trials

Once a species has progressed through screening, research into the biology and husbandry of the animal should be undertaken. Research provides the information required for the development of appropriate culture systems and husbandry techniques. During the research trials it is necessary to perform controlled studies to investigate:

- environmental requirements, including water quality parameters (pH, salinity, nitrogenous wastes, alkalinity etc.), temperature and photoperiod; it is important that the ranges of these parameters providing optimal and significant growth, as well as lethal levels, are determined
- optimal stocking densities
- reproductive physiology and broodstock husbandry
- nutrition and growth of larvae, juveniles and adults. The nutritional requirements for both live and artificial diets, feeding techniques and expected growth rates should be determined.

The species may prove to be undesirable on the basis of one or more of the above areas and be rejected.

Pilot trial

Using the information derived from this research, a pilot trial can be conducted. Farmers consider this as taking culture from the laboratory into the 'real world'. In this trial, the culture structure, e.g. tanks, ponds, cages, raceways, is small and a low number of animals is used. This trial will provide semi-commercial values for and or information on:

- survival
- potential yields
- food conversion ratios
- water quality problems
- handling difficulties
- economics of culture.

It is important at this stage to determine impediments to larger-scale culture. If problems are indicated, the farmer should either perform more research or modify techniques and conduct further pilot trials.

Commercial trial

The commercial trial is a scale-up of the pilot trial. During commercial trials the farmer should use full-sized culture units and larger numbers of animals. At this stage a site should be secured and some capital invested on infrastructure development.

The commercial trial provides information on production costs and profits, and husbandry of large numbers of animals. The commercial trial may identify difficulties in animal husbandry or cost dynamics associated with the larger culture systems.

At this time some preliminary market development should be undertaken. However, it should be noted that it is difficult to establish firm markets before production is established.

Full-scale production

Full-scale production involves the farmer developing the full number of culture units. It is important that other aspects of the culture units, such as structure size and shape, flow dynamics and husbandry, are not modified from those used in the commercial trial. Full production requires substantial capital and infrastructure development and a suitable site. Undercapitalisation has been a major problem in developing aquaculture industries.

As production expands, firm markets should be established.

2.9.2 Time scales for development

The time scale for development of a culture species is important to all prospective farmers. In the above protocol the length of time spent at each stage depends on many factors including:

- how much information is available on the target species
- husbandry difficulties
- the existence of markets
- available capital.

As a rule of thumb the following should be expected:

(1) screening: at least 2 years
(2) research: at least 5 years for species previously not studied; often longer, particularly if information is scarce and the species is difficult to culture
(3) pilot trial: over at least two growing seasons
(4) commercial trial: over at least two growing seasons.

Failure to observe this protocol is a common mistake in aquaculture development. It is often costly and results in the failure of many new aquaculture ventures. Prospectuses for new companies often do not include a development protocol, and many ventures proceed with insufficient information about culture requirements of the target species. Many ventures targeting new species move directly to commercial trials or attempt full-scale production. Omitting pilot trials, even if the species is well known, or not correcting major problems before moving to commercial trials often results in the construction of inappropriate aquaculture facilities, leading to economic failure. However, sequential development, as outlined above, provides a sound basis for decision-making and appropriate progression of aquaculture ventures.

References

Avault, J. W. Jr (1996). *Fundamentals of Aquaculture: A Step-by-Step Guide to Commercial Aquaculture*. AVA Publishing Co., Baton Rouge, LA.

Avault, J. W. Jr (2001). Selecting new species for commercial aquaculture. *Aquaculture Magazine*, **27**(30), 55–7.

Beveridge, M. C. M. (1996). *Cage Aquaculture*, 2nd edn. Fishing News Books, Oxford.

Billard, R. (1999) Biological diversity in pond fish culture. In: *Proceedings of the 5th Indo-Pacific Fish Conference, Noumea, New Caledonia, 3–8 November 1997* (Ed. by B. Seret & J. Y. Sire), pp. 471–9. Societe Francaise d'Ichthyologie, Paris.

Boyd, C. E. (1991). *Water Quality in Ponds for Aquaculture*. Auburn University, Alabama.

Cohen, D. (1997). Integration of aquaculture and irrigation: rationale, principles and its practice in Israel. *International Water and Irrigation Review*, **17**, 8–18.

Edwards, P. (1998). A system approach for the promotion of integrated aquaculture. *Aquaculture Economics and Management*, **2**, 1–12.

FAO (1995). *1993 Fisheries Statistics: Capture Production*, Vol. 82/1. Food and Agriculture Organization of the United Nations, Rome.

FAO (2001). *1999 Fisheries Statistics: Capture Production*, Vol. 88/1. Food and Agriculture Organization of the United Nations, Rome.

Huazhu, Y., Yingxue, F. & Zhonglin, C. (1994). Description of integrated fish farming systems in China and the allocation of resources. In: *Proceedings on the Basic Theories of the Integrated Fish Farming and Aquaculture Bio-economic Studies*, pp. 1–33. Freshwater Fisheries Resources Centre, Chinese Academy of Fisheries Sciences. Science Publishing House, Beijing.

Huguenini, J. E. & Colt, J. (1989). *Design and Operating Guide for Aquaculture Seawater Systems*. Elsevier Science Publishers, Amsterdam.

Landau, M. (1992). *Introduction to Aquaculture*. John Wiley & Sons Inc., New York.

Lawson, T. B. (Ed.) (1994). *Fundamentals of Aquaculture Engineering*. Kluwer Academic Publishers, Boston.

Losordo, T. M. (1998a). Recirculating aquaculture production systems: the status and future. *Aquaculture Magazine*, **24**(1), 38–45.

Losordo, T. M. (1998b). Recirculating aquaculture production systems: the status and future, part II. *Aquaculture Magazine*, **24**(2), 45–53.

Losordo, T. M., Masser, M. P. & Rakocy, J. (2001). Recirculating aquaculture tank production systems. An overview of critical considerations. *World Aquaculture*, **32**(1), 18–22.

Quayle, D. B. & Newkirk, G. F. (1989). *Farming Bivalve Molluscs: Methods for Study and Development. Advances in World Aquaculture,* Vol. 1. The World Aquaculture Society, Baton Rouge, LA.

Tacon, A. G. J. (1996) Feeding tomorrow's fish. *World Aquaculture*, **27**(3), 20–32.

Wang, A. W. & Fast, A. W. (1991). Shrimp pond engineering considerations. In: *Culture of Marine Shrimp: Principles and Practices* (Ed. by A. Fast & L. J. Lester), pp. 415–30. Elsevier Scientific Publications, Amsterdam.

Wheaton, F. W. (1977). *Aquaculture Engineering*. Wiley, New York (reissued, 1993).

Wyban, J. A. & Sweeney, J. N. (1989). Intensive shrimp growout trial in a round pond. *Aquaculture*, **76**, 215–25.

Yoo, K. H. & Boyd, C. E. (1994). *Hydrology and Water Supply for Pond Aquaculture*. Chapman & Hall, New York.

Zweig, R. D. (1985). Freshwater aquaculture in management for survival. *Ambio,* **14**, 66–74.

3

Water Quality

Michael Poxton

3.1 Introduction

The provision of water of adequate quality and quantity is a primary consideration in both site selection and aquaculture production management.

There is a growing body of information on water quality, particularly with respect to pollution and the necessity to protect inland and coastal environments (Chapter 4). The science and technology of water quality management for aquaculture can be identified in three major areas:

(1) water quality criteria of resource systems potentially suitable for aquaculture development
(2) water quality requirements within culture systems
(3) external impacts caused by aquaculture effluent water.

The last topic is dealt with in Chapter 4, and topics (1) and (2) are treated here. This chapter focuses on key aspects of water quality, their influence on aquatic organisms and their maintenance within appropriate levels in aquaculture systems.

3.2 Natural water quality and basic water chemistry

All natural waters contain varying amounts of a wide variety of substances, and water, the universal solvent, contains at least trace amounts of almost every substance it happens to come into contact with. Indeed, traces of all known natural elements occur in seawater.

3.2.1 Salinity and the major elements in seawater

Nine elements constitute more than 99% of natural sea salts. The major constituents of seawater are shown in Table 3.1. Although this table gives a fair idea of the major constituents, it does not give a complete picture of the chemical composition of seawater. This is because the great numbers of minor elements, dissolved gases and organic matter occurring in seawater as anions, cations or molecules in dynamic equilibrium, make it difficult to study its chemical composition.

As a supporter of life, natural seawater is a highly complex system. Some of its constituent elements such as carbon (C), nitrogen (N), phosphorus (P) and silicon (Si) are involved in biochemical cycles. Furthermore, when seawater is allowed to evaporate to dryness and then reconstituted with freshwater, it does not sustain life as it originally did. This is because of the lack of life-supporting micro-organisms, which tend to balance and recondition natural seawater. Seawater improves with age, a point to remember when making it up from artificial salt mixtures.

Table 3.1 The major constituents of seawater (35‰ salinity)

Element	g/kg	mequiv./kg
Sodium	10.752	467.56
Magnesium	1.295	106.50
Calcium	0.416	20.76
Potassium	0.375	10.10
Strontium	0.008	0.18
Total cations		605.10
Chlorine	19.345	545.59
Sulphate	2.701	56.23
Bromine	0.066	0.83
Fluorine	0.0013	0.07
Bicarbonate*	0.145	–
Boric acid	0.027	–
Total anions		602.72
Alkalinity (cations minus anions)		2.38

*Bicarbonate and carbonate vary according to pH (see text).
Reprinted from Poxton & Allouse (1982) with permission from Elsevier Science.

Salinity is defined as the weight of dissolved substances left from 1 kg of seawater after complete oxidation of organic matter, complete replacement of carbonates by an equivalent quantity of oxides, and the replacement of bromide and iodide by an equivalent amount of chloride. Consequently, an empirical relationship exists between salinity and chlorinity. The latter is defined as the chlorine equivalent of the total halide concentration:

$$\text{Salinity (S) (‰)} = 1.80655 \times \text{chlorinity (‰)}$$

The largest variation in the composition of sea salts occurs in their calcium and carbonate contents as a result of their use in biological processes. The salinity range in open oceans is only 33–37‰, with an average of 35‰. Over continental shelves, however, salinity can range between 0‰ and 41‰ at the surface. The variation is usually due to rainfall, river inflow and evaporation.

3.2.2 Minor elements in seawater

Minor or trace elements, by definition, occur in only small concentrations in seawater (Table 3.2). The term 'minor' refers to their quantity, not their importance, for they are essential to the biota, either as nutrients for growth (i.e. P- and N-containing substances) or for their important metabolic roles in aquatic organisms (i.e. iron and copper).

Table 3.2 Some minor elements of 19‰ chlorinity (~34‰ salinity) seawater

Element	mg/kg
Boron	5.0
Silicon	0.010–7.000
Nitrogen	0.001–0.700
Iron	0.001–0.290
Copper	0.010–0.024
Phosphorus	0.001–0.170
Molybdenum	0.0003–0.016
Zinc	0.005–0.014
Manganese	0.001–0.010
Vanadium	0.0002–0.007
Chromium	0.001–0.003
Cobalt	0.0001–0.0005

Reprinted from Poxton & Allouse (1982) with permission from Elsevier Science.

There are about 61 minor elements in seawater. In contrast to the major constituents, trace elements show very marked fluctuations as a result of their involvement in biological processes. This is especially true for P and N. A variation of an order of magnitude in the concentrations of minor elements may occur only under extreme conditions, but any change in their concentration may be significant in its effect on marine life (Poxton & Allouse, 1982).

3.2.3 Dissolved oxygen

In water, the dissolved gases that are of particular biological and ecological importance are oxygen (O_2) and carbon dioxide (CO_2). They either originate from various life processes or are dissolved from the atmosphere at the surface. Gases dissolve in water according to temperature, salinity and their individual partial pressure (pp) gradients across the surface.

The amount of dissolved oxygen (DO) in seawater varies from 0 to 12 mg/L O_2, varying geographically and seasonally, and with depth and environment. Variations in DO result from photosynthesis and from free exchange with the atmosphere at the surface. The latter usually ensures that surface layers contain high DO levels, whereas deep waters, which obtain DO through mixing, wind action and diffusion, often have low DO levels. In addition to reduced atmospheric exchange, deep marine waters have low photosynthetic activity and may reach a condition where the respiratory demand for DO exceeds that supplied through photosynthesis and mixing. Eutrophic water (i.e. water with high concentrations of nutrients, which promote algal growth) may face dangerously low DO levels at night, when the algae switch from photosynthesis to respiration. Nevertheless, complete deoxygenation occurs only in isolated water masses with poor circulation or unusually high rates of oxygen utilisation.

Although it is appropriate to describe waters as containing higher DO levels, this is only relative to the solubility of O_2 in water. Air contains about 210 000 mg/L O_2 while freshwater rarely contains more than 10 mg/L and seawater usually has less than this. Thus, aquatic environments at best contain about 0.005% of the concentration of O_2 in air.

Seasonally low levels of DO may occur in winter at high latitudes as a result of the decay of organic matter under ice cover, which results in rates of O_2 replenishment from the atmosphere being reduced. High temperatures also reduce DO by reducing the solubility of O_2 in water while increasing the O_2 demand from respiration. At ocean surfaces, DO may reach 100% and generally varies only slightly. Up to 120% saturation is common in areas of high photosynthetic activity during long sunny days.

3.2.4 Dissolved carbon dioxide, pH, alkalinity, hardness and buffering capacity

Dissolved CO_2, pH, alkalinity and hardness are dynamically inter-related. In freshwater, elevated levels of CO_2 may be present in spring or bore-hole supplies or may originate as a result of plant growth in surface waters or from metabolic activity of animals or bacteria, or both, in recycled waters. Elevated CO_2 levels have a dramatic effect in lowering pH as the percentage of each species in the carbonate system (free CO_2, bicarbonates and carbonate) depends on H^+ concentration (Table 3.3). Free CO_2 varies in seawater in the range 67–111 mg/L. Thus, it is much more soluble than O_2, for which the equivalent range is 0–12 mg/L (section 3.2.3).

The capacity of seawater to accept H^+ ions is known as its alkalinity, which is defined as the number of milliequivalents of H^+ that are neutralised by 1 kg of seawater when an excess of acid is added.

Table 3.3 Variations in the carbonate system according to temperature, pH and chlorinity (19‰ = 34.325‰ S)

pH	Temp. (°C)	Percentage of molar fractions		
		H_2CO_3	HCO_3^-	CO_3^{2-}
Seawater				
7.5	8	3.9	94.0	2.1
8.0	8	1.2	92.2	6.6
7.5	24	2.9	93.9	3.2
8.0	24	0.9	90.7	8.4
Freshwater				
7.5	8	8.8	91.2	0.0
8.0	8	3.0	96.7	0.3
7.5	24	6.9	92.9	0.2
8.0	24	2.3	97.3	0.4

The milliequivalent amounts of cations and anions in seawater are not balanced against each other, the cations exceeding the anions by approximately 2.38 mequiv. (Table 3.1). This results in seawater being slightly alkaline and having a strong buffering capacity, allowing it to resist changes in its pH value. In general, the pH of seawater ranges from 7.5 to 8.4, although exceptions commonly occur. Normally, the pH is at its maximum within 100 m of the surface, falling to a minimum at depths of 200–1200 m. Freshwater is less buffered: the amount of CO_2 added and removed during plant growth, and added by animal and plant respiration, has a more dramatic effect on pH. It is the higher alkalinity of seawater compared with freshwater that provides greater protection against the effects of a build-up of CO_2.

The alkalinity of seawater is mainly due to the presence of the carbonate and borate systems. When the amount of H^+ needed to convert all the anions of the weak acids to their unionised acids is known, the term total alkalinity (TA) is used:

$$TA = HCO_3^- + 2CO_3^{2-} + B(OH)_4^-$$

Carbonate alkalinity, however, contributes only to the alkalinity attributable to the bicarbonate (HCO_3^-) and carbonate (CO_3^{2-}) ions. The borate [$B(OH)_4^-$] contribution to alkalinity arises from the ionisation of boric acid:

$$H_3BO_3 + H_2O \rightleftharpoons B(OH)_4^- + H^+$$

The $B(OH)_4^-$ contribution to TA can be obtained by subtracting the contribution of the carbonates from the measured TA. At the normal pH of seawater, the borate system is not as important as the carbonate system.

The carbonate system in water follows certain steps, each of which is characterised by an equilibrium constant:

$$CO_2 \text{ (gas)} \rightleftharpoons CO_2 \text{ (aqueous)}$$

$$H_2O + CO_2 \text{ (aqueous)} \rightleftharpoons H_2CO_3$$

$$H_2CO_3 \rightleftharpoons H^+ + HCO_3^-$$

$$HCO_3^- \rightleftharpoons H^+ + CO_3^{2-}$$

The proportion of each species of the carbonate system (free carbon dioxide, bicarbonates and carbonate) depends on the H^+ concentration of the water and, conversely, a change in these values will affect the pH. Below pH 4.0, only free CO_2 occurs in seawater of 35‰; at pH 7.5, the proportion of bicarbonate reaches its maximum; and at pH values > 7.5 bicarbonate is increasingly replaced by carbonate ions. Hence, at the normal pH range of seawater the bicarbonate ion is the most important species in the carbonate system (about 90%), the remainder being free CO_2 near the lower end of the pH range, or carbonate ion near the upper end. Table 3.3 indicates how the system varies with the temperature and pH in both fresh and marine waters.

Hardness has been confused in the literature, perhaps because of familiar use in the home and in the wastewater engineering industry, where the degree of hardness is determined from the amount of soap required to form suds. It is not correct to apply the term to waters having a high alkalinity, for although most waters of high alkalinity are hard, this is not always the case (Table 3.4). Total hardness is the total concentration of metal ions in water (mg/L) in terms of equivalent calcium carbonate ($CaCO_3$). Total hardness is usually related to total alkalinity because the anions of alkalinity and the cations of hardness are normally predominantly derived from carbonate minerals. The ions are primarily those of Ca^{2+} and Mg^{2+}, but hardness also includes K^+, Na^+, Fe^{2+} and Mn^{2+}.

Table 3.4 Total hardness and total alkalinity of fish pond waters

Type of water	Total alkalinity (mg/L)	Total hardness (mg/L)
Pond on sandy soil	13.2	12.9
Pond on acidic, clay soil	11.6	12.3
Pond on calcareous soil	51.1	55.5
Pond filled with soft but alkaline well water	93.0	15.1
Pond in arid region	346.0	708.0

Reprinted from Boyd (1982) with permission from Elsevier Science.

Calcium concentrations in water generally increase with increasing salinity. Very soft freshwater, usually defined as containing < 20 mg/L $CaCO_3$, may be unsuitable for aquaculture as Ca is required for the formation of exoskeleton and bone in shellfish and fish respectively. Fish grown in very soft water are therefore prone to skeletal deformities, and crustaceans may not be able to moult. Very soft water, with low alkalinity, has very poor buffering capacity, so the pH may fluctuate both quickly and widely. This will be detrimental to fish *per se*, and they will also be more susceptible to toxic pollutants under such conditions.

In fertilised fish ponds with < 20 mg/L of total alkalinity, production increases with increasing alkalinity. This may not result entirely from the increasing alkalinity and hardness, however, as P and other nutrients will typically be increasing as well. Boyd (1982) reported little effect on fish production in fertilised ponds having alkalinities in the range 20–120 mg/L, although he noted that natural freshwaters that contain > 40 mg/L total alkalinity are more productive for aquaculture.

At high temperatures, evaporation may result in concentration and subsequent precipitation of ions responsible for alkalinity, which may therefore be lower than total hardness. Along coastal plains, however, groundwater sometimes has high alkalinity and low hardness (Table 3.4). In the latter case, some of the bicarbonate and carbonate will be associated with Na and K rather than Ca and Mg, but in the former case some of the Ca and Mg is associated with anions other than bicarbonate and carbonate.

The total hardness of seawater is about 6600 mg/L.

3.2.5 Dissolved and particulate organics

Organic substances arise from animal and plant excretion, degradation of dead biota, natural seeps and human waste inputs, which may be direct or indirect from land runoff. As long as there is photosynthetic activity, C is fixed and released as a by-product either during life or after death. In growth periods, plants release minor amounts of C fixed during photosynthesis. Under stress, much larger amounts of C are released, and after death and degradation still greater amounts are liberated. The substances differ in each of these circumstances. Carbohydrates, peptides, fatty acids, toxins, antibacterial materials and vitamin B_{12} have all been reported to be released from cultures of marine phytoplankton. Despite this variety of compounds, dissolved organic carbon (DOC) is often used as an approximation for dissolved organic matter (DOM).

DOC decreases with depth in the sea, values for the upper 100 m usually being in the range 0.6–1.0 mg/L. Annually, the DOC undergoes small fluctuations in surface waters as a result of phytoplankton blooms, but in deeper layers it is constant. A loss of DOC may occur as a result of conversion into particulate organic carbon (POC), absorption on air bubbles or by processes such as decomposition or utilisation. POC includes living (phytoplankton, bacterial aggregates), non-living (detritus) and suspended particles that are larger than 0.5–1.0 μm diameter. Detrital POC often exceeds the living POC, but the total POC is generally only a fraction of the DOC. In general, the quantitative relationship between the amounts of organic matter in sediments, dissolved in seawater, suspended in seawater and in living organisms is 1000:100:10:1.

Direct methods of carbon analysis require expensive equipment, so standardised indirect techniques are generally used. These usually involve the reduction of an oxidising agent such as permanganate, dichromate or persulphate (chemical oxygen demand or COD), or the biological utilisation of DO by bacteria (biological oxygen demand or BOD). Considering the nature of aquaculture, the latter method is generally employed for routine monitoring, especially as it also reflects the self-purification capacity of the water, for the bacteria involved are its current complement. The results are expressed in terms of DO depletion over a period of 5 days at 20°C, under dark conditions (BOD_5).

The measurement of ultraviolet (UV) absorbance is a further, low-cost, popular method of monitoring organic carbon levels in water. Most organic compounds absorb UV between the wavelengths of 200 and 260 nm, while inorganic salts, with the exception of transition metal ions, do not exhibit significant absorbance over 250 nm. Most investigators have therefore used a wavelength of 250–260 nm; 254 nm being a popular choice as it corresponds to a strong and sharp spectral line of the low-pressure mercury lamp. The UV method is often preferred in aquaculture to measure the 'yellowing' of culture water, which is due to excreted organics, but also to leaching from uneaten food (Schuster, 1994) and faeces, and products of bacterial metabolism.

3.2.6 Total solids, suspended inorganic particles and turbidity

Suspended particles can be quantified in terms of their mass, or quantity present or size distribution, using filtration, counting or characterisation techniques respectively. If a water sample is evaporated to dryness, the weight of the residue (mg/L) is referred to as the total solids concentration, and this includes all dissolved and particulate material with the exception of gases. If this residue is then ignited at 550°C and reweighed, then the weight lost represents the total volatile solids (a measure of the total dissolved and particulate organic matter), and the weight remaining is a measure of the dissolved and suspended inorganic particles.

Dissolved and particulate solids are distinguished by passing a water sample through a 0.5- to 1.0-μm filter. If the filtrate is then evaporated to dryness, the weight of the residue represents the concentration of dissolved substances, excluding gases. Allowing the filter paper to dry and then reweighing it yields the total particulate matter. The volatility of both parameters may be determined by ignition at 550°C.

High dissolved solids indicates the presence of large concentrations of solutes and, if volatility is low, then a high mineral content is indicated. Water with a high suspended solids content is associated with a high BOD and is often referred to as being 'turbid'.

Analytically, turbidity is a measure of the penetration of light through water, usually measured by noting the depth at which a black and white (Secchi) disc lowered into the water disappears (Fig. 3.1). Clearly, turbidity will also depend on other parameters, such as the amount of plankton in the water.

Information on particle size distribution can be an aid to their identification and may be used to assess the quantity and distribution of live feed populations (phyto- and zooplankton) and culture vessel hydrodynamics, as well as water quality. Methods include size fractionation using membranes or sieves, laser diffraction or a high-speed particle counter. The particle counter is particularly useful, for it yields data on both particle number and size distribution, and it can be combined with other analytical techniques, such as nutrient concentration studies, which may be used to characterise aquaculture effluents (Cripps, 1993).

Fig. 3.1 A Secchi disc is lowered into pond water to determine the turbidity. Using graduations on the rope (or handle), the turbidity is 'measured' by noting the depth at which the disc is no longer visible.

In aquaculture, solids can be categorised in several ways.

(1) Settleable solids. Conventional gravity settling tanks or ponds may be used to remove these solids, which will usually have a diameter $> 10^{-2}$ mm.
(2) Suspended solids ($> 10^{-3}$ mm) will not generally settle in < 1 h and may pass through settling tanks if the detention time is less than this period.
(3) Colloidal particles (10^{-3}–10^{-6} mm) may be bacteria, viruses or fine clay.
(4) Dissolved particles generally have a diameter of $< 10^{-6}$ mm.

Uneaten food and faeces released from culture facilities may form extensive biodeposits, their distribution depending on the settlement characteristics of the materials and the physical nature of the environment to which they are released (section 4.2.1). For example, if salmon cages are moored in low-energy coastal environments, such as the deep basins of some sea lochs (Scotland), fjords (Norway) or inlets (British Columbia), these wastes will fall to the seabed in close proximity to the cages. Such wastes have a high BOD, resulting in DO depletion at the bottom water, especially as the latter may be retained in deep basins for months if not years.

3.2.7 Nitrogen compounds

Nitrogen occurs in aquatic environments in six forms:

- dissolved gaseous N_2 (from the atmosphere)
- dissolved gaseous NH_3 (largely as waste from animal metabolism)
- ammonium ions, NH_4^+
- nitrite ions, NO_2^-
- nitrate ions, NO_3^-
- in a large variety of organic molecules in solution, living tissue and non-living organic matter.

The term 'ammonia' is often used to include both the dissolved unionised gas (NH_3) and the ionised NH_4^+, and as such is better expressed as total ammonia. The presence of each fraction of ammonia in aqueous solution is governed by the dissociation constant K_b:

$$NH_3 + H_2O \rightleftharpoons NH_4^+ + OH^-$$

$$K_b = \frac{NH_4^+ \cdot OH^-}{NH_3 \cdot H_2O}$$

From the above reaction it is clear that pH plays an important role in determining the NH_3 content of any solution containing ammonia. Thus the percentage of unionised ammonia in an aqueous ammonia solution is as follows:

$$\text{Unionised ammonia} = \frac{100}{1 + \text{antilog}\,(pK_a - pH)}$$

where pK_a is log K_b.

pK_a decreases as the temperature increases, so that for any increase in temperature or pH, the percentage of unionised ammonia increases (equation above). For an increase of one pH unit, the percentage of NH_3 increases by a factor of 10 at the same temperature, whereas a rise in temperature of 10°C doubles the percentage of NH_3 at the same pH (Table 3.5). As pH decreases as the amount of CO_2 increases, this causes a reduction in NH_3 levels. Furthermore, the percentage levels of NH_3 are substantially lower in seawater than in freshwater at equivalent pH and temperature conditions (Table 3.5).

The amounts of N present in solution may be expressed as NH_3-N or NH_4-N (alternatives). The amount given relates to elemental N present, not to the amount of the nitrogenous molecule. It is expressed as a number of milligrams or micrograms per unit volume of water, e.g. mg/L.

Chemical inter-relationships between nitrogenous compounds in water are mediated by heterotrophic and autotrophic bacteria. These bacteria are endemic to water and all surfaces in contact with water, particularly sediments. In intensive aquaculture, particular media and design characteristics are chosen to optimise their density in a variety of biological filters (Lawson, 1995; Midlen & Redding, 1998). Heterotrophic bacteria (e.g. *Bacillus pasteurii*), so called because they derive energy from organic matter, mineralise organic nitrogenous compounds, such as urea, converting them to ammonia:

$$O{=}C{-}2(NH_2) + H_2O \rightarrow CO_2 + 2NH_3$$

Nitrifying bacteria are mainly autotrophs, so called as they derive energy from the oxidation of inorganic substrates and carbon from inorganic sources such as carbon dioxide, carbonates and bicarbonates. They are from the family Nitrobacteraceae, which use inorganic nitrogen as their primary source of energy. Bacteria such as *Nitrosomonas* species and the marine *Nitrosocystis oceanus* are involved in the oxidation of ammonia to nitrite:

$$NH_4^+ + 0.5O_2 \rightarrow NH_2OH\ (\text{hydroxylamine}) + H^+$$

$$NH_2OH + O_2 \rightarrow H^+ + NO_2^- + H_2O$$

In some texts the above two reactions are often summarised as:

$$NH_4^+ + OH^- + 1.5O_2 \rightarrow H^+ + NO_2^- + 2H_2O$$

Oxygen is involved in two ways, being directly incorporated into the substrate and also accepting electrons generated through the cytochrome system. It is difficult to establish whether specific reactions are chemical or biological as many of the proposed nitrogenous intermediates between ammonia and nitrite are unstable. Nitrification may in fact be semi-cyclical:

$$NH_2OH + HNO_2 \rightleftharpoons NO_2NHOH + 2H$$

(hydroxylamine) + (nitrous acid) $\rightleftharpoons$ (nitrohydroxylamine)

In the second step of nitrification, nitrite is oxidised to nitrate by bacteria such as *Nitrobacter* species, the reaction generally being symbolised as:

$$NO_2^- + 0.5O_2 \rightarrow NO_3^-$$

The extra oxygen atom in nitrate is, however, generated from water and not from O_2, which acts only as a terminal electron acceptor.

As nitrifying bacteria require both nutrients and oxygen, they are generally limited, in natural

Table 3.5 Percentage of NH_3-N in the NH_3/NH_4 equilibrium at various levels of pH, temperature and salinity

Temperature (°C)	pH			
	6.0	7.0	8.0	9.0
Freshwater				
5	0.01	0.12	1.2	11.1
15	0.03	0.27	2.7	21.4
25	0.06	0.56	5.4	36.2
35	0.11	1.11	10.1	52.9
Seawater				
5	n/a	0.07	0.68	6.4
15	n/a	0.15	1.5	13.1
25	n/a	0.31	3.1	23.9
35	n/a	0.62	5.9	38.6

Derived in part from Huguenin & Colt (1989) with permission from Elsevier Science; n/a, not available.

environments, to the surface layers of sediments. In aquaculture, this nitrification process is used in biological filters, the design of which is heavily influenced by the need to provide maximum surface area for bacterial attachment. Two examples of designs to achieve this are shown in Figs 3.2 and 3.3, although evidence seems to suggest that many such bacteria exist in flocs held in the voids between the substrate particles. Optimising the surface area to void space ratios for particular flow rates of water through biological filters is a critical design parameter (Lawson, 1995). If sufficient oxygen is present, nitrification rates will be determined by influent ammonia concentration, and nitrate will tend to accumulate. If, however, there is insufficient oxygen to satisfy stoichiometric requirements, nutrient oxidation efficiency will be determined by oxygen availability. *Nitrosomonas* species have a higher affinity for O_2 than *Nitrobacter* species, which will therefore have difficulty in competing in the presence of the former, especially as they are sensitive to ammonia, and nitrite will accumulate.

Complete oxidation of ammonia to nitrite involves the consumption of 3.43 kg O_2 per kg NH_3-N oxidised and produces a cell yield of 147 g of *Nitrosomonas* species. Oxidation of NO_2-N to NO_3-N consumes 1.14 kg O_2 per kg NO_2-N oxidised and produces a cell yield of 20 g of *Nitrobacter* species. These cell yields are low relative to those produced by heterotrophs, so it is important to prevent clogging of biological filters to minimise organic inputs.

The pH optima of *Nitrosomonas* and *Nitrobacter* species are from 8.5 to 8.8, the growth range being 5.1–9.4 and 5.7–10.2 respectively. Nitrifying bacteria can adapt to low pH, the period and their generation times depending on temperature.

Denitrification can occur in both aerobic and anaerobic conditions, and both autotrophic and heterotrophic bacteria are responsible. Denitrification is usually mediated by aerobes using nitrate as an electron acceptor. This allows aerobic (O_2-requiring) bacteria to develop anaerobically:

$$NO_3^- + 2H^+ \rightarrow NO_2^- + H_2O$$

$$NO_2^- + H_2O + 2H^+ \rightarrow [NOH]$$

$$[HO{-}N{=}N{-}OH] \rightarrow N_2O + H_2O$$

and/or

$$[HO{-}N{=}N{-}OH] + 2H^+ \rightarrow N_2 + 2H_2O$$

In deeper layers of natural sediments, anaerobic bacteria are involved in reducing the end products of nitrification back to lower oxidation states. For example, *Pseudomonas* species can reduce nitrate ions. The exact end product (nitrite, ammonia, nitrous oxide or molecular nitrogen) depends on the

Fig. 3.2 Biological filtration using a rotating drum with closely placed fibreglass plates providing a large surface area for nitrifying bacteria. At any time, 80% of the drum is submerged (photograph by Dr Sagiv Kolkovski).

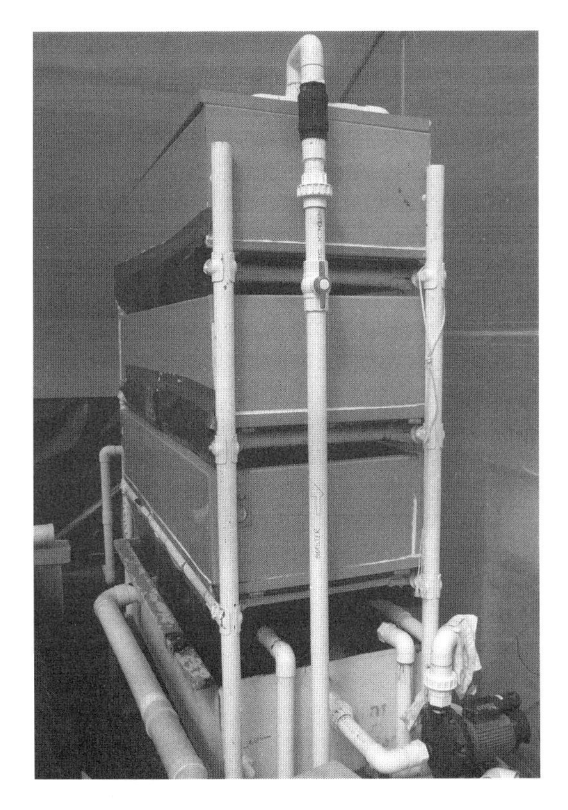

Fig. 3.3 Biological filtration unit in which water trickles from the top, through three containers filled with plastic medium, to a sump at the bottom. The plastic medium provides a large surface area for nitrifying bacteria (photograph by Dr Paul Southgate).

species of bacteria involved. Complete dissimilation involves the production and release of nitrogen and is called denitrification. For example, *Micrococcus denitrificans* uses hydrogen to reduce nitrate:

$$5H_2 + 2NO_3^- + 2H^+ \rightarrow N_2 + 6H_2O$$

Thiobacillus denitrificans is interesting in that, while reducing nitrate, it can also oxidise S, H_2S or thiosulphate:

$$5S + 6NO_3^- + 2H_2O \rightarrow 5SO_4^{2-} + 3N_2 + 4H^+$$

As many of the bacteria involved are heterotrophs, this process can be enhanced in aquaculture by providing a carbon source. Obviously, nitrification and denitrification cannot occur optimally in the same place and at the same time. This, however, does not exclude both processes occurring within close proximity of each other at suboptimal rates. Clearly, denitrification rates depend on the nitrate concentration and therefore on the rate of nitrification. As many denitrifying bacteria will use O_2 in preference to nitrate in aerobic conditions, denitrification will be inhibited. DO is also important in determining whether any intermediate products accumulate. Critical DO levels may be 0.2 mg/L, above which denitrification does not occur, while nitrification is inhibited below 0.6 mg/L.

The optimum temperature for nitrification is 30–36°C, while that for denitrification is 65–75°C. The reduction of nitrate is most sensitive to low temperatures (15°C) and ceases completely at 5°C. Clearly, then, in temperate or cold waters both processes will occur at suboptimal rates, and this should be taken into consideration in the design of recirculating systems. Substrate inhibition also needs consideration. High ammonia concentrations inhibit *Nitrobacter* species, as does nitrate. Organic nitrogen compounds such as urea have also been described as having an inhibitory effect, although in part at least this will result from competition with heterotrophs.

3.2.8 Phosphorus compounds

The productivity of most surface freshwaters is limited by phosphorus, N/P ratios of 10:1 being common. In addition, many species of blue-green algae can fix N biologically. The concentration of P, however, is not nearly as important as its rate of supply, which is conditioned by its rate of uptake, use and release by phytoplankton. Kinetics is therefore replacing concentration when considering the role of P in aquatic ecosystems.

Phosphorus is usually detected in water as soluble orthophosphate, of which a number of different forms exist in equilibrium ($H_2PO_4^-$, HPO_4^{2-}, PO_4^{3-}) depending on pH, although in fish ponds soluble organic P and particulate P also occur. These compounds are, however, mineralised by bacteria to soluble orthophosphate. For practical purposes, the difference between total P and soluble orthophosphate, which is usually < 10% of total P, is an indication of the amount of P contained in phytoplankton and detritus.

Phosphates in aerobic sediments are generally those of Ca, Fe and Al. As these are all slightly soluble they are in dynamic equilibrium across the sediment–water interface. Consequently, as plants absorb P, further phosphate will be released from the sediment. This equilibrium is, however, highly skewed in favour of the sediment (water–sediment ratios are usually < 1:> 99 P), which most authors regard as a P 'sink' (Boyd, 1995). Pond soil rich in organic matter has a reduced capacity for P adsorption and, if heavily fertilised, may become saturated.

P is strongly bound to mineral particles only in well-aerated sediments. Under anaerobic conditions, concentrations of phosphate in the interstitial water increase, as the low redox potential (E_h) favours the solubility of iron and aluminium phosphates (e.g. insoluble Fe^{3+} is converted to soluble Fe^{2+} phosphate). Rooted macrophytes can use these sources of P in the sediment, but they are not available to phytoplankton. Instead they precipitate at the interface between the anaerobic and aerobic layers in the sediment (Boyd, 1982). Similarly, ferrous phosphate diffusing from the pond soil at the sediment–water interface will quickly precipitate and fall back to the soil surface as ferric phosphate if the overlying water is aerobic. If it is anoxic, however, the ferrous phosphate will remain in solution; but, as this circumstance typically occurs in thermally stratified ponds and lakes, it will not be available to phytoplankton until mixing occurs on destratification.

The presence of N and P compounds as nutrients is important in extensive culture and in the intensive pond cultivation of herbivores, such as mullet (*Mugil* species) and milkfish (*Chanos chanos*), in that they support the growth of phytoplankton and blue-green algae. Phytoplankton reduces the sunlight reaching the bottom of ponds by shading. This may reduce cannibalism and extend the hours of feeding in benthic crustaceans such as shrimp and freshwater crayfish.

The N and P levels in the freshwater supplies to 22 rainbow trout (*Oncorhynchus mykiss*) farms in the UK are shown in Table 3.6. The farms produced trout in earth ponds (14 farms) or circular tanks (eight farms), using water from a variety of sources (13 rivers, seven springs, one bore-hole, one lake). A general indication of the water quality required for success in freshwater rainbow trout production in this region may be inferred from these data.

Table 3.6 Mean and range of nutrient concentrations (mg/L) in the intake freshwater of 22 UK rainbow trout production units

	NH_4-N	NO_2-N	NO_3-N	PO_4-P	Suspended solids
Usual range					
Minimum	0.02	0.00	0.1	0.01	1.0
Maximum	0.05	0.02	1.9	0.05	9.9
Mean	0.05	0.02*	1.0	0.03	5.5
SD	0.03	0.02*	0.7	0.03	3.7
Worst site overall	0.12	0.06	3.2	0.07	12.7

*A single value of 2.45 mg/L NO_2-N was recorded at one site, but was not used in the calculations.

Reproduced from Poxton (1990) with permission of the European Aquaculture Society.

3.2.9 Sediment quality and redox potential

Sediment quality in aquaculture ponds is particularly important for benthic shellfish and bottom-feeding fish. In intensive and semi-intensive culture, faeces, uneaten food and other organic matter may accumulate. The majority of aquatic bacteria are heterotrophs and their numbers are largely determined by the concentration of organic matter, so that any enrichment is likely to result in the selective enhancement of specific bacterial groups. Under aerobic conditions, mineralisation of organic material produces CO_2 and ammonia, and nitrification will also occur (section 3.2.7). Under anaerobic conditions, however, nitrogen, sulphides and methane are produced and diffuse into the overlying water. Sediment quality monitoring is therefore important in aquaculture, for poor sediment quality precedes and often generates poor water quality. Sediment quality will be largely determined by stocking density and the intensity of feeding and fertilisation, but it will also be substantially influenced by sediment characteristics such as the grain size distribution (porosity), the exchange rate of water over the sediment and whether or not aerators are used (Kochba *et al.*, 1994; Hussenot & Martin 1995).

Hussenot & Martin (1995) evaluated the use of four parameters to measure sediment quality:

- redox potential (E_h)
- acidity/alkalinity (pH)
- hydrogen sulphide potential (pH_2S)
- NH_4-N.

E_h is a measure of the flow of electrons as oxidation or reduction reactions occur, and it is easily and rapidly monitored using electrodes (the higher the value the more oxidised is the substance). A highly reduced sediment usually gives a negative value, although different electrodes vary. E_h is also affected by pH and temperature so calibration is necessary, usually using a buffer of known potential to pH 7.0 and 25°C. As microbial populations increase in enriched sediments, their demand for O_2 will increase proportionately. Reducing conditions will occur as available DO is consumed. For example, sulphate is reduced to potentially toxic sulphide at an E_h of –200 mV, while methanogenesis occurs at –250 mV.

Fig. 3.4 shows selected depth profiles of E_h in the sediments of some different shrimp ponds. Hussenot & Martin (1995) presented them as representing three types of E_h profiles in sediments:

(a) The profile of a well-oxidised (aerobic) sediment, showing a high E_h value down through 20+ cm depth. This corresponds to a new or well-managed pond.
(b) A profile showing good surface oxidation and then reduced conditions below 5 cm depth. This corresponds to ponds with intensive culture and aeration of ponds in which there has been poor soil preparation between the crops, e.g. no drying out of the pond bottom or removal of sludge.
(c) A sediment profile showing poor oxidation throughout the range of soil depths. This condition prevails as a further step in deterioration from type (b). It occurs where there is inadequate aeration and in confined ponds where there are a series of successive crops without soil treatment between them. Some E_h values in type (c) are less than –200 mV, the level at which sulphate is reduced to the potentially toxic sulphide. Even if these conditions do not cause heavy mortality of the cultured shrimp, they will adversely affect production.

Sediment pH is usually measured on the same

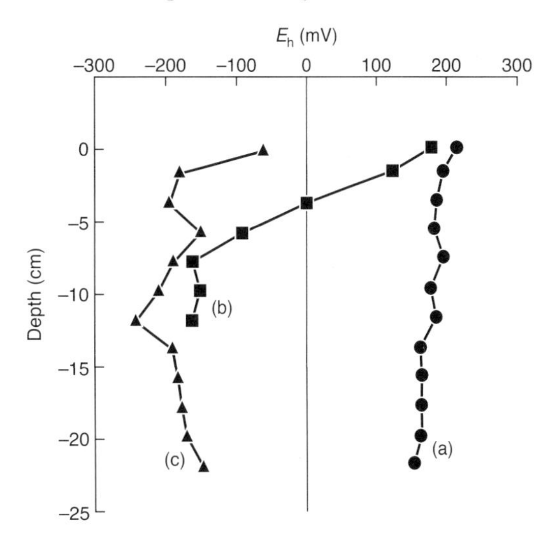

Fig. 3.4 Depth profiles of the redox potential (E_h) in three kinds of shrimp pond sediment. See the text for explanations of lines a, b and c. Reproduced from Hussenot & Martin (1995) with permission from Kluwer Academic Publishers.

core sample used for E_h, and for convenience a combined electrode is often used. Clifford (1992) considered that pH 7.0 was the lowest pH value not to have adverse effects on shrimp. In marine sediments having a salinity of 35‰, non-ionic hydrogen sulphide concentration (mg/L) is given by:

$$[H_2S] = 2.91 \times 10^{(4-pH_2S)}$$

where pH_2S is related to the activity a of the non-ionic form of H_2S:

$$pH_2S = -\log a_{H_2S}$$

Total hydrogen sulphide is the sum of unionised H_2S and the ionised forms HS^- and S^{2-}, although the latter is generally negligible in seawater and marine sediments. The concentration of HS^- (mg/L) can be obtained from:

$$[HS^-] = 4.59 \times 10^{(pH - pH_2S - 2.98)}$$

Sulphide oxidation is mediated by micro-organisms in the sediment, although it can also occur by purely chemical processes (Boyd, 1995):

$$H_2SO_4 + 2H^+ \rightarrow H_2SO_3 \text{ (sulphite)} + H_2O$$

$$H_2SO_3 + 2H^+ \rightarrow H_2SO_2 \text{ (sulphoxylate)} + H_2O$$

$$H_2SO_2 + 2H^+ \rightarrow H_4SO_2 \text{ (sulphur hydrate)}$$

$$H_4SO_2 + 2H^+ \rightarrow H_2S \text{ (hydrogen sulphide)} + 2H_2O$$

3.3 Effects of water quality parameters on animals

There has been major progress in aquaculture technology in the past three decades, although a number of expensive failures by commercial companies have demonstrated that there are weaknesses in our knowledge of water quality. Thus, this plays a key role in total system performance. Huge investments in large-scale, modern production systems have often failed after a number of years of operation because of unknown water quality requirements under a given set of operational circumstances. Indeed, many of the recent successes in aquaculture development have related to technologies deployed in open water areas where extensive supplies of natural water have been available.

Water quality criteria in European freshwaters have been studied by the European Inland Fisheries Advisory Commission (EIFAC), and are summarised by Alabaster & Lloyd (1980). Although there is no equivalent data bank for marine waters, Poxton & Allouse (1982) identified the conditions that are both completely safe and directly harmful to natural populations. Water quality criteria considered to be appropriate for intensive fish culture are summarised in Table 3.7. Similar criteria for cultured crustaceans and molluscs are presented in Tables 3.8 and 3.9 respectively.

The effects of these water quality parameters tend to follow three patterns (Fig. 3.5). In Fig. 3.5 'Performance' would most obviously be reflected in rates of growth and survival, but it may also be reflected in other factors, such as fecundity.

(1) There is an optimum level of the water quality parameter for rates of growth and survival, and this may be a very narrow or a broad range of optimum conditions. Environmental parameters that influence aquatic animals in this way are, for example, salinity, pH and temperature (Tables 3.7–3.9).
(2) Performance declines progressively with increases in the parameter until there is no survival. This is characteristic of toxic substances, such as nitrogenous compounds (note 'less than' values for NH_3, NO_2^- and NO_3^- in Tables 3.7–3.9), heavy metals, insecticides and other pollutants.
(3) Performance increases progressively with increases in the parameter until it reaches an upper level. There is little effect of further increase in the parameter, except at extremely high levels, when there may be an adverse effect. Examples of this pattern are for parameters DO (note 'greater than' values in Tables 3.7–3.9), food availability and dissolved nutrient concentrations.

The other pattern that is evident from Tables 3.7–3.9 is that the tolerances and optimum conditions for these animals are related to their normal environments. For example:

Table 3.7 Water quality criteria for intensive fish culture

Parameter	Species	Concentration (mg/L)
Oxygen (DO)	Salmonids	> 5.0–6.0
	Most finfish	> 5.0 (90% sat)
	European eel	> 4.0
	Common carp	> 3.0
	Channel catfish	> 3.0
	Tilapia	> 3.0
Carbon dioxide	Most fish	< 6.0 as CO_2
Ammonia	Rainbow trout	< 0.04 as NH_3-N
Nitrite	Salmonids	< 0.01 as NO_2-N
	Most freshwater fish	< 0.02 as NO_2-N
	Most marine fish	1.0 as NO_2-N
Nitrate	Most fish	< 50 as NO_3-N
Suspended solids	Most fish	< 15 (dry wt)
		Temp. (°C)
Temperature	Salmonids	12–17
	European eel	22–26
	Common carp	25–30
	Channel catfish	28–30
	Tilapia	29–32
pH	Freshwater fish	6.0–9.0
	Most marine fish	6.5–8.5

Reproduced from Poxton (1991b) with permission of the European Aquaculture Society.

(1) Some freshwater fish and shellfish live in conditions where low DO values occur and they may be reared without adverse effects at lower DO levels than most other species.
(2) Tropical fish and shellfish have higher optimum temperature ranges than temperate species.
(3) The penaeid shrimps, which live in brackish waters, may be cultured successfully over a broader salinity range than marine and freshwater species of decapod crustaceans.
(4) Marine fish, living in an environment where pH changes are well buffered, tolerate a narrower pH range than freshwater species, which live in an environment that is less buffered.

It is possible that the stressful environment created in culture systems will be overcome by domestication, in that, after successive generations, cultivated organisms may become adapted by selection to their new environment. Since, however, the nutritional and disease statuses of cultured animals are also important, in that poor feeds and infections can reduce their ability to withstand deteriorations in water quality, there may in practice be no absolute ‘safe’ levels (Poxton, 1990). Hence, all that can reasonably be expected at present is the identification of water quality conditions most likely to lead to success.

3.3.1 Temperature

In common terminology, cultured aquatic animals are ‘cold-blooded’. In biological terminology, they are described as

- poikilothermic (with fluctuating body temperature)
- ectothermic (with body temperature determined by the environment).

Table 3.8 Water quality criteria for culture of crustaceans

Parameter	Species	Concentration (mg/L)
Oxygen (DO)	Freshwater prawn (*Macrobrachium* species)	> 5.0
	American crayfish (*Procambarus* species)	> 3.0
	Australian crayfish (*Cherax* species)	> 5.0
	Marine lobster (*Homarus* species)	> 6.0
Carbon dioxide	Marine shrimp	> 12 as inorganic C
Ammonia	Marine shrimp	< 0.10 as NH_3-N
	Homarus species	< 0.12 as NH_3-N
Nitrite	Marine shrimp	< 0.50 as NO_2-N
	M. rosenbergii	< 0.61 as NO_2-N
Nitrate	Marine shrimp	< 50 as NO_3-N
Salinity	*Penaeus monodon*	20–33 g/L
	Homarus species	30–33 g/L
	Cherax species	< 12 g/L
	Procambarus species	< 5
Total hardness	Freshwater crayfish	> 50
		Temp. (°C)
Temperature	*M. rosenbergii*	29–34
	Penaeus monodon	13–33
	Cherax species	20–33
pH	Marine shrimp	6.5–9.0
	Procambarus species	6.5–8.5

Fish and shellfish therefore lack the means of controlling body temperature other than by behaviour, such as seeking warmer or colder water conditions when these are available within their environment. As the temperature increases, their metabolic rate and oxygen consumption increase. They become more active, and their rates of feed intake and digestion increase. The metabolic rate often increases by a factor of up to 2–3 for a 10°C rise. Consequently, they use more DO and produce more CO_2 and other excretory products, such as ammonia, as temperature increases. As growth and food conversion efficiency (FCE) are bound to metabolic rate through energy expenditure or storage, activity or stress may affect growth at different temperatures as a result of increased energy demand. Conversely, as the temperature decreases, the metabolic rate and all the other processes that are influenced directly by the temperature or by the metabolic rate decline. Ectothermic animals can be likened to a test tube filled with a multitude of chemical reactions that increase and decrease in rate with temperature changes.

Some consequences of the temperature for aquaculture are listed below.

- The feeding regime must be adjusted according to the prevailing temperature regime.
- It is important to avoid subjecting animals to abrupt temperature changes.
- The water temperature can be lowered during transport of stock to reduce activity, toxic waste accumulation and stress.
- It is very important to know the temperature tolerances of a species and to relate these to local water temperatures and climate when selecting culture sites.
- The prevailing environmental temperatures will strongly affect grow-out periods.

Growth rates of fish and shellfish increase up to

Table 3.9 Water quality criteria for mollusc culture

Parameter	Species	Concentration (mg/L)
Oxygen (DO)	Surf clam (*Spisula solidissima*)	> 2.3
Ammonia	Surf clam (*Mercenaria mercenaria*)	< 0.0014 as NH_3-N
	American oyster (*Crassostrea virginica*)	< 0.0014 as NH_3-N
Nitrite	*M. mercenaria*	< 0.14 as NO_2-N
	C. virginica	< 0.14 as NO_2-N
Nitrate	*M. mercenaria*	< 50 as NO_3-N
		Temp. (°C)
Temperature	European oyster (*Ostrea edulis*)	5–18°C
	S. solidissima	5–25°C
	C. virginica	8–36°C
	M. mercenaria	5–33°C
pH	*S. solidissima*	> 7

an optimum level, above which the increased energy requirements for feed conversion and other metabolic processes ensure that the 'law of diminishing returns' applies. Lower FCEs also occur, partially due to increased wastage, so that optimum temperatures for cultivation are determined by economic as well as biological considerations. The temperature range for growth is narrower than that for survival, while that for spawning, egg survival and hatching is narrower than for other life stages. In general, the egg is the stage most vulnerable to the effects of thermal stress.

Optimum temperatures for FCE, growth and reproduction are more useful to culturists than are lethal limits. They are not easy to identify, as they vary with age, and many factors such as light, salinity, diet, activity and crowding, which may interfere and interact significantly.

Egg incubation rates, larval size at hatching and rates of energy reserve utilisation are related to temperature (Poxton, 1991a). For example, no viable Dover sole (*Solea solea*) larvae hatch at 22°C, whereas ~10% will hatch at 19°C, but many will be abnormal. At 16°C and 13°C, survival to hatching is high, reaching 100% at 10°C, but larval growth rate is very poor at 10°C and reaches a maximum between 23°C and 24°C. As the temperature increases, the time to metamorphosis decreases and occurs at a smaller size. The time to metamorphosis is 2 weeks at 22°C, when the larvae are 9 mm; at 13°C it takes 4 weeks, but the larvae are 10 mm. Growth is very poor at 10°C and ceases altogether at 7°C (Fonds, 1979).

The temperature tolerance is also affected by past thermal history (i.e. temperature acclimation). The previous temperature range of an animal, especially fish, influences heat and cold tolerance when the animal is moved to new temperatures. As acclimation temperatures increase, so optimum and lethal temperatures also progressively increase. Acclimation to increased temperature is a rapid process, occurring within a day at temperatures above 20°C. The loss in this increased tolerance and the gain in resistance to low temperatures takes 2–4 weeks. The relationship between the temperature requirements of fish and their final preferred temperature and lethal limits, and how their thermal responses vary with acclimation temperature, are shown in Fig. 3.6. It is important to appreciate that for most species the final preference is several degrees higher than the

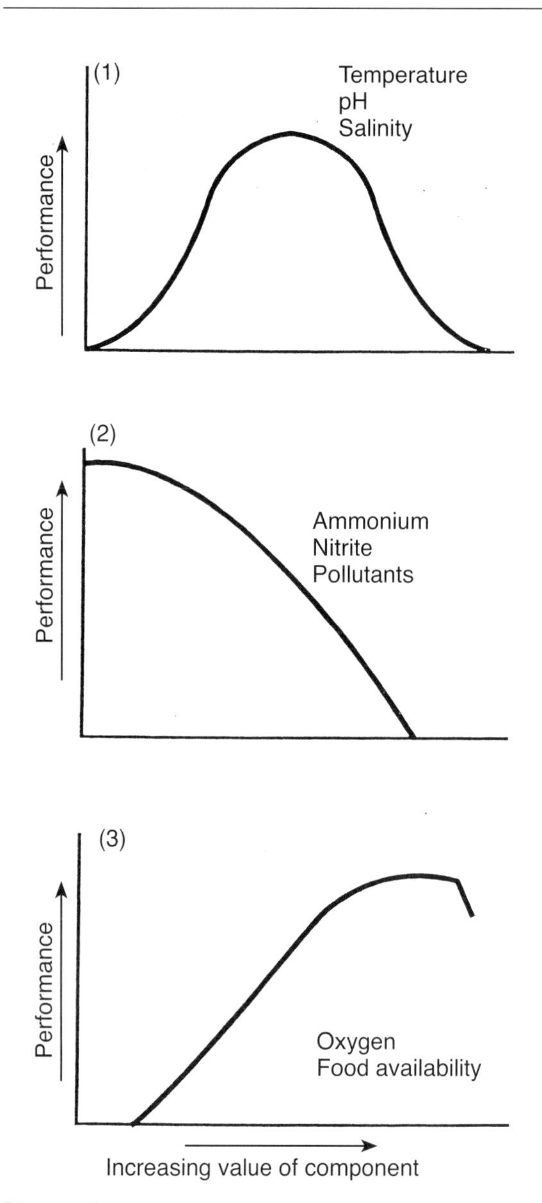

Fig. 3.5 Three curves showing different patterns for the influence of environmental parameters on the performance of cultured animals. After Tomasso (1996) with permission of the World Aquaculture Society.

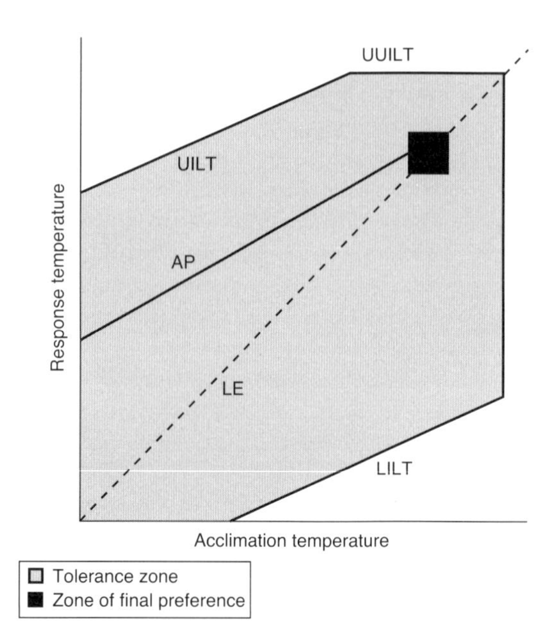

Fig. 3.6 The temperature relations of fish. AP, acute thermal preference; LE, line of equality; LILT, lower incipient lethal temperature; UUILT, ultimate upper incipient lethal temperature; UILT, upper incipient lethal temperature. Reproduced from Poxton (1990) with permission from the European Aquaculture Society, after Jobling (1981) with permission from Elsevier Science.

optimum temperature at which growth rate is highest when fish are given excess feed (Fig. 3.7). The strong correlations shown in Figs 3.6 and 3.7 are important in that determinations of preferred temperatures are much easier to perform than long-term growth trials. This allows aquaculturists to follow changes in the preferred temperature of their animals with age or other factors (McCauley & Casselman, 1981) and so adjust water temperature when this is a management option. Early aquaculturists sought to maximise growth, but it is far better to maximise economic return, which will also reduce wastage and environmental pollution.

No single water quality parameter should be considered in isolation, and this applies even, for instance, to the role of temperature in egg incubation (Poxton, 1991a). For example, yolk utilisation efficiency in larval turbot (*Scophthalmus maximus*) is higher at 15°C in seawater of 32‰ S than at 21°C in brackish water of 17‰ S, but declines with time under all experimental conditions.

Both turbot and Atlantic halibut (*Hippoglossus hippoglossus*) display an ontogenetic change in optimum temperature for growth, with lower temperature optima as fish grow bigger. Extended photoperiods or continuous light can enhance the growth rate in these juveniles. Both growth rate and growth efficiency of juvenile turbot and halibut vary among

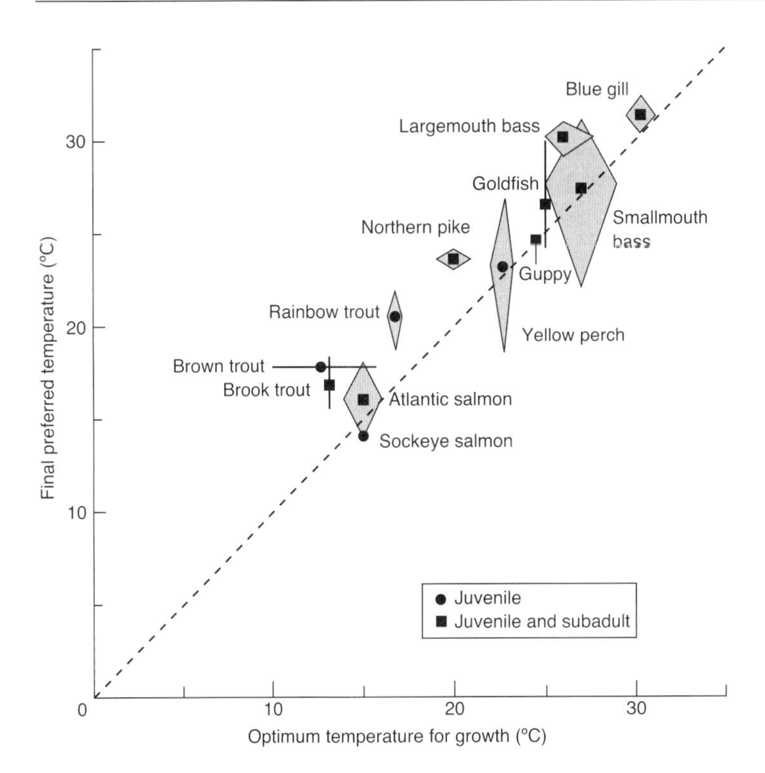

Fig. 3.7 The relationship between final preference and optimum temperature for growth in 12 selected fish species. Reproduced from Poxton (1990) (after McCauley & Casselman 1981) with permission of the European Aquaculture Society.

population groups along a latitudinal gradient, suggesting that a shorter growing season for populations at higher latitudes is compensated by better growth performance. This effect seems to be mixed with temperature adaptation, with growth performance increasing with latitude and temperature optima being related inversely with latitude (Imsland & Jonassen, 2001).

3.3.2 Salinity

The direct influence of salinity on aquatic animals is through its effects on the concentration, and hence osmotic pressure of the animals' body fluids. In lethal salinities, the animal is unable to maintain its body fluid and ion concentrations within tolerated ranges.

In many cases, the animal's body fluid concentration is not usually equal to the environment. Animals in freshwater have hypertonic body fluids (greater than the osmotic pressure of the environment). At higher salinities, molluscs conform to the environment (isotonic body fluids) but fish go from being strongly hypertonic in freshwater species to being strongly hypotonic in marine species (body fluids greater than osmotic pressure of the environment). Decapod crustaceans tend to be like fish, but show more variability. The body fluid/environment differences in osmotic pressure (and ion concentrations, even when the body fluids are isotonic) involve energy-requiring transport of ions such as Na^+ and K^+ across the animal's body surface. It is when these ion transport mechanisms cannot cope with the gradient between the environment and the body fluids that lethal changes can occur in the latter.

There are major changes in metabolic rate in aquatic animals when they live in different salinities. These are more than would be predicted purely on the basis of regulatory energetics at different salinities. Factors such as behaviour must be also involved. This means that for aquaculture the feeding regime may need to change with the salinity regime, and that variations in salinity will be a factor in growth rate. Furthermore, salinity may alter thermal responses by lowering the optimum temperature at lower salinities.

The majority of aquatic organisms are limited to marine or freshwater habitats and have limited tolerance of salinity variations (they are stenohaline). Other aquatic organisms inhabit environments of varying salinity, e.g. estuaries, mangroves and hypersaline lakes, and these of necessity tolerate varying degrees of salinity variation (they are euryhaline). Some other species migrate between different salinity environments at different phases in their life cycle or in different seasons, e.g. in relation to breeding. Thus, anadromous Atlantic salmon ascend rivers from the sea to breed in freshwater (section 15.2.2), whereas catadromous barramundi descend rivers to high-salinity conditions to breed. The gonads of barramundi do not mature in low salinities (section 18.2.2). The salinity requirements of the larval and the juvenile stages may be quite different to those of the adult stages. For example, the larval stages of the freshwater prawn, *Macrobrachium rosenbergii,* cannot tolerate freshwater and must be reared in more saline conditions (section 20.3.5).

When considering the importance of salinity for aquaculture, it is relevant to note that aquaculture organisms tend to be relatively euryhaline. This is partly because some species, e.g. tilapias and shrimp, live in brackish habitats, but for many species their degree of salinity tolerance reflects the general 'robustness' required in the selection of an aquaculture species. Thus, carps can be reared in low-salinity water (section 5.3) and rainbow trout can be reared in seawater. The subtropical–tropical marine species used in aquaculture tend to be inshore species that live in habitats where they are naturally subject to salinity variations in rainy seasons. They must have a natural degree of euryhaline tolerance.

3.3.3 Oxygen

The physical factors influencing oxygen levels and respiration in water and in air are overwhelmingly in favour of air (Schmidt-Neilson, 1997).

(1) As described previously, water contains at best ~0.005% of the oxygen in air per unit volume. Thus, aquatic animals require vastly more medium to pass across their respiratory surfaces than terrestrial animals to extract the same amount of oxygen.
(2) The density and viscosity of water are magnitudes greater than air, so the energy required to ventilate the respiratory surfaces is much greater in water.
(3) The rate of diffusion of oxygen in water is very low compared with the rate in air and quite inadequate for passive oxygen diffusion from the environment to supply the needs of all but the most minute organisms.

All these factors work against respiration in aquatic environments. This means that aquatic animals must have very effective respiratory systems for obtaining oxygen. Also, oxygen levels in aquaculture systems with a high biomass of cultured animals per unit volume of water must have mechanisms to sustain acceptable levels of DO and minimise potentially disastrous fluctuations.

Some aquatic animals, such as some bivalve molluscs, can switch to anaerobic metabolism for long periods when DO levels are low. Others, such as many crustaceans and fish, have little capacity for anaerobic metabolism and rapidly die in low DO levels. Therefore, it is important to monitor DO levels in aquaculture ponds and tanks, avoiding crowding and excessive phytoplankton blooms, and providing aeration when needed to sustain DO levels.

Rates of growth, activity and survival are limited by DO supply, which is the primary water quality parameter to be assessed in farm site selection. As surface water supplies are subject to natural fluctuations and can rarely be maintained at 100% saturation, even having chosen a 'good site', many freshwater farmers find their production limited by DO.

When fish are exposed to low DO, they react immediately. Resistance depends on temperature, activity level, feeding rate, species and age. Reports on the behavioural responses of fish exposed to low DO are often contradictory and include a decrease in activity, an increase in activity and avoidance. Elevated heart rates and respiration rates are also a kind of increased activity, but are usually temporary effects. Acclimation to reduced DO increases the ability of fish to tolerate still lower DO (Alabaster & Lloyd, 1980). For example, trout acclimated to 3 mg/L DO subsequently tolerated DO levels between 1.0 and 1.2 mg/L, whereas fish acclimated to 10–11 mg/L subsequently tolerated DO levels no lower than 1.8 mg/L.

Oxygen is needed as well as food for metabolism and growth. Accordingly, both energy requirements and food availability will determine the limiting level of DO. In nature, a balance is maintained between food eaten and the O_2 needed for its metabolism, but this condition cannot necessarily be reached in culture. The inability to take up sufficient O_2 may limit the growth rate, even at or above saturation levels when food is abundant. Any marked reduction in DO from full saturation can, at moderately high temperatures, impair growth. For example, the growth rate of silver bream (*Sparus sarba*) decreases when DO is less than 60% of saturation and FCE decreases markedly at 40% DO. Reported critical DO levels influencing growth rate are difficult to compile owing to differences in their definition.

A minimum constant DO level of 5 mg/L should be satisfactory for most stages and activities of freshwater fish and shellfish provided that other environmental factors are favourable. A similar level should be sufficient for most marine species. Hardy species such as common carp (*Cyprinus carpio*), channel catfish (*Ictalurus punctatus*) and tilapia (*Sarotherodon* and *Oreochromis* species) can tolerate 3 mg/L for short periods, whereas 4 mg/L is acceptable for the European eel (*Anguilla anguilla*) (Table 3.7). By statistical treatment of the available data, Davis (1975) derived DO criteria based on percentage saturation at three levels of protection. A, B and C are high, medium and low levels of protection:

(1) A is the safest level and is derived from the mean (level B) threshold of DO for the group +1 SD (standard deviation) above the mean.
(2) B represents the level at which the average members of a fish population of the same species start to show symptoms of distress as a result of low DO (the mean threshold).
(3) C is a rather dangerous level that should normally be avoided, being derived from the mean (level B) –1 SD below the mean.

This seems a very reasonable approach because it is neither realistic nor meaningful to use a defined DO level for all water bodies regardless of salinity, ionic composition or natural fluctuations (Poxton, 1991b).

Some shellfish, such as freshwater crayfish and table oysters, can tolerate DO levels well below 3 mg/L for brief periods.

3.3.4 pH

The most important mechanism for acid–base regulation in fish utilises the transfer of appropriate ions between intra- and extracellular body fluid compartments and the surrounding water. Although some transfer takes place across the skin, with mucus forming an important barrier, by far the most important site is the gill surface. Ionic transfer is facilitated by the large surface areas of the gills, high water flow rates and the semi-permeable nature of the branchial epithelium.

Regulation is mainly achieved by dynamic manipulation of the net transepithelial fluxes of Na^+ and Cl^- (Schmidt-Nielsen, 1997). The dominant mechanism for correcting internal acidosis is the simultaneous reduction of Cl^- uptake and, to a lesser extent, increased Na^+ uptake. These responses transfer basic equivalents from the water to the fish, a process commensurate to the transfer of acidic equivalents from the fish to the water. It metabolically corrects the acid–base disturbance. The reduction of Cl^- uptake reduces HCO_3^- excretion, and the increase in Na^+ uptake enhances H^+ excretion (Rankin & Jensen, 1993). The ionic content of the surrounding water has a major modifying effect. Of most importance is the availability of appropriate counter-ions, but other factors such as $[Ca^{2+}]$ are also important in limiting transfer rates and plasma $[HCO_3^-]$.

H^+ excretion normally occurs down a concentration gradient, but in acidified water this gradient is reduced and at low pH may even be reversed. A pH of 6.0–9.0 is regarded as giving reasonable protection to freshwater fish (EIFAC, 1969), whereas a pH of 6.5–8.5 is appropriate for most marine species (Table 3.10). Fish can be expected to die below pH 5.0, although some may be acclimated to low values near pH 4.0. Lethal effects become almost inevitable at pH 4.0 and 9.5. The pH limits of embryonic and larval fish may differ markedly from those of fully developed young and adults.

The toxicity of several common pollutants, such as ammonia and cyanides, is markedly affected by pH changes within the normal range. pH toxicity also depends on the mineral content and buffering capacity of the water. The presence of metals such as iron can raise the danger point of low pH because of the precipitation of ferric hydroxide on the gills

Table 3.10 Effect of various pH ranges on various species of fish

pH range	Effect on fish
3.0–3.5	Rapidly lethal to most species
4.0–4.5	Likely to be harmful to most species that have not been acclimated. Fish resistance increases with age and size
5.0–6.0	Unlikely to be harmful unless free CO_2 is > 20 mg/L or iron salts are present. Can reduce feeding of some marine species and may cause mortality
6.0–6.5	Unlikely to be harmful to fish unless free CO_2 is >100 mg/L
6.5–8.0	Harmless, although changes within this range may have an indirect effect, by changing the toxicity of other poisons
8.0–9.0	Feeding may be affected in marine fish larvae although juveniles may be acclimated
9.0–9.5	Likely to be harmful to marine fish larvae
9.5–10.5	Lethal to marine fish with prolonged exposure, but can be withstood for short periods
10.5–11.0	Prolonged exposure to the upper limits of this range is lethal to cyprinids (carps, etc.)
11.0–11.5	Rapidly lethal to all species of fish

Reprinted from Poxton & Allouse (1982) with permission from Elsevier Science. Modified from EIFAC (1969) with permission from Elsevier Science.

(EIFAC, 1969). For example, fish that tolerated a pH of 4.8 were all dead at pH 5.6 in the presence of 0.09 g/L iron. Aluminium in acid water damages fish gills, which become covered in mucus to such an extent that fry may suffocate. In many streams that are high in organic matter, however, the potential toxicity of metals may be low due to extensive chelation by organic acids. This would not necessarily lower toxicity if uptake of the metal was facilitated by chelation and, in any event, lethal effects can still occur due to high H^+ and low Ca^{2+} concentrations occurring concurrently. Freshwater samples, in particular, can have the same pH but differing kinds and amounts of dissolved materials. Lethal pH levels are affected by other variables such as water hardness, the concentration of free CO_2 and the presence of other toxic substances, as well as the animal's size, age, acclimation and duration of exposure (Poxton, 1990).

The effects of varying pH on the growth and survival of shrimp depend on the inorganic carbon contained in the water. In high-alkalinity water (inorganic C > 15 mg/L) shrimp can withstand a wider pH range than in low-alkalinity water (Wickins, 1976).

The interactions of many environmental and individual variables make it difficult to separate pH from other factors that may also contribute to mortality. Mortalities as a result of factors other than extreme pH can occur when the pH is approaching its lethal limits. Thus, the data in Table 3.10 must be regarded as tentative criteria for general predictions of the effects of pH on fish in culture.

3.4 The toxicity of metabolites

3.4.1 Carbon dioxide

In poorly buffered waters (e.g. 'soft' freshwater), relatively small amounts of free CO_2 from metabolism, e.g. from phytoplankton in a pond at night, can cause considerable changes in pH, which may be toxic. In addition, discharge of acid pollutants to waters with a high bicarbonate alkalinity will liberate enough CO_2 to be directly lethal to fish, even though the pH of the water at the culture site does not fall to a harmful level. In this situation, the animals' metabolites are adversely affecting them, as normal diffusion gradients are reduced, destroyed or, in extreme cases, reversed. A correlation between high CO_2 levels and nephrocalcinosis (white calcareous deposits of calcium phosphate in the kidneys) has been demonstrated in a number of rainbow trout farms (Poxton, 1990). As is typical in aquatic toxicology, the severity of the problem depends both on concentration and on exposure time, so broodstock are most likely to be affected.

3.4.2 Nitrogenous wastes

The end products of carbohydrate and lipid metabolism

are mainly H_2O and CO_2, which are excreted as such. The end products of protein metabolism are, however, excreted as a variety of compounds:

- ammonia
- urea
- uric acid
- trimethylamine oxide
- creatine
- creatinine
- other minor substances, including several nucleic acids.

Fish and shellfish excrete mainly NH_3, and fish also excrete urea. In fish, the NH_3 produced from the deamination of glutamine from the kidney is excreted as NH_4^+ in the urine and across the gills.

In general, factors that tend to increase stress or activity levels of fish increase rates of waste elimination. High temperatures, increased feeding rates and stocking densities are among the factors governing the metabolic rate and thus the excretion of ammonia. Ammonia production is directly proportional to feeding rates, but inversely proportional to fish size, stocking density and water flow rates. A rise of 8°C in ambient temperature may cause an order of magnitude increase in the rate of ammonia excretion in rainbow trout. This response may be reduced for less active fish. Rates of ammonia excretion are also a function of ambient ammonia concentration. For example, rainbow trout in freshwater containing no ammonia excreted about 130 mg/kg/day of NH_3-N, and in ambient ammonia of 5 mg/L NH_3-N decreased the excretion rate to about 100 mg/kg/day of NH_3-N. This occurs because NH_3, which can easily pass through the lipid membranes of the gills, is exchanged down a diffusion gradient, and any increase in ambient ammonia decreases the rate of ammonia excretion. Indeed, an influx of NH_3 can occur if the external concentration is sufficiently high.

NH_3/NH_4 toxicity

The accumulation of ammonia in water is known to be one of the major causes of functional and structural disorders in aquatic animals. As described in section 3.2.7, the term ammonia is often used to include both the unionised gas (NH_3) in solution and the ionised NH_4^+, and as such is often expressed as total ammonia. Only unionised ammonia (NH_3) is very toxic in that it can readily diffuse across the gill membrane into the circulation, whereas the ionised form cannot. Haemoglobin in the blood loses the ability to combine with O_2 or to liberate CO_2 when fish are exposed to high NH_3.

pH plays an important role in determining the unionised NH_3 vs. NH_4^+ content of any water containing ammonia (section 3.2.7), thus modifying its toxicity to fish. Increases in the percentage of NH_3 are directly related to pH and temperature. The effect of free CO_2 in the water is related to this pH effect by decreasing the concentration of NH_3, and causing a reduction in NH_3 toxicity until the level of CO_2 itself starts to become toxic. Ammonia toxicity can also be influenced by salinity (Table 3.11). For example, rainbow trout are more resistant to ammonium chloride in seawater than in freshwater. Resistance increases threefold up to a maximum in 30% seawater concentration and then decreases at higher salinities. The effect of salinity in reducing NH_3 toxicity is related to the influence of salinity in lowering the percentage of NH_3 under equivalent pH and temperature conditions (Table 3.5; section 3.2.7).

NO_2 toxicity

Nitrite enters fish through the chloride cells across their gills, and uptake from the environment is so active that blood concentrations in the range of 10–70 times greater than that of their medium have been recorded. Once in the blood, it combines with haemoglobin to form methaemoglobin, which is unable to bind with, and therefore transport, oxygen. Chronic toxicity symptoms are those of hypoxia, caused by a decrease in the oxygen-carrying capacity of the blood, which can be seen to be chocolate brown rather than red. It is probably for this reason that a rise in temperature of 10°C may lead to a doubling in toxicity. Fish blood commonly contains 10% methaemoglobin, and even when levels approach 50% there is usually little or no mortality. Above this, problems are likely to occur. Toxic effects of nitrite have also been recorded in the absence of pronounced methaemoglobinaemia, but other causal mechanisms of toxicity are unclear (Jobling, 1994).

Table 3.11 Comparison of the toxicity of unionised ammonia to fish

Species	Salinity (‰)	Unionised ammonia (mg/L NH_3-N)	Notes
Acute			
Chinook salmon	0.0	0.30 (24 h)*	DO 8.2–9.8 mg/L, 12–13°C
(parr)	5.2	0.72	
	9.6	1.80	
	16.9	1.14	
	27.6	0.95	
Rainbow trout	0.0	0.46 (24 h)	pH 7.45, 13.6°C
(yearlings)	5.0	0.66	
	10.0	1.06	
	36.0	0.59	
Atlantic salmon	0.0	0.23 (24 h)	pH 7.51–7.81
(smolts)	18.0	0.29	
	27.0	0.27	
Atlantic salmon	0.0	0.12 (24 h)	DO 9.6 mg/L
(smolts)	10.2	0.23	DO 9.5 mg/L
	0.0	0.07	DO 3.5 mg/L
	10.2	0.09	DO 3.1 mg/L
Striped bass	12.0	0.87 (96 h)	pH 7.5–8.0, 15°C
	12.0	1.12	
	35.0	0.66	
	35.0	0.87	
Sea bream	40.5	1.59 (96 h)	DO 93‰
(juveniles)	40.5	1.05	DO 61‰
	40.5	0.80	DO 33‰
	40.5	0.34	DO 26‰
Sublethal			
Turbot	34.0	0.08	(11 days) Threshold for no effect on (Juveniles) growth
	34.0	0.30	No growth; pH 6.8
	34.0	0.90	no growth; pH 7.9
Dover sole	34.0	0.064	(14 days) Threshold for no effect on (Juveniles) growth
	34.0	0.38	No growth; pH 6.9
	34.0	0.59	No growth; pH 8.1

*Data in parentheses indicate duration of acute LC_{50} tests (lowest concentration killing 50% of the fish) or exposure duration in sublethal tests.

Reproduced from Handy & Poxton (1993) with permission from Kluwer Academic Publishers.

Fish in brackish water are more resistant to NO_2-N than fish in freshwater, e.g. 12 mg/L NO_2-N proved acutely toxic to milkfish juveniles in freshwater, but at 16‰ salinity tolerance increased to 675 mg/L NO_2-N.

As the NO_2^- transport system is the same as that of other monovalent ions, such as Cl^-, bromine (Br^-) and HCO_3^-, the presence of these ions reduces nitrite toxicity. Indeed, the relationship between NO_2 toxicity and Cl^- concentration is linear and for maximum protection a weight ratio (mg/L Cl^- to mg/L NO_2-N) of about 17 is necessary for trout in freshwater, while a ratio of about 8 is recommended for coarse fish (non-salmonids). The strength of the chloride effect is greatest for the least sensitive species and smallest for the most sensitive species such as salmonids.

HCO_3^- is only 1% as effective as Cl^- and Br^- in protecting against toxicity, while divalent and trivalent anions have very little effect (Jobling, 1994).

Other ameliorating effects include those of Ca^{2+}, which is thought to induce changes in gill permeability. For example, the acute toxicity of NO_2^- to steelhead trout (*O. mykiss*) in freshwater was reduced by a factor of about 24 for 5-g fish and 13 for 10-g fish when the total water hardness was increased from 25 to 300 mg/L $CaCO_3$. Consequently, caution must be applied to any data on nitrite toxicity that fail to take in to account or specify the ionic content of the water, together with other factors such as temperature, DO, pH and size of the fish. The susceptibility of different species to nitrite toxicity appears to be related to the extent to which they either concentrate nitrite in (salmonids and ictalurids) or exclude it from the plasma (centrarchids) (Jobling, 1994).

In natural seawater, containing 416 mg/kg Ca^{2+} and 19 g/kg Cl^- (Table 3.1), the toxicity of nitrite to marine and estuarine fish may be low. Considering the absence of significant quantities of data for chronic toxicity effects on marine fish, nitrite concentrations of 10% of those known to produce methaemoglobinaemia should be considered potentially toxic (Handy & Poxton, 1993).

Nitrification and denitrification may occur during toxicity studies, leading to mixtures of ammonia, nitrite and nitrate being present together. These poisons often exert greater toxic effects when more than one substance is present. It is therefore unfortunate that long-term, sublethal effects of such mixtures have not been extensively studied, especially as such conditions commonly occur in recirculating systems during the establishment of new biological filters (sections 2.4.4 and 3.2.7).

When natural waters contain NO_2 that is derived from sewage, they may also contain amines. These could form nitrosamines, which have carcinogenic or mutagenic effects, in the presence of bacterial contamination. Such circumstances are much more likely to occur in farm water than in drinking water, in which bacterial numbers are controlled.

NO_3 toxicity

The increased use of fertilisers on arable farmland has led to increased NO_3^- levels in many water bodies. In recirculating culture systems, NO_3^-, which is the end product of nitrification in biological filters, will accumulate unless a denitrification or plant filter is installed (Fig. 3.8). Although NO_3^- is not acutely toxic to fish, it must not be left to accumulate, as phytoplankton blooms, inhibition of nitrification and chronic toxicity may eventually result. Adverse effects on osmoregulation and oxygen transport have been recorded, and dietary aspects must also be considered.

Freshwater fish such as channel catfish and largemouth bass (*Micropterus salmoides*) are tolerant of concentrations as high as 400 mg/L NO_3-N, as are turbot in seawater (Poxton & Allouse, 1982). One must, however, consider the possibility of denitrification occurring in the gut, particularly in herbivorous freshwater species. Indeed, the European Commission (EC) stipulated that drinking water must not contain > 50 mg/L NO_3-N, as blue baby syndrome has been linked to methaemoglobinaemia. Furthermore, the presence of nitrosamines and nitrosamides in foodstuffs, including fish, has been linked to stomach cancer in humans (Garrett, 1993). It would therefore seem prudent to also set this as the limit for fish culture in order to limit human dietary intake of nitrate (Poxton, 1990). This is particularly so as the EC target is to reduce this standard to

Fig. 3.8 'Algal scrubber' units within a recirculating seawater system. Water from the system passes in a shallow layer over a mesh screen densely covered with macroalgae, which removes N compounds from the water. The algae are periodically harvested to remove the N from the system. Such units must be exposed to high light levels (photograph by Dr Paul Southgate).

25 mg/L NO_3-N, and the World Health Organization (WHO) guideline has been set at 10 mg/L NO_3-N.

3.4.3 Hydrogen sulphide

Chien (1992) estimated that the toxicity of H_2S to the black tiger shrimp (*Penaeus monodon*) was 0.05 mg/L (96 h LC_{50}, i.e. the lowest concentration at which there was > 50% survival after 96 h of exposure). This corresponded to a pH_2S of 5.75. For Kuruma shrimp (*Marsupenaeus japonicus*), growth was arrested when pH_2S concentrations of 5–6 were recorded 1 cm below the surface of the sediment (Hussenot & Martin, 1995). Chronic toxicity to bluegill sunfish (*Lepomis macrochirus*), recorded as a lowering of fecundity, was noted at only 0.001 mg/L H_2S, corresponding to a pH_2S of 7.45 (Smith *et al.*, 1976). Hussenot & Martin (1995) indicated that a value of 7.5 for pH_2S should be considered as the danger limit for shrimps or molluscs living in contact with the sediment. H_2S produced in bottom sediments beneath cages of salmonids has been implicated in gill damage and mortality. H_2S production will, however, be at least an order of magnitude less under freshwater cages than in seawater as a result of the relative scarcity of sulphur in freshwater. Under most conditions H_2S will be quickly oxidised to non-toxic SO_4^{2-} ions.

3.5 Maintenance of water quality

Apart from temperature control in hatcheries and the precaution of using a sub-sand extraction system, the management of water quality in flow-through systems is generally limited to aeration and pH control. The control of suspended solids, BOD and COD is mainly aimed at achieving effluent standards of typically 10, 3 and 2 mg/L respectively. It should be noted, however, that farm influents may be in excess of 10 mg/L for suspended solids (Table 3.6) and these would clearly benefit from a sub-sand extraction system, or a settlement pond/tank to remove suspended solids before use in culture.

Means of removing particulate matter from water in closed or recirculating aquaculture systems are discussed in section 2.4.4. Suspended solids are generally removed from such systems using self-cleaning screen filters (Fig. 3.9).

3.5.1 Temperature control

Many hatcheries, including bivalve, shrimp and fish hatcheries, use large temperature-controlled rooms to control the air temperature and hence the water temperature in their culture tanks. The water temperature is often raised to increase the growth rate of larvae, but the temperature may have to be lowered when ambient temperatures outside the hatchery building are above the optimum. In addition, when intake water differs markedly from the required temperature, it may be heated or cooled, usually with a heat exchanger. Care must be taken to avoid gas supersaturation (Lawson, 1995; Midlen & Redding, 1998).

Air-conditioning large hatchery rooms and heating or cooling the large volumes of water used are expensive operations, and some methods have been developed for reducing these costs. In some cases involving freshwater fish, subterranean sources of heated water are used for both hatchery and later development. Heated effluents from industry, e.g. large electrical power generators, can be used directly when they are unpolluted or used as a heating source. Counter-current heat exchange systems are another means for conserving energy; the cooled or heated water being discharged from the hatchery tanks and the intake water are passed in opposite directions in close proximity over a large surface area within the exchanger, promoting heat transfer. The result is that

Fig. 3.9 A rotating drum filter removes solids by passing culture water through fine-mesh screens (photograph by Dr Sagiv Kolkovski).

the intake water is heated or cooled, as appropriate, to approach the temperature of the discarded water and considerable energy is saved. Temperature and light (photoperiod) manipulations are also frequently coordinated.

Use of heat exchange and greenhouses to control water temperature is described in sections 5.3.1, 5.3.2 and 5.5.

3.5.2 Aeration and ozonisation

An aeration system has five main uses in aquaculture:

(1) re-aeration of waters when DO is low
(2) emergency provision of DO in the case of pump failure or in low-flow, high-temperature situations
(3) control of CO_2 levels (this may be particularly important in broodstock maintenance)
(4) preventing mortalities in hatchery tanks by ensuring that eggs or larvae remain suspended in the water column
(5) degassing and problems related to Oversaturation of water with gases.

Oversaturation can occur naturally or as a result of poor design, and even municipal supply systems cannot be relied upon to be unaffected. Oversaturation is a particular problem in freshwater fish hatcheries, where it can cause gas bubble trauma (Lawson, 1995).

Paddle-wheel aerators are generally the method of choice in pond cultivation (Fig. 3.10). Other types of aeration devices used in the pond culture of shrimp are described in section 19.8.3, and the means of maintaining DO levels in recirculating systems are outlined in section 2.4.4.

The emergency provision of DO is an important fail-safe mechanism that must not be overlooked in the more intensive forms of aquaculture. There is an important distinction between the production of batches of fish and shellfish and the production of batches of whisky, beer or chemicals: one cannot usually start with a new batch immediately, and sometimes the next breeding season must be awaited. A whole production cycle may be lost as a result of a system failure leading to lethal DO levels (or other lethal conditions), and this can have a major impact on cash flow as well as on profits.

Fig. 3.10 Paddles wheels are a commonly used form of aerating system in ponds (photograph by Dr Peter Appleford).

Ozonation is often used, despite its cost, in research facilities and in commercial recirculation systems for disinfection purposes, and to oxidise otherwise intransigent organic compounds that contribute to the yellowing of culture water (Christensen *et al.*, 2000). Ozonation can improve the efficiency of biological filters, by reducing suspended solids and total organic carbon (TOC), and therefore COD and usually BOD. The oxidised products will also serve to feed biological filters, and the number of bacteria suspended in culture water will be reduced (Summerfelt & Hochheimer, 1997). However, as ozone is toxic to cultured animals and to humans, it must always be carried out as a side-stream operation with great care, preferably using automatic systems that function at night in the absence of operators. It must never be used to excess: dosages as low as 0.05–0.30 mg/L O_3 with contact times as short as 0.3–2.0 min usually prove effective. A counter-current contactor, similar to a foam fractionator, is generally used (Lawson, 1995; Midlen & Redding, 1998).

3.5.3 pH control

The effects of low pH in ponds can be mitigated by the direct addition of basic substances such as limestone ($CaCO_3$), lime [CaO and $Ca(OH)_2$], soda ash (Na_2CO_3), olivine [$(Mg,Fe)_2SiO_4$], fly ash and industrial slag as neutralising agents. There is, however, no single accepted method of neutralisation. Furthermore, the effects of acidification on cultured animals are also partly a function of elevated levels of metals such as aluminium.

The lime requirement for a pond will vary with the particle size and organic content of the substrate, water and soil chemistry and water depth. Details of liming shrimp ponds are given in section 19.4.4.

For freshwater recirculating systems stocked at high density, a sodium bicarbonate ($NaHCO_3$) dosing methodology was developed to control pH, because nitrifying bacteria consume alkalinity at a rate of about 7 mg (expressed as $CaCO_3$) per 1 mg total ammonia nitrogen oxidised to nitrate (Loyless & Malone, 1997). Large amounts of CO_2 were generated and were controlled by air-stripping using simple airlift pumps.

3.6 Summary

Under natural conditions, aquatic animals are often subjected to a variety of changing water quality fluctuations that may or may not be of biological significance. Many parameters are inter-related in a way that precludes considering one parameter without considering the other possible interacting variables. Furthermore, there are variations in the adaptive response of the animals to one or multi-variants. Fish and shellfish are genetically adapted or capable of adaptation through acclimation or both.

Despite its deficiencies, research has provided information on the essential characteristics of each environmental factor and at least drawn attention to the complex inter-relationships. For example, the water quality levels that are completely safe or directly lethal to some of the commonest temperate freshwater fish are well identified for each parameter. Unfortunately, much less work has been done on many of the marine species now being cultured. There is danger in relying on information relevant for freshwater fish in the design of marine culture systems.

Many more studies of long-term chronic toxicity tests under aquaculture conditions are required before realistic criteria can be specified. The empirical approach of basing water management on conditions experienced on farms is inadequate, and useful only in the short term in the absence of specifically designed studies. This is due not only to differences in natural water quality on various farms, but also to differences between aquaculture species and their physiological state under test. For example, the effects of complex mixtures of, say, elevated levels of ammonia and nitrite need to be investigated while simultaneously considering the effects of low DO, high temperatures and high stocking densities. Information on safety margins in the event of an accident, such as a failure of the water or air supply, or an additional stress, such as grading, transportation or a disease challenge, is also required. Risks due to the effects of pollutants, such as toxic organic compounds, heavy metals and antifouling or disease treatments, whether these are caused by others using the water or by fish farmers themselves, also need to be evaluated.

References

Ahmad, T. & Boyd, C. E. (1988). Design and performance of paddle wheel aerators. *Aquacultural Engineering*, **7**, 39–62.

Alabaster, J. S. & Lloyd, R. (1980). *Water Quality Criteria for Freshwater Fish*. Butterworths, London.

Boyd, C. E. (1982) *Water Quality Management for Pond Fish Culture*, Vol. 9. Developments in Aquaculture and Fisheries Science. Elsevier, New York.

Boyd, C. E. (1995). *Bottom Soils, Sediment and Pond Aquaculture*. Chapman & Hall, New York.

Boyd, C. E., Ahmad, T. & La-fa, Z. (1988). Evaluation of plastic pipe, paddle wheel aerators. *Aquacultural Engineering*, **7**, 63–72.

Chien, Y.-H. (1992). Water quality requirements and management for marine shrimp culture. In: *Proceedings of the Special Session on Shrimp Farming* (Ed. by J. Wyban), pp. 144–56. World Aquaculture Society, Baton Rouge, LA.

Christensen, J. M., Rusch, K. A. & Malone, R. F. (2000). Development of a model for describing accumulation of color and subsequent destruction by ozone in a freshwater recirculating aquaculture system. *Journal of the World Aquaculture Society*, **31**, 167–74.

Clifford, H. C. (1992). Marine shrimp pond management: a review. In: *Proceedings of the Special Session on Shrimp Farming* (Ed. by J. Wyban), pp. 110–37. World Aquaculture Society, Baton Rouge, LA.

Cripps, S. J. (1993). The application of suspended particle size characterization techniques to aquaculture systems. In: *Techniques for Modern Aquaculture*, ASAE Conference, Spokane, Washington.

Davis, J. C. (1975). Minimal dissolved oxygen requirements of aquatic life with emphasis on Canadian species: a review. *Journal of the Fisheries Research Board of Canada*, **32**, 2295–332.

EIFAC (1969). Water quality criteria for European freshwater fish: extreme pH values and inland fisheries. *Water Research*, **3**, 593–611.

Fonds, M. (1979). Laboratory observations on the influence of temperature and salinity on development of the eggs and growth of the larvae of *Solea solea* (Pisces). *Marine Ecology Progress Series*, **1**, 91–9.

Garrett, M. K. (1993). Nitrogen losses – form, origin and impact. In: *Developments and Ethical Considerations in Toxicology* (Ed. by M. I. Weitzner), pp. 110–28. Royal Society of Chemistry, London.

Handy, R. D. & Poxton, M. G. (1993). Nitrogen pollution in mariculture: toxicity and excretion of nitrogenous compounds by marine fish. *Reviews in Fish Biology and Fisheries*, **3**, 205–41.

Huguenin, J. E. & Colt, J. (1989). *Design and Operating Guide for Aquaculture Seawater Systems*. Elsevier Science Publishers, Amsterdam.

Hussenot, J. & Martin, J.-L. M. (1995). Assessment of the quality of pond sediment in aquaculture using simple, rapid techniques. *Aquaculture International*, **3**, 123–33.

Imsland, A. K. & Jonassen, T. M. (2001). Regulation of growth in turbot (*Scophthalmus maximus* Rafinesque) and Atlantic halibut (*Hippoglossus hippoglossus* L.): aspects of environment × genotype interactions. *Reviews in Fish Biology and Fisheries*, **11**, 71–90.

Jobling, M. (1981). Temperature tolerance and the final preferendum – rapid methods for the assessment of optimum growth temperatures. *Journal of Fish Biology*, **19**, 439–55.

Jobling, M. (1994). *Fish Bioenergetics*. Chapman & Hall, London.

Jobling, M. (1995). *Environmental Biology of Fishes*. Chapman & Hall, London.

Kochba, M., Diab, S. & Avnimelech, Y. (1994). Modelling of nitrogen transformation in intensively aerated fish ponds. *Aquaculture*, **120**, 95–104.

Lawson, T. B. (1995). *Fundamentals of Aquaculture Engineering*. Chapman & Hall, New York.

Loyless, J. C. & Malone, R. F. (1997). A sodium bicarbonate dosing methodology for pH management in freshwater-recirculating aquaculture systems. *Progressive Fish Culturist*, **59**, 198–205.

McCauley, R. W. & Casselman, J. M. (1981). The final preferendum as an index of the temperature for optimum growth in fish. In: *Aquaculture in Heated Effluents and Recirculation Systems*, Vol. 2 (Ed. by K. Tiews), pp. 81–93. Heenemann Verlagsgesellschaft, Berlin.

Midlen, A. & Redding, T. A. (1998). *Environmental Management for Aquaculture*. Kluwer Academic Publishers, Dordrecht.

Poxton, M. G. (1990). A review of water quality for intensive fish culture. In: *Aquaculture Europe '89: Business Joins Science* (Ed. by N. de Pauw & R. Billard), pp. 285–303. Special Publication 12. European Aquaculture Society, Bredene, Belgium.

Poxton, M. G. (1991a). Incubation of salmon eggs and rearing of alevins: natural temperature fluctuations and their influence on hatchery requirements. *Aquacultural Engineering*, **10**, 31–53.

Poxton, M. G. (1991b). Water quality fluctuations and monitoring in intensive fish culture. In: *Aquaculture and the Environment* (Ed. by N. de Pauw & J. Joyce), pp. 121–43. Special Publication 16. European Aquaculture Society, Ghent.

Poxton, M. G. & Allouse, S. B. (1982). Water quality criteria for marine fisheries. *Aquacultural Engineering*, **1**, 153–91.

Rankin, J. C. & Jensen, F. B. (1993). *Fish Ecophysiology*. Chapman & Hall, London.

Schmidt-Nielsen, K. (1997). *Animal Physiology: Adaptation and Environment*, 5th edn. Cambridge University Press, New York.

Schuster, C. (1994). The effect of fish meal content in trout food on water colour in a closed recirculating aquaculture system. *Aquaculture International*, **2**, 266–9.

Smith, L. L., Oseid, P. M., Kimball, L. L. & El-Kaudelgy, S. M. (1976). Toxicity of hydrogen sulfide to various life history stages of bluegill (*Lepomis macrochirus*). *Transactions of the American Fisheries Society*, **105**, 442–9.

Summerfelt, S. T. & Hochheimer, J. N. (1997). Review of ozone processes and applications as an oxidising agent in aquaculture. *Progressive Fish Culturist*, **59**, 94–105.

Tomasso, J. R. (1996). Environmental requirements of aquaculture animals – a conceptual summary. *World Aquaculture*, **27**(2), 27–31.

Wickins, J. F. (1976). The tolerance of warm-water prawns to recirculated water. *Aquaculture*, **9**, 19–37.

4

Environmental and Other Impacts of Aquaculture

Simon Cripps and Martin Kumar*

4.1 Introduction

> Every person has a general environmental duty to take all reasonable and practical measures to prevent or minimise environmental harm.
>
> Dallas J. Donovan (1999)

The intensification of aquaculture and rapid growth of the industry over the past decade has had environmental impacts and has raised concerns for governments. After a period of often uncontrolled activity, concerns about the environmental implications of intensive aquaculture have increased significantly. Environmental impacts are now more frequently taken into account when aquaculture developments are undertaken. Aquaculturists must be environmentally aware to survive. The negative environmental impacts attributed to aquaculture have most often resulted from:

- poor planning
- inappropriate site selection
- inappropriate management procedures
- lack of attention to environmental protection.

There are legitimate environmental concerns about aquaculture, particularly because past errors in the design and management of aquaculture facilities have been made at the expense of the environment. Much publicity has surrounded the adverse impacts of aquaculture on the environment, and some has been justified. Now, however, the modern industry is maturing and adapting to current values and philosophies, and is responding to environmental concerns. This chapter describes the various interactions between the aquaculture industry and the environment, with emphasis on the means by which adverse impacts can be avoided or reduced.

There are also positive impacts of aquaculture on the environment. Aquaculture can be used as a tool to treat wastewater (Kumar *et al.*, 2000). Water and nutrients can be recycled through an aquaculture operation. When industries must comply with environmental standards that require treatment of effluent, this often constitutes an added operational cost. If, however, the treatment itself produces income, minimises pollution and complies with environmental standards, it not only increases profitability, but also enhances the sustainability of the industry.

*The views expressed in this chapter are those of the authors and are not the policy of WWF International, Gland, Switzerland, nor of SARDI.

The 'waste', which provides income by producing a valuable product, effectively becomes a 'resource'. A number of by-products, such as bio-energy (gas and heat), aquaculture products (fish) and aquatic plant and agricultural products, can be produced while organic waste is treated. Thus, this chapter also deals with the positive environmental impacts of aquaculture.

4.2 Aquatic pollution from land-based aquaculture

In general, environmental effects associated with extensive aquaculture systems are considered minimal. As the production and intensity of aquaculture increase, the potential impacts of pollutants from aquaculture also increase. A serious environmental impact of land-based aquaculture (including coastal areas) occurs when untreated effluent is discharged into surrounding bodies of water. Major environmental concerns about pond, tank, raceways and coastal farming include:

- destruction of natural habitats (natural vegetation, such as forests and mangroves)
- eutrophication and sedimentation in natural bodies of water caused by effluent
- excessive use of resources such as water, food stuffs and electricity
- negative effects on native fisheries and biodiversity.

A number of measures are currently used to minimise the negative impacts of land-based aquaculture. Although aquaculturists are progressing towards a system in which the discharge of effluent is minimised, the problem of effluent disposal continues to challenge the aquaculture industry.

4.2.1 Effluent

Sources of pollutants in aquaculture effluent

The environmental impact of aquaculture from effluent discharge depends strongly on species, culture methods, stocking density, food composition, feeding techniques and hydrography of the site. Nutrient-rich effluent discharged from aquaculture systems can cause reduced dissolved oxygen concentration (DO), raised biological oxygen demand (BOD) and increased suspended solids (SS), phosphorus (P) and nitrogen (N) compounds. The concentrations of P in freshwater and N in marine water can be considered as factors limiting the growth of aquatic plants. However, they are also important nutrients for cultured organisms, and enter farms in the form of plant fertiliser or animal feed. Nutrient compounds have been shown to be associated with suspended solids in effluent water. Thus, the removal of suspended particles from aquaculture effluents is advantageous in reducing its nutrient content.

Nutrients and SS in aquaculture effluent come from a variety of sources (Fig. 4.1). Particles are carried into the farm with the water supply and also produced within the culture system. In flow-through systems, particles consist mainly of uneaten or regurgitated food, faeces and fragmented tissue. In recirculation systems, additional inputs may come from water-conditioning devices, such as filter media escaping from retainment vessels or bacterial flocs breaking away from tertiary filter media (Cripps, 1993).

Depending on the species and culture techniques, up to 85% of the P and 52–95% of the N entering a fish culture system as feed may be lost into the environment through feed wastage, fish excretion, faeces production and respiration (Wu, 1995). Some feed may be lost from the tank or pond outlet or through the bottom of a cage before the animals have eaten it. This loss is increased with over-feeding. Nutrients also pass through the animals and reach

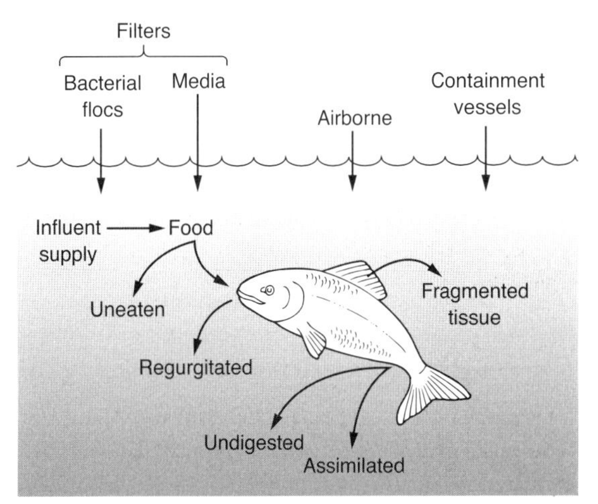

Fig. 4.1 Suspended particle sources in an aquaculture system.

the environment as faeces. If feeding management is incorrect, the food conversion ratio (FCR) will be poor, resulting in increased wastage of feed. In high-density cultures, or those in large volumes, it can be difficult to assess the amount of food required. In such cases, it is normal to err on the side of caution and to over-feed to ensure good growth. The FCR has, however, improved, and N and P levels in feeds have decreased as the aquaculture industry has developed (Figs 4.2 and 4.3). Similar feed wastage is evident in the farming of shrimp. In the Philippines, for example, it has been estimated that 4500 ha of intensive shrimp culture ponds used 43 200–108 000 mt of formulated feeds in 1992 and that ~15% of this (6480–16 200 mt) was not consumed (Primavera, 1994). Thus, careful feed management is an important aspect in limiting the environmental impacts of aquaculture, and an important economic consideration for aquaculturists.

A number of studies have estimated the total N and P discharged into receiving waters from aquaculture of various species. It has been estimated that for each metric ton of channel catfish (*Ictalurus punctantus*) produced, an average of 9.2 kg of N and 0.57 kg of P is discharged (Schwartz & Boyd, 1994). Similarly, for each metric ton of juvenile Atlantic salmon (*Salmo salar* L.) cultured in freshwater, environmental loadings of 71 kg of N and 10.9–11 kg of P have been reported (Kelly *et al.*, 1996). A recent study conducted in an intensive system for the culture of seabream (*Sparus aurata*) in earthen ponds estimated the direct consequences of discharge into receiving waters. The study estimated that 9105 kg of total suspended solids (TSS), 843 kg of particulate organic matter (POM), 235 kg of BOD, 36 kg of NH_4^+-N, 5 kg of NO_2^--N, 7 kg of NO_3^--N and 3 kg of PO_4^{3-}-P were discharged into the environment for each metric ton of fish cultured (Tovar *et al.*, 2000).

A further source of potential pollutants in aquaculture effluents is that from undigested food, which

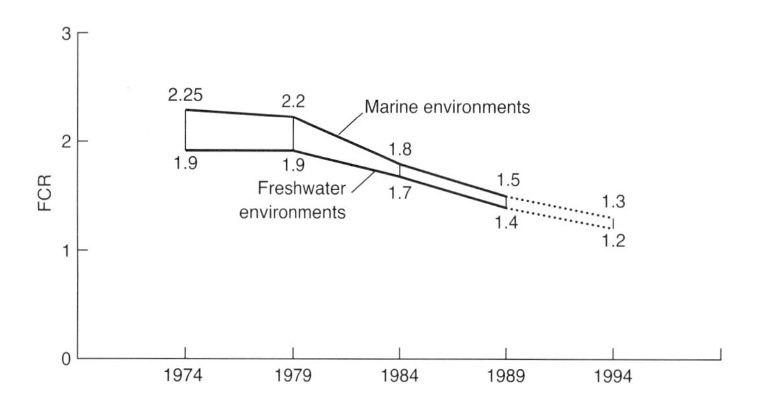

Fig. 4.2 Changes in the food conversion ratio (FCR) in Nordic fish feeds. After Ackefors & Enell (1994).

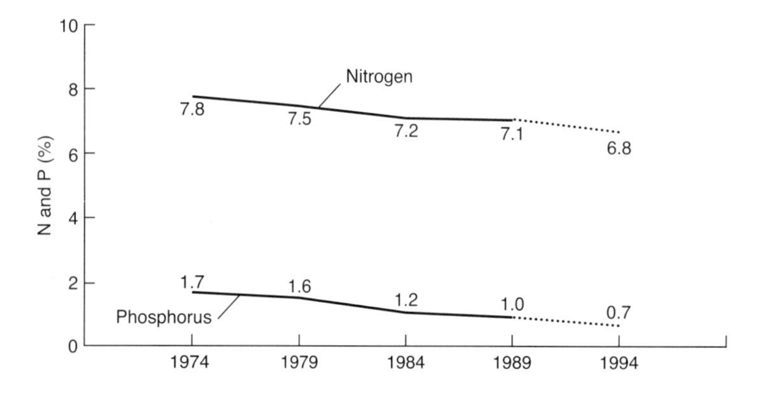

Fig. 4.3 Changes in the nitrogen (N) and phosphorus (P) content in Nordic fish feeds. After Ackefors & Enell (1994).

is eaten by the animal but not assimilated and passed out as faeces. In a balanced diet, 15–20% of the ingested dry matter may not be digestible. Using the figures for Norwegian salmonid culture as an example, it is possible that approximately 47 000 mt of feed was lost to the environment as faeces in Norway in 1989. Advances in feed composition and feeding practices over recent years have decreased the figure considerably and are partly responsible for the economic success of Norwegian fish farming in the twenty-first century.

Similarly, it has been estimated that in intensive shrimp ponds in Asia, only 17% of the food presented is converted to shrimp biomass. About 15% of the feed remains uneaten or is lost through nutrient leaching, 20% is egested as faeces and a large part of the remainder (48%) is lost to the environment as excreted metabolites and moults (Primavera, 1994).

High nutrient content feeds can also result in losses to the recipient water. Fish utilise protein N for growth, as an energy source and for metabolic functions. It is more cost-effective (excluding flesh quality considerations) and better environmentally if diets contain only sufficient protein to support growth and metabolic requirements so that energy requirements are met from dietary carbohydrates and lipids ('high-energy diets'). These protein- or energy-sparing diets have become popular in recent years and, in conjunction with improved feeding techniques, have resulted in substantial reductions in the wastage of N into the environment.

Table 4.1 shows the concentrations of suspended solids, total phosphorus (TP), total nitrogen (TN) and BOD in effluents from farms in Europe. There is surprisingly close agreement between the studies in the concentrations of all four parameters. This is despite the range of geographical locations and variations in farm type and farm management practices between studies. Table 4.2 indicates typical industrial effluent contaminant concentrations for comparison. When placed in the context of other industries and typical municipal wastewater, it can be seen that the concentrations (as opposed to mass flow) of these four contaminants in aquaculture effluents are relatively low. Separation and concentration of the contaminants from aquaculture wastewater for subsequent disposal are therefore difficult because of their low concentrations.

Impacts of effluent on aquatic systems

Eutrophication can be translated from Greek as 'nutrient (food) rich'. Nutrient enrichment in both fresh- and saltwater is a natural process. Nutrients can be carried into low-nutrient (oligotrophic) areas, which then become mesotrophic (intermediate nutrient content) and ultimately become eutrophic when the nutrient level becomes high. This natural process can be greatly accelerated by additions of nutrients from untreated sewage and from discharges from agriculture, aquaculture and other industries (Pillay, 1992; Welch & Lindell, 1992). As the concentration of organic matter increases, so also do the populations of aerobic bacteria, which use oxygen to break down organic matter. The DO content of the water then decreases, resulting in stress, death or exclusion of other aquatic species. The point at which the DO concentration is at its lowest is usually some distance downstream from the point at which discharge water enters the recipient body of water. Thus, in a lake or river, this 'critical point' may be the site of considerable damage to the environment, some distance from the farm. In contrast, settlement from cage culture often occurs below the cages and the harmful effects are close to the cultured organisms.

According to Boyd (2001), levels of suspended solids in effluents must be limited for several reasons.

(1) The suspended solids may form a plume of discoloured water in the discharge area, and the plume may reduce light penetration and phytoplankton productivity.
(2) Restricted light penetration by turbidity affects sea grasses, coral reefs and other sensitive underwater habitats.
(3) Sedimentation in shallow water may interfere with local navigation, fishing and other beneficial uses.
(4) Excessive sedimentation can stress or kill trees where effluent is directed into mangroves.
(5) Sediment accumulation can bury and smother benthic organisms.
(6) A high sediment oxygen demand can produce undesirable anaerobic conditions, and toxic metabolites (e.g. ammonium, methane and hydrogen sulphide) may enter the water from anaerobic sediment and harm sensitive aquatic animals.

Table 4.1 Reported contaminant concentrations in aquaculture effluent in Europe

Location	SS (mg/L)	TP (mg/L)	TN (mg/L)	BOD (mg/L)
21 EIFAC farms	9	–	–	5
Typical in Norway	3	0.100	0.5	–
Rogaland, Norway	1.6–14.1	0.08–0.27	0.43–0.70	–
Northern Sweden	[6.9]	0.11	0.70	–
Northern Ireland	–	0.11	0.531	–
Typical range	5–50	0.05–0.26	0.5–5.0	5–20
Finland	–	0.055	–	–
> 31 UK farms	11.1	0.082	–	4.00
Typical in Denmark	5–50	0.05–0.15	0.5–4.0	3–20
Typical values	14	0.125	1.4	8

BOD, biological oxygen demand; EIFAC, European Inland Fisheries Advisory Committee; SS, suspended solids; TN, total nitrogen; TP, total phosphorus.
Reproduced from Cripps & Kelly (1996).

Table 4.2 Reported domestic and industrial wastewater contaminant concentrations

Facility	SS (mg/L)	TP (mg/L)	TN (mg/L)	BOD (mg/L)
Aquaculture (typical)	14	0.125	1.4	8
Domestic wastewater				
Weak	350	4	20	110
Medium	720	8	40	220
Strong	1200	15	85	400
Stormwater runoff	170	0.350	3.5	14
Meat processing	300	–	3	640
Paper pulp mill	–	–	–	1800

BOD, biological oxygen demand; EIFAC, European Inland Fisheries Advisory Committee; SS, suspended solids; TN, total nitrogen; TP, total phosphorus.
Reproduced from Cripps & Kelly (1996).

In addition, large quantities of suspended particles may cause clogging of the respiratory apparatus of aquatic animals and result in poor feed uptake in visual feeders.

The deposition of suspended solids is pronounced in slow-moving or shallow water. In such locations, the majority of benthic species can be excluded, and only those that are opportunistic survive. The correlation between pollution and biomass (or diversity) is so strong that indices have been devised that can be used to assess pollution levels by calculating the biomass, species diversity or presence of sensitive species intolerant to pollution.

An increase in dissolved nutrients can greatly increase primary production, resulting in large concentrations of phytoplankton or algal 'blooms'. If a bloom contains a species of toxic algae, then this can be dangerous for both humans and aquatic animals. Such blooms have occurred in many parts of the world and have affected bivalve aquaculture in particular (section 21.5.3); however, it appears that aquaculture has often been the victim of these problems rather than the cause.

4.2.2 Effect of land-based aquaculture on natural habitat

The negative environmental implications of aquaculture on the habitat can be alleviated through the use of good operational techniques and better farm management practices. Land-based aquaculture, which includes aquaculture in the coastal regions, has had negative impacts on the natural habitat. Large-scale removal of mangrove forest for shrimp culture and pressure on forest resources in mountain areas for freshwater aquaculture have been well documented (Kumar & Hiep, 1995).

Coastline degradation

The large-scale exploitation of coastal regions for aquaculture is a major concern in some developing countries. Environmental issues may, out of necessity, take second place to food production or economic development in these countries. In Indonesia and the Philippines, it is estimated that 300 000 ha and 200 000 ha of mangroves, respectively, have been converted into shrimp ponds (Fig. 4.4). The Philippines has lost more than 66% of its mangroves since 1920 (Primavera, 1994). Similarly, substantial mangrove losses have occurred in other SE Asian countries. For example, Vietnam cleared 102 000 ha of mangroves for shrimp farming between 1983 and 1987, and approximately 65 200 ha of mangroves were converted into shrimp farms in Thailand between 1961 and 1993 (Primavera, 1998). Similarly in South America, around 21 600 ha of shrimp ponds in Ecuador have been developed in mangrove areas. The processes of mangrove forest clearance are continuing in these regions.

Fig. 4.4 Banks being built for a shrimp farm near a coastal village on Los Negros, Philippines. The natural wetland will be destroyed and the mangroves in the distance lost (photograph by Professor C. Tisdell).

This development is a major concern for several reasons. Mangroves are the main factor stabilising many coastal and estuarine regions. Without the extensive root systems of mangroves, the mud flats on which they grow can easily be washed away, particularly during tropical storms. Mangrove clearance also influences the quality of water available to agriculture and may cause changes in stream flow, siltation and coastal sedimentation characteristics. In estuaries where mangroves have been cleared for aquaculture ponds, flooding may result because the water volume retained by the ponds exceeds the carrying capacity of the estuary. Without the stabilising effects of mangroves, the land is gradually eroded despite the building of pond walls.

Mangroves are a major source of coastal productivity, adding huge amounts of organic detritus to coastal ecosystems. Mangroves are also known to be important nursery grounds for a wide range of species and, in particular, species that support coastal fisheries. Removal of mangroves can result in depletion of a wide range of native species and can lead to serious reductions in stocks of fished and non-commercial species. In South America, for example, young shrimp are hand-netted from beaches and grown on in coastal farms. Serious mangrove depletion has resulted in a sharp decline in shrimp stocks, which has affected the viability of the shrimp farms themselves.

Similarly, temperate marshes can be highly productive regions containing vegetation and organic detritus, which is a food source for many organisms. They also prevent coastal erosion, buffer flood damage, act as over-wintering and nesting grounds for birds, and contain many species requiring conservation (Pillay, 1992).

Clearing mangroves and coastal marshes for aquaculture development is not sustainable because of the large-scale environmental and social degradation, and the depletion of fisheries stocks that may result. Despite this, large amounts of aid money are still directed towards this type of development, with only the promise of short-term profits.

Several means are available to reduce these impacts.

(1) Legislation against further facilities in sensitive coastal areas, using zonation plans, such as those described by Black (1992) for temperate regions, or by the total prohibition of new developments without independent impact assessment prior to granting aid.
(2) Intensification of farming practices (Kapetsky, 1987).
(3) Sustainable use of mangroves, summarised by the following guidelines (Nor, 1984).
    — do not alter the physical structure, biological activity or chemical properties of the substrate
    — do not alter the hydrology
    — do not alter the tidal flooding and surface circulation patterns.

Impacts on other resources

In some developing countries, such as Vietnam, farmers are clearing forest to accommodate agriculture, including aquaculture. However, some of these developing nations have recognised the negative environmental impact of such practices and are taking positive steps towards environmental protection for the long-term sustainability of the farming sectors.

Use of grass carp as a major aquaculture species has had a significant negative impact on forests. Grass carp require large amounts of grass or green vegetation as food (40:1 FCR). They are cultured predominantly in Vietnam (Figs 4.5 and 4.6) in small-scale operations where farmers often obtain this foliage from the forest. A large number of farmers looking for green leaves in the forest to feed cultured grass carp in their ponds can have major impacts on forests. Recognising the long-term environmental impacts of undesirable aquaculture practice, the government is taking measures to protect forest and is providing assistance for more sensitive fish culture practice. Also, the government is investigating alternative species composition for better economic return without including grass carp in freshwater polyculture.

Fig. 4.5 Grass carp (*Ctenopharyngodon idella*).

Fig. 4.6 Farmers collect large quantities of fresh grass or green foliage for feeding grass carp in Vietnam.

Some freshwater aquaculture practices use excessive water. The diversion of streams, the addition of canals and reservoir construction has negatively affected the environment in some areas. Freshwater resources are becoming even more limited and require conservation. Freshwater aquaculture ponds must be operated using appropriate water management techniques, which minimise water exchange.

4.3 Impacts of aquaculture within large water bodies

The culture of aquatic animals within large water

bodies has the potential to cause both onshore and offshore impacts on the surrounding environment. Large water bodies including oceans, rivers and lakes are widely used for aquaculture purposes. Commonly used methods for aquaculture include:

- cage or pen culture for fish (both marine and freshwater fish)
- culture of bivalve molluscs within a few kilometres of the shoreline
- ranching activity (stock enhancement programmes).

Fig. 4.7 Seafloor beneath a seacage for fish culture showing accumulation of organic matter.

4.3.1 Pollutants from cage culture and its impact on the environment

Dense populations of animals such as farmed fish in cages (section 2.2.6) may be a major source of organic and inorganic compounds in water. Such high densities of animals may produce marked changes in the chemical composition of the water (e.g. DO components and N/P ratio), which may subsequently affect populations of phytoplankton. The main sources of pollution are animal faeces and excretions and feed waste. Unlike its land-based counterpart, cage culture relies upon natural water movement to deliver water and oxygen to sustain production and remove wastes. The impact of aquaculture on the surrounding biota has been a growing concern because of the rapid expansion of cage culture in the past few decades. Uneaten feed and faeces contribute significantly to the overall solid-waste production from cage systems (Fig. 4.7). The limited potential that exists for treatment of the waste material produced is a key issue in the environmental concerns raised against cage aquaculture.

The effects on the benthic community structure of organic loading originating from fish farms are most pronounced under and in the immediate vicinity of fish cages. They are less so at increasing distances from the farming operation. Accumulation of waste products in farm sediments usually results in dramatic changes in sediment chemistry and also the macro- and meiobenthic communities (Karakassis *et al.*, 1999). If the sedimentation rates from cages are high, the fauna may disappear completely. The community becomes more diverse and normal with increasing distance from the cages. The benthic fauna play an important role in the supply and mineralisation of organic matter and affect nutrient cycling in the sediment. Nitrogen cycling is faster, and both nitrification and denitrification are enhanced compared with azoic sediments. Various studies have shown that microbiological metabolism in sediments is stimulated by organic wastes from cages. In a shallow marine cage farm in Kolding fjord, Denmark, microbiological metabolism in the sediments was measured at ~10 times (525–619 mmol/m^2/day of CO_2) that at a reference site, most of which could be accounted for by sulphate reduction (Holmer & Kristensen, 1992). Studies on the production of macrobenthic populations in the vicinity of cage farms in western Scotland (Pearson & Black, 2001) suggested that the production of infaunal benthos close to the cages is 4–6 times greater than background levels.

Site rotation and 'fallowing' are methods used by cage culture industries to minimise impacts in the vicinity of fish cages (section 15.4.2). The fallowing of sites and rotation of cages has now become recommended practice in many areas where hypertrophic sedimentary conditions are a problem. A study undertaken in western Scotland on the benthic faunal succession and sedimentary conditions during the fallowing periods indicated that communities adjacent to fish cages returned to normal 21–24 months after destocking (Pearson & Black, 2001). The switch from highly enriched to moderately enriched communities took 9 months, followed by a further 9 months to achieve a lightly enriched status.

The rapid expansion of cage culture in the last

decade, particularly in the salmon industry, has increased the use of many chemicals which may have direct environmental impacts. Poor water quality and crowded conditions induce stress in caged fish, contribute to impaired growth and predispose them to disease. This in turn necessitates increased use of medicinal solutions. Pesticides are being used to combat sea-lice infestations, and disinfectants help to prevent the spread of viral infections. These substances add to the chemical wastes going into the environment and may have a negative impact on it (Haya *et al.*, 2001).

4.3.2 *Cage farming impacts on wild fauna and flora*

Cage culture also has direct impacts on wild fish populations by supplying food, either directly or through an increase in algal and zooplankton biomass, and by providing refuge in fish cage structures. Escaped fish may also have ecological impacts. A review undertaken by Phillips *et al.* (1985) of the freshwater sector showed that populations of species including roach and native brown trout have increased as a consequence of rainbow trout cage culture.

Meadows of seagrass, *Posidonia* species, cover vast areas in shallow regions of the Mediterranean. They are regarded as a cornerstone of the littoral ecosystem, providing a wide variety of niches, accounting for the high diversity of these areas. A recent study concluded that, although nutrient input from fish cages resulted in an increase in leaf length and increase in the biomass of epiphytes and ichthyofauna, there was evidence of decreased meadow density and total disappearance beneath cages (Pergent *et al.*, 1999). The study suggested that *Posidonia* species is a useful bioindicator for monitoring fish farm impacts in these environments.

4.3.3 *Impact of bivalve culture*

Bivalves filter feed on phytoplankton and other fine organic particles (Chapter 21). Bivalve culture usually requires the introduction of structures into the water body, from which the bivalves are either supported, protected or suspended. The introduction of such structures has an immediate effect on local hydrography and provides a substratum upon which other epibiota can settle and grow. The introduction of high densities of cultured organisms increases local oxygen demand and elevates the input of organic matter into the immediate environment.

The effects of biodeposits from suspended mussel culture on the local benthic environment have been considered in a number of studies. Heavy sedimentation of faeces and pseudo-faeces beneath mussel farms leads to organic enrichment and thus alters macrofaunal communities (section 21.6). Sedimentation rates up to three times higher than at reference sites were recorded in several studies, suggesting that average biodeposits in suspended culture could reach quantities up to 345 $kg/m^2/year$ (Chamberlain *et al.*, 2001). A study on biodeposits conducted in Ireland indicated that, in general, the effect of sedimentation was restricted to a radius of 40 m around the farm. There is increasing awareness of the potential environmental effects of bivalve culture and a number of measures can be undertaken to minimise these impacts (section 21.6).

4.4 Aquaculture's impact on the environment in general

There are general environmental impacts that are applicable to both land-based and water-based aquaculture. National development plans and policies for aquaculture must account for the general and specific impacts of aquaculture, and approval processes must include stringent measures to prevent or minimise such impacts.

4.4.1 *Disease transfer*

The transfer of disease between farms and the environment has long been a contentious issue. Farmed animals may infect wild animals but, also, indigenous organisms may carry endemic pathogens to and between farms. There is disagreement about which of these two vectors is the more common or important.

In terms of environmental impacts, two main types of pathogens can be identified:

- obligate pathogens, which occur only when there is a suitable host as a substrate for growth
- opportunistic pathogens, which are widespread, almost ubiquitous, and cause disease when the resistance of the host is reduced.

Stress on fish resulting from high-density, intensive cultivation is often sufficient to allow opportunistic pathogens to take hold and form a disease reservoir capable of infecting other culture stocks or the surrounding environment (Chapter 10). Alternatively, natural populations may also be under considerable stress as a result of sub-optimal conditions, such as a poor food supply or chronic obligate disease. They can then form a disease reservoir, with the potential to disperse over a much larger area than farmed species.

Several vectors for the transfer of diseases between cultured animals and the environment have been determined. Escaped animals have received the highest media profile. Intermediate hosts can be important vectors. Wild fish, for example, may move in and out of floating cage systems. Bivalves are known to contain and concentrate pathogens in addition to various chemicals and heavy metals. Although the presence of pathogens in such intermediate hosts may have originated from culture facilities, the hosts themselves do not necessarily suffer. In the case of bivalves, the organisms can be depurated with clean water to ensure that diseases and contaminants are flushed away. Effluents can both carry pathogens from a culture facility and predispose the surrounding environment to disease outbreaks by chemically stressing indigenous fauna or by improving the environment for the growth of pathogens (e.g. increasing the organic matter). There is no firm evidence that predators of cultured organisms carry pathogens out of culture facilities, although pathogens are found in faeces and regurgitated food from birds that feed at aquaculture farms. Also, parts of the cultured organism may be lost or discarded after predation, allowing them to decompose away from the farm. The practice of removing dead animals from farms and dumping them at sea is now illegal in most countries. Solid or liquid wastes from processing facilities, such as blood or offal, can also contain high concentrations of pathogens. Animals cultured for restocking purposes may carry disease with them when they are stocked into a fishery.

There are several ways of reducing the impact of diseases on wild stocks and the environment. Great care must be taken when introducing exotic culture species into a location or during the relocation of a species within its normal range, either as eggs, fry or adults. There are several documented accounts of pathogens entering a new and possibly highly favourable environment in this way, and affecting a host population that has little resistance to rarely occurring or exotic diseases. The diversity of a pathogen often increases when introduced into a new environment. To reduce these possibilities, heavy restrictions have been placed by environmental protection authorities on the movement and introduction of culture organisms. Depuration or quarantining may be necessary. Care must be taken to ensure that the water and containers in which the animals are transferred are also adequately sterilised prior to contact with a new environment.

All sensible measures must be taken to ensure that escapes and predator attacks are reduced. Treatment of wastewater can reduce the incidence of pathogens and prevent the build-up of conditions that predispose an area to disease. Dead animals from the culture population must be removed rapidly and regularly to ensure that pathogens do not build up. They must be disposed of with care, possibly at a land tip or by some form of ensilation.

The use of unprocessed whole fish by-catch ('trash fish') from the fishing industry as an aquaculture feed has greatly decreased in recent years and is illegal in some countries. Such feed, particularly if ungutted and not fresh, can harbour disease.

The use of prophylactic drugs in intensive fish culture has greatly increased in recent years. Although this helps to reduce the reservoir of pathogens that could be transferred to the environment, their widespread use has other detrimental environmental effects, including the build-up of drug-resistant strains in wild and cultured animals. Management strategies that are less aggressive and more sustainable are showing great success in reducing disease transfer. Maintaining only one year-class of animals in a region at one time has prevented the transfer of disease throughout the life cycle of the culture animals. In Scotland, neighbouring companies have agreed on a joint strategy in which a region can be left fallow between generations, thus reducing direct disease transfer or transfer via secondary hosts and the sediments. Adequate spacing between farms is also important, so that the carrying capacity (the ability of the environment to accept effluent) is not exceeded, and to prevent direct transfer of pathogens.

The presence of a disease in the environment may not be the result of farming operations. This link is often difficult to prove because so many contributing factors are involved. Also, farm diseases do not always disseminate in the environment if stress levels are low. Microbial activities may well increase in sediments, but they are not necessarily pathogenic.

Diseases in culture organisms, and the transfer of disease to the environment, can be greatly reduced by careful management (Chapter 10). Some of the management practices used to minimise the occurrence and transfer of diseases in aquaculture systems are discussed in more detail in the later chapters of this book, which outline specific culture methods for various species/groups (Chapters 13–22).

4.4.2 Impact on genetics

The introduction of exotic species or strains into a new region is not confined to aquaculture, nor is the introduction of animals and plants through natural, accidental or deliberate causes a new phenomenon. For example, terrestrial animals have been deliberately introduced for public health reasons (e.g. pathogen competitors), for reduction of pests (e.g. a new predator) or for improved farming efficiency (e.g. a faster-growing stock).

Introduction of exotics

Within the aquatic environment, natural movement of stocks into a new area or region can occur as a result of climatic events, such as glaciation at high latitudes or river flooding in the tropics. There are examples of accidental introduction from the aquarium fish trade and the discharge of ballast water from shipping, the latter vector being a particularly serious problem in California, Australia and New Zealand. Deliberate introductions include predatory fish brought into a catchment to improve the vitality of a managed stock and also ranching.

Introductions and transfers have been defined as follows:

Transfers take place within the present geographical range of a species and are intended to support stressed populations, enhance genetic characteristics or re-establish a species that has failed locally. Introductions are movements beyond the present geographical range of a species and are intended to insert totally new taxa into the flora and fauna.

(GESAMP, 1991)

The escape of some species of cultured animals is not covered well within the above definition. At many, if not the majority of, locations, cultured animals are grown within their natural range; however, the cultured stock may be a different strain from local wild individuals of the same species. Escapes may then be classified as threatening introductions rather than transfers. Despite precautions, escapes from farms are common. The breaching of containment vessels, ponds, cages, tanks, etc., during storms or floods is common. Predators such as birds and marine mammals take and then drop live prey. Damage to pond walls by burrowing mammals and to seacages by seals occurs. Operational losses may occur during grading, transferring between vessels and harvesting. The potential for losses because of poaching and vandalism must also be considered.

The scale of this problem varies with location. In some lakes in which cages have been broken, the majority of a class of animals, e.g. salmonid fishes, may be of cultured origin. An example of this is the presence of rainbow trout (*Oncorhynchus mykiss*) in some Scottish freshwater lochs. At more open marine locations in which few escape incidents have occurred, there may be insignificant introduced populations. Mills (1989) estimated that the number of escaped salmonids in Britain was 15% of the farmed production at that time. The World Wildlife Fund (WWF, 2001) concluded that in Norway alone at least half a million farmed salmon escape from aquaculture facilities annually, so that 30–50% of coastal catches may be farmed fish.

Escaped animals may have the following impacts:

- interbreeding with wild stock, i.e. genotype modification
- competition for space and food – niche competition
- transfer of disease
- changes in the local ecology by changing trophic interaction
- habitat modification.

Although precautions can be taken, escapes are bound to occur. Both the frequency and magnitude of such losses must however be minimised, and in many countries this is becoming a legal requirement. The following procedures will help to reduce losses:

- regular inspection and maintenance of pond walls, weir gates and farm effluent gratings
- secondary back-up containment facilities
- predator deterrence: nets, scarers, etc.
- underwater predator netting around net cages
- site surveys, prior to farm construction, covering flood frequency, height above water table, etc.
- adequate security or circumspect location to avoid or deter poaching and vandalism
- good public relations to reduce vandalism
- all in-farm animal handling operations to be conducted within confined areas.

The introduction of non-native species into a new region must be considered with great caution. Once in a new location, the ecology of the species may alter, or unforeseen consequences, such as unpredicted inter-specific competition, may occur. For example, introduced brown trout are thought to be a major factor in the decline of several species of freshwater fish in south-eastern Australia. The major contributing factors are thought to include competition for habitat, food resources and predation.

Exotic species have been introduced to facilitate the development of aquaculture in a number of cases. For example, a number of species of shrimp, including *Marsupenaeus japonicus, Litopenaeus vannamei, L. styirostris* and *Fenneropenaeus chinensis*, have been introduced into the Philippines since the 1960s. The effects of these introductions on endemic species and the local environment is unknown.

Care must be taken not just with the species concerned, but also with associated organisms, which may accidentally be imported with it. For example, some species have invaded an area after being transported there in holding water, in the body of an introduced culture animal or in packaging. This is a particular problem if the introduced organism is pathogenic or noxious. The risk of introducing these organisms can be reduced if suitable quarantine procedures are adopted. The culture of giant clam spat for reintroduction to depleted areas is a good example of this. Giant clam seed were routinely transferred to Pacific nations in the 1980s and 1990s and stringent quarantine procedures were developed (Lucas, 1994).

Genetic interaction

There is increasing concern about the interbreeding of wild and cultured stock (WWF, 2001). Controversy concerns not only the effects, but also the extent to which this occurs. ‘Genetic pollution’ can be associated with the introduction of new species into an area or the escape of organisms from a culture facility. Rather than just outcompeting the native species, or transferring diseases, escaped animals may interbreed with wild stocks. The mode and scale of possible genetic interaction from introductions and escapees has been described in the previous section.

Cultured species, in which the life cycle has been closed in captivity, have been bred to be suitable for farming. Such adaptations, e.g. reduced stress in high stocking density situations, may not necessarily be suitable for success in the wild. There is some evidence to suggest that hatchery-reared salmonids do not compete well in the wild with corresponding native stocks. By implication then, interbreeding between wild and cultured stocks may produce less fit individuals. The following list indicates other possible impacts of interbreeding.

(1) Loss of genetic diversity by interbreeding of wild stocks with relatively uniform culture stocks (Crozier, 1993).
(2) Loss of distinct stock integrity. Some native strains, although not necessarily suitable for culture directly, may form a store of a particular genotype, e.g. resistance to a particular disease, or fast growth in warm, low-oxygen conditions. By back-crossing these genes, the genotype can be incorporated into a culture or a threatened species. Interbreeding with these distinct stocks can dilute such characteristics and result in loss of the strain.
(3) Cross-bred hybrids can be sterile, so the native stock becomes less viable and may ultimately be lost.

The following actions can be taken to reduce the probability of interactions and their impacts occurring:

- restrict the movement of exotic genotypes capable of interbreeding with native stock
- reduce escapes
- only permit the use of sterile stock within sensitive regions (methods of inducing sterility are not, however, 100% effective, so that the escape or introduction of a large number of animals capable of breeding can result in a viable population)
- enforce exclusion zones to protect particularly rare or vulnerable stocks
- do not locate aquaculture facilities in areas where there are valuable or sensitive native stocks, e.g. at the mouths of important spawning rivers.

A particular problem exists when animals are cultured as part of a restocking programme in which individuals will eventually mature and may reproduce with existing members of a wild population. As mentioned above, giant clams are a good example of animals that were cultured for this purpose and which were transferred between Pacific nations. Because of concern for the potential genetic implications of these transfers, guidelines were drawn up that outlined preferred practices for hatchery production of spat and for translocation of spat (Lucas, 1994).

4.4.3 Chemical additions

A wide range of chemicals are currently used in the aquaculture industry. These have been summarised by Beveridge *et al.* (1993) as follows:

- chemotherapeutants, e.g. parasiticides, antimicrobials
- vaccines
- hormones, e.g. growth stimulants, sex control
- food additives, e.g. vitamins, pigments
- anaesthetics
- disinfectants
- water treatment compounds, e.g. pH regulators, herbicide, pesticide and piscicides
- antifoulants, e.g. tin and copper compounds
- construction additives, e.g. stabilisers, pigments.

Compared with the public health or general veterinary fields, relatively few chemotherapeutants are suitable for use with aquaculture species for human consumption. This is because of problems of strain resistance and because few pharmaceutical companies consider aquaculture to be a sufficiently large market to target, primarily because of licensing costs. Pressure from environmentalists also limits marketing. The main classes of therapeutants are antibiotics, nitrofurans, sulphonamides, acriflavine, copper sulphate/potassium permanganate stains, formalin/formaldehyde, oxolinic acid, iodine and di-*n*-butyl tin oxide.

The use of antibiotics in aquaculture is widespread. Although the timing, quantity, concentration and type is carefully controlled in developed countries, they have been subject to misuse, especially in Asian shrimp culture. Antibiotics can be added to the feed to control bacterial diseases or used prophylactically to prevent the outbreak of diseases. Considerable quantities can be used.

Most antibiotics are associated with particulate matter and leave the farm on particles in the effluent. They will therefore have a large impact on sediments. The period of activity or persistence of antibiotics in the environment varies with environmental conditions, such as temperature and substrate, and according to the particular antibiotic used.

The major environmental impacts of antibiotics are on non-target species, including humans, and on the resistance of the target organisms. For example, resistant strains of *Vibrio* bacteria have developed in shrimp ponds in SE Asia as a result of the widespread prophylactic use of antibiotics (Pillay, 1992). The main impacts on the environment have yet to be definitively quantified, but possible effects may include:

(1) resistance of the target organisms to antibiotics, leading to the occurrence of more virulent strains
(2) antibiotic resistance of non-target pathogens of other species, such as human pathogens, resulting in the development of increased disease incidence in non-culture species
(3) reduction in general microbial action within the sediments leading to rapid degradation of site quality

(4) persistence within the cultured organism, which can result in allergic reactions in humans if the organisms are consumed too soon after antibiotic treatment.

To reduce these impacts, the following measures are being taken:

- improved aquaculture management to reduce stress and the occurrence of diseases (Smith, 1992)
- implementation of antibiotic withdrawal periods prior to harvesting
- effluent screening, using 40 μm or finer mesh sizes, to reduce the concentrations of suspended solids acting as pathogen substrates (section 19.4.5)
- mandatory dose rates and periods
- optimisation of antibiotic concentrations in the feed
- restrictions on prophylactic use
- vaccination (although this is not effective for all diseases).

Some chemical treatments for sea-lice (*Lepeophtheirus salmonis*) contain dichlorvos, an organophosphate (Jackson & Costello, 1992). Concentrated dichlorvos is a highly toxic neurotoxin that is dangerous when inhaled, absorbed across the skin or ingested. Treatments are stressful to the cultured fish, and toxic to various molluscs and crustaceans (Jackson & Costello, 1992). The withdrawal period is 7 days at temperatures greater than 10°C and the half-life in seawater is 4–7 days. This short half-life and its breakdown to harmless products reduces its capacity to bioaccumulate; however, because of the serious negative impacts, alternatives are now being widely used. For example, wrasse, which act as cleaner fish, are proving successful, and vaccines are also being developed. Pyrethrum, a plant extract, is effective against sea-lice but its administration is difficult. A fish pump that counts and grades the fish at the same time as bathing them in recycled treatment water appears promising. Parasite incidence can also be reduced, but not eradicated, through appropriate management practices such as faster water flow rates in cages, attention to net fouling and fallow periods.

Steroid hormones are used for inducing maturation and spawning, and for sex reversal. These chemicals are used commercially only for broodstock and not for marketed animals (section 7.5).

Cultured animals may not have access to naturally occurring pigments, so any carotenoids required must be added to the diet (e.g. in salmonid culture). Although 88% of the pigments may not be retained by the cultured organism, they oxidise rapidly in the environment. There are no data to indicate a significant environmental impact from carotenoids.

The use of herbicides, pesticides and piscicides was reviewed by Pillay (1992). For example, teaseed cake, derris powder and rotenone are commonly used to remove pests from aquaculture ponds. The toxicity of these compounds appears to last for 2–4 days, although 20 mg/L rotenone concentrations can remain toxic to juvenile fish in tropical shrimp ponds for 8–12 days.

A good flow of water is required through floating net cages at all times. They must therefore be kept clean from fouling organisms such as seaweeds and bivalve molluscs. Similarly, pipes and pumps in land-based systems must be kept clean. It is now common practice to impregnate the netting with some form of antifoulant, either at the factory or on site. The most commonly used antifoulants are either organotin [e.g. tributyl tin oxide (TBT)] or copper-based compounds. These compounds are highly toxic to bivalves, and there is growing evidence that some fish species may also be affected. Research is continuing to find less toxic biocide-free antifoulants. Again, suitable management practices can reduce the amount of antifoulant required and therefore the effect on the environment. Cleaning nets from seacages on land, after drying, is labour intensive, but it reduces the chemical requirement for antifouling (section 15.4.5).

4.4.4 Predator conflicts

Prevention of predation on cultured stocks by native animals is a day-to-day problem for many aquaculturists. This impacts directly on the native predatory species. The concentration of potential prey animals, waste food and opportunistic non-cultured species within and around aquaculture facilities is attractive to predators. Aquaculture facilities are inherently insecure with regard to predation because access to ponds and cages is required regularly, and flow-through of water, which is as unrestricted as possible, is beneficial.

There are six major groups of predators in aquaculture: mammals, birds, fishes, molluscs, crabs and echinoderms. Each group has its own mode of operation, scale of impact and control measures.

Mammalian predators include seals, otters, mink and water rats. Like much information regarding predation losses, data on mammalian activities are somewhat anecdotal. A survey of salmon farms by Ross (1988) found that 82% of respondents considered seals to cause damage or loss on their farms (section 15.4.6). Seals attack fish in cages, pushing up under the net to bite and suck pieces out through the net (Howell & Munford, 1992). Uneaten fish can be damaged, or the net perforated, leading to stock losses. Mink (often mistaken for otters) and otters usually attack by climbing over enclosures or making access holes. Water rats are a problem for freshwater crayfish culture because they dig under and climb over perimeter fences.

Birds take cultured animals by wading or perching, surface feeding, aerial swoops, surface dives and plunge dives (Howell & Munford, 1992). In addition to the loss or damage of stock, birds may be a vector for the transmission of disease both to and from farms and, if the prey is dropped, can cause escapes. Faeces from gulls feeding on harvested Dutch mussels have increased bacterial concentrations in mussel flesh to levels that are unfit for human consumption (Anon, 1993). Major bird predators include ducks, cormorants, oystercatchers, gulls, pelicans and herons.

Bottom-browsing fish such as plaice and flounder can damage the siphons of sediment-dwelling bivalves or eat the whole animal. Scavengers, such as dogfish and eels, and parasitic lampreys, may be attracted to culture sites. Various species of crabs and starfish are a particular problem for producers of commercial bivalves around the world (section 21.5.1). A major reason for the failure of dammed enclosures for the culture of flatfish was predation by crabs.

Measures used to control these predators that have an impact on the environment include:

- deterrents, such as human presence, loud noises ('bird guns') and flashing lights
- underwater acoustic scarers to protect against mammals
- biological scarers, e.g. simulation of higher predators
- heavy exclusion netting around seacages, which may cause entanglement and drowning
- netting or lines above the surface of facilities, and fences and lines at pond margins, which may cause entanglement
- shooting, poisoning and trapping, with obvious ecological consequences.

Further impacts of aquaculture may, in extreme circumstances, include an increase in the predator population and possible destabilisation of the local ecology. Deterrents can also cause habitat disruption. Four main strategies to reduce the environmental impact of predator deterrence are:

- careful site selection, avoiding feeding grounds and regions containing high predator densities
- careful selection of culture techniques, such as off-bottom culture of molluscs
- prevention of feed losses likely to attract predators
- sacrifice of a proportion of the culture stock in the hope of protecting the majority.

Consultation between nature conservation and aquaculture groups can help to minimise serious impacts and breaches of legislation.

4.4.5 Social aspects

Socioeconomic impacts

The development of aquaculture has a number of advantageous impacts, particularly in rural areas. These include employment, the stabilisation of rural populations, improved economy, improved infrastructure (communications, roads, etc.) and, most importantly, in developing countries, the production of a fresh supply of protein-rich food. Even in industrialised countries, a fresh, local supply of quality fish or shellfish is desirable. However, these developments can result in social problems in some cases:

- displacement of traditional local industries such as capture fisheries

- reduction in employment because of intensive and relatively skilled aquaculture activities
- depletion in the seed required for capture fisheries (e.g. shrimp postlarvae)
- conversion, and therefore degradation, of a traditional multi-use subsistence habitat into a monoculture facility (Pollnac, 1992)
- changes in traditional water use and consumption;
- increased farm intensity, which may lead to increases in income differential and social stratification (Pollnac, 1992)
- overexploitation of the carrying capacity of a region, reducing the possibility of other uses.

Though rarely considered, the socioeconomic advantages and problems must be assessed and weighted before any major new aquaculture industry or expansion of an existing industry is undertaken in a region.

Aesthetics

The location of farms in areas of outstanding natural beauty may be considered a problem (Fig. 4.8). Aquaculture farms are often situated in remote areas with little other development. Very often these areas are also considered to be in need of protection. A conflict of interest then arises between those who require jobs and income in remote communities and those who wish to keep the area unspoilt.

There is little doubt that the tactless location of an aquaculture facility in a scenic area will arouse understandable controversy, and development of this sort has been restricted in some locations. Often, where development restrictions are imposed in a particular area, another area will be 'sacrificed' for development. Little research has been conducted to assess the importance of this parameter on the local community, although in some locations opposition based on subjective, but nevertheless important, suppositions and issues can restrict or prevent aquaculture development. The lack of survey data indicates the lack of importance that is attached to this issue of aesthetics.

There are ways of minimising the visual impact of aquaculture facilities in scenic and recreational

Fig. 4.8 Long-lines for mussel culture in a loch in northern UK.

regions. If possible, buildings must be kept low lying or placed away from the shoreline, and their colour must blend in with the surroundings. Tanks and cages must be similarly unobtrusive. Brightly coloured buoys for floatation purposes are particularly noticeable and must be avoided, although navigation buoys must of course remain visible. Equipment not actually in use, such as nets, tanks, forklift trucks, etc., must not be abandoned or stored in obvious sites.

Noise

The noise from aquaculture facilities in remote areas is often a problem. In the worst cases, constant disturbance and noise from farm boat motors and vehicles can scare off shy animals and annoy local residents. Automatic feeding devices can fire food into tanks or cages on a regular basis, perhaps every 5 min, throughout the day. Although such noise would scarcely register on a sound meter, the annoyance value is similar to that of a dripping tap at night. Thus, effort must be made to reduce the noise emitted from aquaculture facilities, for example pumps must be sound-proofed.

Other social impacts

The following list summarises some of the main social issues associated with aquaculture developments (Gowen, 1992):

- competition for space
- traditional fishing
- navigation
- anchorages and marinas for recreation boating
- different forms of aquaculture and between aquaculture and other industries (e.g. wood pulp)
- amenity, recreation and tourism
- visual impact and loss of wilderness aspects of the countryside
- restrictions of access to land, foreshore and inshore areas, which may affect outdoor activities (water sports and harvesting of shellfish for non-commercial purposes)
- reduction in the amenity value of freshwater for recreational fishing
- reduction in property values.

Many of these conflicts can be resolved at the planning stage by adequate discussion and legislation. The fact that many such conflicts are not resolved amicably indicates that in many cases planning consent procedures and legislation are inadequate. It is in the interest of all parties to ensure that well-defined impact assessment and planning regulations exist and are implemented and that protected areas are not compromised. This will help ensure that environmental impacts are adequately considered and developments are appropriate.

4.4.6 Public health issues

Aquaculture operations can have several impacts directly on humans, in their roles as the operators of aquaculture facilities and consumers of aquaculture products.

Safety at work

Aquaculture is an industry that is frequently located both onshore and in deep water, contains heavy and electrical equipment in proximity to water, and has to operate in all weather. This leads to a variety of potential hazards:

- electrical and mechanical hazards, e.g. electrocution
- hazardous chemicals, e.g. organochlorine sea-lice treatments
- boating accidents
- drowning
- muscular injuries
- diving accidents and practices (Buchanan, 1992), e.g. multiple 'bounce' dives for the removal of fish mortalities from cages.

These potential hazards may be reduced by the employer by providing the following facilities and knowledge (Wall, 1992):

- a safe place of work
- safe means of access and exit
- safe plant and machinery
- safe working practices
- information, instruction, training and supervision
- suitable protective clothing
- preparation of emergency plans
- arrangements for use of any article
- provision of welfare facilities
- employment of competent persons.

Product safety

In addition to obvious public health considerations, it is particularly important for the aquaculture industry, which frequently markets its products as clean and healthy, that contamination of products is minimised. Legal regulations also usually require some level of security. Aquaculture facilities sited inland or inshore are particularly susceptible to product contamination. This can range from slight tainting to greater levels of contamination by bioaccumulating or biomagnifying substances. Serious contamination can have an important impact on consumers. Substances that can cause serious contamination of aquaculture products include heavy metals, such as mercury; petroleum compounds from oil spills; microbes, such as those resulting from sewage (particularly important in the bivalve mollusc industry); organochlorines entering the farm as contamination or as pesticide treatment; and algal toxins resulting from algal blooms, such as 'red tides'. Careful site selection can play an important part in reducing contamination from particular sources. Good management will monitor the prevailing ambient conditions, and so specific pollution incidents may be avoided, e.g. closure of pond intakes during toxic algal blooms. The controlled purification of bivalves in clean water (depuration) for a period prior to marketing is often a legal requirement and greatly reduces the potential threat to human health. In areas where a serious pollution incident has occurred, an exclusion zone, in which culture animals cannot be sold for consumption, may be required to safeguard not just the consumer but also the reputation of the product and the local industry. This occurred following oil contamination of a small number of salmon farms in Scotland.

4.4.7 Miscellaneous impacts

There are several other possible impacts from aquaculture that have not been described above. At present their incidence is infrequent, their effect is small or current social thinking considers them to be unimportant.

The relationship between aquaculture and capture fisheries is important to consider. For example, the collection of seed from the wild (such as young shrimp and juvenile fish) for aquaculture could detrimentally influence capture fisheries. Alternatively, production of juveniles that have been cultured through their early life stages (during which mortality in the wild is high), for release to supplement natural stocks, may have a beneficial impact on capture fisheries. Cultured animals can also be used to enhance or replace overfished wild stocks, augment sports-fishing stocks and replace populations depleted through industrial developments, such as the damming of rivers used by migrating fish.

As capture fisheries decline and aquaculture expands, competition is likely to increase between the two in terms of product price and quality. This may ultimately result in a shift in labour, first on a local basis and then nationally. A benefit of this is that natural stocks may have a better chance of recovery, although market forces will probably ensure that further stock protection legislation is still required.

The use of 'trash fish' to provide fish meal for the production of aquaculture feeds is not an efficient use of global resources. It is more efficient to use plant-based nutrient sources, although, from a nutritional perspective, 100% replacement is not yet possible (section 8.7.3). The use of fish meal means that, although aquaculture production may replace the dwindling capture fisheries of species for human consumption, non-target species will still be impacted as a source of cheap fish meal. Advances in the use of alternative feeds, such as soya meal for some cultured species, are reducing this dependence on wild-caught protein sources.

Increases in the number of cultured species deliberately released, or which escape into the wild, may result in the impacts described in sections 4.3.2–4.4.2, but angling catches may increase as a result. There is currently a heated debate about whether cultured fish, such as rainbow trout, provide sport that is as good as wild fish.

The various methods for processing aquaculture products are covered in Chapter 11. Consideration must be given to the containment and disposal of blood and offal that contain high levels of polluting nutrients and BOD. For example, the procedure of bleeding salmonids at the culture site (sections 11.9 and 15.9) can cause pollution.

Refuse must be disposed of so that it does not represent a public health hazard or an eyesore. Similarly, aquaculture feed stores must be secure against rodents, which can be a serious source of disease. For example, deaths from Weil's disease, which is transferred to cuts by handling material contaminated with rat urine, are increasing among aquaculture workers. Rodents can also damage equipment by biting through cables, etc. Insecure storage of feed and inadequate protection of automatic feeders may therefore increase local rodent populations. Rodent control measures include restricting the food supply, filling burrows, reducing the number of possible nest sites, capture and poisoning.

Another consideration of the impact of aquaculture activity is the welfare of the cultured species itself. Although animal welfare has received little consideration in the past, concern is increasing. The welfare of animals can be summarised by 'five freedoms':

- freedom from thirst, hunger or malnutrition
- freedom from thermal or physical discomfort
- freedom from pain, injury or disease
- freedom from fear and distress
- freedom to express a normal pattern of behaviour.

There is currently a lack of knowledge about the

impact of culture on aquatic vertebrates (and squid and octopus). The welfare of the animals at various stages in the culture system may require further attention if public opinion requires it. Parameters to consider include water quality, stocking density, culture environment (e.g. shade, tank or net dimensions), physical handling, disease control, harvesting, transport and slaughter. Care must be taken not to attribute human feelings and preferences to the culture species, but clearly, for ethical and economic reasons, animal welfare must be considered and openly discussed.

4.5 Impact assessment

An environmental impact assessment (EIA) is the gathering of information required to estimate the possible environmental impacts resulting from a development. The results of the study are presented in an environmental impact statement (EIS). The EIA is usually conducted by the developer. Regulating authorities normally inspect the EIA protocol and the result is the EIS. These authorities ensure the correct monitoring of the predicted impacts during and after development. The EIS can also be made available to other relevant government authorities, financiers and the local community. The main stages of the EIA, conducted with the aid of specialists, are:

- scoping and impact identification
- baseline studies
- prediction and evaluation
- identification and mitigation measures
- assessment or comparison of alternatives
- impact monitoring/auditing (from Pillay, 1992).

Assessments must be restricted to the most serious and likely impacts to focus efforts most efficiently. The EIS must include the following information described in detail by O'Sullivan (1992):

- a description of the proposed development
- the data necessary to identify and assess the main effects which the development is likely to have on the environment
- a description of the measures envisaged in order to mitigate these effects
- a non-technical summary.

EIA guidelines vary with local legislation and the nature of the development. Protocols have been presented by a number of workers. One such method is shown in Fig.4.9.

4.6 Technical solutions

The ecological effects of aquaculture operations can be reduced considerably with careful management. Careful feed management and feed composition have been discussed above. Site location is also an important consideration. In addition to other environmental, economic and operational considerations, site surveys for proposed aquaculture developments must examine the following effluent impact parameters:

- water exchange
- effluent dilution and dispersion
- depth under cages
- poor water quality near the culture species
- carrying capacity
- density of organisms
- concentration of culture vessels – ponds, tanks, cages
- proximity of other farms
- proximity of other loading industries
- residence time of receiving water body
- protection of sensitive sites
- proximity to sensitive ecosystems, such as coral reefs and mangroves
- multi-use conflicts (e.g. tourism, fisheries)
- treatment requirements
- legal requirements
- ability to retain the effluent
- suitable treatment method
- consent regulations
- allowable discharges
- costs.

The removal of suspended particles from aquaculture effluents is commonly used for primary treatment, because the low contaminant concentrations are difficult to treat using other technology. Several methods for reducing or removing suspended particles from aquaculture effluents are currently available, including particle screens, sedimentation, media

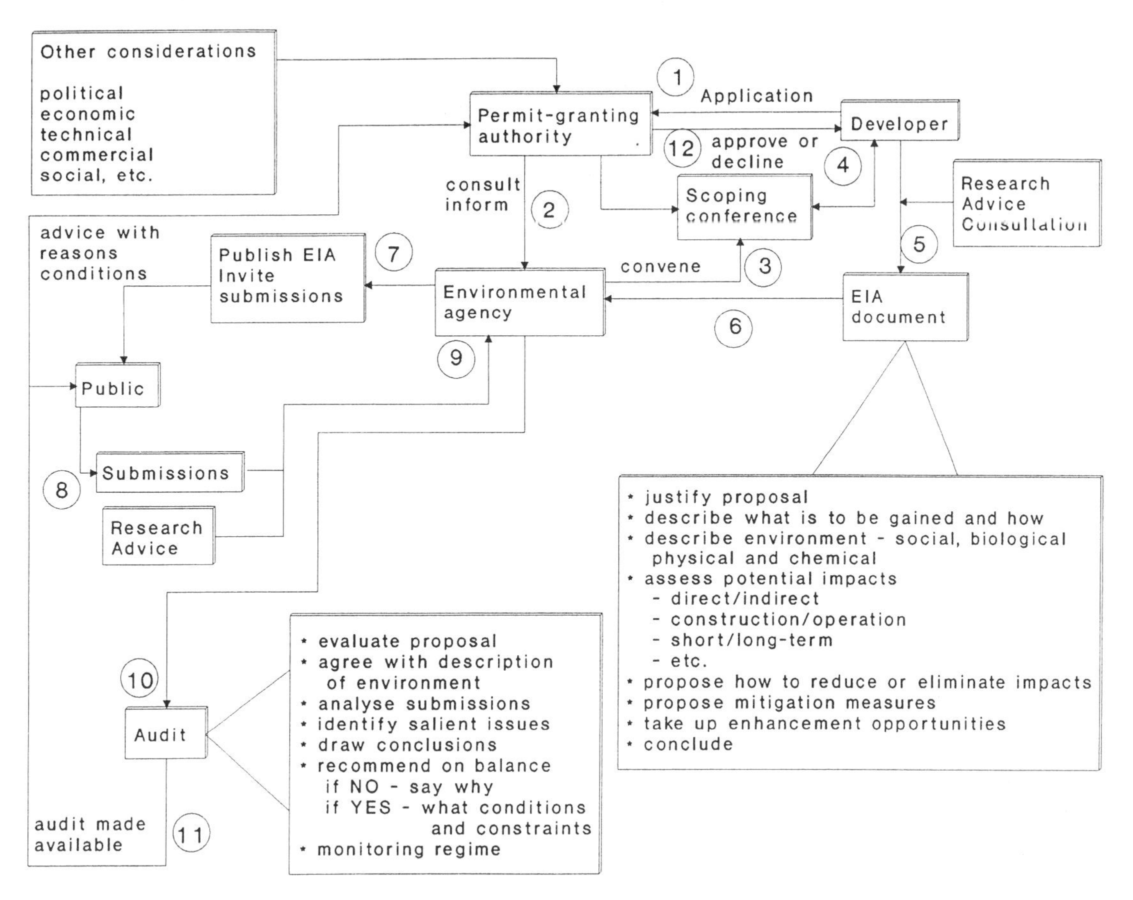

Fig. 4.9 Example of an environmental impact procedure. Based on GESAMP (1991) with permission.

filters and biological removal of nutrients (Wheaton, 1977; Cripps & Bergheim, 2000). The effectiveness of these methods depends on the particular application. Several procedures that can be used to reduce the impact of aquaculture pond effluent on coastal waters were discussed in detail by Boyd (1998).

It is impossible to collect water from floating cage farms. Care then must be taken to locate cage farms where the impact will be minimised. Large cages have been developed to be strong enough to withstand conditions in the open ocean. Clearly, in these locations, the water flow-through and the depth under the cage are greater than at inshore locations. For similar reasons, culture on offshore barges and ships is becoming more common. Although suspended solids and nutrient levels are not reduced by this practice, wastes are dispersed over a much wider area, and this reduces their most acute effects. Closed cages also need to be developed and they will become more popular as costs decrease. Particle traps under cages have been developed, and are proving suitable for commercial applications.

Recirculation systems, in which the water is used multiple times, offer the best protection against wastewater loadings. The small volume of water that passes out from the farm can be intensively treated, but it may have a reduced impact because of the low effluent volume. Such high-technology farms are currently operating with up to 100% recirculation (make-up water is used only for replacing evaporation and intermittent losses), but they have a high capital cost and require skilled management. As such, they are usually confined to high cost/unit biomass stock, such as hatchery production.

4.7 Integrated wastewater treatment and aquaculture

Aquaculture can have positive impacts on the environment. Integrated farming has significantly enhanced agricultural production and sustainability in many parts of the world, including parts of Asia, where integrated aquaculture has been practised for many centuries (section 2.3.6). An underlying process in integrated farming is recovering resources such as nutrients and water for re-use. This improves the sustainability of the system and minimises environmental pollution. Integrated farming and wastewater treatment embrace a diverse set of technologies that link fish culture to terrestrial farming. By incorporating aquaculture into waste treatment, the incentive to reclaim nutrients from wastewater, release clean effluent and simultaneously produce fish has proved successful in many parts of the world. Purification levels have reached those attained by the best alternative treatment methods.

It has increasingly been recognised that organic waste including sewage is not necessarily a pollutant but a nutrient resource that can be recycled through integrating farming practices. Extensions of the traditional practices of recycling domestic sewage through agriculture, horticulture and aquaculture, being basically biological processes, have been in vogue in several countries. The sewage-fed fish culture of Munich, Germany, and 'bheries' in Calcutta, India, are world famous. The emphasis in these practices has been on the recovery of nutrients from the wastewaters. Taking a cue from these practices and deriving information from the new databases in different disciplines of wastewater management, aquaculture is recognised as an important tool in many developing countries and is being adapted as a standardised technology for treatment of domestic sewage. Waste-fed aquaculture is a proven nutrient-recycling technology that is practised successfully in many countries. Properly designed and managed sewage-fed fish ponds offer a viable low-cost wastewater treatment cum usage opportunity. Several studies have clearly demonstrated that fish production of 5–7 mt/ha/year is achievable in tropical climates where year-round growth occurs.

4.7.1 Waste-fed aquaculture in Asia and Oceania

The concept of using aquaculture as a tool for wastewater treatment has been evaluated through a systematic research programme carried out over a period of 5 years by the Central Institute of Freshwater Aquaculture, Bhubaneswar, India. In collaboration with the Public Health Engineering Department, Government of Orissa, the Indian Aquaculture Sewage Treatment Plant (ASTP), comprising duckweed and fish culture, was designed. Its field facility has been developed under a project on 'Aquaculture as a tool for utilisation and treatment of domestic sewage'. The ASTP comprises a set of duckweed ponds where algae and duckweeds are utilised in the removal of the nutrients and the reduction of BODs and CODs (chemical oxygen demands). These are complemented by fish ponds and marketing (holding) ponds (CIFA, 1998). The system can receive primary-treated sewage after the removal of solids. The intake BOD levels for the ASTP are in the range of 100–150 mg/L and, consequently, it may be necessary to incorporate an anaerobic unit in which the organic load and BOD levels are very high. Duckweed culture, before the fish ponds, aids in the removal of excessive nutrient concentration and residues. The waste contains BOD_5 levels of about 100 mg/L. After treatment in the system with a total retention period of 5 days, the final effluent BOD_5 is brought down to 18–22 mg/L (section 3.2.5). Fish ponds are stocked with five carp species: catla (*Catla catla*), rohu (*Labeo rohita*), mrigal (*Cirrhinus migrila*), silver carp (*Hypophthalmichthys molitrix*) and common carp (*Cyprinus carpio*). The production levels recorded from the fish ponds are in the ranges of 3–4 mt/ha/year.

The Research Institute for Aquaculture No 1, the Fisheries University and Hanoi University, Vietnam, has carried out a research programme on the characteristics of wastewater, and its reclamation and re-use for fish culture. Wastewater has been used in aquaculture and agriculture in areas near Hanoi for many decades. Tuan & Trac (1990) described the three systems involving waste-fed fish culture that are used:

- fish culture
- fish–rice rotation
- fish–rice–vegetable rotation.

The average net yield from fish culture is about 2.1 mt/ha/year. Increased gross yield has also been achieved (4–7 mt/ha/year) by controlling the sewage flow and thereby adjusting the N/P ratio along with the organic load. Tilapia (*Oreochromis mossambicus* and *O. niloticus*), silver carp and common carp are the main species utilised in the waste-fed culture system in Hanoi.

Luu & Kumar (2000) presented the national status of sewage-fed aquaculture systems in Vietnam. In medium-sized towns and cities, where the sewage output is relatively high and the catchment area is suitably designed for aquaculture, fish culture is a common practice. In larger cities, where sewage is available around the year, the intermediate catchment areas are always used for aquaculture. Aquaculture is also widely practised in most of the sewage lakes. In Hanoi, there is a daily discharge of 320 000 m^3 of sewage that flows by gravity to the flood plains of Thanh Tri district, where it is used and treated by the agricultural–aquacultural systems. Other lakes, such as Truc Bach and West Lake, Bay Mau Lake, are simple catchment areas for domestic sewage and are used for aquaculture.

The sewage lakes or ponds are usually stocked with fingerlings of Chinese major carp (silver and grass), Indian major carp (rohu and mrigal), tilapia and common carp. In the intermediate catchment lakes, where the water level is less exchangeable, algal blooms develop quickly, sometimes resulting in sudden planktonic collapse and DO depletion to critically low limits. Experience shows that in such lakes the percentage of fish species such as silver carp and tilapia, which feed on phytoplankton, algae and detritus, can be increased to 50–60%. On the other hand, in sewage lakes and ponds where water is periodically pumped to balance the nitrogen content, the fish species thriving predominantly on detritus, zooplankton and zoobenthos can be stocked at a higher ratio. Stocking density depends on the quality of sewage. However, the commonly followed stocking density is up to four fingerlings per square metre within the size range 30–70 g. With this stocking density, fish productivity of these ponds/lakes reaches 5–7 mt/ha/year without other inputs, such as feed, fertilisers and chemicals.

According to Zhang (1990), the use of municipal wastewater in China has developed rapidly since the 1950s. In 1985 the total area of wastewater-fed aquaculture involving more than 30 sites was 8000 ha, with a total fish production of 30 000 mt. In the 1970s, people in China, especially environmentalists, started to review the positive and negative sides of waste-fed aquaculture. After the review, the government continued its development of waste-fed aquaculture. Most of the waste-fed aquaculture in China is located near cities, and it contributes a large part of commercial fish products to city markets. Silver carp and bighead carp are the main species stocked, along with common carp (*Cyprinus carpio*) and crucian carp (*Carassius auratus*). The stocking density is usually 15 000 per hectare of 20- to 30-g fingerlings. In general, fish yields from wastewater-fed ponds are 2–4 times higher than those from ordinary fish farms. The production ranges from 1.5 to 11 mt/ha/year. In China, polyculture systems, using Chinese major carp species with common carp, are common. The efficient usage of food resources coupled with year-round growing conditions (temperature) allows the systems to obtain high production rates. The major problem in waste-fed aquaculture in China is that some industrial toxic waste is mixed with municipal sewage. A great deal of effort has gone into separating industrial pollutants from wastewater.

4.7.2 Waste-fed aquaculture in Europe

Waste-fed aquaculture dates back more than a century in Germany. According to Prein (1990), two groups of systems can be distinguished:

- polishing
- wastewater-fed fish ponds.

The former receive well-treated effluent from wastewater purification systems and are subdivided into:

- ponds that receive drainage effluents from sewage fields
- ponds that receive effluents from other biological treatment systems.

The latter are designed to purify raw wastewaters, which have only been mechanically pretreated. Net freshwater fish production from waste-fed aquaculture averages 500 kg/ha per 7 months (860 kg/ha/year), with loading rates equivalent to 2000 persons/ha/year. Prein (1990) recorded over 90 installations across the country, ranging from small, single ponds, receiving wastewater amounts equivalent to the treatment of only a few dozen people, up to large systems, e.g. Munich, with 233 ha designed to treat the wastewater from 500 000 people and to produce a gross fish yield of 100–150 mt/year. The main fish species used are common carp (*Cyprinus carpio*) and tench (*Tinca tinca*). The German system was designed to operate in temperate latitudes with low stocking densities. Thus, the fish yields are lower than those of tropical waste-fed aquaculture systems where year-round optimum growth is possible because of tropical temperature conditions.

In Hungary, the first experimental sewage-fed fish culture trial was carried out in the city of Fonyod on a total water surface area of 211 ha. A 5-year research programme followed by a commercial operation along with a monitoring programme was implemented for the next 5 years (1979–83). Hungarian technical guidelines for domestic sewage-fed fish ponds were elaborated and approved by the government in 1982 (Olah, 1990). The technological package developed in Hungary was the culmination of the 5-year research project plus the 5 years of monitoring a commercial operation system. The sewage-fed fish pond technology in Hungary can receive, process, utilise and purify domestic sewage and produce 12–20 kg/ha/day. It applies basic principles of complete grazing pressure on both planktonic and benthic communities by implementing polyculture methods using silver carp and common carp, which are able to utilise the planktonic and benthic fish food resources completely (Fig. 2.8). The results from the Hungarian farm indicate that nutrient levels in the sewage are substantially reduced, e.g. ammonia from 48–50 mg/L to 0.3–0.5 mg/L, total N from 50–55 mg/L to 2–3 mg/L and total P from 10–12 mg/L to 0.70–1 mg/L. At the same time, the DO level has been increased to 8.3 mg/L.

4.7.3 Integrated farming

Modern integrated farming is an approach to farming that incorporates the concept of integrated resource management (IRM). This approach integrates the management of additional enterprises, particularly aquaculture, with those of the existing farming system and with their natural resource systems, so that opportunities for rehabilitation and synergism can be exploited (section 2.3.6). IRM systems with aquaculture as a major or minor component differ from traditional farming systems. The basic concept is to use aquaculture as a tool for recycling wastewater and nutrients without creating adverse environmental impact. In other words, IRM can complement and improve the overall recycling and efficient use of many types of farm and other wastewater in a sustainable way (Fig. 4.10).

In the system shown in Fig. 4.10, animal wastes are collected in a reception pit and then fed semi-continuously into an aerobic reactor, which is aerated through a membrane diffuser. Microbial populations in the reactor degrade organic material to CO_2 and microbial biomass. The aerobic process significantly reduces the organic content (BOD and COD) of the slurry, and transforms N and P. The slurry is separated into solid and liquid phases, with the solid phase subsequently composted. The liquid phase is diluted and passed into shallow microalgal ponds, which are mixed slowly using paddle-wheels. Prolific algal growth occurs, with nutrient removal via biotic and abiotic processes. Effluent from the ponds is fed into fishponds to produce zooplankton and aquatic plants on which the fish feed. The final effluent from fishponds is passed to constructed wetland, from which there may be recycling, and is also used for irrigation.

Overcoming constraints of enhancing food and economic security while minimising environmental impact is a prime concern of many developing nations in SE Asia. Vietnamese rural families are required to make better use of the available land, particularly within the highland region. About 80% of the total land area of Vietnam is mountainous with depleted forest. The Vietnamese government

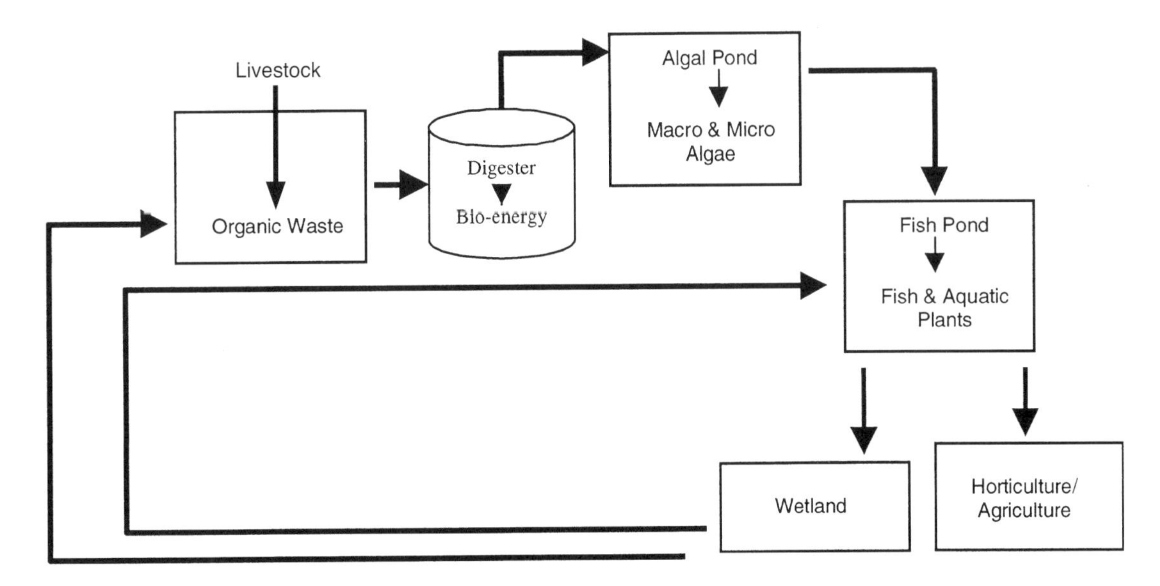

Fig. 4.10 Integrated wastewater treatment model at Urrbrae, South Australia.

is trying to address this problem through promoting an integrated farming model. The integrated farming system (section 2.3.6) practised in Vietnam has provided a significant boost to the income levels of farmers in lowland areas. Research is being carried out to maximise the production and sustainability of integrated farming in the mountain region.

4.8 Conclusion

Some of the negative impacts of aquaculture cannot be regarded as serious, particularly in comparison with other industries, including many forms of agriculture. Nevertheless, impacts such as nutrient discharges and coastal degradation are very serious. For the sake of the future expansion of the aquaculture industry, it is important to minimise environmental impacts and minimise negative public perception of the industry. Attention also needs to be paid to those impacts that the public (i.e. consumers) consider to be important, even if scientific evidence suggests otherwise.

Serious impacts are often an indication of bad management, wrong site location or poor planning, which will also be detrimental to the culture economics. In such an ecologically based industry, minimal impacts go hand in hand with good business. Aquaculturists have to work with and within the environment to succeed.

References

Ackefors, H. & Enell, M. (1994). The release of nutrients and organic matter from aquaculture systems in Nordic countries. *Journal of Applied Ichthyology*, **10**, 225–41.

Anon. (1993). Seagulls do the dirty on Dutch shellfish trade. *New Scientist*, **1890**, 8.

Beveridge, M. C. M, Ross, L. G. & Kelly, L. A. (1993) Environmental impacts of aquaculture – disturbances of biotopes and influence of exotic or local species new to aquaculture. In: *World Aquaculture '93 Conference, Torremolinos, Spain, May 26–28, 1993*, p. 494. Special Publication No. 19. European Aquaculture Society, Ostend, Belgium,

Black, K. D. (1992). Coastal resource inventories: a Pacific coast strategy for aquaculture development. In: *Aquaculture and the Environment. Reviews of the International Conference Aquaculture Europe '91, Dublin, Ireland, June 10–12, 1991* (Ed. by N. De Pauw & J. Joyce), pp. 441–60. Special Publication. European Aquaculture Society, Gent, Belgium.

Boyd, C. E. (1998). *Codes of Practice for Responsible Shrimp Farming*. Global Aquaculture Alliance, St Louis, MO.

Boyd, C. E. (2001). Water quality standards: total suspended solids. *Global Aquaculture Advocate*, **4**(1), 70–1.

Buchanan, J. S. (1992). Health and safety in the work environment: the view from the industry. In: *Aquaculture and the Environment. Reviews of the International Conference Aquaculture Europe '91, Dublin, Ireland, June 10–12, 1991* (Ed. by N. De Pauw & J. Joyce), pp. 381–92. Special Publication. European Aquaculture Society, Gent, Belgium.

Chamberlain, J., Fernandez, T. F., Read, P., Nickell, T. D. & Davies, I. M. (2001). Impacts of biodeposits from suspended mussel (*Mytilus edulis* L.) culture on the surrounding surficial sediments. *ICES Journal of Marine Science*, **58,** 411–16.

CIFA (1998). *Sewage Treatment Through Aquaculture*. Central Institute of Freshwater Aquaculture, Bhubaneswar, India.

Cripps, S. J. (1993). The application of suspended particle characterisation techniques to aquaculture systems. In: *Techniques for Modern Aquaculture. Proceedings of an Aquacultural Engineering Conference, 21–23 June 1993, Spokane, Washington, USA* (Ed. by J. Wang), pp. 26–34. American Society of Agricultural Engineers, St Joseph, MI.

Cripps, S. J. & Bergheim, A. (2000). Solids management and removal for intensive land-based aquaculture production systems. *Aquacultural Engineering*, **22,** 33–56.

Cripps, S. J. & Kelly, L. (1996). Reduction of wastes from aquaculture. In: *Aquaculture and Water Resource Management* (Ed. by D. G. Baird, M. Beveridge, G. A. Kelly & B. F. Muir), pp. 166–201. Oxford, Blackwell Scientific Publications.

Crozier, W. W. (1993). Evidence of genetic interaction between escaped farmed salmon and wild Atlantic salmon (*Salmo salar* L.) in a Northern Irish river. *Aquaculture*, **113,** 19–29.

Donovan, D. J. (1999). *Industry Environmental Code of Best Practice for Freshwater Finfish Aquaculture*. Report prepared for the Department of Primaries Industries and the Queensland Finfish Aquaculture Industry. Brisbane, Queensland.

GESAMP (IMO/FAO/UNESCO/WMO/WHO/IAEA/UN/UNEP (1991). Joint Group of Experts on the Scientific Aspects of Marine Pollution. Reducing Environmental Impacts of Coastal Aquaculture. *Reports and Studies of GESAMP*, Vol. 47. Food and Agriculture Organization of the United Nations, Rome

Gowen (1992). Aquaculture and the environment. In: *Aquaculture and the Environment: Reviews of the International Conference Aquaculture Europe '91, Dublin, Ireland, June 10–12, 1991* (Ed. by N. De Pauw & J. Joyce), p. 23. Special Publication. European Aquaculture Society, Gent, Belgium.

Haya, K., Burridge, L. E. & Chang, B. D. (2001). Environmental impacts of chemical wastes produced by the salmon aquaculture industry. *ICES Journal of Marine Science*, **58,** 492–6.

Holmer, M. & Kistensen, E. (1992). Impact of marine fish cage farming on metabolism and sulphate reduction of underlying sediments. *Marine Ecology Progress Series*, **80,** 191–201.

Howell, D. L. & Munford, J. G. (1992). Predator control on finfish farms. In: *Aquaculture and the Environment: Reviews of the International Conference Aquaculture Europe '91, Dublin, Ireland, June 10–12, 1991* (Ed. by N. De Pauw & J. Joyce), pp. 339–64. Special Publication. European Aquaculture Society, Gent, Belgium.

Jackson, D. & Costello, M. J. (1992). Dichlorvos and alternative sealice treatments. In: *Aquaculture and the Environment: Reviews of the International Conference Aquaculture Europe '91, Dublin, Ireland, June 10–12, 1991* (Ed. by N. De Pauw & J. Joyce), pp. 215–21. Special Publication. European Aquaculture Society, Ostend, Belgium.

Kapetsky, J. M. (1987). Conversion of mangroves for pond aquaculture: some short-term and long-term remedies. In: *Papers contributed to the Workshop on Strategies for the Management of Fisheries and Aquaculture in Mangrove Ecosystems, Bangkok, Thailand, June 1986* (Ed. by R. H. Mepham). FAO Fisheries Reports. (370) Suppl., pp. 129–41.

Karakassis, I., Hatziyanni, E., Tsapakis, M. & Plaiti, W. (1999). Benthic recovery following cessation of fish farming: a series of successes and catastrophes. *Marine Ecology Progress Series*, **162,** 243–52.

Kelly, L. A., Stellwagen, J. & Bergheim, A. (1996). Waste loadings from a fresh water Atlantic salmon farm in Scotland. *Water Resources Bulletin*, **32,** 1017–25.

Kumar, M., & Hiep, D. (1995). An investigation on issues and constraints of aquaculture development in the buffer zone of Ba Vi National Park, Viet Nam. In: *Proceedings of Fish Asia '95 conference, 19–21 September, Singapore*, pp. 23–6. Primary Production Department, Ministry of National Development, Singapore

Kumar, M., Clarke, S & Sierp, M. (2000). Linkage between wastewater treatment and aquaculture: initiatives by the South Australian Research Development Institute (SARDI). In: *National Workshop on Wastewater Treatment and Integrated Aquaculture Production, 17–19 September 1999* (Ed. by M. S. Kumar), pp. 153–9. SARDI Aquatic Sciences, Adelaide, South Australia.

Lucas, J. S. (1994). The biology, exploitation, and mariculture of giant clams (Tridacnidae). *Reviews in Fisheries Science*, **2,** 181–223.

Luu, L. & Kumar, D. (2000). Aquaculture – an effective biological approach for recycling of organic waste into high quality protein food, In: *National Workshop on Wastewater Treatment and Integrated Aquaculture Production, 17–19 September 1999* (Ed. by M. S. Kumar), pp. 49–53. SARDI Aquatic Sciences, South Australia.

Mills, S. (1989). Salmon farming's unsavoury side. *New Scientist*, **29**(April), 40–2.

Nor, S. M. (1984). Major threats to the mangroves of Asia and Oceania. In: *Proceedings of the Workshop on Productivity of the Mangrove Ecosystem: Management Implications* (Ed. by J.-E. Ong & W.-K. Gong), pp. 69–78. Universiti Sains Malaysia, Penang.

Olah, J. (1990). Wastewater-fed fishculture in Hungary. In: *Wastewater-fed aquaculture, Proceedings of the International Seminar on Wastewater Reclamation and Reuse for Aquaculture, Calcutta, India, 6–9 December 1988* (Ed. by P. Edwards & R. S. V. Pullin), pp. 79–89.

Environmental Sanitation Information Centre, Asian Institute of Technology, Bangkok, Thailand.
O'Sullivan, M. (1992). Environmental impact assessment and aquaculture in Ireland. In: *Aquaculture and the Environment: Reviews of the International Conference Aquaculture Europe '91, Dublin, Ireland, June 10–12, 1991* (Ed. by N. De Pauw & J. Joyce), pp. 103–8. Special Publication. European Aquaculture Society, Gent, Belgium.
Pearson,T. H. & Black, K. D. (2001). The environmental impacts of cage culture. In: *Environmental Impacts of Aquaculture* (Ed. by K. D. Black), pp. 1–32. Sheffield Academic Press, Sheffield.
Pergent, G., Mendez, S., Pergent-Martini, C. & Pasqualini, V. (1999). Preliminary data on the impact of fish farming facilities on *Posidonia oceanica* meadows in the Mediterranean. *Oceanologica Acta*, **22,** 95–107.
Phillips, M.J., Beveridge, M. C. M. & Ross, L. G. (1985). The environmental impact of salmonid cage culture on inland fisheries: present status and future trends. *Journal of Fish Biology*, **27,** 123–37.
Pillay, T. V. R. (1992). *Aquaculture and the Environment.* Fishing News Books, Oxford.
Pollnac, R. B. (1992). Multiuse conflicts in aquaculture – sociocultural aspects. *World Aquaculture*, **23,** 16–19.
Prein, M. (1990). Wastewater-fed culture in Germany. In: *Wastewater-fed Aquaculture. Proceedings of the International Seminar on Wastewater Reclamation and Reuse for Aquaculture, Calcutta, India, 6–9 December 1988* (Ed. by P. Edwards & R. S. V. Pullin), pp. 13–47. Environmental Sanitation Information Centre, Asian Institute of Technology, Bangkok, Thailand.
Primavera, J. H. (1994). Environmental and socioeconomic effects of shrimp farming: the Philippine experience. *Infofish International*, **1**(94), 44–9.
Primavera, J. H. (1998). Tropical shrimp farming. In: *Tropical Mariculture* (Ed. by S. S. De Silva), pp. 257–89. Academic Press, London.
Ross, A. (1988). *Controlling Nature's Predators on Fish Farms.* Marine Conservation Society, Ross on Wye, UK.
Schwartz, M. F. & Boyd, C. E. (1994) Effluent quality during harvest of Channel catfish from watershed ponds. *Progressive Fish-Culturist*, **56,** 25–32.
Smith, P. (1992). Antibiotics and the alternatives. In: *Aquaculture and the Environment: Reviews of the International Conference Aquaculture Europe '91, Dublin, Ireland, June 10–12, 1991* (Ed. by N. De Pauw & J. Joyce), pp. 223–34. Special Publication. European Aquaculture Society, Gent, Belgium.
Tovar, A., Moreno, C., Manuel-Vez, M. P. & Garcia-Vargas, M. (2000). Environmental implications of intensive marine aquaculture in earthen ponds. *Marine Pollution Bulletin*, **40,** 981–8.
Tuan, P. A. & Trac, V. V. (1990). Reuse of wastewater for fish culture in Hanoi, Vietnam. In: *Wastewater-fed Aquaculture, Proceedings of the International Seminar on Wastewater Reclamation and Reuse for Aquaculture, Calcutta, India, 6–9 December 1988* (Ed. by P. Edwards & R. S. V. Pullin), pp. 69–71. Environmental Sanitation Information Centre, Asian Institute of Technology, Bangkok, Thailand.
Wall, V. (1992). Aquaculture, relevant EC and national legislation on safety, health and welfare. In: *Aquaculture and the Environment: Reviews of the International Conference Aquaculture Europe '91, Dublin, Ireland, June 10–12, 1991* (Ed. by N. De Pauw & J. Joyce), pp. 367–79. Special Publication. European Aquaculture Society, Gent, Belgium.
Welch, E. B & Lindell, T. (1992). *Ecological Effects of Wastewater: Applied Limnology and Pollutant Effects.* Chapman & Hall, London.
Wheaton, F. W. (1977). *Aquacultural Engineering.* Chichester, John Wiley & Sons.
Wu, R. S. S. (1995). The environmental impact of marine fish culture. Towards a sustainable future. *Marine Pollution Bulletin*, **31,** 159–66.
WWF (2001). Threats to salmon populations. In: *The Status of Wild Atlantic Salmon: A River by River Assessment*, pp. 45–56. World Wildlife Fund, Oslo,
Zhang, Z. S. (1990). Wastewater-fed fish culture in China. In: *Wastewater-fed Aquaculture. Proceedings of the International Seminar on Wastewater Reclamation and Reuse for Aquaculture, Calcutta, India, 6–9 December 1988* (Ed. by P. Edwards & R. S. V. Pullin). pp. 3–12. Environmental Sanitation Information Centre, Asian Institute of Technology, Bangkok, Thailand.

5

Desert Aquaculture

Sagiv Kolkovski, Gideon Hulata, Yitzhak Simon and Avi Koren

5.1 Introduction

The desert and the parched land will be glad;
the wilderness will rejoice and blossom.

Isaiah 35:1

This chapter is based on developments in Israel. The experiences in Israel are applicable to other countries that have similar water and weather conditions in at least some parts of their geography. These countries or regions include the interiors of Central Africa, Eastern Asia, North America and Australia, and most of the countries in the Middle East. The developments are also applicable to countries with limited or no coastline and to heavily populated countries with limited land areas available for aquaculture. Although the chapter discusses 'desert aquaculture', the systems and methods described in it may be applied to any area where aquaculture depends on limited water resources from terrestrial sources.

Israel is located in a semi-arid zone, with distinct winter (wet) and summer (dry) seasons, and a low annual rainfall of around 500 mm. The only large, inland water body is the Lake of Galilee, which mainly supplies freshwater for human consumption. Moreover, in the central-north areas of Israel, where the majority of the rainfall occurs, the hilly and mountainous land cannot naturally hold water. In spite of the obvious climatic constraints and overall shortage of water, both agriculture and aquaculture are highly developed in Israel.

To deal with these impediments, different solutions and methods to maximise water use were developed. Some of these solutions are listed below.

(1) Reservoirs to store rainwater during the wet season. Israeli agriculture is now largely intensive and depends on irrigation from these reservoirs during the dry summer. Recently, it has become common to use irrigation reservoirs for fish culture in integrated farming systems (section 2.3.6). These integrated agriculture–aquaculture systems use the water twice:
   - within an aquaculture production system
   - subsequently to supply irrigated agriculture systems.

   This system, now a few decades old, was a significant step in the intensification of inland fish culture in Israel (Hepher, 1985; Sarig, 1988).
(2) Large-scale recirculating systems, in which water from fish ponds, raceways and tanks, which are usually inside greenhouses, is passed into sediment ponds to remove the solids. The water is then passed through a water treatment system, which may include different filters and biological filters, and is then returned to the fish rearing systems.
(3) Super-biomass-intensive recirculating systems that incorporate water filtration systems, such as drum filters, biological filters, protein skimmers and oxygen injection systems. Super-biomass systems may support up to 150 kg of fish per cubic metre of water, with an average

level for these systems of ~100 kg/m^3, i.e. 10% of the water volume is fish biomass. Culture is intensive, as the stock is entirely dependent on a comprehensive artificial diet (see definitions of extensive, semi-intensive and intensive culture in Chapter 2). These systems are usually compact, take up a relatively small area and are extremely efficient with water usage.

(4) Greenhouse technology was adopted from desert agriculture and includes environmental control, i.e. humidity, temperature, light and radiation. These conditions are important in arid areas, which have large temperature changes between day and night and summer and winter.

5.2 Regional variation in Israel

Israel can be divided into two climatic regions.

(1) The southern arid/semi-arid areas have very low annual precipitation (< 100 mm) and consist of the Negev Desert and the Arava Valley. These areas are part of the Surian–African Break. The Arava Valley sits between two mountain systems: the Jordan in the east and the Negev in the west. It starts at the southern end of the Dead Sea, some 400 m below sea level, and finishes at the Gulf of Eilat (an extension of the Red Sea). The Negev Desert lies west of the Arava Valley and on most of the southern part of Israel.

(2) The central-north of the country has a relatively high annual rainfall (> 800 mm). This region is the most populated area of Israel and the competition for resources is intense. These areas include the flats near the Mediterranean sea, the hilly area towards the east that borders the Lake of Galilee, and the Galil and Golan mountains in the north.

The two areas differ in their sources of water, its use and the types of integrated aquaculture systems employed. In the southern, more arid, region water is pumped year-round from geothermal bores. It is used for super-biomass-intensive aquaculture in a very highly integrated system of water management from the bore to the end user. In the central-north region, irrigation reservoirs storing winter rainfall also operate as large fish ponds with semi-intensive culture. As noted above, although this central-north region has moderate rainfall, compared with the southern region, it potentially suffers from water shortages throughout much of the year as a result of its topography.

5.3 Desert aquaculture

The term 'desert aquaculture' refers to aquaculture production of fish, shellfish and aquatic plants in arid areas or in areas where water supply is restricted. It sounds paradoxical given the obvious lack of suitable surface waters for such a purpose in these areas. However, during the last two decades, the commercial 'desert aquaculture' sector has thrived in southern Israel, utilising the vast groundwater resource in the southern arid region. This development also results from the constant freshwater shortage in the central-north region of Israel during recent years, which in turn has limited further development of aquaculture in that region.

'Desert aquaculture' in Israel, as in many places around the world, offers many specific advantages in regions where:

- Large quantities of brackish groundwater can only partially be used for agriculture.
- There is a warm ambient climate.
- Geothermal bores can maintain high temperatures in the winter through greenhouse use.
- The dry climate allows water cooling in summer.
- There is inexpensive land.
- Geographic isolation provides a natural quarantine.
- There are minimal ecological risks.
- Year-round production can be achieved.

These advantages have attracted large investment over recent years in a diversity of fish and shellfish species, including striped bass, barramundi, carp, mullets, tilapias, redclaw crayfish, brackish water shrimps (*Litopenaeus vannamei*) and ornamental fish species.

5.3.1 Aquaculture in geothermal water

Arid aquaculture in southern Israel began in 1979 with the discovery of locally available geothermal water (at 60°C) near Moshav Faran, a settlement in the Arava Valley. The idea of using hot groundwater for super-biomass aquaculture, to achieve maximum growth throughout the year, has subsequently been developed commercially. Combined heating of greenhouse-covered microalgae cultures (*Spirulina* and *Dunalliela* species) and fish ponds has also been successfully trialled.

For both economic (the cost of 1 m^3 of water in Israel is about US$0.25) and ecological reasons, the design of integrated aquaculture projects with agriculture areas as end-users is essential in arid areas. In contrast to the central-north areas of Israel, integrated aquaculture in the southern, more arid, areas is based on super-biomass-intensive systems with very tight water budgets. Water loss is minimal and is predominantly due to evaporation. However, even when there is no need for heating during the summer, most of the fish farms have water exchange of at least 10% per day to maintain water quality. A small fish farm of 2000 m^3 will therefore use about 200 m^3 of water per day, which in turn will irrigate about 4 ha of crops in the desert summer. In winter, when a large amount of water is needed to supply the heat energy to the fish ponds in the aquaculture system, there is a need to find a solution for all the output water or effluent (Fig. 5.1).

There are two options for transferring heat energy to the fish ponds in these production systems.

(1) A closed system using heat exchangers. When using a closed system, the geothermal water is used for heating the fish pond via a heat exchanger.
(2) Direct supply of water to the fish pond. When a direct supply of geothermal water to the fish pond is used for heating, the water is also used for flushing the organic matter from the pond and to contribute overall to the water quality of the pond. Accordingly, the outlet water is loaded with suspended solids, micro-organisms, algae and plankton as a result of the high nutrient loading on the intensive rearing system (Zoran *et al.*, 1994; Milstein *et al.*, 1995).

When the end-user of the effluent is drip irrigation, the water needs to be filtered or otherwise treated prior to being distributed under pressure through the irrigation system (Fig. 5.1). Usually, a small reservoir (0.1–1 ha surface area) is attached to the fish farm for this purpose. This reservoir, together with water treatment facilities, is used to provide a buffer between the agriculture project (e.g. greenhouse or open field) and the aquaculture system (Fig. 5.2). Fish are also reared in this reservoir, but at relatively low biomass/unit volume or area. The water treatment facilities typically include high-pressure pumps, a chlorine injection system (or other form of disinfection) and an automatic filtration system. Secondary filtration is undertaken at each irrigation head to ensure good water quality for final reticulation and to prevent drippers from clogging with particulate waste matter.

Knowing the bore water salinity is crucial for any agricultural crop, with 0–5‰ salinity being an acceptable concentration in most cases in Israel. Most of the geothermal water available in Israel is considered too saline (8–12‰), especially if increased salinity occurs as a result of evaporation in fish ponds. Rearing sensitive crops is not feasible at these higher salinities, although other crops, e.g. watermelons, alfalfa and tomatoes, are highly successful. 'Desert sweet tomatoes', a brand name for a very sweet variety of tomato that was developed in Israel and is produced in saline groundwater, is extremely successful in both local and European markets. Higher salinities have also been successfully used in production of olives and date palms in an integrated agri-aquaculture system.

Of five model pilot-scale farms established during the 1980–90s, two expanded to a full commercial scale of 200–400 mt/year of aquaculture production. These farms were built from modular units of 8 × 300 m^3 capacity ponds (Fig. 5.3). The ponds are connected to a water treatment unit that includes a settlement pond (100–200 m^3 capacity) together with an 'activated suspension' method for nitrification of nitrogen into protein by bacteria (Avinmelech *et al.*, 1989, 1994; Avinmelech, 1998) (section 5.3.2).

Two different types of water use systems were developed for these fish ponds.

(1) Autonomic pond systems that include the ponds, settlement pond and water treatment system. These systems function primarily as aquaculture systems with only small daily amounts of water discharge.
(2) Pond systems that are connected to an irrigation reservoir, exchanging water with it. A water treatment system may be included for treating the

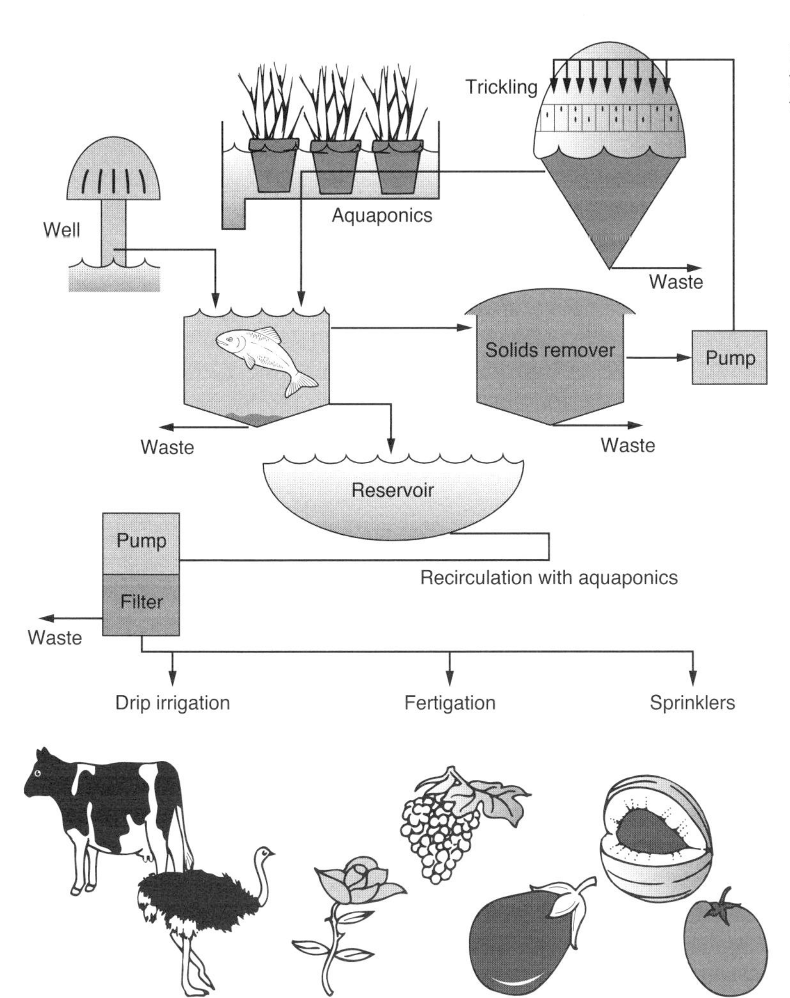

Fig. 5.1 Schematic description of a form of desert aquaculture (see text for a description of processes).

Fig. 5.2 Interior of greenhouse containing large tanks for intensive fish culture.

Fig. 5.3 Pilot-scale farms established during the 1980s and 1990s are built from modular units of 8 × 300 m³ capacity ponds.

pond effluent before it returns to the reservoir. These systems have a dual function of irrigation and aquaculture.

Ein Tamar fish farm on the banks of the Dead Sea is an example of the first type of pond system. The farm includes a large group of greenhouses with 16 × 300 m³ circular concrete ponds. The ponds are drained three times per day to remove solid waste via a central standpipe. A drum filter (Fig. 3.9) removes the solids and the water is passed through a trickle biological filter system, which sprays the water over a large surface area of trickle filter medium (section 3.2.7). The water exchange of the pond system is 5–10% (of total pond volume) per day, with exchange water obtained from a nearby bore. The pond exchange rate (recirculated through the filtration system) is 300–400% per day. The total pond volume is ~5000 m³ with annual production of 150 mt. The fish species reared in the farm are carp, tilapia and barramundi (the last of which was imported in the early 1990s from Queensland, Australia).

The second system is based on using a large irrigation reservoir as a biological filter and water source. Kibbutz Neve Etan fish farm includes 8 × 200 m³ circular concrete ponds. Water is pumped from the fish ponds to a settlement pond for solids removal and then pumped to the main irrigation reservoir. In most cases, the reservoir is very large with a capacity of several million cubic metres of water. The effluent water from the fish farm is 'diluted' in the reservoir, which acts as a biological filter, so the water that is pumped back into the pond systems is relatively clean. Some of the irrigation reservoirs are also stocked with fish that can maintain the water quality, i.e. by controlling the algae and other aquatic organisms. This method (Levanter, 1987) involves a range of fish species that are usually stocked in these reservoirs to control the aquatic environment. These are usually a variety of carp species that feed at different levels of the food chain. For example, silver carp filter unicellular algae, bighead carp feed on zooplankton, grass carp eliminate vegetation and black carp control aquatic mollusc populations (section 2.3.5 and Fig. 2.8). The fish are not considered an environmental threat (because of wild spawning) since they can be stocked as sterile triploids. In Israel there are centres and hatcheries that specialise in producing these species (for example, Gan Shmuel Fish Breeding Centre). Since these reservoirs are used for a dual propose, i.e. agriculture and aquaculture, synchronisation between fish and crop production needs to be managed so that at the end of the summer, when the reservoir is near empty or very low, the fish will be harvested. Stratification of water in these reservoirs may occur during the rearing season and can cause mortalities. Different methods were developed to break the stratification, mainly using aeration (Zoran & Milstain, 1998; Milstein *et al.*, 2000).

5.3.2 *Aquaponics: integration of aquaculture and hydroponics*

Different methods of 'aquaponics' have been developed and tested in Israel over many years at a pilot scale on desert aquaculture farms (Fig. 5.1). 'Fill and flush' is one of the common methods used, in which aquatic plants are housed in buckets or containers filled with lava gravel substrate (0.4–1 cm grade). The containers (with 0.5-cm holes) are placed in a shallow, plant pond bed (1 m × 10–15 m). Inlet water, obtained from the sedimentation basin/settlement pond or biological filtration unit of an aquaculture system, fills the plant pond bed to the top of the gravel. The water is then drained, by siphoning or automatic valve, to a small reservoir at the edge of the pond bed system. From this reservoir the water is then pumped back to the fish ponds. Owing to high temperature and evaporation, during the summer months, additional water is required (up to 8–10 m^3/0.1 ha/day).

Efficient integration of hydroponics and fish ponds allows a ratio of hydroponics to fish pond land usage (area) of about 1:1. Water quality monitoring is essential, and addition of potassium, iron and calcium carbonate is desirable. In recent aquaponics field experiments, crop production from this type of system was double that of conventional (field produced) crops. Furthermore, this was achieved with no addition of pesticides or insecticides, possibly making the crops more marketable as 'organic' or 'environmentally safe'. Aquaponics crops tested included leafy plants and herbs such as basil, mint, chives, salvia, rosemary and lettuce. Many other field crops such as honey melons, watermelons, cucumbers, tomatoes, peppers and fruit trees were also successfully trialled.

A pilot-scale operation is currently being tested in the northern part of Israel in Kibbutz Ain Shemer. Although this project may not be economically viable, it demonstrates some of the principles for agri-aquaculture projects. The project includes fish-rearing sections of 150 m^3 plastic-lined elliptic ponds, a heat exchange system and a nursery with different plants. The plant nursery and the fish ponds are built inside a greenhouse. The temperature in the greenhouse is controlled by heat exchange, fans and windows. The heat exchanger is placed 5 m above the ponds and includes a polyethylene pipe (110 mm i.d.) connected to 150 polyethylene pipes (25 mm i.d.). Each of the pipes has 10 spray nozzles that produce water spray (40 mm i.d.). The device is activated at intervals and creates a fine spray of water that reduces the temperature (Fig. 5.4). Effluent water is pumped from the fish pond through a drum filter.

The water treatment unit includes a conical separator that, while pumping the pond water at a rate of 11 m^3/h, creates a vortex action and removes the solids. The biological filter includes a rotating drum with fibreglass plates. Eighty per cent of the drum surface is submerged and 20% of the surface area is exposed to the air. By rotating the drum, air is captured and 'locked' into the space between the plates. The rotating movement ensures that the nitrifying bacteria growing on the drum plates (as on any biological filter medium) have constant interaction with organic matter and dissolved oxygen. The organic matter within the ponds is treated using the 'activated suspension' method (Avinmelech *et al.*, 1989, 1994). This method is based on the concept that the organic particles in the pond itself are used as 'carriers' for bacteria. The bacteria change the excess inorganic and organic N, by nitrification and carbon metabolism, to harmless microbial protein that may be used as a food source for plankton in the pond itself.

The main fish species reared in the fish ponds is mirror carp; however, other freshwater fish can be reared in such a system.

5.4 Water-limited aquaculture

With water limited throughout much of the year and lack of natural water containments, the development of aquaculture in the central-northern region of Israel also required specific systems to be developed in which dual-purpose reservoirs are used for irrigation and fish culture.

5.4.1 *The annual cycle*

Fish are reared in the reservoirs during the irrigation period from spring to autumn (April–November). Fish harvesting is carried out in autumn, when the reservoirs are nearly empty and awaiting refill over

Fig 5.4 Fish ponds and plant nurseries inside a greenhouse. Water temperature is controlled using a heat exchanger placed 5 m above the ponds. Water temperature control includes polyethylene pipes with spray nozzles that produce water spray to cool the water.

winter. Reservoirs begin to fill during late autumn with a combination of ground (spring) water and early winter rains. When the water column reaches a height of 1 m (normally around December) the reservoir can be stocked with fingerlings. These fish reach market size within a year. The reservoir continues to fill until January or early spring, depending on its geographical location and holding capacity. In late spring or early summer (May–June), the stored reservoir waters are pumped out for crop irrigation and the water level in the reservoir drops gradually while the fish are growing. By September–October, when only 1–2 m depth of water remains, the farmers progressively harvest the fish and prepare for final draining of the reservoir and harvesting of the remaining fish.

In the 1960s, the first reservoirs were constructed solely for irrigation in the Harod Valley to serve the needs of the kibbutz settlements in that region. Five reservoirs were constructed, with a total surface area of 90 ha. These rather shallow reservoirs collected and stored brackish ground (spring) water, flowing year round, for use during the dry summer. The reservoirs were subsequently deepened to increase their storage capacity for winter rains. The farmers of these communal settlements then decided to use them for fish culture also. In a few years it became apparent that rearing fish in such reservoirs was profitable, although professional and technological expertise were lacking.

This secondary use of water for fish culture, by introducing aquaculture into irrigation reservoirs, improved the efficiency of water usage and reduced the cost of water needed for fish culture in conventional earthen ponds. However, the main drawback was harvesting the fish from these reservoirs, as the engineers planning their construction did not consider such activities in their initial design. This led to dramatic technological developments during the late 1970s, when many new reservoirs were constructed and were specifically planned for dual-purpose use, i.e. they were equipped for efficient harvesting of the fish. This development, in turn, changed the emphasis on water usage in this region. In the newly constructed reservoirs, fish culture became the primary activity and crop irrigation a 'by-product'. Most fish farms in the central-northern region of Israel now operate such reservoirs, which have proved to be an efficient and profitable tool for fish culture.

In Israel, both in conventional earthen fish ponds and in reservoirs, freshwater fish are typically cultured in a polyculture system, stocked with different species of fish (Hepher, 1985). Most reservoirs are stocked with:

- 80% common carp and tilapia (at various percentage combinations)
- 20% accompanying species, such as grey mullet, grass carp, red drum and the silver carp × bighead carp hybrids.

Only about 30% of reservoirs are used for monoculture of either carp or tilapia. The main advantages of the monoculture system relate to easier harvesting operations and storage of recovered fish.

Juvenile fish to be stocked into reservoirs are initially reared to an appropriate size in conventional fish ponds that are shallow and earthen. None of the target species can grow from a 1-g juvenile (at first stocking) to market size from spring to autumn (~8 months), which is the usual operational period of a reservoir. Thus, the farm must have additional rearing ponds in order to efficiently operate a reservoir for fish culture. These ponds will also be used at harvest time to hold fish until they are marketed (all through the year), as a farm cannot market the whole harvest of a reservoir at once. Thus, the ponds are operated all year round: rearing juvenile fish in the spring and summer, and holding fish for market or for stocking the reservoirs (in the following spring) during the winter. The stored fish are fed maintenance rations during winter, and this increases operating costs.

5.4.2 Economics of reservoirs vs. conventional fish ponds

Detailed examples of production figures for dual-purpose reservoirs are presented in Hepher (1985) and Sarig (1988). Production inputs and associated capital depreciation costs per unit weight of fish output in such reservoirs compare favourably with those in conventional earthen fish ponds (Table 5.1).

To compare water use in particular, production of 1 kg of tilapia requires:

- 7.4 m^3 of water in conventional earthen ponds with semi-intensive (medium) biomass
- 4.6 m^3 in concrete ponds with high biomass
- 4.0 m^3 when cultured with low biomass in reservoirs
- 1.4 m^3 in an industrial culture system that is indoors and super-biomass intensive.

Production costs of the major species are rather similar for the two culture systems (stand-alone and dual-purpose reservoir), with those in dual-purpose reservoirs being slightly lower (Table 5.2). Most of this difference is due to the different feed conversion ratios (FCRs) obtained in the two culture systems. Current farm-gate prices (to the farmer) are:

- 13–14 NIS*/kg for carp
- 12–13 NIS/kg for tilapia
- 5.7 NIS/kg for silver carp (and its hybrid)
- 17–20 NIS/kg for mullet and red drum.

A detailed breakdown of production costs of fish in dual-purpose reservoirs is shown in Table 5.3.

5.5 Greenhouse technology

Greenhouse technology for aquaculture has developed rapidly during the past decade to accommodate the need for an appropriate environment for recirculating systems (Fig. 5.1). Greenhouse technology was originally developed for agriculture crops and in recent years has been adapted to the aquaculture industry. Owing to the extreme changes of temperature between day and night in desert areas, there is a need to cool down the greenhouse during the day

Table 5.1 Production inputs and associated capital depreciation costs for producing 1000 kg of fish in ponds and reservoirs in Israel

Data	Dual-purpose reservoirs	Earthen ponds	Comments
Water	–	50 000 m^3/ha	Reservoir water price is charged to irrigated field crop
Feed	1300 kg	2200 kg	Ponds are used for holding fish during winter, so more feed is required overall
Labour	5 days	6 days	
Seed	4000 per ha	5000 per ha	
Energy	5000 kW	6000 kW	
Depreciation	2000 NIS*	2500 NIS	

*1 NIS is approximately US$0.25.

Table 5.2 Comparative costs of rearing carp and tilapia to commercial size in reservoirs and conventional earthen ponds

Reservoirs	Conventional earthen ponds
8 NIS per kg of carp*	9 NIS per kg of carp
11 NIS per kg of tilapia	12 NIS per kg of tilapia

*1 NIS is approximately US$0.25.

Table 5.3 Itemised costs of producing 1 kg of fish in dual-purpose reservoirs

Item	Unit	Quantity	Unit price	Cost
Water	m^3	3.5	0.10	0.35
Feed	kg	2	1.20	2.40
Fingerlings (50 g)	Number	3.02	0.65	1.96
Energy (pumping, aeration)	kW/h	5	0.30	1.50
Maintenance, machinery		1	0.30	0.30
Marketing		1	0.80	0.80
Labour, management	Days/mt	3	300.0	0.90
Financing	%	5		0.41
Depreciation		1	1.44	1.44
Total				10.06

Price and cost figures in NIS per kg (1 NIS is approximately US$0.25).

and to heat or maintain the temperature at night. Radiation also plays an important role in generating heat, as well as high light intensity, and needs to be controlled on a daily cycle.

The greenhouses are typically modular so that they can be built and expanded to specific requirements. Most greenhouses are constructed from an arc or square steel skeleton with shade cover. In most of these greenhouse systems, algae growth within the tanks is restricted or inhibited by using a green plastic shade material that was developed in Israel and can be tailor-made to precise specifications (percentage radiation and light intensity). There are also pond/shade cover systems in which a circular concrete pond has an 'igloo' construction to support the shade cover, creating a greenhouse effect under it.

Often, a reflective thermal screen is installed inside the greenhouse. This helps to decrease heat stress during periods of high radiation by shading and dispersing light evenly. It then aids in reducing energy loss in the heated greenhouse during cold weather periods. The installation of a mobile screen enables it to close or open, which gives the grower substantial control over air temperature, humidity, condensation, etc., in the greenhouse. The thermal screen is made of aluminised synthetic fibre netting. The aluminium is incorporated into the synthetic fibre before weaving, thus ensuring a longer life. The netting gives an effective shading of 40%. Owing to the reflective powers of the aluminium, the temperature loading is decreased in the summer months and increased during winter. For colder climates, screens with higher energy-saving properties can be designed. The screens are suspended on a system of cables and can be opened or closed manually or computer controlled.

The most cost-effective way to lower the temperature in greenhouses is evaporative cooling. Air is expelled from the closed environment by means of fans, which tends to create a vacuum inside. This vacuum is then filled with outside (fresh) air, flowing in through specially corrugated cardboard pads, kept constantly moist. The degree of the lowering of the inside temperature depends on the temperature and humidity of the outside air (Table 5.4).

The design of the system is based on the desired temperature and required humidity and is directly related to the location of the greenhouse, environment conditions, size of the greenhouse, and the cooling pad. The placement of the cooling system depends on the direction of the light and wind. The system is based on pumping water from a water tank through a dripping system onto the cardboard pads. The air (wind or fans or both) passes through the wet cardboard into the greenhouse. Gutters at the bottom of the cardboard collect the excess water and it is pumped back to the water tank.

5.6 Species for water-limited aquaculture

It is not a simple matter to give a list of species that are suitable for use in aquaculture systems in conditions of water shortage. There are several reasons for this.

(1) It is difficult to characterise these culture systems. As described in this chapter, there are many different systems, from low-biomass reservoirs to high-biomass intensive systems. They all work on the principle of 'smart usage' of the water available.
(2) A decision on the 'best' cultured fish or shellfish species should take into account many local and non-biological factors, such as market value and volume and the cost of all the operation segments (for example labour, land and electricity).

In addition to the general considerations for choosing the best species in any aquaculture project (section 2.8), there are specific biological and other attributes required for a species to be suitable for water-limited aquaculture. In general, aquaculture species reared in water-limited aquaculture systems need to be more stress tolerant than species reared in flow-through systems or sea cages. The former are more likely to have broader tolerances of adverse water quality parameters associated with crowding: low dissolved oxygen (DO), low pH, high N compounds and high suspended solids. Depending on the environment or ability to economically control these parameters, they may be required to tolerate substantial temperature and salinity variations.

Various fish species are currently reared around the world in water-limited systems, including recirculating systems, pond systems and others. Probably the most common fish species for water-limited systems are tilapias. They may be reared intensively or semi-intensively in polyculture systems with other fish species, such as carps and mullets. In many countries, tilapias are considered to be the 'best' candidate for water-limited systems. They have very high stress tolerance and can be reared in a range of salinities, from fresh to almost seawater salinity (Chapter 16). Tilapia can be reared in very high densities of up to 150 kg/m^3. The rearing period is relatively short (6 months to market size), and in many countries there are very good market values and demand. Other fish species suitable for water-limited aquaculture systems include carps, mullet and barramundi.

Table 5.4 Outdoor/indoor temperature and humidity in greenhouses using cooling pads (Dagan Agriculture Automation, Israel)

Outdoor dry bulb temperature (°C)	Indoor dry bulb temperature (°C) at outdoor relative humidity (%) of										
	10.0	15.0	20.0	25.0	30.0	35.0	40.0	45.0	50.0	55.0	60.0
30	15.5	17.5	19.0	20.0	20.5	21.0	22.0	23.0	24.0	24.5	25.0
35	20.0	21.0	22.0	23.5	24.0	25.0	26.0	27.0	28.0	29.0	29.5
40	23.0	24.5	25.5	27.0	28.0	29.0	30.5	31.0	32.5	33.0	34.0
45	26.0	28.0	29.5	30.5	32.0	33.5	35.0	36.0	37.0	37.5	38.5
50	29.0	31.0	33.0	34.5	36.0	37.0	39.0	40.0	41.0	42.0	43.0

5.7 Conclusions and future directions

During the past few years, Israeli restrictions on the use of water for agriculture and aquaculture, largely aggravated by prevailing drought conditions, have focused attention on the development of 'water-smart' culture systems. More productive use of saline groundwater in desert areas is increasingly important. Very high-density and high-yield systems that can maximise the use of the water are being developed by:

- recirculation through water treatment systems
- recirculation through large reservoirs that can also be used for irrigation.

Different integrated agri-aquaculture systems are also being tested. Accordingly, Israel is currently investing in the development of salt-tolerant crops that are suitable as the end-water users after first use by aquaculture. Integrated aquaculture systems with these crops, including water-efficient pond/reservoir and recirculating/greenhouse systems, are being further developed and improved. Other priority issues for industry in Israel are the evaluation of new aquaculture species that suit these systems and which have local and export market potential.

Many countries face the problem of increasing salination, where soil becomes increasingly saline, as a result of salty groundwater rising, which renders soil unsuitable for agriculture. For example, Western Australia has already lost 30% of the land used for agriculture as a result of rising groundwater. This problem needs to be addressed by finding salt-resistant crops and, in the longer term, by rehabilitation of the soil. In Israel, various exotic plants have been imported from Africa, and others have been genetically improved to cope with the high salinity.

Use of the water for agri-aquaculture can provide a solution when the water is in the range of 0–8‰ salinity. This salinity range allows production of various crops such as date palms, olives and varieties of green vegetables.

The smart usage of water is sometimes needed where it is impossible to dump vast amounts of saline water into the environment. In this situation, fully or partially recirculated systems are the best solution. This is also the case whenever water quality is a problem and there is a need for water treatment, prior to use, to address problems such as high iron and low pH.

Finally, even after finding the right fish species for the local conditions and markets, and the optimal rearing system, the most important aspect is the aquaculturist behind the system. Trained and knowledgeable personnel are the key to any successful aquaculture venture.

Acknowledgement

We acknowledge, with gratitude, technical information and advice received in the preparation of this chapter from Mr Cobi Levanon, project manager, Midge 2000; Hamama 2000, Kibbutz Ain Shemer; and DAGAN Agriculture Automation Ltd.

References

Avnimelech, Y. (1998). Minimal discharge from intensive fish ponds. *World Aquaculture*, **21**, 32–7.

Avnimelech, Y., Kochva, M. & Diab, S. (1994). Development of controlled intensive aquaculture systems with a limited water exchange and adjusted carbon to nitrogen ratio. *The Israeli Journal of Aquaculture – Bamidgeh*, **46**, 119–31.

Avnimelech, Y., Mokady, S. & Schroeder, G. L. (1989). Circulated ponds as efficient bioreactors for single-cell protein production. *The Israeli Journal of Aquaculture – Bamidgeh*, **41**, 58–66.

Hepher, B. (1985). Aquaculture intensification under land and water limitations. *GeoJournal*, **10**, 253–59.

Laventer, H. (1987). *Contribution of Silver Carp (Hypophthalmichthys molitrix) to the Biological Control of Reservoirs*. Mekoroth Water Co., Israel.

Milstein, A., Zoran, M. & Krambeck, H-J. (1995). Seasonal stratification in fish culture and irrigation reservoirs: potential dangers for fish culture. *Aquaculture International*, **3**, 1–7.

Sarig, S. (1988). The development of polyculture in Israel: a model of intensification. In: *Intensive Fish Farming* (Ed. by C. J. Shepherd & N. R. Bromage). pp. 302–32. Blackwell Scientific Publications, Oxford.

Zoran, M. & Milstein, A. (1998). A device for gradual breaking of stratification in deep fish culture reservoirs. *International Reviews in Hydrobiology*, **83**, 673–80.

Zoran, M., Milstein, A. & Krambeck, H.-J. (1994). Limnology of dual purpose reservoirs in the coastal area and Jordan Valley of Israel. *The Israeli Journal of Aquaculture – Bamidgeh*, **46**, 64–75.

6
Reproduction, Life Cycles and Growth

Paul Southgate and John Lucas

6.1 Introduction

> [W]hereas the organs of reproduction, with their product the seed and embryo, are of paramount importance!
>
> Charles Darwin (1859)

The three major groups of aquaculture animals will be considered in this chapter: fish, bivalve molluscs and decapod crustaceans (the last two are the major shellfish groups). The morphologies of the species in these three groups are extremely different (see invertebrate and vertebrate textbooks), as are their reproductive systems, life cycles and patterns of growth. However, although the details and requirements of the various life stages of aquaculture species vary enormously, there are some general patterns.

6.2 Reproductive physiology

6.2.1 Fish

The sexes in cultured fish are separate and their paired gonads are located dorsolaterally in the body cavity. Reproductive activity is confined to a particular season of the year. Reproduction is usually triggered by environmental cues, such as increase in day length or water temperature (in temperate and tropical species), or changes in salinity or turbidity (in tropical species). These cues trigger hormonal changes within the animal brought about by stimulation of the pituitary gland.

The pituitary gland contains three major hormones concerned with reproduction.

(1) Gonadotrophin-release hormone (GnRH), which controls the release of gonadotrophin from the pituitary.
(2) Gonadotrophin-release-inhibitory factors (GnRIF, primarily dopamine), which inhibits the release of gonadotrophin from the pituitary.
(3) Gonadotrophins (GtHs), which regulate the release of gonadal steroid from the gonad. Gonadotrophins are composed of GtH I (follicle-stimulating hormone, FSH) and GtH II (luteinising hormone, LH).

The major male and female gonadal steroids are 11α-ketotestosterone and 17β-oestradiol respectively. They control the major aspects of reproduction such as reproductive behaviour, oocyte maturation, spermatogenesis and ovulation. These steroids also have a negative-feedback influence on GtH production from the pituitary. This hormonal system is shown in Fig. 6.1. Although low levels of GtH are found in the blood throughout the year, final maturation of gametes and ovulation is brought about by a surge in GtH levels in response to final environmental cues.

Knowledge of this system allows aquaculturists to control reproduction in captivity and obtain spawnings of high-quality eggs when required. Control of reproduction also allows hatchery managers to plan

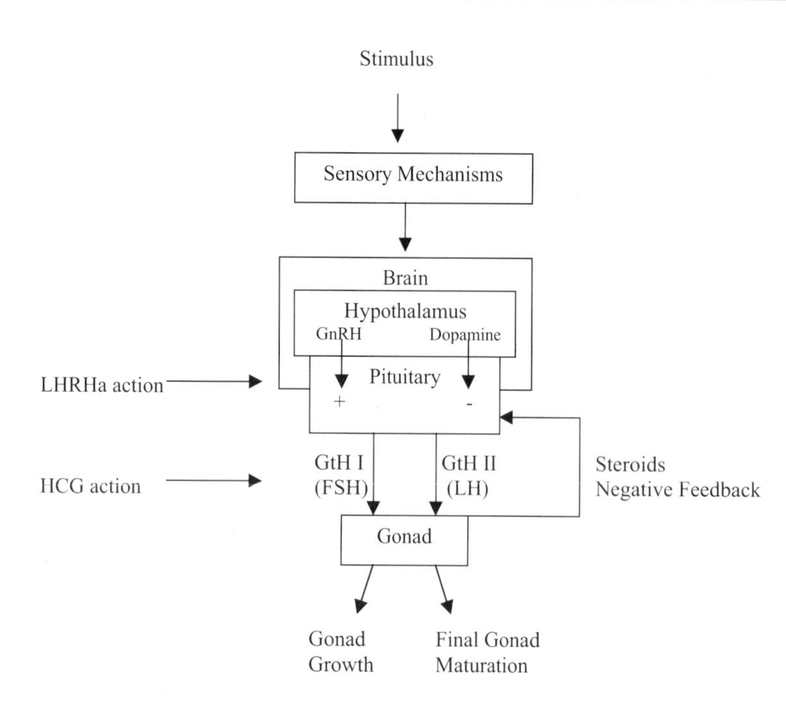

Fig. 6.1. The reproductive endocrine physiology of fish showing the sites of action of LHRHa and HCG used to artificially induce maturation.

for maximum food production for larvae and juveniles when needed. In captivity, however, the final environmental cue for reproduction is often lacking, in which case eggs do not undergo final maturation and fish do not ovulate or spawn because of the lack of a surge in GtH levels. This problem is usually overcome in one of two ways.

(1) Environmental manipulation. The environmental cues necessary for gamete maturation and spawning are provided. This method requires precise knowledge of the factors governing reproduction in a particular species. Important cues include water temperature, salinity, photoperiod and food availability.
(2) Hormonal manipulation. The final GtH surge is artificially attained by injection of appropriate hormones into the fish. Although crude pituitary extract can be used, human chorionic gonadotrophin (HCG) and luteinising hormone-releasing hormone analogue (LHRHa) are more commonly used.

The actions of these hormones are detailed in Table 6.1 and their points of action within the maturation sequence of fish are shown in Fig. 6.1.

Successful spawning induction depends on a number of factors, which are outlined in the following sections.

Stage of maturity of brood fish

Artificial surges in GtH will be an ineffective spawning inducer unless female oocytes have previously reached a certain stage of development. To determine this, a small sample of developing eggs is removed from the female and observed microscopically. Cannulation is used, in which a fine plastic tube is passed up through the oviduct to the ovary. Gentle suction by the mouth at other end of the tube provides a small sample of eggs for inspection. Eggs must possess yolk globules and be of a certain size. The required size can only be determined experimentally for each species, but, as a rule of thumb,

Table 6.1 The actions of human chorionic gonadotrophin (HCG) and luteinising hormone-releasing hormone analogue (LHRHa) in influencing reproduction in fish

Hormone	Action
HCG	Acts directly on the gonad to induce the release of gonadal steroids (sex hormones)
LHRHa	Acts on the pituitary gland to stimulate the release of gonadotrophins (GtH)

they should be at least half the diameter of eggs at spawning. Ripe males should exude milt (sperm) from the genital opening when pressure is applied to the abdomen.

The correct hormone to use and hormone dosage

HCG and LHRHa are the preferred hormones for spawning induction in fish; both are available commercially and available in purified form.

(1) LHRHa is more effective in bringing about oocyte maturation. It is usually used at a dosage of 10–50 μg per kg fish.
(2) HCG is usually used at dosages of 250–2000 IU (international units of activity) per kg of fish.

Initially, a mid-range dose is used, and the optimum dose is determined by trial and error. If the dose is too low, it will fail to induce a spawning. Too high a dose will cause final oocyte maturation to occur too rapidly and will result in poor egg quality. A very low dose will be effective if the oocytes are very mature (determined by cannulation).

Method of hormone administration

Hormones are administered either intramuscularly (IM) or intraperitoneally (IP) (into the body cavity). IM injections have the disadvantage of hormone loss when the needle is withdrawn; however, minimising the injection volume can reduce this. IP injections avoid the problem of hormone loss, but can result in damage to internal organs or injection into the intestine, where the hormone will be ineffective.

Hormones can be administered by injection of liquid or as a pellet containing the hormone. Usually, the hormone is mixed with cholesterol and compressed to form a pellet, which is injected into the muscle. The advantage of using pellets is that the release of the hormone into the blood system occurs more evenly and does not result in a sudden increase in hormone levels as does liquid injection. Cellulose can also be incorporated into the pellet to help regulate the rate of hormone release: the more cellulose, the slower the rate of release.

Timing of hormone administration

Hormonal induction of spawning is more successful if the hormone is administered so that spawning occurs at a time of day when it would occur naturally. To achieve this, knowledge of the time of natural spawnings and the length of time taken between hormone administration and ovulation ('latent period') is required. For example, barramundi spawn naturally after dark and have a latent period of ~36 h. Therefore, the hormone is administered at around 07.00 h to induce a spawning after dark the following night.

As detailed above, GnRIF inhibits the release of GtH within the pituitary (Fig. 6.1). GnRIF is actually dopamine, and dopamine antagonists such as pimozide and domperidone can block its inhibitory action. Thus, spawning may be facilitated by administering a dopamine antagonist to brood fish. The use of a combination of GnRF analogue (e.g. LHRHa) and a dopamine antagonist (e.g. domperidone) to induce ovulation and spawning of cultured fish is known as the 'Linpe method'. A mix of LHRHa (stimulating GtH) and pimozide (inhibiting GnRIF) is now available commercially and spawning kits are available for the major groups of cultured fish such as carps and salmonids.

6.2.2 Decapod crustaceans

The sexes in cultured decapod crustaceans are separate, and the paired gonads are located dorsally and laterally to the gut. Their reproduction and gonad maturation are hormonally regulated. Synthesis and release of crustacean reproductive hormones occur in response to both exogenous and endogenous factors and, in the wild, reproduction is closely related to seasonal cues. These seasonal cues are similar to those for fish, e.g. photoperiod, water temperature and food availability.

Reproduction in decapod crustaceans is controlled by hormones released from the sinus gland and associated centres (X-organ and Y-organ) within the eyestalks. These hormones are gonad-inhibiting hormone (GIH), moult-inhibiting hormone (MIH) and several other hormones. A commonly used technique to induce reproductive maturation in shrimp is eyestalk ablation. This removes the eye, together with the eyestalk and its source of hormones. Eyestalk ablation is usually unilateral (applied to one eye only) and is achieved by:

- cutting off the eyestalk
- cauterising, crushing or ligating the eyestalk.

In females, ablation results in an increase in total ovarian mass, owing to the acceleration of primary vitellogenesis and the onset of secondary vitellogenesis. In males, ablation induces spermatogenesis, enlargement of the vas deferens and hypersecretion in the androgenic gland. Eyestalk ablation also removes the source of MIH and other compounds that control moulting (ecdysis). Reproduction in decapod crustaceans is usually characterised by a pre-copulation/pre-spawning moult.

Gonad maturation after ablation can be very rapid in shrimp, and females can develop full ovaries within 3–4 days. At mating, males insert a spermatophore (a large bundle of non-motile spermatozoa) into the thelycum (receptacle for the spermatophore) on the ventral surface of a female shrimp. Fertilisation occurs externally upon ovulation and passage of the oocytes through the gonadophore. The eggs are shed by the female shrimp into the water column. Spermatophores may remain implanted through several ovarian maturation cycles and may fertilise as many as six spawnings within one moult cycle.

Copulation in decapod crustaceans typically involves a soft-shelled female that has just moulted and a hard-shelled male. As decapod crustaceans moult regularly (section 6.4.2), it means that fertilisations can be reliably obtained by keeping female and male broodstock together in appropriate conditions. This is used for hatchery production in freshwater prawns, freshwater crayfish, crabs, lobsters and some shrimp species. In the groups other than shrimp, the fertilised eggs are not shed into the water column, but are retained on the female's abdominal appendages (pleopods) until the larvae or juveniles hatch (Fig. 20.2). Females brooding eggs on their pleopods are known as 'berried' or ovigerous females.

6.2.3 *Bivalves*

Most species of cultured bivalves have separate sexes, but the total situation within the group is more complex than in fish and decapod crustaceans. Some cultured bivalves are hermaphrodites, e.g. some scallops, with eggs and sperm produced simultaneously. Other bivalves change sex during development. Usually these are protandrous, with younger individuals being male and older individuals being female. Some bivalves may undergo more than one change (often annual) in functional sexuality and are said to show rhythmical consecutive hermaphroditism (e.g. *Ostrea* species). Although the causes of sex change in bivalves are poorly understood, gene-activated components that respond to environmental factors, sex-determining genes and food supply have all been suggested.

As in fish and decapod crustaceans, gamete maturation in bivalves is influenced by a number of exogenous factors, including water temperature, food availability, light intensity and lunar periodicity. It is also influenced by endogenous factors including hormones, genetic factors and levels of endogenous energy reserves. Gonad maturation in bivalves relies on both direct food intake and utilisation of endogenous energy reserves. Water temperature is perhaps the major influence on gonad maturation in bivalves. However, the relationship between increased water temperature and increased natural phytoplankton production (food availability) is also very important.

Major spawning cues for bivalves include a change in water temperature, a change in salinity, lunar periodicity and chemicals (pheromones) associated with water-borne gametes from other individuals. With the exception of lunar periodicity, these are also commonly used spawning cues in bivalve hatcheries (see section 6.3.2). Gametes are generally liberated into the surrounding water where they are fertilised. In some species (e.g. *Ostrea edulis*), fertilised eggs are retained in the mantle cavity, where they develop into swimming larvae before being released.

6.3 Life cycles

At breeding, mature adults shed their gametes (eggs and sperm) freely into the water or the male impregnates the female. Fertilisation results in a zygote and subsequently an embryo, which develops within the egg. Embryonic development occurs over a period of hours or days, depending on the species and temperature. In a generalised life cycle for a species with planktonic larvae, the embryo emerges from the egg as a swimming larva that develops into a juvenile over a period of days, weeks or even, in some species, months. This may be a progressive process, as in shrimp and most fish; or the larva may abruptly settle out of the plankton onto an appropriate substrate and metamorphose into a juvenile, as in bivalves and other benthic invertebrates. The juvenile then grows progressively over a period of months or years until it reaches sexual maturity and the life cycle is completed (Fig. 6.2).

From the viewpoint of aquaculture, these life cycles can be divided into the following sequence of stages:

(1) broodstock conditioning to produce ripe adults for spawning
(2) spawning, either naturally or induced (the latter being more common in aquaculture)
(3) egg fertilisation
(4) larval rearing
(5) postlarval and juvenile rearing
(6) grow-out rearing to commercial size.

These phases of aquaculture are outlined in the following sections.

In intensive aquaculture of a species, many of these phases in the life cycle require different culture techniques. For example, intensive fish culture uses different culture methods for larvae, juveniles and grow-out stock. In extensive aquaculture, however, there may be very little change in culture techniques throughout all phases of the life cycle, e.g. extensive culture of tilapia and freshwater crayfish.

6.3.1 Broodstock selection and conditioning

This involves selecting appropriate animals to serve as sources of gametes. It may involve selecting stock with particular genetic traits (Chapter 7). In many cases the broodstock are simply obtained from the field. Most aquatic animals become sexually mature and reproduce during the warmer months of the year in response to increasing photoperiod, food availability and rising water temperature. Knowing the major cues for gonad maturation, conditions in an aquaculture hatchery can be manipulated to bring on sexual maturity outside natural reproductive seasons. This is known as broodstock conditioning, and it allows hatchery production on a year-round basis.

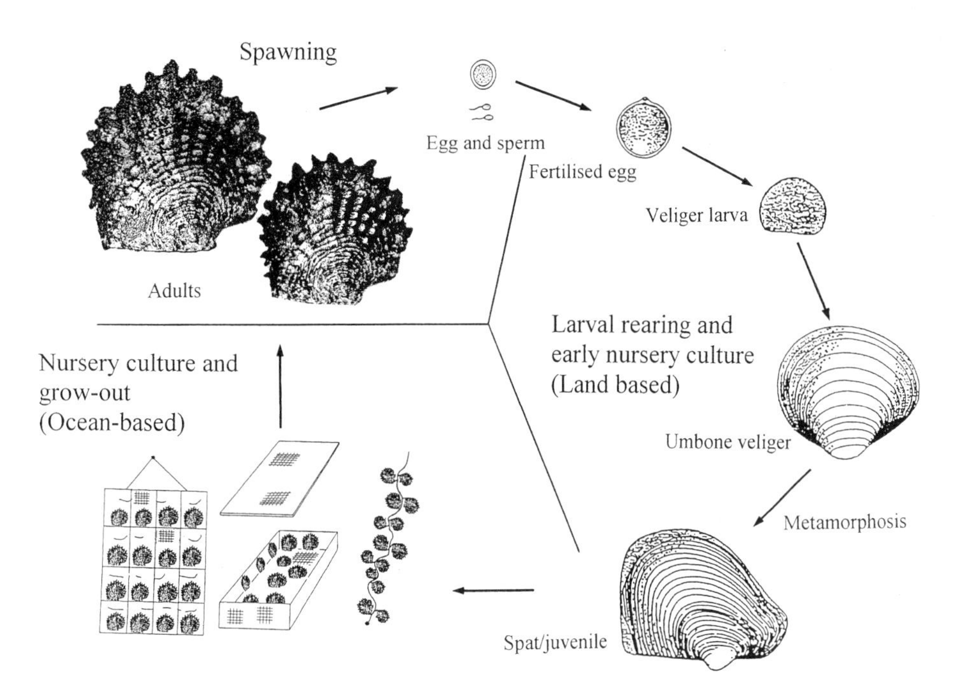

Fig. 6.2 Generalised life cycle of a pearl oyster (*Pinctada* species) showing the phases of culture.

6.3.2 Spawning

Methods of inducing spawning in individual species and groups are described in detail in other chapters. However, it is appropriate to summarise these methods in five categories.

(1) Mild stress, such as abrupt temperature increases, e.g. with bivalves.
(2) Manipulating the body's reproductive hormones by injection or implantation of reproductive hormones into the body in fish or by eyestalk ablation in shrimps.
(3) Transferring gametes from a container where animals have spawned to one where they have not and thus allowing the pheromones associated with gametes to trigger responses from the recipients, e.g. bivalves.
(4) Gonad stripping, where the broodstock are known to have ripe gametes. The body is appropriately massaged to expel gametes, e.g. fish, or gonads are removed from a dead animal and lacerated to release gametes, e.g. bivalves.
(5) Injecting a neurotransmitter substance directly into the gonad of a ripe animal, presumably to cause gonad contractions, e.g. some bivalves.

It is evident that the greatest range of induced spawning techniques is used with bivalves (section 21.4.2). To an extent this reflects the fact that they are more readily induced to spawn than fish or shrimp.

6.3.3 Egg fertilisation

Where the broodstock sexes are segregated, fertilisation involves mixing an appropriate amount of sperm suspension with a suspension of eggs. The objective is to run a course between low fertilisation rates, as a result of insufficient sperm, and polyspermy, where eggs are penetrated by a number of sperm (Stephano & Gould, 1988). When an egg is penetrated by a sperm, its surface often changes to prevent further sperm penetration. If, however, sperm are very abundant, other penetrations may occur before the barrier is established, and the egg then contains two or more haploid nuclei from sperm. These result in a polyploid (2+ n) zygote and the subsequent embryo develops abnormally.

6.3.4 Larval rearing

Embryonic development is usually a brief process, from a few hours' to several days' duration. It requires the developing embryos to be cleaned of any debris and chemicals from spawning induction and fertilisation. The embryos need to be maintained in finely filtered (e.g. 0.2–2 μm) water to reduce levels of bacteria, which may potentially invade the surfaces of the eggs. Developing embryos must also be provided with adequate levels of dissolved oxygen.

The larval stages of most fish, crustaceans and bivalves are planktotrophic, that is dependent on exogenous food supplies for much of their development. When larvae first hatch, however, they usually have sufficient energy reserves in the yolk to support development for a day or more before they need to feed. These reserves were present in the egg and were passed on from the mother. Fish larvae, for example, do not develop a functional gut until some time after hatching: they depend on endogenous reserves until they can support themselves by feeding on appropriate foods in their environment (Fig. 6.3). The embryos and early larval stages of most aquatic animals utilise lipids and proteins from yolk reserves to fuel development to this point. It has been estimated, for example, that embryonic development in some bivalves utilises about 70% of the lipid reserves of the egg (Gallager & Mann, 1986). Bivalve larvae must again rely on endogenous energy reserves at the end of larval development, during metamorphosis, when they temporarily lose the ability to feed. Unlike fish and crustacean larvae, which retain the ability to feed during the transition from larva to juvenile, bivalve larvae must accumulate substantial energy reserves during larval development. In some species (e.g. abalone and trochus), the eggs contain sufficient nutrient reserves to support larval development through to the juvenile phase without the larva having to feed. This lecithotrophic development is usually relatively brief and associated with relatively large and yolky eggs.

The life cycles of freshwater animals often differ from this general pattern in that planktonic larval stages are omitted or much abbreviated. This is known as direct development. Freshwater species with direct development typically produce larger and yolkier eggs than related marine animals. These

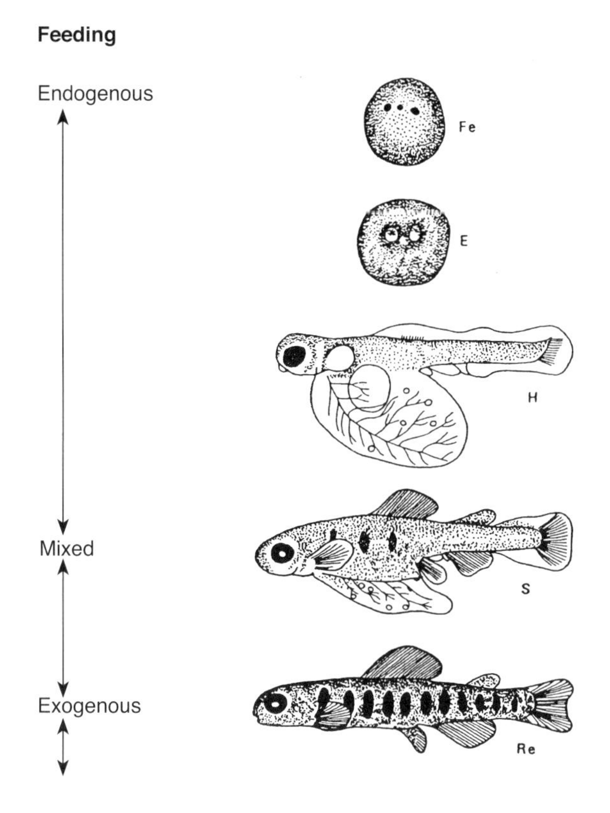

Fig. 6.3 Diagram showing early development in salmonids. E, eye development of embryo inside the egg; Fe, fertilised egg; H, hatching with large yolk sac; S, larva beginning to seek external food; Re, completion of yolk sac resorption and total reliance on exogenous nutrition. Reproduced from Kamler (1992) modified from Raciborski (1987) with permission from the Polish Academy of Sciences.

large eggs support a comparatively longer embryonic period to result in the hatching of juveniles or near-juveniles. The eggs are usually protected during this long embryonic development and the adult may carry them, e.g. on the female crayfish's pleopods or in the mouths of some fish, or there may be 'nests' and patterns of reproductive behaviour to protect broods of eggs. (Egg-protecting behaviour also occurs in some freshwater and marine fish that have normal larval stages.)

Planktotrophic larval development is often the most demanding phase of the life cycle for aquaculture, and it can last from several days to many weeks. The larvae are physically and physiologically fragile. They require well-controlled environmental parameters, such as dissolved oxygen (DO), pH, nitrogen waste levels, bacterial populations, levels of organic and inorganic particles, and temperature. They are often reared in large tanks to buffer these environmental parameters; but tank culture also has hazards, such as contamination and pathogenic bacterial blooms.

As the larvae switch from relying on their endogenous yolk reserves to feeding, providing appropriate food becomes a major aspect of larval culture. There are two main forms of food:

- microalgae for the filter-feeding larvae of molluscs and shrimps
- rotifers and brine shrimp nauplii for the larvae of fish and older larvae of shrimp (Chapter 9).

Larval density is important in hatchery culture. The advantage of high larval density is that more postlarvae can be produced per unit volume of culture tank. However, if larval density is too high, it may compromise water quality and affect larval growth and survival. The culture tank water is usually changed at regular intervals to maintain water quality. The density of larvae is reduced as they grow, despite inevitable attrition that reduces numbers. If 1 larva per millilitre of culture water completes development, this corresponds to one million larvae per 1000 litres of culture water. Aquaculture hatcheries use larval rearing tanks up to 20 000 litres and produce tens of millions of postlarvae per rearing.

6.3.5 *Postlarval and juvenile rearing*

It is necessary to provide appropriate substrates in the culture system to induce bivalve larvae to settle. The larvae are selective because, at settlement and metamorphosis, they attach and make a habitat commitment for the remainder of their lives. For fish and shrimp larvae there is no abrupt metamorphosis. There is, however, a progressive change in behaviour of shrimp postlarvae as they become more substrate oriented. The same pattern occurs in benthic fish.

Marine and brackish water crabs, such as the mudcrab, are another example of an abrupt metamorphosis from planktonic larval stages to the first benthic juvenile stage. In this case it is via a transitional megalopa stage.

The early part of this period is often a particularly difficult one for rearing fish, as the postmetamorphic juveniles must be progressively weaned off live feeds on to artificial diets. Standard procedures have been developed for the major culture species (e.g. section 9.5.1), but this is often one of the particular points of difficulty in developing the culture process for a new fish species.

Postmetamorphic juveniles of fish, decapod crustaceans and bivalves are very small, physiologically fragile and vulnerable to predation. Over a period of days to months or more than a year, they are grown in protected conditions, before they are put out into the grow-out environment. This may involve a period of culture in tanks at the hatchery and an onshore nursery, followed by a period of protected culture in the field. Care is required to minimise mortality when transferring juveniles from the relative comfort of the hatchery (e.g. adequate feed and filtered water, which may be heated above ambient temperature) to field-based nursery conditions with cooler water, fluctuating food supply, predators, disease and fouling.

6.3.6 *Grow-out rearing*

Grow-out rearing is the final phase, during which the juveniles are put out into the adult environment and reared until being harvested. They are not treated in the same manner throughout this phase; factors such as food pellet size, pond size, and mesh sizes of protective or enclosing nets, are varied as they grow.

6.3.7 *Other considerations*

In some cases, the whole culture process from spawning broodstock to final marketing takes place on one farm. In other cases, the cultured organisms may be sold several times during the culture process. Broodstock may be captured by fishing or reared on a specialist farm and sold to the hatchery. Late larval stages or early juveniles are often sold from a specialist hatchery to nursery growers. Juveniles ready to be grown out may be sold from a specialist nursery to grow-out farms. This is because these different phases of culture require particular expertise and facilities. Most typically, the culture process involves a hatchery or hatchery/nursery selling juveniles to farmers, because the hatchery aspect is most technically demanding and because one hatchery can supply many farms. In some Asian countries, such as Taiwan, this process is taken further. There is a series of specialised facilities with expert technical staff who culture specific early-development stages of fish. During development, the fish may be sold on to other facilities a number of times. Each facility may culture equivalent stages of a number of species.

The above describes general methods for culturing stock from gametes to harvesting. In many cases, however, only the later part of the life cycle is cultured on a particular farm and the difficult early stages are omitted. Many aquaculture industries rely on natural recruitment of juveniles, thereby omitting the more technically demanding phases of aquaculture outlined in sections 6.3.1–6.3.4. Relying on natural recruitment is possible by knowing the times and localities of this recruitment and other biological information about the recruits, such as settlement substrate. It may be possible to stock coastal ponds with recruits, e.g. juvenile fish, shrimps or crabs, by filling the ponds at high tides when juveniles are abundant in the adjacent inshore waters. Alternatively, the recruits may be netted from coastal water when they are abundant and then stocked into ponds. In the case of bivalve larvae, it is a matter of providing appropriate substrates, e.g. wooden battens for oyster larvae or raked particulate sediment for cockle or clam larvae, to attract late-stage larvae to settle from the water column. In freshwater ponds, the adults may be allowed to mate and care for their eggs, and then the juveniles are removed at an appropriate stage. There may still be several commercial stages in this process, whereby juveniles from the field, obtained from privately owned settlement substrates or by fishing, are sold to farmers for grow-out.

Natural recruitment is often the start of an extensive or semi-intensive culture process, i.e. where the stocking rate is low to moderate and capital is limited. It is the basis of many major aquaculture industries, but there are the inherent problems, as in fisheries, that natural recruitment levels vary from year to year, and are not completely predictable in time. Fishing pressure on the wild adult stock may reduce its reproductive output and hence natural recruitment. Another disadvantage of relying on natural recruitment is that there is no opportunity for stock improvement. It is impossible to select for desirable traits for aquaculture.

6.4 Growth

6.4.1 Size vs. age

There are two patterns of growth occurring in the developing organism.

(1) Absolute growth rate, e.g. rate of increase in length or mass per unit time, is quite insignificant initially, when the animal is very small (Fig. 6.4). Then, as the animal grows, its capacity to feed and assimilate food progressively increases. The absolute growth rate increases and the size vs. age curve becomes steeper. This is known as the exponential phase of growth.
(2) Relative growth rate, e.g. growth increment per unit body mass per unit time, is very rapid during the early phases of an animal's life cycle (Fig. 6.4). That is, growth relative to size is very rapid. It may be 200% and more during the first week or so of development and remain at a high level during the early months.

The size vs. age relationship reflects the changing pattern of absolute growth, producing a sinusoidal curve (Fig. 6.4). There is an upper inflection in the size vs. age curve, and the upper curve typically slopes gently towards a final theoretical size, L_∞, which it would reach if the animal lived indefinitely.

Some complex logistic equations, such as the von Bertalanffy growth function (VBGF), are used to describe the relationship between size and age. One form of the VBGF is:

$$L_t = L_\infty (1 - be^{-K_t})^{-1}$$

where L_t is size at time t; L_∞ is maximum size; b and K are growth coefficients; b is related to the ratio of maximum size to initial size; and K is the growth coefficient, often used in comparisons of growth rates between populations and species.

However, over short periods of measurement (months), when the animals are in the exponential (rapid) phase of growth, a relative growth coefficient can be obtained from the following equation:

$$k = \frac{(L_2 - \ln L_1)}{(t_2 - t_1)}$$

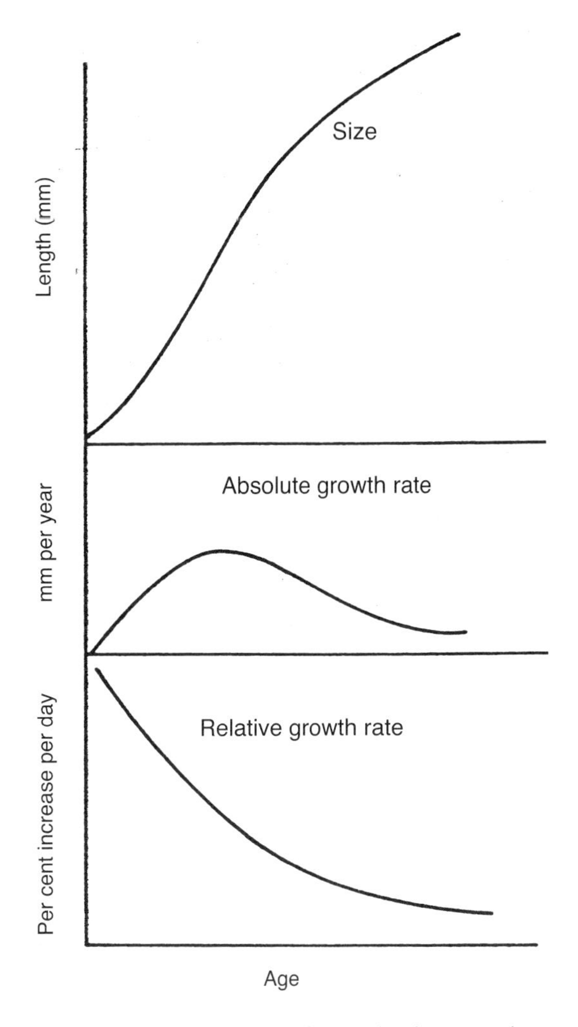

Fig. 6.4 Generalised curves of size, absolute growth rate and relative growth rate vs. age. Reprinted from Malouf & Bricelj (1989) with permission from Elsevier Science.

where L_1 and L_2 are sizes at initial time, t_1, and final time, t_2, respectively. This growth coefficient is simple to calculate and is widely used in comparative studies.

6.4.2 Growth in decapod crustaceans

Growth in decapod crustaceans basically follows the pattern outlined in section 6.4.1. The pattern of growth is, however, complicated by moulting, whereby the animal sheds its old exoskeleton and expands rapidly during a short period before a new exoskeleton hardens. Thus, growth appears to occur

Table 6.2 Events of a moult cycle in a decapod crustacean, from one inter-moult period (instar X) to the next (instar X + 1)

Phase	Duration	External body size	Organic tissue mass	Behaviour	Shell
Inter-moult (instar X)	Days–weeks	Fixed	Increasing	Feeding; locomotion	Hard
Pre-moult	Hours–several days	Little change	Decreasing to increase organic content of body fluids	Ceasing feeding; may seek cover	Shell decalcifying
Moult	Minutes–hours	Rapid increase	Low	Rapidly shedding old shell; no feeding; may be concealed	Old shell discarded; new soft shell
Post-moult	Hours–days	Little change	Recommences increasing	Recommences feeding	Hardening
Inter-moult (instar X + 1)	Days–weeks (months*)	Fixed	Increasing	Feeding, locomotion	Hard

*Last instars of large species, e.g. marine lobsters, mudcrabs.

in a series of abrupt steps rather than as a continuous process. The details of the moult cycle are outlined in Table 6.2. The decapod crustacean's life is divided into a series of intermoult periods or instars, which may be numbered consecutively for convenience. These are punctuated by moults (or ecdyses) and the animal continues to moult, unless it goes into a final terminal instar without moulting (terminal anecdysis). The moults are more frequent earlier in life and the size increments at moults are relatively larger, but absolutely smaller. The result is a size vs. age relationship that fundamentally looks like the relationship in Fig. 6.4, but with a stepwise pattern instead of a smooth curve.

The processes of

- moulting to expand body dimensions
- growth, in terms of adding organic tissue and energy content

are completely independent (Table 6.2). The decapod crustacean feeds and builds up organic tissue during the intermoult period. This is the critical period of growth, although there is no change in external dimensions: body size is constrained by the exoskeleton. The animal does not feed during the periods immediately before, during and after moulting. Its shell is too soft for feeding and it is vulnerable. This is the period of non-growth, although it is the period when the body expands rapidly by absorbing water.

6.4.3 Energetics of growth

The growth rate depends on the extent to which net energy intake exceeds metabolic rate, and to what extent energy is diverted from growth to reproduction in mature animals (section 8.5). Energy committed to gametogenesis is a major component of the body's resources in many aquaculture animals. Hence gametogenesis is at least part of the cause for growth tapering off with age.

The energy for growth may be expressed as a summation of input and outputs.

$$G = I - M - E - F - R$$

where G is the growth rate; I is the food ingestion and absorption rate; M is the metabolic rate; E is the excretion rate of waste molecules from metabolism; F is faeces production; R is gametogenesis; and $[I - (E + F)]$ is the net rate of energy intake, i.e. food intake – (excreted wastes and faeces). The equation can be reorganised:

$$G = [I - (E + F)] - M - R$$

or

$$G = [I - (E + F)] - M \text{ during immaturity.}$$

6.4.4 Measuring growth

General

A very important aspect of growth is to measure it accurately in the organisms being cultured. This is a means of predicting when they will be ready for harvest. Evidence of suboptimum growth rates draws attention to factors such as health, adequacy of feeds and nutrition, environmental quality, stock source, etc., that may be adversely influencing performance.

The change in size of a substantial representative sample of the cultured organisms must be measured to determine the growth rate. For example, an appropriate sample from each pond on a shrimp farm may be at least 100 individuals. The size of aquaculture organisms may be measured in various ways, but different methods are more appropriate for seaweeds, fish, bivalves and crustaceans. The four main measurements of size used in aquaculture are listed below.

(1) A linear dimension measured with callipers or a ruler, according to size. The advantages of this method are that it is quick and easy, and the animal is not unduly harmed. The disadvantages are that it cannot be used for flexible organisms (e.g. seaweeds) and it does not give an indication of the condition (fat/thin) of the animal.
(2) Wet weight of the whole living organism, weighed on scales. The advantage of this method is that it is quick and easy, and the animal is not unduly harmed.
(3) Dry weight. The dead organism, or a representative part of it, is dried in an oven at ~50°C until all the water has evaporated. The advantages of this method are that it gives a tissue weight value that is not complicated by water content and the value is relatively quickly and easily obtained. With any assessment of animal tissue growth, whether it is fish, bivalves or crustaceans, dry tissue weight is the most accurate. In times of physiological or nutritional stress, the water content of animal tissues may increase, leading to an increase in wet tissue weight and giving a false impression of tissue growth.
(4) Ash-free dry weight (AFDW). Once the dry weight of an organism (or part thereof) is determined, it is heated in a furnace at ~500°C for ~24 h. All organic matter is burnt off in the furnace, leaving only inorganic ash. The weight of the ash is calculated and subtracted from the dry weight. This gives the AFDW, which was the dry weight of organic matter in the animal. The particular advantage of AFDW determinations is that it is more accurate for animals with very large inorganic components, such as shells (e.g. bivalves), for which it is difficult to estimate organic tissue content because it is a relatively small.

Determining dry weight and AFDW requires specialised equipment such as a drying oven, 500°C furnace and a sensitive balance. As such, linear measurement and wet weight determination are used most routinely as means of assessing size or growth of aquaculture organisms.

Frequently it is sufficient for the aquaculturist to observe measured growth, but in some circumstances there may be a need to calculate growth rates, e.g. for comparisons of growth rate through the seasons or between batches. The growth rate may calculated on the assumption of linear growth, i.e. change in size over period of measurement. However, as illustrated in Fig. 6.4, growth is not linear. The growth coefficient provided in section 6.4.1 can be used as a more accurate value for comparisons of relative growth rate, provided the animals are in the appropriate phase of growth.

Fish

Fish length has long been used as an assessment of size and growth. Two measurements are typically used.

(1) Standard length (SL). SL is measured from the tip of the snout to the base of the caudal fin. In fish larvae, however, SL is from the tip of the snout to the tip of the notochord.
(2) Total length (TL). TL is measured from the tip of the snout to the tip of the caudal fin. However, measuring TL may be inaccurate if there is damage to the caudal fin, and it is difficult to measure accurately in larvae and juveniles.

As noted previously, length takes no account of fatness or condition of a fish and may therefore be a poor guide to tissue growth. Consequently, wet weight is also frequently used for size and growth assessment, usually in combination with length. Although wet weight is most commonly used as a measure of condition in fish, there are advantages of using dry weight for fish (see above).

Bivalves

Bivalves are usually assessed in terms of shell dimensions, e.g. shell length (SL), shell height (SH) and shell width (SW). If only one dimension is used, as is often the case, it is the largest dimension of the shell. This is not the same in all bivalves, e.g. the largest shell dimension is SH in oysters and SW in clams.

The major problem with shell dimensions as a measure of growth is that they take no account of tissue growth. The bivalve shell tends to grow, regardless of the condition of the animal. It is not unusual for the shell to increase in dimensions at an abnormally slow rate while tissue mass declines. This can happen during periods of inadequate phytoplankton abundance or in response to adverse environmental conditions or disease. Clearly, shell increments in these circumstances do not represent growth. They show only apparent growth. This is where AFDW determinations would indicate the true situation with regard to the state of the tissue.

There is also a simpler method for determining the relative amount of tissue in relation to the shell. This is called the condition index (CI) and the method for determining it is described in section 21.2.5. The CI is used for table oysters and is more concerned with gonad condition and how 'fat' the oysters are than with tissue growth *per se*.

Decapod crustaceans

Decapod crustaceans tend to be measured by a linear measurement of their carapace. Crabs, with a carapace that is broader than it is long, have their maximum carapace width (CW) measured. Other decapod crustaceans, shrimp, freshwater crayfish, prawns, etc., with carapaces that are longer than they are broad, have their carapace lengths (CL) measured. The last group of crustaceans have a well-developed tail (abdomen). This tends to flex vigorously when they are held, hindering total length measurements, which are not usually used.

Because of the moult cycle, the water content of decapod crustacean tissues is very variable and wet weight measurements of individuals are dubious. Wet weight measurements of large groups of individuals are more reliable as they accommodate the range of moult conditions in the group. The total wet weight is then divided by the number of individuals. This technique has the advantage of being more rapid than taking individual carapace measurements, but it only gives a mean size.

Frequency of moulting is a fair indication of relative growth rate in a cultured population of decapod crustaceans but it is difficult to assess. It is no substitute for size measurements.

References

Darwin, C. (1859). *On the Origin of Species by Means of Natural Selection, or the Preservation of Favourable Races in the Struggle for Life*. John Murray, London.

Gallager, S. M. & Mann, R. (1986). Growth and survival of larvae of *Mercenaria mercenaria* (L.) and *Crassostrea virginica* (Gmelin) relative to broodstock conditioning and lipid content of eggs. *Aquaculture*, **56**, 105–21.

Kamler, E. (1992). *Early Life History of Fish: An Energetics Approach*. Chapman & Hall, London.

Malouf, R. E & Bricelj, V. M. (1989). Comparative biology of clams: environmental tolerances, feeding and growth. In: *Clam Mariculture in North America* (Ed. by J. J. Manzi & M. Castagna), pp. 23–73. Elsevier, Amsterdam.

Merrick, J. R & Lambert, C. N. (1991). *The Yabby, Marron and Red Claw: Production and Marketing*. J.R. Merrick Publications, Artarmon, NSW.

Raciborski, K. (1987). Energy and protein transformation in sea trout (*Salmo trutta* L.) larvae during transition from yolk to external food. *Polskie Archiwum Hydrobiologii*, **34**, 437–502.

Stephano, J. L. & Gould, M. (1988). Avoiding polyspermy in the oyster (*Crassostrea gigas*). *Aquaculture* **73,** 295–307.

7
Genetics and Stock Improvement

Douglas Tave

7.1 Introduction

> [G]enetically, most marine fish production remains equivalent to the use of undomesticated wild ancestral cattle, chicken, etc. in ancient terrestrial agriculture.
>
> Knibb *et al.* (1998)

Although most aquaculturists feel that nutritional, health management or water quality management aspects of fish farming are the most important facets of productivity and profit, the genetic aspects are equally, if not more, important. Farming is, first and foremost, the management and husbandry of a population, and its genes determine the biological potential of that population. Consequently, it is not possible to maximise productivity if the population's genes are managed improperly.

Unfortunately, there is no simple genetic formula or recipe that an aquaculturist can follow to enable the best management of the population's genes. The programme that a farmer or hatchery manager needs is determined by his or her specific goals:

- Some farmers may want to improve productivity by selecting for larger or heavier animals.
- Some farmers may wish to improve growth via hybridisation.
- Some may wish to produce a sterile or monosex population by sex reversal or chromosomal manipulation.
- Those who produce animals that will be stocked in natural bodies of water can maximise survival by minimising inbreeding and genetic drift.

To date, most genetic research in aquaculture has been conducted with fish. As a result, this chapter deals predominantly with genetic aspects of fish culture (see also Purdom 1993; Tave, 1993). Purdom (1993) and Tave (1993) consider mainly freshwater and andromous fish, and Knibb *et al.* (1998) provide some case studies of genetic improvement in cultured marine fish.

7.2 Genetics and breeding for qualitative phenotypes

Phenotypes can be divided into two major categories: qualitative and quantitative.

(1) Qualitative traits are those that are described, such as eye colour and body colour.
(2) Quantitative traits are those that are measured, such as weight and eggs per kilogram of female.

Although both are controlled by genes that follow Mendelian laws, there are genetic differences in the way that the genes produce the phenotypes. This means that different types of breeding programmes are needed to exploit the genes that control these phenotypes in order to improve the population and to improve profits.

In general, qualitative phenotypes are controlled by one, two or three genes and can be understood and managed using the Mendelian genetics typically taught in introductory biology courses. These phenotypes are usually easy to manipulate, and the breeding programmes needed to fix (frequency goes to 100%) or to eliminate a phenotype are relatively simple. These traits are often cosmetic, but if consumers are willing to pay more for a more attractive body colour, a farmer would be foolish not to oblige them. For example, virtually all chickens and turkeys that are farmed have white feathers because dark feathers leave melanistic pigment spots in the skin when they are plucked. Consumers feel that the dark spots indicate spoilage or unwholesomeness, so poultry breeders have selected for and fixed white feathers to improve marketability.

Occasionally, an abnormal phenotype can improve the value of the crop. Much of the wheat that is grown has been selected to fix the dwarf phenotype because it has a stronger stalk. Goldfish (*Carassius auratus*) farmers can select for and produce varieties with telescope eyes and can sell what is a bizarre deformity to fish fanciers.

Although selection can be used to select for or against various qualitative phenotypes, the genetics behind the traits should be understood before selection begins in order to make the programme efficient. Many abnormal phenotypes are birth or congenital defects that occur because of developmental mistakes, so selection will have no effect on their frequencies. Even if a trait is heritable, selection for or against the trait will not always improve the population; the mode of gene action determines whether selection or another breeding programme will be needed to either fix or eliminate a trait.

The basic genetic rules that determine the breeding programmes that can be used to exploit qualitative traits are listed below.

(1) If the desired phenotype is controlled only by a homozygous genotype, selection will fix the trait and create a true-breeding population in a single generation.
(2) If the desired phenotype is controlled by more than one genotype, progeny testing must be used to fix the trait and create a true-breeding population.
(3) If the desired phenotype is controlled by the heterozygous genotype, selection cannot fix the trait. A population containing only the desired trait can be produced only by crossing the two homozygous phenotypes.

7.2.1 Autosomal phenotypes

The genes that control qualitative phenotypes either reside on an autosome, which means that they are inherited, and usually expressed identically, in males and females, or they reside on a sex chromosome, which means that they are differentially inherited and expressed in the sexes.

Most qualitative phenotypes that have been deciphered in fish are controlled by single autosomal genes. In general, autosomal genes express themselves in either an additive or a non-additive manner. Golden ($G'G'$), palomino ($G'G$) and normal body pigmentation (GG) in rainbow trout (*Oncorhynchus mykiss*) are the only phenotypes discovered in fish that are controlled by a single autosomal gene with additive gene action. All other phenotypes that are controlled by single autosomal genes are produced by genes with either complete dominance (Table 7.1) or incomplete dominance (Table 7.2). An example of a phenotype controlled by a single autosomal gene with complete dominance is albino and normally pigmented channel catfish (*Ictalurus punctatus*) (Fig. 17.1). Normal pigmentation is the dominant trait and is produced by the $++$ and $+a$ genotypes. Albinism is the recessive trait and is produced by the aa genotype. An example of a phenotype controlled by a single autosomal gene with incomplete dominance is shown by black, bronze and gold Mozambique tilapia (*Oreochromis mossambicus*). Black is produced by the GG genotype, bronze is produce by the Gg genotype and gold is produced by the gg genotype.

Two or more genes control some phenotypes and, when this occurs, epistasis is usually involved. Epistasis occurs when one gene influences or alters the phenotypes produced by a second gene. Other types of gene interactions can also occur, such as 'additive interaction' or the 'interaction of complementary genes', in which the simultaneous expression of two phenotypes produces a unique

Table 7.1 Examples of phenotypes controlled by single autosomal genes with complete dominance in fish

Species	Phenotype
	Recessive phenotype
Channel catfish (*Ictalarus punctatus*)	Albinism
Common carp (*Cyprinus carpio*)	Blue; gold
Rainbow trout (*Oncorhynchus mykiss*)	Albinism
Tilapia (*Oreochromis niloticus*)	Blond, syrup; light coloured (pink); caudal deformity syndrome
Goldfish (*Carassius auratus*)	Blue; telescope eyes
Grass carp (*Ctenopharyngodon idella*)	Albinism
	Dominant phenotype
Tilapia (*Oreochromis niloticus*)	Red
Common carp (*Cyprinus carpio*)	Light yellow band on dorsal fin, yellow on head

Table 7.2 Examples of phenotypes controlled by single autosomal genes with incomplete dominance in fish

Species	Dominant phenotype	Heterozygous phenotype	Recessive phenotype
Common carp (*Cyprinus carpio*)	Death	Light coloured	Normally pigmented
Blue tilapia (*Oreochromis aureus*)	Death	Saddleback (abnormal dorsal fin)	Normal
Mozambique tilapia (*Oreochromis mossambicus*)	Black (normally pigmented)	Bronze	Gold
Goldfish (*Carassius auratus*)	Transparent scales	Calico	Normal scales

phenotype. Table 7.3 gives some examples of phenotypes that are controlled by the epistatic interaction of two genes in fish. Fig. 7.1 illustrates the genetics of scale pattern in common carp (*Cyprinus carpio*), arguably the most important phenotypes controlled by epistasis in fish. The four scale phenotypes in common carp are controlled by dominant epistatic interaction between the *S* and *N* genes. The *S* gene controls the scale pattern and the *N* gene modifies it. The scaled (normal) phenotype is the dominant phenotype and is produced by the *SS* and *Ss* genotypes. The recessive trait is mirror (the number of scales is reduced and those that remain are large) and is produced by the *ss* genotype. The *NN* genotype is lethal, regardless of the genotype at the *S* locus. The *Nn* genotype changes the scale phenotype into the line phenotype (*SSNn* and *SsNn*) and changes the mirror phenotype into the leather phenotype (*ssNn*). Line fish have large scales along the lateral line and the dorsal and ventral margins. Leather fish have few or no scales. The *nn* genotype does not alter the scale pattern produced by the *S* gene.

Phenotypes can be controlled by more than two genes, and genes can have far more than two alleles. For example, major body and fin colours in the Siamese fighting fish (*Betta splendens*) are controlled by the epistatic interactions among four genes (Wallbrunn, 1958), and the gene that controls spot pattern in the platyfish (*Xiphophorus maculatus*) has nine alleles (Gordon, 1956).

Table 7.3 Examples of phenotypes controlled by the epistatic interaction of two genes in fish

Species	Phenotype	Type of epistasis
Common carp (*Cyprinus carpio*)	Scale pattern	Dominant epistasis
Chinook salmon (*Oncorhynchus tshawytscha*)	Flesh colour	Duplicate recessive gene interaction
Goldfish (*Carassius auratus*)	Albinism	Dominant epistasis
	Transparent scales	Duplicate dominant gene interaction
Mexican cave characin (*Astyanax fasciatus*)	Eye colour	Recessive epistasis
Sumatran tiger barb (*Barbus tetrazona*)	Trunk striping	Duplicate genes with cumulative effects

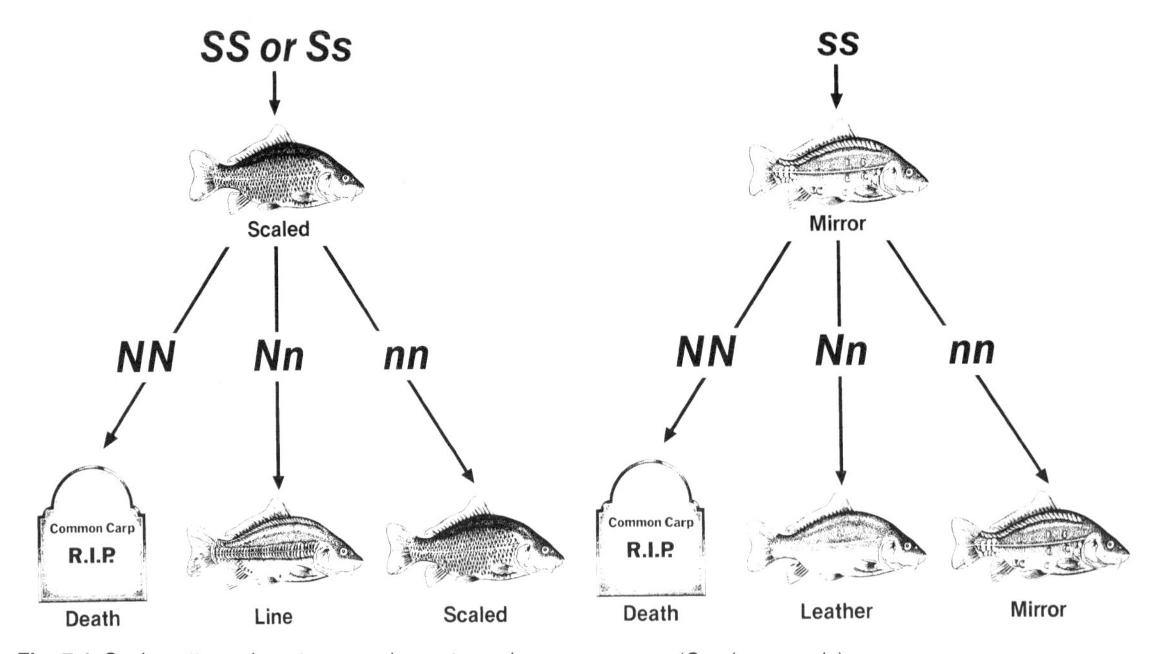

Fig. 7.1 Scale pattern phenotypes and genotypes in common carp (*Cyprinus carpio*).

Selection for phenotypes produced only by homozygous genotypes

In some instances, fixing a recessive trait can improve the market value of a fish. For example, a farmer may wish to produce a true-breeding population of albino channel catfish or light-coloured (pink) Nile tilapia (*Oreochromis niloticus*) if local consumers are willing to pay premium prices for a more attractive fish. A true-breeding population is one that produces only the desired phenotype because the allele that controls the phenotype has been fixed.

When the desired phenotype is recessive, a true-breeding population can be created in a single generation (a single act of selection), because the desired phenotype is controlled only by a homozygous genotype (homozygous recessive). The culling (removal: culled fish are not allowed to reproduce) of fish with the undesired dominant phenotype will remove all copies of the undesired dominant allele from the population. For example, a farmer who wants to produce a true-breeding population of pink Nile tilapia only has to cull all normally pigmented ones. Pink is the recessive phenotype and it is produced by the recessive *bb* genotype. Normal pigmentation is the dominant phenotype and it is produced by the *BB* and *Bb* genotypes (Table 7.4).

Culling all normally pigmented fish removes all

Table 7.4 The BB and Bb genotypes

Phenotype	Normally pigmented	Pink
Genotype	*BB* and *Bb* ↓ Cull	*bb* ↓ No *B* alleles The *b* allele has been fixed The population will breed true and produce only pink fish

B alleles from the population. The only fish that remain are pink, and they possess only the recessive *b* allele (all are *bb*). Consequently, a single generation of selection creates a true-breeding population of pink fish.

This type of breeding programme can also be used to fix a dominant phenotype if the homozygous dominant genotype produces a phenotype that is different from that produced by the heterozygous genotype. This occurs when the mode of gene action is incomplete dominance (Table 7.2). For example, black (*GG*), bronze (*Gg*) and gold (*gg*) body colours in Mozambique tilapia are controlled by incomplete dominance (Tave *et al.*, 1989). A farmer can create true-breeding populations of both black and gold fish in a single generation by culling the bronze fish and by isolating the black and gold individuals (Table 7.5).

The same concept applies when the desired phenotype is controlled by more than one gene. If the desired phenotype is controlled only by a homozygous genotype, selection will produce a true-breeding population in a single generation. Perhaps the most important qualitative phenotype that is controlled by two genes is scale pattern in common carp (Fig. 7.1) (Wohlfarth *et al.*, 1963; Kirpichnikov, 1981). Depending on locality, various scale patterns bring different prices in local markets, so the production of fish with the desired scale pattern brings greater profits. A form of dominant epistasis controls scale pattern when the epistatic gene (the gene that modifies the phenotypes produced by the other gene) is lethal in the homozygous dominant state.

In Europe, consumers like common carp with a reduced scale pattern because they are easier to clean. The mirror scale pattern is the reduced scale pattern that is produced only by a homozygous genotype (*ssnn*). Because of this, if all other scale phenotypes are culled, the dominant *S* and *N* alleles that produce the undesired scale patterns (as well as the lethal phenotype) will be culled. The population will contain fish with only *s* and *n* alleles (only mirror fish will remain and they are *ssnn*), which means that the population will breed true and produce only common carp with the mirror scale pattern.

Selection for phenotypes controlled by more than one genotype

The most common type of selection programme is one in which a farmer selects against an undesired recessive phenotype in an effort to eliminate it because it is subviable or abnormal, i.e. selection for the dominant phenotype in an effort to fix it. Occasionally, this is done because the dominant phenotype is better or more attractive than the recessive phenotype. Although this type of selection is practised routinely, it will not work. Selection cannot eliminate a recessive allele and the recessive phenotype, and thus fix the dominant allele and dominant phenotype to produce a true-breeding population. This breeding programme cannot succeed because many recessive alleles will be masked by a dominant allele in heterozygous fish. Since the homozygous

Table 7.5 The black and gold genotypes

Phenotype	Black	Bronze	Gold
Genotype	*GG* ↓ All g alleles culled; the *G* allele has been fixed; the population will breed true and produce only black fish	*Gg* ↓ cull	*gg* ↓ All *G* alleles culled; the g allele has been fixed; the population will breed true and produce only gold fish

dominant and heterozygous genotypes produce identical phenotypes, selection cannot cull all recessive alleles. Selection can decrease the recessive allele's frequency, but selection cannot reduce it to 0%.

For example, a tilapia farmer who wants to produce a true-breeding population of red Nile tilapia will find that this goal cannot be accomplish by selection. Red is the dominant phenotype, and it is controlled by the *RR* and *Rr* genotypes. Normal pigmentation is the recessive phenotype and it is produced by the *rr* genotype (McAndrew *et al.*, 1988) (Table 7.6).

Progeny testing is the breeding programme that is needed to cull all recessive alleles. Progeny testing is a breeding programme that deciphers a parent's genotype by examining its offspring's phenotypes. To do this, parents with the desired dominant phenotype are paired and mated to fish with the recessive phenotype. Fish with the recessive phenotype can produce gametes with only the recessive allele, so the dominant parent will be the one that determines the offspring's phenotypes: a homozygous dominant fish will produce 100% dominant offspring, while a heterozygous dominant parent will produce offspring with a 50:50 phenotypic ratio.

The production of a single offspring with the recessive phenotype is reason enough to condemn and cull the dominant parent, because this automatically means that it carries the undesired recessive allele. If no individuals with the recessive phenotype are observed in a random sample of 20–50 offspring, the dominant parent can be considered to be homozygous dominant and should be saved for breeding purposes. The greater the sample size, the greater the guarantee that a correct decision will be made. After a sufficient number of fish has been progeny tested, there will be a group of selected dominant fish that are homozygous and contain only the dominant allele, and a true-breeding population has been created.

Table 7.6 The *rr* genotype

Phenotype	Red	Normally pigmented
Genotype	*RR* and *Rr*	*rr*
	↓	↓
	Both *R* and *r* alleles remain; red and normally pigmented fish will be produced	Cull

The same principle applies for phenotypes controlled by more than one gene. If one of the genes responsible for the production of the desired phenotype can produce this phenotype in the heterozygous state, progeny testing must be used to fix the desired alleles and create a true-breeding population. For example, in Asia, most consumers prefer scaled common carp (scaled fish have either the *SSnn* or *Ssnn* genotypes). The only way to fix the *S* allele and create a true-breeding scaled population is by progeny testing to identify and cull *Ssnn* fish and to save *SSnn* individuals for broodstock. It is easy to fix the *n* allele; this is accomplished by culling fish with the line and leather phenotypes (Fig. 7.1).

Selection for phenotypes controlled only by heterozygous genotypes

Selection cannot create a true-breeding population if the desired phenotype is controlled only by the heterozygous genotype. For example, a rainbow trout farmer who wants to create a true-breeding population of palominos will find that this cannot be accomplished by selection (Table 7.7). However, a farmer can produce 100% palomino rainbow trout by mating golden females to normally pigmented males, or vice versa (Table 7.8).

The same principle applies to phenotypes controlled by two or more genes. If one, or more, gene that must be in the heterozygous state controls the desired phenotype, selection cannot produce a true-breeding population. The only way to produce a population with only the desired phenotype is to mate fish with the homozygous genotypes.

7.2.2 Sex-linked phenotypes

Sex-linked genes are known only for species with the XY sex-determining system (Table 7.9), and virtually all information about this form of inheritance comes from the guppy (*Lebistes reticulatus*) and platyfish.

It is easy to fix Y-linked phenotypes in males and create true-breeding populations, because only one copy of a Y-linked gene exists on the single Y chromosome of a normal male. Y-linked phenotypes cannot be fixed in normal females, which do not possess a Y chromosome. Because males have only one Y chromosome, a Y-linked allele's phenotype is always

Table 7.7 A true-breeding population of palominos cannot be accomplished by selection

Phenotype	Normal pigmentation	Palomino	Golden
Genotype	*GG*	*GG′*	*G′G′*
	↓	↓	↓
	Cull	Both *G* and *G′* alleles remain; all three phenotypes will be produced	Cull

Table 7.8 Palomino rainbow trout produced by mating golden females to normally pigmented males

Normally pigmented female (*GG*)	Golden male (*G′G′*)
↓	
All offspring palominos (*GG′*)	

Table 7.9 Sex-determining system in some cultured fish

XY Sex-determining system	WZ Sex-determining system
Channel catfish (*Ictalurus punctatus*)	Blue tilapia (*Oreochromis aureus*)
Rainbow trout (*Oncorhynchus mykiss*)	Wami tilapia (*Oreochromis urolepis*)
Coho salmon (*Oreochromis kisutch*)	Japanese eel (*Anguilla japonica*)
Nile tilapia (*Oreochromis niloticus*)	
Mozambique tilapia (*Oreochromis mossambicus*)	
Common carp (*Cyprinus carpio*)	
Silver carp (*Hypophthalmichthys miolitrix*)	
Grass carp (*Ctenopharyngodon idella*)	
Goldfish (*Carassius auratus*)	

expressed. This makes selection relatively simple. A single generation of selection can eliminate the undesired Y-linked phenotype and the Y-linked allele that produces it and can fix the desired Y-linked phenotype and its Y-linked allele, as was described in the section 'Selection for phenotypes controlled by more than one genotype' for autosomal phenotypes controlled only by recessive genotypes.

X-linked genes are expressed differently in normal males (single copy in XY) and normal females (two copies in XX). In females, most X-linked genes exhibit complete dominance. Because males have only a single X chromosome, an X-linked allele is always expressed in males. The fixation of recessive X-linked alleles can be accomplished in a single generation, as was described in the section 'Selection for phenotypes produced only by homozygous genotypes' for autosomal phenotypes controlled only by recessive genotypes. The fixation of a dominant X-linked allele is a two-step process. First, the recessive X-linked allele must be culled in males, as was described in the section 'Selection for phenotypes produced only by homozygous genotypes'. Then the recessive X-linked allele must be culled in females by progeny testing, as was described in the section 'Selection for phenotypes controlled by more than one genotype' for autosomal genes.

7.2.3 Pleiotropy

Fixing a phenotype by selection should only be carried out after it has been studied to determine its effect on growth, fecundity, disease resistance, etc. Many alleles carry excess baggage, in that they depress the growth rate or food metabolism. These secondary effects are called 'pleiotropic effects'. Studies have shown that many mutant colours in fish have negative pleiotropic effects, that is they depress the growth rate or decrease viability. If negative pleiotropic effects exist, selection to fix a given trait may be counter-productive. Consequently, a trait should be thoroughly studied before selection begins. Fig. 7.2 shows the negative pleiotropic effect that red body colour has on viability in blue tilapia (*Oreochromis aureus*). Variability of red siblings (A × R-R) was compared with that of normally pigmented siblings (A × R-NP) and that of normally pigmented fish from another population (A × A). The variability under four stages of production is illustrated.

7.3 Genetics and breeding for quantitative phenotypes

7.3.1 Quantitative phenotypes

Most of the important production traits are quantitative phenotypes: length, weight, number of eggs per kilogram of females and the food conversion ratio

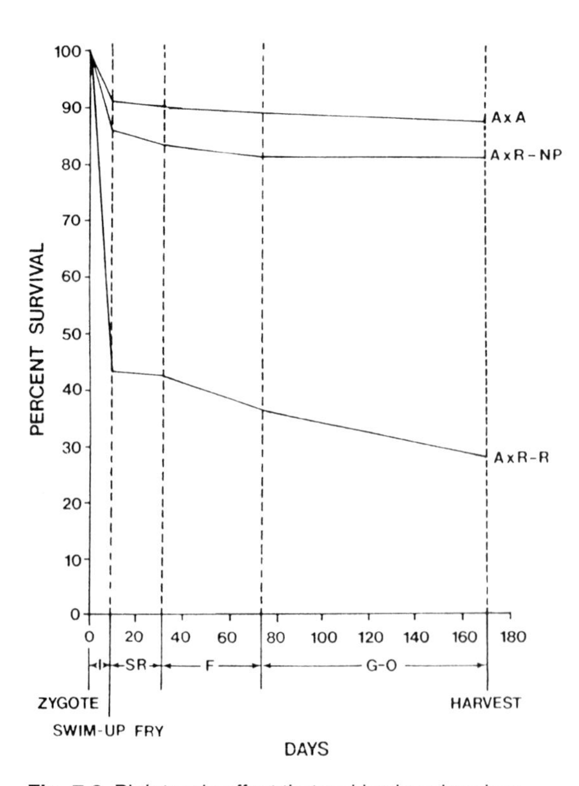

Fig. 7.2 Pleiotropic effect that red body colour has on viability in blue tilapia (*Oreochromis aureus*). F, fingerling production; G-O, grow-out; I, artificial incubation of eggs; SR, sex reversal (after El Gamal, unpublished thesis, used with permission).

(FCR). These traits are measured (e.g. the weight in grams or kilograms) rather than described (e.g. body colour), so phenotypic differences among fish are matters of degrees rather than of kind. Because these phenotypes are measured, they do not fall into discrete categories that form ratios; instead, quantitative traits are described by their central tendency (the mean) and the distribution around the mean (variance and standard deviation).

Quantitative phenotypes form continuous distributions in a population, rather than discrete non-overlapping categories, because they are more complicated genetically. These traits are usually controlled by dozens to hundreds of genes. The exact number is usually never known. The simultaneous segregation of these genes and the different modes of gene action at each individual locus, coupled with the interactions between, among and within loci, means that the overall genetic expression for the trait is variable, which in turn means that phenotypic expression will also be variable. In addition, environmental variables play a major role in the production of quantitative phenotypes, and these variables do not impact upon all individuals identically. This too results in variable phenotypic expression. Finally, there is an interaction between genotype and environment that also helps produce the continuous distributions of phenotypic expression.

Because quantitative phenotypes exhibit continuous distributions, the only way to analyse the phenotypes and work with the underlying genetics in order to improve productivity is to partition the phenotypic variance (V_P) into its component parts:

- genetic variance (V_G)
- environmental variance (V_E)
- genetic–environmental interaction variance (V_{G-E}).

$$V_P = V_G + V_E + V_{G-E}$$

The correct breeding programme that should be used to exploit V_G (the component that is of most interest to breeders) is determined by its subcomponents:

- additive genetic variance (V_A)
- dominance genetic variance (V_D)
- epistatic genetic variance (V_I):

$$V_G = V_A + V_D + V_I$$

These subcomponents of V_G do not refer to additive, dominance and epistatic gene action, but to their variance. The two important subcomponents are V_A and V_D. Most breeders ignore V_I because it is comparatively small and it is usually difficult to exploit. Additive genetic variance is the subcomponent that is the result of the additive effect of the genes, i.e. it is the sum of all alleles across all loci taken separately. Dominance genetic variance is due to the interactions of alleles at each locus, i.e. the interactions between the paired alleles at each locus. Epistatic genetic variance is due to the interactions of alleles across loci, i.e. with alleles other than its homologue.

Additive genetic variance and V_D are essentially opposites. Additive genetic variance is the sum of alleles taken independently and does not depend on any interaction, so it is transmitted from a parent to its offspring in a predictable and reliable manner. Dominance genetic variance, on the other hand, is the interactions of the paired alleles at each locus and it is not transmitted from a parent to its offspring. Additive genetic variance is a function of the gametes, which is a random sample of an animal's alleles, whereas V_D is a function of an animal's genotype, which is a function of the diploid state. Because V_A is a function of the alleles, and thus the gametes, it is transmitted from parents to offspring. Because V_D is a function of the diploid state, it cannot be transmitted from a parent to its offspring; instead, it is created anew and in different combinations each generation.

The differences between V_A and V_D mean that different types of breeding programmes are needed to exploit them. Selection is used to exploit V_A and improve a trait. Hybridisation is used to exploit V_D.

Additive genetic variance is the most important genetic subcomponent, and the percentage of V_P that is due to V_A is one of the most important pieces of information about a quantitative phenotype. The proportion is called heritability (h^2):

$$h^2 = V_A/V_P$$

The reason that it is important to know a trait's h^2 is that, once known, it can be used to predict the response to selection by using the following formula:

$$R = Sh^2$$

where R is the response to selection and S is the selection differential or reach (the superiority of the select broodstock over the population average).

Thus, if h^2 is known, it is possible to predict whether selection will be an efficient way of improving the trait. In general, traits with h^2 about or < 0.15 are difficult to improve by selection, whereas traits with h^2 about or > 0.25 can be improved efficiently by selection. Heritabilities for various production traits that have been determined in five species of fish are listed in Table 7.10. It is not mandatory that the h^2 be known prior to initiating a selection programme, but if it is known the type of selection can be tailored to maximise improvement. Conversely, a small h^2 can generally be used to predict that selection will not work and this can prevent several years of expensive breeding work that would end in frustration.

7.3.2 Selection

In general, there are two types of selection programmes:

(1) individual selection
(2) family selection.

Fig. 7.3 illustrates the differences between the two and is a representation of (a) individual selection; (b) between-family selection; (c) within-family selection; (d) a combination of between- and within-family selection.

There are five families (the vertical lines) and individual fish within each family are arranged along these lines. The position of each fish represents its phenotypic value, e.g. the highest in each family line has the highest phenotypic values. The small horizontal line for each family represents the family mean and the large horizontal line (X) in Fig. 7.3a represents the population mean. Black fish are selected and white fish are culled.

In individual selection (Fig. 7.3a), family means are ignored and all fish within a population are compared; the best are chosen to become the select broodstock, and the remainder are culled. The demarcation that delineates the best is called the 'cut-off value' (CO). This value is often predetermined, but it can

Table 7.10 Examples of heritabilities (h^2) for some production phenotypes in five important cultured fish

Phenotype	h^2
Channel catfish (*Ictalurus punctatus*)	
18-month weight	0.33
Dressing weight	0.43
Percentage of fat	0.61
Rainbow trout (*Oncorhynchus mykiss*)	
147-day weight	0.52
1-year weight	0.38
2-year weight	0.32
Dressing percentage	0.36
Abdominal fat	0.25
Female spawning age	0.55
Egg size	0.28
Egg number	0.32
Atlantic salmon (*Salmo salar*)	
10-month weight	0.33
2-year weight	0.38
3-year weight	0.38
Percentage of dead alevins	0.04
Egg size	0.44
Egg number	0.30
Dressing percentage	0.03
Common carp (*Cyprinus carpio*)	
1-year weight	0.49
2-year weight	0.50
Fat content	0.15
Nile tilapia (*Oreochromis niloticus*)	
90-day weight	0.04
Weight at first spawning	0.46
Fecundity at first spawning	0.09
Female 136-day GSI	0.28

GSI, gonad/somatic index, i.e. gonad weight as a proportion of body weight.

also be created after the fish, or a random sample, are measured. Fish that meet or exceed the cut-off value become the select broodstock. The cut-off value is usually expressed as a percentage (e.g. the top 10% or top 5%).

When determining the cut-off value, sexual dimorphism must be taken into account. In many species, one sex grows faster. If this occurs, separate cut-off values must be made for each sex or the select fish may be of only one sex.

In between-family selection (Fig. 7.3b), family means are determined and ranked and a cut-off value selects the top families; all fish from the remaining families are culled. Either every fish from the select families is retained as select broodstock or random and equal samples from each family are retained.

In within-family selection (Fig. 7.3c), family means are determined and fish within each family are compared only with their family mean. After individual rankings within each family are determined, a universal cut-off value in terms of percentage or number of fish, not the absolute phenotypic value, is established and the top fish from each family are chosen to become the select broodstock. All fish that fall below the cut-off value within each family are culled.

The two types of family selection can be combined (Fig. 7.3d). The first step is to use between-family selection to choose the best families. Then, within-family selection is used to choose the best fish from each of the select families.

Individual selection is usually used when h^2 is about or > 0.25. When h^2 is about or < 0.15, family selection is usually used. When h^2 is small, family selection is more efficient because little heritable variance (V_A) exists and it can be more accurately assessed by selecting at the family level. Family selection is also preferred if there are environmental variables that can strongly influence the trait, such as spawning date, maternal care, mother's age and family size. Individual selection is easier and cheaper because all fish can be grown together. When family selection is used, families must be cultured separately until they can be marked.

A selective breeding programme is essentially an experiment (even on commercial farms) and the only way to assess progress is to compare the select offspring with a control population. If no control is maintained, it is not possible to determine how much of the improvement was due to genetic improvement and how much was due to environmental factors. New feed formulations, better water quality management techniques, new advances in health care and the acquisition of better fish-handling skills can produce tremendous improvements in growth rate or other traits. A control is needed to show how much of the improvement in the select population is due to these environmental factors. The control can be a random sample from the original population that is maintained at the hatchery, or it can be a population that is purchased from another farm and raised every year.

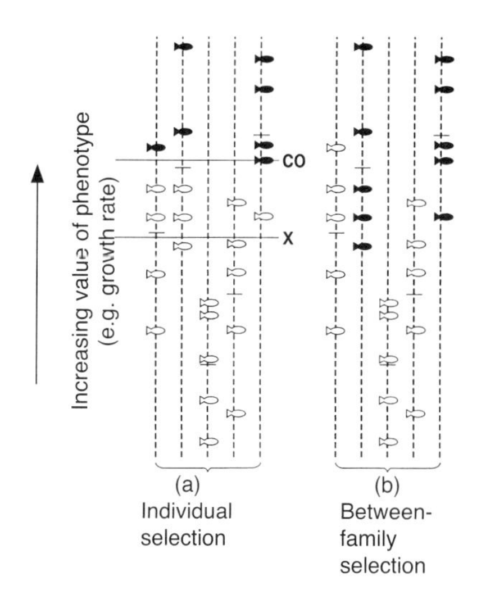

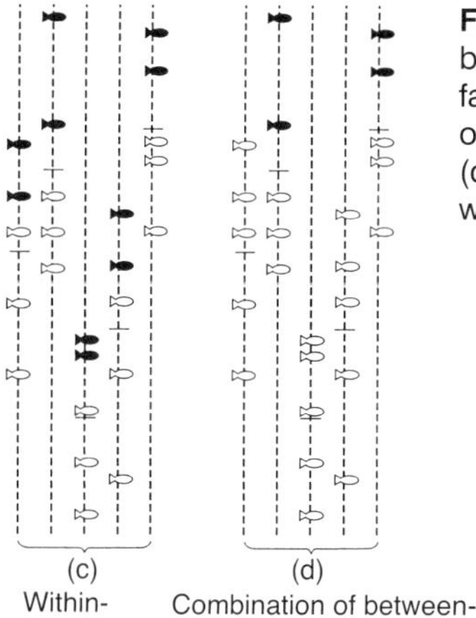

Fig. 7.3 Individual selection (a), between-family selection (b), within-family selection (c) and a combination of between- and within-family selection (d). Reproduced from Falconer (1957) with permission.

Fig. 7.4 is a diagram of individual selection for weight. Mean weight in the parental generation was 1.25 kg. The cut-off value was 1.65 kg. A random sample of fish from around the mean was used to choose the control broodfish. Mean weights in the F_1 select and F_1 control generations were 1.88 kg and 1.39 kg respectively. The weight improved by 36% over that of the control line. Without a control, it would have been impossible to determine genetic gain. Improvements in management improved the average weight by 0.14 kg.

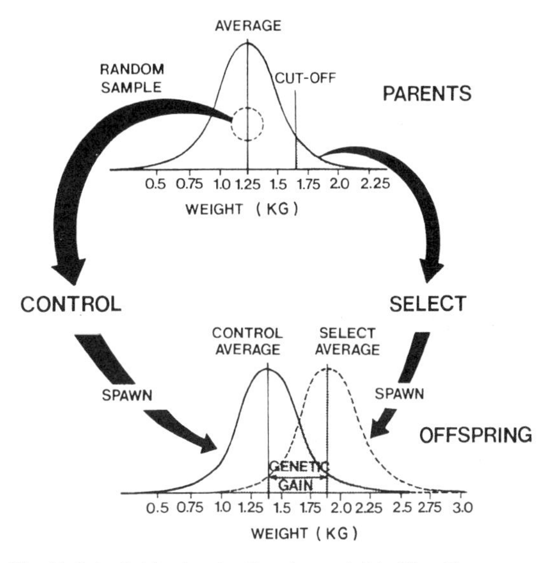

Fig. 7.4 Individual selection for weight. After Tave (1987) Improving productivity in catfish farming by selection. *Aquaculture* **13** (5), 53–5.

A number of selective breeding programmes have been conducted with important cultured species, and traits such as weight, length, percentage of smoltification, age at reproduction and disease resistance have been improved. In some cases, the improvement has exceeded 15% per generation. An example of the improvement that can be made via selective breeding is shown by the change in the spawning date in rainbow trout (*Oncorhynchus mykiss*). After six generations of selection, the spawning date changed by 69 days in even years and by 67 days in odd years.

A selective breeding programme was used to improve body weight in coho salmon (*Oncorhynchus kisutch*) being raised for the 'pan-size' market (300–350 g). After four generations of selection, the average weight at the end of the freshwater phase was 14.6–19.2 g and 17.6–23.8 g in the odd- and even-year lines respectively. The weight after 8 months in saltwater increased from 239 g to 430 g and from 296 g to 406 g in odd- and even-year lines respectively. The time needed to produce a marketable fish decreased from 11 months to 6–8 months in the selected lines, but decreased by only 1 month in controls (Hershberger *et al.*, 1990). This programme was conducted at a commercial facility, and the fish that were culled during the saltwater sampling were retained as broodstock and used to produce the fish that were raised for market. This enabled the farmers to exploit some of the genetic improvement immediately and did not compromise the breeding programme.

When conducting a selective breeding programme, it is crucial to control environmental variables. If this component of V_P is not controlled, it can be confounded with V_A and can also make assessment difficult. Environmental variables such as spawning time, female age, female size, stocking rate, feeding practices and food particle size have been shown to affect growth. If they are not controlled they can confound data for length or weight, which will make it difficult, if not impossible, to determine which fish are genetically superior. Fish that are superior simply because of an environmental variable obviously cannot transmit this superiority to their offspring and are of little value in a selective breeding programme. Although it is impossible to equalise all environmental vectors across all fish, the results of selection will be improved the more they are managed and controlled.

7.3.3 Hybridisation

Hybridisation (also called crossbreeding) is the other major traditional breeding programme that is used to improve productivity. In general, crossbreeding is used to produce superior animals for grow-out. Selection is used to produce superior broodstock. In most cases, crossbred animals are sold to consumers. Crossbreeding improves productivity by exploiting V_D. Because V_D is not transmitted from a parent to its offspring, the creation of superior hybrids is a hit-or-miss proposition.

The most common type of hybrid that is created is called an 'F$_1$ hybrid'. This is created by mating fish from one population to those from another. If the mating is made between two strains within a species, it is called an 'intraspecific hybrid'; if it is made between two species, it is called an 'interspecific hybrid'. The results of hybridisation (measured and defined as heterosis = 'hybrid vigour' when there is positive heterosis) depend on the fortuitous combination of alleles; the results cannot be predicted. Research has shown that better results have been obtained by hybridising cultured strains than by hybridising cultured and wild strains.

Interspecific hybridisation is usually not as successful as intraspecific hybridisation, because chromosomal and behavioural differences can either prevent or preclude reproduction or the production of viable fry. However, some promising interspecific hybrids have been created and are being cultured or developed for food or sport:

- striped bass (*Morone saxatilis*) × white bass (*Morone chrysops*)
- lake trout (*Salvelinus namaycush*) × brook trout (*Salvelinus fontinalis*)
- channel catfish × blue catfish (*Ictalurus furcatus*).

Some of the more interesting interspecific hybrids that have been produced are among the tilapia (sections 16.2.4 and 16.4.1). Tilapias are cultured worldwide and have many desirable qualities, but they have one undesirable trait that has severely curtailed the expansion of tilapia farming: they reproduce when young and are much smaller than the desirable market size. This diverts feed energy from growth to reproduction and also produces a large biomass of unmarketable fish. It was accidentally discovered that interspecific hybridisation among tilapia can produce an all-male population (section 16.4.1).

This occurs because some species have the XY sex-determining system, whereas others have the WZ sex-determining system. In the XY sex-determining system, females are the homogametic sex and are XX, while males are XY (the heterogametic sex). In the WZ sex-determining system, males are the homogametic sex and are ZZ, and females are WZ. The mating of a homogametic female (XX) with a homogametic male (ZZ) produces XZ offspring; because the X chromosome is a female-producing chromosome and the Z chromosome is a male-producing chromosome, XZ offspring are males (Table 7.11).

Unfortunately, only about 90–95% males are produced in many cases. Possible reasons for this are contaminated broodstock and autosomal sex-modifying genes. Impure stocks of tilapia can contain more than one sex-determining system, and broodstock that contain 'incorrect' sex chromosomes will not breed as expected. An autosomal sex-determining (sex-reversing) gene has been identified in common carp (Komen *et al.*, 1992), and research suggests that

Table 7.11 The mating of a homogametic female (XX) with a homogametic male (ZZ)

XX female		ZZ male
Nile tilapia		Wami tilapia
Mozambique tilapia		blue tilapia
(XX)	×	(ZZ)
	↓	
	All offspring males (XZ)	

one or two exist in tilapia (Avtalion & Hammerman, 1978; Hammerman & Avtalion, 1979). Factors other than sex-determining genes that influence the sex of tilapia are outlined in section 16.4.1.

Because the results of hybridisation are inconsistent, there is no pattern in the literature that can be used to create a set of guidelines for developing a crossbreeding programme. Even though some F_1 hybrids have been failures, crossbreeding has been used to produce F_1 hybrids that have improved growth rate, lower FCRs, improved dressing percentage, improved disease resistance and improved harvestability.

Hybridisation does not have to end with the production of F_1 hybrids. Three-way or four-way hybrids can be produced. Although F_1 hybrids are usually produced just for grow-out, research with Nile tilapia has shown that F_1 intraspecific hybrid females produced faster-growing offspring than pure-strain females (Tave *et al.*, 1990). This is called 'maternal heterosis'. This has not been tested in other species, so it is not possible to generalise these results.

Hybridisation can also be used to produce synthetic strains for fish farming. The best way to do this is to find two strains that have many desirable characteristics (e.g. fast growth rates) and discover whether they produce F_1 hybrids that exhibit hybrid vigour. Once this combination has been discovered, F_1 hybrids are created, and when sexually mature they are mated among themselves to produce F_2 hybrids. Selection is then carried out in the F_2 hybrids to choose the fastest-growing fish, etc. This creates a synthetic strain and the F_3 generation will be the first that is grown commercially. This can even be done with two species to create a synthetic hybrid species.

7.3.4 Inbreeding

General aspects

Inbreeding is the third major traditional breeding approach that can be used to improve productivity. Although many associate inbreeding with detrimental results, if used properly it can be a powerful tool to improve a population. Inbreeding is the mating of relatives. This increases homozygosity. Inbreeding is quantified as the percentage increase in homozygosity over the population average. Inbreeding can be used to fix certain desirable traits in a population. This occurs because the population becomes homozygous at certain loci. The creation of inbred lines to fix certain genotypes followed by their crossbreeding to create a population in which every individual has only the desired genotypes (and thus phenotypes) is the classic way that some important varieties of crops are produced.

Inbred lines can be crossed with non-inbred lines. This is called 'top-crossing' and is one way to produce outstanding F_1 hybrids, e.g. top-crossed rainbow trout, brook trout and brown trout (*Salmo trutta*) exhibit hybrid vigour.

Linebreeding is another form of inbreeding that can be used to produce outstanding broodstock. Linebreeding is a mating programme in which an outstanding individual (usually a male) is mated to one or more descendants. This is done to increase the percentage of his descendants' genes. If done properly, linebreeding can produce near clones of the animal in question.

Although inbreeding can be a powerful tool, if used improperly it can ruin a population. Inbreeding can lower growth rates, viability and fecundity, and can simultaneously increase the percentage of abnormalities. This occurs because detrimental alleles are paired when relatives mate. Such pairings occur when unrelated individuals mate, but the mating of relatives increases the likelihood that such pairings will occur, and the closer the relationship the greater the likelihood.

As there have been few inbreeding studies in fish, the levels of inbreeding that cause problems have not been well quantified. Even if extensive studies had been conducted with one species, it would not be possible to extrapolate the results for all species. In other animals, low levels of inbreeding have been shown to improve some traits and to make others worse. The results from studies in rainbow trout are illustrated in Fig. 7.5. In this figure, a value of 0 means that the inbred fish and the controls are the same. A positive value means that inbred fish are better than the controls, whereas a negative value means that they are worse. Inbreeding values ranged from 12.5% to 59% (one generation of brother–sister mating produces inbreeding of 25% and two generations produce inbreeding of 37.5%). The only way

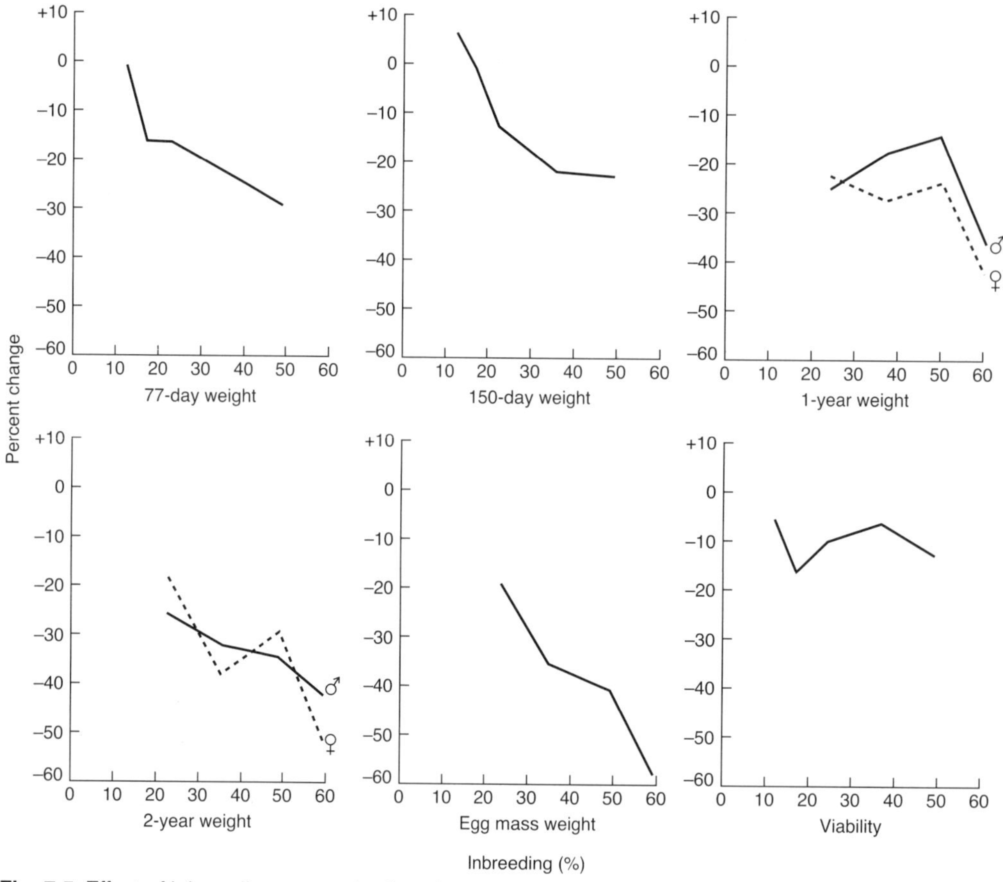

Fig. 7.5 Effect of inbreeding on production phenotypes in rainbow trout (*Oncorhynchus mykiss*). Reproduced from Tave (1988) with permission of the British Columbia Salmon Farmers Association.

to calculate individual inbreeding values is to know individual pedigrees. This is usually impossible with fish. Although inbreeding cannot be determined, this does not mean that it should be ignored. Uncontrolled inbreeding can quickly ruin a population.

Managing a population's effective breeding number

Unintentional inbreeding and genetic drift occur in broodstock populations because they are small. Hatchery managers usually try to prevent the importation of new fish once the broodstock is established, in order to prevent the importation of diseases. Small, closed populations can produce high levels of inbreeding and lose much of the original genetic variance through genetic drift.

Small, closed populations should not be described by the number of fish or by the number of broodstock. Instead, they should be described by something called the 'effective breeding number' (N_e). The N_e is determined by

- the number of males that produce viable offspring
- the number of females that produce viable offspring
- the sex ratio (N_e will approximate the rarer sex).

It is also influenced by the variance in family size, so unequal family size can have a negative impact on N_e. Furthermore, the N_e over a series of generations is the harmonic mean of the N_e for each generation, which means that the generation with the smallest N_e has a disproportionate effect on overall N_e.

The reason it is important to know and manage a population's N_e is that both inbreeding and genetic drift are inversely related to N_e. The relationship between inbreeding and N_e is illustrated in Fig. 7.6. This figure shows that small N_e values can produce high levels of inbreeding. Effective breeding numbers of less than 50 can rapidly produce detrimental levels of inbreeding. Table 7.12 gives minimal values that should be maintained to prevent inbreeding exceeding 5%.

Because small N_e values can produce high levels of inbreeding, a major goal of broodstock management should be to manage the population's N_e. Unfortunately, there is no single N_e that can be used to manage every hatchery population. The N_e that is desired can be determined only after the goals are examined. In general, N_e values for populations that are cultured for fisheries management or for conservation biology programmes must be larger than those needed for populations that are farmed for food. Recommended N_e values for these programmes are presented in Table 7.12. These values are not absolute; they are valid only for the assumptions that were used in their calculation.

7.4 Strain evaluations

One way to improve productivity is to evaluate a number of strains and discover which one performs best. In most cases, this would be the strain that has the best growth rate but, if diseases are a major problem, it may be the most disease-resistant strain. This information is generally site or region specific. If the results of a yield trial demonstrate that a particular strain is the best, this does not mean that the strain is the best under all culture conditions.

Because of genetic–environmental interactions (V_{G-E}), regional yield trials should be conducted. A number of studies have shown that V_{G-E} can be a major factor influencing the growth rate in strains of fish. This is a well-established fact in agronomy and horticulture. That is why there are so many varieties and cultivars of fruits, vegetables and grains.

7.5 Production of monosex populations by sex reversal

7.5.1 *Rationale*

Sex reversal is a breeding programme that can be used to produce monosex populations. Monosex populations are desirable:

- if one sex grows faster than the other
- if one sex is burdened with undesirable baggage, such as precocious sexual maturity followed by death
- to prevent unwanted reproduction during grow-out.

In all cases, the object is to increase marketable yield. For example, channel catfish males grow faster than females, and the production of a monosex male population will increase the yield by about 7%. Many

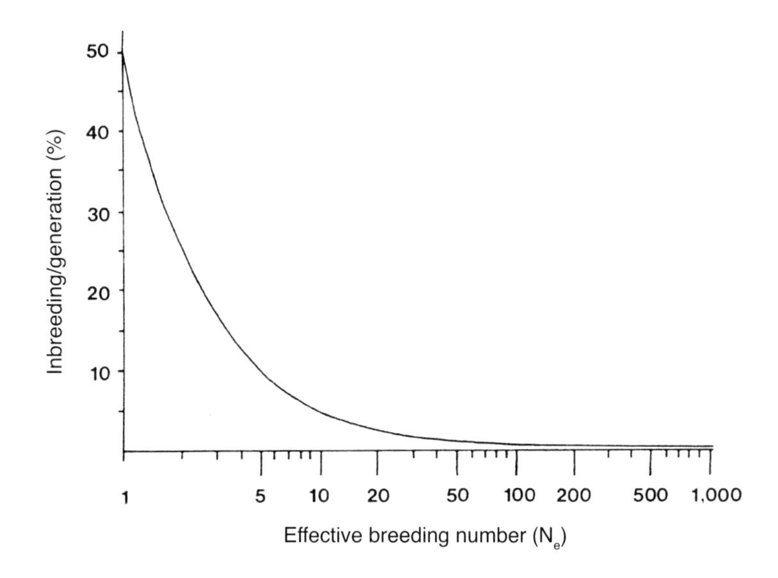

Fig. 7.6 The relationship between effective breeding number and inbreeding (from Tave, 1990a).

Table 7.12 Minimum effective breeding numbers (N_e) for hatchery populations when fish are raised as a farmed commodity and when they are cultured for fisheries management programmes

Number of generations	Food fish farming	Fisheries management
1	45	–
5	61	–
10	100	344
15	150	364
20	200	378
25	250	390
50	–	500
60	–	600
70	–	700
80	–	800
90	–	900
100	–	1000

Farmed fish N_e values were generated using the following assumptions: keeping inbreeding ≤ 5% and giving a 99% guarantee of keeping an allele whose frequency is 0.05. N_e values for fisheries management programmes were generated using the following assumptions: keeping inbreeding ≤ 5% and giving a 99% guarantee of keeping an allele whose frequency is 0.01.
After Tave (1993) with permission from Kluwer Academic Publishers.

Atlantic salmon (*Salmo salar*) males become sexually mature precociously and die before they reach marketable size. This can occur in up to 25% of the males at a farm, so the production of an all-female population will improve survival by up to 12.5% and will also increase yields and profits (section 15.7). In the tropics, tilapia can reproduce when only 3 months old, so the culture of a mixed-sex population will result in a population of few marketable-size fish, as most of the feed energy goes to reproduction and production of fingerlings, not to the growth of food fish (Chapter 16). A monosex male population of tilapia is desirable because this will eliminate reproduction; additionally, the yield will increase because males can grow twice a fast as females.

Sex reversal can be divided into two components:

(1) Direct sex reversal of fry. This is primarily used to produce animals for grow-out and, as such, is not genetics but an aspect of hatchery management and fingerling production.
(2) Production of sex-reversed broodstock. The basics behind direct sex reversal must be understood in order to produce sex-reversed broodstock.

7.5.2 Direct sex reversal

Because the fry stage of a fish's life cycle takes place when the fish is free-swimming, it is possible to alter development and to direct it to some extent. This is possible with a fish's sex. Until a specific moment during development, fry are neither males nor females (the exact moment is species specific). Genetically, each fish's sex is determined at fertilisation, but phenotypically fry are totipotent (able to develop either sexual phenotype). At the biologically prescribed moment, one or more genes fire a chemical signal that directs the totipotent gonadal tissue to develop into either male or female organs. It is at that point that phenotypic sex is determined.

The normal genetic signal can be overwhelmed by the external application of anabolic steroids, which, if done properly, will direct phenotypic sex and sex reverse half the population. The efficiency of sex reversal depends upon:

- the species
- the hormone used and dosage
- the method used to administer the hormone
- the temperature
- the beginning and ending dates
- the culture system used during treatment.

In general, sex-reversed males are produced by feeding fry rations containing 17α-methyltestosterone at a rate of 60 mg/kg for 30–60 days. If done properly, half the fish will be normal males, and the other half will be sex-reversed males (genetic females that are phenotypic males). Sex-reversed females are produced by feeding fry rations containing 17β-oestradiol at a rate of 20 mg/kg for 30–60 days. If done properly, half the fish produced will be normal females, and the other half will be sex-reversed females (genetic males that are phenotypic females).

Direct sex reversal is relatively simple and is a form of technology that is available even to farmers in developing countries. However, the use of anabolic steroids can create marketing problems because more and more consumers are demanding that their food be free from pesticides and antibiotics, and they do not want hormonal residues in their meat. Although the amount of hormone needed to sex reverse a fish is quite small, and although studies have shown that virtually all of the hormone is gone 3 weeks after it is removed from the diet, an industry that depends on its use could collapse from adverse publicity in a health-conscious society.

7.5.3 Sex-reversed broodstock

To avoid this problem, sex reversal can be used to create sex-reversed broodstock that are capable of producing monosex offspring. Most of the work in this area has been done with tilapia and salmonids in an attempt either to eliminate reproduction during grow-out or to avoid sexual precocity in males.

Most important cultured species of fish have the XY sex-determining system (Table 7.9). A three-generation breeding programme is generally needed to create broodstock that can produce an all-male population. The first step (generation) is to use oestrogens to create sex-reversed XY 'females'. These females look the same as XX females, and they can be distinguished only by progeny testing, which is the second generation of breeding (75% of the offspring produced by XY 'females' will be males, not the normal 50%). The male offspring produced by sex-reversed XY 'females' are either normal XY males or YY super-males. To differentiate the two, they are progeny tested (super-males will produce only sons), which is the third generation of breeding. Once identified, the super-males can be used to produce all-male offspring. Fig. 7.7 shows the sex-reversal breeding programme that can be used to create YY super-males for species with the XY sex-determining system.

If a monosex female population is desired for species with the XY sex-determining system, a simpler two-generation programme is needed. In the first generation, fry are sex-reversed with androgens to create XX 'males' (Fig. 7.8). Because XX and XY males are indistinguishable, they must be progeny tested when mature. They are easy to identify because XX 'males' will produce only daughters, whereas XY males will produce 50% males and females. These XY males are culled. Once XX 'males' are identified, they are bred to produce an all-female population.

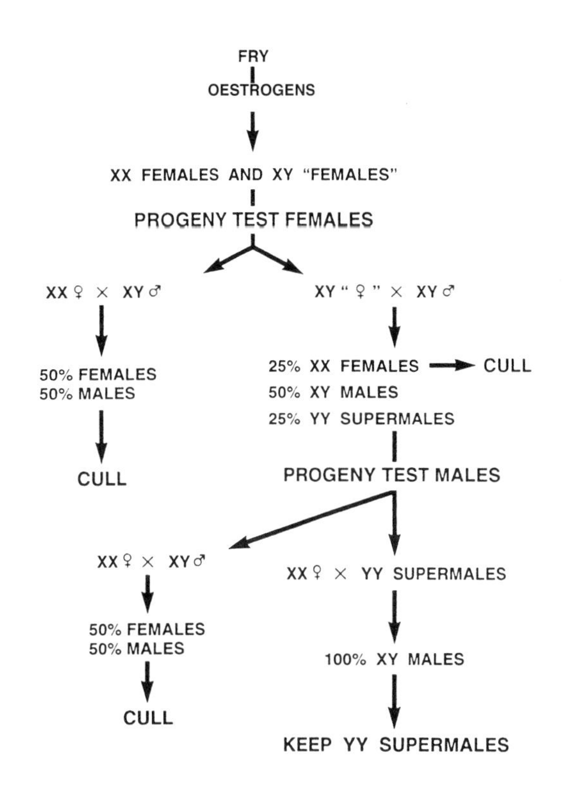

Fig. 7.7 The sex-reversal breeding programme to create YY super-males. Reproduced from Tave (1993) with permission from Kluwer Academic Publishers.

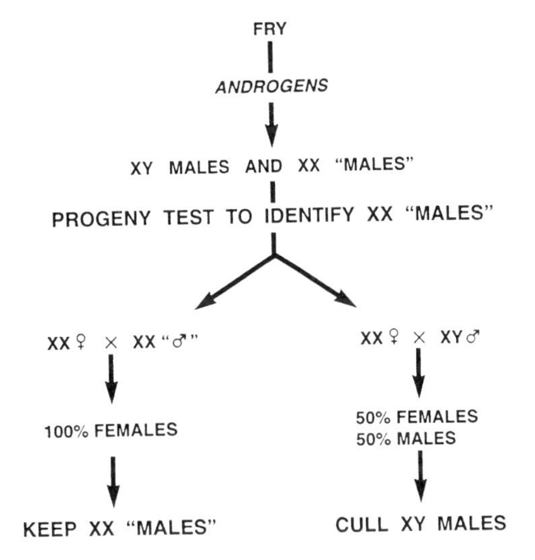

Fig. 7.8 The sex-reversal breeding programme to create sex-reversed XX 'male' broodstock. Reproduced from Tave (1993) with permission from Kluwer Academic Publishers.

This is the basis for the production of all-female populations of salmonids (section 15.7.2).

Although these protocols are not that daunting, and despite the fact that they have been shown to produce monosex populations in small-scale experiments, they are only 90–95% successful at the commercial level. Reasons for this may be autosomal sex-reversing or sex-modifying genes or impure broodstock (section 7.3.2).

This technique is being used commercially to produce monosex female rainbow trout in North America and Europe. Many of the rainbow trout farmed in the UK and chinook salmon farmed in Canada are all-female progeny produced by sex-reversed broodstock. Interestingly, rainbow trout do not need to be progeny tested, because sex-reversed XX 'males' have incomplete sperm ducts. If all males are examined internally before the milt is extracted, sex-reversed XX 'males' can be identified; this breeding programme is therefore only one generation long.

Blue tilapia is one of the few important fish that has the WZ sex-determining system (females are WZ and males are ZZ) (Table 7.9). The breeding goal for a farmer who raises this species is to produce an all-male population. This can be accomplished with a two-generation sex-reversal breeding programme (Fig. 7.9). In the first generation, oestrogens are used to create sex-reversed ZZ 'females'. Because the sex-reversed females are indistinguishable from their normal counterparts, the females must be progeny tested to identify the sex-reversed ones. This procedure is relatively simple because the sex-reversed females produce all-male offspring. Once identified, they can be bred to produce a monosex male population. Again, although this breeding programme has been shown to work on an experimental basis, it is generally only 90–95% successful at a commercial level.

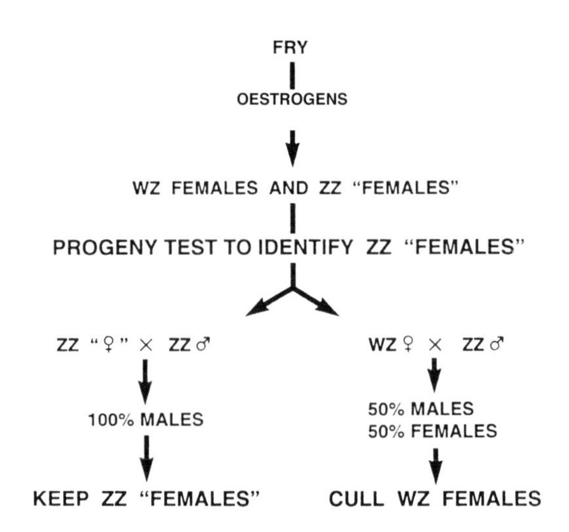

Fig. 7.9 The sex reversal breeding programme to create sex-reversed ZZ 'female' broodstock. Reproduced from Tave (1993) with permission from Kluwer Academic Publishers.

7.6 Production of sterile or monosex populations by chromosomal manipulation

Fertilisation and embryological development occur externally in most cultured species. Because of this, it is possible to manipulate gametes and zygotes in order to alter the chromosomal set number or to create animals that have only a single parent. This is usually done to produce sterile animals or to produce genetically unique animals for breeding purposes.

7.6.1 *Triploids*

Triploids are individuals that have three sets of chromosomes ($3n$). There are naturally occurring triploids, but such species are rare. For practical purposes, most cultured fish and shellfish species are diploids ($2n$). Some species are natural tetraploids ($4n$), but they behave like diploids and can be considered to be diploids for practical breeding purposes.

Triploids are created primarily to produce sterile animals for grow-out. In theory, triploids should grow faster than diploids because they have 33% more genes, have larger nuclei and do not divert energy to gamete production, reproduction or parental care. However, triploid fish have seldom outperformed diploids in yield trials; when they did, it was often for only short periods. Under culture conditions, pre-maturation growth of triploid fish is generally considered to be inferior to that of diploids (Arai, 2001). Growth and survival of triploid tilapia, for example, is generally inferior to that of diploids during early life stages (section 16.5.2). Triploids may, however, show better growth, survival and meat quality than diploids during the period of final maturation (Arai,

2001). All female triploids are commonly used in the salmonid industry to avoid pseudo-maturation characters seen in triploid males (section 15.7). However, triploid rainbow trout and brown trout have been reported to show slower growth than diploids to commercial size (Bonnet *et al.*, 1999).

In contrast to fish, very clear benefits of triploidy have been reported for some species of bivalve mollusc, with triploid oysters showing considerably better growth rates than diploids to market size (Nell *et al.*, 1994; section 21.4.2).

The exact reason why triploids are sterile is not known, but it is generally attributed to 'gametic incompatibility', in that the chromosomes cannot properly align themselves during meiosis, because it is impossible to divide three sets of chromosomes into two equal sets. Gametes that are produced are usually aneuploid in that they contain an unbalanced chromosome set, and this usually produces abnormal or sub-viable offspring.

Triploids are generally produced by shocking eggs (E) shortly after fertilisation to prevent the second polar body (SPB) from leaving the egg (Fig. 7.10). This produces a fertilised egg that has a haploid (n) egg nucleus (EN), a haploid sperm nucleus (S) and a haploid second polar body nucleus. These three haploid nuclei fuse to form a triploid zygote nucleus (Fig. 7.10).

The type of shock and the exact moment that the shock has to start and stop in order to produce triploids are species specific and must be determined experimentally. Temperature, pressure and chemical shocks can be used, but temperature is preferred, because of safety and because it is inexpensive and easy to control. Coldwater fish eggs are usually shocked in warm water, whereas warmwater fish eggs are shocked with cold water. The exact temperature needed to produce optimal results must be determined experimentally, but it is usually within a degree or two of the lethal limit for that species. The shock usually begins after fertilisation but before the second polar body is extruded, and it continues until it is unlikely that the second polar body will be extruded. Suitable methods for triploid induction in salmonids are outlined in section 15.7.3. Triploidy in bivalves is usually induced by exposing eggs to the chemical cytochalasin B (CB) (Nell *et al.*, 1994). However, CB and other chemicals are not commonly used for triploid induction in fish (Arai, 2001).

Even though triploid fish may not grow faster than diploids, they are created in order to produce sterile animals for grow-out. For example, in the USA many farmers and natural resource agencies want to use grass carp (*Ctenopharyngodon idella*) for aquatic weed control. However, because it is an exotic species, most states ban its importation and use. Many of these states will allow triploid grass carp because they have been shown to be sterile. Consequently, states that allow diploid grass carp have created an aquaculture industry in which farmers produce triploid grass carp for states that ban the diploids.

Some farmers do not want their animals to reproduce during grow-out because the fish die after they spawn or because the market price is greater for larger fish. Both problems can be avoided by raising sterile triploids. Oyster (*Crassostrea gigas*) farmers in the USA have discovered that triploid oysters can be harvested year-round. When oysters produce gametes, they become less tasty because glycogen reserves are used for gamete production; in addition, some consumers do not like the texture of these oysters because they feel 'mushy'. Because triploid oysters do not produce gametes, they do not deplete glycogen reserves, so triploid oysters are always 'sweet', and they always have the desired texture. The preference for unripe oysters is related to particular markets. Ripe oysters are preferred in other countries and triploids are a disadvantage.

7.6.2 *Tetraploids*

The major reason for producing this type of genetically engineered animal is that they can be bred to produce triploids (Chourrout *et al.*, 1986; Blanc *et al.*, 1987). Because triploids are sterile, they must be produced every year. The creation of tetraploids enables fish farmers to mate tetraploids ($4n$) with diploids ($2n$) to produce interploid triploids ($3n$). In addition, the tetraploids can be used to produce successive generations of tetraploids.

Tetraploids are produced by shocking normal diploid embryos during first cleavage (Fig. 7.10). The shock usually begins after the nucleus has begun to divide but before the zygote has divided. The shock

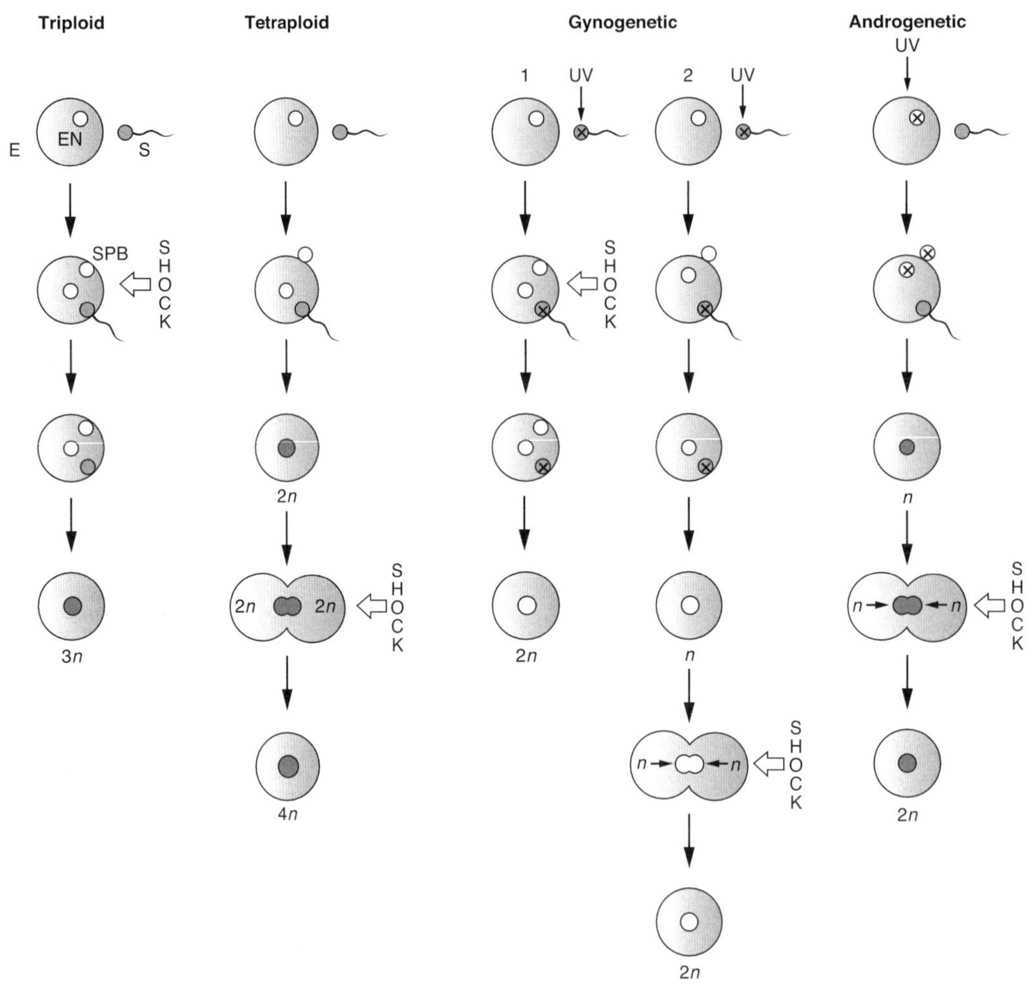

Fig. 7.10 The timing of the shocks that are needed to create triploids, tetraploids, gynogens and androgens. See the text for the abbreviations and explanations. E, egg; S, sperm; EN, egg nucleus. After Tave (1990b) Chromosomal manipulation. *Aquaculture* **16** (1), 62–5.

prevents cell and nuclear division. However, since it is initiated after the chromosomes have replicated, the nucleus contains four sets of chromosomes and the zygote is a tetraploid.

The best way to use tetraploids is to mate tetraploid females to diploid males to produce interploid triploid offspring. If diploid females are mated to tetraploid males, the ovum micropyle may be too narrow to admit the large nucleus of a diploid sperm ($4n$ to $2n$ after meiosis). Tetraploids have been created in a number of species, but the most advanced work has been undertaken with rainbow trout.

7.6.3 Gynogens and androgens

Gynogens and androgens are animals that have only a single parent:

- Gynogens have only a mother and all genes come from her.
- Androgens have only a father and, similarly, all genes come from him.

Like tetraploids, such fish are not produced for grow-out, but are produced for breeding programmes.

Gynogens are produced by 'fertilising' eggs with irradiated sperm. Because the sperm's DNA has been destroyed, the eggs are not actually fertilised; instead, they are activated. Because the egg contains only a haploid nucleus, the egg will produce a haploid zygote. Haploids can develop for a short period, but few will develop normally or hatch. Diploid gynogens are produced by shocking the eggs or zygotes in one of two ways (Fig. 7.10).

(1) They can be shocked as was described for the production of triploids. This prevents the second polar body from leaving the egg and produces what is called a 'meiotic gynogen'.
(2) Shocking the haploid zygote to produce a diploid zygote as was described for the production of tetraploids can also produce them. This produces what is called a 'mitotic gynogen'.

In both cases, this produces an animal that has only a mother; and, in both cases, all chromosomes come from the mother.

Androgens are usually produced by using normal sperm to fertilise eggs whose DNA has been destroyed by irradiation (Fig. 7.10). To restore the normal diploid complement, the developing haploid embryo is shocked as was described for the production of tetraploids. This changes a haploid androgen into a diploid androgen and produces a diploid in which all chromosomes come from its father.

Androgenesis can be used to produce YY super-males for species that have the XY sex-determining system. Half the sperm produced by normal XY males contain a Y chromosome, so androgenesis will produce 50% XX androgenetic females and 50% YY androgenetic super-males. YY super-males can then be used to produce all-male offspring (Fig. 7.7).

Gynogenesis can also be used to produce YY super-males. Eggs from sex-reversed XY 'females' (half the eggs that they produce will contain a Y chromosome) are activated with irradiated sperm. When the haploid zygotes are shocked to produce diploid gynogens, half will be YY gynogenetic super-males.

7.7 Genetic engineering

Genetic engineering has great potential to create important new and improved varieties and cultivars of animals and plants. It is a high-technology molecular process whereby one or a few genes are transferred from one organism to another. The transferred gene can be within a species or it can be across kingdoms, e.g. from viruses and bacteria to higher organisms, or it can be a 'fabricated' gene. If the gene, together with promoter genes (which influence its expression), is incorporated into the genome of the target organism, inherited and expressed, then this transgenic organism will have a new genotype and a new phenotype. The phenotype will depend on the nature of the inserted gene and the strength of promoter genes.

This breeding technique has been used for some commercial plants such as cotton, corn and sorghum to produce insect- and herbicide-resistant strains. In these cases the sources of genes for producing transgenic plants are bacteria and other plants. Transgenic mammals have also been developed, but they have not been used in commercial animal husbandry. In animals, the growth hormone genes have been the frequent targets for transgenic studies in order to produce transgenic animals with higher growth rates.

The technology for this type of genetics is very sophisticated and expensive. Consequently, research institutions and large agribusinesses carry out this type of research. In addition, it is highly regulated by governments because of concerns about potential adverse environmental and public health effects. This very cautious approach to the implementation of commercial-scale production with transgenic animals applies to aquaculture (Entis, 1997).

Transgenic fish have been produced, and the potential benefits for fish and shellfish culture are outlined by Hew & Fletcher (2001):

- increased growth rates
- improved feed conversion rates (FCRs)
- controlled reproduction (maturation, sterility and sexual expression)
- greater disease resistance
- increased tolerance of environmental extremes
- improved flesh quality.

These are also the same objectives as other stock improvement programmes.

So far, the aquaculture-related transgenic

Table 7.13 Growth enhancement in fish with growth hormone gene inserted

Fish	Growth rate enhancement	Source
Tilapia	1.81	Martinez *et al.* (1996)
	2.00	Maclean *et al.* (1995)
Atlantic salmon	3–10	Du *et al.* (1992)
	3–10	Hew *et al.* (1995)
Pacific salmon	3–10	Devlin *et al.* (1995)
	6–11	Devlin *et al.* (1994)

Reprinted in part from Hew & Fletcher (2001) with permission from Elsevier Science.

Table 7.14 Transgenic technologies that may be used in developing disease resistance in fish

Experimental approach	Principles
Antisense technology	Complementary RNA to complex with viral or foreign RNA
Ribosome	Specific RNA-based enzymes to destroy viral or foreign RNA
Expression of viral coat protein	Viral coat protein to occupy the receptor binding sites, thus competing with normal viral binding
Expression of antibacterial, antimicrobial substances and peptides	Lysozyme and other cationic peptides against a broad spectrum of pathogens
Expression of cytokines, interferons and other host genes involved in immune defence	To boost the host's general immune systems

Reprinted in part from Hew & Fletcher (2001) with permission from Elsevier Science.

programmes have largely been conducted with freshwater and andromous fish and there have been a number of successful gene transfers. The effects of the transfer of a growth hormone gene on growth rate in some fish species are shown in Table 7.13. There were outstanding tenfold increases in growth rate in some fish. Some of the fastest-growing transgenic Pacific salmon showed abnormalities, but the moderately fast-growing fish were normal and fertile, and they have produced second- and third-generation offspring from the original transgenic broodstock (Hew & Fletcher, 2001).

With the intensification of aquaculture in recirculating systems, the potential for catastrophic disease outbreaks is greatly increased (Chapters 2 and 10). Hence, genetic engineering may make a great contribution to modern aquaculture technology with mechanisms to control bacterial, viral and parasitic diseases (Table 7.14).

References

Arai, K. (2001). Genetic improvements of aquaculture finfish species by chromosome manipulation techniques in Japan. *Aquaculture*, **197**, 205–28.

Avtalion, R. R. & Hammerman, I. S. (1978). Sex determination in sarotherodon (*Tilapia*). I. Introduction to a theory of autosomal influence. *Bamidgeh*, **30,** 110–15.

Blanc, J. M., Chourrout, D. & Krieg, F. (1987). Evaluation of juvenile rainbow trout survival and growth in half-sib families from diploid and tetraploid sires. *Aquaculture*, **65**, 215–20.

Bonnet, S., Haffray, P., Blanc, J. M., Vallee, F., Vauchez, C., *et al.* (1999). Genetic variation in growth parameters until commercial size in diploid and triploid freshwater rainbow trout (*Oncorhynchus mykiss*) and seawater brown trout (*Salmo trutta*). *Aquaculture*, **173**, 359–75.

Chourrout, D., Chevassus, B., Krieg, F., Happe, A., Burger, G. & Renard, P. (1986). Production of second generation triploid and tetraploid rainbow trout by mating tetraploid males and diploid females. Potential of tetraploid fish. *Theoretical and Applied Genetics*, **72**, 193–206.

Devlin, R. H., Yesaki, T. Y., Biagi, C. A., Donaldson, E. M., Swanson, P. & Chan, W.-K. (1994). Extraordinary salmon growth. *Nature*, **371**, 209–10

Devlin, R. H., Yesaki, T. Y., Donaldson, E. M., Du, S. J, & Hew, C. L. (1995). Production of germline transgenic Pacific salmonids with dramatically increased growth performance. *Canadian Journal of Fisheries & Aquatic Sciences*, **52**, 1376–84

Du, S. J., Gong, Z., Tan, C. H., Fletcher, G. L. & Hew, C. L. (1992). Development of an 'all fish' gene casette for gene transfer in aquaculture. *Molecular Marine Biology and Biotechnology*, **1**, 290–300.

Entis, E. (1997). Aquabiotech: a blue revolution? *World Aquaculture*, **28**, 12–15.

Falconer, D. S. (1957). Breeding methods. I. Genetic considerations. In: *The UFAW Handbook on the Care and Management of Laboratory Animals*, 2nd edn (Ed. by A. N. Worden & W. Lane-Petter), pp. 85–107. Universities Federation for Animal Welfare, London.

Gordon, M. (1956). An intricate genetic system that controls nine pigment cell patterns in the platyfish. *Zoologica*, **41**, 153–62.

Hammerman, I. S. & Avtalion, R. R. (1979). Sex determination in *Sarotherodon (Tilapia)*. Part 2. The sex ratio as a tool for the determination of genotype – a model of autosomal and gonosomal influence. *Theoretical and Applied Genetics*, **55**, 177–187.

Hershberger, W. K., Myers, J. M., Iwamoto, R. N., McAuley, W. C. & Saxton, A. M. (1990). Genetic changes in the growth of coho salmon (*Oncorhynchus kisutch*) in marine net-pens, produced by ten years of selection. *Aquaculture*, **85**, 187–97.

Hew, C. L. & Fletcher, G. L. (2001). The role of aquatic biotechnology in aquaculture. *Aquaculture*, **197**, 191–204.

Hew, C.L., Fletcher, G. L. & Davies, P. L. (1995). Transgenic salmon: tailoring the genome for food production. *Journal of Fish Biology*, **47**, 1–19

Kirpichnikov, V. S. (1981). *The Genetic Bases of Fish Selection*. Springer-Verlag, New York.

Knibb, W., Gorshkova, G. & Gorshkov, S. (1998). Genetic improvement in culture marine finfish: case studies. In: *Tropical Mariculture* (Ed. by S. S. de Silva), pp. 111–49. Academic Press, London.

Komen, J., de Boer, P. & Richter, C. J. J. (1992). Male sex reversal in gynogenetic XX females of common carp (*Cyprinus carpio* L.) by a recessive mutation in a sex-determining gene. *Journal of Heredity*, **83**, 431–4.

McAndrew, B. J., Roubal, F. R., Roberts, R. J., Bullock, A. M. & McEwen, I. M. (1988). The genetics and histology of red, blond and associated colour variants in *Oreochromis niloticus*. *Genetica*, **76**, 127–37.

Maclean, N., Nam, S., Williams, D. & Lavender, L. (1995) The use of transient expression of reporter genes to test promoter efficiency in tilapia and zebrafish. *Molecular Biology in Fish, Fisheries and Aquaculture. An International Symposium*. p. 118 (Abstract). Fish Society of British Isles, Plymouth.

Martinez, R., Estrada, M. P., Berlanga, J., Guillen, I., Hernandez, O., Cabrera, E., *et al.* (1996). Growth enhancement of intransgenic tilapia by ectopic expression of tilapia growth hormone. *Molecular Marine Biology and Biotechnology*, **5**, 62–70

Nell, J. A., Cox, E., Smith, I. R. & Maguire, G. B. (1994). Studies on triploid oysters in Australia. I. The farming potential of triploid Sydney rock oysters *Saccostrea commercialis* (Iredale & Roughley). *Aquaculture*, **126,** 243–55.

Purdom, C.E. (1993). *Genetics and Fish Breeding*. Chapman & Hall, London.

Siitonen, L. & Gall, G. A. E. (1989). Response to selection for early spawn date in rainbow trout, *Salmo gairdneri*. *Aquaculture*, **78**(2), 153–61.

Tave, D. (1987). Improving productivity in catfish farming by selection. *Aquaculture Magazine*, **13**, 53–5.

Tave, D. (1988). Effective breeding number and broodstock management. In: *Genetics, Breeding and Domestication of Farmed Salmon Workshop* (Ed. by E. A. Kenney), pp. 46–57. Ministry of Agriculture and Fisheries and BC Salmon Farmers' Association, Vancouver, BC.

Tave, D. (1990a). Effective breeding number and broodstock management. I. How to minimise inbreeding. In: *Proceedings Auburn Symposium on Fisheries and Aquaculture* (Ed. by R. O. Smitherman & D. Tave), pp. 27–38. Alabama Agricultural Experiment Station, Auburn University, Auburn, AL.

Tave, D. (1990b). Chromosomal manipulation. *Aquaculture Magazine*, **16**, 62–5.

Tave, D. (1993). *Genetics for Fish Hatchery Managers*, 2nd edn. Van Nostrand Reinhold, New York.

Tave, D., Rezk, M. & Smitherman, R. O. (1989). Genetics of body color in *Tilapia mossambica*. *Journal of the World Aquaculture Society*, **20**, 214–22.

Tave, D., Smitherman, R. O., Jayaprakas, V. & Kuhlers, D. L. (1990). Estimates of additive genetic effects, maternal genetic effects, individual heterosis, maternal heterosis, and egg cytoplasmic effects for growth in *Tilapia nilotica*. *Journal of the World Aquaculture Society*, **21**, 263–70.

Wallbrunn, H. M. (1958). Genetics of the Siamese fighting fish, *Betta splendens*. *Genetics*, **43**, 289–98.

Wohlfarth, G., Lahman, M., & Moav, R. (1963). Genetic improvement of carp. IV. Leather and line carp in fish ponds of Israel. *Bamidgeh*, **15**, 3–8.

8
Nutrition

Trevor Anderson and Sena De Silva

8.1 Introduction

[A]dequate nutritional practices have an ever increasing determinant role, not only in terms of economic optimisation, but also in terms of maintenance of good health and improvement of reproductive and growth performance.

Sadasivam Kaushik (1993)

Nutrition is the science of feeding the body to ensure its optimal development, health and maintenance. In all forms of animal husbandry, providing a supply of nutrients to match the requirements of the cultured animal is fundamental to achieving optimal growth and production efficiency and hence maximising economic return (Fig. 8.1). For nutritionists to deliver the optimal balance of nutrients to an animal, they must understand the processes of ingestion, digestion, absorption and metabolism. Although some of these processes vary greatly between individual species, many are consistent and allow general description.

The mechanism of ingestion largely impacts upon the form in which feed is delivered. Ingestion mechanisms show enormous variety, e.g.:

- Molluscs are largely filter feeders.
- Crustaceans generally manipulate food to their mouthparts with chelipeds.
- Fish have the usual vertebrate ingestion apparatus, albeit in huge variety, with an equally diverse range of digestive tract morphology.

Individual fish species are often categorised by feeding types.

- Herbivores feed largely on living plant material.
- Carnivores feed on animal matter.
- Detritivores feed on detritus.
- Omnivores consume a mixed plant and animal diet.

Most cultured crustaceans are omnivorous, and filter-feeding molluscs generally consume phytoplankton and other small particulate organic matter (POM). Fish may also be broadly categorised into ecological groups according their mode of feeding, i.e. pelagic plankton feeders, benthos feeders, etc. Pelagic plankton feeders can be further subdivided into surface and columnar feeders. These divisions are important in certain types of aquaculture, in particular in warm-water polyculture, in which a combination of species is used, each occupying a different ecological niche (Fig. 2.8). In some cases these divisions are relevant to understanding the nutrient requirements of a species, such as herbivorous fish, which are more likely

Fig. 8.1 Using a water cannon to feed pellets to Atlantic salmon at a farm in Tasmania, Australia (photograph by Dr John Purser).

to be able to use carbohydrate as an energy source than carnivorous fish (see below).

The bulk of food consumed is made up of three major nutrients: proteins, lipids and carbohydrates. Proteins are digested to release their component amino acids, which subsequently are either used for synthesis of new proteins or catabolised for energy. Lipids are broken down to constituent fatty acids in the gut, absorbed and then re-esterified into glycolipids. These coalesce in the blood to form micelles or droplets that move through the circulation. In order to be used by the organism, they must again be degraded to their constituent fatty acids. Fatty acids are important structural components of cell membranes and also act as precursors to other compounds. Fatty acids are also degraded for energy. Carbohydrates are generally consumed as complex molecules, the most common forms in aquaculture diets being starch and cellulose. Although evidence is emerging that some cellulose is digested by fish, in general it is considered indigestible. Starch is broken down to produce glucose, which again is further degraded to provide energy. There are other nutrients required for optimal growth of aquatic organisms. These include minerals, vitamins, purines and pyrimidines, which generally act as cofactors in enzyme-mediated reactions and/or can be incorporated into tissues (e.g. calcium in bone and iron in haemoglobin).

The degradation of amino acids, fatty acids and simple sugars is mediated by catabolic pathways, each of which is regulated by a variety of mechanisms. These pathways were initially described in studies of mammalian metabolism, but most have been confirmed in studies of other species. Some interesting and significant differences between the metabolism of fish, crustaceans and molluscs which have important consequences for their nutrient requirements are discussed below.

Animals in different parts of their life cycle may require different nutrients, but the principles of nutrition are the same no matter how old the animal. Specialist areas of study exist for broodstock and larval nutrition, but most research has investigated the nutrition of aquatic animals in the pre-reproductive phase of their life. It is known that the composition of the broodstock diet is a major influence on gamete viability (Izquierdo *et al.*, 2001). Developing embryos and pre-feeding larvae rely on maternally derived energy and nutrient reserves for their requirements (section 6.3.4). Demand for essential nutrients is very high during periods of rapid tissue growth, such as during embryonic and early larval development. If maternally derived sources of a particular essential nutrient are insufficient to meet the demand for growth, the viability of the developing larvae is likely to be reduced. The presence or absence of these essential nutrients in broodstock diets will influence their presence in the gametes produced and their availability to the developing embryos or larvae (Izquierdo *et al.*, 2001). When larvae begin to feed they usually have a period of transition from endogenous to exogenous nutrient supply and, once the endogenous reserves have been used, they must rely totally on exogenous nutrients (Fig. 6.3). As the animals grow, and for the rest of their lives, the food available to them must continue to satisfy their nutritional requirements. Appropriate nutrition of culture animals is a key factor in determining the productivity and economic success of aquaculture ventures.

8.2 Digestion

Digestion is performed primarily by digestive enzymes, which are secreted into the lumen of the alimentary canal. Enzymes may originate from the oesophageal, gastric, pyloric caecal or intestinal mucosa, or from the hepatopancreas. Production of acidic gastric fluid probably occurs in most fish, except in those without a stomach as neither hydrochloric acid (HCl) nor pepsin is found in the gut. The fluids and enzymes secreted in teleost fish are summarised in Table 8.1. Digestive enzymes are water-soluble proteins that can be divided into proteases, lipases and esterases, and carbohydrases, depending upon the macronutrient they digest. Enzymes present in food organisms (such as rotifers or brine shrimp) may significantly enhance endogenous enzyme activity (section 9.7.2). This is particularly the case in fish larvae, which do not produce enzymes until some days after hatching. Most animals do not feed continuously. Generally, secretion of digestive juices and enzymes is elicited by factors such as feeding stimulants and time of day, so that their presence will correspond with that of food in the alimentary canal. The presence of food in the tract also stimulates enzyme and acid/alkaline juice secretion in regions not immediately adjacent to the food.

In fish, digestion is predominantly extracellular and takes place in the lumen of the alimentary canal, whereas mollusc digestion often occurs intracellularly (i.e. following absorption). The rate at which food is digested by an animal depends upon the specific enzyme activity, the volume of the digestive juices, the rate at which food fills the gut and the time of transit. Factors that influence the rate at which food moves through the alimentary canal and impact on the efficiency with which a diet is used are:

- species
- age
- size
- sex
- temperature
- stocking density
- meal size
- type of food and its presentation
- feeding rate.

The rate of digestion is important in aquaculture: rapidly digested food may provide a short burst of nutrients to the animal, which are metabolised inefficiently, whereas poorly digested food will result in nutrients being voided before they can be utilised by the animal.

8.3 Nutrient absorption

Products of digestion are absorbed by active transport and diffusion. An example of active transport is the uptake of glucose involving a carrier. It is an energy-requiring process that moves glucose across a membrane into the epithelial cells even if there is already a high concentration of glucose within that cell. Diffusion can occur either as facilitated or as simple diffusion. Facilitated diffusion occurs when there is a carrier system that allows the compound to move across an otherwise impermeable membrane. Facilitated diffusion does not require energy and will not move the compound up a concentration gradient. Simple diffusion does not require a carrier or energy. Fatty acids are examples of compounds that are absorbed into intestinal epithelium by simple diffusion. As the micelles or droplets of fatty acids and bile salts approach the epithelial cell surface, lipids are released to pass through the cell membrane. Once inside, the lipids reform as droplets, this time called chylomicrons, which have a very thin protein coat to make them water soluble. The chylomicrons are moved through the cell to be released into the circulation and transported to the liver for further processing. Pinocytosis may be used to absorb macromolecular protein in the hindgut (Anderson, 1986). As stated above, the processes of absorption are common to all animals with only minor variations, many of which have little relevance to aquaculture.

8.4 Digestibility

Only a proportion of ingested food is digested and its nutrients absorbed. The rest is voided as faeces. Digestibility is a relative measure of the extent to which ingested food and its nutrient components have been digested and absorbed by the animal. It is necessary to know the digestibility of a feedstuff in order to evaluate its quality as a source of nutrients.

Total and dry matter digestibility refers to the degree of digestibility of the complete diet and the corresponding ingredient in question. Nutrient digestibility refers to the degree with which a specified nutrient, such as protein, lipid, amino acid, carbohydrate, etc., is digestible. When providing a cultured organism with a cost-effective diet that is well assimilated, it is not sufficient to consider only the nutrient requirements of the organism. A nutritionally balanced diet will only be a good diet if it is also easily digested and effectively utilised. The digestibility of a diet depends on the nature and type of ingredients that go into the diet and the final

Table 8.1 Digestive fluids and enzymes secreted in teleosts

Site/type	Fluid/enzyme	Function
Stomach (gastric glands)	Zymogens	Proteolytic enzymes; pepsinogen
	HCl	Reduces gut pH and prepares pepsinogens to act. Cleaves peptide links and other structures such as cell walls by hydrolysis
	Pepsin	Attacks most proteins
	Amylase	Acts on carbohydrates
	Lipase	Acts on lipids
	Esterases	Act on esters (class of lipids)
	Chitinase	Acts on chitin
Pancreas	Enzymes	Enzymes are stored as zymogens. Proteases produced by intestine transform trypsinogen into trypsin, which in turn activates others
	HCO_3^-	Neutralises HCl entering intestine, and prepares intestine for alkali digestion
	Proteases (trypsin, chymotrypsin, carboxypeptidases and elastase)	Optimal action of enzymes at pH 7.0
	Trypsin	Cleaves peptide linkages at carboxy groups of lysine or arginine
	Chymotrypsin	Attacks peptide with carbonyls from aromatic side-chains
	Elastase	Attacks peptide bonds on elastin
	Carboxy-peptidases	Hydrolyse the terminal peptide bond of their substrates
	Amylase	Carbohydrate digestion at non-acidic pH
	Chitinase (+ NAGase)	Splits chitin into dimers and trimers of *N*-acetyl-D-glucosamine (NAG), which is further broken down by NAGase
	Lipases	Hydrolyse triglycerides, fats, phospholipids and wax esters
Liver/anterior intestine	Bile (bile salts, organic anions, cholesterol, phospholipids, inorganic ions)	Makes the intestinal medium alkaline; most bile salts are reabsorbed from the intestine and returned to the liver in enterohepatic circulation
Brush border of the intestinal epithelium but could be partly pancreatic in origin	Amino-peptidases (alkaline and acid)	Split nucleosides
	Polynucleotidase	Splits nucleic acids
	Lecithinase	Splits phospholipids into glycerol, fatty acids
	Various carbohydrate-digesting enzymes	

physical form (such as hardness, palatability, stability in water) of the product. From the point of view of the aquaculturist, a thorough knowledge of digestibility is more important than knowledge of digestive physiology, even though digestibility is dependent upon and determined by physiology.

Digestibility values are expressed as percentages of the original ingested material. Generally, fish meal is digested efficiently, with digestion often exceeding 90%, but the apparent protein digestibility of various plant products is extremely variable. Equally, the digestibility of macrophytes by different species is variable. These variations have a number of practical implications. If, for example, it is necessary to formulate a diet that contains 30% protein by dry weight, the required protein can be supplied by a variety of ingredients ranging from animal products, such as fish meal or feather meal, to plant products, such as soybean meal, oilseed cakes, pulses, etc. Diets incorporating these different ingredients, although formulated to have the same protein content, are unlikely to be equally digestible with respect to protein or to the total dry matter of the diet. Accordingly, the performance of the cultured organism fed on different diets containing the same amount of protein, but made up of different ingredients, will be different.

8.4.1 *Determining digestibility*

Digestibility of a diet or feed ingredient can be determined directly or indirectly. Unlike comparable studies with terrestrial animals, those with aquatic animals have an inherent difficulty because of the medium in which they live. Faecal traps, for example, are impossible to use, and the voided faeces lose nutrients immediately on discharge. Therefore, all digestibility estimations on aquatic animals, whichever method one chooses, are subject to some degree of error. Because of hindgut absorption by pinocytosis, this is the case even when the animal is killed and rectal contents are dissected out.

In the direct method, the quantity ingested (total or nutrient) and the quantity of faecal matter voided are determined. The ratio gives the percentage digestibility of the feed or the nutrient under consideration. The indirect method of estimating digestibility relies on the use of markers. A marker is usually an indigestible material introduced in small quantities and distributed evenly in the test diet, or it may be an indigestible component of the diet itself. These are known as external and internal markers respectively. Since it is indigestible, the marker will concentrate in the faeces relative to the digestible material and the relative quantities will provide a measure of the digestibility of the diet or its nutrient components.

Clearly, a marker should not influence the digestive physiology of the experimental animal; it should move along the gut at the same rate as the rest of the food material and it should not be toxic. Digestibility is estimated from the determined values as follows:

% DMD or % TD =
100 – 100(% marker in diet/% marker in feed)

% ND (protein, lipid, etc.) =
100 – [100 × (% marker in diet/% marker in faeces) × (% nutrient in faeces/% nutrient in feed)]

where DMD is the dry matter digestibility, TD is total digestibility and ND is nutrient digestibility.

These digestibility estimates are only apparent digestibility estimations since the faeces are contaminated with endogenous material. Unless a correction factor is introduced and/or allowance made for such material, the true digestibility of a feed and/or an ingredient is not being estimated. Therefore, they are correctly expressed as percentage apparent dry matter and/or total digestibility and percentage apparent nutrient digestibility. In normal practice, for the evaluation of diet and ingredient suitability, the estimation of apparent total and/or nutrient digestibility is sufficient.

8.4.2 *Ingredient digestibility*

In feed formulation and manufacture, it is essential to have knowledge of the digestibility of the main ingredients of a diet, as well as of the whole diet. It may appear that the ingredient in question could be treated as a 'diet' and the method described above used to determine its digestibility. However, this is not desirable for many reasons. For example, the ingredient may behave differently according to whether it is provided alone or as a component of a compound diet. Alternatively, the ingredient may not be accepted (ingested) by itself.

The procedure that is presently accepted for estimating digestibility of an ingredient is to use a diet for which apparent total and nutrient digestibilities are known for the particular species being investigated. This diet is known as the reference diet. A test diet is prepared by mixing 15–30% of the ingredient to be tested with the reference diet and apparent total and nutrient digestibility of the test diet is determined (De Silva & Anderson, 1995). The apparent total and/or nutrient digestibility of the ingredient is estimated using the following equation [assuming that the test diet (T) has been made up in a ratio of 8: 2 of the reference diet (R) to the ingredient].

Apparent total digestibility of the ingredient = [(100/20) × DMD of T] – [(80/100) × DMD of R]

Apparent total digestibility of the ingredient = [(100/20) × ND of T] – [(80/100) × ND of R]

Commonly used external markers (i.e. those introduced into the diet) are chromic oxide (Cr_2O_3), ferrous oxide (FeO), silicon dioxide (SiO_2), polypropylene and ytterbium chloride (YbCl). By far the most commonly used of these in fish is Cr_2O_3. YbCl provides an interesting alternative as it can be included at very low levels in the diet (0.01%) compared with Cr_2O_3 (0.1%) and is therefore less likely to interfere with other digestive processes. However, analysis of YbCl requires mass spectrometry, which increases expense. Differential movement of Cr_2O_3 has been observed in crustaceans, and so external markers may be inappropriate for digestibility studies in those animals.

Endogenous or internal markers commonly utilised for digestibility estimations are crude fibre (CF), hydrolysis-resistant organic matter (HROM), and hydrolysis-resistant ash (HRA). Occasionally ash has been used, particularly with respect to digestibility estimations of naturally ingested food material. CF and HROM refer to the same group of materials; cellulose and chitin (when present) are the chief constituents of HROM, and cellulose and lignin are the major components of the CF fraction. HRA is the fraction of mineral ash resistant to acid digestion.

Having determined the digestibility of a feedstuff, it is then possible to consider how it can satisfy the energetic and nutritional requirements of the culture species.

8.5 Energy

A major function of animal feeds is to provide energy to the animal. Energy is required for the chemical reactions that build new tissues, maintaining osmotic balance, moving food through the digestive tract, respiration, reproduction, locomotion, etc. Animals obtain the energy they require from food or, in periods when they are deprived of food, from body stores.

Energy is not a nutrient itself, but is present in the chemical bonds that hold the molecules in the nutrients together. There are many different types of bonds, each containing a different amount of energy. Accordingly, the amounts of energy in the various nutrients that make up a feed are of great importance. In addition, the capacity of different species to utilise the energy contained in different nutrients varies considerably. For example, some species are able to use carbohydrates as a significant energy source, whereas others utilise carbohydrates poorly and rely to a greater extent on protein for energy.

A number of terms are commonly used in discussions of energetics and it is necessary to define these.

(1) Gross energy (*E*) is the energy that is released as heat when a substance is completely oxidised to carbon dioxide, nitrous oxide or water.
(2) Consumed energy or intake energy (*C*) is the gross energy consumed by an animal in its food. The majority of intake energy is present in the form of carbohydrate, protein or lipid.
(3) Retained energy (*P*) is that portion of the energy contained in the food that is retained as part of the body or voided as a useful product such as gametes.
(4) Faecal energy (*F*) is the gross energy of the faeces. Faeces consist of undigested food and metabolic products that may include sloughed

gut epithelial cells, digestive enzymes and excretory products.

(5) Excretory energy (*U*) is the energy contained in all excretory products, including compounds absorbed from the food but not utilised, the energy of waste products of metabolic processes such as ammonia, the compounds excreted through the gills and the energy lost from the surface or skin.

(6) Respiratory or metabolic energy (*R*) is all energy associated with the various metabolic processes. This includes basal metabolism required for those activities that are necessary to maintain the life of the animal (i.e. cellular activity, respiration and blood circulation), the energy of locomotion, metabolism associated with waste formation and excretion, and feeding metabolism associated with the processing of food. The last parameter is termed heat increment (H_iE) or specific dynamic action (SDA) and is appreciable when expressed in terms of the energy content of the food ingested, being about 6% for sucrose, 13% for fats and 30% for proteins (Gordon *et al.*, 1977). This is particularly relevant in the case of fish and crustaceans, which rely heavily on protein as an energy source.

The major aim of nutrition in aquaculture is to maximise *P* and minimise all other energy losses in a cost-effective manner while maintaining an acceptable body composition of the end product.

The energy balance of an organism is described as:

$$C = F + U + R + P \quad (1)$$

This expression contains the parameters that can be readily and accurately measured by experimentation. The values of *C* and *F* are determined directly by bomb calorimetry. The value of *U* is obtained either by measuring ammonia and urea excreted and converting to equivalent energy values using 23.05 J/mg of ammonia and 24.85 J/mg of urea (De Silva & Anderson, 1995) or by using the model of Brett & Groves (1979) relating energy lost in excretion to energy consumed as:

$$U = 0.07C$$

The value of *R* is determined using a respirometer and converting to an equivalent energy value using 13.56 J/mg of oxygen (Elliott & Davison, 1975). These values, however, have been determined in studies on salmonids and may be inaccurate for other species of cultured animals, particularly crustaceans and molluscs.

Additional parameters are commonly used in the analysis of the energy flows in aquaculture species. The energy within a food that is digested (the apparent digested energy, ADE) is determined as the energy in the feed minus the energy in faeces (i.e. ADE = $C - F$). It is generally assumed that the fish absorbs all ADE.

The partitioning of energy between the various parameters described by the terms on the right-hand side of equation (1) is affected by a variety of environmental and metabolic processes. A number of factors affect basal metabolic rate (*R*), most of which are related to an animal's environment. Maintaining these parameters at the optimal level for the target species ensures minimal energy expenditure for maintenance. These are issues of system management and animal husbandry rather than nutrition (section 3.3).

8.6 Energetics and feeding

Food intake is regulated by dietary energy. Feeding adequate amounts of energy is necessary for the economical production of aquatic organisms, while feeding excess energy will result in obesity and deterioration of flesh quality. Growth involves the laying down of muscle, fat, and epithelial and connective tissue. The proportion of protein or fat deposited in these tissues is highly dependent upon the composition of the diet and the presence of essential nutrients. Essential nutrients are those that are required for metabolic processes but which cannot be synthesised by the animal from ordinarily available dietary precursors or which are synthesised at an insufficient rate. By definition, essential nutrients must be supplied in the diet. For example, for protein synthesis to occur, the correct ratio of essential amino acids (EAAs) must be provided. The EAA requirements of aquatic animals vary between species. For any given protein source, essential amino acids will be limiting (i.e. available in insufficient quantities from the diet to satisfy the requirement for protein synthesis). The most deficient amino acid is known as the first limiting amino acid. Protein synthesis will continue until the first limiting amino acid is completely consumed. Use by the animals of limiting amino acids is not 100% efficient, for some will be used to provide energy. The process of protein synthesis has an energy demand on top of the basal metabolic rate. Unless a form of non-protein energy is provided in the diet, an animal has to direct amino acids into energy-liberating pathways in order to provide the energy needed for basal metabolism and growth, and this will limit the scope for growth. Alternative energy sources that can be included in the diet to meet these needs are carbohydrate and lipid.

Inclusion of non-protein energy sources in diets allows the formulator to reduce the protein content of the diet; however, the ability to utilise carbohydrate and lipid for energy varies between species. Lipids contain more energy per unit weight than any other dietary component (Table 8.2) and they are used efficiently by fish as energy sources. Lipids increase the palatability of feeds (up to a point), and assist in reducing dust and stabilising the pellets during manufacture, transportation and storage. As a result, lipid is an excellent source of non-protein energy for fish. However, this is not the case for crustaceans, which are unable to efficiently utilise diets with more than about 10% lipid. As a result of their inability to utilise lipid, carbohydrate is the major alternative source of energy (Cuzon & Guillaume, 1997). Carbohydrate is utilised less efficiently than lipid as an energy source by fish, but it is much less expensive than alternative sources of energy for fish diets. There is considerable variation in the recommended optimal levels of carbohydrate that can be included in an aquatic animal feed. This level varies according to species, from less than 12% in rainbow trout (Phillips *et al.*, 1948) to as high as 33% for channel catfish (Wilson, 1991). Similar variations are seen in carbohydrate utilisation by different species of crustaceans (Shiau, 1997).

Table 8.2 Energy content of the major dietary nutrients

Nutrient	Gross energy (MJ/kg)	Digestible energy (MJ/kg)
Carbohydrate	17.15	8.37–12.55
Protein	23.01	15.90–17.78
Lipid	38.07	33.47

8.7 Protein

8.7.1 Protein digestion

Proteolytic enzymes (proteases) digest proteins and arise from inactive precursors known as zymogens. Proteases break down the peptide links of protein molecules. Different enzymes are capable of acting on peptide bonds.

- Exopeptidases act on peptide bonds at the end of a protein molecule.
- Endopeptidases act at a point within the molecule.

Endopeptidases are very specific in their action and exert their effect at a particular point within the protein molecule. Most protein digestion in fish occurs in the stomach as a result of the action of pepsin, which is most active in acidic solutions. It is believed that stomach acidity also causes lysis of plant cell walls in macrophyte-feeding fish such as tilapia (Bowen, 1976). In stomachless animals, the role of pepsin is taken by alkaline proteases, which are most active in alkaline environments. Activity of digestive enzymes has been shown to be correlated to nutrient composition. For example, reduced protease activity has been reported to result from a decrease in the proportion of fish meal (protein) in the diet (Kawai & Ikeda, 1972).

8.7.2 Protein as an energy source

Although there is considerable interaction between protein, carbohydrate and lipid as energy sources, each is dealt with separately here, as there are some important differences between them. For the purposes of this discussion, it is assumed that the composition of protein provides a balanced mix of the EAAs. This assumption makes it possible to ignore any growth-limiting effects associated with a restricted supply of EAAs.

Since dietary protein is the largest and most expensive component of aquaculture diets, nutritionists aim to formulate diets in which the energy required by the animal is provided by non-protein sources to the maximum extent. This will allow the majority of the protein in the diet to be directed towards protein synthesis. The ability to spare proteins for growth and utilise other nutrients for energy is commonly referred to as the 'protein-sparing capability'. The protein-sparing capability differs between species in both the extent to which protein is spared and the energy source used for the sparing. For example, lipid is the energy source used for sparing in salmonids (Hardy, 2000), whereas carbohydrate is used in anguillid eels (Hidalgo *et al.*, 1993). An imbalance in the ratio of dietary protein to other energy sources will, however, lead to either wasted protein or the production of lower-value fatty animals. Optimal dietary protein–energy (P/E) ratios provide a measure of the amount of energy required by an animal to allow most efficient use of protein for growth. The values that have been reported for fish average about 22 mg/kJ protein. However, values range from 17 mg/kJ for tilapia (De Silva *et al.*, 1989) to 28.7 mg/kJ for channel cat fish (Reis *et al.*, 1989). Optimum dietary protein–energy ratios also vary between species of crustaceans and with energy source. For example, the optimum P/E ratio varies from 20 mg/kJ to 36 mg/kJ for *Penaeus monodon,* depending upon the source of carbohydrate, and from 21 mg/kJ to 38 mg/kJ for other penaeids (Cuzon & Guillaume, 1997).

The optimum dietary P/E ratio also varies within a species according to the digestibility and amino acid composition of the protein source. Optimal P/E ratios are also affected by water temperature (Hidalgo & Alliot, 1988) and are likely to be affected by other environmental parameters that affect the partitioning of energy.

8.7.3 Protein sources in the diet

The assumption, made earlier, that all protein is balanced for amino acids is a considerable simplification. Although many workers discuss protein requirements, it is more appropriate to consider that aquatic animals have a requirement for a well-balanced mixture of essential and non-essential amino acids. Table 8.3 shows the estimated protein requirements for juvenile fish of a variety of species, and the growth responses of cultured carps to a range of dietary protein levels is shown in Fig. 8.2. The protein requirements range from 30% to 56% of the diet. These values can be considered to be estimated levels of protein required in diets, for most studies, sometimes incorrectly, have assumed that all amino acids are available for metabolism and protein synthesis.

Fish meal has been the traditional protein source in diets for aquaculture species. Fish meal is expensive, and efforts to replace it with cheaper protein sources (e.g. soybean meal and cottonseed meal) have provided a large contribution to the literature in recent times. In general, most alternative protein sources can replace fish meal to some extent. A number of factors affect the proportion of fish meal that can be replaced, and these depend upon the nature of the protein source. In some grain meals, the essential amino acid composition is not adequate, while many others contain anti-nutritional factors (section 9.10.5). Despite these limitations, there can often be considerable cost advantage in replacing some fish meal with alternative protein sources. In spite of considerable research effort, with the exception of soybean meal, which is often incorporated into pelleted feeds, other agricultural by-products are rarely used in commercial feeds.

In the light of recent findings on bovine spongiform encephalopathy (BSE) and other transmittable diseases from animal meals, the use of rendered meat products, such as blood meal, bone meal and meat meal, in aquaculture has been prohibited in the European Union and the USA. Consequently, there has been an increase in attempts to explore the use of aquatic food product waste in aquaculture feeds, and the relevant aspects have been discussed by Gunasekera *et al.* (2002).

Another factor influencing selection of protein sources in aquatic animal feeds is the release of nutrients from uneaten food and faeces into the environment. These nutrients, particularly nitrogen and phosphorus, stimulate natural productivity and may lead to eutrophication (section 4.2.1). To prevent this, the nutritionist seeks to formulate feeds that are well utilised by the animal. Thus, efficient protein utilisation (to limit nitrogen in the effluent) and phosphorus utilisation (discussed further below) are important. Efficient protein utilisation is best achieved by a clear understanding of the amino acid requirements of the target species.

Table 8.3 Estimated protein requirements of juvenile fish

Species	Protein source(s)	Estimated requirements (%)
Channel catfish (*Ictalurus punctatus*)	Whole egg protein	32–36
Common carp (*Cyprinus carpio*)	Casein	38
Estuary grouper (*Epinephelus salmoides*)	Tuna muscle meal	40–50
Gilthead bream (*Pagrus aurata*)	Casein, FPC* and amino acids	40
Grass carp (*Ctenopharyngodon idella*)	Casein	41–43
Japanese eel (*Anguilla japonica*)	Casein and amino acids	44.5
Largemouth bass (*Micropterus salmoides*)	Casein and FPC	40
Milkfish (fry) (*Chanos chanos*)	Casein	40
Plaice (*Pleuronectes platessa*)	Cod muscle	50
Puffer fish (*Fugu rubripes*)	Casein	50
Red sea bream (*Chrysophrys major*)	Casein	55
Salmonids		
Chinook salmon (*Oncorhynchus tshawytscha*)	Casein, gelatin, and amino acids	40
Coho salmon (*Oncorhynchus kisutch*)	Casein	40
Rainbow trout (*Oncorhynchus mykiss*)	Fish meal	40
	Casein and gelatin	40
	Casein, gelatin and amino acids	45
Sockeye salmon (*Oncorhynchus nerka*)	Casein, gelatin and amino acids	45
Smallmouth bass (*Micropterus dolomieui*)	Casein and FPC	45
Snakehead (*Channa micropeltes*)	Fish meal	52
Striped bass (*Morone saxatilis*)	Fish meal and SP†	47
Tilapias		
Tilapia aurea (fry)	Casein and egg albumin	56
Tilapia aurea (juvenile)	Casein and egg albumin	34
Tilapia mossambica	White fish meal	40
Tilapia nilotica	Casein	30
Tilapia zillii	Casein	35
Yellowtail (*Seriola quinqueradiata*)	Sand eel and fish meal	55

*Fish protein concentrate.
†Soy proteinate.
Reproduced from Wilson (1989) with permission from Academic Press.

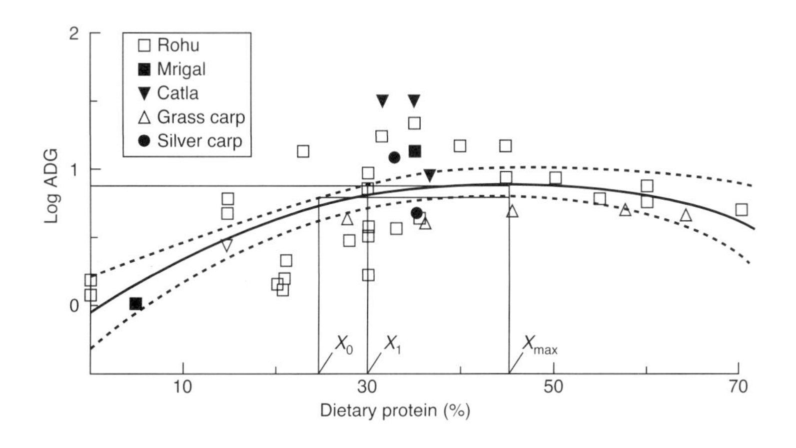

Fig. 8.2 Growth response of cultured major carps to dietary protein level and the line of best fit depicting the relationship between these parameters. X_{max}, the dietary protein level when maximum growth occurs; X_1, optimal dietary protein level; X_0, economically optimal dietary protein level. ADG, average daily gain. Reprinted from De Silva & Gunasekera (1991) with permission from Elsevier Science.

8.7.4 Amino acids

Amino acids conform to a particular general structure:

$$-NH_2CHRCOOH-$$

where R is any one of a number of organic side-chains having some or all of carbon, hydrogen, oxygen, nitrogen and sulphur atoms (Table 8.4). Some of the side-chains are quite complex, forming ring structures. Amino acids can be divided into two major groups: those that are used in protein synthesis and those that have other functions. The 20 standard amino acids found in proteins can be divided further into the essential and the non-essential amino acids (Table 8.5).

Amino acids suffer one of three fates in an animal:

- used for protein synthesis
- subjected to structural change to produce other compounds
- degraded for energy production by deamination, resulting in the production of ammonia (Fig. 8.3).

8.7.5 Amino acid synthesis

Many amino acids are synthesised by pathways that are present only in plants and micro-organisms. The EAAs required by animals for protein synthesis must therefore be obtained in their diets. The remaining amino acids, those that can be synthesised by the animal, are non-essential amino acids (Table 8.5). These requirements appear to be common to fish and crustaceans. Molluscs, however, require proline as an essential amino acid (Knauer & Southgate, 1999). Meeting the requirement for EAAs is not straightforward; some EAAs are readily converted to non-essential acids if these are lacking. As a result, the dietary requirement for an EAA often reflects the requirement for it and at least part of the requirement for its non-essential relatives. Two good examples of this are methionine and cysteine, and phenylalanine and tyrosine. Part of the degradative pathway of methionine involves the production of cysteine. Therefore, when considering the dietary requirement for methionine, the requirement for cysteine must also be considered. The inclusion of cysteine in the diet can alleviate some, but not all, of the methionine requirement. Similarly, tyrosine is readily produced from phenylalanine and the dietary requirement for phenylalanine may include an amount that can be used to produce tyrosine. As was the case above, inclusion of tyrosine in the diet can alleviate some, but not all, of the requirement for phenylalanine. The relationship between phenylalanine and tyrosine is closer than that between methionine and cysteine, for cysteine can be synthesised from molecules other than methionine, whereas tyrosine can only be produced by modification of phenylalanine.

All the non-essential amino acids except tyrosine can be synthesised by simple pathways leading from common metabolic intermediates that are readily produced by the citric acid cycle.

8.7.6 Amino acid requirements

A reliable technique for determining the essentiality of amino acids uses a ^{14}C-labelled precursor (e.g. glucose), which is injected into the target animal. The radiolabel is incorporated into those amino acids that the animal is able to synthesise. All other amino acids (not radiolabelled) are considered essential and are therefore required in the diet. This method identifies the essentiality of an amino acid, but not the dietary level required.

A number of techniques have been used to determine the dietary EAA requirements of aquatic animals. The most common method measures the growth of animals fed diets containing graded levels of a given EAA ranging from deficiency to excess. The diet with the lowest level of the EAA, but producing the maximum or near-maximum level of growth, is considered to contain the minimum requirement for that particular EAA.

Another common method determines the dietary level of a given amino acid at which the blood concentration of that amino acid starts to rise. With increasing levels of the EAA up to sufficiency, no change in the plasma level of that amino acid is observed. However, with greater than sufficient levels of the amino acid, there is a proportional increase in plasma concentrations. This method is accurate because most enzymes responsible for catabolising amino acids are not increased by increasing dietary amino acid concentrations (Wilson, 1989).

Table 8.4 The structures of the 20 amino acids found in proteins

Name	Structural formula*	Name	Structural formula*
Alanine	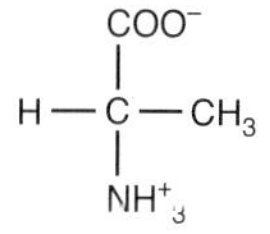 COO^- $H-C-CH_3$ NH^+_3	Leucine	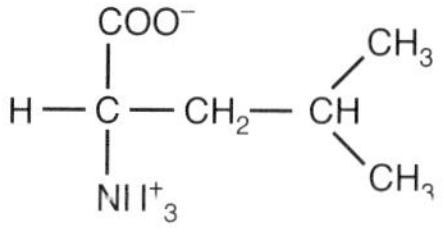 COO^- CH_3 $H-C-CH_2-CH$ NH^+_3 CH_3
Arginine	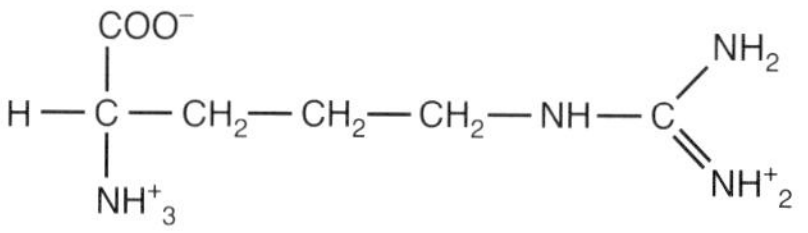 COO^- NH_2 $H-C-CH_2-CH_2-CH_2-NH-C$ NH^+_3 NH^+_2	Lysine	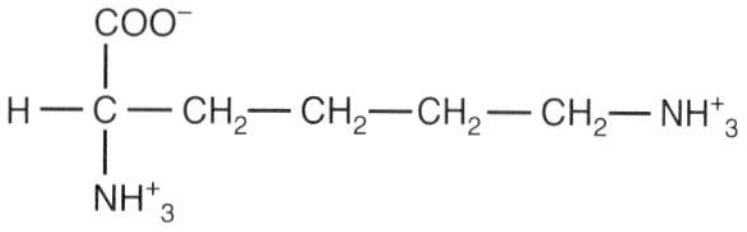 COO^- $H-C-CH_2-CH_2-CH_2-CH_2-NH^+_3$ NH^+_3
Asparagine	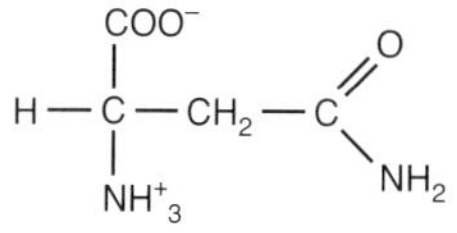 COO^- O $H-C-CH_2-C$ NH^+_3 NH_2	Methionine	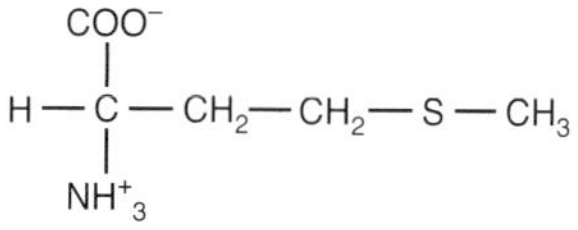 COO^- $H-C-CH_2-CH_2-S-CH_3$ NH^+_3
Aspartic acid	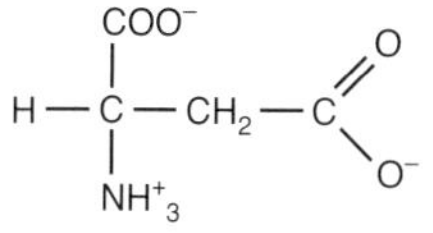 COO^- O $H-C-CH_2-C$ NH^+_3 O^-	Phenylalanine	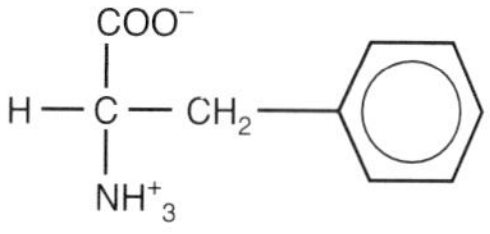 COO^- $H-C-CH_2-$(benzene ring) NH^+_3
Cysteine	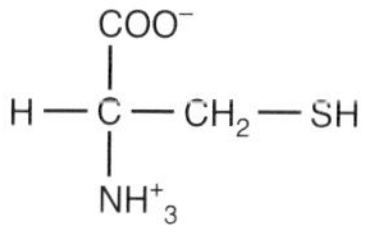 COO^- $H-C-CH_2-SH$ NH^+_3	Proline	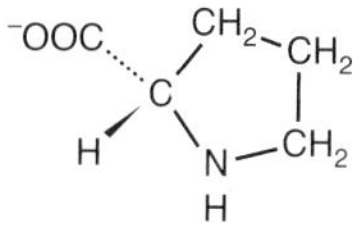 ^-OOC CH_2-CH_2 C H N CH_2 H
Glutamic acid	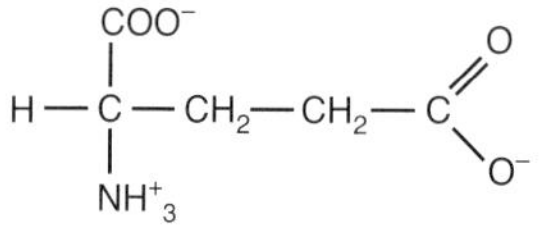 COO^- O $H-C-CH_2-CH_2-C$ NH^+_3 O^-	Serine	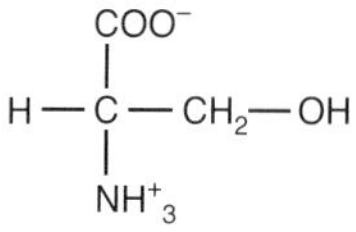 COO^- $H-C-CH_2-OH$ NH^+_3
Glutamine	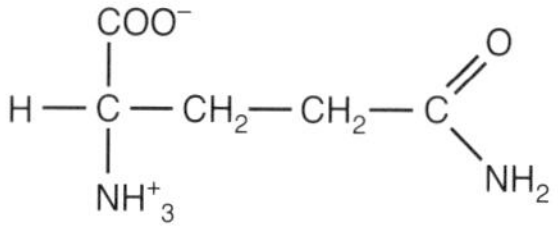 COO^- O $H-C-CH_2-CH_2-C$ NH^+_3 NH_2	Threonine	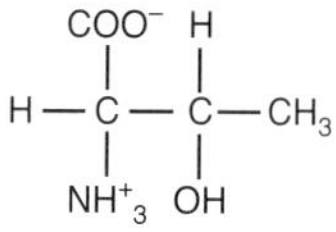 COO^- H $H-C-C-CH_3$ NH^+_3 OH
Glycine	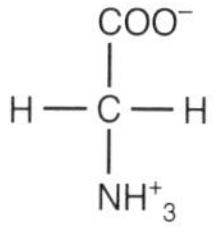 COO^- $H-C-H$ NH^+_3	Tryptophan	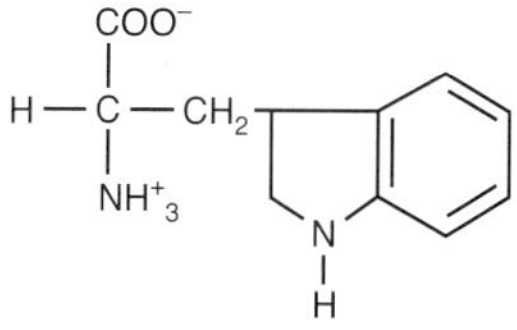 COO^- $H-C-CH_2-$(indole ring) NH^+_3 N H
Histidine	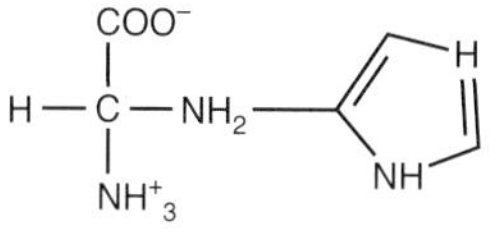 COO^- H $H-C-NH_2-$(imidazole ring) NH^+_3 NH	Tyrosine	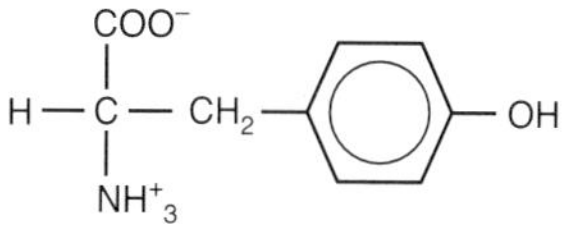 COO^- $H-C-CH_2-$(benzene ring)$-OH$ NH^+_3
Isoleucine	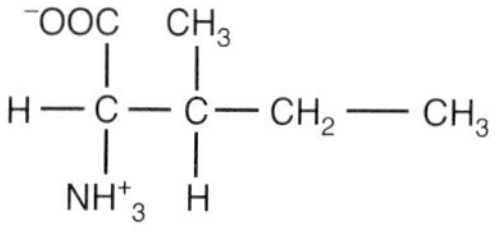 ^-OOC CH_3 $H-C-C-CH_2-CH_3$ NH^+_3 H	Valine	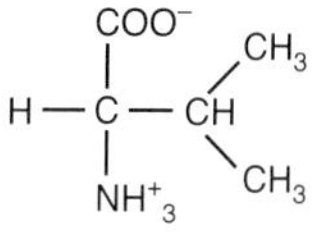 COO^- CH_3 $H-C-CH$ NH^+_3 CH_3

*The ionic forms shown are those predominating at pH 7.0.
Adapted from De Silva & Anderson (1995) with permission from Kluwer Academic Publishers.

Table 8.5 Essential and non-essential amino acids in aquatic animals

Essential	Non-essential
Arginine	Alanine
Histidine	Asparagine
Isoleucine	Aspartate
Leucine	Cysteine
Lysine	Glutamate
Methionine	Glutamine
Phenylalanine	Glycine
Threonine	Proline*
Tryptophan	Serine
Valine	Tyrosine

*Proline is an essential amino acid for molluscs.

Table 8.6 Essential amino acid requirements of fish and crustaceans

Amino acid	Requirement (g per 100 g of protein)
Arginine	3.3–5.9
Histidine	1.3–2.1
Isoleucine	2.0–4.0
Leucine	2.8–5.4
Lysine	4.1–6.1
Methionine*	2.2–6.5
Phenylalanine†	5.0–6.5
Threonine	2.0–4.0
Tryptophan	0.3–1.4
Valine	2.3–4.0

*Requirement varies depending upon the amount of cysteine in the diet.
†Requirement varies depending upon the amount of tyrosine in the diet.
Data from De Silva & Anderson (1995) and Guillaume (1997) with permission from the World Aquaculture Society.

The detailed and accurate determination of amino acid requirements for any species is a lengthy procedure and requires considerable experimentation. A technique that has been applied to assist in the initial formulation of diets is to identify the proportion of each EAA in the body tissue of the species in question, commonly referred to as the 'essential amino acid index'. Comparison with the requirements for EAAs indicates that this is generally a good measure of the relative requirements for EAAs. Table 8.6 summarises the range of EAA requirements that have been determined for a number of fish and crustaceans. The requirements for individual amino acids are fairly consistent between species, although variability is apparent between both species and studies on the same species. A large part of the variability may be explained by differences in the methods used. An example of this is the arginine requirement reported for rainbow trout, which has been determined at 3.3% of the protein (Kaushik, 1979) and 5.9% of the protein (Ketola, 1983). Both studies used semi-purified diets, which contained a mixture of purified (e.g. an amino acid) and whole ingredients (e.g. fish meal). Kaushik (1979) used a diet containing zein, fish meal and free amino acids as the amino acid source, whereas Ketola (1983) used a diet containing corn gluten meal and free amino acids as the amino acid source. Failure to determine the comparative digestibility of these protein sources or a variation in the amino acid balance might have resulted in the variable data obtained. Although there is no alternative, the use of free amino acids in test diets may confound results in such studies as they may leach from the diet between feeding and ingestion. There is also the possibility of differential uptake of free amino acids, which appear in the blood sooner than those bound in proteins. This leads to differential metabolism between free and protein-bound amino acids.

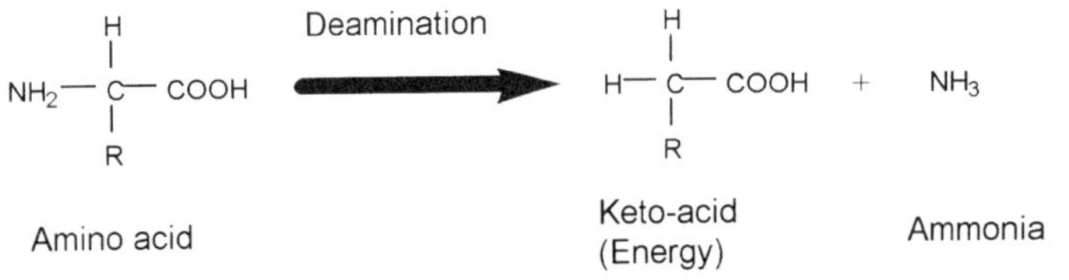

$$NH_2\text{–CH(R)–COOH} \xrightarrow{\text{Deamination}} H\text{–CH(R)–COOH} + NH_3$$

Amino acid → Keto-acid (Energy) + Ammonia

Fig. 8.3 Deamination of an amino acid produces ammonia, which is excreted, and a carbon skeleton (keto acid) subsequently used to provide energy.

8.8 Carbohydrates

8.8.1 Carbohydrate structure

Carbohydrates are essential components of all living organisms, with roles as readily metabolised energy stores, as molecules which facilitate transfer of energy throughout the organism, and as structural components. The basic units of carbohydrates are monosaccharides (e.g. glucose), which are synthesised as a result of a process called gluconeogenesis or are the products of photosynthesis. Monosaccharides are also found linked together in chains to form polymer complexes, which fall into two distinct groups.

- Oligosaccharides consist of only a few linked monosaccharide units.
- Polysaccharides consist of many covalently linked monosaccharide units and are extremely large.

Oligosaccharides are often associated with proteins, forming glycoproteins, and with lipids, forming glycolipids. Polysaccharides usually function as structural molecules (e.g. cellulose in the cell walls of plants or chitin in the exoskeleton of invertebrates) or as dense high-energy storage molecules such as starch in plants and glycogen in animals. The physical characteristics of a polysaccharide are determined by the monosaccharides it is composed of and the way that they are joined together. The links are described as either α, which result in a loose structure that is easily degraded by enzymes, or as β, which produce a very flat, layered structure resistant to degradation. There are numerous examples in all organisms of carbohydrate-digesting enzymes (carbohydrases) capable of hydrolysing α-linked carbohydrates (α-carbohydrases). However, carbohydrases capable of hydrolysing the β-linkage in cellulose occur naturally only in bacteria. Consequently, α-linked sugars are much more useful as carbohydrate sources for aquaculture diets than β-linked sugars.

The most important carbohydrates in aquaculture nutrition are starch, chitin, sucrose and cellulose, and these are found in the greatest amounts in aquaculture feeds.

Starch is a complex of many α-linked glucose molecules and is a useful component of aquaculture diets since it is generally readily available to most aquatic animals. The accessibility of starch does vary between species, however, and the maximum inclusion level needs to be determined for each. Chitin is common in the exoskeleton of invertebrates. It is particularly important in the diets of larval and juvenile fish, and it is usual practice to feed prey such as brine shrimp (*Artemia* species), *Daphnia* and other small crustaceans to larval fish (section 9.4). Chitinase, the enzyme that degrades chitin, has been described in the digestive tracts of a number of fish (Lindsay & Gooday, 1985) and so, presumably, chitin is available to these animals. The efficiency, however, with which it is digested is unknown. Glycogen is an important polysaccharide in metabolism and similar in structure to starch. Both function as energy stores: glycogen in animals and starch in plants.

8.8.2 Carbohydrate digestion

A large number of different carbohydrases, each with very specific actions, are present in the gut of aquatic animals. Carbohydrases have been found in pancreatic juice, in the stomach, in the intestine and in the bile, but not necessarily at all sites in all species investigated. In most species, however, the pancreas is the main producer of carbohydrases. Carbohydrase activity responds to the level of dietary carbohydrate. In general, carbohydrase, and in particular amylase, activity differs from species to species, and appears to be related to feeding habits.

8.8.3 Carbohydrate metabolism

After digestion, carbohydrates are absorbed through the wall of the digestive tract. They are absorbed and enter the bloodstream as monosaccharides (i.e. glucose, fructose, galactose, etc.). In vertebrates, the absorbed nutrients are transported to the liver, where the initial processes of metabolism occur. However, all the processes of metabolism that are to be described here occur in all tissues. Of the monosaccharides, glucose is the most important, acting as a major metabolic energy source circulating in all vertebrates. Some tissues (e.g. brain) use only glucose as an energy source, and so maintenance of blood glucose levels is a very important process.

Four reaction pathways exist:

- glycolysis – the breakdown of glucose-liberating energy

- gluconeogenesis – the synthesis of glucose from other molecules
- glycogen synthesis – the storage of excess glucose in glycogen molecules
- glycogenolysis – the breakdown of glycogen to provide free glucose.

There is an apparent relationship between the natural diet of a species in the wild and its capacity to deal with dietary carbohydrate. Carnivorous fish such as yellowtail (*Seriola quinqueradiata*) respond less well to glucose tolerance tests than omnivores, such as common carp (*Cyprinus carpio*), which clear blood glucose more rapidly (Furuichi & Yone, 1981) but slower than mammals. This information indicates that fish are unable to metabolise glucose quickly. When fish are presented with diets high in carbohydrate, the excess glucose appears to be used to synthesise glycogen (Palmer & Ryman, 1972).

Although the enzymes required for glycogen degradation are present in at least some fish species, their role in maintaining a constant blood glucose level between bouts of feeding seems to be minimal in both eels (*Anguilla anguilla*) (Dave *et al.*, 1975) and Pacific salmon (*Oncorhynchus nerka*) (French *et al.*, 1983). Furthermore, extensive periods of starvation did not result in the depletion of muscle glycogen in these species. Glycogen has been shown to decline during prolonged starvation in the hepatopancreatic tissues of carp (Nagai & Ikeda, 1971) and in the liver of golden perch (*Macquaria ambigua*) (Collins & Anderson, 1995).

Murat *et al.* (1978) showed that the gluconeogenic pathway is more important in maintaining blood glucose levels in carp than the breakdown of glycogen. From this information it is apparent that carbohydrate in aquaculture diets has to be carefully controlled, for the excess deposited as glycogen is subsequently less readily available to the fish as an energy source than other stores. Feeding studies by Furuichi & Yone (1980) showed that optimal levels of dietary carbohydrate are:

- 30–40% for carp
- ~20% for red sea bream
- 10% for yellowtail.

When incorporating carbohydrate into diets for aquatic animals, it is preferable to utilise a carbohydrate that requires some degree of digestion, such as starch, rather than monosaccharides, such as glucose. This will at least allow a time lag between consumption of the carbohydrate and the appearance of glucose and other monosaccharides in the blood. The resultant slower increase in the plasma of the animal will result in a greater degree of catabolism of these substrates for energy. This preference for complex carbohydrates is one of the reasons that sucrose has not been widely used in aquatic animal diets.

8.9 Lipids

The main function of lipids in animals is either as high-energy storage molecules or as components of cell membranes. There are five major classes of lipids:

- fatty acids
- triglycerides
- phospholipids
- sterols
- sphingolipids.

8.9.1 Lipid digestion

There are two types of secretions that are important in lipid digestion. These are bile and lipases. Bile is produced in the liver, stored in the gall bladder (if present) and released when food arrives in the intestine. Bile emulsifies the fats, breaking large fat droplets into very small droplets, thereby increasing the surface area and making them more accessible to fat-splitting enzymes. All fat-digesting enzymes are classed as lipolytic enzymes or lipases.

The origin of lipase activity varies between species. In fish, it has been found in extracts of the pancreas and upper digestive tract. A major difference between fat hydrolysis and that of proteins or carbohydrates is that lipases show relatively little substrate specificity. Thus, the progressive breakdown of fat through various intermediate stages is often catalysed by a single lipase, and there is not a precise succession of different enzymes as there is in proteolysis. All fat-digesting enzymes act in alkaline media, the pH optimum being slightly variable from group to group.

Table 8.7 Some of the common biological fatty acids

Symbol*	Common name	Systematic name	Structure
Saturated fatty acids			
16:0	Palmitic acid	Hexadecanoic acid	$CH_3(CH_2)_{14}COOH$
18:0	Stearic acid	Octadecanoic acid	$CH_3(CH_2)_{16}COOH$
20:0	Arachidic acid	Eicosanoic acid	$CH_3(CH_2)_{18}COOH$
22:0	Behenic acid	Docosanoic acid	$CH_3(CH_2)_{20}COOH$
24:0	Lignoceric acid	Tetracosanoic acid	$CH_3(CH_2)_{22}COOH$
Unsaturated fatty acids			
16:1	Palmitoleic acid	9-Hexadecenoic acid	$CH_3(CH_2)_5CH{=}CH(CH_2)_7COOH$
18:1	Oleic acid	9-Octadecenoic acid	$CH_3(CH_2)_7CH{=}CH(CH_2)_7COOH$
18:2	Linoleic acid	9,12-Octadecadienoic acid	$CH_3(CH_2)_4(CH{=}CHCH_2)_2(CH_2)_6COOH$
18:3	α-Linolenic acid	9,12,15-Octadecatrienoic acid	$CH_3CH_2(CH{=}CHCH_2)_3(CH_2)_6COOH$
18:3	γ-Linolenic acid	6,9,12-Octadecatrienoic acid	$CH_3(CH_2)_4(CH{=}CHCH_2)_3(CH_2)_3COOH$
20:4	Arachidonic acid	5,8,11,14-Eicosatetraenoic acid	$CH_3(CH_2)_4(CH{=}CHCH_2)_4(CH_2)_2COOH$
20:5	EPA	5,8,11,14,17-Eicosapentaenoic acid	$CH_3CH_2(CH{=}CHCH_2)_5(CH_2)_2COOH$
24:1	Nervonic acid	15-Tetracosenoic acid	$CH_3(CH_2)_7CH{=}CH(CH_2)_{13}COOH$

*Ratio of the number of carbon atoms to the number of double bonds.

8.9.2 Lipid metabolism

Fatty acids are carboxylic acids with long-chain hydrocarbon side-groups. The more common biological fatty acids are listed in Table 8.7. The nomenclature of fatty acids follows a particular convention. This is:

$$C_{x:yn\text{-}z}$$

where x denotes the number of carbon atoms, y denotes the number of double bonds in the chain and z denotes the carbon atom at which the first double bond appears numbering from the non-carboxyl (COOH) end of the molecule. For example, α-linolenic acid is 18:3n-3 and γ-linolenic acid is 18:3n-6 (Table 8.7). In popular jargon n is often replaced by ω (omega): hence ω-3 and ω-6 fatty acids.

Saturated fatty acids are those without double bonds, whereas monounsaturated fatty acids have one double bond and polyunsaturated fatty acids (PUFAs) have more than one double bond. PUFAs with four or more double bonds are sometimes referred to as highly unsaturated fatty acids or HUFAs. In higher plants and animals, the predominant fatty acid residues are those of the C_{16} and C_{18} species, of which half are unsaturated and are often polyunsaturated.

Fatty acids rarely occur free in nature, but more generally occur in an esterified form, called triacylglycerols or triglycerides, so named as they are triesters of glycerol composed of three fatty acids joined to a glycerol 'backbone' (Fig. 8.4). Triacylglycerols are an efficient form in which to store metabolic energy, mainly because they are less oxidised than carbohydrates or proteins and hence yield significantly more energy on oxidation. As they are not soluble in water, they are stored in an anhydrous form, but glycogen binds about twice its weight of water under physiological conditions. Hence, a given weight of lipid (mixture of triacylglycerols) provides about six times the metabolic energy of an equal weight of hydrated glycogen.

Another class of lipids is the phospholipids (also called glycerophospholipids), which are the major lipid components of biological membranes. These molecules have a similar structure to triacylglycerols, except that one of the fatty acid chains is replaced by a phosphate attached to another organic molecule. The fourth major class of lipids is the sterols, identified by their four fused carbon rings to which a variety of side-chains are attached to give them their individual characteristics. The most notable sterol is cholesterol, which is a major component of animal

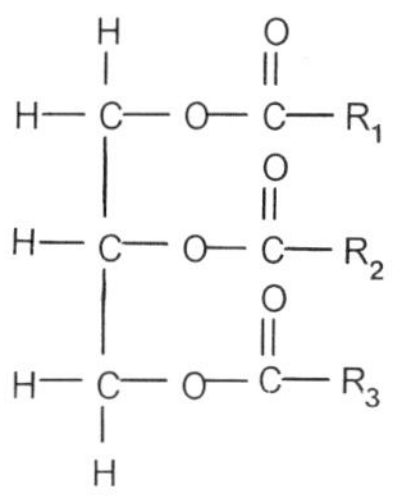

```
    H       O
    |       ||
H — C — O — C — R1
    |
    |       O
    |       ||
H — C — O — C — R2
    |
    |       O
    |       ||
H — C — O — C — R3
    |
    H
```

Fig. 8.4 General structure of triacylglycerol, where R_1, R_2 and R_3 are fatty acids.

cell membranes and a precursor of steroid hormones and bile acids.

8.9.3 Lipid as an energy source

Lipid is digested and metabolised with greater relative ease than carbohydrate and so serves as a much better source of energy for protein sparing. Again, though, too much lipid can be included in the diet, which results in production of 'fatty' fish and in growth retardation in crustaceans. The protein-sparing effect of lipid varies between species but appears to be optimal at about 15–18% of the diet in fish (Lie *et al.*, 1988; De Silva *et al.*, 1991), particularly when coupled with low dietary protein levels (De Silva *et al.*, 1991). The optimal dietary lipid level for crustaceans appears to be less than 10% (D'Abramo, 1997).

8.9.4 Lipid catabolism

Lipids are transported in the bloodstream either as lipoprotein complexes called very low-density lipoproteins (VLDLs) or as very small droplets called chylomicrons. Complexing the lipids to proteins allows otherwise insoluble lipid components to be maintained in aqueous solution. The triacyglycerol components of VLDLs and chylomicrons are hydrolysed to free fatty acids and glycerol in the target tissues, generally adipose tissue and skeletal muscle. Fatty acid oxidation then liberates the energy contained in the fatty acid.

8.9.5 Lipid biosynthesis

Fatty acid biosynthesis occurs through the linkage of C_2 units. The end product is palmitic acid (16:0), which is the precursor of longer-chain saturated and unsaturated fatty acids produced through the actions of enzymes called elongases and desaturases. Elongases lengthen the fatty acid in a step-wise manner, adding two carbon units at a time. Unsaturated fatty acids are produced by terminal desaturases. A variety of unsaturated fatty acids may be synthesised by combinations of elongations and desaturation reactions. It is generally considered that there are at least three terminal desaturases in vertebrates. Desaturation in animals does not occur at positions beyond C9. Since palmitic acid is the shortest available fatty acid in animals, the formation of the double bond at C12 of linoleic acid (18:2*n*-6), or at C15 of α-linolenic acid (18:3*n*-3), requires an enzyme found only in plants and some micro-organisms. Therefore *n*-3 or *n*-6 fatty acids required by animals must be obtained from the diet and are termed essential fatty acids. The *n*-3 and *n*-6 fatty acids play very important roles in the proper functioning of animals, particularly in providing membrane fluidity and acting as precursors of some important hormones.

The regulation of fatty acid metabolism is controlled by two peptide hormones:

- insulin, which stimulates the synthesis of fatty acids
- glucagon, which stimulates their degradation.

These hormones are secreted by the pancreatic tissue, depending upon the levels of circulating metabolites and hence the overall energy status of the animal.

Animals can synthesise most other lipid classes from precursors found in the cell. Cholesterol is synthesised by fish and molluscs by a process that utilises acetate as a substrate. Because of this, there is no dietary requirement for cholesterol in these animals. Crustaceans, however, do not possess the enzymes to conduct these reactions and therefore require cholesterol in their diet. Other lipid classes, triglycerides, phospholipids and sphingolipids, utilise fatty acids and various other molecules as substrates and thus the dietary requirement is for fatty acids.

8.9.6 Dietary lipid requirements

Apart from satisfying the requirements of an aquatic animal for essential lipids (fatty acids and sterols), dietary lipid acts as a source of energy. In general, a 10–20% level of lipid in fish diets gives optimal growth rates without producing an excessively fatty carcass (Cowey & Sargent, 1979). Interspecific variation in the ability of different species to utilise lipid as a source of energy is prevalent. For example, when rainbow trout were fed diets with lipid levels from 5% to 20% and protein contents of 16% to 48%, the optimum ratio of protein to lipid was found to be 35% protein to 18% lipid (Takeuchi *et al.*, 1978a,b). However, carp fed diets with a fixed protein level of 32%, with lipid varying from 5% to 15% and with corresponding decreases in carbohydrate, did not show increased growth or food conversion rates (Takeuchi *et al.*, 1979).

Fasting fish often utilise lipid reserves as an energy source in preference to protein and carbohydrate. A study of Coho salmon indicated that this occurred initially as an increase in the rate of lipid breakdown, since the activities of lipid-synthesising enzymes remained unchanged for a period of 2 days (Lin *et al.*, 1977a,b). However, significant decreases in lipid synthesis were noticeable after 23 days of food deprivation. There is also evidence that lipid mobilisation in starved fish is selective, with shorter (C_{18} and C_{16}) and less saturated fatty acids being mobilised first (Sargent *et al.*, 1989). However, again interspecific variation predominates.

8.9.7 Essential fatty acid requirements

Aquatic animals possess a range of elongase and desaturase enzymes capable of modifying dietary fatty acids and the products of endogenous fatty acid biosynthesis. They do not possess the desaturases necessary for converting 18:3*n*-3 and 18:2*n*-6 from 18:1*n*-9. Consequently, all *n*-3 and *n*-6 PUFAs in aquatic animal lipids originate from *n*-3 and *n*-6 PUFAs consumed in the diet and are therefore essential fatty acids (EFAs). All fish studied to date appear to require 18:3*n*-3 at about 1–2% of the diet by dry weight. This requirement can be reduced by feeding longer-chain *n*-3 PUFAs, such as 20:5*n*-3, 22:5*n*-3 or 22:6*n*-3, to approximately 0.5% of the diet by dry weight. Larval fish, crustaceans and marine fish species are unable to efficiently or sufficiently elongate C_{18} fatty acids and thus require long-chain 20(*n*-3) or 22(*n*-3) fatty acids as essential inclusions in their diet (Sargent *et al.*, 1989). In these cases, diets should include 1–2% long-chain PUFAs.

The requirements for *n*-6 PUFAs have been less well studied, but these fatty acids are also required in the diet. Generally, dietary *n*-3:*n*-6 ratios should be between 1 and 2.5, depending upon the species. Again, in crustaceans and marine fish species that have limited capacity to elongate and desaturate, these fatty acids should be provided as long-chain fatty acids. In contrast, most freshwater species are capable of chain elongation and desaturation of 18:3*n*-3 and 18:2*n*-6, the two biologically active base fatty acids, to the corresponding HUFAs. Therefore HUFAs need not be provided in the diet.

Determining the requirement for individual fatty acids is difficult because aquatic animals require very small amounts of essential fatty acids, and the likelihood of other ingredients used in test diets containing PUFAs of unknown composition is high. It is therefore extremely difficult to be certain that a known amount of PUFA is being fed. Care must be taken not to provide excess short-chain PUFAs.

These requirements have significance to the preparation of aquaculture diets. As meat meals from terrestrial herbivorous ruminant contain *n*-3 PUFAs in very low amounts and the seed oils of terrestrial plants are rich in short-chain *n*-3 and *n*-6 PUFAs, they do not satisfy the general requirement of cultured marine species for long-chain *n*-3 PUFAs. As such, aquaculture is reliant upon the lipids found in those aquatic organisms with a high proportion of long-chain *n*-3 PUFAs for provision of the *n*-3 PUFA requirements of cultured animals.

8.9.8 Phospholipids and sterols

Crustaceans are unique in aquatic animal nutrition in that they are unable to synthesise sterols. Sterols are the substrates for the synthesis of cholesterol and many hormones, and are therefore required in the diets of crustaceans. This sterol requirement is usually satisfied by cholesterol in aquaculture diets, because it is nutritionally superior to other sterols (Table 9.11). Cholesterol is required at levels ranging from 0.12% for *Macrobrachium rosenbergii* to 2.0% for *Homarus americanus* juveniles (Teshima, 1997).

Both larval fish and crustaceans show improved growth and survival when fed a diet incorporating phospholipids. Experiments to determine phospholipid requirements usually use a product such as soybean lecithin, which is a mixture of neutral and polar phospholipids. Additional phospholipids normally improve growth and survival in larval fish, and improve growth and enhance rates of metamorphosis in larval crustaceans. Inclusion of refined soybean lecithin at between 0.68% and 3% has been described (Teshima, 1997).

8.10 Vitamins

Vitamins are organic molecules that act as cofactors or substrates in some metabolic reactions. They are generally required in relatively small amounts in the diet and, if present in inadequate amounts, may result in nutrition-related disease, poor growth or increased susceptibility to infections (Fig. 8.5). Vitamins can be divided into two groups: the fat-soluble vitamins, A, D, E and K, and the water-soluble vitamins, which are the B group, vitamin C and some specific cofactors.

The vitamin requirements of the vast majority of fish and shellfish in culture have not been determined. Therefore, the data obtained from studies of salmonids, carp or catfish are usually applied to other species of fish, and those of penaeid shrimp applied to all crustaceans. Most practical diets (diets formulated and manufactured from ingredients that are readily available) include vitamins at the levels published by the National Research Council (1983) for warm-water fish, unless there is evidence of a different requirement that has been obtained from specific studies. Measured vitamin requirements of different species vary considerably (Table 8.8). Too much vitamin in the diet not only increases the cost of the feed unnecessarily, but may also result in vitaminosis. Vitaminosis is usually caused by excess amounts of fat-soluble vitamins, which are difficult for an animal to excrete.

Fig. 8.5 A radiograph of a juvenile barramundi (*Lates calcarifer*) showing spinal deformity resulting from dietary vitamin deficiency (photograph by Ms Helen Clifton).

Determining vitamin requirements in fish, as in other animals, is difficult, because many vitamins are produced by micro-organisms in the gut. Depending upon the activity of an animal's gut microflora, the requirement for dietary vitamins can vary. It is likely that all vitamins are required by all aquatic animals, but whether it is necessary to include them in a formulated diet is a different matter. A species such as common carp, *Cyprinus carpio*, can obtain many nutrients from decaying organic matter, which it will consume in addition to any artificial diet in all but the most sterile of conditions. As such, the requirements for vitamins in supplementary feeds for this species are less than for a species that has a carnivorous feeding habit and relies entirely upon the components of an artificial diet for its vitamin intake. Similarly, animals held in tanks have less opportunity to consume natural sources of vitamins than those held in ponds and so require greater levels of vitamins in any artificial feed that they may be given.

As stated above, the requirement for vitamins by different species varies greatly according to their usual feeding habit and their capacity to synthesise them. The vitamin requirements for some of the fish that have been studied are shown in Table 8.8. Variations in the vitamins required and the dietary requirement are apparent from Table 8.8. This variation may be due to differences in experimental design, but is more likely to be due to variations between species. Vitamins are contained in the fresh products from which aquaculture diets are made. However, the processing method and time can destroy these compounds, a problem not easily resolved (section 9.10.4, Table 9.12). It is therefore very difficult to determine a generally recommended level of vitamin supplementation that will satisfy all species.

Inclusion of vitamins in aquaculture diets is complicated further since most of them are highly labile molecules and are readily destroyed during processing. It is easy to overcome this problem for the water-soluble vitamins by adding excess amounts,

Table 8.8 Summary of the vitamin requirements of various species of fish

	Sea bass	Atlantic salmon	Ayu	Channel catfish	Common carp	Eel	Turbot	Pacific salmon	Rainbow trout	Red drum	Sturgeon	Yellowtail	Tilapia
Thiamine	R		12	1	R	R	0.6–2.6	10–15	40			1[illegible].2	
Riboflavin	R	R	40	9	7–14	R		20–25	9			1[illegible].0	
Pyridoxine	5	5	12	3	5–6	R	1.0–2.5	15–20	9			1[illegible].7	
Pantothenic acid	R	R	50	15	30–50	R		40–50	40			35.9	
Nicotinic acid			100	14	28	R		150–200	300			12.0	
Biotin			0.3	R	1	R		1–1.5	0.4			0.67	
Inositol	R		400	NR	440	R		300–400	510			423	
Choline			350	400	4000	R		600–800	11 100		1700–3200	2920	
Folic acid			3	R		R		6–10	21			1.2	
Ascorbic acid	700	50	300	60	R	R	R	100–150	400	60–75	NR	122	R
Vitamin A			10 000 IU	1000–2000 IU	10 000 IU			2000–2500 IU	7000 IU			5.68	
Vitamin D			2000 IU	250–500 IU					3000 IU			NR	
Vitamin E	R	35	100 IU	50 IU	100 IU	200		30 IU	200			119.0	50–100
Vitamin K		R	10 IU	R					50			NR	
B_{12}				R		R		0.015–0.02	0.21				NR

Values are mg/kg diet unless stated otherwise.
IU, international units; NR, not required; R, required.
Reproduced from De Silva & Anderson (1995) with permission from Kluwer Academic Publishers.

because any excess consumed is readily excreted. However, excessive fat-soluble vitamins accumulate in the body and may cause vitaminosis or vitamin poisoning. Adding large amounts of vitamins, even those that are water soluble, has the additional disadvantage of the extra expense, which is undesirable. The common practice of basing inclusion of vitamins on known requirements of other species generally means using data from rainbow trout for coldwater fish, from catfish or carp for warmwater fish and from penaeids for crustaceans. This strategy probably provides a compromise between determining the exact requirement experimentally and adding vitamins to excess.

8.11 Minerals

Minerals, or inorganic elements, are needed by animals to maintain many of their metabolic processes and to provide material for major structural elements (e.g. skeleton). Not all elements that are used in metabolism are required in an animal's diet. Aquatic animals, particularly marine animals, have an advantage in that their surrounding medium contains many of the elements needed for growth and survival.

The minerals required for normal metabolism can be divided into two groups:

(1) Major minerals are required in large quantities and include calcium, phosphorus, magnesium, sodium, potassium, chlorine and sulphur.
(2) Trace minerals are those required in trace amounts and include iron, iodine, manganese, copper, cobalt, zinc, selenium, molybdenum, fluorine, aluminium, nickel, vanadium, silicon, tin and chromium.

The roles of some important minerals are shown in Table 8.9.

In general, the dietary requirements for minerals are poorly understood because of the difficulty in devising mineral-deficient diets and the need to deplete tissue mineral stores. Often, generalisations from studies of higher vertebrates must be made. The known mineral requirements of fish are shown in Table 8.10. Animal meals are rich sources of iron, magnesium, zinc, iodine, selenium, calcium

Table 8.9 Some important minerals and their functions

Mineral	Functions
Calcium	Component of skeleton, scales, teeth, exoskeleton, etc. Roles in physiological processes including metabolism, nerve and muscle function and osmoregulation
Phosphorus	Component of bones and scales of fish and exoskeleton of crustaceans Roles in metabolic reactions Constituent of many important molecules such as ATP and phospholipids
Sodium, potassium, chlorine	Appropriate levels of these ions needed for proper functioning of cells, in maintaining ion gradients between the inside and outside of cells and for maintaining nerve function
Iron	Constituent of haemoglobin and cytochromes (proteins) important in energy metabolism
Magnesium	Component of skeletal tissue Important cofactor in a number of metabolic reactions Important in maintaining muscle tone
Manganese	Important cofactor in a number of metabolic reactions Important in maintaining proper nerve cell function
Copper, zinc	Important components of a number of metalloenzymes involved in a wide variety of metabolic processes. Approximately 20 different enzymes have been found to contain zinc
Iodine	Important component of thyroid hormones; important in growth regulation
Selenium	Constitutes an integral part of the enzyme glutathione peroxidase Imparts protective effect against toxicity of heavy metals
Cobalt	Component of vitamin B_{12}
Chromium	Important in normal carbohydrate and lipid metabolism
Sulphur	Required for the synthesis of the amino acid cysteine

Table 8.10 Summary of the mineral requirements of various species of fish

	Sea bass	Atlantic salmon	Channel catfish	Common carp	Eel	Sole	Pacific salmon	Rainbow trout	Red drum	Tilapia
General	2% UPS XII*									
Phosphorus		13 000	4500	6000–7000	2500–3200	7000	R	6000	8600	9000
Manganese		20	2.40	13			R	20		12
Copper		6	5	3				5		3–4
Iron		73	30	150	170			60		
Selenium		R	0.25				R	0.3		
Iodine and Fluorine		R					0.6–1.1 mg/kg			
Potassium		NR					8 000			
Keto-carotenoids		R								
Calcium			4500		2700					6500
Magnesium			400	400–500	400			500		590–770
Zinc			20	15–30			R	30	20–25	10
Cobalt				0.1						

*Proprietary mineral mix.
Values are mg/kg diet unless stated otherwise.
R, required, NR, not required. Reprinted from De Silva & Anderson (1995) with permission from Kluwer Academic Publishers.

and phosphorus, and plant meals contain adequate amounts of manganese and copper. Selenium, sodium, potassium, chlorine and magnesium are also commonly found dissolved in water and can be absorbed by animals through gills and other body surfaces.

8.12 Determining the quality of a diet

The end point of the efforts of nutritionists is the development of a suitable feed formulation (Fig. 8.6). The quality of a feed is determined by how well it satisfies the nutrient requirements of the target species. Not only must the feed contain the correct proportions of nutrients but the nutrients must be digested and absorbed in a form that makes them available to provide energy and substrates for growth. This is termed bioavailability. Digestibility of food is the primary determinant of bioavailability. However, digestibility varies with species, the source of the nutrient, the temperature at which it is evaluated and often between two samples of exactly the same feedstuff that are treated in different ways (e.g. different drying heats) (Pfeffer *et al.*, 1991). These factors make it difficult to relate the data obtained by separate groups of workers, or even by one group of workers at different times. Nevertheless, digestibility remains the most widely used method of determining how much of a given food component is bioavailable.

Fig. 8.6 Aquaculture feeds are formulated to satisfy the nutritional requirements of the target species and maximise growth rates for the lowest cost. Commercially produced pelleted aquaculture feeds with species-specific formulations are now widely used (photograph by Dr Darryl E. Jory).

It is important to note that the digestibility of carbohydrates, lipids and proteins contained within a single food source, for instance soya meal, vary from one another (Table 8.11). This influences the proportion of the energy requirements of a culture animal that are met by the different components of a feed (DE). An alternative term, metabolisable energy (ME), more closely estimates the energy available to the culture animal for growth and metabolism, because it allows for energy lost in urine and gill excretions. Retained energy (*P*) is easily measured by determining the energy in the whole carcass of the culture animal. This term takes account of the digestibility of the diet and is equal to metabolisable energy minus energy lost during metabolism and activity. Retained energy is a measure of the energy used to increase body mass. Fig. 8.7 shows the relationship between DE, ME and *P*.

There is some debate about the relative value of DE, ME and *P* for the evaluation of the quality of aquaculture feeds. In practice, and despite its relatively simple determination, *P* is rarely used as it gives little information about why one feed may be better than another. ME is also rarely used because determination of the energy contained in urine and gill excretions involves technical difficulties and assumptions, and so DE serves as the main determinant of the value of feeds.

Once formulated, the nutritionist also needs to be able to evaluate how the formulation performs in providing the necessary nutrients and energy for growth. A number of parameters are routinely determined by aquaculturists to determine feed quality. These include food conversion ratios (FCRs), also known as utilisation efficiency, and protein efficiency ratios (PERs).

FCR = mass of food consumed (dry)/increase in mass of animal produced (wet)

Generally, FCRs range between 1.2 and 1.5 for animals fed carefully prepared diets, but values as low as 0.8 have been reported in fish.

PER = increase in mass of animal produced (wet wt)/mass of protein in fed (dry wt)

Table 8.11 Digestibility of different components of feed by the African catfish, *Clarius gariepinus*

Energy in diet (kJ/g)	Apparent protein digestibility (%)	Apparent fat digestibility (%)	Apparent energy digestibility (%)
8.4	94.0	90.5	50.0
12.4	95.5	94.0	65.0
16.8	86.5	73.0	60.0

Reprinted from Machiels & Henken (1985) with permission from Elsevier Science.

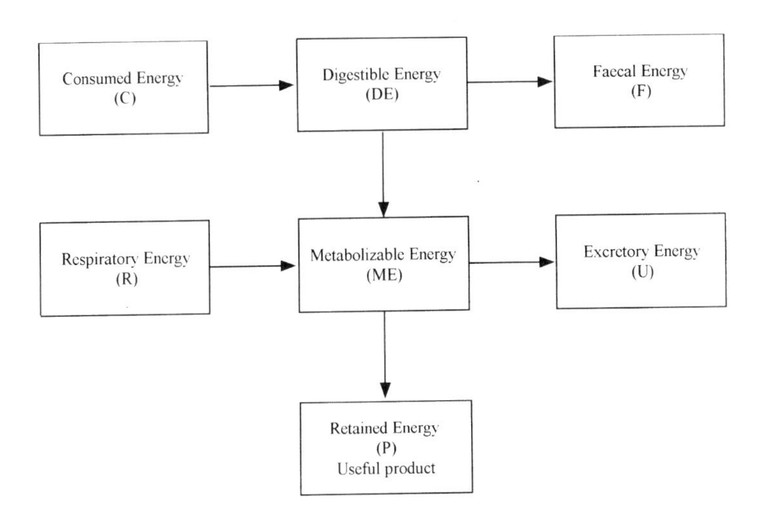

Fig. 8.7 Energy flow in an aquatic organism showing the relationship between digestible energy (DE), metabolisable energy (ME) and retained energy (*P*).

PER gives a measure of how well the protein source in the diet provides for the essential amino acid requirement of the animal and how well the diet is balanced for energy and protein. The protein efficiency ratio measures the deposition of fat as well as protein, which means that diets producing fatty fish can be associated with high PER values. Because of this, net protein utilisation (NPU), also known as protein retention (% PPV), is a better measure of the feed quality than protein efficiency ratios (Lie, *et al.*, 1988). It is calculated by the formula:

%NPU = [protein gain in fish (g)/protein intake in food (g)] × 100

8.13 Future directions

Our understanding of aquatic animal nutrition has come a long way in the last 25 years, but much remains to be discovered. Three significant gaps in our knowledge can be readily identified: broodstock nutrition, larval nutrition and the requirements for 'environmentally friendly' aquaculture diets.

Many aquaculture hatcheries continue to rely on 'trash fish' for broodstock nutrition. This may be crudely supplemented with a vitamin mix but our understanding of the nutrient requirements of broodstock is limited (Izquierdo *et al.*, 2001). Improved understanding of the relationships between nutrient supply, gonad maturation and the quality of the gametes and larvae produced will allow us to formulate specialist broodstock diets that improve hatchery productivity. Such diets already exist for some well-established species (e.g. salmonids), for which we have a good understanding of broodstock nutrition.

Similarly, aquaculture hatcheries still rely on live larval feeds such as rotifers and *Artemia* (section 9.4). However, these add significantly to the cost of operations and the size of hatcheries. The development of successful artificial diets to replace live feeds will require a greater understanding of the nutritional requirements of the larval stages of culture species. Achieving this will also require improved production methods for larval diets to maximise bioavailibility (section 9.7).

Many of the feed formulations used in

aquaculture have high levels of fish meal, which is rich in protein. Fish meal also contains phosphorus and this limits biovailability. Environmental concerns arise from this (Thoman *et al.*, 1999). The first is that faeces and uneaten food used in aquaculture often release large amounts of nitrogen and phosphorus into the environment. These nutrients may lead to eutrophication of the environment and consequent environmental degradation (section 4.2.1). Second, the continued use of fish meal in animal feeds is unsustainable in the medium to long term as the supply of these materials is limited. Thus, research to identify alternative protein sources, to develop methods to improve bioavailability of nutrients and to accurately describe the protein requirements of each species is necessary to reduce environmental impacts and improve the sustainability of aquaculture.

These areas provide significant new challenges to aquatic animal nutritionists and exciting opportunities for careers in aquaculture nutrition.

References

Anderson, T. A. (1986). The histological and cytological structure of the gastrointestinal tract of luderick, *Girella tricuspidata* (Pisces: Kyphosidae), in relation to diet. *Journal of Morphology*, **90**, 109–19.

Bowen, S. H. (1976). Mechanisms for digestion of detrital bacteria by the cichlid fish *Sarotherodon mossambicus* (Peters). *Nature*, **260**, 137–8.

Brett, J. R. & Groves T. D. D. (1979). Physiological energetics. In: *Fish Physiology*, Vol. 8 (Ed. by W. S. Hoar, D. J. Randall & J. R. Brett), pp. 279–352. Academic Press, New York.

Collins, A. L. and Anderson, T. A. (1995). The regulation of endogenous energy stores during starvation and refeeding in the somatic tissues of the golden perch, *Macquaria ambigua* (Percichthyidae) Richardson. *Journal of Fish Biology*, **47**, 1004–15.

Cowey, C. B. & Sargent, J. R. (1979). Nutrition. In: *Fish Physiology*, Vol. 8 (Ed. by W. S. Hoar, D. J. Randall, and J. R. Brett), pp. 1–69. Academic Press, New York.

Cuzon, G. & Guillaume, J. (1997). Energy and protein: energy ratio. In: *Crustacean Nutrition. Advances in World Aquaculture*, Vol. 6 (Ed. by L. R. D'Abramo, D. Conklin, & D. M. Akiyama), pp. 51–70. World Aquaculture Society, Baton Rouge, LA.

D'Abramo, L. R. (1997). Triacylglycerols and fatty acids. In: *Crustacean Nutrition. Advances in World Aquaculture*, Vol. 6 (Ed. by L. R. D'Abramo, D. Conklin, & D. M. Akiyama), pp. 71–84. World Aquaculture Society, Baton Rouge, LA.

Dave, G., Johanssen-Sjobeck, M. L., Larson, A., Lewander, K. & Lidman, U. (1975). Metabolic and hematological effects of starvation in the European eel, *Anguilla anguilla* L. 1. Carbohydrate, lipid, protein and inorganic ion metabolism. *Comparative Biochemistry and Physiology*, **52A**, 423–30.

De Silva, S. S. & Anderson, T. A. (1995). *Nutrition in Aquaculture*. Chapman & Hall, London.

De Silva, S. S. & Gunasekera, R. M. (1991). An evaluation of the growth of Indian and Chinese major carps in relation to dietary protein content. *Aquaculture*, **92**, 237–41.

De Silva, S. S., Gunasekera, R. M. & Atapattu, D. (1989). The dietary protein requirements of young tilapia and an evaluation of the least cost dietary protein levels. *Aquaculture*, **80**, 271–84.

De Silva, S. S., Gunasekera, R. M. & Shim, K. F. (1991). Interactions of varying dietary protein and lipid levels in young red tilapia: evidence of protein sparing. *Aquaculture*, **95**, 305–18.

Elliott, J. M. & Davison, W. (1975). Energy equivalents of oxygen consumption in animal energetics. *Oecologia*, **19**, 195–201.

French, C. J., Hochachka, P. W., & Mommsen, T. P. (1983). Metabolic organisation of liver during spawning migration of sockeye salmon. *American Journal of Physiology*, **245**, 827–30.

Furuichi, M. & Yone, Y. (1980). Effect of dietary dextrin levels on the growth and feed efficiency, chemical composition of liver and also muscle and the absorption of dietary protein and dextrin in fishes. *Bulletin of the Japanese Society of Scientific Fisheries*, **46**, 225–9.

Furuichi, M. & Yone, Y. (1981). Change of blood sugar and plasma insulin levels of fishes in glucose tolerance tests. *Bulletin of the Japanese Society of Scientific Fisheries*, **47**, 761–4.

Gordon, M. S., Bartholomew, C. A., Grinnell, A. D., Jorgensen, C. B. & White, F. N. (1977). *Animal Physiology: Principles and Adaptations*, 3rd edn. Macmillan, New York.

Guillaume, J. (1997). Protein and amino acids. In: *Crustacean Nutrition. Advances in World Aquaculture*, Vol. 6 (Ed. by L. R. D'Abramo, D. Conklin, & D. M. Akiyama), pp. 26–50. World Aquaculture Society, Baton Rouge, LA.

Gunasekera, R.M., Turoczy, N.J., De Silva, S.S., Gavine, F. & Gooley, G.J. (2002). An evaluation of the potential suitability of selected aquatic food processing industry waste products in feeds for three fish species based on similarity of chemical constituents. *Journal of Aquatic Food Product Technology*, **11**, 57–78.

Hardy, R. W. (2000). Advances in the development of low-pollution feeds for salmonids. *Global Aquaculture Advocate*, **3**, 63–7.

Hidalgo, F. & Alliot, E. (1988). Influence of water temperature on protein requirement and protein utilization in juvenile sea bass, *Dicentrarchus labrax*. *Aquaculture*, **72**, 115–29.

Hidalgo, M. C., Sanz, A., Garcia-Gallego, M. G., Suarez, M. D. & de la Higuera, M. (1993). Feeding of the European eel (*Anguilla anguilla*). I. Influence of dietary carbohydrate level. *Comparative Biochemistry and Physiology*, **105A**, 165–9.

Izquierdo, M. S, Fernandez-Palacios, H. & Tacon, A. G. J. (2001). Effect of broodstock nutrition on reproductive performance of fish. *Aquaculture*, **197**, 25–42.

Kaushik, S. (1979). Application of a biochemical method for the estimation of amino acid needs in fish: quantitative arginine requirements of rainbow trout in different salinities. In: *Proceedings of World Symposium on Finfish Nutrition Technology*, 20–23 June, Hamburg Vol. 1. pp. 197–207. Heenemann, Berlin.

Kaushik, S. J.(1993) Preface. In: *Fish Nutrition in Practice. Proceedings of the IV International Symposium on Fish Nutrition and Feeding*, 24–27 June, 1991, Biarritz, France. (Ed. by Kaushik, S.J. & Luquet, P.), pp. 15–16. INRA Les Colloques, n° 61. Institut National De La Recherche Agronomique, Paris.

Kawai, S. & Ikeda, S. (1972). Effects of dietary change on the activities of digestive enzymes in carp intestine. *Bulletin of Japanese Society of Scientific Fisheries*, **38**, 265–70.

Ketola, H. G. (1983). Requirement for dietary lysine and arginine by fry of rainbow trout. *Journal of Animal Science*, **56**, 101–7.

Knauer, J. & Southgate, P. C. (1999). A review of the nutritional requirements of bivalves and the development of alternative and artificial diets for bivalve aquaculture. *Reviews in Fisheries Science*, **7**, 241–80.

Lie, O., Lied, E. & Lambertsen, G. (1988). Feed optimization in Atlantic cod (*Gadus morhua*): fat versus protein content in the feed. *Aquaculture*, **69**, 333–41.

Lin, H., Romsos, D. R., Tack, P. I. & Leveille, G. A. (1977a). Influence of dietary lipid on lipogenic enzyme activities in coho salmon, *Oncorhynchus kisutch* (Walbaum). *Journal of Nutrition*, **107**, 846–54.

Lin, H., Romsos, D. R., Tack, P. I. & Leveille, G. A. (1977b). Effects of fasting and feeding various diets on hepatic lipogenic enzyme activities in coho salmon (*Oncorhynchus kisutch* (Walbaum)). *Journal of Nutrition*, **107**, 1477–83.

Lindsay, G. J. H. & Gooday, G. W. (1985). Chitinolytic enzymes and the bacterial microflora in the digestive tract of cod, *Gadus morhua*. *Journal of Fish Biology*, **26**, 255–65.

Machiels, M. A. M. & Henken, A. M. (1985). Growth rate, feed utilization and energy metabolism of the African catfish, *Clarias gariepinus* (Burchell 1822), as affected by the dietary protein and energy content. *Aquaculture*, **44**, 271–84.

Murat, J .C., Castilla, C. & Paris, H. (1978). Inhibition of gluconeogenesis and glucagon-induced hypoglycemia in carp (*Cyprinus carpio* L.). *General and Comparative Endocrinology*, **34**, 243–50.

Nagai, M. & Ikeda, S. (1971). Carbohydrate metabolism in fish. I. Effects of starvation and dietary composition on the blood glucose level and the hepatopancreatic glycogen and lipid contents in carp. *Bulletin of Japanese Society of Scientific Fisheries*, **37**, 404–9.

National Research Council (1983). *Nutrient Requirements of Warmwater Fishes and Shellfishes*. National Academy Press, Washington, DC.

Palmer, T. N. & Ryman, B. E. (1972). Studies on oral glucose intolerance in fish. *Journal of Fish Biology*, **4**, 311–19.

Pfeffer, E., Beckmann-Toussaint, J., Henrichfreise, B. & Jansen, H. D. (1991) Effect of extrusion on efficiency of utilization of maize starch by rainbow trout (*Oncorhynchus mykiss*). *Aquaculture*, **96**, 293–303.

Phillips, A. M., Tunison, A. V. & Brockway, D. R. (1948). *The Utilization of Carbohydrate by Trout*. Fisheries Research Bulletin, No. 11, Conservation Department Bureau of Fish Culture, New York.

Reis, L. M., Reutebuch, E. M. & Lovell, R. T. (1989). Protein-to-energy ratios in production diets and growth, feed conversion and body composition of channel catfish, *Ictalurus punctatus*. *Aquaculture*, **77**, 21–7.

Sargent, J., Henderson, R. J. & Tocher, D. R. (1989). The lipids. In: *Fish Nutrition* (Ed. by J. E. Halver), pp. 154–219. Academic Press, London.

Shiau, S-Y. (1997). Carbohydrates and fibre. In: *Crustacean Nutrition. Advances in World Aquaculture*, Vol. 6 (Ed. by D'Abramo, L. R., Conklin, D. & Akiyama, D. M.), pp. 108–22. World Aquaculture Society, Baton Rouge, LA.

Takeuchi, T., Watanabe, T. & Ogino, C. (1978a). Supplementary effect of lipids in a high protein diet of rainbow trout. *Bulletin of Japanese Society of Scientific Fisheries*, **44**, 677–81.

Takeuchi, T., Watanabe, T. & Ogino, C. (1978b). Optimum ratio of protein to lipid in diets of rainbow trout. *Bulletin of Japanese Society of Scientific Fisheries*, **44**, 683–8.

Takeuchi, T., Watanabe, T. & Ogino, C. (1979). Availability of carbohydrate and lipid as dietary energy source for carp. *Bulletin of Japanese Society of Scientific Fisheries*, **45**, 977–82.

Teshima, S. (1997). Phospholipids and sterols. In: *Crustacean Nutrition. Advances in World Aquaculture*, Vol. 6 (Ed. by L. R. D'Abramo, D. E. Conklin, & D. M. Akiyama), pp. 85–107. World Aquaculture Society, Baton Rouge, LA.

Thoman, E. S., Davis, D. A. & Arnold, C. R. (1999). Evaluation of growout diets with varying protein and energy levels for red drum (*Sciaenops ocellatus*). *Aquaculture*, **176**, 343–53.

Wilson, R. P. (1989). Amino acids and proteins. In: *Fish Nutrition* (Ed. by Halver, J. E.), pp. 112–51. Academic Press, New York.

Wilson, R. P. (1991). *Handbook of Nutrient Requirements of Finfish*. CRC Press, Boca Raton, FL.

9

Feeds and Feed Production

Paul Southgate

9.1 Introduction

> Among the most important considerations in aquaculture management are improving growth rate and feeding efficiencies to the extent possible.
>
> Kiyoshi Kakazu (1987)

Feed is one of the major costs for aquaculture operations, typically making up between 30% and 60% of the total operating costs, depending on the intensity of the operation. Clearly, the suitability of the food used, the efficiency with which it is utilised for growth by the culture animals and the feeding practices used are major factors determining the profitability of an aquaculture operation.

There are two general types of aquaculture feeds:

(1) Hatchery feeds usually consist of live organisms such as microalgae, rotifers and brine shrimp, which need to be cultured on-site at the hatchery.
(2) Nursery and grow-out feeds are artificially formulated to satisfy the known nutritional requirement of the culture animal.

Grow-out feeds are usually manufactured as pellets at specialised facilities away from the grow-out site. Clearly, the development of effective aquaculture feeds for hatchery and grow-out requires an understanding of the nutritional requirements of cultured aquatic animals (Chapter 8). As more information on the specific nutritional requirements of aquaculture species becomes available through research, more efficient aquaculture feeds can be developed and feed costs reduced.

9.2 Foods for hatchery culture systems

Intensive rearing of the larval stages of fish and shellfish currently relies on the availability of live food organisms. The three major types of live foods used for larval rearing are:

- microalgae
- rotifers
- brine shrimp (*Artemia* species).

Much less use is made of other zooplankton, such as copepods and freshwater *Daphnia* species.

9.3 Microalgae

Cultured microalgae have a central role as a food source in aquaculture. Microalgae are used directly as a food source for larval, juvenile and adult bivalves, and for early larval stages of some crustaceans and fish. They are also very important as a food source for rearing zooplankters, such as rotifers and brine shrimp, which, in turn, are used to feed crustacean and fish larvae.

The golden-brown flagellates (Prymnesiophyceae and Chrysophyceae), the green flagellates (Prasinophyceae and Chlorophyceae) and the diatoms (Bacillariophyceae) (Fig. 9.1) are the most widely used microalgae in aquaculture. As their name suggests, the flagellates possess one or more flagellae, which give the cells motility, whereas diatoms lack a flagellum and are non-motile. Diatoms contain silica in their cell walls and may possess siliceous spines. Some species of diatoms exist as single cells (e.g. *Chaetoceros gracilis*), and other species have cells joined to form chains (e.g. *Skeletonema costatum*). The generalised morphology of microalgae used in aquaculture is shown in Fig. 9.1.

9.3.1 Culture methods

The simplest method of microalgal production is to 'bloom' local species of phytoplankton in ponds or tanks. This is achieved by filling a pool or pond with local water, which is filtered to remove zooplankton, detritus and other unwanted particulates while retaining the smaller phytoplankton. With the addition of fertiliser (usually an inorganic fertiliser), adequate light and aeration, blooms of natural phytoplankton will develop. This is known as the 'Wells–Glancy' method. Although inexpensive, this method can be an unreliable method of food production: the bloom is not guaranteed, and there is little control over the species composition of the bloom. As such, the nutritional value of microalgae produced in this way is unpredictable.

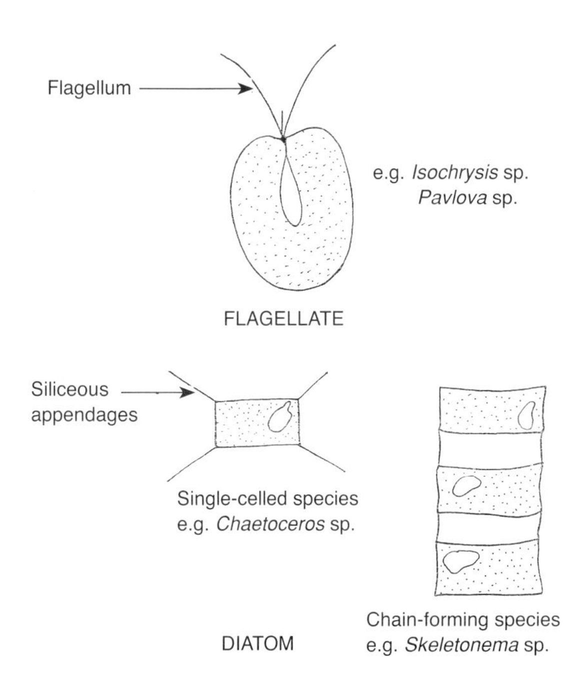

Fig. 9.1 Diagram showing general morphology of a flagellate and diatoms.

More commonly in aquaculture, monospecific cultures of microalgae are intensively reared in systems in which efforts are made to minimise or exclude bacterial contamination. Monospecific axenic (bacteria-free) 'starter' cultures of many species of microalgae are available to the aquaculture industry from specialised laboratories. These are the basis for microalgae production, which involves scaling up the culture volume and the density of algal cells by maintaining favourable conditions for algal growth. An example of a suitable scale-up procedure is shown in Fig. 9.2. Stock cultures are maintained under controlled conditions of temperature and light. During the scaling-up process, microalgae are usually transferred from container to container under axenic conditions using a laminar-flow cabinet. Air entering the cabinet is filtered to remove bacteria and the flow of air within the sterilised cabinet prevents non-filtered air from entering. Culture vessels receiving the inoculant contain seawater that has been filtered (usually to 0.2–0.45 μm) and sterilised by autoclaving. Inoculating microalgae cultures under these conditions reduces the possibility of bacterial contamination. However, it is impracticable to use this method for culture vessels with volumes greater than ~20 L. In larger bag and tank cultures (Fig. 9.3), efforts are made to reduce the bacterial population in the culture water by fine filtration (0.2–0.45 μm), which may be followed by passage of the water through an ultraviolet (UV) steriliser.

Flasks, bags or cylinders of microalgae may finally be transferred to ponds or pools. In large aquaculture operations, where large volumes of microalgae are required, algae may be cultured in tanks or pools either outside or inside.

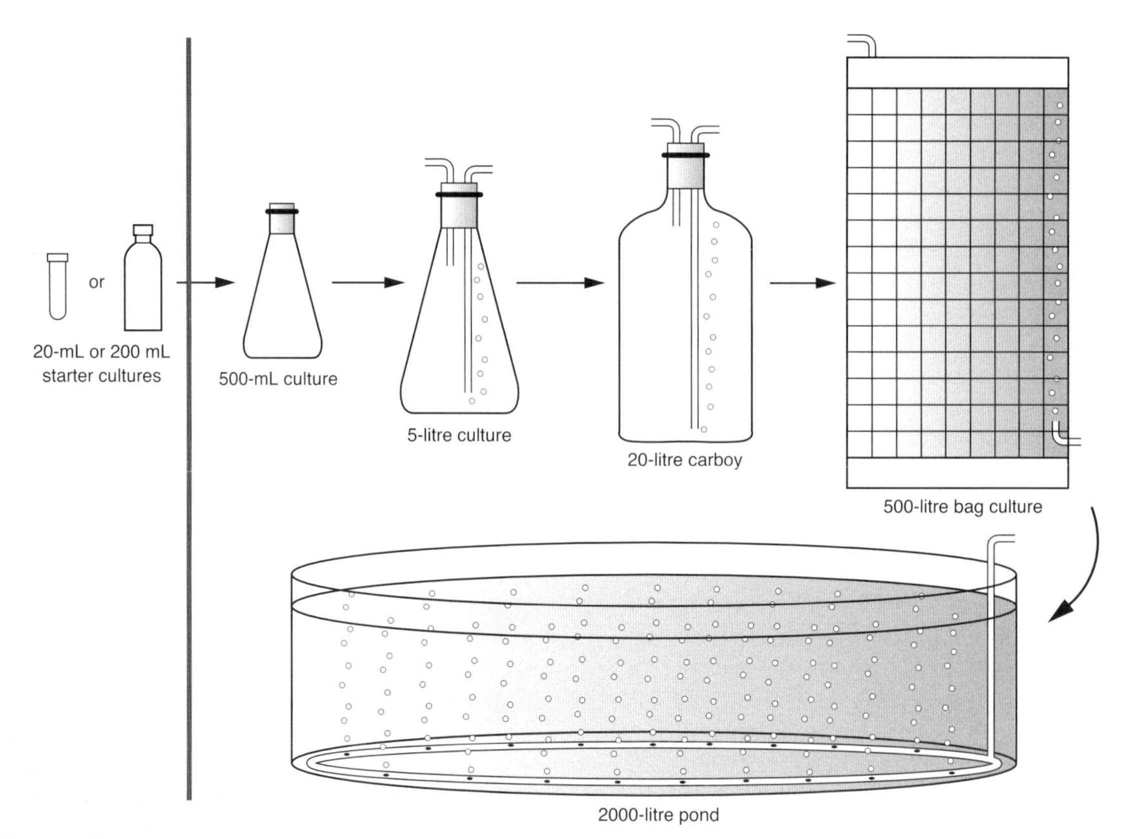

Fig. 9.2 Typical scale-up of microalgae cultures from starter cultures. Reproduced from Brown *et al*. (1989) CSIRO Marine Laboratories Report 205 – *Nutritional Aspects of Microalgae Used in Mariculture: A Literature Review*, with permission from CSIRO Publishing.

Fig. 9.3 Microalgae culture in 200-L bags and 1000-L tanks.

For growth, microalgal cultures require:

- aeration
- a suitable nutrient medium
- light.

At each transfer, a clean vessel containing filtered, preferably sterile, seawater and culture medium is inoculated with microalgae. The newly inoculated vessel is provided with filtered air (0.2–0.45 μm) to maintain microalgae cells in suspension and to supply sufficient carbon dioxide (CO_2) for their growth. The air supply may also be supplemented directly with CO_2 gas to further stimulate growth. Microalgae cultures are maintained under a controlled light and temperature regime, e.g. suitable conditions for most species of microalgae are provided by a photoperiod of 12–16 h, providing irradiance of 70–80 $mE/m^2/s$ at 20–25°C. The growth of microalgae follows a distinct pattern and consists of a number of different phases (Fig. 9.4).

(1) A lag phase occurs following inoculation and is characterised by a steady cell density.
(2) The exponential or log phase is marked by a

rapid increase in cell density within the culture. This is the time when microalgae have optimal nutritional value.

(3) The stationary phase is reached as nutrients start to become limiting, and increasing cell density results in reduced light intensity within the culture, the rate of cell division slows and cell density reaches a plateau.
(4) The death phase is reached as, eventually, cells within the culture begin to die as the nutrients become exhausted and the culture enters the phase characterised by declining cell density.

The densities of microalgae cultures are usually determined using haemocytometer counts or high-speed electronic particle counters. The volume of microalgae culture that needs to be added to larval rearing tanks is calculated using the equation described in section 9.5.1.

There are three main methods for culturing microalgae:

(1) Batch culture. This is when a microalgae culture is grown to a point at which it is completely harvested.
(2) Semi-continuous culture. This is when partial harvesting of culture vessels is conducted periodically and culture vessels are 'topped up' with new water and fresh nutrient medium.
(3) Continuous culture. This is when microalgae are harvested on a continuous basis and the volume removed from the culture vessel is continually replaced by new water and fresh nutrient medium.

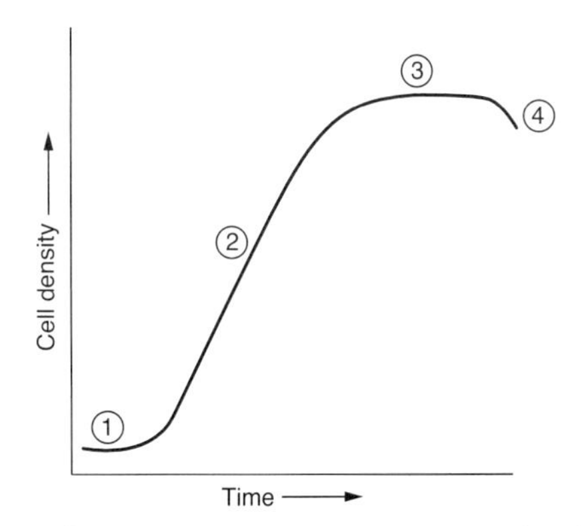

Fig. 9.4 General pattern of changes in cell density over time in microalgae batch cultures. 1, Lag phase; 2, exponential or log phase; 3, stationary phase; and 4, death phase.

The objective of continuous and semi-continuous cultures is to maintain the cultures at maximal growth rate (exponential phase). This maximises microalgae production and reduces variability in the biochemical composition (and nutritional value) of the algae. In batch cultures, the biochemical composition can vary widely according to the growth phase and age of the culture.

9.3.2 Nutrient media

Many nutrient media have been developed for rearing microalgae. They generally contain macronutrients to provide nitrogen and phosphorus (e.g. sodium nitrate, sodium glycerophosphate), trace metals and vitamins. A commonly used medium is the 'f/2' medium of Guillard (1972), whose composition is shown in Table 9.1. Nutrient media are made up from distilled water to which nutrients are added. It is convenient to make up concentrated standard stock solutions of media, which are then added to microalgae culture vessels to provide appropriate nutrient levels. In general, 1–3 mL of stock nutrient solutions is added to each litre of culture water. Given the structural importance of silica in diatoms, a source of silica (usually sodium metasilicate) must be provided to diatom cultures for optimal growth. A stock solution is prepared by dissolving sodium metasilicate (40 g) in 1 L of distilled water, and 0.2–0.4 mL of the resulting solution is added to culture vessels per litre of culture water. Nutrient media are generally added to smaller culture containers (e.g. glass flasks) before they are sterilised by autoclave. Once cooled, containers are inoculated to begin new microalgae cultures.

9.3.3 Nutritional value of microalgae

When considering the suitability of various species of microalgae as a hatchery feed, the first concern is their physical characteristics. Factors such as:

- cell size
- thickness of cell wall
- digestibility

Table 9.1 The composition of 'f/2' medium for microalgae culture

Nutrient	Concentration/L
$NaNO_3$	75 mg
NaH_2PO_4	5 mg
*Na_2SiO_3	15–30 mg
Trace metals	
Na_2EDTA	4.36 mg
$FeCl_3 \cdot 6H_2O$	3.15 mg
$CuSO_4 \cdot 5H_2O$	0.01 mg
$ZnSO_4 \cdot 7H_2O$	0.022 mg
$CoCl_2 \cdot 6H_2O$	0.01 mg
$MnCl_2 \cdot 4H_2O$	0.18 mg
$Na_2MoO_4 \cdot 2H_2O$	0.006 mg
Vitamins	
Cyanocobalamin	0.5 μg
Biotin	0.5 μg
Thiamine HCl	100 μg

*Required for diatom cultures only.
Reproduced from Guillard (1972) with permission from Kluwer Academic Publishers.

- presence of spiny appendages
- chain formation (e.g. diatoms)

influence the nutritional value of a particular species. Clearly, microalgae must have suitable physical characteristics to enable ingestion and, once ingested, must be digestible. Cultured invertebrate larvae vary in their feeding and digestive mechanisms, and this greatly influences the sizes and kinds of microalgae that can be ingested and digested. For example, shrimp larvae have a complete set of setous mouthparts adapted to feeding on chain diatoms. However, such diatoms cannot be captured and ingested by the ciliated feeding structures of bivalve larvae.

Assuming suitable physical characteristics, the nutritional value of a given microalga is determined by its biochemical composition. Biochemical composition varies greatly between species (Table 9.2) and according to growth phase. Composition is also influenced by abiotic factors such as:

- light (photoperiod, intensity and wavelength)
- temperature
- nutrient medium (composition and concentration)
- salinity
- nitrogen availability
- CO_2 availability.

For example, protein levels decrease while lipid and carbohydrate levels typically increase during the stationary phase of a culture. Similarly, the protein content of microalgae is greatly influenced by the nitrogen content of the culture medium. Culture conditions also influence levels of micronutrients such as fatty acids and vitamins in the algae. As such, the conditions under which microalgae are grown and the stage at which they are harvested may greatly influence their nutritional value.

The nutritional requirements of cultured aquatic organisms were discussed in Chapter 8. Numerous growth trials using different species of microalgae as food have shown that differences in the food value of microalgae are related primarily to their fatty acid and carbohydrate compositions. As detailed in section 8.9.7, marine fish and shellfish larvae have an essential dietary requirement for *n*-3 highly unsaturated fatty acids (HUFAs). As such, *n*-3 HUFA content is an important factor in determining the nutritional value of microalgae, and it is generally accepted that species containing the essential fatty acids eicosapentaenoic acid (EPA, 20:5*n*-3) and docosahexaenoic acid (DHA, 22:6*n*-3) will be of high nutritional value for cultured animals.

Table 9.3 shows the *n*-3 fatty acid compositions of various species of microalgae, which vary widely between species. Golden-brown flagellates and diatoms generally contain relatively high levels of essential fatty acids (EFAs), whereas others, notably species of green algae, contain low levels of EFAs or none at all.

Differences in the carbohydrate compositions of microalgae are another important factor in determining their nutritional value. Recent studies have shown that, assuming EFA requirements are met, growth and condition of bivalve larvae are correlated with dietary carbohydrate content. Dietary carbohydrate is used primarily as an energy source and is considered to spare dietary protein and lipid, which can then be utilised for tissue growth (section 8.7.2). Microalgal diets are generally composed of a mixture of species. Diets consisting of more than one species are considered nutritionally superior to a single-species diet and are thought to provide a better balance of nutrients by minimising any nutritional

Table 9.2 The gross nutritional composition of microalgae commonly used in aquaculture*

Species	Composition (%)		
	Protein	Carbohydrate	Lipid
Golden-brown flagellates			
Isochrysis clone I-ISO	44	9	25
Isochrysis galbana	41	5	21
Pavlova lutheri	49	31	12
Diatoms			
Chaetoceros calcitrans	33	17	10
Phaeodactylum tricornutum	33	24	10
Skelotenema costatum	37	21	7
Green flagellates			
Dunaliella salina	57	32	9
Tetraselmis suecica	39	8	7

*Data compiled from Parsons *et al.* (1961), Utting (1986) and Whyte (1987).

Table 9.3 The *n*-3 fatty acid compositions (% total fatty acids) of selected species of microalgae used in aquaculture

Species	Fatty acid		
	18:3*n*-3	20:5*n*-3	22:6*n*-3
Golden-brown flagellates			
Isochrysis clone T-ISO	3.6	0.2	8.3
Pavlova lutheri	1.8	19.7	9.4
Diatoms			
Chaetoceros gracilis	–	5.7	0.4
Chaetoceros calcitrans	Trace	11.1	0.8
Thalassiosira pseudonana	0.1	19.3	3.9
Green flagellates			
Tetraselmis suecica	11.1	4.3	Trace
Dunaliella tertiolecta	43.5	–	–
Nannochloris atomus	21.7	3.2	Trace

Reproduced from Volkman *et al.* (1989) with permission from Elsevier Science.

deficiencies present in any of the component species.

The choice of species of microalgae to be used in an aquaculture hatchery requires careful consideration of their suitability for culture and use under local conditions. This is particularly important when microalgae are cultured in outdoor tanks. The three most important factors to consider for outdoor culture are temperature, salinity and light intensity. For example, in the tropics, the ability of microalgae to tolerate fluctuating salinity and temperature is particularly important in areas where cultures may be subjected to periods of high rainfall and high temperatures. Microalgae vary in their optimal temperature and salinity ranges (Table 9.4). Light intensity also affects the growth rates of microalgae and may alter their biochemical composition and therefore their nutritional value (Jeffrey *et al.*, 1992).

Again, this is of particular importance in areas of high natural light intensity such as the tropics.

On the basis of their known tolerance ranges, microalgae can be categorised into species suitable for culture and use in different environments. For example, Jeffrey *et al.* (1992) divided a range of microalgae species according to their temperature tolerances into:

(1) excellent universal species, which show good growth at 10–30°C (e.g. *Tetraselmis suecica*, *T. chuii*, *Nannochloris atomus*)
(2) excellent tropical and sub-tropical species, which show good growth at 15–30°C (e.g. *Isochrysis* clone T-ISO, *Chaetoceros gracilis*, *Pavlova salina*)
(3) good temperate species, which show good growth at 10–20°C (e.g. *Chroomonas salina*, *Skeletonema costatum*, *Thalassiosira pseudonana*).

9.3.4 Recent developments in microalgae production

The suitability of microalgae as an aquaculture feed has been demonstrated by its use as a food source for cultured organisms for at least 50 years. However, cultured microalgae have a number of disadvantages for aquaculture hatcheries:

(1) On-site production of microalgae is labour intensive and is associated with high running costs (up to 30–50% of hatchery operating costs).
(2) On-site production of microalgae requires specialised facilities and dedicated personnel. There are substantial establishment costs, and microalgae culture requires significant hatchery space that could otherwise be devoted to larval production.
(3) Microalgae cultures can 'crash' through failure of the culture system or become infected with contaminant or pathogenic organisms. Either case may result in a shortage of food for the culture organisms.

In an effort to overcome some of these problems, there have been two relatively recent developments:

- production of dried microalgae
- production of microalgae concentrates.

Both allow microalgae to be stored in concentrated form until required, thereby alleviating the need for on-site microalgae culture.

Microalgae concentrates are prepared by removing the culture medium from the microalgae culture to produce a thick paste of concentrated algal cells. The medium is usually removed by centrifugation,

Table 9.4 Salinity and temperature tolerances of microalgae used in aquaculture

Species	Salinity tolerance (‰)	Temperature tolerance (°C)
Golden-brown flagellates		
Isochrysis sp. (T-ISO)	7–35	15–30
Pavlova salina	21–35	15–30
Pavlova lutheri	7–35	10–25
Diatoms		
Chaetoceros calcitrans	7–35	10–30
Chaetoceros gracilis	7–35	15–30
Thalassiosira pseudonana	–	10–20
Skeletonema costatum	14–35	10–20
Green flagellates		
Tetraselmis suecica	7–35	10–30
Dunaliella tertiolecta	7–35	10–30
Nannochloris atomus	7–35	10–30
Nanochloropsis oculata	7–35	10–30

Reproduced from Jeffrey *et al.* (1992) with permission from NSW Fisheries.

and the resulting concentrates can be stored for many weeks in a refrigerator. Some species are better suited for this process than others, success being largely determined by the integrity of the cell wall during centrifugation. Promising results have been achieved with microalgae concentrates when used as a food for bivalve larvae (Nell & O'Connor, 1991). A number of species of microalgae are now available as concentrates from commercial suppliers. Many can be stored for weeks or months under appropriate conditions (fridge/freezer).

Dried microalgae preparations have been produced from microalgae grown heterotrophically. This technique involves growing microalgae in the dark, using sugars rather than light as an energy source (section 13.3.4). Growth under these conditions produces microalgae with a considerably different biochemical composition from that of the same species grown using conventional methods. Although the number of microalgae species that can be produced in this manner is limited, some have been produced commercially. Whether produced heterotrophically or by conventional means, dried microalgae are likely to have a promising future as a hatchery food. A number of studies have shown the value of dried microalgae as a food source for crustacean and fish larvae (Biedenbach *et al.*, 1990; Navarro & Sarasquete, 1998; Cañavate & Fernández-Díaz, 2001) and for the larvae and spat of bivalves (Laing & Millican, 1992; Knauer & Southgate, 1999).

The major advantage of both these developments is that microalgae can be cultured at a central facility and distributed to hatcheries. This system eliminates the need for hatcheries to have microalgae culture facilities and could result in considerable cost savings.

9.4 Zooplankton

Hatcheries that culture fish and crustacean larvae also rely on zooplankton as a larval food source. The two major organisms cultured for this purpose are rotifers and brine shrimp. However, recent years have seen considerable research effort directed towards the development of mass culture techniques for copepods and their use as live feeds in aquaculture.

9.4.1 Rotifers

Rotifers (*Brachionus* species) are widely used in aquaculture as a food for fish and crustacean larvae, and their use in aquaculture has been reviewed by Lubzens *et al.* (1987). Two species are typically used in aquaculture hatcheries:

- the smaller *Brachionus rotundiformis* (known as S-type; Segers, 1995), which is 90–190 μm in length
- the larger *B. plicatilis* (known as L-type; Segers, 1995), which is 120–300 μm in length.

Clearly, an important consideration in rotifer culture is selection of a strain most suited to the mouth size of the prey.

Rotifers consist of a lorica or body shell from which the foot extends ventrally and the head extends dorsally (Fig. 9.5). The head has two bands of cilia used for the capture of food particles and for locomotion. The life cycle of the rotifer (egg–juvenile–adult) takes 7–10 days. Cultures may contain both males and females, but males are rare and considerably smaller than females. Rotifers reproduce sexually or asexually depending on culture conditions. Under favourable conditions, reproduction is asexual and the female produces diploid amictic eggs, which she carries until they hatch into females:

Female rotifer → amictic egg ($2n$) → female rotifer → amictic egg ($2n$) → female rotifer, etc.

Most reproduction in cultured rotifer populations occurs by this method.

Under unfavourable conditions, reproduction occurs sexually; females produce smaller haploid mictic eggs that hatch into males if not fertilised. If fertilised, they become resting eggs that have a dehydration-resistant outer shell. They can remain dormant for several years and hatch into females when conditions become favourable. On this basis, the presence of males in rotifer cultures indicates poor culture conditions.

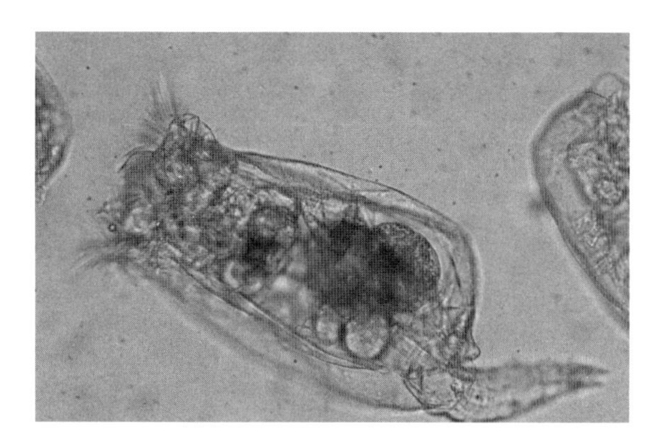

Fig. 9.5 A female rotifer (*Brachionus plicatilis*). The ciliated head is to the left and the flexible foot extends ventrally from the rotifer to the bottom right of the photograph. The body is enclosed within a shell or lorica.

9.4.2 Rotifer culture

Rotifers are hardy and are easily mass cultured on a wide variety of foods. Mass culture of rotifers is usually initiated by inoculating a culture of microalgae with rotifers. Under suitable conditions, the rotifers consume the microalgae and their population rapidly increases. Consumption of microalgae must be monitored regularly and more microalgae added when required; it is important to ensure constant food availability. A portion of the culture water is usually removed from the rearing vessels on a daily basis and replaced with a similar volume of microalgae culture. Water can be removed by siphoning through a 60-μm sieve, which prevents removal of rotifers. Bakers' yeast (*Saccharomyces cerevisiae*) and commercially produced modified yeast are also commonly used as a food for rotifers, either singly or in combination with microalgae. Various species of microalgae are used to culture rotifers, including *Nannochloropsis*, *Tetraselmis*, *Isochrysis* and *Pavlova*. The microalga *Nannochloropsis oculata* and bakers' yeast are considered excellent foods for maintaining rotifer cultures. Rotifers are generally cultured using either batch, semi-continuous or continuous methods (Lubzens *et al.*, 1987).

Both *B. rotundiformis* and *B. plicatilis* are euryhaline and productive at salinities between 4‰ and 35‰. Optimal water temperature varies between species, with *B. rotundiformis* and *B. plicatilis* most productive at high (30–35°C) and low (15–25°C) water temperatures respectively. Rotifer cultures are generally maintained at a salinity of 20–35‰, within a temperature range of 20–30°C and with gentle aeration. Successful rotifer culture requires the maintenance of constant conditions. Water quality must be maintained by regular cleaning to prevent the build-up of detritus and faecal matter. Conventional rotifer cultures can be very productive and may reach densities of 700–1000 individuals per millilitre; however, ultra-high-density culture methods with 10 000–30 000 rotifers/mL have been developed in Japan (Yoshimura *et al.*, 1996).

The health of rotifer cultures can be assessed by monitoring the swimming activity and the number of eggs present. Healthy cultures will contain females that are active and rapidly swimming, with many carrying more than one egg. The presence of male rotifers indicates imminent production problems as sexual reproduction occurs only when environmental conditions become unfavourable.

The nutritional value of cultured rotifers is largely determined by their food. For example, rotifers reared on bakers' yeast, which is deficient in EFAs, are themselves deficient in these fatty acids. For this reason, rotifers reared on yeasts or other foods with low levels of EFAs are usually fed microalgae or artificial feeds high in EFAs prior to feeding to fish larvae. This process is generally known as enrichment and is outlined in section 9.4.7.

9.4.3 Brine shrimp

Brine shrimp (*Artemia* species) are found worldwide in salt lakes and similar habitats. Their inactive dry cysts can be harvested in large quantities and stored in a dry state for many years. When immersed in saline water, the cysts rehydrate and become spherical, and the embryo inside begins to metabolise. The cyst ruptures after ~24 h and a free-swimming nauplius emerges (Fig. 9.6). This first larval stage (instar I) is generally 400–500 μm long and brown-orange in colour. It has a single red eye and three sets of appendages, which have sensory, locomotory and feeding functions. Instar I larvae do not feed as their digestive tract is not yet functional. After ~12 h, the nauplius moults to the instar II stage, which has a functional gut and begins to ingest small particles such as microalgae.

Brine shrimp undergo ~15 moults over 8–14 days to produce mature adults 10–20 mm in length. Like rotifers, they can reproduce sexually or asexually. Under favourable conditions, females produce free-swimming nauplii (ovoviviparous reproduction); however, under unfavourable conditions, such as high salinity and low oxygen, the shell glands of the female become active and secrete a thick shell around the developing gastrula, which enters a dormant state (diapause). These embryos are released by the female as cysts. Under optimal conditions, brine shrimp can reproduce at a rate of 300 nauplii or cysts every 4 days. Production of cysts has obvious advantages for aquaculture. Dry cysts can be easily stored and live feed (in the form of nauplii) can be produced when required.

9.4.4 Hatching brine shrimp cysts

Although cysts can be successfully incubated in full-strength (35‰) seawater, the hatching rate is generally superior at low salinities and a salinity of 5‰ is optimal. Cysts are incubated at densities up to 5 g/L culture medium, which is maintained at 25–30°C with vigorous aeration. Dissolved oxygen content must be maintained above 2 mg/L and, to facilitate good aeration and water movement, culture vessels are usually V-shaped or conically based. Cultures require a pH of 8–9 and constant illumination at the water surface. Culture conditions must be constant during incubation. Within 24 h, the majority of cysts will have hatched.

To harvest hatched nauplii, aeration is stopped, causing the cyst shells to float to the top of the culture vessel. Nauplii are positively phototaxic, and this behaviour can be used to concentrate them prior to harvesting by siphon. It is important that the number of cyst shells accompanying the nauplii is limited. Cyst shells have the potential to introduce disease and bacteria into larval cultures and can cause digestive disorders in fish larvae. Contamination with cyst shells can be minimised if the cysts are decapsulated prior to incubation.

9.4.5 Decapsulation of cysts

The process in which the outer shell or chorion is removed from hydrated brine shrimp cysts is decapsulation. This is achieved by treating hydrated cysts with hypochlorite solution, which dissolves the chorion without damaging the embryo inside. Prior to decapsulation, dried cysts are rehydrated in freshwater for 60–90 min at the rate of 1 g of cysts per 30 mL of water. Approximately 20–30 mL of liquid bleach

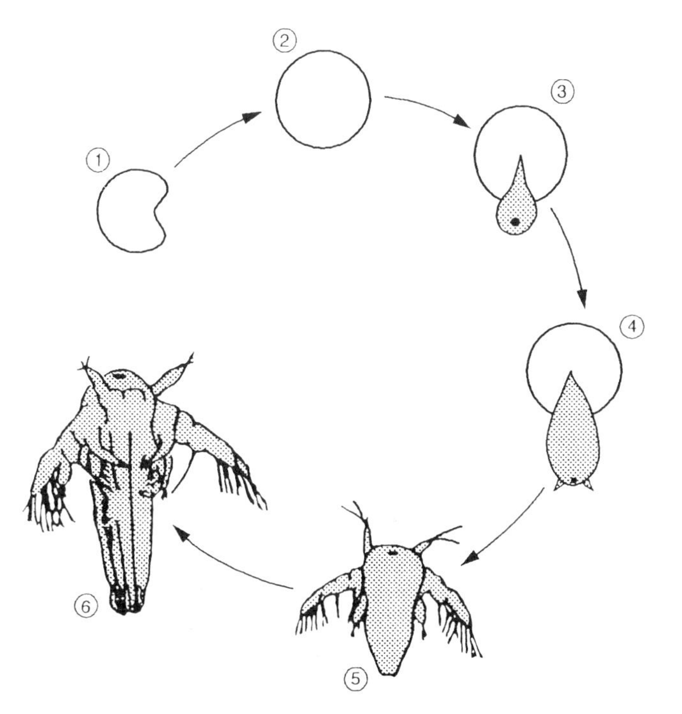

Fig. 9.6 Hatching and development of brine shrimp (*Artemia* species). 1, Dry cyst; 2, hydrated cyst; 3, breaking; 4, hatching; 5, nauplius; 6, larger metanauplius.

(sodium hypochlorite – NaOCl) is added per gram of cysts, and the solution stirred continuously. The colour of the solution changes as the chorion is dissolved, and decapsulation is complete within 2–4 min when the solution becomes orange in colour. The solution is then poured through a sieve to remove the chlorine solution. The decapsulated cysts retained on the sieve are washed thoroughly with seawater or freshwater until no further chlorine smell can be detected. Any residual chlorine can be removed from decapsulated cysts by washing in 0.1% sodium thiosulphate solution for 1 min. The decapsulated cysts are then washed and placed into a medium for hatching, or they may be stored at 4°C for a short period before hatching. Decapsulation offers major advantages in limiting potential digestive and disease problems caused by cyst shells; it disinfects brine shrimp embryos and improves hatch rate. Decapsulated cysts can also be offered directly as a larval food source, with a major advantage being that, prior to hatching, embryos have their maximum energy content.

9.4.6 *Culturing brine shrimp*

For the production of large or adult brine shrimp, nauplii are reared in tanks at an initial stocking density of ~1000–3000 per litre. Nauplii are initially fed cultured microalgae at a density of approximately 5×10^5 to 1×10^6 cells/mL; the feeding rate is adjusted as the brine shrimp grow and more food is required. Best growth of brine shrimp cultures occurs with:

- good aeration
- good water quality
- a readily available food supply
- low light conditions
- 25–30°C
- 30–35‰ salinity.

Culture tanks must be cleaned regularly to remove detritus and faecal matter to maintain water quality. Under suitable conditions, production rates in the order of 5–7 kg (wet weight) of brine shrimp per cubic metre are achievable using batch culture techniques (Sorgeloos *et al.*, 1986).

Brine shrimp nauplii have their greatest energy content at hatching. There is a substantial decrease in the nutritional value of nauplii between instar I and instar II, with a reduction in organic content of 24% and a decrease in lipid and energy content of 27% (Sorgeloos *et al.*, 1986). Although the lipid and fatty acid content of brine shrimp varies according to geographical origin of the cysts (Webster & Lovell, 1991), they are generally considered to be deficient in essential fatty acids. The nutritional value of instar II nauplii, particularly their fatty acid content, can be significantly improved using an appropriate enrichment procedure.

9.4.7 *Enrichment of rotifers and brine shrimp*

Rotifers and many strains of brine shrimp have very low levels (or a total lack) of certain EFAs that are required for normal growth and development of marine larvae (section 8.9.7). To overcome these deficiencies, the EFA content of rotifers and brine shrimp has to be manipulated using fatty acid enrichment techniques. This process involves feeding a nutrient source rich in EFAs to the rotifers or brine shrimp prior to feeding them to the cultured larvae (Fig. 9.7). Various materials can be used for enrichment, including microalgae, oil suspensions, microencapsulated diets and yeasts (Leger *et al.*, 1986), and a number of enrichment preparations are available commercially. The benefit of enrichment is clearly shown by increases in the EFA content of the live food organism (Table 9.5). Enrichment of rotifers and brine shrimp boosts dietary EFA intake, resulting in improved survival and growth of larvae.

The degree to which EFAs are incorporated into rotifers and brine shrimp during enrichment is influenced by the duration of the enrichment procedure, the density of rotifers or brine shrimp, and the density and EFA content of the enrichment material. Although enrichment procedures were developed to improve the fatty acid composition of live food organisms, this process is also used to improve levels of other important nutrients such as vitamins, and as a means of presenting therapeutic compounds.

9.4.8 *Copepods*

Copepods occur in all aquatic systems and are natural prey for virtually all fish larvae. There are over 10 000 known species, with most planktonic forms ranging between 0.5 mm and 2.5 mm in size. Given their

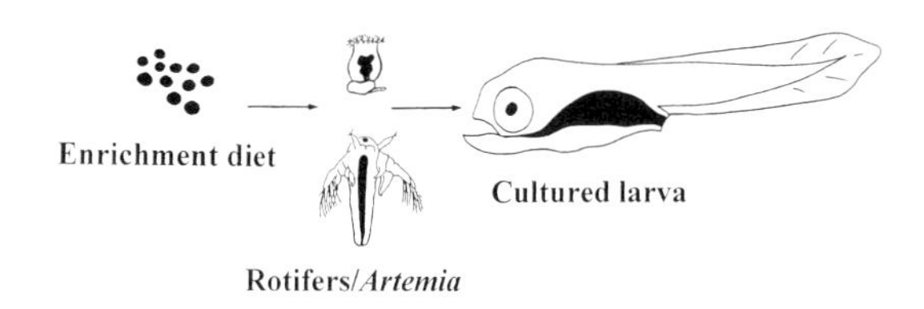

Fig. 9.7 The process of enrichment when a diet rich in the required nutrient(s) (e.g. essential fatty acids) is fed to rotifers or brine shrimp prior to feeding them to cultured larvae.

Table 9.5 The fatty acid composition of unenriched and enriched brine shrimp nauplii

	Composition (% total fatty acids)	
Fatty acid	Unenriched	Enriched
18:2*n*-6	5.0	4.5
18:3*n*-3	33.1	25.4
18:4*n*-3	6.6	5.4
20:5*n*-3 (EPA)	2.3	4.7
22:6*n*-3 (DHA)	0	2.5
Total EPA + DHA	2.3	7.2

Reproduced from Southgate & Lou (1995) with permission from Elsevier Science.

importance as prey for wild fish larvae, there is clear potential for the use of copepods in aquaculture, and developments in this field were reviewed by Nellen (1986). Research has focused primarily on the harpacticoid copepods, *Tisbe*, *Tigriopus* and *Euterpina* species, and the calanoids, *Eurytemora* and *Acartia* species. Research with fish larvae has shown that provision of copepods as food can improve larval survival, growth rates, pigmentation and gut development.

Rotifers and brine shrimp are extensively used in aquaculture, primarily because they are amenable to mass culture, not because they are an ideal food source (Table 9.6). Copepods are readily digested by fish larvae and are superior to rotifers and brine shrimp in terms of nutritional value. In particular, they contain high levels of *n*-3 HUFAs. The small size of copepod nauplii makes them an ideal food for the early larvae of species with a small mouth gape, such as the groupers.

Despite these favourable characteristics, copepods are not yet widely used as a food source in fish hatcheries. Copepods have generally proved to be difficult to mass culture, often with variable and unreliable production. However, intensive culture techniques have been reported for some copepod species (e.g. Payne & Rippingale, 2001). It is likely that copepods will assume increasing importance as a food source in fish hatcheries as more reliable mass-culture techniques are developed and a greater number of species are investigated for their culture potential.

9.5 Feeding strategy for larval culture

9.5.1 Feeding protocols

A generalised feeding protocol for marine fish larvae begins with rotifers at first feeding followed by brine shrimp nauplii and larger brine shrimp as larvae increase in size (Fig. 9.8). Artificial (formulated) diets are then introduced and larvae are weaned from live food organisms. Fish hatcheries also culture microalgae as a food source for rotifers and brine shrimp and, as such, they generally culture three different live foods to feed the larvae of a single target species. Shrimp hatcheries generally begin feeding with microalgae (usually a diatom such as *Chaetoceros* species), which are usually followed by rotifers and brine shrimp or just brine shrimp as the larvae grow. Bivalve hatcheries rely exclusively on cultured microalgae as a larval food source.

The volume of a microalgae, rotifer or brine shrimp culture that has to be added to larval rearing tanks to obtain the desired density of food organisms is calculated as:

$$A = (B \times C)/D$$

where A is the required volume (L) of the live food culture; D is the density of the live food culture (number/mL); B is the required density of micro-algae, rotifers or brine shrimp in the larval tank (number/mL); and C is the volume of the larval tank (L).

The feeding regimen is an important aspect of hatchery management. Overfeeding is wasteful and expensive. It also compromises water quality, which can lead to disease and affect larval growth.

Table 9.6 Some potential problems associated with use of rotifers and brine shrimp

Disadvantage	Comments
Nutritional deficiency	Both brine shrimp and rotifers have inadequate fatty acid compositions for marine larvae. They have to be artificially 'enriched' prior to their use as food, which adds to the expense of food production
Nutritional inconsistency	Live feed organisms may vary in their nutritional composition according to the source of stock (i.e. source of *Artemia* cysts) and the nutritional composition of their food
Reliability of supply	Hatcheries in most parts of the world rely on a supply of brine shrimp cysts from North America. Availability may fluctuate from year to year (Lavens & Sorgeloos, 2000) and imports may be subject to quarantine problems

Underfeeding reduces growth rates, thereby increasing hatchery running costs. It is important to monitor the presence of food in larval tanks to avoid these problems.

9.5.2 *Some disadvantages of live feeds organisms*

A number of disadvantages are common to intensive culture of microalgae, rotifers and brine shrimp. Some of these have been outlined for microalgae production in section 9.3.4, but they also apply to rotifer and brine shrimp production. Other potential problems relating specifically to rotifers and brine shrimp are listed in Table 9.6.

In response to these problems and, in particular, because of the high costs associated with live food culture, there has been considerable interest over recent years in developing artificial hatchery feeds as alternatives to live foods.

9.6 Artificial and alternative hatchery foods

9.6.1 *Advantages*

The high cost of live food production in aquaculture hatcheries could be reduced by cheaper production of live foods and earlier weaning to formulated feeds in the case of crustaceans and fish. However, complete or significant replacement of live foods is the goal of research in this field. Perhaps the most significant advantage of artificial diets is that, unlike live foods, the size of the food particle and diet composition can be adjusted to suit the exact nutritional requirements of the larvae. Artificial diets offer the advantages of nutritional consistency and off-the-

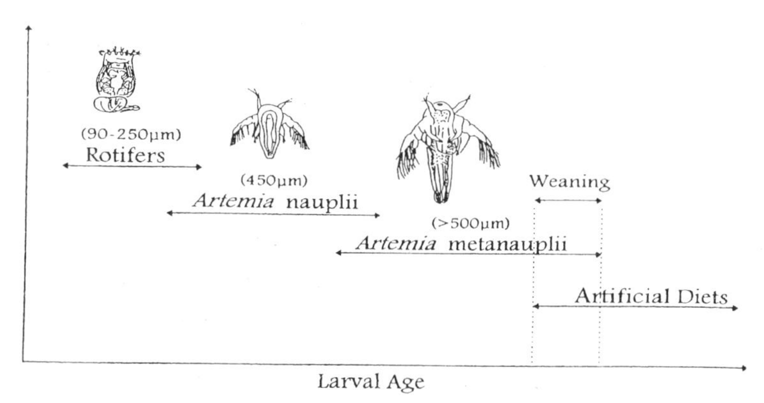

Fig. 9.8 A generalised feeding protocol for marine fish larvae begins with rotifers at first feeding followed by brine shrimp nauplii and larger brine shrimp as larvae increase in size. Larvae are then weaned to artificial formulated feeds. Reproduced from Southgate & Partridge (1998) with permission from Academic Press.

shelf convenience; however, they must satisfy a number of criteria (Table 9.7).

Various materials have been assessed for their potential to replace live microalgae as a feed for bivalves. These include dried and concentrated microalgae (section 9.3.4), dried and pulverised macroalgae, yeasts and cereal products (Knauer & Southgate, 1999). They also include formulated diets, such as microbound and microencapsulated diets, which are generally known as microdiets. Although dried microalgae have also been used in the culture of crustacean and fish larvae (Biedenbach *et al.*, 1990; Cañavate & Fernández-Díaz, 2001), much of the research to develop artificial diets for aquatic larvae has focused on microdiets.

9.6.2 *Microbound diets*

In microbound diets (MBDs), nutrients (both particulate and dissolved) are bound within a particle matrix consisting of a binding material such as agar, gelatin, alginate or carrageenan. Dietary ingredients are mixed with the binder to form a slurry, which is then dried, ground and sieved to produce food particles of the desired size. MBDs allow precise manipulation of dietary contents and, for this reason, have been used extensively in research with larvae (particularly crustaceans and fish) to determine nutritional requirements. However, because MBDs have no barrier between dietary ingredients and the culture water, there is potential for nutrient leaching and they are susceptible to direct bacterial attack.

9.6.3 *Microencapsulated diets*

Microencapsulated diets (MEDs) consist of dietary materials enclosed within a microcapsule wall or membrane. This greatly reduces nutrient leaching and the susceptibility of the diet to bacterial attack. MEDs have been used with some success as a replacement for microalgae for bivalve spat and larvae (Knauer & Southgate, 1999). MEDs have been commercially available for shrimp larvae for a number of years and are widely used in hatcheries (section 19.6.3). It is generally accepted that a combination of artificial diets and live feeds supports superior growth and survival of shrimp larvae than either feed alone. This practice is known as co-feeding. It is very likely that in the near future artificial diets will completely replace live feeds in penaeid shrimp hatcheries (section 19.6.3). However, research in this field has been less successful with the larvae of prawns (caridean) and other crustaceans. The development and use of artificial diets for crustaceans was reviewed by Jones (1998). Despite the development of successful artificial diets for shrimp larvae and their routine use in shrimp hatcheries, similar success has not been achieved with fish larvae.

9.7 Development of artificial diets for fish larvae

9.7.1 *Limited success*

Numerous studies have been conducted to assess the nutritional value of microdiets for marine fish larvae

Table 9.7 Desired characteristics of artificial diets for aquatic larvae

Characteristic	Comments
Acceptability	Artificial diets must be attractive and readily ingested. Diet particles must be of suitable size for ingestion and must elicit a feeding response from the larvae. Diet particles must remain available in the water column
Stability	Artificial diet particles must maintain integrity in aqueous suspension and nutrient leaching must be minimal. Some nutrient leaching may be beneficial in enhancing diet attractability
Digestibility	Artificial diets must be digestible and their nutrients easily assimilated
Nutrient composition	Artificial diets must have an appropriate nutritional composition. Material added to the diet as binders or the components of microcapsule walls must have some nutritional value
Storage	Artificial diets must be suitable for long-term (6–12 months) storage with nutrient composition and particle integrity remaining stable

(see Southgate & Partridge, 1998; Koven *et al.*, 2001). In general, they have resulted in lower survival and poorer growth of larvae than those fed live foods and they often lead to a higher incidence of deformity. These results indicate that total replacement of live prey with artificial diets is still not possible for the larvae of most marine fish. Despite this, partial replacement of live foods can result in cost savings, and some studies have shown that between 50% and 80% of a live feed ration can be replaced with a microdiet without affecting larval growth (Kanazawa *et al.*, 1989; Koven *et al.*, 2001). Weaning fish larvae to artificial diets at the earliest possible age is another means of reducing feed costs, and it has been estimated that weaning European bass 15 days earlier enables savings in brine shrimp production of up to 80%.

9.7.2 Constraints to developing artificial diets for marine fish larvae

The relatively poor performance of artificial diets in studies with marine fish larvae is thought to result primarily from reduced rates of ingestion and poor digestion.

Successful artificial diets must be ingested at a similar rate to live foods. This is a particular problem with carnivorous fish larvae, which require the visual stimulus of moving prey to initiate a prey capture response. In efforts to overcome this problem

- various chemicals (using light refraction) have been included in artificial diets to impart a sense of motion
- food dyes have been incorporated into diets to simulate the colour of brine shrimp nauplii
- amino acids that naturally emanate from live food organisms have been used to enhance larval feeding response and may be incorporated into artificial diets to improve attractability.

Most marine fish larvae are poorly developed at hatch, and in many species the digestive tract does not develop fully until after 'metamorphosis'. Marine fish larvae also have low gut enzyme activity compared to adult fish and, again, secretion of some enzymes begins only after metamorphosis when a functional stomach is present. Marine fish larvae generally improve in their ability to digest artificial food particles with age. Live food organisms consumed by larvae assist digestion by 'donating' their digestive enzymes either by autolysis or as zymogens, which activate endogenous digestive enzymes within the larval gut. The digestive enzymes contributed to fish larvae by live food organisms and the implications for artificial diet development were recently outlined by Kolkovski (2001). The inclusion of digestive enzymes (particularly proteases) in artificial diets has been shown to improve nutrient assimilation by up to 30%, resulting in superior larval growth. Similarly, inclusion of digestive system neuropeptides in microdiets may also improve nutrient assimilation and growth.

9.7.3 Weaning diets

Larvae reared on live feeds during the hatchery phase require weaning to artificial feeds towards the end of the larval period (Fig. 9.8). The weaning process usually involves feeding live feeds together with artificial feeds over a period during which the live feed component of the diet is gradually reduced and the artificial component is increased. The duration of the weaning process varies, but weaning is usually completed within 30 days. A wide variety of weaning diets are available commercially in the form of MBDs, MEDs, flake diets, crushed pellets (crumbles) and yeast-based diets. Development of successful artificial diets for fish larvae would eliminate the need to wean larvae from live to artificial foods: larvae could simply be fed larger food particles as they grow.

9.7.4 Practical problems with artificial hatchery feeds

Artificial hatchery feeds are negatively buoyant and this presents problems maintaining food particles in suspension. This may reduce the availability of food to the larvae, and food particles that settle at the bottom of larval culture tanks may pollute culture water and enhance bacterial activity. In contrast, living foods generally remain motile in larval culture tanks, and this maximizes their availability to larvae and reduces contamination of the culture water from uneaten food. Tank design and aeration systems are important in maximizing particle buoyancy and for maintaining particle movement. The use of artificial hatchery feeds requires careful consideration of tank design and aeration, and regular monitoring of feeding rates. Adding small quantities of food a number of times per day optimises water quality and maximises the food available to the larvae.

9.7.5 *Further development of artificial hatchery feeds*

As described above, use of commercially produced artificial hatchery feeds is standard practice in shrimp hatcheries. However, the situation is not as good for marine fish and bivalve hatcheries. Although metamorphosis of fish and bivalve larvae fed artificial diets has been achieved in the laboratory, commercial fish and bivalve hatcheries still rely exclusively on live food production. Development of artificial hatchery feeds for fish and bivalves has been hindered by lack of knowledge of their nutritional requirements. It has also been hindered by problems relating to the attractability and digestion of artificial food particles and their use in culture systems. Development of more suitable artificial diets for marine fish larvae will require further research in the following areas:

- improved ingestion and digestion of artificial diets
- greater understanding of the nutritional requirements of larval stages
- development of more appropriate culture system designs.

The potential cost savings offered by the use of suitable artificial diets will ensure that research in this field is ongoing.

9.8 Harvesting natural plankton

Natural sources of zooplankton represent a large, relatively untapped, potential food source for aquaculture hatcheries. A large portion of this plankton is composed of copepods, which can occur naturally at densities up to 10 000 per cubic metre. Utilising this potential food source requires efficient extraction of zooplankton from very large volumes of water. This can be achieved by pumping water through sieves that collect zooplankton. Plankton harvesting machines have been developed that harvest and grade plankton by size. One of the drawbacks of harvesting from natural waters is that plankton densities may vary between locations and, as such, the reliability of the food supply is questionable. However, this may be overcome by harvesting from dedicated plankton ponds in which high plankton loads can be encouraged by fertilisation. The use of harvested natural zooplankton as a food source for aquaculture is not widespread; however, the variety of organisms present in natural zooplankton would undoubtedly provide a nutritionally superior diet to the standard rotifers/brine shrimp diets used routinely in aquaculture hatcheries.

9.9 Pond fertilisation as a food source for aquaculture

Extensive and semi-intensive pond culture of herbivorous and omnivorous species is usually based on food production through pond fertilisation (section 2.3.4). Fertilisation of ponds for semi-intensive culture of tilapia, for example, is outlined in detail in section 16.8.2. Although this system is more commonly used for grow-out, pond fertilisation has also been used successfully for larval fish culture (section 18.3.4). The fertilisers used for this purpose may be inorganic or organic in nature, or a combination of both.

9.9.1 *Fertilisers*

Inorganic fertilisers are chemical fertilisers that contain at least one of the primary nutrients nitrogen (N), phosphorous (P) and potassium (K). Commercially available agricultural fertilisers such as ammonium sulphate and superphosphate are widely used in aquaculture. Animal manures are probably the commonest organic fertilisers used in aquaculture, although decomposed plant materials are also widely used. Use of organic fertilisers in aquaculture is an ancient practice and is an economical means of increasing production in aquaculture ponds. There is greater reliance on organic fertilisers in developing countries as they are more readily available than chemical fertilisers. They are also more economical to use and more efficient if pond culture is integrated with crop or animal production. In developing countries, terrestrial and aquatic animals (usually fish) are often reared together in integrated systems (section 2.3.6).

9.9.2 Production in fertilised ponds

Fertilisation encourages primary productivity and promotes a succession of organisms within the pond. Initially, fertilisation results in blooms of protozoa and bacteria, which are generally followed by blooms of algae and then zooplankton. The natural food organisms present in ponds can be divided into a number of categories:

- bacteria and protozoans
- plants (phytoplankton, periphyton, macrophytes)
- animals (mainly invertebrates: zooplankton, zoobenthos, small nekton)
- fish.

Fertilisation increases the biomass of potential food organisms present in a pond. For example, Schroeder (1974) reported zooplankton levels of < 0.055 g/m^3 and 3.3–424 g/m^3 in non-manured and manured ponds respectively.

Ecological conditions within a pond determine which organisms are present, the proportion of each and their abundance. All the listed organisms in the pond form the biocenosis (self-regulating ecological community) of the pond, which therefore contains all potential food sources for cultured organisms. However, a given species will only feed on a certain portion of the biocenosis and this is dictated by its feeding habit (e.g. carp species; Fig. 2.8). The specific portion of the biocenosis consumed by an organism is its trophic basis. The trophic basis of a particular species may differ during its life cycle, e.g. the larvae of many herbivorous fish eat zooplankton.

The finite biomass of natural food in a pond can only support a finite standing crop of animals under culture. When the standing crop is low, the amount of available food exceeds the requirements of the culture population and so each animal is able to find sufficient food to support its energy requirement for maintenance and maximum growth. However, an increase in the culture population brings about an increase in their food requirement. At a certain population density, the amount of food required by the culture population for maintenance and growth exceeds that available in the pond. Since the maintenance requirement must be satisfied, the amount of food that can be utilised for growth is reduced and growth rates decline as a result. This standing crop is termed the critical standing crop (CSC). A continued increase in the standing crop further limits the amount of food that can be utilised for growth, and the point at which the natural food available in the pond is sufficient to support only maintenance requirements is known as the carrying capacity of the pond. Fertilisation increases the CSC and carrying capacity of a pond, and growth rates of the culture population can only be increased above these levels if supplementary feed is added to the pond (Table 9.8).

Supplementary feeds are generally classified into simple feeds and compound (or compounded) feeds. Simple feeds may be of animal origin (e.g. trash fish, slaughterhouse waste and fish meal), or of plant origin (e.g. forage, oil meals, rice bran and sorghum). Many simple feeds are relatively cheap agricultural by-products. The simple feeds used by a given aquafarm are dictated by what is locally available. As would be expected, the nutritional composition of simple feeds varies greatly. Those of plant origin are usually rich in carbohydrate, whereas others, such as oil meals and animal products, are rich in protein. Compound feeds consist of mixtures of ingredients that are commonly bound together to form doughs or pellets.

An aquaculture system that utilises supplementary feeds is classed as semi-intensive (Chapter 2), and animal densities are higher than in equivalent extensive systems that rely on natural foods alone. In many developing countries, however, the use of artificial feeds may not be feasible because it raises the cost of production. These countries may also have a limited capability to manufacture or import compound feeds.

9.9.3 Pond culture of fish larvae

Freshwater and marine/estuarine species can be reared extensively in earthen ponds with blooms of natural plankton as their food source. The larvae of barramundi (*Lates calcarifer*) are a good example of larvae reared in this manner. This method has proved to be very successful, and achieves ~50% survival through larval rearing and growth rates greater than those achieved using intensive culture systems (section 18.3.4).

Table 9.8 The effect of fertilisation and supplemental feeding on critical standing crop and carrying capacity of freshwater ponds

Treatment	Critical standing crop (kg/ha)	Carrying capacity (kg/ha)
No feeding, no fertilisation	65	130
No feeding, fertilisation	140	480
Feeding and fertilisation	550	2500

Reproduced from Hepher (1988) with permission from Cambridge University Press.

9.10 Compound feeds

9.10.1 Formulation of compound feeds

Semi-intensive and intensive aquaculture, especially the latter, require artificial feeds, which are usually obtained as uniform particle or pellet sizes from a commercial source. These feeds are typically used for rearing carnivorous and omnivorous fish and crustaceans, i.e. the highly valuable products of aquaculture in developed countries (Chapter 1). As previously indicated, the use of these compound feeds commences for many culture species when the larval stages have been weaned off the initial live feeds.

The major aim in developing commercial diets for aquaculture is to formulate a diet that satisfies the nutritional requirements of the target species for the minimum possible cost. This is known as least-cost formulation. The information required before diet formulation can begin includes:

- a list of available raw materials and information on their compositions and costs
- knowledge of the nutritional requirements of the target species
- the specifications of the diet to be made (i.e. desired levels of protein, lipid, amino acids, fatty acids, etc.)
- knowledge of the special suitability of available raw materials for the target animal.

Feed composition tables provide data on proximate composition (see following) of feed ingredients, and they are useful for the initial choice of potential raw materials for aquaculture diets. Information on the proximate composition of feed ingredients is given according to six major components.

- Moisture is a measure of the water content of the feed or feed ingredient.
- Crude protein (CP) is a measure of the protein content, estimated after total nitrogen analysis.
- Ether extract (EE) is a measure of the lipid- and fat-soluble vitamin content.
- Crude fibre (CF) is a measure of the insoluble polysaccharide content (e.g. cellulose).
- Nitrogen-free extract (NFE) is equivalent to the carbohydrate content.
- Ash is a measure of the inorganic content.

Other compositional data important in feed formulation are the calcium–phosphorus ratio, the available phosphorus content, the amino acid content (particularly lysine, methionine and cysteine) and the polyunsaturated fatty acid contents. An example of a feed composition table is shown in Table 9.9. The information from feed composition tables can be used to construct a table detailing the proximate compositions and costs of complete diets following formulation. The list of potential ingredients often reflects what is locally available.

Protein is usually the largest and most expensive part of an aquaculture diet and, as such, the protein composition of the diet is usually considered first when formulating a new diet. The crude protein level in a feed is most commonly balanced using the 'square method'. A worked example of this method using two ingredients is shown in Fig. 9.9. In this example, fish meal, with a protein content of 54.7%, and maize meal, with a protein content of 10.9%, are to be included in a diet in which the desired protein level is 23%. This desired value is inserted into the middle of the square and the protein content of the two protein sources placed at the left-hand corners

Table 9.9 Feed composition table showing the composition of selected feed ingredients

	Composition (%)						
Ingredient	CP	EE	CF	NFE	Ash	DE (kcal/kg)	Cost ($/mt)
Fish meal	54.7	5.3	4.0	6.1	29.9	3121	580
Shrimp meal	30.6	9.7	0.3	2.3	57.1	2150	400
Groundnut meal	46.9	7.7	6.4	31.7	7.3	3032	330
Maize meal	10.9	5.0	3.1	76.9	4.1	3119	185
Rice bran	13.7	5.4	18.1	48.9	13.9	2417	130

CF, crude fibre; CP, crude protein; DE, digestible energy; EE, ether extract; NFE, nitrogen-free extract.

of the square. The latter values are subtracted from the value in the centre of the square (ignoring minus signs) and the values obtained are placed on the diagonally opposite right-hand corners of the square.

The values obtained allow us to calculate the proportions of each of the protein sources required to obtain the desired protein level. This is calculated on a percentage basis by dividing the value obtained for each protein source by the sum of both values and multiplying by 100. Thus, a combination of 27.6% fish meal and 72.4% maize meal would provide the desired protein level of 23% (Fig. 9.9).

In practice, aquaculture diets are usually formulated with more than two ingredients. A more complicated example in which four nutrient sources are included in the formulation is shown in Fig. 9.10. In this example, a mean protein content is calculated for combined nutrient sources. From the example, the protein content of a 1:1 mixture of fish meal (54.7% protein) and groundnut (peanut) meal (46.9% protein) is calculated as (54.7 + 46.9)/2. The resulting mixture contains 50.8% protein. If the desired proportion of groundnut meal to fish meal was 2:1, the protein content of the mixture would be calculated as (54.7 + 46.9 + 46.9)/3; the resulting mixture would contain 49.5% protein.

Using the example shown in Fig. 9.10, two other combined ingredients, maize meal and rice bran, are also to be included with the fish meal and

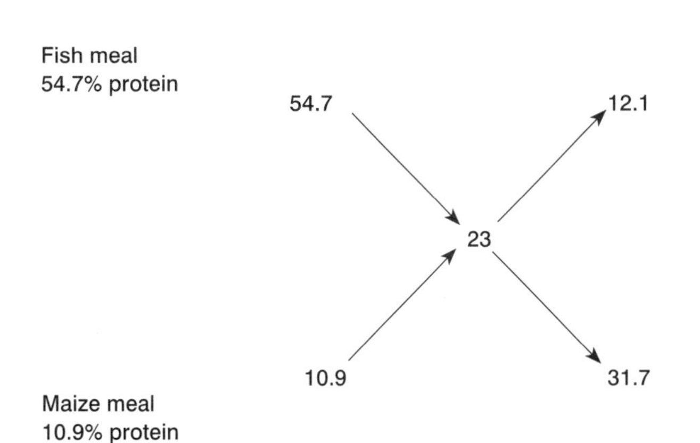

Fig. 9.9 Balancing the crude protein level in a feed using the square method. In this example, fish meal (54.7% protein) and maize meal (10.9% protein) are to be included in a diet with a desired protein level of 23%. This is achieved with a combination of 27.6% fish meal and 72.4% maize meal.

Required proportions

Fish meal	$12.1 \div (12.1 + 31.7) \times 100 = 27.6\%$
Maize meal	$31.7 \div (12.1 + 31.7) \times 100 = 72.4\%$

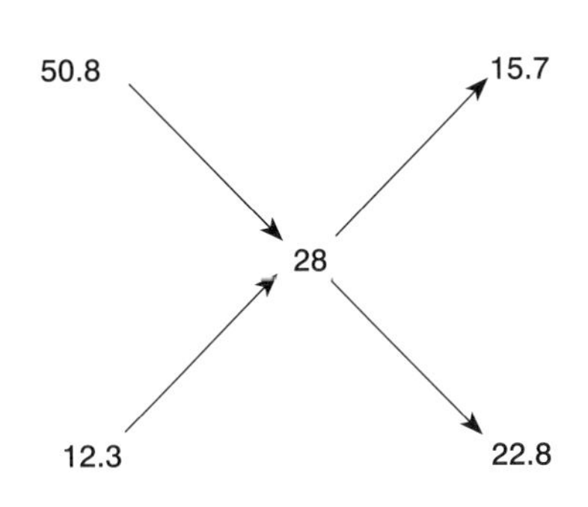

Fig. 9.10 Balancing the crude protein level in a feed containing: (1) a 1:1 mixture of fish meal (54.7% protein) and groundnut meal (46.9% protein) with a protein content of 54.7 + 46.9)/2 = 50.8%; and (2) a 1:1 mixture of maize meal (10.9% protein) and rice bran (13.7% protein) with a protein content of 10.9 + 13.7/2 = 12.3% protein to achieve a desired protein content of 28%. The final formulation is shown in Table 9.10.

Required proportions

Fish meal/groundnut meal $15.7 \div (15.7 + 22.8) \times 100 = 40.8\%$

Maize meal/rice bran $22.8 \div (15.7 + 22.8) \times 100 = 59.2\%$

Final combination is

Fish meal 20.4%
Groundnut meal 20.4%
Maize meal 29.6%
Rice bran 29.6%

Table 9.10 Composition of the diet formulated in Fig. 9.10

Ingredient	Inclusion (%)	Protein	Lipid	Inclusion cost ($/mt)
Fish meal	20.4	11.16	1.08	118.3
Groundnut meal	20.4	9.56	1.57	67.3
Maize meal	29.6	3.23	1.48	54.8
Rice bran	29.6	4.05	1.59	38.5
Total	100.0	28.0	5.72	278.9

groundnut meal in a diet containing 28% protein. A 1:1 mixture of maize meal (10.9% protein) and rice bran (13.7% protein) has a protein content of 12.3%. A combination of 20.4% fish meal, 20.4% groundnut meal, 29.6% maize meal and 29.6% rice bran will give the desired level of 28% protein (Fig. 9.10). The formulation can now be examined in terms of its cost and also with regard to its content of other nutrients, such as lipid (Table 9.10). This information can be calculated from data in the feed composition table (Table 9.9). The method outlined above is clearly not restricted to balancing the protein content of a diet. It can be use to calculate the combination of ingredients required to provide a desired level of other nutrients (e.g. lipid).

In practice, diet formulation may be far more complex than the examples shown above. Clearly, ingredients must be chosen for inclusion in aquaculture diets so that the completed diet provides the target animal with the required nutrient balance. The completed diet must provide all essential nutrients; dietary protein must provide essential amino acids (EAAs) and dietary fat must provide essential lipids. Other considerations in diet formulation are energy content, a suitable mineral balance and vitamin content. When selecting ingredients from a wide range of available raw materials, it is useful to make a crude comparison of them on a best-value basis. For example, which of the potential ingredients offers the best value per unit protein (cost per unit protein)? Their amino acid compositions must also be considered; for example, fish meal may have a better EAA profile than plant proteins, but the latter may have higher protein contents. The majority of pelleted feeds for marine fish and crustaceans are based on marine protein meals such as fish meal, squid meal and shrimp meal. Although of high nutritional value, marine meals are expensive and may be difficult to obtain in a regular supply. Increasingly, plant meals (in particular soybean meal) are being studied as a partial replacement for marine meals in aquaculture feeds. Plant meals are considerably cheaper than marine meals, although they are generally deficient in the sulphur amino acids (methionine and cysteine) and they may contain anti-nutritional factors such as a trypsin inhibitor in soybean meal.

Aquaculture species differ greatly in their nutritional requirements. Clearly, aquaculture feeds must cater for these differences, and commercial diets are usually formulated on a species-specific basis. The compositions and ingredients of two commercial pelleted aquaculture feeds, one for marine shrimp and the other for a marine fish, are shown in Table 9.11. Their compositions reflect the nutritional requirements of the target species. For example, unlike fish, shrimp are unable to synthesise sterols, which must be supplied in the diet (section 8.9.8). A source of sterols (usually cholesterol) and phospholipids (e.g. lecithin) is usually included in diets for shrimp (Table 9.11). These are not usually supplied as discrete nutrients in diets for fish, although sterols and phospholipids are present as components of other ingredients.

9.10.2 Pellet manufacture

Compressed pellets

There are two steps involved in the manufacture of compressed pelleted aquaculture feeds.

(1) Grinding involves a reduction in the particle size of coarse ingredients using a hammer mill or roller. This process helps efficient mixing of ingredients and aids the pelleting process. It also increases pellet stability and digestibility of the final product. Once weighed, dietary ingredients are thoroughly mixed to produce a homogeneous blend prior to pelleting.
(2) Pelleting converts this mixture into a compressed water-stable form and involves forcing the feed mixture through holes in a metal die plate. Exposure to steam during the process increases the temperature of the diet mixture to ~85°C and its moisture content to ~16%. This process causes gelatinisation of some of the starch in the diet, which helps pellet formation and nutrient binding. Pellets of the desired length are cut off as they emerge from the die. Newly manufactured pellets are cooled using a cooler–drier, in which the pellets are moved on a belt through a stream of cold air. The resulting pellets have a moisture content of ~10%.

Pellet quality is greatly influenced by the fat level of the diet. Ideally, the fat content of the dietary mixture must be between 2% and 10%. Fat in the mixture aids lubrication of the diet through the die and helps reduce dustiness. However, excess fat in the mixture can result in insufficient compression and a reduction in pellet hardness. The fat content of pelleted feeds can be increased by top-dressing. This involves spraying fat onto the surface of the pellet after manufacture. Top-dressing can also be used to apply heat-sensitive nutrients, such as vitamins, with the fat.

Extruded pellets

Extrusion is a process whereby the dietary mixture is exposed to controlled conditions of high temperature (125–150°C), increased moisture content (20–24%) and high pressure. This is achieved by forcing the diet through a die plate under high pressure accompanied by an injection of steam. This process causes the water in the feed to evaporate, which leaves air spaces in the resulting pellets. This can be controlled to regulate the bulk density of the pellet, enabling production of floating or sinking pellets, depending on design. The benefits and disadvantages of the extrusion process are shown in Table 9.12.

Extruded pellets are superior in nutritional value to pellets of the same formula made by conventional steam pelleting. Extruded feeds have a better food

Table 9.11 Examples of the typical ingredients and general compositions of commercial aquaculture feeds for grow-out of a carnivorous marine fish and shrimp

		Composition (%)		
Animal	Ingredients	Crude protein	Fat (lipid)	Fibre
Fish	Fish meal, cereals, oilseed meals, protein meals, fish oil, vitamin, minerals	43	10	5
Shrimp	Fish meal, shrimp meal, squid meal, fish/squid oil, wheat flour, soybean meal, broken rice, cholesterol, phospholipid (lecithin), vitamins, minerals	37	2.8	3

Table 9.12 Some benefits and drawbacks of extrusion processing

Benefits	Drawbacks
Starches are gelatinised rendering them more digestible	Increased production costs
Gelatinisation also aids pellet binding	Destruction of heat-labile ingredients such as vitamins. Compensated for by over-fortifying with these ingredients, which adds to production costs
Proteins are rendered more digestible	Reduced availability of certain nutrients (e.g. amino acids)
Anti-nutritional factors are eliminated	
Control over pellet buoyancy	

conversion ratio and protein efficiency ratio (ratio of wet weight gain to amount of protein consumed) than conventional pellets (i.e. less feed is required for the same growth rate). This has been estimated to allow a 7–8% reduction in production costs for rainbow trout. However, the extrusion process also has some drawbacks (Table 9.12).

9.10.3 Storing pelleted feeds

Aquaculture feeds are used in large amounts in aquafarms and this necessitates on-site storage (Fig. 9.11). Storage conditions can have considerable influence on the nutritional value of pelleted feeds and a number of factors are important in this regard:

Fig. 9.11 A feed storage shed at a large shrimp farm.

(1) Age of the feed. Feed must be as fresh as possible when it arrives at the aquaculture site for storage.
(2) Environmental factors such as:
   - moisture (of feed and humidity of storage area)
   - temperature
   - light
   - oxygen.

These affect the stored feed directly or influence other factors causing deterioration, such as fungal contamination and insect infestation.

Regardless of the initial moisture content of pelleted feeds, they will gradually absorb moisture and achieve equilibrium with atmospheric humidity. Because of this, feeds cannot be stored as long in tropical areas because of the high humidity. Ideally, feeds must be stored with a 10–12% moisture content.

Atmospheric oxygen is required for the development of oxidative rancidity of dietary fats and for growth of fungi and insects. However, it is impracticable to exclude oxygen from storage areas. Although most fungi are killed during feed processing (by heat), their spores may remain viable and reinfect the feed if conditions become suitable. Fungal damage is a major concern in the tropics, where high temperature and humidity encourage their growth. Fungal infection causes increase in temperature and moisture, off-flavour and discolouration, and may also produce mycotoxins. These include aflatoxins, which are known to be toxic to fish. Plant products in feeds are particularly prone to contamination with mycotoxins.

9.10.4 Chemical changes during food storage

The major chemical changes that can occur in aquaculture feeds during storage include:

- oxidation of lipids
- fermentation of carbohydrates
- loss of vitamin activity
- loss of pigment colour.

Lipids may undergo irreversible autoxidation when exposed to air, causing rancidity. Rancidity results in a number of compounds that have toxic properties and the peroxides are probably the better known of these. Feeds containing ingredients with high levels of polyunsaturated fatty acids (PUFAs), such as fish meals and oil cakes, have a greater risk of oxidation. Grains contain natural antioxidants that protect them from rapid deterioration. Antioxidants may also be added to feeds to slow deterioration. Commonly used antioxidants include butylated hydroxyanisole (BHA), butylated hydroxytoluene (BHT) and ethoxyquin (EQ). Fermentation of carbohydrates in aquaculture feeds produces alcohols and volatile fatty acids. Oxidation and fermentation reduce the nutritional value and the palatability of aquaculture feeds and can produce toxic chemicals that may depress growth.

The potency of vitamins may decrease during storage, with vitamin C (ascorbic acid) and vitamin B_1 (thiamine) being particularly susceptible. More stable forms of ascorbic acid, such as calcium ascorbate, L-ascorbyl phosphate and ascorbyl palmitate, are now available and are more suited for use in aquaculture feeds. The vitamin content of pelleted aquaculture feeds can be improved immediately prior to use by coating or dipping the feed in a vitamin-rich material.

In order to ensure the maximum nutritional value of stored aquaculture feeds a number of guidelines must be followed:

(1) The storage time must be minimised. Food must be used as quickly as possible using the oldest feed first. The feed must be ordered in appropriate quantities.
(2) The feed must be protected from the elements (rain, light, high temperature). Efforts must be made to reduce humidity and allow adequate ventilation.
(3) The storage area must be kept clean to minimise pest populations.
(4) Storage under refrigeration or air conditioning is ideal.

9.10.5 Other components of aquaculture feeds

A number of non-nutritional or undesirable materials are either deliberately incorporated into aquaculture feeds (e.g. to improve feed utilisation and prevent deterioration) or are present incidentally as components of diet ingredients.

The materials present incidentally may seriously affect the growth and survival of culture animals (e.g. toxins), and some may be passed on to humans through consumption of contaminated aquaculture products. Many of these factors are naturally occurring and are associated with protein-rich plant materials, fungi and bacteria.

(1) Protease inhibitors have been isolated from a number of feed ingredients from plants including soybean meal and wheat flour. Inhibition of enzyme activity (e.g. trypsin) results in poor feed utilisation and depressed growth rates.
(2) Gossypol is a toxic pigment present in cottonseed meal. It binds to protein, thereby reducing amino acid availability and causing reduced growth rates and poor utilisation of feed.
(3) Saponins are water-soluble plant toxins. Toxicity is believed to result from destruction of erythrocytes. Aquatic animals vary in their sensitivity to saponins, with fish generally being

more sensitive than crustaceans (which do not have erythrocytes). This is used to advantage in the culture of shrimp when saponin-containing material is added to ponds as a piscicide to eliminate fish predators.

(4) Cyanogens are plant materials containing cyanide. Their toxicity results from the release of free cyanide following enzymic hydrolysis.
(5) Mycotoxins are metabolites produced by fungal contamination of feeds or feed ingredients.
(6) Aflatoxins are mycotoxins produced by *Aspergillus flavus* that can be extremely toxic to aquatic organisms. Aflatoxins have been shown to cause tumours and death in fish at very low levels, although sensitivity varies greatly between species. Levels of aflatoxin in feeds lower than 1 mg/kg have been shown to cause death in rainbow trout within days.
(7) Bacterial contamination of aquaculture feeds can cause disease and mortality of aquaculture animals. For example, *Mycobacterium* species associated with trash fish fed to cultured Pacific salmon have been reported to cause disease. A bacterium of particular note is *Clostridium botulinum*, which produces a toxin (botulinum toxin) causing death in fish.
(8) Chemical contaminants may be present in aquaculture feed ingredients in many forms. The most common contaminants are residues of agricultural chemicals such as organic pesticides and herbicides, and heavy metals from industrial wastes. These materials can have a direct toxic effect on culture animals, but may also accumulate in their tissues, rendering the animals unsuitable for human consumption.

As well as these undesirable and incidental factors that may occur in feeds, some non-nutritional additives may be deliberately included in feeds. The various reasons for including these are outlined in Table 9.13. Some of the additives commonly included in shrimp feeds are outlined in section 19.9.1.

Table 9.13 Some non-nutritional additives to aquaculture feeds

Additive	Function
Antioxidants	Oxidation leads to losses of vitamins, amino acids and fatty acids from the feed, and to the production of toxic by-products. Natural antioxidants such as α-tocopherol and lecithin or artificial antioxidants such as BHA, BHT and EQ are often included in the vitamin or mineral premix used in aquaculture feeds
Antifungal agents	Fungal growth leads to production of toxins such as aflatoxins. Examples of antifungal agents used in aquaculture feeds are potassium sorbate, which is a preventative only and does not kill existing fungi, and polypropylene glycol, which prevents fungal growth and kills existing fungi. Neither will remove aflatoxins already present in the feed
Binders	Binders increase the water stability of the feed and reduce nutrient leaching. Examples of binders used in aquaculture feeds are alginate, guar gum, gelatin and starches
Antibiotics	Antibiotics may be used at high levels for short-term treatment of disease. Used at low levels, antibiotics may improve growth and feed conversion in terrestrial animals by facilitating nutrient uptake
Hormones	Hormones influence the growth rate of an animal and hormone supplements in feeds can improve growth and feed utilisation. Major problems may be the accumulation of residues in tissues
Pigments	Chemicals included in aquaculture feeds to influence tissue colour (e.g. carotenoid pigments in salmonid feeds)
Flavourings	Gustatory stimulants such as animal extracts and certain amino acids may increase feed consumption and palatability of aquaculture feeds

BHA, butylated hydroxyanisole; BHT, butylated hydroxytoluene; EQ, ethoxyquin.

9.11 Dispensing aquaculture feeds

9.11.1 Feed ration

The amount of food presented to culture animals is a major influence on the productivity and running costs of an aquaculture venture. Overfeeding wastes expensive feed and promotes water quality problems, whereas underfeeding reduces growth rates, overall yield and profitability. Species-specific feeding tables have been developed as a guide to the appropriate rations to feed culture stock (Table 9.14). They are available to farmers from commercial feed manufacturers. Feeding tables take into account the size of culture animals and the water temperature; both influence optimal ration size. Effective use of feeding tables requires knowledge of animal size, the number of animals in a culture unit and the water temperature. However, feeding tables should be used in conjunction with observation of factors such as feeding activity, water quality and uneaten food. In pond culture of shrimp for example, in which feeding activity is difficult to observe directly, feeding trays are commonly used to monitor food consumption (section 19.9.8). Some potential problems with reliance on feeding tables are outlined in section 19.9.7. Feeding tables should be used as a guide only, and rations should be adjusted on the basis of observation if required.

Dividing the daily ration over a number of separate feeds results in more efficient use of feeds and reduces waste. A successful feeding protocol should also account for the natural feeding habits of the culture species. For example, shrimp farms in Asia provide the majority of the daily feed ration for *Penaeus monodon* at night, to better suit the feeding activity of this species (section 19.9.8).

9.11.2 Methods of feeding

Aquaculture feeds are commonly dispensed to tanks and ponds by hand. This allows the farmer to observe feeding activity and inspect culture stock on a regular basis. On larger land-based farms (e.g. shrimp farms), the feed is usually transported to tanks or ponds using trucks or tractors. The feed can then be thrown or shovelled into ponds or can be distributed using machines such as air-blowers, which can propel feed a considerable distance from the pond bank (Fig. 19.10). In ocean-based aquaculture (e.g. cage culture of fish), the feed can be distributed from air-blowers mounted on boats or from water cannons (Fig. 8.1).

However, manual distribution of feeds (whether machine assisted or not) is costly, labour intensive and inappropriate on large farms. In these cases, feeders can be mounted 'permanently' over a pond, tank or cage and can operate as demand feeders or use timers to control feed release (Fig. 9.12). All have hoppers to hold large amounts of food (Fig. 15.8). Demand feeders have a rod or feeder pendulum, which runs down from the hopper to a depth of ~20–30 cm below the water surface.

Table 9.14 An example of a feeding table for a cultured tropical fish

	Water temperature (°C)								
	20	21	22	23	24	25	26	27	28
Fish weight (g)	Daily food allowance (kg/1000 fish)								
50	0.51	0.72	0.95	1.12	1.30	1.48	1.60	1.75	1.84
60	0.65	0.87	1.06	1.29	1.44	1.62	1.78	1.89	2.01
70	0.76	1.01	1.20	1.41	1.59	1.75	1.90	2.03	2.13
80	0.88	1.33	1.52	1.70	1.87	2.03	2.13	2.24	2.34
90	0.99	1.22	1.44	1.63	1.82	1.97	2.11	2.24	2.36
100	1.09	1.32	1.54	1.73	1.92	2.07	2.22	2.35	2.45
150	1.52	1.75	1.97	2.17	2.35	2.52	2.65	2.79	2.89
200	1.89	2.12	2.33	2.53	2.71	2.87	3.02	3.15	3.24
250	2.20	2.44	2.65	2.85	3.02	3.19	3.33	3.46	3.56
300	2.48	2.72	2.93	3.13	3.30	3.47	3.61	3.74	3.83
400	2.97	3.21	3.42	3.63	3.8	3.96	4.10	4.23	4.33

Fig. 9.12 Automatic feeding devices are commonly used in tank-based nursery culture of fish. This type of feeder allows the amount of food dispensed and the frequency and time(s) of feeding to be programmed.

Movement of the pendulum when touched by fish brings about the release of a small quantity of food. Computerised control of feeders is now common on large aquaculture farms. This allows automation of the amount of feed dispensed, the number of feeds per day and the timing of feeds. To improve the efficiency of automated feeders, new technologies are being developed to more closely match the supply of feed to demand (section 15.5.2). Modern 'adaptive feeders' often include some method of feedback by which the amount of feed dispensed is adjusted in response to the feeding activity of the fish. This approach minimises wasted feed and has obvious environmental benefits (section 4.2.1).

References

Biedenbach, J. M., Smith, L. L. & Lawrence, A. L. (1990). Use of a new spray-dried algal product in penaeid larval culture. *Aquaculture*, **86**, 249–57.

Brown, M. R., Jeffrey, S. W. & Garland, C. D. (1989). *Nutritional Aspects of Microalgae Used in Mariculture: A Literature Review*. CSIRO Marine Laboratories report 205. CSIRO, Hobart.

Cañavate, J. P. & Fernández-Díaz, C. (2001). Pilot evaluation of freeze-dried microalgae in the mass rearing of gilthead seabream (*Sparus aurata*) larvae. *Aquaculture*, **193**, 257–69.

Guillard, R. R. L. (1972). Culture of phytoplankton for feeding marine invertebrates. In: *Culture of Marine Invertebrate Animals* (Ed. by W. L. Smith & M. H. Chanley), pp. 29–60. Plenum Press, New York.

Hepher, B. (1988). *Nutrition of Pond Fishes*. Cambridge University Press, Cambridge.

Jeffrey, S. W., Leroi, J. M & Brown, M. R. (1992). Characteristics of microalgal species needed for Australian mariculture. In: *Proceedings of the Aquaculture Nutrition Workshop* (Ed. by G.L. Allan & W. Dall), pp. 164–73. NSW Fisheries, Australia.

Jones, D. A. (1998). Crustacean larval microparticulate diets. *Reviews in Fisheries Science*, **6**, 41–54.

Kakazu, K. (1987). Tilapias. In: *Aquaculture in Tropical Areas* (Ed. by S. Shokita, K. Kakazu, A. Tomori & T. Toma), Midori Shobo, Japan (in Japanese) (pp. 127–36 in English edition prepared by M. Yamaguchi, 1991).

Kanazawa, A., Koshio, S. & Teshima, S. (1989). Growth and survival of larval red sea bream *Pagrus major* and Japanese flounder *Paralichthys olivaceus* fed microbound diets. *Journal of the World Aquaculture Society*, **20**, 31–7.

Knauer, J. & Southgate, P. C. (1999) A review of the nutritional requirements of bivalves and the development of alternative and artificial diets for bivalve aquaculture. *Reviews in Fisheries Science*, **7**, 241–80.

Kolkovski, S. (2001). Digestive enzymes in fish larvae and juveniles – implications and applications to formulated diets. *Aquaculture*, **200**, 181–201.

Koven, W., Kolkovski, S., Hadas, E., Gamsiz, K. & Tandler, A. (2001). Advances in the development of microdiets for gilthead seabream, *Sparus aurata*: a review. *Aquaculture*, **194**, 107–21.

Laing, I. & Millican, P. F. (1992). Indoor nursery cultivation of juvenile bivalve molluscs using diets of dried algae. *Aquaculture*, **102**, 231–43.

Lavens, P. & Sorgeloos, P. (2000). The history, present status and prospects of the availability of *Artemia* cysts for aquaculture. *Aquaculture*, **181**, 397–403.

Leger, P., Bengtson, D. A., Simpson, K. L. & Sorgeloos, P. (1986). The use and nutritional value of *Artemia* as a food source. *Oceanography and Marine Biology Annual Reviews*, **24**, 521–623.

Lubzens, E., Tandler, A. & Minkoff, G. (1989). Rotifers as food in aquaculture. *Hydrobiologia*, **186/187**, 387–400

Navarro, N. & Sarasquete, C. (1998). Use of freeze-dried microalgae for rearing gilthead seabream, *Sparus aurata*, larvae. I. Growth, histology and water quality. *Aquaculture*, **167**, 179–93.

Nell, J.A. & O'Connor, W. (1991). The evaluation of fresh and stored algal concentrates as a food source for Sydney rock oyster, *Saccostrea commercialis* (Iredale & Roughley) larvae. *Aquaculture*, **99**, 277–84.

Nellen, W. (1986). Live animal food for larval rearing in aquaculture: non-*Artemia* organisms. In: *Realism in Aquaculture: Achievements, Constraints and Perspectives* (Ed. by M. Bilio, H. Rosenthal & C. J. Sindermann), pp. 215–50. European Aquaculture Society, Bredene, Belgium.

Parsons, T. R., Stephens, K. & Strickland, J. D. H. (1961). On the chemical composition of eleven species of marine phytoplankters. *Journal of the Fisheries Research Board of Canada*, **18**, 1001–16.

Payne, M. F. & Rippingale, R. J. (2001). Intensive cultivation

of the calanoid copepod *Gladioferens imparipes*. *Aquaculture*, **201**, 329–42.
Schroeder, G. L. (1974). Use of cowshed manure in fish ponds. *Bamidgeh*, **26**, 84–6.
Segers, H. (1995). Nomenclatural consequences of some recent studies on *Brachionus plicatilis* (Rotifera, Brachionidae). *Hydrobiologia*, **313**, 121–2.
Sorgeloos, P., Lavens, P., Leger, P., Tackaert, W. & Versichele, D. (1986). *Manual for the Culture and use of Brine Shrimp Artemia in aquaculture*. Artemia Reference Centre, Sate University of Ghent, Ghent, Belgium.
Southgate, P. C. & Lou, D. C. (1995). Improving the *n*3-HUFA composition of *Artemia* using microcapsules containing marine oils. *Aquaculture*, **134**, 91–9.
Southgate, P. C. & Partridge, G. J. (1998). Development of artificial diets for marine finfish larvae: problems and prospects. In: *Tropical Mariculture* (Ed. by S. DeSilva), pp. 151–69. Academic Press, London.
Utting, S. D. (1986). A preliminary study on growth of *Crassostrea gigas* larvae and spat in relation to dietary protein. *Aquaculture*, **58**, 123–38
Volkman, J. K., Jeffrey, S. W., Nichols, P. D., Rogers, G. I. & Garland, C. D. (1989). Fatty acid and lipid composition of 10 species of microalgae used in mariculture. *Journal of Experimental Marine Biology and Ecology*, **128**, 219–40.
Webster, C. D. & Lovell, R. T. (1991). Lipid composition of three geographical sources of brine shrimp nauplii (*Artemia* sp.). *Comparative Biochemistry and Physiology*, **100B**, 555–9.
Whyte, J. N. C. (1987). Biochemical composition and energy content of six species of phytoplankton used in mariculture of bivalves. *Aquaculture*, **60**, 231–41.
Yoshimura, K., Hagiwara, A., Yoshimatsu, T. & Kitajima, C. (1996). Culture technology of marine rotifers and the implications for intensive culture of marine fish in Japan. *Marine and Freshwater Research*, **47**, 217–22.

10
Diseases

Leigh Owens

10.1 Introduction

Diseases: The most common cause of bankruptcy in aquaculture.

Disease can be defined as 'any process that limits the productivity of a system' and is one of the most seriously limiting factors in aquaculture (Kabata, 1985). Economic evaluations of disease problems are relatively rare, but when they have been conducted the calculated costs are staggering. The best-known example is the US$3 billion loss of cultured shrimp in 1994 due to a combination of whitespot syndrome virus (mostly) and Taura virus (Lundin, 1997). Similar losses include:

- the US$0.5 billion crash of cultured black tiger shrimp (*Penaeus monodon*) in Taiwan in 1987–88, mainly due to viral infections
- the protozoan infections in French oysters (*Ostrea edulis*), which caused a loss of FFr1.8 billion over an 8-year period and a further FFr1.3 billion lost in associated industries
- US$33 million loss caused by Hitra disease (*Vibrio salmonicida*) in Norwegian salmon farms in 1985.

Diseases include both infectious (Table 10.1) and non-infectious (environmental, nutritional and genetic) problems. The non-infectious diseases are solely due to management practices and are often limited to particular farms. However, infectious diseases have the potential to threaten whole industries and will therefore form the basis of this chapter.

10.2 General principles of diseases in aquaculture

10.2.1 Interaction between host, pathogen and environment

The Sneizko three-ring model of the interactions between host (the aquacultured species), pathogen and environment is well known (Fig. 10.1). It illustrates the fact that most infectious disease is a three-way interaction needing all these components:

- pathogen
- host
- environment.

Various modifications of this model have been made to illustrate specific points. Non-infectious disease is an interaction only between the host and environment. The area of overlap between pathogen and host represents obligate pathogens, which are the most threatening group, as they do not need environmental stress to cause clinical disease. Epizootic haematopoietic necrosis virus in redfin perch and crayfish plague (*Aphanomyces astaci*) in signal crayfish are two such examples of obligate pathogens producing disease in the most pristine conditions.

Table 10.1 Some major pathogens of aquaculture species

Group	Genera, etc.
Viruses	Bacilliform viruses, herpesvirus, iridovirus, nodavirus, rhabdovirus, coronavirus, birnavirus
Bacteria	Rickettsiales, *Aeromonas*, *Enterococcus*, *Flavobacterium*, *Flexibacter*, *Pseudoalteromonas*, *Pseudomonas*, *Streptococcus*, *Vibrio*
Fungi	*Aphanomyces*, *Branchiomyces*, *Lagenidium*, *Saprolegnia*, *Sirolipidium*
Protozoa	Amoebae: *Neoparamoeba* Flagellates: *Hexamita*, *Ichthyobodo* Ciliates: *Ichthyophthirius*, *Trichodina* Sporozoans: *Bonamia*, *Loma*, *Marteilia*, *Perkinsus*
Helminths	*Dactylogyrus*
Nematodes	
Annelids	*Polydora*
Crustaceans	Fish 'lice': Isopods: Fish 'lice': Branchiura: *Argulus* Copepods: *Lernaea*, *Ergasilus*, *Mytilicola* Crabs: Pinnotherids
Gastropods	Pyramidellids

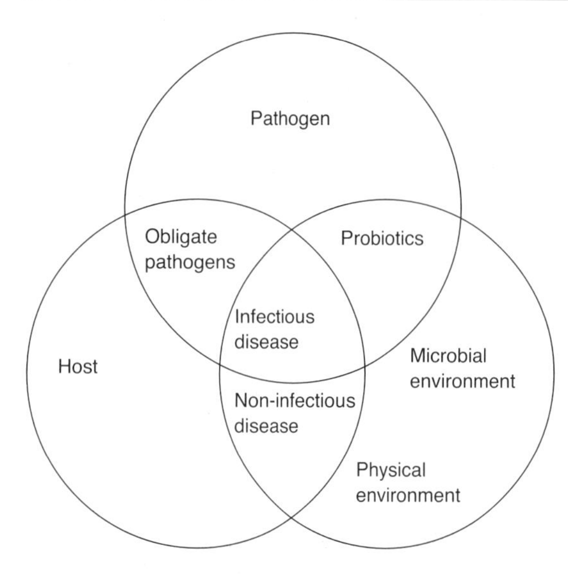

Fig. 10.1 A modified Sneizko three-ring model depicting the interaction between host, pathogen and the environment.

For specific diseases of cultured species, the three-ring model can be modified by changing the size of the rings to reflect the relative importance of the various components.

10.2.2 Density and disease

The spread of pathogens is a density-dependent process and is therefore affected by stocking rates; the higher the density, the smaller the distance between neighbours. This leads to a higher likelihood of pathogens crossing the distance between hosts in a viable state. Immobile pathogens such as

- viruses
- non-motile bacteria
- sporozoans
- parasite eggs

basically follow the diffusion laws and, therefore, in still water conditions a concentration gradient of the pathogens will be formed around an infected individual. Other pathogens such as

- bacteria
- fungal zoospores
- protozoa
- metazoans

generally have active but variable dispersal capabilities. As the distance increases, fewer pathogens will be able to reach susceptible hosts to establish or continue a disease epizootic (outbreak). As there is natural attrition of pathogens in the environment, if the pathogen does not reach a susceptible host within a defined period of time, the chance of establishing a new infection is almost zero.

Higher densities lead to genetic selection of mutant pathogens that are virulent.

(1) In the natural environment, a newly mutated pathogen does not 'know' where the next susceptible host will be encountered. In such a case, pathogens that slowly release progeny are selected for because this will give a random-moving infected host a chance to spread the pathogen in the environment and to the next host. That is, a pathogen 'wants' to destroy its current host, but rather allow it to function as close to normal as possible, thus increasing the chance of an encounter with another susceptible host.
(2) However, under culture conditions the next host is very close. A mutant that uses all the available resources of a host to produce pathogenic progeny, no matter what the consequences to the host, will be selected for (i.e. a virulent organism). One of its large number of progeny is the most likely to infect the next host, as opposed to a conservative, slow-release adapted pathogen. Once a virulent, resource-utilising pathogen arises by mutation, it is very quickly selected for in an aquaculture environment. A single mutation in a nucleotide that begins a codon for a single amino acid can change virulence from non-virulent to highly virulent. This may contribute to the sequential rise of pathogen problems that beset so many aquaculture industries (Fig. 10.2).

10.2.3 The effect of aquaculture on life cycles of pathogens

By stocking facilities with monocultures, aquaculturalists exclude both predators and competitors of the species being cropped. A large number of the prey items of the cultured species are also excluded. Exclusion of cohabiting animals results in removal of intermediate hosts and definitive hosts from the aquacultural ecosystem. This effectively breaks the life cycle of many of the multi-host helminths (flatworms and nematodes), which consequently have less of a role in disease in aquaculture than in wild populations. Seacages are much less effective at breaking these life cycles than are ponds or recirculation systems.

10.3 The philosophy of disease control

Disease control in aquaculture is usually attempted on the assumption that an absence of pathogens is the desired state. However, the chance of beginning an aquaculture venture without any potential pathogens in the system are very slim and the question arises whether it is cost-effective to reach the pathogen-free state. This 'total elimination of pathogens' strategy is the classic approach to disease control: the pathogenocentric approach. This may not always be the best strategy. If, for instance, the chance of reinfection from the local environment is very high, the infection starts to re-establish immediately after

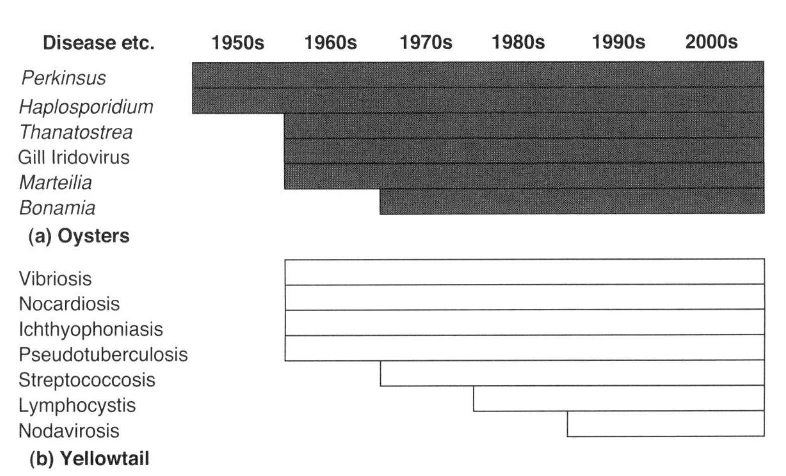

Fig. 10.2 Sequential appearance of microbial diseases of cultured animals. The dates are approximate. (a) Atlantic oysters (*Crassostrea virginica* and *Ostrea edulis*); (b) Japanese yellowtail (*Seriola quinqueradiatus*).

control is completed. Clearly, the eradication philosophy is untenable in such areas.

There are a number of factors to consider when deciding on control measures.

(1) The cost of the control measure. Some pathogens make culture uneconomical in their presence and must be totally removed from the culture system (e.g. highly pathogenic organisms such as crayfish plague). Others might be self-limiting and reduce standing stocks by only a small percentage (e.g. peritrichous ciliates on crustaceans). In this case, living with the pathogen is more cost-effective than trying to eradicate it. In most cases, however, there are no accurate cost estimates of losses due to a pathogen. Rather, a general and often very inaccurate 'feel' for the costs associated with a disease agent are the best estimates available. However, this should not stop the aquaculturist from trying to economically evaluate the cost of the control method and the downstream benefits.
(2) The likelihood of reinfection. Ideally, there should be almost no chance of the pathogen being reacquired from the environment or from wild stocks in the vicinity. Alternatively, infection with a pathogen and subsequent treatment will often allow the vertebrate immune system to be primed and thus further infections are limited (e.g. white spot, *Ichthyophthirius multifiliis* on fish). However, sometimes the environment is supersaturated, with the pathogen from adjoining farms suffering the same problem, and unless the farm can be isolated treatment may be almost useless (e.g. vibriosis in South-East Asian shrimp farms). In seacage situations, wild stock will often congregate outside the cages, where they can easily recontaminate caged stock that have been treated. Therefore, an understanding of the probability of reinfection is needed to assess the control strategy correctly. Again this information is usually lacking.
(3) An adequate assay for the pathogen. It must be possible to accurately identify the pathogen to be able to assess the effect of the control measures on the pathogen. In the first instance this relies on an accurate diagnosis and later on a sampling regime that will determine true positives (sensitivity) and true negatives (specificity) (Table 10.2). The sampling regime is constructed by assuming a certain prevalence which is the maximum allowable for that pathogen and the confidence level that is acceptable by the aquaculturist. If the test is insensitive, the sample size or frequency of sampling must be increased to compensate for the low discriminatory ability of the test. If the test is non-specific, then animals will be assessed as infected when they are not, and a pathogen control or treatment regime may be discounted when it had worked at the level required. The numbers needed for accurate sampling of low prevalences are large, and few commercial ventures willingly part with the necessary numbers without monetary compensation or it being mandatory compliance to legislation.

10.4 Generalised disease management techniques

The most important factor for the movement and introduction of pathogens to farms and, indeed on any geographical scale, is the movement of animals. This includes:

- live broodstock in particular
- live larval forms for stocking
- live alternative hosts, such as hobby fish
- frozen carcasses for human consumption
- aquaculture feeds
- bait.

Probably 99% of new introductions of pathogens to uninfected systems is due to the unrestrained movement of contaminated animals. Sometimes this is unavoidable as aquaculture does not exist

Table 10.2 Relationship between a diagnostic test and the real presence of a pathogen

Result of test	Pathogen present	Pathogen absent
Positive	A	B
Negative	C	D
	A + C	B + D

Sensitivity (%) = A/(A + C) × 100.
Specificity (%) = D/(B + D) × 100.

without either broodstock or live juveniles for stocking. However the biosecurity of broodstock, the number 1 contaminator, has been greatly neglected and should be the first point considered, i.e. are pathogen-free broodstock available? If not, what is the pathogen status of the broodstock that is being used? For example, the major scourges of marine fish aquaculture and shrimp aquaculture, viral encephalopathy and retinopathy, and whitespot syndrome virus, respectively, are both spread vertically from broodstock to larvae and then distributed through infected postlarvae and juveniles to farms.

Although it is impossible to have strategies that will work for all pathogens, there are a number of procedures that can help limit pathogens within culture systems.

10.4.1 Batch culture

Batch culture works on the 'all in, all out' principle. Continuous culture eventually suffers from the early batches acting as pathogen 'factories' that contaminate the environment to such a level that later batches cannot be raised (e.g. viral encephalopathy and retinopathy barramundi nodavirus). Furthermore, young animals are often susceptible to levels of a pathogen that only mildly affect older stages. This is due to a naive immune response in young animals and, having fewer cells in a target organ, the animal is more compromised when cells lose their function because of a pathogen. Drying out and sterilisation of the culture system and associated items between batches stops the magnification in numbers of pathogens for later batches. This is widely practised in hatcheries, but it is often not used in grow-out systems, where individuals from a number of spawnings and of different ages may be mixed, to stock up to an optimum density. However, batch culture during grow-out and techniques to reduce pathogens between batches, such as drying-out ponds and leaving ponds or seacage areas fallow, are employed by the salmonid and marine shrimp industries (Chapters 15 and 19).

10.4.2 Incoming water treatment

Treatment of incoming water is essential in recirculating culture systems and more useful in hatcheries than in grow-out situations because of the sheer volume of water involved in the latter. Water treatment includes chemical sterilisation (chlorine, iodophores, ozone) and physical sterilisation (e.g. UV light) (Chapters 2 and 3). All are greatly enhanced by the inclusion of good settlement ponds preceding the treatment. Particulate matter, producing high turbidity, offers a substrate, nutrition and protection for pathogens, particularly bacteria, which prefer to be benthic rather than pelagic. It has been shown that such settlement for 9 days causes *Vibrio* bacteria to die out and be replaced by oligotrophic, less pathogenic, species. The bodies of each generation of bacteria are used as a nutrient for a progressively smaller biomass of bacteria, so that the numbers and biomass spiral downwards. This phenomenon is called 'self-cure'. Settlement ponds reduce the particulate load of incoming water. Sterilisation is most effective against obligate pathogens that cannot use an alternative life cycle to build up their numbers (e.g. viruses, rickettsia, chlamydia, sporozoans). However, aerosols arising from vigorous aeration are very common in hatcheries. They allow facultative pathogens to by-pass the sterilisation processes and build up to threshold numbers, when they then cause problems. For example, bacteria in shrimp hatcheries have been shown to travel 8 m in aerosols to contaminate other tanks.

Chlorine is very dangerous to use as a sterilising agent in organically polluted waters because of the formation of chloramines. These are highly reactive, have a long half-life and are not neutralised by chemicals used to remove free chlorine. Many aquaculturists have neutralised free chlorine and subsequently watched in horror as their fry proceeded to die from chloramine toxicity when the water was used.

Ozone is a particularly useful sterilising chemical, especially for recirculation, where its application can be tied into reduction–oxidation (redox) measuring devices. When reduction potential is high, anaerobic bacteria easily produce electrons from electron-donating substances and electron acceptors are relatively rare. These electrons are highly reactive and can cause cellular damage as a means of stabilising. At high oxidation potential, electron acceptors are abundant and free electron damage is low. For aquaculturists, oxidative conditions are preferable and ozone produces such conditions. Ozone is also safe as it and its ozide derivatives have a very short half-life and the final product, oxygen, is very useful.

10.4.3 Lower stocking density

By lowering the stocking density, the average 'inter-organism' distance is increased and the probability of a pathogen reaching the next host is reduced on an exponential scale. On theoretical grounds, epizootics (disease outbreaks) will fall to extinction unless a threshold number of hosts is present in a given area. Simplistically, each infected host must infect at least two other hosts as it succumbs or the epizootic will not propagate. Furthermore, lowering stocking densities will also decrease the level of sibling-interaction-induced stress and competition for space and food.

10.4.4 Single-spawning stockings

Differential growth is a good indicator of poor health in a captive population. Runts are very useful to screen for diseases as they are either stunted by pathogens or behaviourally and nutritionally stressed by being at the bottom of a pecking order (Fig. 10.3). Such stressed animals will also express pathogens. If a mixed spawning population is used to stock a culture system, the differential growth as a result of age, genetics or variations in hatching conditions will obscure pathogen-caused, differential growth. Thus, stocking with a single spawning is of particular benefit for an aquatic pathobiologist. This technique is not quite as useful in many fish species for which size grading is a normal part of culture (e.g. eels, salmon and trout), but it works well for invertebrates (e.g. freshwater crayfish). This technique also highlights the problem of a very common practice among fish farmers. At harvest, most fish farmers will put the runts that are too small to meet market needs into a pond to allow them to grow to market size. This overlooks the most likely reasons for their failure to reach market size: they are compromised by having a disease. Therefore, the farmer in reality, is keeping a reservoir of diseased individuals on the farm to potentially infect the next stocking.

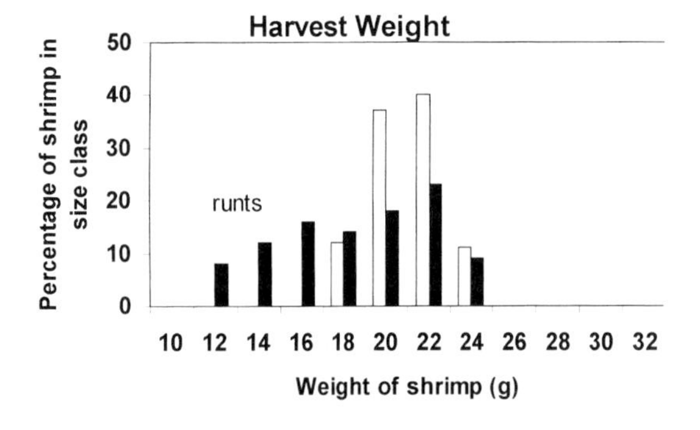

Fig. 10.3 A typical proportion of runts in a population of shrimp diagnosed with midcrop mortality syndrome (MCMS) (■) compared with uninfected shrimp (□). The diseased population shows a spread across more size classes than the uninfected shrimp.

10.4.5 Specific pathogen-free broodstock

Most pathogens are more virulent to the younger stages of a host. By producing offspring from broodstock free of specific pathogens, the offspring have a good chance of growing to a non-susceptible size before being infected, and thus a crop can be produced even in an enzootic area. This can also work if all life stages are equally susceptible. By late infection of the host, the crop can be grown to harvest before the epizootic has a chance to establish. This is the approach taken for infectious hypodermal and haematopoietic necrosis virus in shrimp (*Litopenaeus vannamei*) culture.

10.4.6 Vaccination

Vaccination basically works on the premise that an immunological memory exists and that prior exposure to a pathogen will allow a better and quicker immune response. Vaccination works reasonably well in vertebrates with both specific and cross-reactive antigens (Ellis, 1988), but immunological memory has not been demonstrated in invertebrates. However, evidence is accruing that adaptive immunity does exist in invertebrates. Enhanced stimulation of invertebrate haemocytes (blood cells) by beta-glucans or killed *Vibrio* cells in the diet has been shown to confer some protection. This is thought to work either by increasing the production and maturation rate of haemocytes or by blocking the binding sites of the pathogens. On theoretical grounds, not every fish has to be fully vaccinated to achieve protection of the entire population. If about half the population in each pond is protected, this is equivalent to halving the stocking density of susceptible fish and, in many cases, this will stop epizootics from progressing. Vaccination success against bacteria in Norwegian salmon was almost solely responsible for the increase in production during the early 1990s.

In most cases, the only practical method of delivery of a vaccine is via immersion at the nursery stage (Table 10.3). Furthermore, many factors affect the success of vaccination (Table 10.4).

10.4.7 Stress reduction

Stress is often used as an excuse for problems when no other logical explanation is available. Despite this nebulous usage of the concept of stress, it does have a real physiological basis and real consequences (Pickering, 1981). Unfavourable conditions lead to an adaptive response and a new level of homeostasis is achieved. If this is not achieved, then exhaustion follows and the overproduction of the stress hormones (including, in fish, the corticosteroids cortisol, cortisone, corticosterone, 11-deoxycortisol and adrenocorticotrophic hormones). High plasma levels of corticosteroids:

- induce lymphopenia
- reduce phagocytosis
- reduce access of lymphocytes to inflammatory sites
- deplete vitamin C reserves and consequently reduce wound repair
- increase protein catabolism (gluconeogenesis), leading to muscle wasting and low antibody and collagen synthesis, which again restricts wound repair.

Levels of the stress hormones do not correlate well with levels of stress as some fish pass through the high levels of secretion to a new, highly stressed state with no secretion of the hormones. The two most practicable ways of limiting stress are to double the aeration, thus alleviating any oxygen stress that may be occurring, particularly during hot summers, and to lower stocking density as mentioned above.

10.5 Major diseases

This topic is impossible to cover adequately in a general text and the reader is directed to the more detailed accounts of aquaculture diseases listed in the Reference section of this chapter. Many of these deal with diseases of specific groups of culture animals (Pickering, 1981; Kabata, 1985; Ellis, 1988; Sindermann & Lightner, 1988; Elston, 1990; Kent & Margolis, 1995; Lundin, 1997; Austin & Austin, 1999; Hoole *et al.*, 2001). The following rendition gives a general account of the more common diseases in the various aquaculture industries.

10.5.1 Molluscs

Edible oysters

Worldwide, protozoan parasites are the most significant cause of losses to bivalve industries. This predominance of protozoan parasites is reflected in a guide to diseases for the mollusc farmer (Elston, 1990). Of the 11 'Notable Oyster Diseases' described in this guide, protozoans are responsible for seven:

Table 10.3 The effect of the route of vaccination on the immune response

Route	Systemic antibody response	Problems
Immersion	None	Only the integument is protected by monomeric IgM. Systemic infections can still occur through skin breaches. Antigen might possibly enter the rectum, and give some systemic protection
Oral	None	The acid stomach often denatures the antigen, so large doses are often needed to reach the lower intestine 'Peyer's patch-like' accumulations of macrophages where stimulation will occur. Memory cells are sensitised
Inoculation	Yes	Best protection as both memory cells and antibodies are produced, but practically it can be very costly to handle fish individually. Handling stress is increased. Swim bladder inoculation is superior to muscle, which is better than intraperitoneal

IgM, immunoglobulin M.

Table 10.4 Factors affecting vaccination responses

Factor	Response
Temperature	The higher the temperature the better the response, within physiological limits
Size	The larger the animal the better the response, independent of age. Earliest size is 0.5 g but 10 g is better
Diet	Vitamins C and E are most important, and high-protein diets increase response
Stress	High stress can prevent an effective response
Dose of antigen	Larger is not necessarily better. A small primary dose followed by a larger secondary dose is most often the best
Timing of booster vaccination	If the booster is given when the primary response is waning, the secondary response is better
Age	The older the better. Ova may have maternal antibody which will neutralise the antigen and prevent an immune response. Non-specific immunity (interferon, transferrin and lysozymes) is present in the very young. Cell-mediated immunity is present within 2 weeks Antibacterial antibodies within 4 weeks and antiprotein antibodies within 8 weeks Vaccines have not been extensively tested with larval invertebrates
Antibiotics	Antibiotics can suppress the immune system by 90%, particularly oxytetracycline, which lowers protein (antibody) production at the ribosomal level

- *Perkinsus marina*
- *Haplosporidium nelsoni*
- *H. costalis*
- *Mikrocytos mackini*
- *Bonamia ostrea*
- *Marteilia refringens*
- *Hexamita nelsoni.*

Although this guide to diseases is based primarily on the experiences of the North American and European cultured bivalve industries, similar diseases affect bivalve industries around the world.

For example, *M. sydneyi* (Acetospora) causes summer mortality in Sydney rock oysters (*Saccostrea glomerata = commercialis*) in Australia. A flushing of rivers usually brings on an outbreak of the disease, with deaths in about 6 weeks. The digestive gland is attacked, first in the Leydig tissue and later in the epithelium. The gland becomes pale yellow and watery and the gonad condition may be greatly reduced. No control is known, but growers try not to have oysters on their leases over the wet summer months. *M. refringens* causes similar problems in *Ostrea edulis* in France and Spain, with infection occurring first in summer and most deaths occurring in winter.

A group of intracellular parasites causes problems worldwide, with *Mikrocytos* species infecting the cupped oysters and *Bonamia* species infecting the flat oysters. They cause mortality in oysters 3 years or older, when the oysters enter the female stage of their life cycle. *Mikrocytos roughleyi* causes winter mortality in Sydney rock oysters and requires high salinity (30–35‰). Small yellow to brown pustules occur on the gills, palps and mantle. *M. mackini* occurs in North America and causes similar problems in the Pacific oyster (*Crassostrea gigas*).

Bonamia ostreae is a major problem in the cultured flat oysters of western Europe. It appears that it was originally taken to France in shipments of oysters from western USA, similar to the route of the iridovirus, which causes oyster velum viral disease (OVVD). *Bonamia* has also caused local problems in New Zealand bluff oysters (*Ostrea chiliensis*), among which it destroyed over 80% of the industry over a 6-year period. In Australia in 1991, the Victorian flat oyster industry also lost 90% of its production to *Bonamia*. Early this century, boats took live bluff oysters from New Zealand to Australia because the local fisheries had collapsed. It has been suggested that this spread the *Bonamia* from New Zealand to Australia, but the question of what first caused the collapse of these oyster industries needs to be answered. Furthermore, the occurrence of *Bonamia* in Western Australia supports the premise that the parasite is native to Australia.

Bonamia first infects the tissues below the gut and later invades the haemocytes that are involved in resorbing unspent gonadal material, especially the ovaries. *Bonamia* is a density-dependent disease. Reducing the stocking density and growing the oysters on hanging lines to keep them off the bottom should allow marketable crops to be grown in areas where *Bonamia* is endemic.

It is not just protozoans that cause disease in molluscs however; viruses and bacteria cause problems in bivalve aquaculture.

A herpesvirus (131 nm diameter) has been associated with larval mortalities in Pacific oysters in northern New Zealand. Feeding ceased between days 3 and 4, and 60–100% mortality occurred in days 7–11. The cells infected were the fibroblasts and presumptive phagocyte precursors, which displayed enlarged, marginated nuclei with both intranuclear and cytoplasmic inclusion material. Predisposing conditions included elevated temperatures and crowding. Herpesviruses have been reported from oysters in the north-eastern USA, northern Wales and northern France.

Iridoviral infections cause disease of the larval velum in Pacific oysters in the western USA (OVVD), and gill necrosis and haemocyte diseases in adult *Crassostrea angulata* and, lately, Pacific oysters in France. There has been speculation that the same iridovirus is responsible for all diseases and was imported into France from the USA with Pacific oysters.

Bacterial infections have from time to time caused mortality in Tasmanian oyster hatcheries. *Vibrio* (13 strains), *Alteromonas* (10 strains), *Pseudomonas* (eight strains) and *Flavobacterium* (three strains) were all involved. Five of the *Vibrio* and two of the *Alteromonas* strains caused mortality at 10^7 and 10^5 bacteria/mL, respectively, with the lower doses taking longer. All strains were lethal at 10^8 bacteria/mL. Bacterial virulence for oysters has been strongly associated with a low molecular weight (500–1000 kDa), heat-stable toxin. The toxin stops cilia movement (ciliostasis), which in turn reduces feeding and swimming in larvae and inhibits the cleaning function of the gills of older oysters.

Some diseases affect the oyster's shell and not the soft tissues. The shell may be very weakened structurally in extreme cases, thus exposing the oyster to predation. In lesser infections there is malformation of the shell, making the oyster less presentable or unacceptable to the market, which is as good as killing it from the viewpoint of the industry.

There is a shell disease caused by a fungus that grows as filaments through the shell, weakening the shell and causing dark raised 'warts' on the inner surface of the shell. It occurs in the European flat oyster, in which it has caused heavy mortalities in some regions. Similarly, boring sponges, *Cliona* species, riddle the shells of some bivalves, including clams and pearl oysters, with very deleterious effects on marketability. However, *Cliona* is apparently not a problem with table oysters.

Boring polychaete worms (*Polydora* species) invade the shells of oysters, some other benthic bivalves (e.g. scallops) and even abalone. They have a widespread distribution. *Polydora* is responsible for the use of intertidal stick culture of Sydney rock oysters in northern New South Wales. These 'mudworms' bore through the shell, causing blisters on the inner surface. By culturing oysters in the intertidal zone, the racks dry out during low tides and newly settled polychaete larvae are killed. *Polydora* also affects abalone and benthic mussels in southern Australia. However, rafted mussels are generally free of infection.

Other molluscs

There is a large family of ectoparasitic gastropods, the Pyramidellidae, and some parasitise oysters and other bivalves. Pyramidellids are small gastropods that suck the blood or other tissue fluids of the host by using a long, penetrating proboscis. The host has less energy for growth in mild infections. In heavy infections, the host may die of tissue and fluid loss. The pyramidellid *Boonea impressa* is an important parasite of the American oyster.

Epizootics have been observed in cultured pearl industries. There were considerable long-term mortalities in blacklip pearl oyster (*Pinctada margaritifera*) in French Polynesia during the late 1980s, but no causative agent was identified. In the silverlip pearl oyster (*P. maxima*) industry in Australia, a bacterium, *Vibrio harveyi*, was isolated from the haemolymph of dying oysters. Coldwater temperatures (19°C) apparently predisposed the pearl oysters to infection and crowded transportation with very little water circulation allowed the bacteria to establish. A protozoan parasite, *Perkinsus*, has been associated with mortalities of pearl oysters in north Queensland, and a haplosporidian has been found in Western Australian pearl oysters.

Die-back of abalone in South Australia has been associated with *Perkinsus olseni*. When water temperatures are elevated, the cellular immune system of the abalone cannot encapsulate and destroy the *Perkinsus*. Treatment consists of lowering the water temperature by 8°C in a holding tank. This will stop even a progressing disease outbreak in 2 days. A related species, *P. marinus*, has caused kills of table oyster in the south-eastern USA during summer periods.

There are several families of decapod crustaceans which include species that live in the mantle cavities of large bivalves, such as large clams, mussels and oysters. These are species of the shrimp family Pontoniidae and the crab family Pinnotheridae. These crustaceans often live in pairs within the host. They usually occur on the gills, where they feed from the host's food grooves. They cause mechanical damage to the host's gill tissue. There is no evidence that these parasites cause mortality, but, presumably, they have some minor adverse effect on the host.

Some of the more important afflictions of abalone are described in section 22.2.3.

10.5.2 Crustaceans

Shrimp

Viruses have caused hatchery mortalities and considerable grow-out problems in marine shrimp. The most devastating virus known to date is whitespot syndrome virus (WSSV). It started in 1993 in China, where it destroyed almost US$1 billion worth of *Fenneropenaeus chinensis*. When infected kuruma shrimp (*Marsupenaeus japonicus*) were imported from China and Korea to Japan, the virus was spread with devastating consequences. Further spread of WSSV to Taiwan, then to Thailand and the rest of South-East and Central Asia ensued. Later, the virus was moved to Texas, USA, presumably via frozen commodity shrimp. Central and South America followed, so by 1999 only the Philippines, Australia and Oceania were known to be free of this virus. Unfortunately, illegal movement of broodstock from the Indo-Malaysian archipelago to the Philippines infected that country. On a regional scale, the spread of this virus was always with live animals or frozen commodity shrimp. Interestingly, other rod-shaped viruses have been described from crabs, yet crabs do not seem to be as susceptible to the virus as shrimp. The original outbreak of this virus coincided with the upsurge in the practice of feeding raw crustaceans, in particular mud crabs (*Scylla* species), to broodstock shrimp as a maturation diet.

Shrimp stocks have become partially resistant to WSSV so that carrier animals are common. Carriers can be detected biochemically using nested polymerase chain reactions. The most highly infected animals are detected at the first stage of the reaction and are discarded. If these progeny are stocked, the risk of crop failure is 95%. Progeny of broodstock that are positive at the second stage of the reaction and those that are negative (ideally) are used to stock shrimp ponds. The risk of crop failure from these shrimp is 31%. The risk of crop failure is enhanced by stresses (osmotic, pH, oxygen) associated with the wet season. The failure rate might be around 19% during the dry season, but it may leap to 70% with the onset of the wet season. These same crop failures, which are enhanced by the wet season, occurred in Australia, where they were called midcrop mortality syndrome (MCMS). Stresses from infections with other pathogens also greatly affect the outcome of an infection with WSSV.

Taura syndrome virus has been responsible for widespread mortalities in *Litopenaeus vannamei* in the Americas after initial outbreaks near the Taura River, Equador. Taura syndrome was believed to be due to the toxicity of fungicides used to control black sigatoka disease on banana crops. The washing of the fungicides into the waterways allowed it to accumulate in the shrimp ponds. The fungicides were very similar to crustacean moult inhibitors, and this is believed to be the mode of action. A massive lawsuit against the European chemical companies that produced the fungicides was launched for compensation. Subsequently, cell-free bioassays and electron microscopy unequivocally demonstrated the presence of a picornavirus, which produced the characteristic hypodermal, buck-shot lesions and mortalities. The disease seems to be most aggressive in *L. vannamei* so that *L. stylirostris* has become an alternative crop in infected areas. Unfortunately, survivors of the epizootic are chronic carriers and infected *L. vannamei* have been introduced to Taiwan, China and Indonesia.

Monodon-type baculoviruses (MBVs) can infect all species of shrimp, although their ability to produce disease differs between shrimp species. MBVs have been implicated as causative agents of the mass mortalities that swept through Taiwanese shrimp farms in 1987/88 and resulted in the crash of that industry. This may not be correct, as yellow-head ronivirus (initially called a baculovirus, then a rhabdovirus and then a coronavirus in the literature) (YHV), which was undiagnosed at the time, seems to have had a very prominent role. All life stages of the host after mysis 1 are susceptible to MBVs, but they are diseases of hatcheries rather than later stages, which are largely asymptomatic. In an Australian study, MBVs were found in 8 of a total of 13 monitored hatcheries for *Penaeus monodon*. They have now been largely managed out of existence in *P. monodon*, but have been a major problem with the establishment for aquaculture of other species. The method of control, developed in Tahiti, involves separating floating eggs from the spawning female, surface sterilising them, and then using the phototactic response of the nauplii to separate the larvae from the egg shells and moribund larvae.

Infectious hypodermal and haematopoietic necrosis virus (IHHNV) has all but destroyed the aquaculture industry based on *Litopenaeus stylirostris* in the Americas. Interestingly, as *L. stylirostris* is not susceptible to Taura virus, IHHNV-resistant strains have made a comeback as an alternative crop to *L. vannamei*. Intranuclear inclusions were prominent throughout all tissues except hepatocytes. However, the inclusions from Australian shrimp did not cross-hybridise with an IHHNV gene probe that was 90% specific for the American IHHNV strain. A Philippine IHHNV strain did cross-hybridise with the American gene probe, suggesting that the Australian IHHNV is a different strain. Nor did the gene probe react with lymphoidal parvovirus, which may be a variant in expression of IHHNV. Similarly, an American-produced hepatopancreatic parvovirus (HPV) gene probe based on a virus from Korean *Fenneropenaeus chinensis* did not cross-react well with the Indonesian and Australian shrimp strains nor with HPV from Malaysian freshwater prawns (*Macrobrachium rosenbergii*). HPV has not been found to be a problem in the aquaculture of Australian *Penaeus monodon*, even though it is common in wild stocks of other shrimp species in Australia. However, recent information about *P. monodon* in Thailand demonstrated HPV to cause severe stunting and therefore considerable loss of income to farmers.

The newly diagnosed spawner mortality virus (SMV) (a parvo-like virus) in conjunction with gill-associated virus (a ronivirus) have caused major losses in *P. monodon* broodstock and grow-out. This mid-crop mortality syndrome (MCMS) has become a major grow-out problem in northern Australia, with most farms losing 50% of their stock. A total of A$44 million was estimated to have been lost over the years 1995–97 as a result of MCMS. The disease outbreak is slow in its onset and shrimp grow to around 12–15 g before dying. Much time, labour, feed and money are thus invested in a crop that will be devastated if it is not harvested. A symptom of MCMS is the presence of sick, weak and reddened shrimp at the edge of grow-out ponds.

Bacteria are still a major constraint to hatchery production, with very few batches of postlarvae being able to be produced without antibiotics. Strategic use of antibiotics during larval and early postlarval development is usual. The Vibrionaceae predominate in isolation from shrimp hatcheries, with 8 of 37 strains producing significant mortality in larval bioassays. *Vibrio* species, including *V. harveyi* and *V. tubiashi*, and *Photobacterium damsela* are involved. On a worldwide basis, the non-sucrose-utilising bacteria (those producing green colonies on thiocitrate bile salt agar or TCBS) are often involved in disease in crustaceans. Application of sucrose can alter the bacterial balance, favouring the less pathogenic bacteria, and hence reduce losses.

V. harveyi has been involved in massive mortalities in Ecuador and major mortalities in two grow-out farms in Australia. In one study in Australia, over 50% of the bacteria at the completion of grow-out were *V. harveyi* and *V. alginolyticus*. Strains of *V. harveyi* in Australia and the Philippines have been shown to have very strong antibiotic resistance, with one virulent Australian strain having a plasmid coding for four different antibiotic resistances. *V. penaecida* has been associated with mortalities in ponds in Japan (*Marsupenaeus japonicus*) and New Caledonia (*Litopenaeus stylirostris*); and *V. nigripulchritudo* has been associated with mortalities in ponds in New Caledonia.

In the early years of shrimp culture, the fungi *Lagenidium callinectes* and *Sirolipidium* were frequently involved in causing larval mortality, especially in hatcheries where there was excessive use of antibiotics. Lobster eggs have been shown to have a surface commensal *Pseudoalteromonas* species that protects against fungi. Antibiotics inhibit this commensal, allowing the fungi to become established. A *Pseudoalteromonas* species has been found associated with shrimp eggs, and so the mechanism is presumed to be similar. Use of the herbicide trifluralin has effectively controlled fungi in the hatchery.

Although the fungus *Fusarium solangi* is present in grow-out ponds in Australia, it is only seen when moulting frequency is slowed because of cool water temperatures or another disease problem. It is thus a good indicator of retarded growth. Similarly, the peritrich ciliate protozoans also indicate suboptimal conditions and, at their highest fouling rates on the shrimp gills, can cause death by suffocation if any further stress is added. Peritrich ciliates need high organic loads to feed. So by lowering the amount of uneaten food with water changes or by restricting supplemental feeding, the peritrich numbers can be reduced within a week.

Further information relating to disease and biosecurity in shrimp culture is given in section 19.8.4.

Freshwater crayfish

The European noble crayfish (*Astacus astacus*) industry, which was a major aquaculture industry, has been decimated by a fungus, *Aphanomyces astaci*. Furthermore, in 1986/87 the crayfish industry in Turkey was also devastated by the introduction of this fungus across the natural barrier of the Bosphorus Straits. This has led to the introduction into Europe of the signal crayfish (*Pacifastacus leniusculus*) from the USA, because this species is one freshwater crayfish that is highly resistant to the fungus (section 20.2.2).

With the exception of the devastating crayfish 'plague' in Europe, freshwater crayfish are generally considered to be relatively disease free. This reflects the extensive culture methods used for crayfish, whereby only very low stocking densities (5 per m^2) are used. With the development of more intensive culture methods, disease will become more of a problem. Table 10.5 lists the pathogens already documented from farmed crayfish in Australia. The most important causes of mortality so far are:

- the rickettsia *Coxiella cheraxi*
- *Cherax* baculovirus/ bacteraemia
- bacterial erosion of the eyeballs associated with *Aeromonas hydrophila* and *A. sobria*.

The rickettsia and the baculovirus have been introduced into at least Equador and the baculovirus into the USA with trans-shipments of crayfish from Australia. Emerging pathogens include *Microsporidium* in Western Australia and a systemic parvovirus in the redclaw crayfish, *Cherax quadricarinatus*.

10.5.3 Fish

Carp

As outlined in Chapter 14, the carp and their relatives are the largest group of fish or shellfish produced from aquaculture. In 1998, they made up nearly 50% of all fish and shellfish production, and 80% of all fish and shellfish from freshwater. Section 14.6 on diseases refers the reader to a recent monographic treatment of diseases in carps and their relatives by Hoole *et al.* (2001).

Salmonids

The group of fish that dominates the aquacultural literature on a worldwide basis is the salmonids because, unlike carp, they are cultured in so many countries around the world. Over 20 viral diseases are recognised in salmonids and approximately half are considered to have moderate to high virulence.

Table 10.5 Pathogens found in farmed Australian freshwater crayfish

Type	Pathogen
Viruses	*Cherax* bacilliform virus, *Cherax* giardiavirus, parvovirus, reovirus, picornavirus
Bacteria	*Coxiella cheraxi*, *Aeromonas hydrophila*, *A. sobria*, *Citrobacter freundii*, *Klebsiella pneumoniae*, *Plesiomonas shigelloides*, *Pseudomonas*, *Shewenella putrifaciens*, *Streptococcus*
Microsporidia	*Agmasoma* (= *Thelohania*), *Vavraia parastacida*
Fungi	*Achlya*, *Mucor*, *Psorospermium*
Temnocephalids	*Craspedella spenceri*, *Diceratocephala*, *Didymorchis*, *Notadactylus*, *Temnocephala dendyi*, *T. minor*
Ciliates	*Corthunia*, *Epistylus*, *Lagenophrys*, *Tetrahymena*, *Vorticella*, *Zoothamnium*

(1) Infectious haematopoietic necrosis virus is one of the most studied as it infects a range of salmon species, including Atlantic salmon, rainbow trout, sockeye salmon, chinook salmon and coho salmon. This rhabdovirus is endemic to the north-eastern Pacific, but it has been spread across the northern Pacific as far south as China and Taiwan, the eastern USA, France and Italy. It primarily affects fish less than 6 months old and, as it is found in the sexual secretions of both male and female broodstock, it has ready access to young fish for infection.

(2) Viral haemorrhagic septicaemia (VHS) of rainbow trout was once restricted to Europe but has been introduced to the USA. It grows only at temperatures below 16°C, so its geographic range is limited and losses can be ameliorated by increasing temperature. Recently, VHS has been found in wild southern Californian pilchards and mackerel that are used extensively for bait and fodder in other aquaculture industries such as bluefin tuna culture.

(3) The aquabirnaviruses, e.g. infectious pancreatic necrosis virus, and similar viruses are found worldwide. These viruses have a great propensity for replicating or surviving in hosts other than fish. These other hosts include bivalves, crayfish, shrimp and the intestines of fish-eating birds. Therefore, it is very difficult to eradicate. Luckily, it is easily managed by recognising that there will be fairly predictable losses to the virus. Up to 40% excess of eyed-ova, above what is believed to be necessary for that year's stocking, may be purchased in anticipation of this 40% being killed by the virus.

Bacterial diseases of fish were discussed in detail by Austin & Austin (1999). Bacteria are major pathogens in salmonid culture (Fig. 10.4). For example, *Vibrio anguillarum* serovar C or 01 has been isolated from Tasmanian rainbow trout and Atlantic salmon. Clinical signs include hyperaemic prolapsed rectum, congestion at the bases of the fins and reddened ulcers on the flanks. The muscles may contain enclosed lesions, and severe peritonitis can occur. Antibiotics have been widely used to control outbreaks, but they are becoming less and less acceptable in food animals because of concerns about antibiotic resistance spreading to pathogenic bacteria of humans. Bacteria in the sediments of abandoned aquaculture sites have been assayed for antibiotic resistance 10 years after the culture operation has ceased and were found to still have major resistance against oxytetracycline. This type of problem has led to the production of effective vaccines. The 30% rise in Atlantic salmon production in Norway is due to control of *Vibrio salmonicida* and *V. anguillarum*.

Yersinia ruckeri serovar III (the Australian strain) causes some problems, mainly in young Atlantic salmon, and particularly when water temperatures are above 15°C. Some fish may develop pathognomic haemorrhages in the eye and, in chronic cases, bilateral exophthalmia and anorexia may predominate. The bacteria can be spread in the faeces of carrion-feeding raptors. Non-specific stimulation of the immune system using yeast cell wall β-1,3 and

Fig. 10.4 Wounds on market-sized rainbow trout (2–3 kg) infected with *Vibrio* and *Flexibacter* species bacteria (photograph by Dr John Purser).

β-1,6 M-glucans against *Y. ruckeri* has gained popularity. The transfer of *Aeromonas salmonicida* from Scotland to Norway led to a massive slaughter campaign to control the bacterium. This cost $US100 million in losses. In fact, *A. salmonicida* is probably the most important fish pathogen in all marine aquaculture industries.

Most culinary experts prefer rainbow trout to Atlantic salmon, and accordingly the Australian seacage industry began with rainbow trout. *Enterococcus* (previously *Streptococcus* biovar I) killed 30% of rainbow trout in freshwater and as high as 60% in marine waters associated with the acclimatisation stress. Therefore, the industry swapped *en masse* to Atlantic salmon, which is highly resistant to this bacterium. *Enterococcus seriolicida* (= *Lactococcus garviae*) is still the major problem of yellowtail culture in Japan despite being identified in the 1960s. It is also a problem of rainbow trout in South Africa and southern Australia. Part of the problem is the bacterial A-layer, which defeats the ability of complement to attach to the bacterial cells. Hence, macrophage engulfment is compromised and host antibody production is slowed. Carrier status in the antibody havens of the brain and gonads ensues, allowing sporadic outbreaks whenever stress eventuates.

Kent & Margolis (1995) outlined the diseases caused in seawater-reared salmonids by parasitic protozoans. There are numerous and diverse protozoans that parasitise aquatic vertebrate hosts in freshwater and marine habitats, and these cause various levels of debilitation. For example, *Neoparamoeba* species, an amoeboid protozoan, is an opportunistic pathogen that infects the gills of fish under some conditions. On a coho salmon farm in Washington State, Kent & Margolis (1995) observed 25% mortality that was attributed to *Neoparamoeba* species. The amoebae occurred on marine farms and were quickly eradicated from the fish gills by exposing them to freshwater, e.g. towing the fish cages from seawater into freshwater or by treating the fish in freshwater baths. *Neoparameoba* also caused problems at the outset of the salmonid seacage industry in Tasmania, Australia. Fish in their first year in seawater develop a proliferative gill disease, especially if seacages are fouled or temperatures are above 15°C. Up to 2% of fish per day will die if untreated. Chemical baths are very ineffective, but freshwater baths for a few hours or moving the whole seacage to freshwater work very well (section 15.8). Once the cycle is broken, the immune system prevents further heavy infestations. Unfortunately, there is strong evidence that *Neoparamoeba* is developing resistance to the freshwater treatment, which is now effective only for about 30 days.

Kent & Margolis (1995) listed 11 parasitic protozoans causing significant disease conditions and mortality in salmonids reared in seawater. Virtually all the fish tissues may be infected by the range of parasites, but there is a tendency for these protozoans to parasitise the gills and skin.

Tropical and subtropical fish

An overview of the parasites and diseases of fish cultured in tropical regions was presented by Kabata (1985). Epizootic ulcerative syndrome (EUS) is caused by the fungus *Aphanamyces invadens*. EUS came to prominence in the 1980s as a devastating pathogen of snakehead and other fish of South-East and Central Asia. Snakehead fish are used for protein supplementation in the diets of the rural poor. Destruction of these fish led to widespread malnutrition, starvation and socioeconomic disruption of subsistence lifestyles. After many competing hypotheses on the aetiological agent, including rhabdoviruses and *Aeromonas hydrophila* bacteria, *A. invadens* was identified by a group of researchers in Australia. In the fullness of time, EUS has been traced back to early outbreaks in the 1970s in Australia and Papua New Guinea. The earliest outbreak was in Japan in 1970. One interesting outbreak occurred only in the fish pond at the Sri Lankan international airport, where presumably a passenger had thrown some fish scraps from a meal into the pond. Lately, EUS has become a problem in the establishment of the jade perch (*Scortum barcoo*) as an aquaculture species.

Barramundi (*Lates calcarifer*) is a widely cultured species in Australia and throughout South-East Asia (Chapter 18). Most farms have intermittent problems from time to time, and the most common pathogens are *Cytophaga johnsonae*, causing saddle back, which may lead to dirty yellow ulcers, and *Cryptocaryon irritans*, which causes white spot in fish in seacages (Fig. 10.5). Problems are associated with the onset of cooler temperatures. *Streptococcus iniae* is emerging as the most important disease in barramundi in estuarine seacage operations. The initial episode starts after displacement of soil in the watershed in which the barramundi are being grown. Losses approach 30% and are not ameliorated by winter water temperatures.

There have also been major problems associated with nodaviruses, which cause viral encephalopathy and retinopathy of larval barramundi and kill young, crowded larvae, especially in later batches from a hatchery. The virus has been definitely associated with vacuolate brain and retinal tissues in larvae from 8 of 111 batches (it was not looked for in each batch), and it has been responsible for the placement of certification procedures for movement of larval barramundi between states in Australia. The growing of barramundi outside its natural range in the Murray–Darling watershed of south-eastern Australia and the potential impact of the nodavirus on the endemic freshwater fish has been hotly debated. A similar, if not identical, virus has been found in barramundi in Tahiti and Singapore and in most fish species (about 30 species at present) around the world. Almost all fish raised in marine hatcheries are usually found to be infected with this virus once the correct tools for investigation have been used.

Other pathogens seen from time to time include lymphocystis virus, which is associated with dirty seacages, *Vibrio* species, *Epitheliocystis*, and *Epieimeria*. Experimental infections with Bohle iridovirus, a systemic ranavirus, have shown barramundi to be acutely susceptible in both fresh and marine waters. Total mortalities usually occur within 10 days. More information about diseases in barramundi is given in section 18.4.3.

Common diseases affecting tilapias and channel catfish are discussed in Chapters 16 and 17 respectively.

Fig. 10.5 Barramundi infected with 'white spot' (*Cryptocaryon irritans*) (photograph by Professor Bob Lester).

10.6 Conclusions

The nature of aquaculture and the best anti-disease technologies are the direct antitheses of each other. Aquaculture is about increasing profits by growing more animals in smaller and smaller volumes of water. Anti-disease methodologies include keeping densities as low as possible to break the fish to fish cycles. As profit is the major goal in aquaculture, densities will always be maximised, and, therefore, disease problems amplified. As such, aquatic pathobiologists will always be in demand, and, as the job is perpetually full of challenges, it makes a very interesting career path for students.

References

Austin, B. & Austin, D. A. (1999). *Bacterial Fish Pathogens: Disease of Farmed and Wild Fish*, 3rd edn. Springer, New York.

Ellis, A. E. (Ed.) (1988). *Fish Vaccination*. Academic Press, London.

Elston, R. A. (1990). *Mollusc Diseases: Guide for the Shellfish*

Farmer. Washington Sea Grant Program, University of Washington Press, Seattle.

Hoole, D., Bucke, D., Burgess, P. & Wellby, I. (2001). *Diseases of Carp and Other Cyprinid Fishes*. Fishing News Books, Oxford.

Kabata, Z. (1985). *Parasites and Diseases of Fish Cultured in the Tropics*. Taylor & Francis, London.

Kent, M. L. & Margolis, L. (1995). Parasitic protozoa of seawater-reared salmonids. *Aquaculture Magazine*, **21**, 64–74.

Lundin, C. G. (1997). Global attempts to address shrimp disease. In: *Proceedings of the 2nd Asia-Pacific Marine Biotechnology Conference and 3rd Asia-Pacific Conference on Algal Biotechnology*, 7–10 May 1997, pp. 1–23. Phuket, Thailand.

Pickering, A. D. (Ed.) (1981). *Stress and Fish*. Academic Press, London.

Sindermann, C. J. & Lightner, D. V. (Eds) (1988) *Disease Diagnosis and Control in North American Marine Aquaculture*. Elsevier, Amsterdam.

11

Post-harvest Technology and Processing

Allan Bremner

11.1 Introduction

> God did not design fish with the express intention that they go through rigor mortis.

Post-harvest technology encompasses areas of harvest, handling, slaughter, processing, packaging, manufacturing, storage, transport, distribution and display, through to the point of sale or the time of consumption (Fig. 11.1). This chapter can therefore only be considered as an introduction to the topic and the reader should look also to monographs and proceedings of conferences, listed in the References, for further depth of detail (Huss *et al.*, 1992; Martin, 1992; Shahidi & Botta, 1994; Sikorski *et al.*, 1994; Bremner *et al.*, 1996; Hall, 1997; Sato *et al.*, 1999; Kestin & Warriss, 2001; Bremner, 2002).

The scope of post-harvest technology is somewhat broader in aquaculture than in capture fisheries, since the selection, breeding, rearing, feeding and farming conditions can all be controlled to affect the properties of the product. The appearance, odour, flavour and texture of aquaculture products have a huge bearing on their marketability and acceptability by the consumer. The odour and flavour of either raw or cooked product is a result of the inherent biochemical composition, the accumulated results of bacterial and enzymic action and the effects of processing.

- Characteristic odours of a product are the result of volatile compounds.
- Flavour is due to both volatile and non-volatile compounds.

Inherent non-volatile compounds of importance in this regard are salts, amino acids and other non-protein nitrogenous compounds and, in particular, the breakdown products of the purine nucleotides. The levels of these factors in aquaculture products can be affected by harvesting and processing. Volatile components arise from feeds; many are lipid based in origin and they are products of either enzymic or chemical oxidation. Hence, both nutritional practices, and processing and storage can have profound effects on the flavour and stability of the product.

Fig. 11.1 The beginning of post-harvest. Atlantic salmon pass into an ice slurry bin for chilling (photograph by Dr John Purser).

The texture of the product is of considerable importance too, and harvest stresses and post-harvest practices are factors that combine to affect textural characteristics.

In this chapter, emphasis is given to the interaction between the pre-harvest practices and the post-harvest product, since this is an area of vital concern to aquaculturists in which considerable research is needed. The material presented relates mostly to aquaculture products grown for human consumption. However, the considerations of harvest practices that involve low stress are relevant to species reared for other purposes such as restocking of waterways for ecological, aesthetic or recreational purposes.

11.2 Basic characteristics

There are several basic characteristics that influence post-harvest technology and processing, and these should be determined for each cultured species. These are:

- the gross proximate (water, crude protein, fat, crude fibre, ash, nitrogen-free extract) composition (section 9.10.1)
- the yields of edible product
- the composition of the edible product
- the gross morphology of the edible parts
- alternative processing forms
- the distribution of by-product and off-cuts and their composition
- the chilled life of the product (0°C)
- the frozen storage characteristics of the product
- the pattern of rigor mortis
- the inherent bacterial flora
- the nature of the spoilage bacterial flora
- the inherent nucleotide composition and its rate of change during storage.

Although all these characteristics are affected to some extent by harvest regimes, selection, husbandry practices, feed and by a host of other factors, it is important that they are determined. It is also important to determine the proportions of red to white muscle in a fish species, the position of lipid deposits, the structure of the muscle and the skeletal structure.

11.3 Safety and health

Aquacultured foods generally have a good safety record (Ahmed, 1991). Most of the major food-poisoning organisms of public health significance, at least in the Western world, do not occur in the marine environment unless it has been polluted by human activity. Since the environment of the aquaculture production cycle is controlled for the most part, it could be expected that the product would be free from harmful organisms. It is the task of the industry to ensure this potentially beneficial situation is not eroded. However, even in the relatively controlled environment of brackish water ponds, cultured shrimp can be contaminated with bacteria such as *Salmonella* species and *Vibrio cholerae*. For example, of 304 samples of shrimp and mud/water samples from 131 shrimp ponds, *Salmonella* species were present in 16% of shrimp and 22% of mud/water samples and *V. cholerae* occurred in 1.5% of shrimp and 3.1% of mud/water samples. As a consequence of this and other studies, it may be necessary to reclassify raw cultured shrimp in the same regulatory health risk category as chicken (rather than with fished marine shrimp) in international regulations (Ahmed, 1991; FAO, 1994; Hocking *et al.*, 1997).

In some instances, aquaculture forms part of an existing farming activity and runoff containing faecal material from farm animals and contamination from waterfowl may introduce unwanted organisms. The practice of fertilising ponds with manures and other organic material (Chapters 2, 4, 9 and 14) can also introduce unwanted organisms. Most manufactured feeds are likely to be free of problems, but intensive culture practices that lead to wasted feed and high faecal loads provide an excellent environment for proliferation of undesirable bacteria.

Failure to remove dead animals can even lead to outbreaks of botulism in fish as a result of cannibalism. The result is that ponds need to be completely drained and the soil totally removed before the area is fit for use again.

Problems of harmful micro-organisms are relatively less important in extensive systems or in situations in which there is a high rate of water exchange. However, a number of problems can occur with seafoods from estuarine environments, which are partly due to the nature of the animals involved. For example, because bivalve molluscs, such as oyster and mussels, are filter feeders, they remove bacterial and viral particles from the surrounding water. This situation is worsened because bivalves are often eaten raw and the whole animal, organs and all, is consumed. Bivalves are readily contaminated by organisms of faecal origin from raw or improperly treated sewage (Hocking *et al.*, 1997, section 21.5.3). Prime agents appear to be Norwalk or Norwalk-like viruses that cause viral gastro-enteritis. *Vibrio* organisms are also involved and one species, *Vibrio vulnificus*, can be particularly virulent. Outbreaks often occur after rainfall, since the resulting decrease in salinity releases organisms from the sediments into the adjacent water column where they can be ingested by the animal.

Another major cause for concern is natural blooms of microalgae. In freshwater, blooms of blue-green algae may cause toxins which can be passed along the food chain. In estuarine and marine environments, paralytic shellfish poisoning (PSP), diarrhoeic shellfish poisoning (DSP) and amnesic shellfish poisoning (ASP) can occur with consumption of shellfish contaminated as a result of algal blooms (Hocking *et al.*, 1997; section 21.5.3). When these occur, there is no option other than to close the site to harvest. Monitoring programmes and closures are now practised worldwide to prevent diseases from this source. Although the toxic algal blooms are 'natural', there is growing evidence for the spread of these species in the ballast-water of cargo ships.

Other 'natural' water-borne bacteria, such as *Aeromonas* species and *Vibrio parahemolyticus*, have been associated with poisoning, particularly in the summer months in temperate areas. The infective doses are not known for each organism but they are small. Consumption of raw shellfish should always be approached with care.

The process of depuration, that is treating the shellfish for several days in clean, running water, has been adopted in some areas (section 21.5.3). This process is partly effective in reducing bacterial numbers, but viruses are not easily removed. Far more effective is a properly designed and enforced quality assurance system in which growing sites are selected to be free from problems in the first instance and in which water quality is continually monitored to ensure that no deleterious changes have occurred.

These 'shellfish sanitation' programmes can be integrated into broader industry programmes involving total quality management (TQM) and the hazard analyses critical control point (HACCP) systems (Ahmed, 1991; Anon, 1994; Martin *et al.*, 1997). Schemes such as these are generally a requirement for export. A general Code of Practice for Aquaculture drawn up by the Codex Alimentarius Commission of the World Health Organization and the Food and Agriculture Organization provides a very good guide to practices required to ensure a safe product (FAO, 1994).

Chemical contamination invariably occurs through exposure to toxic wastes dumped or leached into the aquatic environment. The potential for contamination because of previous activities in the area or from adjacent application must also be considered. However, some natural accumulation of heavy metals, such as mercury, can occur if mercuriferous ores (e.g. cinnabar) are found in the region. Again, considerable attention should be paid to the siting of aquaculture ventures. The application of more advanced methods of analysis for heavy metals has indicated that many of the values for heavy metals recorded in the older literature for fish are in error and are far higher than the levels found by these more accurate methods.

In general, aquacultured fish are less likely to be infected with parasites, but this cannot be taken for granted. Many parasites inhabit portions of the fish, such as the head, guts and belly flap areas, that are discarded during processing, but others such as the anisakids are found in the flesh. These are highly infective but are destroyed on thorough cooking and on freezing. More recently, the practice of eating raw fish or lightly preserved or pickled fish dishes, such as ceviche, has become more common in Western societies. It is therefore very important that the product is monitored for the presence of parasites.

11.4 Nutritional aspects

The products of aquaculture are health giving and nutritious, being good sources of vitamins, minerals and protein with a low total-fat content. Marine products represent by far the best natural source of the highly unsaturated fatty acids (HUFAs) eicosapentaenoic acid (EPA) and docosahexaenoic acid (DHA). These fatty acids have important roles in membrane functioning and as precursors for biologically active compounds. They are considered beneficial in human nutrition in reducing plasma cholesterol and triglyceride levels and reducing the risk of cardiovascular disease (Sargent *et al.*, 2001).

As steps are taken to decrease the proportion of fish meal and fish oil in aquaculture feeds (section 8.7.3), so the levels of EPA and DHA in cultured fish will decrease. The level of these fatty acids is one of the major desirable selling points for aquatic produce. Clearly, a balance needs to be struck between maintaining high levels of EPA and DHA and decreasing product costs by altering feed composition.

11.5 Aquaculture and fisheries products

It is difficult to directly compare aquaculture products with wild-caught products. The variability in genetics, in feed composition and in growth conditions makes comparisons difficult to interpret except in the broadest sense. Preferences for wild or farmed products will vary greatly according to season and location of capture. In general, aquaculture products tend to provide flesh that is, compared with wild-caught products (Haard, 1992):

- softer in texture
- less strong in flavour
- often of different hue
- with a higher but more uniform oil content.

Atlantic cod reared in pens have higher contents of sarcoplasmic proteins and different water-binding capacities compared with wild cod. Both factors affect the texture of the cooked flesh. Wild fish have to swim more than captive fish, and the myoglobin content of their red muscle is often higher. Lower levels of free amino acids have been found in cultured coho salmon, red seabream and ayu than in their wild-caught counterparts. Similarly, other non-protein nitrogen compounds appear to occur in lower concentrations in cultured species. Variations in composition and distribution of tissue constituents occur both with changing seasons and with maturation (Kubota *et al.*, 1999). There are also considerable seasonal variations in the mineral content of both wild and cultured fish (Oehlenschläger, 1997), and textural changes are also apparent (Kubota *et al.*, 1999).

11.6 Harvesting

There are stresses during harvesting that may affect the final product. Crowding of fish in a net or in raceways during harvest leads to:

- rises in blood cortisol
- an increase in metabolic rate
- utilisation of glycogen with resulting:
  - increases in lactate levels
  - decrease in tissue pH.

The provision of good aeration can help relieve this situation, but every effort must be made to reduce obvious stress factors during harvest (Robb, 2001).

In rested fish, cortisol levels in the blood are generally below 10 ng/mL, but values of up to 10 times this level are common in stressed or exercised fish. Similarly, lactate levels in the blood and muscle of rested fish are many times lower than in stressed or exercised fish. There is some evidence that chronic stresses, such as those experienced during harvest because of overcrowding or poor water quality, result in a different mix of biochemical indicators from those that may result from imposition of a sudden stress such as physical exertion due to capture.

In crustaceans and molluscs, the expenditure of energy in struggling for survival is met from arginine phosphate and from glycogen. In crustaceans, the main end product is generally lactate, but octopine is formed from the reaction of pyruvate with arginine in many bivalves and gastropods. In the gastropods, tauropine, alanopine and strombine are similarly formed. Tauropine is formed in abalone, and levels of these compounds have been suggested as indicators of harvest and transport stress.

11.7 Live transport

Many markets prefer their products to arrive live to ensure freshness and to allow consumers to select their own dish or, in some instances, to eat the product within seconds or minutes of death or, as in the case of shellfish, while still live (Fig. 11.2). The returns for these markets are often higher, but the risks are greater since the associated costs of transporting animals out of their natural environment can be considerable.

Live kuruma shrimp (*Marsupenaeus japonicus*) fetch prices of approximately ¥18 000 per kilogram in the Tokyo market. This market is supplied by domestic sources, from Taiwan and, more recently, from Australia, which has the advantage of placing the product on the market in the off-season. Getting the product to the market in a live state requires appropriate harvest practices. The shrimp are active only at night and must be harvested then by the gentlest means possible. Simple tangle nets or tunnel nets set near paddle wheels are appropriate. The shrimp are gently laid on to floating wooden trays that stack one on the other and are transferred in these to a chilling tank to bring their temperature down to ~16°C. The tanks are taken from the ponds to the factory, where the temperature is further reduced nearer to 12°C. At this temperature the shrimp become inactive and can be handled readily without becoming excited. They are then size-graded and packed in chilled damp sawdust in corrugated cardboard boxes. About 10 boxes are placed in an outer insulated cardboard box in a manner that allows airflow between them. Chilled coolant gel packs are placed strategically in the box to absorb ingressing heat during transport and to maintain a uniform temperature in the box during transportation. The weights and positions of the coolants are calculated carefully so that the temperature is stable in transit from the farm to the airport and during the flight. It is timed such that the temperature of the shrimp will rise to about 16°C when the packs are opened for inspection at auction. The shrimp will have revived from their inactive state and will be capable of jumping out of the box when the lid is raised. Thus, there must be adjustments for flight times, cargo hold temperatures and seasonal factors. Low-cost disposable temperature recorders are included in each consignment. High survival rates of over 95% have been achieved using these methods (Bremner *et al.*, 1997).

Fig. 11.2 Table oysters are transported live to the point of sale and shucked.

Other crustaceans such as redclaw crayfish (*Cherax quadracarinatus*) can be transported by similar methods. Other species of shrimp can also be transported in this way over short distances, but no other species seems to show the same characteristics as kuruma shrimp. The black tiger shrimp (*Penaeus monodon*) is excitable at temperatures of 12°C and is not docile to handling. In live kuruma shrimp stored emersed in sawdust at 12°C, adenosine triphosphate (ATP) levels are maintained (presumably from arginine phosphate), whereas in other cultured species, such as *P. monodon*, ATP is rapidly depleted.

Similarly, there are lucrative markets for live fish but, in these instances, the fish must be transported in a minimum amount of water to keep them alive. The systems required for live fish transport must include features to:

- prevent injury to the fish
- prevent fouling of the water
- maintain an even (generally chilled) temperature
- allow for sufficient aeration to achieve at least 80% O_2 saturation and removal of CO_2 produced by the fish.

The water must be circulated to achieve these requirements.

Oxygenation is best achieved by direct blowing of air or even oxygen into the water to maintain sufficiently high levels. This also serves to purge CO_2 from the water. CO_2 must then be vented to prevent re-dissolution. Supersaturation with oxygen must be avoided to prevent high CO_2 levels in the blood, with consequent shift towards an acidic pH. Acidosis is lethal, for the oxygen-carrying capacity of the blood is much lower at low pH.

Low (chill) temperatures are commonly used to reduce metabolic activity, respiration and hence production of ammonia (see section 3.3.1). The optimum temperature varies according to species and their environmental temperatures, e.g. some species of flatfish become so inactive that they can be transported in just sufficient water to keep them wet. Containers for air transport range from the relatively simple bag-in-box type to highly specialised containers with pump-operated recirculating units utilising oxygen production from solid chemical pills. Containers may also include internal spray systems to keep the fish moist. Because of the corrosive potential of saltwater in aircraft, airlines are very strict as to the nature of the containers they will accept. Clearly, the package must not leak.

To maintain fish in a docile and inactive state, and to reduce respiration and production of metabolic by-products, anaesthetics can be used during transport. When the fish are for human consumption, the use of anaesthetic is in question and the search for food-grade anaesthetics has resulted in the use of iso-eugenol, a major constituent of oil of cloves, being incorporated into a harmless dispersing system. The usage rate of active ingredient is low, ~20 ppm, but, although the chemically measured residue is negligible, the faint odour of cloves persists and alters the characteristic smell of the product. It is appropriate to starve the fish prior to transportation to allow their gut to empty and, thus, reduce strain on the transport system.

Although stress will occur in the process of transporting animals out of their natural environment, all steps during handling and transport must be designed to minimise stress. Operations leading to sudden stresses must be avoided if possible since fish are often slow to recover (Åkse & Midling, 1997). Once the technical parameters have been overcome, the logistics of transport, and the match of land-freight and airline schedules with customs clearance must be carefully organised.

11.8 Muscle structure: rigor and texture

11.8.1 Muscle structure

The major edible portions of aquaculture products consist of muscle. The muscles of fish and other marine animals are similar in basic structure to other members of the animal kingdom. The muscles are fibrous in nature, contractile and are attached to the shell or skeleton by connective tissue. This connective tissue, in differentiated form, also surrounds each individual muscle fibre to form an interconnecting network (Bremner, 1999).

Muscle fibres in fish are arranged between connective tissue sheets (myocommata) in a roughly parallel fashion, and rhythmical contractions of portions of the body musculature are transmitted through the myocommata, causing flexure of the skeletal system, resulting in a swimming motion. Each individual fibre is surrounded by a network of connective tissue linked into the myocommata. Within the fibre, the major contractile elements are a system of myofibres in which two major proteins, actin and myosin, form the bulk. Contraction of each muscle segment occurs as a nerve impulse results in release of calcium ions into the cell from the surrounding sarcoplasmic reticulum. Relaxation occurs as the calcium is pumped back into the sarcoplasmic reticulum, a process fuelled by ATP (Ochai & Chow, 2000). In death, ATP is no longer present to pump the calcium ions back into the sarcoplasmic reticulum and the muscle remains in the contracted state and rigor mortis is said to have set in. The recruitment of fibres into the rigor process is not uniform and many fibres undergo passive contraction. Muscle in rigor, with contracted fibres, is much tougher than post- or pre-rigor muscle.

The muscle softens after rigor, but the processes by which this occurs are not well understood. Protease enzymes are thought to act at junction areas within each fibril (Jiang, 2000). The actomyosin complex remains intact. Using electron microscopy, evidence of general disintegration can be seen, mostly as a loss of definition in the structures. In addition, it is now clear that changes in the interstitial connective tissue can precede the resolution of rigor mortis within the muscle in some species. Further breakdown in the fine connective tissues of the extracellular matrix leads to a progressive softening of the texture (Ando, 1999; Bremner, 1999).

11.8.2 Rigor mortis and nucleotides

The extent of rigor mortis and its duration and speed of onset are important factors in determining how quickly processing has to occur. However, these are largely dictated by the conditions existing before, during and after harvest, and are thus under the control of the processor. The speed of onset and duration of rigor at any given temperature are controlled by the amounts of ATP and creatine phosphate or arginine phosphate in the muscle. This depends on the feed and the pre-harvest stress, but the major influence is the degree of stress during the harvesting procedure. As described before, when fish struggle violently, ATP is depleted and lactate levels in the muscle increase. Rigor processes set in quickly and the duration of rigor is less (Gill, 2000).

Fish in rigor are difficult to process. Indeed, they should not be handled in the rigor state, since this results in tearing the muscle and consequent downgrading of the flesh. This is a serious problem with high-priced materials, such as salmon, which are filleted for smoking or for production of gravad lax (in which the raw salmon flesh is salted and lightly pickled by the action of naturally occurring lactobacilli), when perfect flesh is very desirable. It is common practice for fish to be harvested and bled then chilled and stored in ice for at least a day until rigor mortis resolves, when the flesh loses its stiffness and can be handled without tearing. Newer methods of rested harvest and the use of rapid mechanical stunning machines, coupled with bleeding, provide opportunities for pre-rigor processing.

The method of slaughter also affects the rigor process. For example, the use of CO_2 as an anaesthetic for salmon results in shorter times for onset and duration of rigor mortis, and it is probable that this would also hold true for other species. Stunning with CO_2 is common practice, but it is far from ideal since considerable struggling occurs in the stun tank. Whatever method of slaughter is chosen, it is important that the process is accomplished in a manner that is as stress free as possible and that the fish are chilled as soon as possible after slaughter (Robb, 2001).

In the post-mortem state, oxygen is no longer available to the tissue to regenerate ATP by the normal cycle and it is broken down through a sequence of compounds, including inosine monophosphate (IMP), inosine (INO), adenosine (ADO) and hypoxanthine (HX). Anaerobic glycolysis is used to regenerate ATP, leading to production of lactate, which decreases the pH of the tissue. In different orders, genera and species, these processes occur at different rates and through different intermediaries.

The pattern of nucleotide breakdown appears to be consistent for each species, and the relative concentrations of the intermediates are controlled by the activities of the various enzymes. Completely unstressed marine animals commonly have a total adenylate nucleotide pool in the region of 10 μmol/g, of which over 90% is present as ATP. In the post-mortem state, the total adenylate nucleotide pool generally remains at or near this level, even for up to 15–20 days in the case of product chilled to 0°C. This occurs until the hypoxanthine is converted enzymically to xanthine (XA) and uric acid (U) or until bacteria metabolise the inosine or hypoxanthine.

In general, the steps leading to formation of IMP (also called 5′-inosinic acid) occur rapidly and this is an important phenomenon as far as the product is concerned, since IMP belongs to a group of compounds known to enhance flavour. Thus, high concentrations of IMP are desirable. The next major breakdown products are inosine (INO) and hypoxanthine (HX). Inosine is said to have no flavour impact but hypoxanthine is bitter. It has been suggested (Fletcher *et al*., 1990) that the balance between the relative levels of IMP and hypoxanthine in the tissue plays a large role in determining the flavour acceptability of the product. Fish have been classified according to whether they are inosine producers or hypoxanthine producers.

The ratio of the unphosphorylated to the phosphorylated nucleotides has been expressed as a ratio known as the *K*-value.

$$K\text{-value} = \text{INO} + \text{HX}/(\text{ATP} + \text{ADP} + \text{AMP} + \text{IMP} + \text{INO} + \text{HX}) \times 100$$

where ADP is adenosine diphosphatase and AMP is adenosine monophosphatase; the other abbreviations have been defined earlier.

In many species this value has been found useful as a time–temperature indicator that integrates product life and is an index of the state of the product (Ehira & Uchiyama, 1986). Sashimi-grade products have a low *K*-value (below 20%), whereas high *K*-values tend to indicate that the product either has been stored for some time or has experienced a high temperature. This rule cannot be applied across all species, and it is essential to know the pattern and rates of change for the species in question and whether it is an HX or an INO producer.

A fish that maintains a high level of IMP in its tissues is likely to have a higher flavour acceptability for a longer period, whereas one that converts it through to HX is likely to lose consumer acceptability quite rapidly.

In the live state, the adenylate energy charge (AEC) of an animal is an indicator of its vital status.

$$\text{AEC} = (\text{ATP} + 1/2\text{ADP})/(\text{ATP} + \text{ADP} + \text{AMP})$$

Harvest stress can force the tissues of shrimp into having an AEC so low (~0.5) that they would be considered dead by this criterion alone, yet many survive if resuscitated in oxygenated water. If these animals are then killed they have high IMP levels and high *K*-values at the start of their storage life, and these indicators are of little value as estimators of post-mortem storage life.

11.9 Post-mortem processing

With fish, processing can be considered to start at harvest, at the point when they are killed (Rørå *et al*., 2001). This is often done by anaesthetising the fish with CO_2 or by plunging them into iced water (Fig. 11.1). The gills are slit to allow them to bleed as a result of the action of the heart and the contractions of the muscle. After the bleeding period, which lasts about half an hour, the fish are chilled in ice slurry or in ice and transported to the factory for processing. If the factory is near the farm, the fish may be processed straight away before rigor sets in. Otherwise they are generally held in chillers for up to 2 days to allow rigor to resolve.

Processing continues with washing and grading, followed by preliminary separating of prime edible materials, generally the flesh, from inedible materials such as the head and gills, and other by-products such as the gut, part of which may be further processed.

Washing can be carried out in vertical or horizontal machines that use the friction of a water jet, the abrasion between the fish and the action of the machine structure, to remove outer debris and, at times, scales. Scaling by hand or by machine with blunt, rotating blades is the next step before gutting and removal of the fillets. This is commonly done by hand, but machines are available to handle large throughput.

Some machines may gill and de-head the fish in the same operation (Sikorski, 1991). The gut can then be removed by suction in small fish. Filleting can be done by hand or by machines in which the fish is conducted end to end past rotating circular knives that cut the flesh from the backbone. Machines differ in configurations: some species are best cut from the head end, whereas others can be cut from the tail. Several types of mechanical skinners are available. In one type, the fillet passes over a refrigerated drum that freezes the skin layer, allowing it to be cut off with a band knife. In another type, the skin side of the fillet is carried on a belt and pressed by a roller against a stationary flat knife.

In planning a processing operation, careful consideration must be given to the proper structure and flow of the process line to prevent build-up of partly processed product at any stage. A logical sequence of procedures to allow steady flow of product must result. Care must be taken to prevent cross-contamination and to keep the product chilled at every opportunity.

Fish are generally gilled and gutted, and the belly cavity thoroughly cleaned out and inspected for parasites or defects. This may be done before rigor when the process area is adjacent to the farm. The cleaned fish is then ready for market, for processing or for freezing and further processing at a later date. For many fish, such as the salmonids, processing involves filleting and removal of small bones, and trimming the fillet prior to dry salting or brining. Salt is applied for 1 day, then the sides are brushed clean and cold smoked to produce highly desirable smoked salmon. Cold smoking is carried out at a temperature near 22°C. Above 25°C the collagen of the connective tissue begins to decompose and the resulting product does not have sufficient integrity to be sliced thinly. Smoked salmon is sold as whole sides or as slices, generally vacuum packed onto a board or plastic tray, and distributed and sold chilled or frozen. Other products such as terrines and patés are made from the trims and off-cuts of the filleting and smoking processes.

Shrimp are harvested by net or by drainage of the pond into pounds (section 19.8.5). Good practice then dictates that they are chilled in ice or ice slurry and transported to the processor. Either at the farm site or on receipt, they are generally treated with a dip in compounds to prevent melanosis (blackspot) developing on the shell. The most common dip is a form of sulphur dioxide (SO_2) supplied by either sodium metabisulphite or sodium sulphite solution. Dip times and concentrations are calculated to be effective yet leave residual levels to comply with domestic and export/import regulations. In most countries this level is 30 mg SO_2 per kilogram of shrimp, calculated on the weight of whole shrimp. The SO_2 is effective, since it inhibits the polyphenoloxidase enzyme responsible for melanosis (Kim *et al.*, 2000). Other dips consisting of acidifying agents, such as sodium acid pyrophosphate, work by acidifying the tissue out of the active range of the enzyme. Recently, compounds such as 4-hexyl resorcinol have become commercially available and these act at very low levels (e.g. 4 mg/kg of shrimp) as substrate analogues to inactivate the enzyme.

Shrimp may be sold raw in whole or peeled forms, chilled or frozen, or they may be cooked. They are normally cooked in vigorously boiling water, either potable or seawater, or with salt added, for up to 5 min to ensure that the enzymes of the digestive gland are thoroughly denatured by the heat. If denaturation is not thorough, enzymes from the gut can digest the tail flesh, leading to staining and off-flavours. After cooking, the shrimp are chilled in fresh cool water, then in ice slurry, and then packed in ice before grading and packing into containers for market or further storage. Size grading is either by hand or by machines. Grading machines consist of rollers or inclined slats with gaps between; the gaps increase in size along the length of the machine.

11.10 Effects of feed on the product

Aquaculturists can favourably affect the flavour characteristics of the final product through manipulation of feed ingredients. In order to do this it is essential that the industry knows the characteristics required by the market. In general, good husbandry practices and care for the feed will result in better products than if these factors are neglected. Most feeds are relatively high in lipids and these oxidise rapidly if exposed to heat, light and oxygen. Not only does this destroy the nutritional value of the product, but the long-chain fatty acids can be broken down into compounds that impart undesirable 'fishy' taints. Thus, the level of antioxidant in the feed and the antioxidant status of the fish are important factors.

Most of the effects of feed on flavour reported in the literature concern changes in the diet from customary ingredients such as fish meal to protein sources from grains and legumes, which are cheaper to obtain (section 8.7.3). The resulting change in flavour can be subtle but of significance to the consumer. The analogy with wines, with the loss or gain of volatile compounds resulting in lower status, price and marketability, is quite apt.

Recently, the importance of the bromophenols in imparting seafood flavour has become apparent (Fig. 11.3). These compounds occur naturally in marine invertebrates, such as bryozoans, and it has been established that even small amounts of 2,6-dibromophenol in the microgram per kilogram range can have a profound effect on the perception of seafood flavour (Whitfield, 1990). The threshold level at which the flavour of 2,6-dibromophenol can be detected in water is extremely low, at 0.00005 μg/kg, whereas in shrimp flesh it is 0.06 μg/kg. The equivalent threshold for 2,4,6-tribromophenol in water is 0.6 μg/kg.

Similarly, terpene compounds from seeds, grasses and algae move their way up the food chain to impart flavour to the wild product. There would seem to be considerable scope to incorporate premixes or minor ingredients rich in these compounds into aquaculture diets to overcome some of the perceived problems of bland flavour. Synthetic or other natural sources could be used.

11.11 Flavours and taints

Flavours arise from the complex of compounds in the tissue and, within a species, these are controlled by the environment and the feed ingredients (Lindsay, 1990). Wild-caught fish and shrimp have stronger 'natural' flavours than their cultured counterparts because of the greater diversity of their diet. Replacing natural feeds with the blander cereal- and legume-based artificial diets thus perpetuates this lack of diversity.

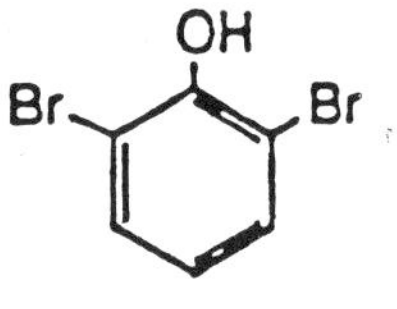

2,6-Dibromophenol

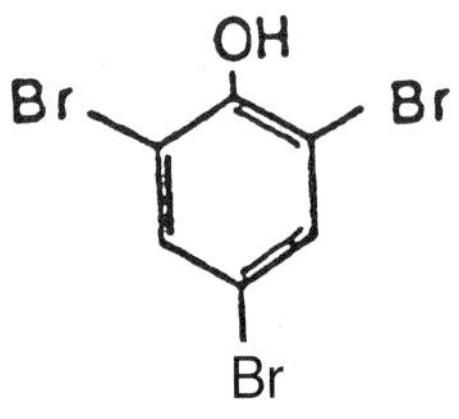

2,4,6-Tribromophenol

Fig. 11.3 Chemical formulae of the two major bromophenols that impart 'seafood' flavour.

One major problem prevalent in freshwater aquaculture is the presence of taints from blue-green algae and actinomycetes, which lead to the compounds geosmin and 2-methylisoborneol accumulating in the lipid portion of the tissue. The taints are variously described as muddy, earthy or musty (Howgate, 2001). Most of the work on these problems has been carried out in the USA on channel catfish, *Ictalurus punctatus*. These taints can be removed by purging the fish in clean water for several days. The time of purge obviously is dependent on the degree of contamination, and each situation needs to be determined on its merits. Exposure of catfish to a concentration of 0.5 μg/kg methylisoborneol in the water leads to rapid uptake of musty flavour within 1–2 h, as detected by a sensory panel. Significant taint still existed in fish even after 48 h of purging in clean charcoal-filtered water. Purging is currently the only remedy since algal outbreaks will always occur in intensive systems. Steam pre-cooking to leach the volatiles followed by canning has been found to neutralise the flavours to some extent with trout.

Contamination of water from domestic and industrial sources can also occur. Traces of the highly odorous musk xylols used as perfumants in commercial detergents have also been detected in freshwater species in Europe, but breakdown products from petroleum can taint both marine and freshwater species. Other taints may arise from oxidised or mouldy feeds, but these are due to bad management practices (section 9.10.3) and will not be considered further here.

Much of the flavour of aquatic foods, and hence of farmed products, is derived from the non-volatile, non-protein nitrogen compounds, including:

- free amino acids
- urea
- betaines
- peptides
- nucleic acids
- nucleotides
- trimethylamine oxide
- other amines
- salts and minerals.

The concentration of the free amino acids, glycine, taurine and alanine, are important determinants of the flavour of abalone. Glycine, proline, arginine, serine, threonine and alanine are particularly important in determining the flavour of shrimp, and their relative content is affected by the salinity of ponds: higher levels occur at higher salinities. Glycine is quite sweet and the major contributor to the sweet flavour of shrimp and scallops (Komata, 1990). It is also readily soluble in water and can be leached out during processing: raw shrimp have a different, much sweeter, flavour than those that have been boiled. Indeed, cooking the product by different methods can also create a different range of heat-induced volatiles (Suzuki *et al.*, 1990).

The main storage carbohydrate in aquatic products is glycogen, which in it self does not contribute to flavour, but its breakdown products of glucose and glucose 6-phosphate are important. Because of the higher content of carbohydrate in artificial diets, most cultured animals tend to have higher glycogen levels than in the wild. Exercise increases glycogen levels in the muscle. The importance of glycogen in determining flesh properties lies in its function as an energy reserve for anaerobic glycolysis when the fish is stressed or in the post-mortem state.

The volatiles in fresh fish flesh are mostly alcohol and carbonyl compounds, which arise from the action of specific enzymes on individual long-chain fatty acids. These lipoxygenase enzymes are involved in the production of leukotrienes and in osmoregulation at the cellular level. The systems in fresh- and saltwater fish are different but, in general, freshly killed fish have delicate odours, similar to the odours of green vegetables. These are replaced within a few days with other enzymic oxidative products that have a more fishy character. In post-mortem tissues, enzymes specifically attacking fatty acids produce a whole suite of odour compounds (Table 11.1) (Cadwallader, 2000; Pan & Kuo, 2000).

11.12 Texture

After death, glycogen in muscle may be broken down by either phosphorolytic or hydrolytic pathways to result in the production of lactic acid, which in turn determines the ultimate pH of the flesh. pH is the major determinant of texture in flesh, with the inherent buffering capacity of amino acids and peptides playing a role in its control. Low pH normally results in firm textured flesh, but too rapid a reduction in pH tends to cause a soft texture even at a low ultimate pH. Thus, the biological conditions, as reflected in glycogen levels and buffering capacity, dictate the flesh texture within a species.

The texture of flesh in various species is a consequence of its structure. In the raw state, the relative proportion of collagen dictates the resistance of the muscle to shear, since the myofibrillar components are present as a soft gel. Those species with low levels of connective tissue are too soft for sashimi; those with too high a level are too tough. The situation reverses in cooked muscle and the myofibrillar components become tough while the collagen molecules shrink, unfold, gelatinise and soften. As a result, the flesh readily separates into flakes, each of which is a muscle block (Hatae, 1999). Thus, the texture of cooked fish is dependent on the state of the myofibrillar proteins. If the proteins are denatured and have lost their ability to hold moisture then the cooked product may taste dry and stringy, since the moisture will only be loosely held. The oil content and the amount of gelatinised connective tissue contribute to the sensation while chewing.

11.13 Chilled storage life

It is customary to use 0°C, the temperature of melting ice, as a reference for chilled storage work (Regenstein & Regenstein, 1991). This is because no mechanical or electrical devices are required for control of temperature, just the ice itself, and because spoilage at this temperature can be readily related to that at other temperatures (Table 11. 2).

Thus, compared to storage at a temperature of 0°C, spoilage at –1.5°C occurs 30% more slowly, leading to a 40% increase in shelf life. In comparison, temperatures of 4, 10 and 15°C result in increases in the rate of spoilage of two times, four times and over five times, respectively, and a substantial decrease in shelf life (Bremner *et al.*, 1986).

It is essential to have a good estimate of the chilled storage life of an aquaculture product in order to determine the scope for distribution. Products that remain marketable for over a fortnight in ice can

Table 11.1 Typical compounds derived enzymically from polyunsaturated fatty acids

Compound	Aroma
Hexanal	Heavy, green*, aldehyde-like
t2-Hexanal	Green, stinkbug-like
c3-Hexanal	Green, leafy
1-Octen-3-ol	Raw mushrooms
1-Octen-3-one	Cooked mushrooms
1,c5-Octadien-3-ol	Heavy, earthy, green, mushroom
1,c5-Octadien-3-one	Crushed geranium leaves
t2,c6-Nonadienal	Cucumber
t2,c6-Nonadienol	Cucumber, melon
(2Z)-Octen-1-ol	Weak green, fishy
2,5-Octadien-1-ol	Fried fish
(2E)-Pentenal	Green apple
(1,5Z)-Octadienal	Geranium leaves
(1,3E,5Z)-Undecatriene	Balsamic
(1,3E,5Z,8Z)-Undecatetraene	Seaweed

*Green denotes cut grass or herb-like aromas.

Table 11. 2 Relative rate of spoilage and potential change in shelf life at various temperatures with reference to storage at 0°C

Temperature of storage (°C)	Rate of spoilage relative to 0°C	Shelf life (%) relative to 0°C
–1.5	0.7	140
0	1.0	100
4	2.0	50
10	4.0	25
15	5.25	19

obviously be distributed more widely than those in which undesirable changes occur in a week. The criteria by which to measure the marketable shelf life are determined by:

- the market
- the suppliers
- some sector of the industry.

There are markets for produce that is not in optimum fresh condition and the products into which such material can be made. It is generally the case with cultured products that the producer is aiming at the fresher end of the market, where prices are higher. Thus, the pattern of change in the product can be more important in determining its market and distribution than the total shelf life.

The typical rates of change seen in products are shown in Fig. 11.4. In general, the highest quality product is available for only a relatively short period (often only 3–4 days), after which the product slowly declines. It is in these first few days that product could be regarded as of a quality suitable for sashimi, i.e. to be eaten raw. The formation of spoilage products is a consequence of bacterial growth. A typical situation is set out in Table 11.3, in which an initial count of only 10^3 bacteria/cm^2 is assumed to be present on the product surface and the reasonable assumption is made that counts will double each day. Bacterial counts are low in the early stages: in the first 4 days, counts increase by only ~20 000 per cm^2 of surface. However, in the 24 h from day 12 to day 13 the increase is 12 million/cm^2! The methods used to determine the storage life may be either direct

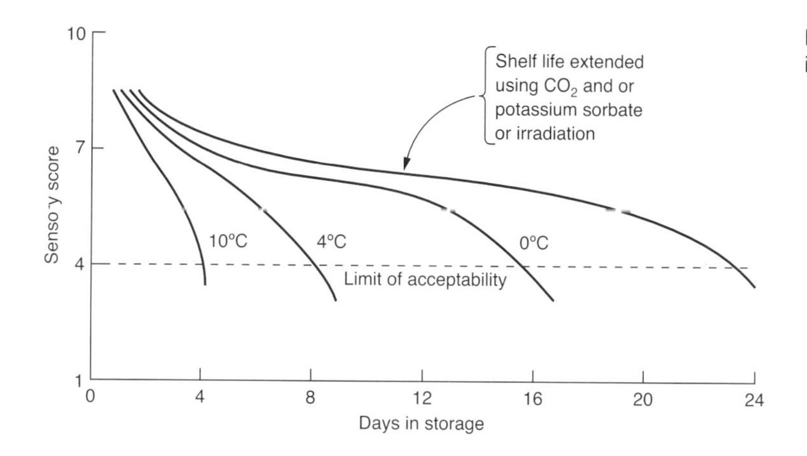

Fig. 11.4 Typical pattern of spoilage in chilled fresh seafood product.

Table 11. 3 Hypothetical examples of calculated bacterial counts on the surface of fish flesh if the bacteria double in number each day. These bacterial counts and percentages can be considered equivalent to the relative concentration of spoilage compounds, assuming that the initial organism has spoilage potential

Storage period in ice (days)	Calculated bacterial counts (thousands of bacteria/cm²)	Bacterial counts as a percentage of count at 13 days*
0	3	0.01
1	6	0.025
2	12	0.05
3	24	0.10
4	48	0.20
5	96	0.39
6	192	0.78
7	384	1.56
8	768	3.13
9	1536	6.25
10	3072	12.5
11	6144	25
12	12288	50
13	24576	100

*End of acceptable shelf life.

or indirect. Direct methods may involve the use of systematic scoring systems or sensory assessment of the cooked or raw material by trained panellists. The systematic scoring system is simple, direct and non-destructive, and has been shown to be capable of integrating time–temperature effects (Bremner *et al.*, 1986). Sensory assessment by trained panellists is very effective, but it requires considerable skill and organisation and suitable facilities.

Sensory methods describe the product when it is eaten and are thus more closely related to consumer appreciation. Full consumer trials must be undertaken to relate the results of trained sensory panels to 'the assessment of the market place'.

Indirect methods include:

- microbiological counts
- measurement of flesh metabolites, such as nucleotides
- measurement of microbial metabolites, such as trimethylamine or products of oxidation.

The microbiology of chilled storage of flesh foods is predominated by the pseudomonad group of bacteria.

In temperate marine situations, the major spoilage organism is a heterotrophic facultative Gram-negative bacterium, *Shewanella putrefaciens* (formerly *Alteromonas putrefaciens*). This can metabolise a number of free amino acids and sugars to produce a range of odorous compounds, including hydrogen sulphide (H_2S). Indeed, this property alone is often linked to spoilage, and enumeration of H_2S-producing bacteria correlates well with spoilage, even though they may represent only a proportion of the total flora. Other pseudomonad bacteria contribute to spoilage and, in tropical and subtropical waters, *Pseudomonas fragi* is often the organism that results in off-odours and off-flavours reminiscent of rotting tropical fruits.

It is important for any given situation to understand the nature of the spoilage microflora and the specific spoilage organism involved so that steps can be taken to inhibit its growth (Huss *et al.*, 1997).

It is common for bacterial counts to decrease in number when warmwater species are placed on ice. For example, tropical shrimp that show a total count of 10^6 organisms per gram when first harvested may only have counts of 10^4/g after storage on ice for a day or two (Fig. 11.5). This is because the microflora (bacteria and fungi) capable of growth at chill temperatures (mainly psychrotrophic organisms) are initially present in only very low numbers and are often introduced with the ice or alternative chilling medium and on boxes or utensils. The native warm-loving (mesophilic) microflora cannot survive the chill conditions and die. For this and other reasons tropical aquaculture products can exhibit a longer chilled shelf life than species from colder waters, in which the psychrophilic microflora are already present in abundance.

11.14 Freezing and frozen storage

Although much aquaculture produce is sold fresh, factors of seasonal production mean that a large proportion is frozen for further processing or as final product for orderly distribution to the market.

In essence, the process of freezing is one of substantially reducing temperature by absorbing heat from the product and immobilising the bulk of the water in the form of ice. The preservative effect of freezing is due to two factors (Regenstein & Regenstein, 1991).

(1) The reduced temperature slows chemical and biochemical reactions, in some cases to imperceptible rates.
(2) Low temperature completely retards microbial and fungal growth.

The reduction of mobile water with freezing reduces the water activity to further retard enzymic and chemical reactions, but it is important to note that not all the water is frozen at temperatures commonly used for commercial storage (–20°C). A concentrated solution of salts, amino acids and other soluble constituents still exists in a mobile phase distributed throughout the tissue (Fig. 11.6), and many biochemical and chemical reactions still continue at appreciable, though reduced, rates (Haard, 1992).

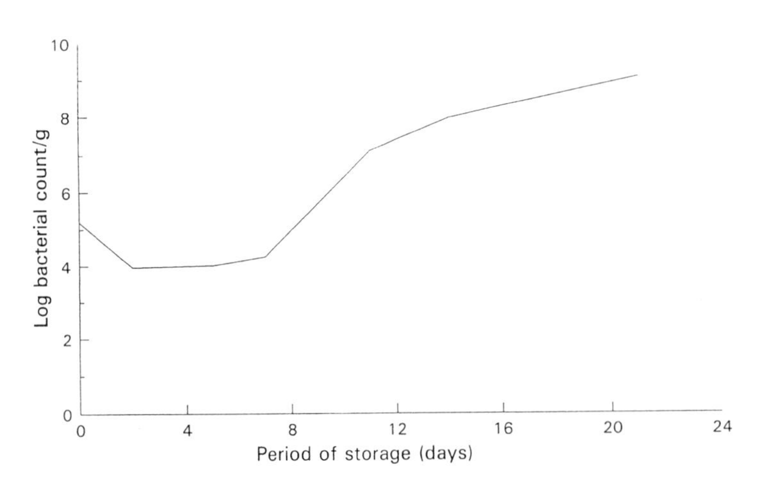

Fig. 11.5 Total bacterial counts of tropical products (e.g. shrimp) often decrease in the initial period of chilled storage.

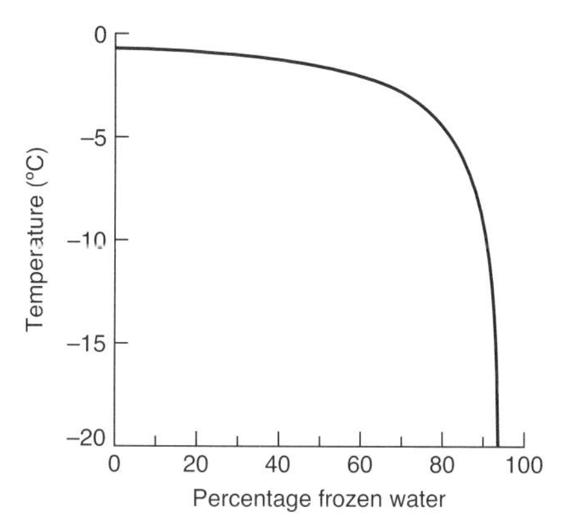

Fig. 11.6 The percentage of water in fish tissue frozen at different temperatures. Note that even at -20°C not all the water is frozen.

The means of freezing depend on the nature of the product.

(1) Blast freezing. Freezing using refrigerated air in a blast freezer is by far the most common and versatile method used, since the same equipment can accommodate different product volumes, shapes and sizes. In addition, continuous blast freezers have been developed to freeze individually quick frozen (IQF) product. In blast freezing, refrigerated air is blown at high speed around the product. Air will take the path of least resistance, and it is important that it passes over the product, not around it in the voids. The velocity must also be sufficient to create a turbulent flow of air at the product surface otherwise heat transfer from the product to the air will not be efficient. Fans with aerofoil sections must be used to achieve sufficient airspeed under the pressure head.
(2) Plate freezing. Horizontal plate freezing, in which the product is frozen between plates arranged in banks, is quite efficient and is commonly used for finished product, e.g. fillets, scallops and shrimp, for which uniform packs are produced. Vertical plate freezing is employed mostly for whole fish or shrimp destined for further processing and is used more at sea than for aquaculture products.
(3) Immersion freezing. Freezing by immersion in refrigerated brine gives efficient heat transfer with consequent rapid freezing times and is best suited to small individual products such as shrimp or scallops. Such products may then be packed as IQF items.
(4) Cryogenic freezing. Freezing with liquid nitrogen tunnels or solid CO_2 'snow' are also used for IQF products. These techniques have the advantage of low capital cost (equipment can be hired), but the disadvantage of high operating costs.

The product is commonly frozen to about –20°C, but –30°C should be used. The rate at which the product reaches this temperature is important since the aim is to produce small ice crystals within the cell structure rather than large jagged crystals, which pierce and disrupt the cell membrane. All frozen product is eventually thawed for consumption, and product in which the structure has been disrupted will be of poorer texture and will release more of the tissue fluids when thawed. These fluids are important contributors to the taste and succulence of the final product and its nutritive value. Furthermore, they represent a considerable loss of revenue if valuable product ends up being discarded.

Much of the heat load on the refrigeration system is due to the latent heat of freezing of water. When a time–temperature profile of freezing is plotted, it shows a plateau area on the curve known as the zone of thermal arrest (Fig. 11.7). In general, this spans the temperature region from –1.5°C to about –5°C,

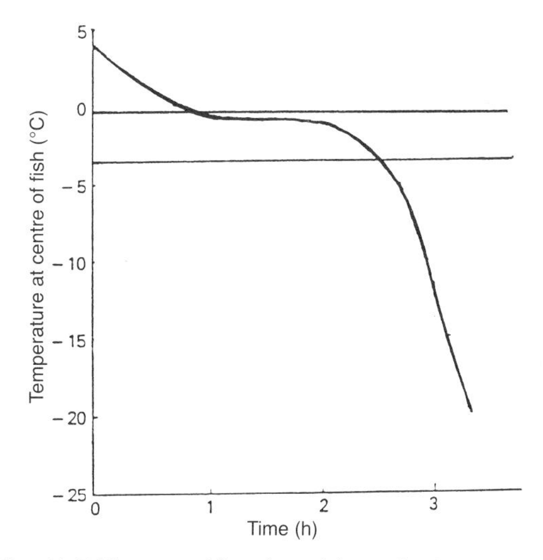

Fig. 11.7 The rate of freezing of tissue is slow as the latent heat of freezing is removed in the 'zone of thermal arrest', which is demarcated by the two horizontal lines.

but the exact values depend on the water-soluble salts and minerals present in the product, which vary according to seasonal and nutritional circumstances. The aim is to drive the temperature of the product through this zone of thermal arrest as rapidly as possible.

The storage temperature of the frozen product is also very important. All products have a finite life that is governed by the temperature of storage and the stability of this temperature. As a very broad guide, a storage temperature of –20°C provides twice the shelf life of –10°C, whereas –40°C provides twice (or more) the shelf life of –20°C. Equally important is the stability of the temperature. If the temperature warms up from –20°C to –10°C, a portion of the ice in the tissue will melt and, if the product is cooled again, this will refreeze to build even bigger ice crystals. Continual temperature cycling will exacerbate this effect, with consequent destructive effects on the product.

Although no bacteria will grow at temperatures lower than about –7°C, enzymes that they have released may still be active if large numbers of bacteria were allowed to proliferate on the product before freezing. Thus, product that is poor due to bacterial growth can deteriorate even further in frozen storage, as undesirable products of bacterial metabolism diffuse through the whole product from the surfaces where the bacteria grew. Inherent enzymic activity also takes place during frozen storage. Lipases are active to produce free fatty acids, which are then prone to oxidation. In marine species, trimethylamine oxidase can split the osmoregulatory compound trimethylamine oxide (TMAO) to dimethylamine and formaldehyde. The formaldehyde is highly reactive and migrates in the aqueous phase either to react with proteins or to destabilise the layer of water molecules that form buttressing structures around the proteins. This results in denatured proteins, which are less able to rehydrate when the tissue is thawed, leading to more fibrous, and often stringy and dry, texture in the product.

The most common cause of deterioration in frozen product is oxidation of the constituent lipids with the oxygen of the air (Table 11.4). Oxidised lipids react with protein, leading to denaturation and textural change. More importantly, oxidation results in off-odours and flavours variously described as straw-like, nutty, paint solvent and stale. Small, fatty fish are notoriously prone to oxidation, but within any species the antioxidant status of the flesh and the lipid composition are critical factors (Undeland, 2001).

Fish oils containing highly unsaturated fatty acids readily react with any available oxygen. The reaction starts with an induction phase during which oxygen is absorbed and free radicals are initiated, but little detectable change is noted. This is followed by the propagative self-catalysed phase, known as autoxidation, in which the reactions are rapid and self-sustaining.

In lipids and oils with low levels of antioxidants, such as vitamin E, the induction period may be slight and the autoxidative phase starts almost immediately (Fig. 11.8). As shown in Table 11.4, peroxides are intermediate products of lipid oxidation. A common method of determining oxidation in aquaculture products is to measure the peroxide content or 'peroxide value'.

The hydroperoxides themselves have little odour or flavour, but they break down to give a wide range of compounds that have quite strong odours and flavours. For example, the compound hept-*cis*-4-enal has been shown to be the main cause of the 'cold storage odour and flavour' found in some species and it contributes nutty 'cardboard' flavours.

Products of oxidation in combination with trimethylamine (produced bacterially in marine species from trimethylamine oxide) are regarded as being responsible for the production of fishy odours that occur during storage.

Thus oxidation in stored products is a problem that causes:

- downgrading of product
- lower acceptability by the consumer
- lower price
- a decrease in the product's nutritional value.

Oxidation can be minimised by ensuring that the product is fresh when frozen and that the precursors of oxidation as a result of the activity of oxidase enzymes or potential exposure to air, light or heat are not allowed to form. Thus, frozen product can be protected either by glazing with water (which may contain antioxidants such as ascorbic acid) or

Table 11.4 A general outline of the process of oxidation

Phase	Reactions	Results
Initiation step	$-RH \Rightarrow R + H^+$ $-O_2 + RH \Rightarrow R + OOH$	Free radical formation
Propagation	$-R + O_2 \Rightarrow RO_2$ $-RO_2 + RH \Rightarrow ROOH + R$	Peroxide and hydroperoxide formation
Termination	$-2R \Rightarrow R\text{-}R$ $-R + RO_2 \Rightarrow ROOR$	Terminal non-propagating products

R is a highly unsaturated fatty acid (HUFA) that may be a free fatty acid (FFA) or attached as part of a phospholipid or a triglyceride.

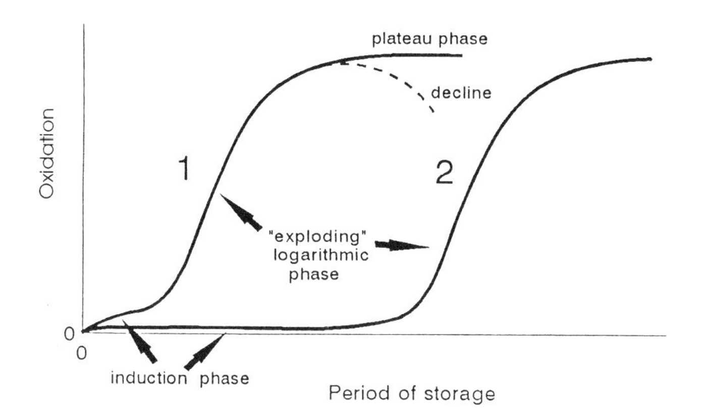

Fig. 11.8 Typical patterns of oxidation in stored fatty fish: 1 represents the situation for a product with little antioxidant protection and 2 represents that for a product with significant antioxidant protection.

by packing in a plastic film with low permeability to oxygen. Vacuum packing is very effective and the inclusion of oxygen-absorbent materials in the packs to harvest any residual oxygen provides even greater protection.

Product can also dehydrate in frozen storage as water evaporates from the surface in the low humidity of the cool store and gets deposited on the evaporator coils. In exposed product, this leads to areas of white, denatured tissue known as 'freezer burn'. Freezer burn is irreversible and causes downgrading of the product.

In packaged products, temperature cycling leads to condensation of water vapour on the inside of packs and its subsequent refreezing into crystals in the bag or box. This represents loss of yield and is an indicator of poor storage conditions and poorer product.

11.15 Packaging

Packaging of the final product is important for its preservation and presentation. The technical requirements of a package for frozen stored product are that :

- oxidation is minimised
- dehydration from the products surface does not occur.

There are a number of options for the chilled product brought about by continual improvements in packaging films and in equipment.

Vacuum-packed products are now common. The product is placed in these packs on a tray or in a pouch (or the tray is placed in the pouch), a vacuum is drawn and the pouch is then sealed by heat, or with a clip or, alternatively, a film is sealed over the edges of the tray. This type of pack provides excellent presentation since the plastic seals in the juices, the product itself can be viewed and the pack can be decorated with brand, advertising and other product information.

Vacuum packing in oxygen-impermeable films is often used for processed materials such as smoked and salted eels and salmonids. A considerable chilled shelf life can be achieved for these products, since the conditions in the pack slow the growth of spoilage bacteria. These conditions of low pH, low oxygen tension, low water activity (due to salt and other curatives) and a low initial microflora (smoke has antibacterial action) all combine to help select a microbial flora that is not spoiling in nature and generally consists of lactobacilli.

Vacuum packing of fresh product in oxygen impermeable films does not improve the shelf life of the product to any worthwhile extent. It does provide more presentable packs, but this can be achieved using less costly oxygen-permeable films. Vacuum packing in films of low oxygen permeability is not necessary for fresh-chilled product.

In modified atmosphere packing (MAP), the product is placed in a tray or pouch, a vacuum is drawn, then the atmosphere is replaced with another gas or gas mixture generally containing at least 40% CO_2 by volume. The pouch is then sealed or a gas-impermeable film sealed over the tray. The gas to product ratio is in the region of 2 or 3 to 1. In these packs, the CO_2 dissolves in the aqueous portion of the flesh of the fish, leading to a reduction in pH due to formation of carbonic acid. The preservative action that occurs is thought to be due to both the lower pH and the amount of undissociated carbonic acid present. This undissociated acid can readily cross bacterial cell membranes, following which it dissociates to release H^+ ions. The decrease in pH and disruption of cell metabolism result in inhibition and death of the bacteria. Thus, spoilage organisms are destroyed or inhibited and can grow only slowly in these packs. The lower pH and low oxygen tension encourage the growth of lactobacilli, which themselves are inhibitors of Gram-negative spoilage bacteria. A doubling of the normal chilled shelf life can be expected for such products, but it should be noted that endogenous enzymes are still at work within the flesh and that deleterious changes still occur slowly.

The degree to which the system works for any particular product is a function of the partial pressure of CO_2 in the pack and the relative volume ratios of product to gas. Fish flesh will absorb more than its own volume of CO_2 so the initial pack must be overinflated to allow for this. The disadvantages of MAP are:

- the bulkiness of the packs
- the fact that the product is loose (unless held by some internal film)
- the pack can fog as a result of condensation
- unsightly drip liquor from the product may be high (due to the low pH), making the product less attractive
- the potential for botulism if low temperature is not maintained.

MAP has been used with part-cooked product and products enrobed in batter to prevent some of these problems.

One of the most serious considerations with MAP is that conditions in the packs can lead to the growth and production of toxin from the botulism bacterium (*Clostridium botulinum* type E) if the temperature is not strictly held at, or below, 3°C. This organism produces the deadliest natural toxin known. The concern is that the longer shelf life obtainable with MAP will provide the opportunity for the organism to produce toxin before there are any overt signs of spoilage which would otherwise warn the consumer not to use the product. Cooking readily destroys the toxin, but it could be inhaled as an aerosol when the pack is opened and smelled. Also, it could contaminate knives, boards and utensils, and be passed on to other products that may not be cooked or to foods that have already been cooked.

MAP is widely used in Europe, but it is viewed as potentially too dangerous in the USA. This probably reflects a difference in culture and approach to safety and food regulatory matters rather than to any real risk. The key to safety is to ensure low product temperatures throughout the distribution system. There are some fears that even vacuum packing could provide conditions suitable for botulism. However, this will not occur if the film used is permeable to oxygen and there is no real need to use oxygen-impermeable films with chilled product, although it is excellent for frozen products.

With smoked salted products, such as those made from trout and salmon, *C. botulinum* cannot grow if the concentration of salt in the water phase of the product is 3%. This level thus provides a critical control point and its measurement should be standard practice in quality control.

11.16 Quality control, quality assurance and HACCP

Quality control (QC) is the practice of determining the properties of products to ensure they are within some agreed range. Quality assurance (QA) is the adoption of practices that will guarantee that the quality attributes of the product lie within that range. Thus, quality assurance is focused on the process, whereas quality control is focused on the product. The two are complementary and QA arises from an understanding of QC and is superior to it.

For processors, QA begins with decisions on the correct site for the factory and its layout, and the processing techniques to be used. A proper QA plan will include schedules for equipment and plant cleaning, the instruction of staff in the use of detergents and sanitisers, and in personal hygiene and food-handling practices. There should be a general maintenance plan and protocols for a full sanitary control programme.

QC procedures generally revolve around measurement taken on raw materials, incoming goods, process operations and finished product. In aquaculture:

- The raw material should always be up to standard since this is controllable.
- Incoming goods, such as packages and cans, must be inspected to ensure that they are of correct specification.
- Process parameters, such as temperatures and times, must be measured and logs kept.
- Finished products must be inspected to ensure they meet company standards.

These standards must be laid down unequivocally. In large operations, QC staff, independent from production operations, must be trained in proper inspection procedures.

Hazard analyses critical control point (HACCP) schemes are mandatory in most countries and are required by importing countries. These schemes are based on:

- evaluation of the hazards involved
- identification of critical points in the process to control the hazards
- adoption of procedures to control the hazards
- recording and verification that these procedures have been carried out correctly (Ahmed, 1991; Anon, 1994; FAO, 1994; Hocking *et al.*, 1997; Martin *et al.*, 1997).

11.17 Canning

The canning of seafood products is well understood, and mostly it is only wild-caught, not cultured, material that is used. The process is one of hermetically sealing the product in a rigid can under partial vacuum and heating the contents sufficiently to ensure destruction of micro-organisms (FAO, 1988). The heat process required is calculated to destroy spores of the botulism bacterium, should they be present, and the times and temperatures required depend on the size of the can and the nature of the pack. For example, a solid tuna pack requires over 60 min at 121°C, whereas shrimp in brine require only 16 min because of the greater convection in the fluid in the can and the more rapid thermal conductivity.

Basic canning operations involve the following stages:

- First preprocessing steps, e.g. cutting, dicing and slicing, are carried out.
- Some species are pre-cooked, then picked over, selected or separated before packing into the cans.
- The seafood is then packed and weighed into cans.
- Vacuum is then drawn either by passing the can through a vacuum chamber or by purging the can headspace with steam before closing of the lid.
- Filled cans are loaded on trolleys or baskets and placed into a steam-heated chamber for processing.
- The air is purged from the retort (a steam-heated horizontal or vertical cylinder) and the temperature rises as steam is introduced.
- Careful monitoring of the temperature is required for each batch.
- The cans are then cooled by the introduction of clean water.
- They are moved from the retort and allowed to stand for their surfaces to dry.
- Labelling occurs several hours later when the can has 'settled'.

11.18 Smoking

The smoking process has been used for many years to preserve and impart a characteristic smoky odour, flavour and texture to fish products (Doe, 1998). Smoke is invariably generated from wood by direct combustion of logs, sticks, chips or sawdust or by friction of grinders on logs. A huge range of organic compounds have been identified in smoke, including:

- phenolics (> 80)
- acids (> 30)
- alcohols (> 10)
- esters (10)
- aldehydes (> 20)
- ketones (> 60)
- hydrocarbons (> 20)
- aromatic hydrocarbons (> 60)
- lactones, ethers, furones and nitrogenous compounds.

These constituents are all volatile in the smoke zone and are carried in the air and deposited on the surface of the product.

It is the phenolics that supply the major characteristic flavour to smoked fish, and the compounds syringol, guaiacol, 4-methylguaiacol, eugenol, 4-methylsyringol and 4-allylsyringol have been reported to be most important (Miler & Sikorski, 1991). These phenolics and the other smoke constituents are antibacterial, so the product surface at the point of exit from the smoker is close to sterile.

There are two main types of smoking, designated cold smoking and hot smoking. Cold smoking is mainly used for fillets, hot smoking for gilled and gutted whole fish. In cold smoking, the product is not cooked and the temperature of the air in contact with the product rarely reaches 30°C. Temperatures in the 22–25°C range for 6–8 h are common. The salt (natural and added) present in the fish, the preservative action of the smoke and the slight drying that occurs give the product microbial stability. It is then ready for eating without the need for cooking. In hot smoking, the product is cooked during smoking. The process lasts only 1–2 h and reaches temperatures above 65°C. The product is eaten either cold or reheated.

It is very important that the product is not smoked at an intermediate temperature that is insufficient to pasteurise it, because in the heating and cooling process bacteria in the interior of the product may well have proliferated to make the product dangerous.

The chill stored life of hot-smoked fish may be up to 2 weeks. Cold-smoked fish are more heavily smoked and salted, and may have a chilled shelf life of up to 6–8 weeks. It depends greatly on the salt content and the degree of smoking, and it is advisable for each producer to check the results from their own system before marketing a product and publicising a use-by date (Fig. 11.9).

11.19 Concluding remarks

Many factors impinge on the attributes of the raw material and finished products from aquaculture, and to further complicate the picture they interact highly with one another.

This has been but a brief introduction to post-harvest technology of aquaculture products. There are many other issues not covered that can properly be considered in this area but are beyond the scope of this chapter. These other important issues involve:

- traceability from harvest to retail sale
- quality chain management
- information technology
- product development
- total utilisation
- genetic and seasonal influences on the properties of the product
- extraction and development of by-products
- biotechnologies for preparation of bioactive materials from aquaculture products.

Fig. 11.9 Vacuum-packed smoked Atlantic salmon.

References

There are many papers specifically dealing with post-harvest technology of aquaculture products in journals such as *Aquaculture*, *Journal of Food Science*, *Journal of Aquatic Food Product Technology*, *Fisheries Science*, *International Journal of Food Science*, *Journal of the Science of Food and Agriculture*, *Journal of Agriculture and Food Chemistry* and *Lebensmittel Wissenschaft und Technologie*. There are, however, few monographs solely devoted to aquaculture products. Consequently, most of these references are from monographs dealing with both captured and cultured species.

Ahmed, E. (Ed.) (1991). *Seafood Safety. National Academies of Science*, National Academies Press, Washington, DC.

Åkse, L. & Midling, K. (1997). Live capture and starvation of capelin cod (*Gadus morhua* L.). In: *Seafood from Producer to Consumer, Integrated Approach to Quality* (Ed. by J. B. Luten, T. Børresen & J. Oehlenschläger), pp. 47–58. Elsevier Science Publishers, Amsterdam.

Ando, M. (1999). Correspondence of collagen to the softening of fish meat during refrigeration. In: *Extracellular Matrix of Fish and Shellfish* (Ed. by K. Sato, M. Sakaguchi & H. A. Bremner), pp. 69–79. Research SignPost, Trivandrum, India.

Anon. (1994). *Fish and Fishery Products. Hazards and Control Guide*. Food and Drug Administration, Washington, DC.

Bremner, H. A. (1999). Gaping in fish flesh. In: *Extracellular Matrix of Fish and Shellfish* (Ed. by K. Sato, M. Sakaguchi & H. A. Bremner), pp. 81–94. Research SignPost, Trivandrum, India.

Bremner, H. A. (Ed.) (2002). *Safety and Quality Issues in Fish Processing*. Woodhead Publishing Limited, Cambridge, UK.

Bremner, A., Davis, C. & Austin, B. (Eds) (1996). *Making the Most of the Catch*. AUSEAS, Brisbane, Australia.

Bremner, H. A, Olley, J. & Vail A. M. A. (1986). Estimating time–temperature effects by a rapid systematic sensory method. In: *Seafood Quality Determination* (Ed. by D. E. Kramer & J. Liston), pp. 413–35. Elsevier Science Publishers, Amsterdam.

Bremner, H. A., Paterson, B. D. & Goodrick, B. (1997). Live transport of crabs and shrimp from Australia – kuruma shrimp and spanner crabs as physiological case studies. In: *Seafood from Producer to Consumer, Integrated Approach to Quality* (Ed. by J. B. Luten, T. Børresen & J. Oehlenschläger), pp. 71–85. Elsevier Science Publishers, Amsterdam.

Cadwallader, K. R. (2000). Enzymes and flavor biogenesis in fish. In: *Seafood Enzymes* (Ed. by N. F. Haard & B. K. Simpson), pp. 365–84. Marcel Dekker, New York.

Doe, P. E. (Ed.) (1998). *Fish Drying & Smoking*. Technomic Publishing, Lancaster, PA.

Ehira, S. & Uchiyama, H. (1986). Determination of fish freshness using the K value and comments on some other biochemical changes in relation to freshness. In: *Seafood Quality Determination* (Ed. by D. E. Kramer & J. Liston), pp. 185–207. Elsevier Science Publishers, Amsterdam.

FAO (1988). *Manual on Fish Canning*. FAO Fisheries Technical Paper, No. 285. FAO, Rome.

FAO (1994). *Joint FAO/WHO Codex Committee on Fish and Fishery Products, Proposed Draft Code of Hygienic Practice for the Products of Aquaculture CX/FFP 94/8*. FAO, Rome.

Fletcher, G. C., Bremner, H. A., Olley, J. & Statham, J. A. (1990). Umami revisited: the relationship between inosine monophosphate, hypoxanthine and Smiley scales for fish flavour. *Food Reviews International* (Special issue. Seafoods: Quality and Evaluation) **6**, 489–503.

Gill, T. (2000). Nucleotide-degrading enzymes. In: *Seafood Enzymes* (Ed. by N. F. Haard & B. K. Simpson), pp. 37–68. Marcel Dekker, New York.

Haard, N. F. (1992). Control of chemical composition and food quality attributes of cultured fish. *Food and Research International*, **25**, 289–307.

Hall, G. M. (Ed.) (1997). *Fish Processing Technology*, 2nd edn. Blackie Academic & Professional, London.

Hatae, K. (1999). Textural changes by cooking. In: *Extracellular Matrix of Fish and Shellfish* (Ed. by K. Sato, M. Sakaguchi & H. A. Bremner), pp. 95–106. Research SignPost, Trivandrum, India.

Hocking, A. D., Arnold, G., Jenson, I., Newton, K. & Sutherland, P. S. (Eds) (1997). *Foodborne Micro-organisms of Public Health Significance*, 5th edn. Australian Institute of Food Science and Technology, Food Microbiology Group, North Ryde, NSW.

Howgate, P (2001). Tainting of aquaculture products by natural and anthropogenic contaminants. In: *Farmed Fish Quality* (Ed. by S. C. Kestin & P. D. Wariss), pp. 192–201. Fishing News Books, Oxford.

Huss, H. H., Dalgaard, P. & Gram, L. (1997). Microbiology of fish and fish products. In: *Seafood from Producer to Consumer, Integrated Approach to Quality* (Ed. by J. B. Luten, T. Børresen & J. Oehlenschläger), pp. 413–30. Elsevier Scientific Publishers, Amsterdam.

Huss, H. H., Jakobsen, M. & Liston, J. (Eds) (1992). *Quality Assurance in the Fish Industry*. Elsevier Science Publishers, New York.

Jiang S. T. (2000). Enzymes and their effect on seafood texture. In: *Seafood Enzymes* (Ed. by N. F. Haard & B. K. Simpson), pp. 411–50. Marcel Dekker, New York.

Kestin, S. C. & Warriss, P. D. (2001). *Farmed Fish Quality*. Fishing News Books, Oxford.

Kim, J., Marshall, M.R. & Wei, C. (2000). Polyphenoloxidase. In: *Seafood Enzymes* (Ed. by N. F. Haard & B. K: Simpson), pp. 271–316. Marcel Dekker, New York.

Komata, Y. (1990). Umami taste of seafoods. *Food Reviews International* (Special issue on Seafoods: Quality and Evaluation) **6**, 457–88

Kubota, S. Yotohara, H. & Sakaguchi, M. (1999). Textural change by maturation. In: *Extracellular Matrix of Fish and*

Shellfish (Ed. by K. Sato, M. Sakaguchi & H. A. Bremner), pp. 107–15 Research SignPost, Trivandrum, India.

Lindsay, R. C. (1990). Fish flavors. *Food Reviews International* (Special issue on Seafoods: Quality and Evaluation) **6**, 437–55.

Martin, A. M. (Ed.) (1992). *Fisheries Processing. Biotechnological Applications*. Chapman & Hall, Oxford.

Martin, R., Collette, R. & Slavin, J. (Eds) (1997). *Fish Inspection, Quality Control, and HACCP: a Global Focus*. Technomic Publishing, Lancaster, PA.

Miler, K. B. M. & Sikorski, Z. E. (1991). Smoking. In: *Seafood: Resources, Nutritional Composition, and Preservation* (Ed. by Z. E. Sikorski), pp. 163–80. CRC Publishers, Boca Raton, FL.

Ochai, Y. & Chow, C-J. (2000). Myosin ATPase. In: *Seafood Enzymes* (Ed. by N. F. Haard & B. K. Simpson), pp. 69–89. Marcel Dekker, New York.

Oehlenschläger, J., (1997). Marine fish – a source for essential elements. In: *Seafood from Producer to Consumer, Integrated Approach to Quality* (Ed. by J. B. Luten, T. Børresen & J. Oehlenschläger), pp. 641–50. Elsevier Scientific Publishers, Amsterdam.

Pan, B.S. & Kuo, J-M (2000). Lipoxygenases. In: *Seafood Enzymes* (Ed. by N.F. Haard & B. K: Simpson), pp. 317–36. Marcel Dekker, New York.

Regenstein, J. M. & Regenstein, C. E. (1991). *Introduction to Fish Technology*. Van Nostrand Reinhold, New York.

Robb, D. (2001). The relationship between killing methods and quality. In: *Farmed Fish Quality* (Ed. by S. C. Kestin & P. D. Wariss), pp. 220–33. Fishing News Books, Oxford.

Rørå, A. M. B., Mørkøre, T. & Einen, O. (2001). Primary processing (evisceration and filleting). In: *Farmed Fish Quality* (Ed. by S. C. Kestin & P. D. Wariss), pp. 249–60. Fishing News Books, Oxford.

Sargent, J. R., Bell, J. G., McGhee, F., McEvoy, J., & Webster, J. L. (2001). The nutritional value of fish. In: *Farmed Fish Quality* (Ed. by S. C. Kestin & P. D. Wariss), pp. 3–12. Fishing News Books, Oxford.

Sato, K., Sakaguchi, M. & Bremner, H. A. (Eds) (1999). *Extracellular Matrix of Fish and Shellfish*. Research SignPost, Trivandrum, India.

Shahidi, F. & Botta, J. R. (Eds) (1994). *Seafoods: Chemistry, Processing Technology and Quality*. Blackie Academic & Professional, London, UK.

Sikorski, Z. E. (Ed.) (1991). *Seafood: Resources, Nutritional Composition, and Preservation*. CRC Publishers, Boca Raton, FL.

Sikorski, Z. E., Pan, B. S. & Shahidi, F. (Eds) (1994). *Seafood Proteins*. Chapman & Hall, New York.

Suzuki, J., Ichimura, N. & Etoh, T. (1990). Volatile components of boiled scallop. *Food Reviews International* (Special issue on Seafoods: Quality and Evaluation) **6**, 537–52.

Undeland, I. (2001). Lipid oxidation in fish during processing and storage. In: *Farmed Fish Quality* (Ed. by S. C. Kestin & P. D. Wariss), pp. 261–75. Fishing News Books, Oxford.

Whitfield, F. B. (1990). Flavour of prawns and lobsters. *Food Reviews International* (Special issue on Seafoods: Quality and Evaluation) **6**, 505–20.

12
Economics and Marketing

Clem Tisdell

12.1 Introduction

> Wealth from get-rich-quick schemes quickly disappears; wealth from hard work grows.
>
> Proverbs 13:11

Economics plays an important role in the survival and development of aquaculture. Technical ability is a precondition for the aquaculture of a given species, but this will fail to develop and survive (in any meaningful sense) if it is commercially uneconomic. Economic failure of an aquaculture project may stem from production, technical or cost problems, or from marketing problems (Fig. 2.1). Therefore, those who want to have a commercially successful aquaculture enterprise must pay considerable attention to economics, including marketing issues.

Aquaculture economics and marketing is now a specialised subject, and its whole range cannot be covered in depth in a single chapter such as this. Therefore, the purpose of this chapter is to highlight, for non-specialists in economics, selected important issues that need consideration:

- in developing aquaculture commercially
- in assessing aquaculture from a community-wide economic perspective.

Readers who require further in-depth coverage of this subject can consult specialised books on aquaculture economics such as Shang (1981), Hatch & Kinnucan (1993) and Jolly & Clonts (1993).

Aquaculture economics can be considered from several different perspectives. It can be considered from the point of view of an individual aquaculture business, from the perspective of the whole industry or from a national standpoint. In western countries, in particular, profitability is likely to be the major economic concern in an individual business, whereas overall net national benefit should be the main focus from the national viewpoint. It should be noted that profitability does not necessarily measure net national economic benefit. For example, if an industry has adverse environmental effects, profits in the industry can overstate its national economic benefit because the social costs of its production may exceed the costs paid by individual firms.

If aquaculture businesses are to make a profit in market economies, they must actively market their product and do so effectively. For established products, this may be relatively easy because it may be possible to tap existing marketing networks, making use of established food processors, transport and distribution channels. Marketing of new aquaculture products can, however, be quite difficult, especially in the absence of appropriate marketing networks (Tisdell, 2001).

Profitability is influenced not only by the market but by the costs of production. The latter will depend on, among other things, the culture techniques used and the costs of inputs to the production process. For example, costs vary according to whether the business is involved in the hatchery phase, the grow-out phase or both, and whether aquaculture is performed in artificial enclosures requiring pumping of water or in natural water bodies.

Modelling the economics of aquaculture is complex, but some insights can be obtained by considering simple economic models. Thus, this chapter will successively consider models analysing:

- the profitability of a business
- the market
- the nature of production costs
- methods by which a firm can deal with business risk and uncertainty
- the social economic evaluation of aquaculture.

Some economic terms used in this chapter, which are not explained in the text, are highlighted (in italics) and explained in Table 12.1.

12.2 Profitability from a business viewpoint

For a single period, say a year, a firm's profits can be obtained by taking the difference between its total revenue and its total costs. Its revenue is equal to its volume of output multiplied by the price at which units of output are sold.

Profits = total revenue (= volume of output × price/unit of output) – total costs

If the market in which the firm sells its product is very competitive, the firm will need to sell at the going market price per unit. This, for example, is likely to be the case for the sale of table oysters and for shrimp on the international market. Clearly, other things being equal, the higher the price for the aquaculture product, the higher will be the profit of the business. In some cases, however, there may be few or virtually no competitors in the market for the cultured product and, up to a point, an aquaculture business selling this product will be a price-maker (as opposed to a price-taker, e.g. the above examples of businesses selling table oysters and shrimp). Price-making has been true of the Japanese cultured pearl industry, but no longer appears to be the case (Tisdell & Poirine, 2000). It is currently true in Australia for producers of pearl-oyster seed.

Note that the economic concept of cost differs from that of accounting costs. The latter considers only actual costs (including purchases, salaries and depreciation). Economic cost takes account of opportunity cost, that is the economic benefit forgone by not choosing the best alternative to the choice actually made.

Economic costs = actual costs + opportunity costs

For example, if family labour is supplied to an aquaculture business free of charge, this would not be included in the accounting cost of the business. However, if that family labour could earn an income if employed elsewhere, the highest income that it can earn elsewhere is its opportunity cost. In order to calculate economic cost, this opportunity cost would be included and ascribed to the family labour employed.

Often, but not always, opportunity costs are zero. Then actual costs and economic costs do not differ. This is likely to be so when all inputs for an aquaculture operation are purchased at market prices, as is frequently the case in western market economies.

In general, we are interested in the profitability of an aquaculture business not only in a single period, but for an interval of time spanning several periods. Economists usually consider the firm's profitability for a planning period covering several time intervals, e.g. for a 10-year period covering 10 annual intervals. The appropriate planning period is likely to vary with the enterprise at hand. However, a very long planning period, say 50 years, is likely to be too long because the *discounted value of future profits* (Table 12.1) and uncertainty will mean that events 50 years hence will have little consequence for current decisions.

The optimal business strategy from the point of view of an aquaculture business is, according to standard economic theory (Tisdell, 1972), that which maximises the business's net present value. It is the *present discounted value* of its stream of profits over

Table 12.1 Explanations of some economic terms

Term	Explanation
Discounted benefits	Future benefits from a project reduced to equivalent present values
Discounted costs	Future costs of a project reduced to equivalent present values
Discounted realisable value	The sum, reduced back to its equivalent present value, that could be realised by a business selling out at the end of its planning period
Discounted value of future profits	This is the sum of profits during the firm's planning period with future profits reduced below their actual future values. The reduction of future profits reflects the fact that a dollar available in the future is worth less than a dollar available now because a dollar available now can be invested at the going *nominal rate of interest* to earn more than a dollar in the future. So a future dollar is equivalent to only a fraction of a dollar now
Equity	The proportion of a firm's assets or capital belonging to the owner(s) of a business
Internal rate of return (IRR)	Indicates the percentage rate of return on funds employed by a business or a project. It is a useful indicator of the degree of profitability of a business or a project. Estimates of IRR take into account the time-pattern of returns
Market transaction costs	Costs involved in arranging market exchanges, e.g. costs involved in searching for potential buyers in arranging contracts, agency costs and so on for a sale of aquaculture products
Nominal rate of interest	This is the rate of interest payable, not adjusted for price inflation. The nominal rate of interest tends to rise with the rate of inflation
Present discounted value	The sum of money that, invested now, would accumulate with the addition of interest to a stated future sum of money
Real rate of interest	This is the rate of interest reduced for price inflation. The greater the rate of inflation, the larger the reduction in the rate of interest needed to obtain the real rate
Spillovers (externalities) from business activity	These are side-effects of the activities of a business or other businesses for which no economic payment (e.g. compensation) or costs are involved. They can be favourable or unfavourable

its planning period plus the *discounted realisable value* of the business.

A unit of currency (e.g. a dollar) available in the future can be expected to be less valuable to a business than a dollar available now. There are two reasons.

(1) If there is price inflation, the purchasing power of a dollar in the future is less than now.
(2) A dollar available now can be invested at the going *nominal rate of interest* with relative safety to earn income from interest and so, in the future, return the initial capital invested plus interest. This makes it more valuable than a future dollar that has not been invested.

Furthermore, the higher the market rate of interest, the lower the net present value of a dollar available in the future.

Usually the market rate of interest on government bonds or similar safe investments are used to take account of the minimum opportunity cost (economic actual costs) of committing funds to a business. This takes account of a relatively safe alternative profit that is forgone in committing funds to the business. For some businesses, however, the opportunity costs will be higher than the returns from this alternative investment, or, if they are borrowing funds, they may also have to pay a higher rate of interest. Therefore a higher rate of discounting of future monetary amounts would be appropriate. Some judgement is required in determining the appropriate discount rate to apply. Here it is only possible to bring attention to this issue, which forms a part of a study of finance. Nevertheless, it should be clear that discounting of future income flows is appropriate when considering the financial returns of an aquaculture business.

If the net present value of a project is positive, it is profitable from an economic point of view because it earns more than the relevant rate of interest.

This principle is applied to cost–benefit analysis. If the net present benefit from investment in a project is positive after being reduced using the appropriate rate of interest, it is economic. This also implies that its *discounted benefits* divided by the *discounted costs* exceed unity, or, in other words, that its benefit–cost ratio exceeds unity (Shang, 1981; Allen *et al.*, 1984).

Alternatively, but subject to some relatively minor qualifications to determine a firm's profitability or the profitability of an investment project, one may calculate its *internal rate of return*. The internal rate of return (IRR) is an excellent measure of the profitability of a project or business. If the IRR of a business exceeds the relevant rate of interest, the business is profitable. It is more profitable as the IRR increases in relation to the rate of interest.

The above types of analysis of profitability often assume that the capital (finance) market is perfect and ignore uncertainties. In practice, a firm may need to give special attention to its liquidity (cash availability) to ensure its continuing viability. It may, therefore, be concerned about how quickly a business enterprise can pay back the investment in it and about how large its debt may become during the planning period: the larger its debt, the greater are its risks.

It is not uncommon for economists to make estimates of internal rates of return for the culture of different species. Treadwell *et al.* (1992) estimated the IRR from cultivating various aquaculture species in Australia. These estimates are set out in Table 12.2. To make these estimates, they considered model (or representative) aquaculture farms and specified annual operating costs and capital cost. These costs, together with predicted levels of revenue, provided the basis for estimating net benefits and subsequently the IRR values for the different types of farms. The aquaculture business's planning interval was assumed to be for a 20-year period.

Table 12.2 sets out the mean probable IRR values for the various model farms. In this case, risk or uncertainty has been specified by the authors as a probability distribution and the ranges of these distributions are shown. For some species, these ranges are very wide, suggesting a considerable degree of risk as far as returns are concerned.

To determine whether the aquaculture of a given species is profitable from an economic viewpoint, we should compare its IRR with the real rate of interest. Treadwell *et al.* (1992) estimated the *real rate of interest* in 1991 to be 6%. That being the case, and using mean IRR values, it follows that species cultured in the following way would potentially be uneconomic:

- barramundi in cages in freshwater
- giant tiger prawns grown in subtropical areas
- rainbow trout grown either in freshwater or in the ocean (model I)
- comparatively small Atlantic salmon farms producing ca. 40 000 smolt/year.

The profitability of culturing the different species varies considerably. The most potentially profitable in decreasing order are:

- Pacific oysters on a good site
- barramundi farming initially involving a nursery phase and then a free-ranging phase
- redclaw freshwater crayfish farming on a 15-ha farm
- shrimp stocked at medium density in north Queensland.

Variations in the range of possible returns between species are considerable. From Table 12.2 it is evident that there is little likelihood of a low return from Pacific oysters on a good site, from freshwater crayfish farming and from large farms growing crocodiles. But there is a risk of low and uneconomic returns from rainbow trout farming, barramundi

Table 12.2 Internal rates of return (IRR) for selected aquaculture species in Australia (as in 1991)

Species and culture	Mean IRR (%)	Possible range of IRR (%)
Barramundi		
Cages in freshwater	3.8	–12.2 to 17.5
Nursery then free ranging	19.6	4.7 to 23.6
Cages in the sea	13.6	–4.8 to 28.6
Crocodiles		
Small breeding farm	6.1	4.4 to 7.4
Small grow-on farm	10.8	9.1 to 12.4
Large breeding farm	8.8	7.4 to 10.2
Large grow-on farm	14.0	12.4 to 15.6
Freshwater crayfish		
Marron	10.3	6.7 to 14.0
Redclaw (15-ha farm)	19.4	15.7 to 23.0
Yabbies	9.3	2.1 to 26.6
Mussels (blue)	12.3	1.0 to 22.7
Pacific oysters		
Good site	31.1	28.4 to 33.8
Poor site	10.8	8.5 to 10.5
Shrimp (mainly giant tiger prawns), medium stocking density (25/m^2)		
North Queensland	17.5	–0.1 to 29.9
Subtropical	5.1	–8.1 to 13.8
Rainbow trout		
Freshwater	5.6	3.4 to 7.8
Ocean model I	0.0	–15.2 to 12.7
Atlantic salmon in 60-m-diameter cages		
40 000 smolt	5.5	–2.7 to 12.3
150 000 smolt	12.5	5.5 to 19.6

Based on Treadwell *et al.* (1992) with permission of ABARE.

cultivation in freshwater in cages and the culture of shrimp in subtropical areas of Australia.

Similar analysis to that used by Treadwell *et al.* (1992) has been applied in the Philippines and by Firdausy & Tisdell (1991) to Indonesia to estimate IRRs for seaweed farming. This was found to be high. Tisdell *et al.* (1993a) estimated the IRR of ocean grow-out of giant clams (*Tridacna gigas*) under Australian conditions, and Leung *et al.* (1994) estimated the net present value of giant clam farming in Palau and various territories in the Pacific.

Although IRR analysis is useful, it requires predictions to be made about future prices for aquaculture products and, therefore, about how markets for these products will behave. For instance, although Treadwell *et al.* (1992) estimated a high economic return from the aquaculture of redclaw freshwater crayfish in Australia, this will not be sustained if many suppliers enter the industry and drive the price of redclaw down. This can happen quickly if the market is thin and difficult to expand. Therefore, let us consider markets and marketing.

12.3 Markets and marketing

The markets for most aquaculture products are influenced by supply and demand conditions in their industry and changes in these (Allen *et al.*, 1984).

For products of aquaculture businesses that are price-takers rather than price-makers, the standard economic analysis of purely competitive markets is relevant (Tisdell, 1982).

The quantity demanded of an aquaculture product depends on many factors. Factors such as its price per unit, the income levels of buyers, price of substitutes, tastes, and so on, can all be expected to have an influence. Usually, as the price of a commodity is reduced, the demand for it increases, and all other factors are held constant. This can be illustrated diagrammatically. However, it should be noted (in advance of the following diagrammatic outline) that, in illustrating market relationships, economists conventionally put the independent variable on the *y*-axis and the dependent variable on the *x*-axis. This convention is followed here and differs from the convention in natural science. So in the discussion which follows, the independent variable, in this case price per unit of the aquacultured product, is shown on the *y*-axis and the market quantity of the product is shown on the *x*-axis.

Normally, the demand curve in a market is downward-sloping (Fig. 12.1, D_1D_1), indicating that buyers purchase more of the product as its price is lowered. The market supply of the product is usually upward-sloping, indicating that greater supplies only become available if producers are paid higher prices (Fig. 12.1, S_1S_1). The quantity demanded of a product as a function of its price represents the market demand curve for a product, all other things being constant. The quantity supplied of a product as a function of its price represents its market supply curve, all things, other than its price, being held constant. The point at which these two curves cross represents the market's equilibrium and the corresponding price is the equilibrium price and the corresponding quantity traded is the market equilibrium quantity. In Fig. 12.1, for instance, the demand curve D_1D_1 might represent the demand for shrimp in Japan in July 2005 and S_1S_1 might represent the supply curve of shrimp. Market equilibrium would be established at point E, with the equilibrium price of shrimp being $\overline{P}$/kg with $\overline{X}$ mt of shrimp being supplied. Supplies may be drawn from cultured shrimp (the supply curve for these may be as indicated by the curve marked S_0S_0) and from captured shrimp, the supply function of which is the difference between curves S_1S_1 and S_0S_0. In the case shown, in market equilibrium $\overline{X}_1$ of supply comes from cultured shrimp and $\overline{X} - \overline{X}_1$ from captured shrimp.

It is clear from Fig. 12.1 that, if the demand curve

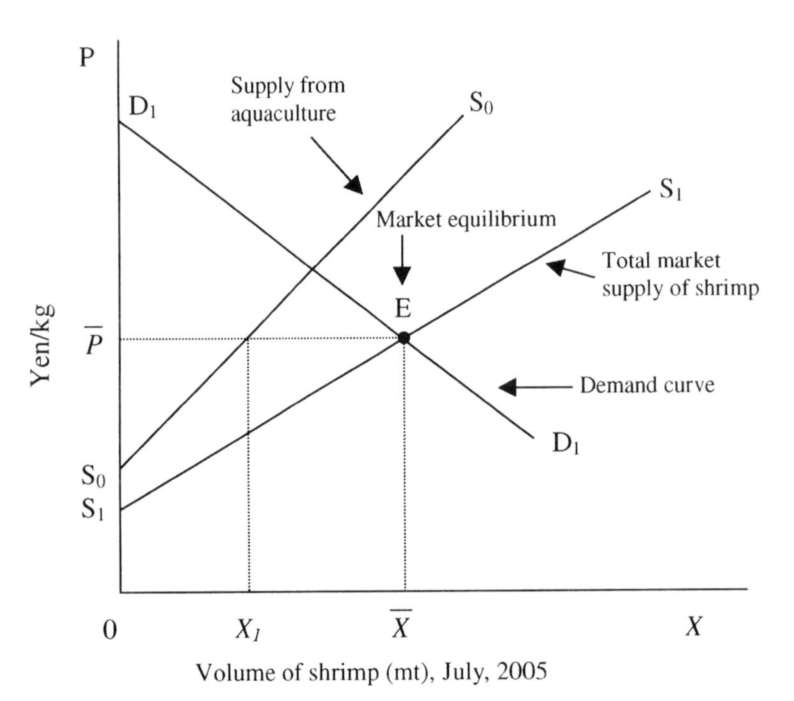

Fig. 12.1 A theoretical market model for shrimp in Japan illustrating market equilibrium and dividing supply into capture and culture components. As mentioned in the text, economists conventionally place the price variable, in this case the independent variable, on the *y*-axis and the dependent variable, in this case the quantity of the product demanded or supplied, on the *x*-axis. This differs from the normal convention in the natural sciences.

for shrimp moves upwards (and everything else remains constant), the equilibrium price and quantity traded will rise. It becomes more profitable for businesses to supply shrimp. Other things held constant, the market demand curve for shrimp may rise, for example, if:

- incomes in Japan rise and, more generally, incomes in market outlets for shrimp rise
- the prices of shrimp substitutes rise
- the human population increases
- tastes alter in favour of the product.

It is important to be able to predict such trends and their influences on demand.

Sometimes the demand curves for aquaculture products are stated in terms of the average consumption per head of population or per household. Miyazawa & Hirasawa (in Liao *et al.*, 1992) reported the relationship between the consumption of shrimp by households in Japan and the price of shrimp (Fig. 12.2) on the basis of annual data for 1980–89. It can be seen that a fall in price led to a substantial increase in the consumption of shrimp by Japanese households. Other data also showed that a rise in Japanese incomes led to a significant rise in the per capita consumption of shrimp in Japan.

A shift downward in the supply curve (that is, increased supply for any given price), other things unchanged, tends to lower the equilibrium price for the aquaculture product, in this case shrimp. Other things constant, the supply curve of an aquaculture

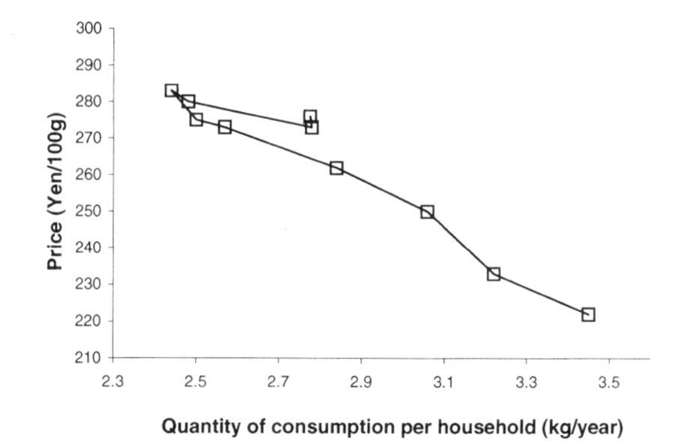

Fig. 12.2 Quantity of consumption of shrimp per Japanese household related to the purchase price of shrimp in the 1980s. Reproduced from Liao *et al.* (1992) with permission from Taiwan Fisheries Research Institute.

product may shift downwards, because, for example:

- The price of one or more inputs falls, e.g. fish food.
- New technologies are discovered that lower production costs, e.g. techniques that greatly reduce food wastage, such as have been developed for the culture of Atlantic salmon (Asche *et al.*, 1999).
- Improved methods may be found to reduce the incidence of pestilence or disease in aquaculture.
- Genetic selection and breeding may raise the productivity of cultured organisms such as tilapias (Dey, 2000) (section 7.3).
- High returns in the industry may result in new businesses entering and investing in the industry, thereby raising supplies.

As indicated above, most aquaculture products compete with supplies of substitutes from the capture fishery. Sometimes these are perfect or near-perfect substitutes. Hence, a reduction in supplies of substitutes from the capture fishery usually raises demand for the farmed product. A rise in supply from the competing capture fishery has the opposite effect. Nevertheless, there is evidence that increased supply of aquaculture products is not completely at the expense of sales of the capture fisheries because market segmentation exists between farmed and wild-caught products (Asche *et al.*, 2001).

Trends or expected variations in relation to all of the above-mentioned demand and supply matters need to be considered in predicting future prices and markets for aquaculture products. To do so accurately can be very difficult, especially if a 20-year planning period is being used.

Of course, there are also marketing decisions to be made at the business level. These include the quality of the product to be supplied and how far to process it. In established industries, middlemen are often present to facilitate marketing and distribution, but one of the difficulties sometimes encountered in developing a market for a new aquaculture product is the absence of suitable networks for its distribution and sales. For example, the sale of giant clam for human consumption in Australia was hampered by the absence of suitable distribution networks for this. On the other hand, the sale of Australian cultured giant clams as aquarium specimens initially progressed quite rapidly because of the existing network of wholesalers and distributors of aquarium specimens. In the absence of suitable distribution and marketing networks, much more of the cost and effort of marketing activities will fall on the aquaculture business. These costs will include advertising the product, its presentation, search for market opportunities and information transfer (Tisdell, 2001).

Many cultured species progress through a typical product cycle (Fig. 12.3 and Table 12.3). In the early stages of this cycle, new production techniques are developed, and to a large extent the market is uncertain. Only innovators or adventurers enter the industry at this stage. At the next stage, sorting of techniques tends to take place, with the least effective ones being discarded, and market penetration may proceed rapidly. The industry may go from a position of earning low and uncertain profits to one of high profit. This induces followers to enter the industry and eventually the industry becomes well established, with 'appropriate' techniques relatively settled and potential markets fully tapped. This is the mature stage in which profitability falls to a level that tends to the average level of business profitability in the economy. Channel catfish culture in the USA is in the mature phase (Chapter 17). Atlantic salmon culture is in the mature phase in Europe, but in Australia is still in a relatively early stage. Redclaw crayfish culture in Australia was also in an early stage in the 1990s and, since returns in 1991 seemed relatively high for little risk, one would have expected considerable entry into the industry, resulting eventually in a fall in returns due to increased supply. However, returns may not fall substantially at first because demand might also expand as consumers become more aware of this product and it gains greater acceptance. There are a number of instances in which this has occurred. For example, when tilapia culture was first introduced to Fiji, local demand for this introduced fish expanded slowly. However, it is now a sought-after fish. Another similar example is that of farming green turtles, *Chelonia mydas*, in the Cayman Islands. American conservationists blocked the sale of cultured green turtle meat in the USA using the argument that it would lead to expanded demand for turtle meat and thereby impact adversely on wild stocks. Their argument is analysed in Tisdell (1986).

When a market needs to be developed or a business plans to supply a new market, a variety of methods may be used to determine the nature of the market and to foster it. These include trials of the product such as taste-testing of a new aquaculture product, pilot or trial marketing, interviews and various types of surveys, and examination of the demand for substitute products (Shang, 1981; Meade, 1989). Because giant clam farming was so new in the 1990s, it was necessary to use all these methods to assess potential demand for cultured giant clams (Tisdell *et*

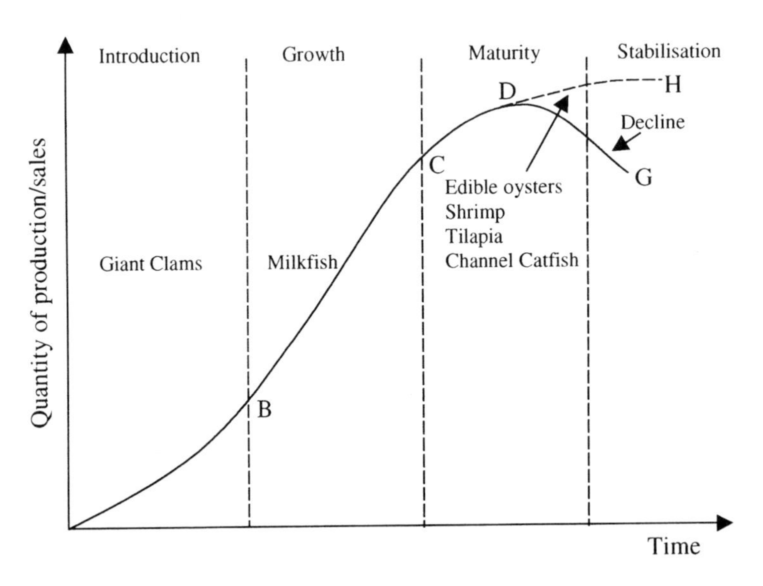

Fig. 12.3 Product cycle showing typical stages which aquaculture industries pass through if they succeed economically and the approximate stage in which some of these industries are now. The demand and volume of supply for all these industries does not decline but stabilises for some as upper limits to production and demand are approached.

Table 12.3 Product cycle stage (see Fig. 12.3) for some European aquaculture industries (information provided by Professor Niall Bromage, University of Stirling, Scotland)

Stage	Aquaculture industry
Introduction	Atlantic cod* Atlantic halibut[†] Sole Haddock Sturgeon Wolffish Dentex Arctic charr Bluefin tuna Striped bass Channel catfish African catfish Abalone Sea urchins Manilla clam Freshwater crayfish
Growth	European seabass[‡] Sea bream[‡] Tilapias Turbot
Maturity	Atlantic salmon
Stabilisation	Rainbow trout Brown trout Carps European eels Mussels Scallops Oysters
Decline	None

*Poised to move into growth phase – will possibly be the next 'salmon'.
[†]Moving into growth phase – but will probably be a niche market.
[‡]Currently experiencing marketing problems after a period of explosive growth, but still growing.

al., 1994). Fig. 12.4 shows a dish using giant clams (one of several) prepared by a Japanese restaurant in Australia for experimental taste testing. An aquaculture business will also have to make economic decisions about how to distribute its product, promote and present it.

As a market expands, it becomes increasingly necessary to standardise the cultured product, or its grades, in order to reduce *market transaction costs* and increase market penetration (see Young, 2001). The industry may itself set standards or a government marketing body may do so. There can be an economic benefit to an aquaculture industry in imposing financial levies on its businesses in order to have its product promoted by a 'government' marketing authority. This is so even though members of the industry as individuals would not be prepared to spend so much on promotion, because others would benefit considerably by their promotion of a relatively generic product, e.g. Atlantic salmon, Pacific oysters, channel catfish.

Fig. 12.4 A dish using cultured giant clam prepared by a Japanese restaurant in Australia for market testing (photograph by Ms Linda Cowan).

12.4 Economies of scale and other factors

An aquaculture business's costs are likely to vary with the size of the undertaking. There are economies of scale or decreasing costs per unit of production for many species, up to some annual volume of output. After this point, costs per unit of production may begin to rise with greater volume of output or rise after remaining stationary over a range (Fig. 12.5).

The scale (volume of annual production) at which a business obtains its minimum cost per unit produced is called its minimum efficient scale. If a business is operating below this level, it will usually be at an economic disadvantage compared with businesses operating at the efficient scale. Consequently, its rate

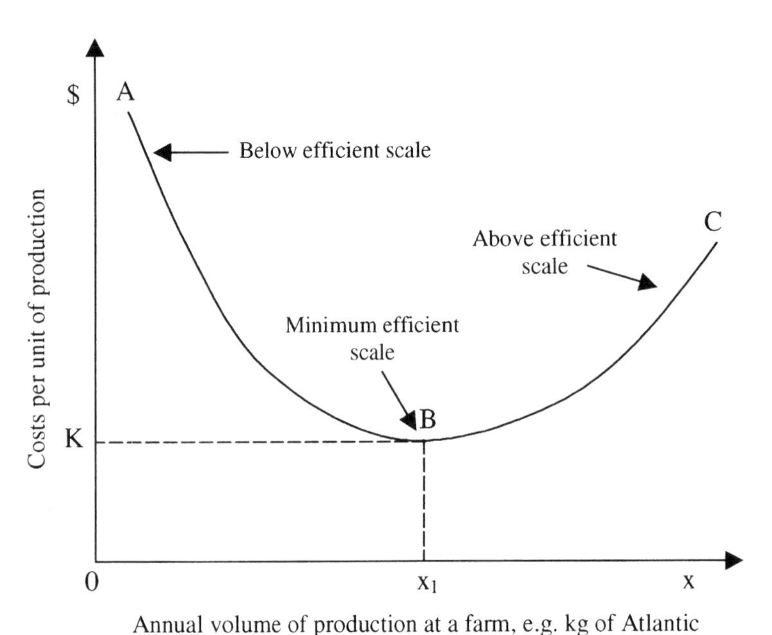

Fig. 12.5 U-shaped average cost of production curve. Businesses having a level of production less than the minimum efficient scale could reduce their costs per unit of production by expanding their level of production.

of profit can be expected to be lower than that for the latter businesses.

From Table 12.2 it can be inferred that economies of scale exist in Australia for model farms producing crocodiles and Atlantic salmon. The IRR for the larger farms can only be explained by lower per unit costs of production for these (e.g. Atlantic salmon) because the businesses are assumed to be price-takers (accepting the available price). Economies of scale also exist in redclaw freshwater crayfish farming. The mean IRR from a 6-ha model farm is estimated to be 9.3% compared with 19.4% for a 15-ha model farm (Treadwell *et al.*, 1992). Available evidence also indicates economies of scale for shrimp farming.

Economies of scale are likely to be significant in land-based aquaculture operations involving the pumping of water to tanks, raceways or ponds and requiring regular water circulation. This is mainly because of engineering relationships, e.g. the volume tends to increase at a faster rate than the circumference of a container, but there may be other economies of scale, for example in being able to use more effectively the services of specialised personnel who can be employed. Economies of scale can also be present for farming *in situ*, e.g. as the case of Atlantic salmon farming indicates. The minimum efficient scale (size of production operations) of an Atlantic salmon farm has tended to increase with the passage of time. Significant economies of scale in production exist for hatchery/nurseries engaged in land-based production, e.g. in the supply of giant clam seed (Fig. 12.6). However, in the case of seaweed production in developing economies, economies of scale do not appear to be significant.

Apart from economies of producing a greater volume of a particular species, other types of economies may exist. These include economies of scope (or diversification) and economies of specialisation. To a large extent, these are the opposite sides of the same coin. To take advantage of economies of scope, if they exist, the firm engages in the supply of multiple products or services and this can include polyculture (section 2.3.5). There may be biological synergies (complementarity) in the production of more than one species so that mixed aquaculture of species is the most profitable. In land-based facilities, it may be possible to spread overheads, e.g. those involved in pumping water, or the employment of specialists, by producing different species in different ponds or containers of various kinds.

Economies of diversification, however, have to be balanced against possible economies from specialisation. Frequently, if specialisation by production

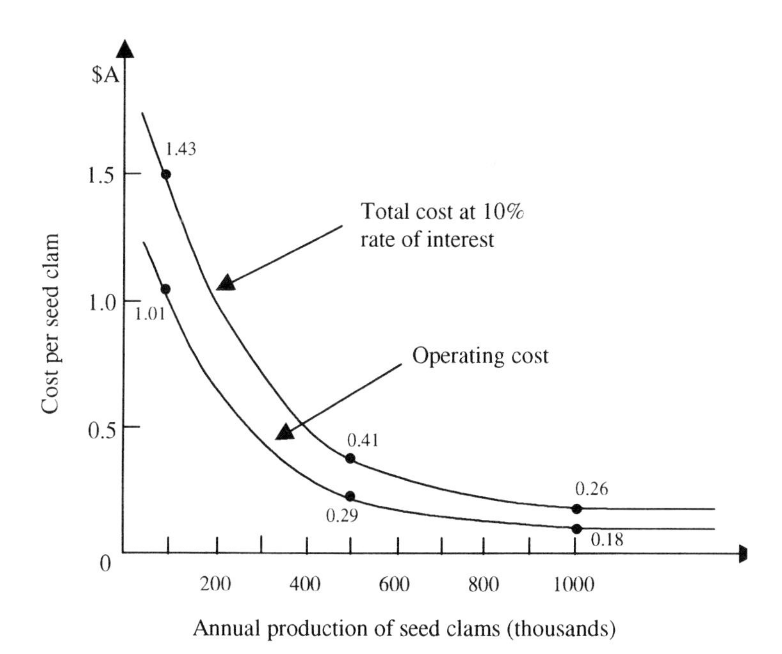

Fig. 12.6 Estimated cost of production curves for a producer of giant clam seed in Australia. Reproduced from Tisdell *et al.* (1993b) with permission from the World Aquaculture Society.

of species is absent, there is often specialisation by stages in the culture of a species. For example, some businesses may specialise in the hatchery/nursery stage culture of a species, whereas others may confine themselves to the grow-out stage, or even just a part of it. This pattern has been developed in Taiwan, with a series of small businesses specialising in successive stages of fish aquaculture. It means that the industry can take advantage of maximum economies of scale at different stages in the culture of a species.

Economies of scope or of diversification seem to be very important at the hatchery/nursery stage from an economic viewpoint. Casual observation indicates that a large number of hatcheries/nurseries supply a range of aquaculture species or varieties of these, even though the range may be restricted to closely related species.

Possibilities for economies of scale, scope and specialisation are limited by the available techniques of production and by those adopted. The appropriate choice of a technique from those available is partly an economic matter. In countries in which labour is cheap relative to capital, from an economic profit standpoint labour-intensive techniques are likely to be more appropriate than capital-intensive ones. However, in developed countries, where labour is relatively expensive, the reverse can be expected.

It should be noted that the location of an aquaculture business is likely to have a significant influence on its cost of production and profitability. The location of an aquaculture business's facilities will affect its cost of access to markets, its availability of inputs and their costs. A good location ecologically may be uneconomic if it is distant from markets and lacks available human resource or services support.

12.5 Dealing with business risk and uncertainty

Aquaculture can be a risky undertaking because of uncertainty about economic variables, such as future prices, and production uncertainties, which can arise from environmental changes, disease, etc. Most aquaculture businesses need to adapt to such uncertainties to survive and minimise possible losses. Some methods of adaptation include:

- product diversification
- diversification in techniques used for production
- incorporation of flexibility into the capital equipment or facilities used to keep options open

- expanding cautiously into a new business area to leave time for learning-by-doing
- making sure that the business has limited liability
- increasing the number of shareholders or partners in the business
- making sure that the business's debt to *equity* (or ownership) ratio does not become so high as to jeopardise its ability to repay loans if its economic performance is below expectation
- ensuring that fixed costs of the business are low so that a substantial economic loss can be avoided if the price of, or demand for, the aquaculture product is low, or if production is below that planned, or if the cost (e.g. price of an important input) is above expected levels.

Fixed costs tend to be high when a production technique is capital intensive, that is, uses a lot of equipment and fixed investment relative to other resources. When capital-intensive aquaculture techniques are adopted by a business, the business must make sure that economic conditions are favourable for this. For example, conditions are more likely to be favourable if the product is of high value or there is a high volume of demand for the business's product or the technique considerably reduces per unit operating costs. Also, the risk of production falling markedly below planned levels should be low, for instance as a result of environmental occurrences.

Fig. 12.7 shows a relatively capital-intensive white eel pond at a cooperative at Shenzhen in China. This type of intensive production is risky because the cooperative relies on high-price fish meal (of good quality) imported from the USA, and the price of this meal fluctuates considerably from year to year. Nevertheless, output is of high value and exported to Japan and California. The operation is profitable on average, even though in some years losses have been recorded due to sudden increases in the price of fish meal.

Fig. 12.7 A pond producing white eels under intensive conditions at Shenzhen, China, for export. Production is risky but the product is of high value.

An intensive shrimp farm on Okinawa, Japan, produces very high-value shrimp and can operate profitably even though its capital, overhead and operating costs are high. A semi-intensive shrimp farm near Shenzhen, China, feeds its shrimp by collecting shellfish from a nearby bay. Both its operating and its capital costs are lower per hectare than in the Japanese case. The shrimp are exported, but the price received is lower than for the Japanese farm. A seasonal extensive prawn farm (*Macrobrachium* species) in Bangladesh requires very little capital investment and has even lower operating costs. The economics of operation of the farms is hampered by the occurrence of typhoons, which result in the escape of shrimp stocks in some years, causing an economic loss. In the Bangladesh case, both capital costs and operating costs are extremely low because the prawns are not given supplementary feeding, but rely on organisms naturally present in the water, which is interchanged with the nearby brackish river system. In this case the business risks are relatively low.

Diversification of production is a common risk aversion strategy. If returns from different products are not perfectly correlated, this will tend to reduce the variability of the business's total returns. The same is true of production using different *techniques*, e.g. juvenile giant clams may be cultured in onshore tanks as well as in floating cages, so reducing the likelihood of a major loss of supply if adverse weather conditions occur. Capital equipment used to farm a range of species needs to be flexible or adaptable. It may be more sensible, taking into account business risks, to use such equipment in culturing a species, than to use equipment specifically designed for the species. Although specific equipment results in lower cultivation costs for a given species it may have little alternative use. Should the culture of the species prove to be uneconomic, flexible equipment can be used to cultivate other species and will have a higher resale value.

Businesses engaging in the culture of a species unfamiliar to them generally go through a period of learning-by-doing. With the passage of time and with the experience gained, their productivity and economic performance in cultivating the species improves. In the early stages, therefore, they might do well to proceed cautiously, e.g. use small-scale or pilot plants, and install flexible or cheap short-lived capital equipment (section 2.9). A late start can be a particular disadvantage for a new entrant to an aquaculture industry in which substantial economies of scale exist. If the new entrant starts on a small scale, this means higher cost for the entrant than for established firms. If the entrant immediately goes at least to the minimum scale of efficient production, this involves considerable risk since it does not allow time for learning.

Institutional arrangements such as the limited liability form of company ownership can reduce personal business risks, and, if risk is shared among a large number of shareholders or partners in a business, losses are easier to bear. In addition, the management needs to give continuing attention to the debt–equity ratio of the business. The higher this ratio is, the greater the risk to the business in the event of unfavourable economic performance. This ratio (debt–equity) is sometimes called the firm's gearing ratio, and if equity is low relative to debt the firm is said to be highly geared. A highly geared business can have a high risk of not surviving. On the other hand, a firm with a high IRR in relation to the rate of interest may be unnecessarily forgoing profitable business opportunities if its equity/debt-gearing ratio is low.

12.6 Economic assessment from a social standpoint

Although an aquaculture business may be privately profitable and an aquaculture industry may be economically thriving, this does not necessarily indicate its value from a social point of view. The social value of production by the industry will, for example, depend upon whether social costs of production are greater than private costs. If they are, private gains overstate social net benefits (Tisdell, 1993).

Social costs will exceed private costs of production by businesses if the aquaculture industry results in unfavourable environmental *spillovers* (externalities) that impose costs on others for which they are not compensated. For example, consider shrimp farming in some less developed countries. In some, e.g. in Ecuador, Thailand, the Philippines and parts of Bangladesh, wetlands are impounded to create ponds for the cultivation of shrimp. Vegetation (such as mangrove trees) is lost and the breeding grounds and food supplies of wild fish stocks are destroyed, with an adverse impact on local fishing communities. When ponds are stocked with captured young shrimp, as in Bangladesh, this may subsequently reduce the population of large shrimp available to the capture shrimp fishery. Furthermore, by converting coastal areas that play an essential role in the life cycle of wild shrimp populations to private shrimp ponds, the aquaculture farms further reduce wild stocks. Eventually, there is a risk of the capture fishery collapsing altogether and supplies of seed shrimp from the wild drying up, as happened in Ecuador (Tisdell, 1991).

By contrast, in some cases aquaculture can give rise to favourable spillovers, and, when this happens, the profits of fish farmers understate the social economic benefits of their activity. The activity might then be on a smaller scale than is socially optimal. Waste from marine fish farms causes nutri ent enrichment of surrounding waters. Up to some level, this may enhance the growth of surrounding wild fish or benefit mollusc production. But, beyond some point, this positive effect can become negative (section 4.2.1).

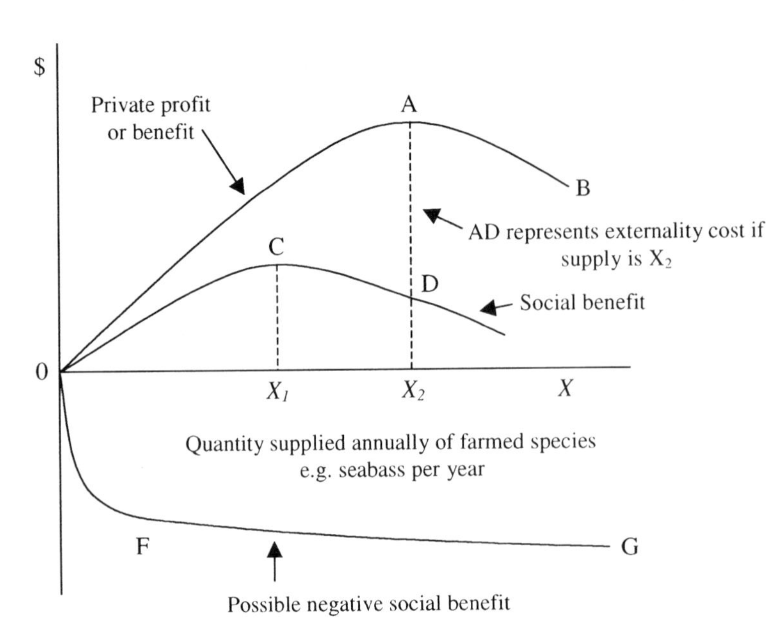

Fig. 12.8 Environmental spillovers from fish farming sometimes result in private decisions being at odds with social economic benefits from these decisions (see text for explanation).

The economic theory underlying this matter is illustrated in Fig. 12.8. Curve OAB represents the profit from farming a fish species, e.g. sea bass, in a region as a function of the quantity produced annually. In this region, however, the farming of the species gives rise to negative environmental effects so the social benefit curve is OCD. This curve is lower than curve OAB and the difference represents environmental costs (curve OFG) not paid for by the sea bass farmers. In order to gain maximum profit, fish farms will produce X_2 mt of the farmed fish in the region annually. This is an excessive amount from a social economic viewpoint. Social net benefit is maximised when only X_1 mt of the species is produced each year. Therefore, because of the adverse environmental effects that are present, the market mechanism fails to ensure a social economic optimum. Hence, it may be desirable for the government to adopt policy measures to restrict production of this farmed species.

The opposite situation can arise if the farming of a species generates favourable environmental effects. In Fig.12.8, the environment curve OFG becomes positive, OCD becomes the profit curve and OAB is the social benefit curve.

It is also important to recognise that in some cases the adverse spillovers generated by the aquaculture of particular species can be so great that its culture should not be tolerated. For example, in Fig. 12.8, although curve OAB may represent the private benefit to producers from farming a species, the social benefit from doing so may be as shown by curve OFG. It is negative. For example, the introduction of a new species to a region can pose significant risks to wild species in the region. Escaped farmed species may compete with other wild species or become predators of them. The risks and potential costs to natural ecosystems of introduction of new species and attendant economic losses may be so great as to make it desirable from a social economic point of view to ban their introduction.

It is also possible for different methods or techniques of aquaculture to give rise to different magnitudes of external costs. It may, therefore, be desirable to introduce public policies that limit the use of some techniques or ban these altogether. The regular feeding, for example, of antibiotics to farmed fish can give rise to a number of serious environmental consequences (section 4.4.3). These include the growing resistance of disease-creating organisms to antibiotics and the reduced natural resistance among the farmed stock, and, possibly where there are escapees, reduced resistance of wild stock to diseases (Tisdell, 1993). Therefore, some governments may consider it to be desirable to ban the use of environmentally 'dangerous' antibiotics in aquaculture or to restrict their use.

Clearly, different methods of husbandry in aquaculture can have significantly different environmental consequences. Nevertheless, it is frequently the case that technological progress reduces the magnitude of environmental effects. For instance, between 1980 and 1997 the average feed conversion ratio in Norwegian salmon aquaculture fell from just under 3 to just over 1 (Asche *et al.*, 1999). This means that less waste per kilogram of fish produced goes into the surrounding environment. The food used nowadays, for example, sinks more slowly through the water, and improved techniques are available to monitor feeding so that the quantity of food supplied to the fish can be adjusted more accurately to consumption (Asche *et al.*, 1999; section 9.11.2). In addition, innovations have resulted in a substantial reduction in use of antibiotics by the Norwegian salmon industry (Asche *et al.*, 1999).

As pointed out by Tisdell (1999), adverse environmental spillovers are often the source of lack of sustainability in production and can result in an activity eventually becoming uneconomic. Although they are not the only source of lack of sustainability in economic production (Tisdell, 1999), they should not be overlooked as a potentially important source. Bardach (1997) gives particular attention to the sustainability of aquaculture, and Shang & Tisdell (1997) concentrate on the economic dimensions involved.

Again, some forms of aquaculture raise income distribution questions. Large-scale aquaculture, which displaces small farmers or adversely impacts on the incomes of poor fishing and subsistence communities, has an adverse income distribution effect. This has happened for shrimp aquaculture in some less-developed countries, for example in Bangladesh (Alauddin & Tisdell, 1998). On the other hand, seaweed farming in Indonesia appears to have reduced rural income inequality, at least in some villages (Firdausy & Tisdell, 1993).

Some forms of crustacean culture raise further sustainability issues (see Be *et al.*, 1999, for some examples) and, in essence, an inter-generational income equity problem. The practice has arisen in some parts of monsoonal Asia of alternating rice and shrimp/prawn production in low-lying estuarine areas, e.g. in the Sunderbarns of Bangladesh. Just before the wet season rice is planted. After the rice is harvested in the dry season, the fields may be flooded with brackish water to create ponds for rearing shrimp or prawns (*Macrobrachium* species). These ponds are drained before the start of the next wet season and the animals are harvested. The land is then prepared for rice and replanted. So the cycle continues. This, however, does not appear to be a sustainable practice. It results in falling rice yields due to rising soil salinity and mineralisation of the soil (Alauddin & Tisdell, 1998).

Various forms of aquaculture can result in a variety of environmental issues including environmental health risks. If, however, aquaculture has adverse environmental impacts, this does not mean that it should be banned from a socioeconomic perspective. Instead, policy measures, such as taxes on effluent, could be adopted to ensure that aquaculture businesses take their external costs into account in their decision-making. (A tax, for example, can result in the firm's private costs of production after tax being brought into line with its social cost.) When this is done, some aquaculture businesses will no longer be economic. Optimal economic policies to control environmental spillovers are outlined in Tisdell (1993), but more attention needs to be given to these specifically in relation to aquaculture. Economic theory indicates that it is not optimal, as a rule, to eliminate all environmental effects, but that government intervention to control them is sometimes justified.

References

Alauddin, M. & Tisdell, C. A. (1998). *The Environment and Economic Development in South Asia*. Macmillan, London.

Allen, P. G., Botsford, L.W., Schuur, A. M. & Johnston, W. E. (1984). *Bioeconomics of Aquaculture*. Elsevier Science Publishers, Amsterdam.

Asche, F., Bjørndal, T. & Young, J. A. (2001). Market interactions for aquaculture products. *Aquaculture Economics and Management*, **5**, 303–18.

Asche, F., Guttormsen, A. G. & Tevterås, R. (1999). Environmental problems, productivity and innovation in Norwegian salmon aquaculture. *Aquaculture Economics and Management*, **3**, 19–29.

Bardach, J. E. (1997). *Sustainable Aquaculture*. John Wiley and Sons, New York.

Be, T. T., Dung, L. C & Brennan, D. (1999). Environmental costs of shrimp culture in the rice-growing regions of the Mekong Delta. *Aquaculture Economics and Management*, **3,** 31–42.

Dey, M. M. (2000). The impact of genetically improved farmed Nile tilapia in Asia. *Aquaculture Economics and Management*, **4,** 109–26.

Firdausy, C. & C. Tisdell (1991). Economic returns from seaweed (*Euchemia cottonii*) farming in Bali, Indonesia. *Asian Fisheries Science*, **4,** 61–73.

Firdausy, C. & Tisdell, C. A. (1993). The effects of innovation on inequality of economic distribution: the case of seaweed cultivation in Bali, Indonesia. *The Asian Profile*, **21,** 393–408.

Hatch, U. & Kinnucan, H. (1993). *Aquaculture: Models and Economics*. Westview Press, Boulder, CO.

Jolly, C. M. & Clonts, H. A. (1993). *Economics of Aquaculture*. Food Products Press, Binghamton, NY.

Leung, P. S., Shang, Y. C., Wanitprapha, K. & Tian, Y. (1994). Production economics of giant clam (Tridacna) culture systems in the U.S. affiliated Pacific islands. In: *Economics of Commercial Giant Clam Mariculture* (Ed. by C. Tisdell, Y. C. Shang & P. S. Leung) pp. 267–91. Australian Centre for International Agricultural Research, Canberra.

Liao, I. Chiu, Chung-Zen Shyu & Nai-Hsien Chao (Eds) (1992). *Aquaculture in Asia*. Taiwan Fisheries Institute, Keelung, Taiwan.

Meade, J. W. (1989). *Aquaculture Management*. Van Nostrand Reinhold, New York.

Shang, Y. C. (1981). *Aquaculture Economics: Basic Concepts and Methods of Analysis*. Westview Press, Boulder, CO.

Shang, Y. C. & Tisdell, C. A. (1997). Economic decision-making in sustainable aquaculture development. In: *Sustainable Aquaculture* (Ed. by J. E. Bardach), pp. 127–48. John Wiley & Sons, New York.

Tisdell, C. A. (1972). *Microeconomics: The Theory of Economic Allocation*. Wiley, Sydney.

Tisdell, C. A. (1982). *Microeconomics of Markets*. Wiley, Brisbane.

Tisdell, C. A. (1986). Conflicts about living-marine resources in Southeast Asian and Australian waters: turtles and dugongs as case. *Marine Resource Economics*, **3,** 89–109.

Tisdell, C. A. (1991). *The Economics of Environmental Conservation*. Elsevier Science Publishers, Amsterdam.

Tisdell, C. A. (1993). *Environmental Economics*. Edward Elgar, Aldershot.

Tisdell, C. A. (1999). Economics, aspects of ecology and sustainable agricultural production. In: *Sustainable Agriculture and Environment* (Ed. by A. K. Dragun & C. Tisdell), pp. 37–56. Edward Elgar, Cheltenham.

Tisdell, C. A. (2001). Externalities, thresholds and marketing of new aquacultural products: theory and examples. *Aquaculture Economics and Management*, **5,** 289–302.

Tisdell, C. A. & Poirine, B. (2000). Socio-economics of pearl culture: industry changes and comparisons focussing on Australian and French Polynesia. *World Aquaculture*, **31,** 30–7, 58–61.

Tisdell, C. A., Tacconi, L., Barker, J. R. & Lucas, J. S. (1993a). Economics of ocean culture of giant clams, *Tridacna gigas*: internal rate of return analysis. *Aquaculture*, **110,** 13–26.

Tisdell, C. A., Thomas, W. R., Tacconi, L. & Lucas, J. S. (1993b). The cost of production of giant clam seed, *Tridacna gigas*. *Journal of the World Aquaculture Society*, **24,** 352–360.

Tisdell, C. A., Shang, Y. C. & Leung, P. S. (Eds) (1994). *Economics of Commercial Giant Clam Mariculture*. Australian Centre for International Agricultural Research, Canberra.

Treadwell, R., McKelvie, L. & Maguire, B. (1992). *Profitability of Selected Aquacultural Species*. Discussion Paper 91.11, Australian Bureau of Agricultural and Resource Economics, Canberra.

Young, J. A. (2001). Communicating with cod and others in some perspectives or promotion for expanding markets for fish. *Aquaculture Economics and Management*, **5,** 241–51.

13
Algae Culture

Macroalgae: C. K. Tseng

Microalgae: Michael Borowitzka

13.1 General introduction

> The sea is 'a field of production' full of unexpected surprises. Therefore, endless efforts to improve the existing techniques [of seaweed culture] are required.
>
> Takeshi Toma (1987)

Macroalgae (seaweeds) and microalgae differ markedly in morphology and life cycles. The methods used for their culture and the purposes for which they are cultured are therefore extremely different.

Commercial cultivation of seaweeds, or phycoculture, has been undertaken in China and Japan for several hundred years. Typically, it involves extensive culture using relatively simple technology. It produces seaweed for direct human consumption, for use as fertilisers and as a source of products such as alginate and carrageenan, both of which are gelling polysaccharides used extensively as thickeners in the food industry (Tables 13.1 and 13.2).

Only in the last 20 years has commercial culture of microalgae come to prominence. The importance of microalgae as a source of certain chemicals (such as β-carotene) has been recognised, and culture systems for microalgae have been developed as a means of commercial production of these chemicals. It often involves advanced biotechnological methods using intensive culture systems.

Because of the very different culture methods, the two kinds of cultured algae will be treated separately in this chapter.

13.2 Macroalgae (seaweeds)

13.2.1 Introduction

World production

Commercial cultivation of macroalgae is exclusively with seaweeds: as yet, no species of macroalgae from fresh water are cultivated. Hence the term 'macroalgae culture' is synonymous with 'seaweed culture'.

Total world seaweed production from aquaculture was estimated to be 9.5 million mt in 1999 (Table 13.3). Seaweed production from aquaculture has averaged an annual growth of around 16% over recent years. At this annual rate of increase, production is now likely to be >10 million mt/year. During the decade 1990–99, production of seaweeds from fisheries remained static, as did other production from fisheries (section 1.3). In 1999, seaweeds from fisheries constituted only 11% of world seaweed production.

Seaweeds constituted 22% of total world aquaculture production by weight in 1999, but only 11% by value (Chapter 1, Fig. 1.2), reflecting its relatively low value per unit weight compared with most other classes of aquaculture products.

Brown seaweeds (Phaeophyta) are the major group of cultured seaweeds, constituting more than

Table 13.1 Some uses and products of seaweeds

Use/product	Seaweed
Dried for human consumption	*Enteromorpha* *Fucus* *Laminaria* *Sargassum* *Undaria* *Eucheuma* *Gelidium* *Gracilaria* *Porphyra*
Medicinal uses	*Dictyota* (antifungal, antibacterial) *Sargassum* and *Ulva* (vermicides) *Chondrus* (blood anticoagulants) *Ascophyllum* (muscle-related problems)
Use as fertilisers	*Macrocystis* *Ascophyllum* *Fucus* *Sargassum* *Laminaria*
Production of gelling colloids	
Agar production	*Gelidium*, *Gracilaria*
Carrageenan production	*Chondrus*, *Eucheuma*
Alginate production	*Ascophyllum, Fucus, Laminaria, Macrocystis, Sargassum*

Table 13.2 Seaweed products and their relative economic values

Product	Value (US$ × 10^6)	Raw material (mt)	Raw material (US$/mt)	Product (mt)	Product (US$/mt)
Carrageenan	240	400 000	600	25 000	9600
Alginate	211	460 000	459	23 000	9174
Agar	132	125 000	1056	7 500	17 600
Soil additives	10	550 000	18	510 000	20
Fertilisers	5	10 000	500	1000	500
Seaweed meal	5	50 000	100	10 000	5000
Totals	603	1 595 000	2733	576 500	41 894

Reproduced from Hanisak (1998) with permission from the World Aquaculture Society.

half the seaweed production in 1999 (Table 13.3), and they are clearly the most important for the industry. However, because a substantial proportion of the Food and Agriculture Organization (FAO) data for aquaculture production of seaweeds falls into a 'miscellaneous aquatic plants' category (i.e. unidentified species), it is not possible to identify the relative importance of the three major groups with any precision. The most widely cultured species of brown algae is *Laminaria japonica*, which made up 93% of the total cultured brown algae in 1999. Reflecting the tradition of macroalgae culture in Asia, by far the largest producer of brown algae is China, which produced 4474 million mt or 85% of total world production in 1999.

Production of red seaweeds (Rhodophyta) made

Table 13.3 World production (thousand mt) of seaweeds from aquaculture and fisheries and its value (million US$) in 1990 and 1999

	1990	1999	Increase/year (%)
Brown seaweeds aquaculture	2521 **1798**	5268 **3217**	10.9
Red seaweeds aquaculture	994 **997**	1782 **1354**	7.9
Green seaweeds aquaculture	30 **20**	10 **3**	–6.7
Total aquaculture production	3572 **2947**	9461 **5688**	16.5
Total fisheries production	1232	1202	–2.4

Data in bold are economic vales (FAO, 2001a,b).

up ~19% of total world seaweed production in 1999 (Table 13.3), the major species in culture being 'laver' or 'nori' (*Porphyra tenera* and other *Porphyra* species), *Gelidium* species, *Eucheuma* species and *Gracilaria* species. China, Japan and the Republic of Korea are the major producers of red seaweeds. World production of cultured green seaweeds (Chlorophyta), although they are considered to be one of the three major groups in culture, is relatively low compared with the other two groups.

Morphology and habitats

The general morphology of seaweeds includes a discrete attachment organ or holdfast, the stem-like stipe and the leaf-like blade or thallus (Fig. 13.1). The blade generally shows little morphological variation, although branches from the main stem, leaf-like organs and floatation structures (vesicles or pneumatocysts) may be present. Blades are generally flattened to facilitate photosynthesis and nutrient absorption from surrounding seawater.

Seaweeds are generally benthic in nature and restricted to solid substrata such as rock. Perhaps the most stressful habitat for seaweeds is the intertidal zone, where they are exposed to the air at low tide and subject to desiccation as well as fluctuations in temperature, light intensity and salinity. Despite this, lush natural beds of seaweeds are found in the intertidal zone, which also supports considerable seaweed-farming activity. Seaweeds require light for photosynthesis and, because light intensity diminishes with depth, their distribution is restricted to a maximum depth of around 170 m. The point at which light intensity no longer supports positive growth of seaweed is the compensation depth or compensation intensity. Light quality also changes with depth. Seaweeds account for this through changes in photosynthetic metabolism relating to the amount and types of chlorophyll they possess and the presence of accessory pigments.

13.2.2 Reproduction and life cycles

A description of the culture methods for seaweeds requires an understanding of their various reproductive strategies, which are often complex. Reproduction may be asexual (vegetative) or sexual, and many species have a life cycle characterised by alternation of generations. These seaweeds have a diploid ($2n$), spore-producing sporophyte stage and a haploid ($1n$) gametophyte stage.

The following sections describe the reproduction and the life cycles of three important culture genera, *Laminaria*, *Porphyra* and *Ulva*, representing the three major culture groups, Phaeophyta, Rhodophyta and Chlorophyta respectively.

Reproduction in Laminaria species (Phaeophyta)

The mature diploid ($2n$) sporophyte of *Laminaria* species bears specialised reproductive organs,

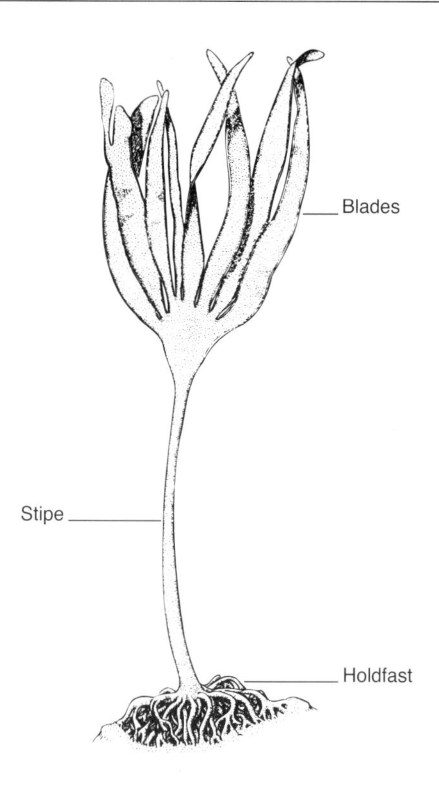

Fig. 13.1 Kelp, *Laminaria andersonii* (Phaeophyta), showing holdfast, stipe and blades. Reproduced from Rost *et al.* (1979) with permission from John Wiley & Sons, Inc.

called sori, which undergo meiotic cell division to produce haploid (n) zoospores (Fig. 13.2). The motile zoospores settle on a suitable substrate and develop into microscopic male or female filamentous gametophytes. Mature female gametophytes release eggs that secrete chemicals to stimulate the release of sperm from the male gametophyte. Motile sperm cells are attracted to the eggs chemotactically. The diploid zygote resulting from fertilisation develops into a mature sporophyte. In many species of Phaeophyta, gamete production is seasonal, indicating that it is controlled by environmental factors such as water temperature, day length and light quality.

Reproduction in Porphyra species (Rhodophyta)

The life history of most red algae (Fig. 13.3) involves male and female gametes developing within the vegetative cells of the haploid (n) gametophyte stage. Spermatangia are formed from mitotic divisions in the protoplast of the male gametophyte, and the female gamete is formed within a carpogonium-like cell on the female gametophyte. The release of non-motile male gametes through wave and current action carry the spermatia to the carpogonium. The two fuse to produce diploid carpospores following fertilisation. These spores are released through an opening in the pericarp and settle onto an appropriate substrate (e.g. mollusc shells) to give rise to a minute filamentous thallus or conchocelis stage. The conchochelis persists through summer and they release conchospores (also called monospores) in autumn (September–November). Conchospores settle and grow into plumules which, in time, produce carpospores to close the life cycle. In some species of Rhodophyta, including *Porphyra* species, neutral spores are released in September–November and these settle to form secondary plumules. Production of neutral spores is replaced by production of carpospores in response to further reduction in water temperature.

The culture of *Porphyra* species initially involves seeding carpospores onto mollusc (e.g. oyster) shells. 'Seeding' of the cultivation nets with conchospores (produced meiotically by the conchocelis) is achieved by placing the nets in cultivation tanks containing the conchocelis-infested shells. The water is agitated in the tanks to enhance the adhesion of the non-motile conchospores to the nets. The conchospores develop into the macroscopic gametophyte generation, which is cultivated for consumption.

Reproduction in Ulva species (Chlorophyta)

The life history of the Chlorophyta is very diverse. These species undergo an alteration of isomorphic generations. That is, the gametophyte and sporophyte are undistinguishable except for their reproductive structures. The diploid ($2n$) sporophyte stage produces haploid zoospores via meiosis from the sporangia of the thallus. These spores germinate into male or female haploid gametophytes. Flagellate gametangia produced from the parent gametophyte are released, and, upon fusion of opposing gametes, a diploid zygote forms that develops into the mature sporophyte.

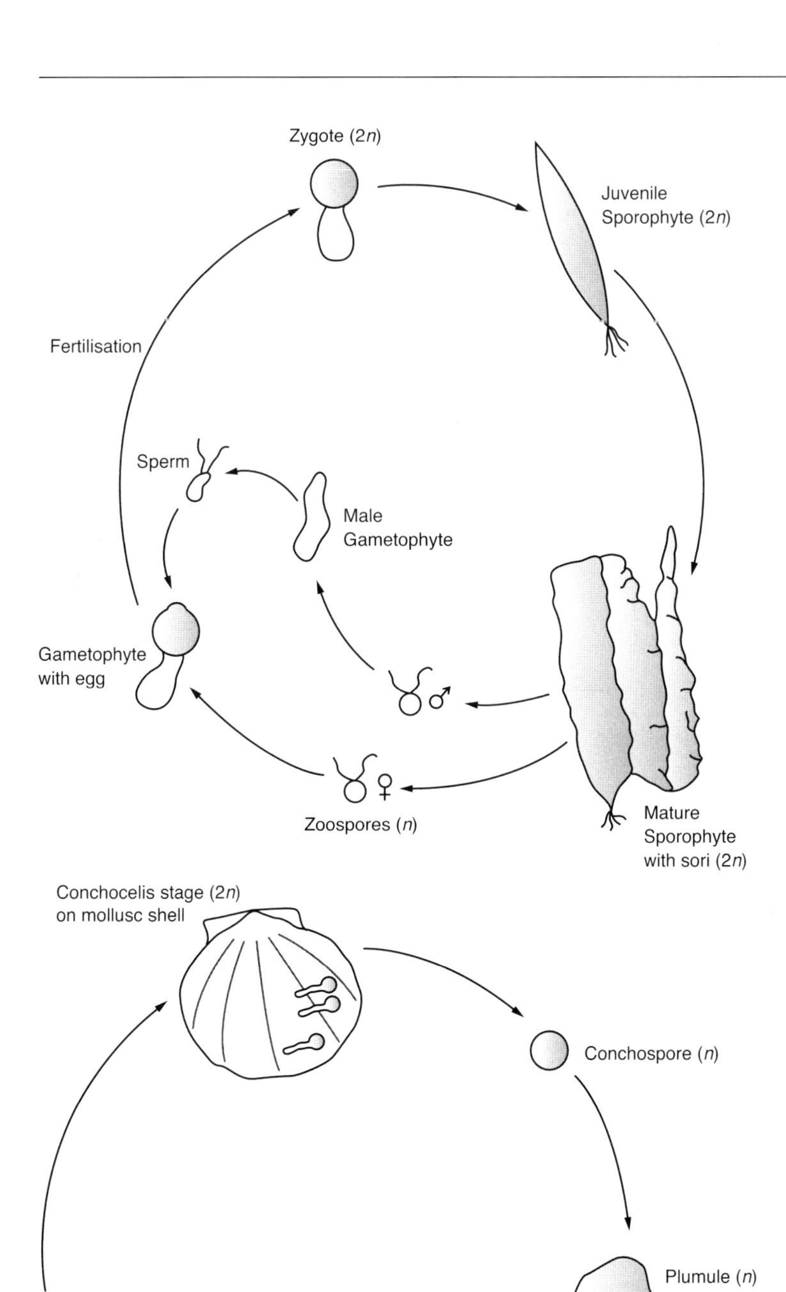

Fig. 13.2 Generalised life cycle of *Lamanaria japonica.*

Fig. 13.3 Generalised life cycle of *Porphyra* species.

13.2.3 Characteristics of seaweed culture

Seaweed culture distinguished from agriculture

Phycoculture has some important characteristics that distinguish it from agriculture.

(1) Aquatic macroalgae do not need a special absorption organ or root, as nutrients in solution can be absorbed by any part of the plant. The holdfast is sufficient to anchor the algae in place. However, the ocean is in constant motion and, as the sea level varies with tide, the amount of light reaching algae on fixed substrates varies. Use of floating rafts in phycoculture, where the depth of cultivation is constant, addresses this problem; raft-based phycoculture also allows the culture depth to be changed as required.
(2) Seaweeds reproduce sexually using spores that are released into open water. The spores cannot be kept alive for a long time after discharge from the parent plants and suitable substrates must be provided for them in the vicinity. This problem can be overcome by using artificial substrates set close to the plants that are discharging spores.
(3) Seaweeds must be attached to substrates. In general, seaweeds grow naturally on rocks or stones that are able to withstand high wave action and remain stable enough to receive adequate light. In the early days of seaweed culture, immobile rocks and stones were regarded as excellent substrates for cultivation. However, several problems associated with the use of rocks and stones are:
    - limited surface area
    - large size and immobility
    - the need to employ divers for harvesting the seaweeds.

    For instance, the Japanese kelp (*Laminaria japonica*) was introduced to China in the 1920s and Chinese production of this species is now many times greater than production in Japan. Although the Chinese coast is a very poor place for growing this plant in terms of natural substrates, large-scale production is possible using floating rafts (Tseng, 1981).
(4) Unlike the application of fertiliser to land plants via moisture in the soil, the application of fertiliser to seaweeds is very difficult given the dynamic nature of the culture medium (the ocean). Fertiliser applied to seaweed in the ocean will be rapidly dispersed. In order to be effective, huge quantities of fertiliser would have to be added, making seaweed culture too costly and having the potential to pollute the coastal environment. Porous containers for fertiliser application have been developed in China, where losses of fertiliser to the open sea are minimised. For large areas cultivating large quick-growing seaweeds, such as the Japanese kelp, the required fertiliser can be sprayed onto the seaweed at low tide.

Problems in common with agriculture

Phycoculture has a number of problems in common with agriculture. Temperature is one of the most important factors in seaweed culture, and optimal and minimal temperature for growth and development may differ within the same species depending on the phase of the life history and growth stage. For example, the sporophyte and gametophyte of *Porphyra tenera* have different optimal water temperatures. Successful phycoculture requires information relating to the optimal culture requirements of all stages of seaweed development and growth.

Like land-based agriculture, phycoculture also has a problem from nuisance or 'weed' algae. In kelp culture, for example, ropes covered with zoospores are placed in the ocean for grow-out in October. Spores of weed algae such as species of *Ectocarpus*, *Enteromorpha* and *Licmorphora* may quickly adhere to the ropes, germinate and grow into macroscopic thalli, which cover the spores or young gametophyte of the kelp. The kelp gametophytes are not exposed to light until December, when the weed algae matures and drops away from the ropes. This delays development to sporophytes for about 2 months (Fig. 13.2). Several methods are employed to combat weed algae. These include the collection of spores in early summer (see section 13.2.4 'Tank culture') and the cultivation of spore-covered ropes in glass-houses with artificially cooled filtered seawater. The young sporelings (called 'summer sporelings') can then be put to sea in autumn, when they are a few centimetres tall. They grow much faster than weed algae, which do not interfere with their growth.

The weed problem in the cultivation of red algae, *Porphyra* species, is more complicated. The predominant weed algae are species of *Monostroma, Entermorpha, Urospora* and *Licmorphora*. Attachment of these algae is discouraged by three methods:

- In the seeding process, the attached conchospores are as densely packed as possible so that there is very little space for the weed spores to attach.
- When manipulating cultivation nets, care is taken not to scrape off *Porphyra* germlings.
- Large thalli are collected and small ones left intact during the harvesting process.

A common way to control the weed algae already present on culture nets is to expose the nets to direct sunlight. Weed algae are generally more susceptible to desiccation than *Porphyra*, and with the correct amount of exposure weed algae can be killed and the *Porphyra* left intact.

13.2.4 Culture methods

The success of commercial cultivation of seaweeds depends on good culture techniques that may differ to some extent according to location. In the 1950s, when red algae cultivation began in China, the Japanese techniques employed were unsuccessful because of the much larger tidal range in China. Subsequently, a semi-floating method of cultivation was developed, which gave much better results than the traditional Japanese method.

There are now five major types of seaweed cultivation:

- natural substrate culture
- long-line culture
- net culture
- pond culture
- tank culture.

Natural substrate culture

This primitive type of phycoculture was developed several hundred years ago in the southern part of Fujian Province, China. It was first employed to enhance the production of the glueweed, *Gloiopeltis furcata*. Natural rock substrates are cleaned by scraping, just before the growing season of the glueweed. The loosened material is carried away by tides and, on the following day, lime is sprayed on the cleaned rocks to destroy other seaweeds and barnacles. The rocks provide a natural substrate for settlement of glueweed spores, which germinate and grow to maturity before being harvested. A similar method has been employed in the production of purple laver/nori (*Porphyra* species) for which, traditionally, seaweed farmers depended on nature for the 'seeds' of *Porphyra*. These days, farmers spray seawater containing conchospores over the cleaned rocks.

Cultivation of *Eucheuma gelatinae* in Hainan is quite simple, with dead coral branches employed as the substrate. First, 'seed' thalli are selected. They must be thick, sturdy and without epiphytes. Seed thalli are tied to a branch of deal coral using rubber rings and scattered over the sea bottom at a density of about 75 000 branches/ha. Arrangement of the seed thalli on the sea bottom is carried out by professional divers, and harvesting is conducted once a year (Tseng, 1981).

In the traditional method of cultivating kelps (*Laminaria* and *Undaria* species), farmers simply throw selected stones into subtidal areas when and where the zoospores of these kelps are released. The stones provide substrate for the zoospores to settle and germinate. This method was widely used to propagate kelps in the 1930s and 1940s in China and is still used today in some areas of Japan and China.

Long-line culture

The long-line culture method (Fig. 13.4) is the usual method for commercial cultivation of kelps (*Laminaria* and *Undaria* species) in China. The long-line for kelp cultivation is composed of a 30- to 60-m-long synthetic fibre rope that is 1–1.5 cm in diameter. The rope is attached to two wooden stakes or suitable anchors by two anchor ropes about 2 cm in diameter. The long-line is supported by a number of floats (15–20 cm in diameter) made of glass or plastic. Attached to the long-line is a series of hanging ropes, which vary in length. Attached to these ropes are cultivation ropes, each about 1.2 m long, having a small stone weight. The distance between two adjacent cultivation ropes is 70–140 cm. Each cultivation rope holds about 30 plants, and the total number of kelp plants growing in one hectare is generally around 150 000–300 000. The distance between two adjacent long-lines is about 6–7 m.

During cultivation there is a difference in growth rate between the upper and lower plants on the same rope. To counteract this, cultivation ropes are regularly inverted. The difference in growth rates can also be minimised by tying adjacent ropes together so that they become oriented more horizontally. Using these methods, the production level and product quality are greatly improved (Tseng, 1981, 1986).

Several other kinds of seaweed are cultivated in China by the long-line method, including *Macrocystis pyrifera* (introduced to China from Mexico) and *Kappaphycus alvarezii,* introduced to China from the Philippines. Local Chinese species, such as species of *Gelidium* and *Gracilaria,* which propagate well asexually, are cultivated from fragments of their thalli using this method. Similarly, *Eucheuma* and *Kappaphycus* species are cultured vegetatively by attaching pieces to a long-line which may be suspended (using floats) or attached to supports driven into the substrate. Plants are generally tied to the long-line with a ribbon. This method is labour intensive and may result in loss of seaweed as a result of improper ties (Zertuche-Gonzalez *et al.*, 1999).

An interesting variation of the long-line culture method was reported from Mexico by Zertuche-Gonzalez *et al.* (1999). In this system, seaweeds (*Chondracanthus pectinatus* and *Eucheuma uncinatum*) are placed into 5-m lengths of plastic mesh tubing (polypropylene), which are then hung horizontally between two parallel long-lines. This not only ensures that each part of the tubing receives adequate light, but reduces the risk of seaweed loss should an attachment to the long-line fail. The size of the seaweed to be cultured determines the diameter of the tubing and size of the mesh. Branches and blades of the seaweed can grow through the mesh that holds the seaweed in place. As well as allowing rapid seeding and harvesting of relatively large amounts of seaweed, this method can be mechanised to reduce labour and cost (Zertuche-Gonzalez *et al.*, 1999).

Net culture

The long-line is used primarily to cultivate large brown algae with zoospores, but net raft culture is better suited for the culture of small and medium-sized red algae with non-motile spores. The best-known seaweed cultivated by the net raft method is purple laver/nori (*Porphyra* species). Net culture methods are illustrated in Fig. 13.5.

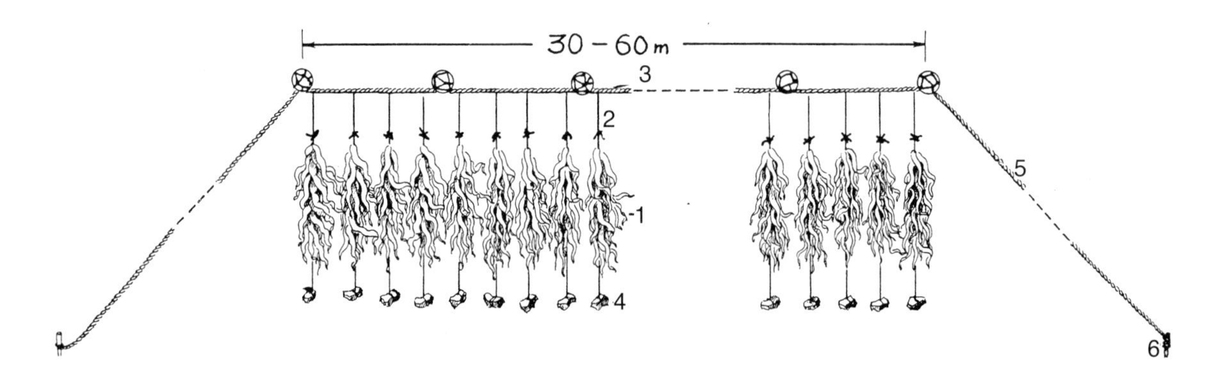

Fig. 13.4 Long-line cultivating *Laminaria* in China. Note that each cultivation rope bearing kelp (1) is attached to a hanging rope (2), which is attached to the long-line (3), and its lower end is tied to a weight (4). The anchor ropes (5) are twice as long as the depth under the long-line and they are anchored to the sea bottom using wooden stakes (6). Reproduced from Tseng (1981).

Nets are first subjected to 'seeding' processes in tanks containing the shells with the conchocelis (see section 13.2.2, 'Reproduction in *Porphyra* species', Fig. 13.3). The cultures are repeatedly examined to determine the progress of conchospore formation. When the number of spores discharged reaches approximately 50 000 spores per shell per day, preparations are made for the seeding process.

Light intensity on the surface of the tank is increased and the water is agitated. When spore discharge reaches 100 000 spores per shell per day, formal seeding begins and net rafts are placed in the tanks. The greatest discharge of spores occurs during mid-morning. A density of 3–5 spores/mm^2 will satisfy the requirement of the seeding process. The net rafts are then taken out to the field for cultivation.

There are three general methods of net cultivation (Fig. 13.5):

- the pillar method
- the semi-floating method
- the floating method.

In the first, pillars are driven into the substrate to serve as vertical supports for the cultivation nets. The nets can be fastened to the pillars by short ropes at a defined level within the tidal range. Alternatively, in places with a greater tidal range, nets with cylindrical floats made of bamboo or plastic pipes on their upper surface can be used. Such nets are fastened to the poles with ropes so that they can move up and down with the change of tide within a defined range (Fig. 13.5) (Kito & Kawamura, 1999).

The intertidal semi-floating net cultivation method is an improvement on the pillar method and is particularly good for cultivation of intertidal seaweed. At high tide, the net floats on the water, maximising

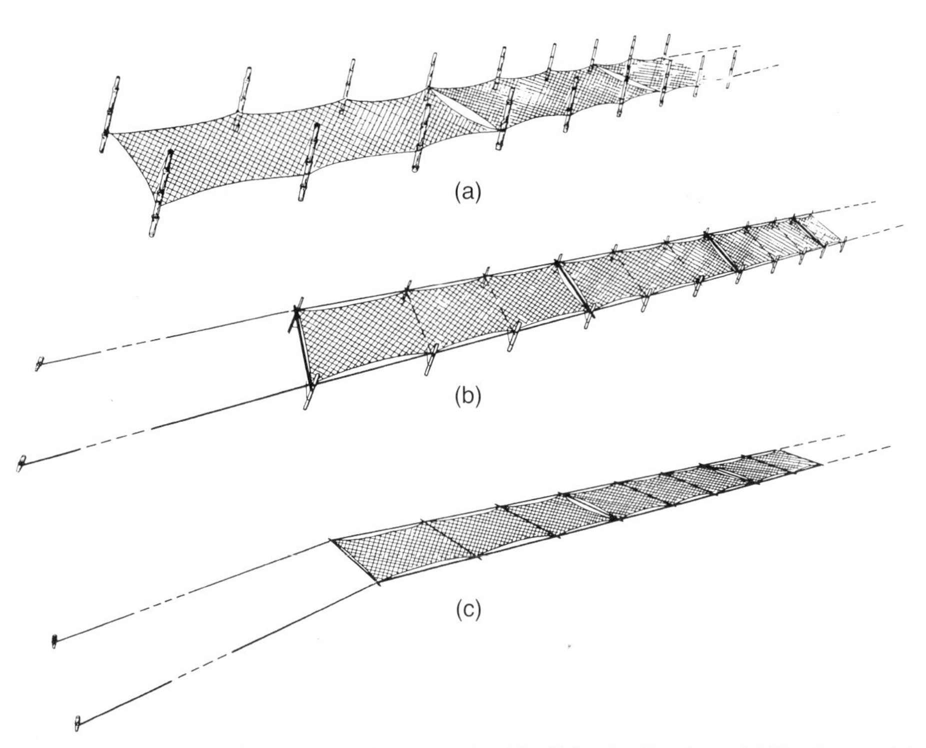

Fig. 13.5 Three methods of net raft seaweed culture practiced in China for *Porphyra*. (a) Fixed type of the pillar method; (b) semi floating method; and (3) floating method. Note the short legs of the semi-floating nets. Reproduced from Tseng (1981).

the light available to the seaweed. At low tide, the net rests on the ground on short legs (Fig. 13.5). Using this method, the sporelings appear earlier and grow better, and production is approximately double that obtained using the pillar method.

The floating net method is used for production of purple laver/nori (*Porphyra* species) in deep water subtidal areas (Fig. 13.5). The floating nets, made of synthetic fibres, are 60 m long and 180 cm broad. They have long anchor ropes so that nets maintain their position on the ocean surface regardless of tidal height. This method is similar to the long-lines used in the cultivation of kelps (see 'Long-line culture', Fig. 13.4).

Pond culture of seaweeds

There are some seaweeds, such as *Gracilaria tenuistipitata* var. *liui*, that multiply quickly in still ponds. Unlike most seaweeds, wave action is not needed for their growth.

Pond cultivation of *G. tenuistipitata* var. *liui* began in Taiwan in 1962, when production was about 10 mt/ha. The polyculture ponds of Taiwan yielded, on average, 9 mt of *Gracilaria* and 6.3 mt/ha of grass shrimp and crab (Shang, 1976). *Gracilaria* grows most rapidly in waters of about 25‰ salinity and at a temperature of 20–25°C. *Gracilaria* begins to die when salinity exceeds 35‰ and freshwater must be available for dilution when pond salinity increases because of evaporation. Cultivation ponds are usually rectangular and about 1 ha in area. They must be 20–30 cm deep during spring to early summer (March–June) and 60–80 cm deep later in summer. Fragments of *Gracilaria* are introduced in spring (April) at a density of 5000 kg/ha. They are strewn evenly and held in place with upright bamboo sticks fixed to the substrate. Alternatively, plants can be kept in place by covering with old fishing nets. The ponds are usually fertilised with urea or fermented pig manure. Harvesting by hand or scoop net takes place every 10 days during summer and autumn (June–November). Harvested plants are washed and sun-dried. Approximately 7 kg of harvested *Gracilaria* yields 1 kg of dry seaweed (Shang, 1976).

Tank culture

Indoor tanks are often used to cultivate juveniles of seaweeds with a biphasic life history in which one phase is microscopic, e.g. the gametophytic phase of kelps and the conchocelis phase of *Porphyra* species (see section 13.2.2, 'Reproduction in *Laminaria* species' and 'Reproduction in *Porphyra* species').

As mentioned in section 13.2.3 in relation to controlling weed algae, the sporelings for commercial cultivation of kelp are cultivated in summer time and hence called 'summer sporelings'. Prior to the seeding process, the parent fronds with abundant sori are cleaned and hung in the air for several hours. When these fronds are placed in seawater, the pressure resulting from the quick absorption of water breaks the sporangial walls and liberates large masses of zoospores. Spore-collectors, in the form of frames with cords, are placed in the spore water and the actively swimming zoospores soon adhere to the collectors to complete the seeding process. These summer sporelings are sold to the kelp farmers, who cultivate them on their own farms (Tseng, 1981).

The seeded frames may be kept in shallow indoor tanks containing seawater previously cooled to 8–10°C and enriched with nutrients. The seeded frames, eventually with gametophytes and young sporophytes, remain in the cool house until autumn, when the juvenile sporophytes are about 1–2 cm high. At this time, when the ambient seawater temperature has dropped to about 20°C, the juveniles are transferred to the farm.

The conchocelis or sporophyte phase of *Porphyra* species is also microscopic and cultivated in indoor tanks in which culture nets are seeded (see 'Net culture'). The tanks vary considerably in size, but they all contain seawater about 20–30 cm deep. The seawater has previously been subject to sedimentation in the dark for a few days, and nutrients added to it. Water temperature is not controlled, but light intensity is controlled by a series of screens to give best growth for conchocelis and to produce the maximum number of conchospores per unit area. For the seeding process, nets are placed in tanks with conchocelis-infested shells (Kito & Kawamura, 1999) and the water is agitated to promote conchospore discharge (Fig. 13.3).

A number of seaweeds (e.g. *Chondrus crispus* and species of *Enteromorpha*, *Gracilaria* and *Porphyra*) are cultured for direct human consumption, in land-based systems in indoor or outdoor tanks. These culture systems require detailed knowledge of the culture requirements of the target species to maximise production and maintain consistent product quality (Craigie *et al.*, 1999). Factors such as irradiance, pH, use of fertilisers and availability of carbon dioxide are of critical importance in such systems (Braud & Amat, 1996; Craigie *et al.*, 1999).

Use of seaweeds in effluents (bioremediation) and polyculture systems

Effluent water leaving fish farms contains high levels of nitrogen excreted by fish into the water. It has been estimated that 13% of the total nitrogen input to fish farms (primarily as protein in feed) is excreted as dissolved inorganic nitrogen (Krom & Van Rijm, 1989). The environmental impacts of aquaculture effluents, such as eutrophication, are discussed in section 4.2.1. One approach to minimising environmental impacts of aquaculture is to use seaweeds for direct removal of dissolved nitrogen from aquaculture effluent (bioremediation). This has the added advantage of diversification (i.e. a second crop) through development of such systems (Chopin & Yarish, 1999).

A number of studies have reported on the polyculture of fish and seaweeds (e.g. *Laminaria* and *Porphyra* species) in open water systems (e.g. Petrell & Alie, 1996; Chopin & Yarish, 1999). The position of the seaweeds relative to fish cages is important in determining the concentration of dissolved nitrogen available to the seaweeds, which, in turn, influences growth rates. Petrell & Alie (1996) noted that technical and economic difficulties with fish/seaweed polyculture systems include:

- marketing and processing two different types of products
- variable nutrient removal efficiencies by seaweeds
- incompatible production rates of fish and seaweeds
- logistical problems resulting from shared space and equipment.

Seaweeds have also been used to remove fish culture waste products in semi-closed aquaculture systems. For example, it has been estimated that a 153-m^2 surface area of tank(s) containing *Ulva rigida* would be required to remove 100% of the dissolved inorganic nitrogen produced by 1 mt of cultured *Sparus aurata* (Jimenez del Rio *et al.*, 1996). In this system, maximum nitrogen uptake was determined as 2.5 g/m^2/day dissolved inorganic nitrogen and the maximum seaweed yield was calculated as 40 g/m^2/day dry weight. In a similar land-based system, *U. lactuca* was reported to remove 74% of ammonia and reduce water use and nitrogenous pollution by half (Neori & Shpigel, 1999). Production of *Gracilaria chilensis* using fish effluents has been estimated at 48 kg/m^2/year; however, fish effluent reduced the agar yield from *G. chilensis* when compared with that grown in clean seawater (Martinez & Buschmann, 1996).

As well as dissolved inorganic nitrogen, fish farm effluent contains particulate matter. A number of studies have described bioremedial systems that incorporate bivalve molluscs and seaweeds to clean fish farm effluent (e.g. Shpigel *et al.*, 1993). In such systems, bivalves remove the particulates and seaweeds remove dissolved nutrients. Seaweed cultured in these systems is not only valuable in cleaning effluent water, and potentially providing a second crop, but may itself provide a food source for other culture organisms such as abalone. A system incorporating fish, seaweeds and abalone has been described by Neori & Shpigel (1999). In such systems (Fig. 13.6), fish effluent drains back to the sea via seaweed 'biofilters' that treat the water and the resulting seaweed is used to feed abalone. A system with 1 mt of fish and a seaweed pond area of 75 m^2 produced 4–5 kg of fish, 20–25 kg of seaweed and 2 kg of abalone per day (Neori & Shpigel, 1999).

13.2.5 Diseases of cultured seaweeds

Cultured seaweeds are affected by both physiological and pathological diseases. The most common physiological diseases in kelp cultivation are 'green rot' and 'white rot', both related to light intensity.

(1) Control of 'green rot' (caused by too little light) is achieved by inverting the cultivation ropes

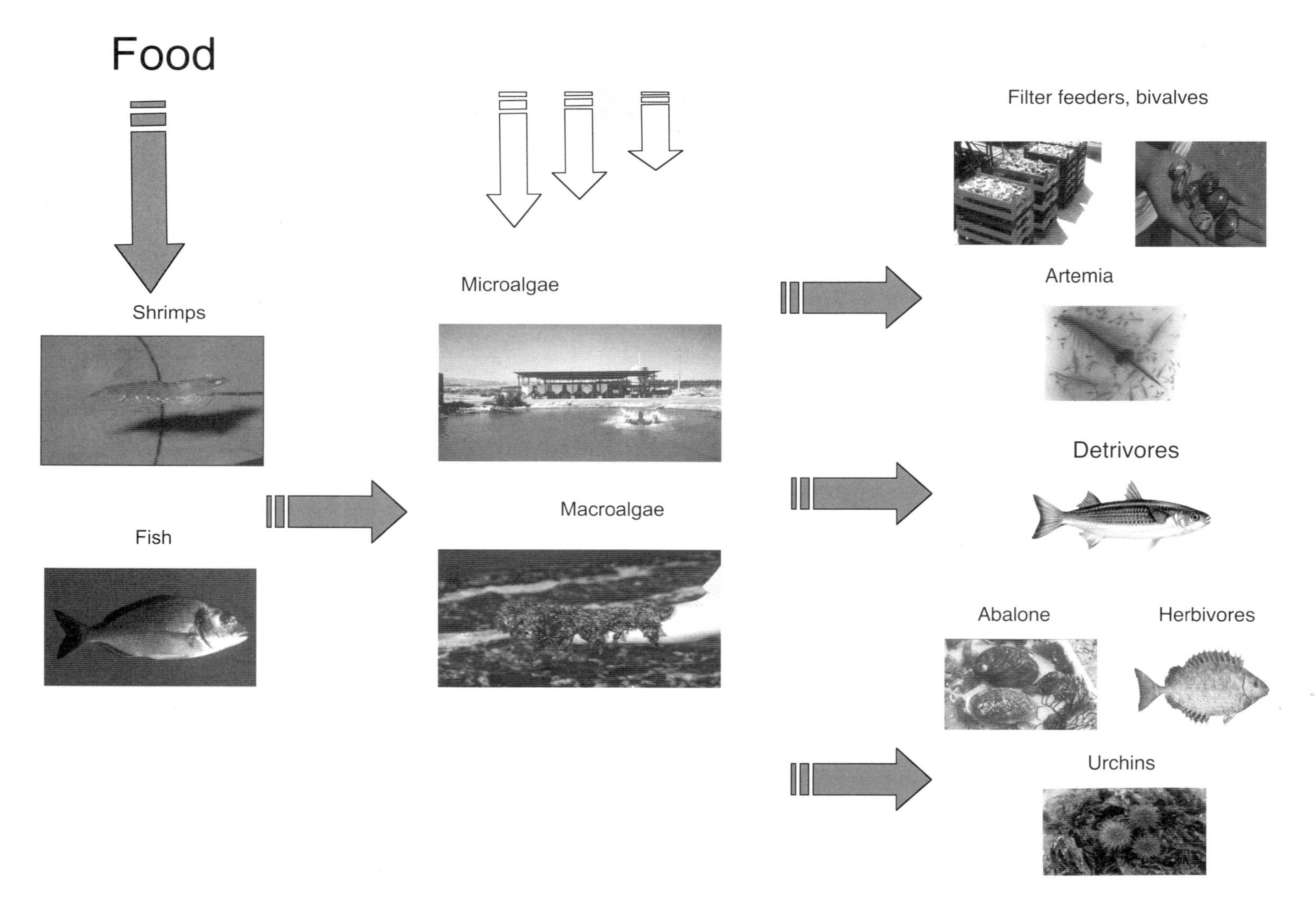

Fig. 13.6 Diagrammatic representation of a pilot mariculture system in Eilat, Israel, which utilises effluent from fish/shrimp culture to provide nutrients for algae culture. In turn, algae are used as a food source for a variety of other culture animals (figure provided by Dr Muki Shpigel).

so that the lower, overshaded, fronds receive sufficient light. If the disease occurs during the fast-growing stage of kelp, tip-cutting may be used to increase light intensity. As much as one-third of the total length of the fronds may be cut off, greatly reducing the overcrowded condition and improving light penetration and frond health. The cut portion has a market in the alginate industry.

(2) 'White rot' disease always occurs in the fronds of the upper part of the cultivation ropes. It is believed that the three factors stimulating this condition are strong light, high water temperature and low nutrient levels. As the principal cause is believed to be strong light, treatment includes reduction in light intensity by lowering the level of cultivated algae in the water column. Fertiliser is also applied.

Several kinds of pathogenic diseases have been recognised in seaweeds, but relatively few have been well documented (Correa, 1996). The sulphate-reducing bacteria and hydrogen sulphide-producing saprophytic bacteria, quite common in cultivation glasshouses for kelp sporelings, were found to be the causal agents in a disease characterised by plasmolysed oogonia and abnormal malformed sporophytes. Prevention measures included separating the sporeling cultivation system from the mature sporophytes and sterilising the water system with chlorine before the seeding process. Rotten and diseased fronds are periodically removed to reduce potential sources of infectious bacteria.

In 'frond-twist' disease of raft-cultivated kelp, the contagious and biotic nature of the disease was confirmed and the causal agent found to be a mycoplasma-like organism. Antibiotics such as tetracycline are effective treatments for this disease. Alginic acid-decomposing bacteria were found to be the causal agent of a disease that causes detachment of summer sporelings. This condition can be effectively controlled using antibiotics.

In his short review of diseases in seaweeds, Correa (1996) pointed out that further development of disease-resistant strains of seaweed will require more information on the mechanisms of pathogenicity and defence and on whether disease susceptibility and resistance are genetically determined traits.

13.2.6 Genetic aspects of seaweed culture

As with all farmed organisms, significant benefits can be gained through appropriate breeding programmes (Chapter 7). Research with seaweeds has sought to enhance characteristics such as yield and growth rates through genetic selection in the sexual phase of the life cycle. For example, in the 1960s and 1970s, superior strains of kelp were developed in China by intensive inbreeding and selection for specific characteristics, such as high productivity, high iodine content and increased thermal tolerance, that better met the demands of industry.

Whereas these developments were generally brought about through breeding programmes and strain selection, more recently major developments in this field have been brought about using modern genetic manipulation techniques or genetic engineering (Cheney, 1999; Minocha, 1999; Qin *et al.*, 1999). Example of some of the modifications made to cultured red seaweeds using these techniques include increased tolerance to higher temperatures (e.g. *Chondrus crispus*, *Kappaphycus alvarezii*), increased agar or carrageenan content (e.g. *Chondrus crispus*, *Kappaphycus alvarezii*, *Gracilaria tikvahiae*) and increased growth rates (e.g. *Kappaphycus alvarezii*, *Eucheuma denticulatum*, *Porphyra yezoensis*) (Cheney, 1999).

13.2.7 Future developments

The high value of seaweed products and their increasing use in industrial processes and as food will ensure continued expansion of this industry. Expansion will occur primarily from improvements to culture techniques and genetically improved culture stock. It will also result from expansion of seaweed culture in countries without a tradition of seaweed culture such as Mexico (Zertuche-Gonzales *et al.*, 1999), Venezuela (Rincones & Rubio, 1999) and Chile (Buschmann *et al.*, 1999), and from the development of culture techniques for new culture species (e.g. Zertuche-Gonzales *et al.*, 1999).

The next decade may see the development of large-scale open-ocean farming of seaweeds. This will require many biological and technical problems to be solved. Furthermore, social and ethical issues will need to be addressed because of the expected increase in the use of genetically modified seaweeds in aquaculture, particularly in the open ocean. In response to growing environmental concerns (Chapter 4), increasing pressure to minimise aquaculture effluent is likely to result in greater use of seaweeds in bioremediation and as major components of new kinds of polyculture systems.

13.3 Microalgae

13.3.1 Introduction

Microalgae are taxonomically diverse and are found in almost every environment in nature. This great taxonomic and environmental diversity is also reflected in the range of metabolites they produce. Several species are grown commercially as sources of high-value, fine chemicals such as carotenoids and fatty acids, and as human food and animal feed (Table 13.4). Others are used in wastewater treatment and in agriculture as soil conditioners (Metting, 1988; Oswald, 1988a). Microalgae are also proving to be excellent sources of bioactive compounds such as antibiotics and anti-cancer drugs and, although some of these compounds can be produced by chemical synthesis, many others will probably have to be produced through microalgae culture (Borowitzka, 1999a). Although commercial culture of microalgae is still a very new industry with only a small number of species and products, global production has grown significantly in the last 20 years. Estimated annual production for the most widely cultured species of microalgae is shown in Table 13.5. The high cost of production (Table 13.6) means that the product must also command a high sale price. It is interesting to note that the most expensive microalgae produced are those grown for use as feed for aquaculture species (Table 13.6). Some of the factors contributing to the high production costs are the high capital costs and high labour costs and, for algae used in aquaculture, the rather small scale of production (section 9.3.1, Fig. 9.2) (Borowitzka, 1997, 1999b).

Since most algae require light for growth a basic feature of all microalgae culture systems is that they are shallow so that light can reach all the cells. Commercial-scale microalgae culture systems may be extensive, semi-intensive or intensive. The cultures may also be either open to the air or closed.

13.3.2 Extensive culture

Extensive culture systems are very large and achieve only low cell densities of 0.1–0.5 g/L dry weight. Extensive microalgal culture is possible only with the small number of algae that live in extreme, highly selective environments. The main microalga grown commercially in extensive culture systems is the chlorophyte, *Dunaliella salina*. *D. salina* is grown in very large shallow ponds in Australia for the production of the carotenoid β-carotene (Borowitzka & Borowitzka, 1989). It grows best at very high salinities (> 25% w/v NaCl) and high temperature (30–40°C). β-Carotene production from *D. salina* is greatest at high salinity and high light levels. The high salinity generally prevents other organisms growing in the ponds and competing with the *D. salina*.

In commercial systems in Australia, *D. salina* is grown in ponds of up to 250 ha in area constructed with earthen walls on the bed of a salt lake. The ponds are about 30–50 cm deep and the only water movement in ponds results from wind or convection. In such a system, the operator has little control over culture conditions other than salinity and nutrient concentrations. The ponds are usually operated in a semi-continuous mode with part of the ponds harvested at regular intervals and with the medium being returned to the ponds after microalgae cells have been harvested. Nutrients are added as required for microalgae growth, and salinity is controlled by the addition of seawater (Borowitzka, 1994). Fig. 13.7 outlines this culture process. Since the algal density achieved in such systems is low, harvesting is very expensive and the final product must have a high value for the overall process to be economical. Despite this, the actual production costs for *D. salina* are among the lowest for any commercially produced microalga.

13.3.3 Semi-intensive culture

Semi-intensive culture systems are still quite large; however, mixing of the cultures and better control of culture conditions results in cell densities of up to about 1 g/L dry weight. The first commercial large-scale cultures of microalgae were developed in Taiwan in the 1950s for culturing the freshwater

Table 13.4 List of the major microalgae grown on a large scale and their applications (the use of algae in wastewater treatment or as biofertiliser is not included)

Alga	Product or application	Status	Countries
Chaetoceras muelleri	Aquaculture feed	Commercial	Global
Chlorella spp.	Health food	Commercial	Japan, Taiwan, Czech Republic, Germany
Crypthocodinium cohnii	Docosahexaenoic acid	Commercial	USA
Dunaliella salina	Beta-carotene	Commercial	Australia, Israel, China, India
Dunaliella tertiolecta	Aquaculture feed	Commercial	Global
Haematococcus pluvialis	Astaxanthin	Commercial	USA, Sweden, (Israel)
Isochrysis spp.	Aquaculture feed	Commercial	Global
Monochrysis lutheri	Aquaculture feed	Commercial	Global
Nannochloropsis spp.	Aquaculture feed	Commercial	Global
Pavlova spp.	Aquaculture feed	Commercial	Global
Porphyridium cruentum	Polysaccharides, pycobilin pigments	R&D	Israel, France
Skeletonema spp.	Aquaculture feed	Commercial	Global
Spirulina platensis	Health food, phycocyanin	Commercial	Thailand, USA, China, India, Vietnam
Tetraselmis suecica	Aquaculture feed	Commercial	Global
Thalassiosira pseudonana	Aquaculture feed	Commercial	Global

R&D, research and development.

Table 13.5 Estimated annual production of microalgae (based on published data, company information and estimated from market value)

Alga	Annual production (dry weight) (mt)
Chlorella	> 2200
Spirulina	> 3000
Dunaliella	> 2000
Haematococcus	~200–300

green alga, *Chlorella*, which is used as a health food. The algae are grown in circular concrete ponds of up to about 500 m^2 surface area. The ponds have a centrally pivoted rotating arm that mixes the culture (Fig. 13.8). This system results in uneven mixing, with the periphery of the pond being mixed much more than the centre because of the higher velocity of the mixing arm at the outer perimeter. The larger the pond, the greater is the difference between the periphery and the centre, and this limits the effective size of the ponds. Because of the inherent instability, the cultures need to be grown in batch mode. Each growth cycle is started as a small (~1 L) laboratory culture (section 9.3.1) and is scaled up by a factor of 10 at each step.

In the 1960s, a better pond design, the 'raceway' pond, was developed. These ponds consist of long channels arranged in single or in multiple loops (Fig. 13.8). Early designs used a configuration consisting of relatively narrow channels with many 180° bends and propeller pumps to produce a channel velocity of about 30 cm/s. In the 1970s, paddle wheel mixers of various designs were introduced and found to be more effective, with reduced energy requirements and reduced shear forces on the microalgae cells. The numerous bends in the channels of the older designs also led to hydraulic losses and problems with solids deposition. These were minimised by using a single loop (raceway) configuration, with baffles at the corners. Simple geometric optimisation has also shown that a large pond with a low length to width (*L*/*W*) ratio gives the largest pond area for the least wall length, and is therefore cheaper to construct. Ponds can be up to about 6 m wide, the width being limited by the paddle wheel design. Pond length is influenced by head loss relative to the mixing velocity and pond depth. Details of the design considerations in such systems were outlined by Oswald (1988b).

Several factors need to be taken into account when designing the optimally sized pond. These include:

Table 13.6 Estimated production costs for microalgae currently grown on a commercial scale

Alga	Estimated production cost (US$/kg dry wt)	Production system
Spirulina (Arthrospira) platensis	8–12	Raceway systems. Raceways of up to about 0.5 ha area
Chlorella spp.	15–18	Centre pivot open ponds
Dunaliella salina	10	Very extensive open ponds up to 250 ha in area
Haematococcus pluvialis	> 40	Closed reactors and open ponds
Algae for aquaculture (e.g. *Isochrysis, Tetraselmis* and *Skeletonema* spp.)	60–1000+	Big bags (lowest cost is for largest aquaculture facility in the USA)
Crypthecodinium cohnii	2	Grown heterotrophically on glucose in fermenters

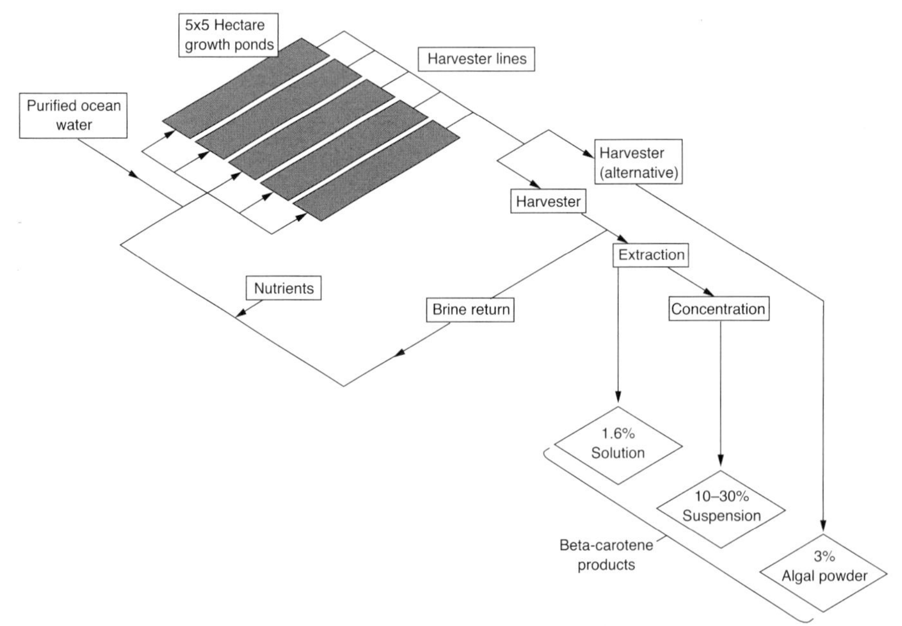

Fig. 13.7 Diagrammatic representation of the operation of the *Dunaliella salina* production plant operated by Western Biotechnology at Hutt Lagoon, Western Australia (note that not all ponds are shown).

- optimal pond depth, taking into account the degree of light penetration
- mixing velocity, which relates to the need to keep the algae in suspension, avoiding any dead spaces and the effects of turbulence on the pond materials
- the energy requirement for mixing
- materials from which the pond is constructed.

This pond design was first developed for high-rate oxidation ponds used in the treatment of wastewater, but was soon also applied to the 'clean' culture of a range of microalgae, especially the blue-green alga, *Spirulina platensis*. It is also used in Israel and India for culturing *D. salina*. The raceway ponds are more efficiently mixed than the circular central pivot ponds and pond size can be up to 1 ha in area.

The ponds are usually constructed of concrete or of concrete walls lined with a plastic liner. The plastic liner is replaced with a concrete bottom in the region of the paddle wheel. Laboratory experiments have shown that microalgae productivity increases with increasing flow rate and that a velocity of at least 10 cm/s is necessary to avoid settlings of the cells;

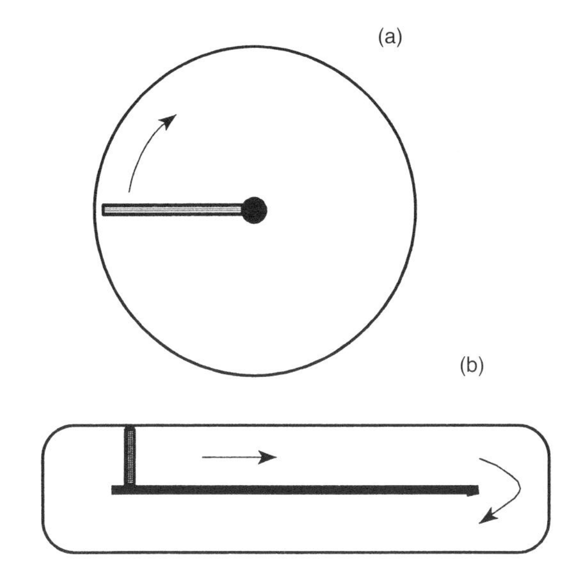

Fig. 13.8 Diagrams of the layout of semi-intensive culture systems. (a) Circular central pivot pond; (b) raceway pond (arrows indicate direction of water flow).

however, practical limitations in pond design mean that velocities in the range of 30 cm/s are optimal. In order to maximise productivity, the pond depth is about 20–30 cm and cell density must be controlled to minimise self-shading by the cells.

Although productivity of up to 30 g/m²/day dry weight has been reported, actual long-term productivity is significantly lower than this. Several attempts have been made to improve the productivity of these ponds. Of these, the introduction of a series of aeroplane-type wings into the water flow seems to be the most promising. The wings introduce extra turbulence into the water flow, exposing the algae to more light and, therefore, increasing the growth rate by 10–20%.

Although the raceway pond design is the main culture system used for the commercial-scale culture of microalgae, its major limitation is that the system is open to the air, which leads to contamination and infection by predators (mainly other algae, protozoa and fungi). These systems are, therefore, only suited to microalgae that grow in a relatively extreme environments such as high pH (e.g. *Spirulina* species) or high salinity (e.g. *Dunaliella* species) or fast-growing algae such as *Chlorella*, *Phaeodactylum* or *Scenedesmus* species, which can outgrow most of their competitors.

13.3.4 Intensive culture

In intensive cultures, the algae are grown under highly controlled optimum conditions in closed reactors, which can result in cell densities of 1–10 g/L dry weight. High cell densities have the advantage of requiring a smaller area and, in addition, harvesting costs are also reduced significantly.

Closed culture systems include:

- bag culture, which is widely used for the culture of algae for aquaculture (section 9.3.1)
- alveolar panels and other flat plate reactors of various designs (Silva & Cortinas, 1994)
- stirred tank reactors with internal illumination (Wohlgeschaffen *et al.*, 1992)
- tower reactors with internal fibre optic illumination (Burgess *et al.*, 1993)
- suspended narrow bags or tubes (Cohen & Arad, 1989)
- tubular reactors (Chaumont *et al.*, 1988; Tredici, 1999).

The predecessors of many of these systems were described for the culture of *Chlorella* species by Davis *et al.* (1953) and by Arthur D. Little, Inc. (1953).

'Big bag' systems

Probably the longest-used closed culture systems for mass culture of microalgae are the 'big bag' systems generally used in aquaculture hatcheries to feed larval fish, crustaceans, molluscs or rotifers (section 9.3.1, Fig. 9.3). Although widely used, these systems are notorious for the instability of the cultures. This instability probably occurs because mixing in these bags is uneven, leading to build-up of cells in unmixed areas, which, in turn, leads to cell death, especially if the culture is not axenic (bacteria free). In order to achieve reasonably reliable cultures, it is essential to maintain axenic conditions, a feature that is not as essential for the tubular photobioreactors.

Tubular photobioreactor

Despite the range of designs listed above, few of these systems are operational at a commercial scale as yet.

The two most promising designs for commercial large-scale culture are the tubular photobioreactors and the flat-panel reactors.

Many tubular reactor designs have been developed to produce cultures of relatively high density (Tredici, 1999). The first large-scale tubular photobioreactor (Chaumont *et al.*, 1988) had a solar receptor constructed of five identical 20-m^2 units made of 25-cm-diameter polyethylene tubes floating on or in a large pool of water. The culture was circulated through these tubes and temperature was controlled by either floating the tubes at the water surface (to heat) or by immersing them in the water (to cool). The water in the pool also provided a convenient support for the long tubes of the solar receptor. At the end of each solar receptor, a gas exchange tower removed photosynthetically produced oxygen and CO_2 could be added. This pilot plant was quite successful in growing the red unicell, *Porphyridium*; however, it was technically complex and expensive, and required a large land area.

A more efficient arrangement for the tubes of the solar receptor is to wind them helically around a tower. This is the design of the 'Biocoil', a system developed in the UK and optimised in Australia. Several pilot scale units of the Biocoil have been in operation in the UK and in Australia, with volumes up to 2000 L, and a very wide range of microalgae including *Chlorella*, *Spirulina*, *Dunaliella*, *Tetraselmis*, *Phaeodactylum*, *Chaetoceras*, *Isochrysis*, *Pavlova*, *Porphyridium*, *Haematococcus* and *Skeletonema* species have been grown successfully. The Biocoil system uses low-density polyethylene or Teflon tubing of 25–30 mm diameter. This narrow diameter has been shown to result in much higher productivities and reduced fouling of the inside of the tubes by the algae. The helical arrangement of the tubing also means that there are no sudden changes in direction of flow, which not only result in significant head losses, but can also lead to undesirable accumulation of algae. The helical design also has the great advantage of good scale-up properties. This means that the results obtained in smaller pilot experiments can be directly related to a full-scale production unit. Systems such as the Biocoil also allow for continuous culture, which results in a more consistent quality of algae produced and is cheaper. In Perth, Western Australia, *Isochrysis* species (T. ISO) has been grown in a 1000-L Biocoil in continuous culture for more than 4 months.

One of the key design features of these systems is the pumping system used to circulate the algal culture. Several types of pumps, including centrifugal, diaphragm, peristaltic and lobe pumps, as well as airlifts, have been used and the choice of pump depends on the degree of fragility of the algae being grown.

There are two other arrangements for the tubes that are being developed for commercial production. In these, the tubes are either arranged vertically in long rows (the 'Biofence') or are laid horizontally on the ground.

Flat-panel photobioreactor

An alternative design for a closed photobioreactor is the flat-panel reactor first described in the 1980s (Samson & Leduy, 1985). This reactor consists of two rectangular panels of glass or Perspex spaced about 25 mm or more apart. Some of the designs have a number of internal baffles. The panels can be inclined to capture the optimum amount of solar irradiation and the algal culture inside is mixed by aeration or circulated by pumping. The aeration not only mixes the culture but also helps to remove photosynthetic O_2, which at high concentrations will limit productivity due to photorespiration. CO_2 can also be added to enhance growth. The temperature is usually controlled by spraying water over the panel surface in order to cool the cultures. Some systems use a heat exchanger; however, this is usually too expensive for large-scale systems. These flat-panel reactors can be very productive, but they are very difficult to scale up to any appreciable size. A large-scale flat-panel reactor for mass production of *Nannochloropsis* was recently described by Cheng-Wu *et al.* (2001) (Figs 13.9 and 13.10). An industrial-scale production plant growing *Chlorella* has been constructed in Germany. This unit consist of a series of large upright plate reactors closely spaced in a glasshouse. The glasshouse is required for temperature control.

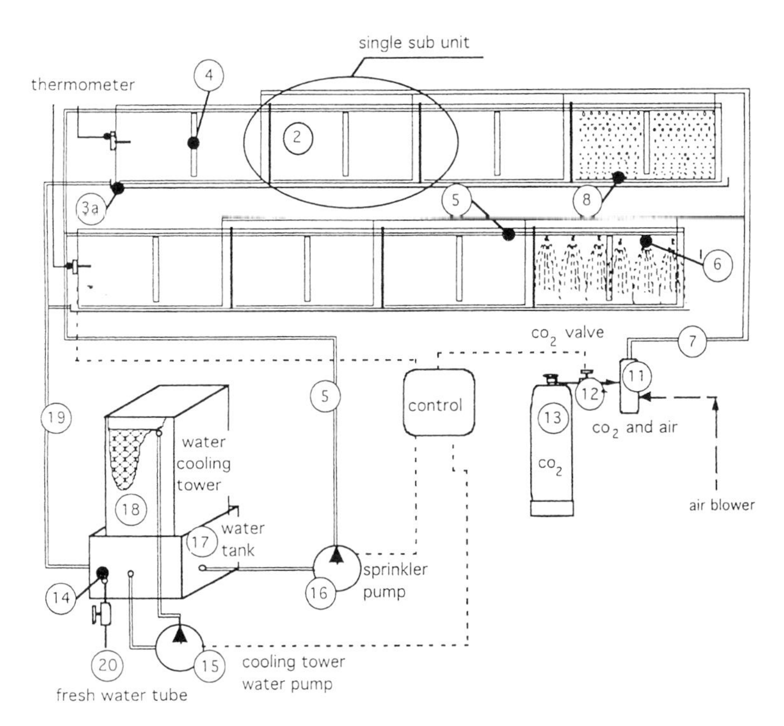

Fig. 13.9 Schematic view of industrial-scale photobioreactor (see legend of Fig. 13.10 for details). Reprinted from Cheng-Wu *et al.* (2001) with permission from Elsevier Science.

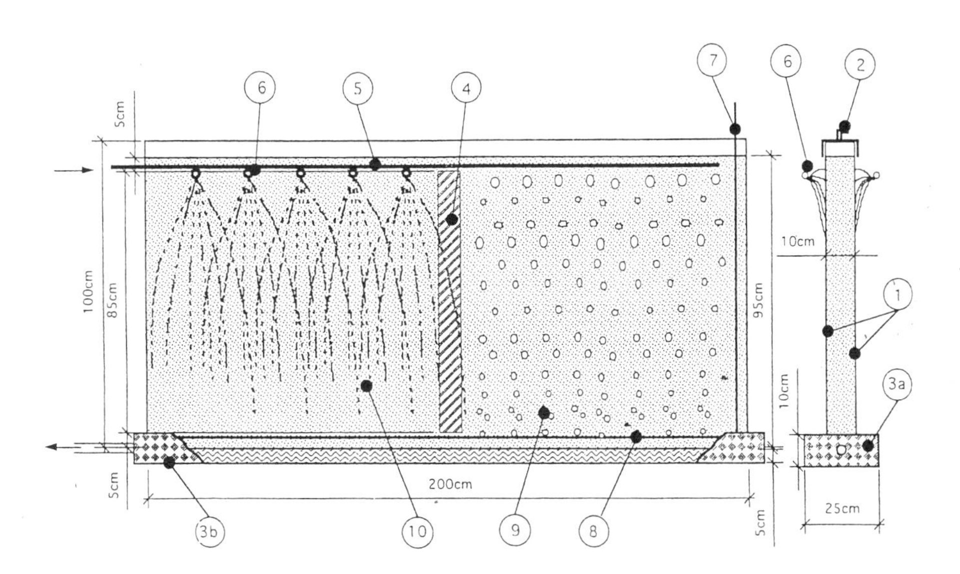

Fig. 13.10 Schematic drawing of one sub-unit from the large-scale flat-plate glass photobioreactor shown in Fig. 13.9. Key for Fig. 13.9 and Fig. 13.10: (1) front and back 10-mm glass plates; (2) glass cover placed over a ceiling film leaving a small opening for out-flowing air; (3a) trough to collect sprayed water, side view; (3b) trough for collecting sprayed water, front view; (4) support connecting front and back glass plates; (5) water tubes; (6) micro-sprinklers; (7) air tubes; (8) air tube with 1-mm perforations placed 5 cm apart; (9) air bubbles; (10) algae culture; (11) mixing tank for air and CO_2; (12) valve for CO_2; (13) CO_2 tank; (14) freshwater floats; (15) cooling tower water pump; (16) sprinkler pump; (17) water tank; (18) water cooling tower; (19) tube to return spray water to water tank; (20) freshwater tube. Reprinted from Cheng-Wu *et al.* (2001) with permission from Elsevier Science.

Table 13.7 Comparison of the properties of different large-scale algal culture systems

Reactor type	Mixing	Light utilisation efficiency	Temperature control	Gas transfer	Hydrodynamic stress	Species control	Sterility	Scale-up
Unstirred shallow ponds	Very poor	Poor	None	Poor	Very low	Difficult	None	Very difficult
Paddle-wheel raceway ponds	Fair–good	Fair–good	None	Poor	Low	Difficult	None	Very difficult
Stirred tank reactor	Largely uniform	Fair–good	Excellent	Low–high	High	Easy	Easily achievable	Difficult
Air-lift reactor	Generally uniform	Good	Excellent	High	Low	Easy	Easily achievable	Difficult
Flat-plate reactor	Uniform	Excellent	Excellent	High	Low–high	Easy	Achievable	Difficult
Tubular reactor (Serpentine type)	Uniform	Excellent	Excellent	Low–high	Low–high	Easy	Achievable	Reasonable
Tubular reactor (Biocoil type)	Uniform	Excellent	Excellent	Low–high	Low–high	Easy	Achievable	Easy

Hybrid systems

The green alga *Haematococcus pluvialis* has proven to be difficult to grow in either open raceway ponds or closed bioreactors. Commercial production of this alga, which is an excellent source of the carotenoid astaxanthin, involves growing the algae initially in a closed photobioreactor. Once sufficient biomass has been produced, the algae are then transferred to open-air raceway ponds, where the cells then accumulate the astaxanthin (Olaizola, 2000).

Heterotrophic culture

Closed fermenters have been developed for the heterotrophic production of long-chain polyunsaturated fatty acids from microalgae (Gladue, 1991; Radmer & Parker, 1994). This technique uses glucose or acetate as the energy and carbon source for the algae and eliminates the need for light, which is a major cost in phototrophic microalgae culture systems. Furthermore, because cell density is not limited by light availability, microalgae can be cultured in relatively high densities with high biomass production. Heterotrophically grown microalgae have been commercially produced and shown to have value as an aquaculture feed (section 9.3.4). However, only a limited number of species can be grown heterotrophically, and therefore systems such as the tubular photobioreactors and flat-panel reactors are likely to be the main systems to be used in the future.

13.3.5 Choice of culture system

Table 13.7 compares some of the main characteristics of the different culture systems used for production microalgae. The choice of culture system depends on many factors and no single system is best for all microalgae. The reliability of the culture is of great importance to commercial operations and, for this reason, closed culture systems are preferred. However, many of these systems are still in the early stages of development and they are generally more expensive to construct and operate. On this basis, the majority of commercial microalgae culture systems are open systems that can only be used with microalgae growing in a highly selective environment. As such, further developments in this field with new species of microalgae will require closed systems.

Large-scale photo-bioreactors require a high surface area to volume ratio to maximise light availability. Although this generally results in higher productivity, greater cell biomass and lower production costs, these systems do have disadvantages. For example, the high surface area to volume ratio provides susceptibility to overheating in outdoor systems. In addition, the high biomass in such systems requires turbulent flow to reduce light limitation and ensure nutrient exchange. On this basis, the number of species that can be cultured in these systems is limited as they are unsuitable for fragile species.

References

Arthur D. Little Inc. (1953). Pilot-plant studies in the production of *Chlorella*. In: *Algal Culture. From Laboratory to Pilot Plant* (Ed. by J. S. Burlew), pp. 232–272. Carnegie Institution, Washington, DC.

Borowitzka, L. J. (1994). Commercial pigment production from algae. In: *Algal Biotechnology in the Asia-Pacific Region* (Ed by S. M. Phang, K. Lee, M. A. Borowitzka & B. Whitton), pp. 82–4. Institute of Advanced Studies, University of Malaya, Kuala Lumpur.

Borowitzka, M. A. (1997). Algae for aquaculture: opportunities and constraints. *Journal of Applied Phycology*, **9**, 393–401.

Borowitzka, M. A. (1999a). Pharmaceuticals and agrochemicals from microalgae. In: *Chemicals from Microalgae* (Ed. by Z. Cohen), pp. 313–52. Taylor & Francis, London.

Borowitzka, M. A. (1999b). Economic evaluation of microalgal processes and products. In: *Chemicals from Microalgae* (Ed. by Z. Cohen), pp. 387–409. Taylor & Francis, London.

Borowitzka, L. J. & Borowitzka, M. A. (1989). β-Carotene (Provitamin A). In: *Biotechnology of Vitamins, Pigments and Growth Factors* (Ed. by E. J. Vandamme), pp. 15–26. Elsevier Applied Science, London.

Braud, J. P. & Amat, M. A. (1996). *Chondrus crispus* (Gigartinaceae, Rhodophyta) tank cultivation: optimising carbon input by a fixed pH and use of a salt water well. *Hydrobiologia*, **326/327**, 335–40.

Burgess, J. G., Iwamoto K., Miura, Y., Takano, H. & Matsunaga, T. (1993). An optical fibre photobioreactor for enhanced production of the marine unicellular alga *Isochrysis* aff. *galbana* T-Iso (UTEX-LB-2307) rich in docosahexaenoic acid. *Applied Microbiology and Biotechnology*, **39**, 456–9.

Buschmann, A. S. H., Correa, J. A., Westermeier, R., Hernandez-Gonzalez, M. & Norambuena, R. (1999). Mariculture of red algae in Chile. *World Aquaculture*, **30**, 41–5.

Chaumont, D., Thepenier, C., Gudin, C. & Junjas, C. (1988). Scaling up a tubular photoreactor for continuous culture of *Porphyridium cruentum* from laboratory to pilot plant (1981–1987). In: *Algal Biotechnology* (Ed by T. Stadler, J. Mollion, M. C. Verdus, Y. Karamanos, H. Morvan, & D. Christiaen), pp. 199–208. Elsevier Applied Science, London.

Cheney, D. P. (1999). Strain improvement of seaweeds through genetic manipulation: current status. *World Aquaculture*, **30**, 55–67.

Cheng-Wu, Z., Zmora, O., Kopel, R. & Richmond, A. (2001). An industrial-size flat plate glass reactor for mass production of *Nannochloropsis* sp. (Eustigmatophyceae). *Aquaculture*, **195**, 35–49.

Chopin, T. & Yarish, C. (1999). Nutrients or not nutrients? *World Aquaculture*, **29**, 31–61.

Cohen, E. & Arad, S. (1989). A closed system for outdoor cultivation of *Porphyridium*. *Biomass*, **18**, 59–67.

Correa, J. A. (1996). Diseases in seaweeds: an introduction. *Hydrobiologia*, **326/327**, 87–8.

Craigie, J. S., Staples, L. S. & Archibald, A. F. (1999). Rapid bioassay of a red food alga: accelerated growth rates of *Chondrus crispus*. *World Aquaculture*, **30**, 26–8.

Davis, E. A., Dedrick, J., French, C. S., Milner H. W., Myers, J., *et al.* (1953). Laboratory experiments on *Chlorella* culture at the Carnegie Institution of Washington Department of Plant Biology. In: *Algal culture. From laboratory to Pilot Plant* (Ed. by J. S. Burlew), pp. 105–53. Carnegie Institution of Washington, Washington, DC.

FAO (2001a). *1999 Fisheries Statistics: Capture Production,* Vol. 88/1. Food and Agriculture Organization of the United Nations, Rome.

FAO (2001b). *1999 Fisheries Statistics Aquaculture Production,* Vol. 88/2. Food and Agriculture Organization of the United Nations, Rome.

Gladue, R. (1991). Heterotrophic microalgae production: Potential for application to aquaculture feeds. In: *Rotifer and Microalgae Culture Systems* (Ed. by W. Fulks & K. L. Main), pp. 275–86. Proceedings of a US–Asia Workshop, The Oceanic Institute, Hawaii.

Hanisak, M. D. (1998). Seaweed cultivation: global trends. *World Aquaculture* **29**, 18–21.

Jimenez del Rio, M., Ramazanov, Z & Garcia-Reina, G. (1996). *Ulva rigida* (Ulvales, Chlorophyta) tank culture as biofilters for dissolved inorganic nitrogen from fishpond effluents. *Hydrobiologia*, **326/327**, 61–6.

Kito, H. & Kawamura, Y. (1999). The cultivation of *Porphyra* (nori) in Japan. *World Aquaculture*, **30**, 35–9.

Krom, M. D. & Van Rijm, J. (1989). Water quality processes in fish culture systems: processes, problem and possible solutions. In: *Aquaculture, A Biotechnology in Progress* (Ed. by N. DePauw, E. Jaspers, H. Ackefors & N. Wikens), pp. 1091–111. European Aquaculture Society, Berdene.

Martinez, L. A. & Buschmann, A. H. (1996). Agar yield and quality of *Gracilaria chilensis* (Gigartinales, Rhodophyta) in tank culture using fish effluent. *Hydrobiologia*, **326/327**, 341–5.

Metting, B. (1988). Micro-algae in agriculture. In: *Micro-algal Biotechnology* (Ed. by M. A. Borowitzka & L. J. Borowitzka), pp. 288–304. Cambridge University Press, Cambridge, UK.

Minocha, S. C. (1999). Genetic engineering of marine macroalgae: current status and future perspectives. *World Aquaculture*, **30**, 29–57.

Neori, A. & Shpigel, M. (1999). Using algae to treat effluents and feed invertebrates in sustainable integrated mariculture. *World Aquaculture*, **30**, 46–51.

Olaizola, M. (2000). Commercial production of astaxanthin from *Haematococcus pluvialis* using 25 000-liter outdoor photobioreactors. *Journal of Applied Phycology,* **12**, 499–506.

Oswald, W. J. (1988a). The role of microalgae in liquid waste treatment and reclamation. In: *Algae and Human Affairs* (Ed. by C. A. Lembi & J. R. Waaland), pp. 255–81. Cambridge University Press, Cambridge, UK.

Oswald, W. J. (1988b). Large-scale algal culture systems (engineering aspects). In: *Micro-Algal Biotechnology* (Ed. by M. A. Borowitzka, & L. J. Borowitzka), pp. 357–94. Cambridge University Press, Cambridge, UK.

Petrell, R. J. & Alie, S. Y. (1996). Integrated cultivation of salmonids and seaweeds in open systems. *Hydrobiologia*, **326/327**, 67–73.

Qin, S. Sun, G., Jiang, P., Zou, L., Wu, Y. & Tseng, C. (1999). Review of genetic engineering of *Laminaria japonica* (Laminariales, Phaeophyta) in China. *Hydrobiologia*, **398/399**, 469–72.

Radmer, R. J. & Parker, B. C. (1994). Commercial applications of algae – opportunities and constraints. *Journal of Applied Phycology,* **6**, 93–8.

Rincones, R. E. & Rubio, J. N. (1999). Introduction and commercial cultivation of the red algae *Eucheuma* in Venezuela for production of phycocolloids. *World Aquaculture*, **30**, 57–61.

Rost, T. L., Barbour, M. G., Thornton, R. M., Weier, T. E. & Stoching, C. R. (1979). *Botany, An Introduction to Plant Biology*, 5th edn. John Wiley & Sons, New York.

Samson, R., & Leduy, A. (1985). Multistage continuous cultivation of blue-green alga *Spirulina maxima* in the flat tank photobioreactors with recycle. *Canadian Journal of Chemical Engineering*, **63**, 105–12.

Shang, Y. C. (1976). Economic aspects of *Gracilaria* culture in Taiwan. *Aquaculture*, **8**, 1–7.

Shpigel, M., Neori, A., Popper, D. M. & Gordin, H. (1993). A proposed model for 'environmentally clean' land-based culture for fish, bivalves and seaweeds. *Aquaculture*, **117**, 115–28.

Silva, H. J. & Cortinas, T. I. (1994). Vertical thin-layer photoreactor for controlled cultivation of cyanobacteria. *World Journal of Microbiology and Biotechnology,* **10**, 145–8.

Toma, T. (1987). Mariculture of seaweeds. In: *Aquaculture in Tropical Areas* (Ed. by S. Shokita, K. Kakazu, A. Tomori & T. Toma). Midori Shobo, Japan. (in Japanese) (pp. 31–55 in English edition prepared by M. Yamaguchi, 1991.)

Tredici, M. R. (1999). Photobioreactors. In: *Encyclopedia of Bioprocess Technology: Fermentation, Biocatalysis and Bioseparation.* (Ed. by C. F. Michael & W. D. Stephen), pp. 395–419. Wiley, New York.

Tseng, C. K. (1981). Commercial cultivation. In: *Botanical Monographs*. Vol. 17. *The Biology of Seaweeds* (Ed. by C. J. Lobban & M. J. Wynne), pp. 680–725. Blackwell Scientific Publications, London.

Tseng, C. K. (1986). *Laminaria* mariculture in China. In: *Case Studies of Seven Commercial Seaweed Resources* (Ed. by M. S. Doty, J. F. Caddy & B. Santelices). FAO Fisheries Technical Paper 281, pp. 239–63.

Wohlgeschaffen, G. D., Rado, D. V. S. & Mann, K. H. (1992). Vat incubator with immersion core illumination – a new, inexpensive setup for mass phytoplankton culture. *Journal of Applied Phycology*, **4**, 25–9.

Zertuche-Gonzalez, J. A., Garcia-Lepe, G., Paceco-Ruiz, I., Chee-Barragan, A. & Gendrop-Funes, V. (1999). A new approach to seaweed cultivation in Mexico. *World Aquaculture*, **30**, 50–66

14

Carps

Sena De Silva

14.1 Introduction

There is no banquet without a fish.

Ancient Chinese proverb

Carps are regarded very differently in different parts of the world. To many people in Western countries, carps are ornamental fish in ornamental ponds and aquaria. In one region, the Murray–Darling River system in south-eastern Australia, feral common carp (*Cyprinus carpio*) are regarded as a pest. Yet carps and related species of the family Cyprinidae are a major source of animal protein for millions of people in many Asian countries. World cyprinid aquaculture production in 1998 was 14 142 298 mt, valued at US$14 281 091 (FAO, 2000). Cultured cyprinid production in 1998 accounted for 30.5% of world aquaculture production and 45.8% of global cultured fish and shellfish production. Furthermore, cultured cyprinid production in 1998 accounted for 78.1% of cultured fish and shellfish production from freshwater environments. As such, cyprinid culture is very important to the world aquaculture industry, outweighing all the other species groups in its contribution to world aquaculture production. The trends in cultured cyprinid production from 1984 to 1997, in comparison with freshwater fish and shellfish production, are shown in Fig. 14.1. It is evident that cyprinids constitute the major group of cultured freshwater fish and shellfish. Moreover, the contribution of cyprinids to world aquaculture production has remained steady throughout the last 15 years, contributing about 30%. Their contributions to total cultured fish and shellfish production (e.g. from marine and freshwater environments), and to total freshwater fish and shellfish production, have ranged from 42% and 48.6%, respectively, in 1987 to 73.0% and 79.8%, respectively, in 1997. The slight decrease in cyprinid contribution in the early 1990s could be partially due to an upsurge in the culture of tilapia in freshwater and in mariculture (Chapter 17).

The family Cyprinidae is a typically freshwater group of fish with a very wide distribution; its members are collectively referred to as carps, barbels and minnows. Carps occur naturally in North America, Africa and Eurasia, but are absent from South America, Australasia and Madagascar. Distinguishing features of the group are the presence of pharyngeal teeth in one to three rows, with not more than eight teeth in any one row, lips that are usually thin and an upper jaw that is usually bordered only by premaxillae. There are about 1600 species in the family Cyprinidae, making it the largest family of fish. The taxonomy of the family is correspondingly

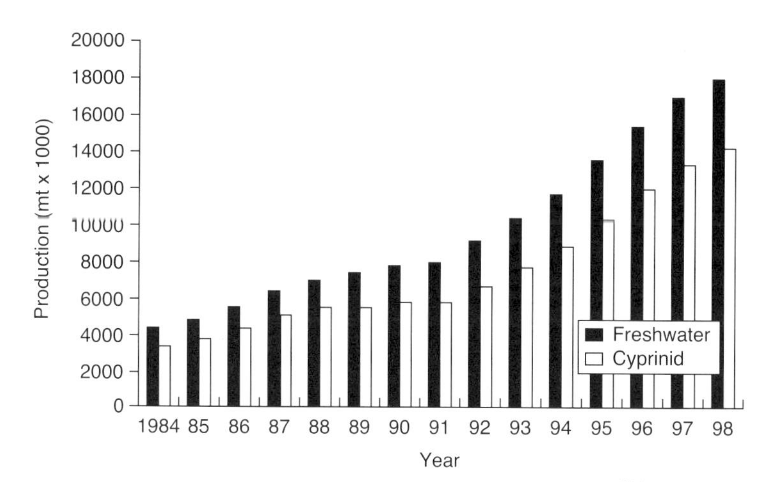

Fig. 14.1 Yearly world production (in thousands of mt) of cultured freshwater fish and shellfish, and cyprinids, 1984–97 (based on data from FAO, 2000).

complex, with about 11 subfamilies and 275 genera. The greatest diversity of the group occurs in Asia.

Despite the large number of species in the family Cyprinidae, only a very small proportion are cultured commercially. Altogether, 29 species of carps are cultured globally (FAO, 2000). However, the predominant species cultured are the Chinese and Indian major carps (Fig. 14.2) and the common carp.

Out of the 29 species of cyprinids that are currently cultured, the production of 11 species exceeds 150 000 mt/year, including five species whose production exceeds 1 million mt/year (Table 14.1). There are only six other fish species or species groups whose production exceeds 100 000 mt/year. More importantly, all species referred to in Table 14.1 have shown a significant increase in production since 1984. The importance of cyprinids in aquaculture is further highlighted by the fact that, of all cultured fish species, the production of only five species exceeds 1 million mt (1998 data) and all of these are cyprinids, i.e. silver carp (*Hypophthalmichthys molitrix*), grass carp (*Ctenopharyngodon idella*), common carp (*Cyprinus carpio*), bighead carp (*Aristichthys nobilis*) and crucian carp (*Carassius carassius*), in that order. In view of the culture potential of the species listed in Table 14.1, most of these species have been introduced to many countries, deliberately and/or inadvertently, much beyond the range of their natural distribution, particularly through Eurasia. For example, species such as the grass carp have been introduced into the USA and New Zealand for the purpose of aquatic weed control. The common carp, on the other hand, was spread across the globe with European colonisation in the eighteenth and nineteenth centuries, in a similar fashion to that witnessed for brown and rainbow trout.

Fig. 14.2 Capture of Chinese carp broodstock in Romania (photograph by Professor R. Billard).

This chapter deals primarily with the culture and those aspects related to culture of the Chinese and Indian major carps. The major carps include three of the Chinese species referred to previously (silver, grass and bighead carps) and three Indian carp species: catla (*Catla catla*), rohu (*Labeo rohita*) and mrigal (*Cirrhinus mrigala*). As the name implies, these species grow to a relatively large size in nature (Table 14.2).

14.2 Aspects of biology

Lin & Peter (1991) reviewed the biology of cyprinids. Cultured carp species are all riverine, typically being found in large river systems. However, the food habits of the species differ from each other

Table 14.1 Cyprinid species that at present contribute more than 150 000 mt/year to world aquaculture production and the country or region that produces most (FAO, 1992, 1999)

Species	Country	Production (mt) 1984	1998
Silver carp (*Hypophthalamichthys molitrix*)	China	842 621	3 308 419
Grass carp (*Ctenopharyngodon idella*)	China	288 423	2 894 017
Common carp (*Cyprinus carpio*)	Eurasia	590 500	2 465 283
Bighead carp (*Aristichthys nobilis*)	China	365 110	1 584 289
Crucian carp (*Carassius carassius*)	Eurasia	58 993	1 036 164
Rohu (*Labeo rohita*)	India	NA	754 677
Catla (*Catla catla*)	India	NA	628 757
Mrigal (*Cirrhinus mrigala*)	India	NA	560 556
Chinese bream (*Parabramis pekinensis*)	China	90 600	449 282
Black carp (*Mylopharyngodon piceus*)	China	36 200	153 633
Mud carp (*Cirrhinus molitorella*)	China	36 382	150 084

NA, data not available.

Table 14.2 Total length (TL) and weight for four species of major carps in natural waters

Age (years)	Silver carp*		Bighead carp*		Rohu†		Mrigal‡	
	TL (cm)	Weight (g)	TL (cm)	Weight (g)	TL (cm)	Weight (g)	TL (cm)	Weight (g)
2	50.0	1803	63.0	3250	50.0	NA	51.1	1512
3	57.6	4650	74.6	10 700	65.0	NA	67.0	3618
4	60.3	5340	75.1	10 900	74.0	NA	79.9	6324
5	63.0	6400	77.8	11 800	80.0	NA	85.8	8030
10					96.0	NA	95.8	11 930

*Chang *et al.* (1983).
†Khan & Jhingran (1975).
‡Jhingran & Khan (1979).
NA, data not available.

(Table 14.3). Food habits also differ between stages of the life cycle of a given species, e.g. the grass carp is zooplanktivorous in the fry and early fingerling stages, and transforms into a macrophyte feeder later on.

Carps are very fecund, and most of the cultured major carp species attain sexual maturity in their third year. In the wild, they spawn once per year, generally with the onset of monsoonal floods. It is generally accepted that the interaction of a large number of factors associated with flooding is responsible for bringing about ovulation and spawning of Indian major carps under natural conditions. The spawning of Indian major carps may be synchronised with the phase of the moon during the floods. In the case of Chinese major carps, it is believed that temperature and photoperiod provide the primary cues for maturation. Chinese and Indian major carps have not been known to spawn naturally in lake waters, nor under captive conditions without hypophysation (section 14.3). All the major carps are single spawners, in that during any one spawning season the female sheds all her mature oocytes within a very short period.

Fertilised eggs of major carps are buoyant, as opposed to the adhesive eggs of common carp and those of most of the smaller cyprinids. The fertilised eggs hatch in 72–96 h and the major carp larvae feed first on microalgae.

Table 14.3 A summary of the predominant feeding habits of Chinese and Indian major carps

Species	Feeding habit(s)
Silver carp	Zoo- and phytoplankton filter feeder; prefers phytoplankton; surface feeder
Grass carp	Omnivorous; prefers higher aquatic plants and submerged grasses
Bighead carp	Predominantly zooplankton filter feeder
Chinese bream	Macrophyte feeder
Black carp	Snails, aquatic insects and crustaceans
Mrigal	Omnivorous, preferring to feed on detritus; predominantly a bottom browser
Rohu	Omnivorous planktophage; predominantly a column feeder
Catla	Plankton feeder; prefers zooplankton and surface feeder

All the major carps grow to about 1 m in length, and generally the Chinese major carps grow to a larger size than the Indian major carps. For example, it is not uncommon for silver and bighead carp to grow up to 1.5 m in length and exceed 10–20 kg in weight. Under culture conditions, however, all except broodstock are harvested in their second or third year, often at a weight approaching 1 kg.

14.3 Artificial propagation

As for most cultured fish species, the single most important breakthrough in carp culture was the development of techniques for artificial propagation of the major species. Before this, carp culture depended on the availability of natural seed for stocking. Specialised fisheries developed in the flood plains of major rivers in mainland China and India to collect the natural seed.

The traditional and most commonly used technique of induced spawning in carps is injection of either:

- crude extract of common carp pituitary gland (or those of other mature fish species that are phylogenetically close to carp)
- partially purified HCG (human chorionic gonadotropin).

Use of hormones and/or analogues for inducing spawning is referred to as hypophysation.

Details of the hormonal control of reproduction in fish and artificial spawning induction are given in Chapter 6 (section 6.2.1). The use of various techniques for spawning induction of Indian and Chinese major carp was discussed by Jhingran & Pullin (1985).

Initially, high efficiency of ovulation was achieved using HCG either alone or in combination with carp pituitary extracts. This treatment was superseded by the use of gonadotropin-releasing hormone analogue (GnRHa) to stimulate reproduction. But GnRHa alone is not entirely effective and has to be accompanied by administration of a dopamine receptor blocker. A better understanding of neuroendocrine regulation of gonadotropin secretion has led to the development of new effective techniques for induced ovulation and spawning of cultured fish species.

Peter *et al.* (1986) dealt with aspects of neuroendocrine regulation of gonadotropin secretion. Basically, modern techniques of inducing ovulation and spawning use a combination of drugs, one of which blocks the inhibiting action of dopamine within the neurohormonal systems (the 'Linpe method'; section 6.2.1). Details of various effective combinations of GnRHa [LHRHa (luteinising hormone-releasing hormone analogue) or sGnRHa (salmon GnRHa)] and a dopamine antagonist (e.g. domperiodone) for induced ovulation and spawning of Chinese major carps are summarised in Table 14.4. The Linpe method is known to be more effective in many ways, ensuring:

- a high rate of ovulation
- consistency between broods
- complete ovulation
- that the time lag between injection and ovulation is short and predictable.

Ovulation and spawning by the Linpe method do

Table 14.4 A summary of the Linpe method of ovulation and spawning of cultured carp in China

	Treatment				
Species	Temperature (°C)	Domperidone (mg/kg)	LHRHa (μg/kg)	sGnRHa (μg/kg)	Time to ovulation (h)
Silver carp	20–30	5	20	–	8–12
		5	–	10	8–12
Mud carp	22–28	5	10	–	6
Grass carp	18–30	5	10	–	8–12
Bighead carp	20–30	5	50	–	6–8
Black carp	20–30	3	10	–	6–8
		7	15	–	6–8

Modified after Lin & Peter (1991) with permission from Kluwer Academic Publishers.

not influence subsequent reproduction cycles of the same brood fish. The Linpe method uses synthetic drugs that are cheaper and more stable and, because only one injection is needed, the brood fish are stressed to a much lesser extent.

Spawning induction of carps in India has been undertaken using commercially available kits (marketed under the trade name 'Ovaparim'), which utilise the Linpe method. Bruzuska (1999) compared the efficacy of LHRHa [(Des-Gly[10])D-Ala[6], LH-RH-ethylamide] and pimozide at 15 μg and 5 mg/kg body weight, respectively, injected simultaneously, vs. carp pituitary suspension on the spawning of grass carp and silver carp. The silver carp responded better than grass carp to both forms of treatment. However, it was stressed that the LHRHa treatment involved less handling of the fish.

Another important advance in induced spawning of the major carps has been the development of the ability to spawn brood fish twice in a calendar year. A second spawning is now achieved for most major carps and is commonly practised. This development has enabled farmers to maintain fewer brood fish and has enhanced seed availability almost year round. This has almost completely eliminated the need for dependence on natural seed.

After hormonal treatment, broodstock are put into spawning ponds in a ratio of three males to two females (Fig. 14.3). They usually spawn at dawn following a second injection of HCG or pituitary extract and are removed from the spawning ponds after spawning. The floating eggs are moved from the spawning pond, by movement of the water through the tank, into a collection box. Alternatively, the brood fish may be stripped and a dry fertilisation performed. The process of stripping gametes from brood fish and dry fertilisation is described in detail in Chapter 15 (section 15.3.3 and Fig. 15.3). This method is becoming increasingly common in major carp culture.

Fig. 14.3 A 'spawning pond' (in Sri Lanka), used for acclimatising selected broodstock before and during hypophysation.

Specially designed spawning tanks are now commonly used for fry production of Chinese carps. They are usually circular or elliptical cement tanks about 1.2–1.5 m deep, containing 50–60 m^3 of water. The tank bottom usually slopes towards the centre, where an outlet leads to an egg collection chamber. Incoming water is directed to create a circular flow within the pool at a rate of 200–400 L/s.

Fertilised eggs are transferred into incubation tanks or hatching pools which are circular,

~3.5–4.0 m in diameter and around 1 m deep. Water flow is maintained at approximately 0.2–0.3 m/s. Eggs are usually incubated at a density of around 700 000–800 000 eggs/m^3. Under these conditions, a hatching rate of about 80% is achieved. In China, 150 000–200 000 eggs are incubated in 150-L clay jars or in funnel-type incubators with vertical water movement. After 4–5 days, when the larvae have resorbed the yolk sac, they are removed to nursery ponds.

Glass hatchery jars are also commonly used for hatching both Indian and Chinese major carp eggs. The jars are generally ~13 cm in diameter and 60 cm long, with conical bases. Each jar is supplied with water up through its conical base to create vertical water movement (Fig. 14.4). The basic concept of all hatchery designs for major carps is to provide a water current of sufficient strength to maintain the eggs in the water column and to remove metabolic waste products.

Fig. 14.4 Fertilised carp eggs being placed into incubation jars in a hatchery in Poland (photograph by Professor R. Billard).

14.4 Nutrient requirements

Of the commonly cultured cyprinid species, the nutrient requirements of common carp are best known (see review by Satoh, 1991). This is to be expected because it was one of the earliest species to be cultured and examined experimentally. The nutrient requirements of most Chinese and Indian major carps are incompletely documented. This is not unexpected, as most cyprinids are cultured extensively or semi-intensively, and are rarely fed commercial feeds that have been compounded and formulated. Because of this, basic nutritional research on the members of the group has lagged behind that of other cultured fish species.

The protein, amino acid and carbohydrate requirements of common carp are known (NRC, 1993). The essential amino acid requirements of the Indian major carp, catla, are also known (Ravi & Devaraj, 1991). Also of some relevance is a study by Ding (1991), who determined the relative proportion of essential amino acids in grass carp muscle and compared it with those of three other cultured species (Table 14.5).

The dietary fatty acid requirements of cultured cyprinids are not well known. Indeed, apart from the early work on common carp (Takeuchi & Watanabe, 1977), when it was demonstrated that this species requires equal amounts of dietary linoleic acid (18:2*n*-6) and linolenic acid (18:3*n*-3), there has not been any study on the fatty acid requirements of carps. It is plausible that this requirement is true for all carps, and conforms to the basic notion that freshwater fish require the two base fatty acids (section 8.9.7). They have the capability to elongate and desaturate these to longer-chain polyunsaturated fatty acids, such as eicosapentaenoic acid (20:5*n*-3), docosahexaenoic acid (22:6*n*-3) and arachidonic acid (20:4*n*-6), among others.

The dietary protein requirement of the major carps, particularly rohu, has been fairly intensively investigated. Most of these investigations have been carried out with fry and fingerling stages, and there is considerable variation in the results of different

Table 14.5 The known essential amino acid requirement of cultured carps. The dietary protein level, where relevant, is given in parentheses

	Common carp*		Catla†	Grass carp‡	Rohu¶
Amino acid	a (38.5%)	b	a (40.0%)	c	a (40.0%)
Arginine	4.3	1.6	4.8	11.1	5.8
Histidine	2.1	0.8	2.5	4.7	2.3
Isoleucine	2.5	0.9	2.4	8.8	NA
Leucine	3.3	1.3	3.7	16.5	NA
Lysine	5.7	2.2	6.2	15.5	NA
Methionine	3.1	1.2	3.6	4.5	2.9
Phenylalanine	6.5	2.5	3.7	7.8	4.0
Threonine	3.9	1.5	5.0	8.5	4.3
Tryptophan	0.8	0.3	1.0	2.4	NA
Valine	3.6	1.4	3.6	11.5	NA

a, Expressed as a percentage of dietary protein (38.5%); b, percentage of dry diet; c, percentage of the total essential amino acids.
*NRC (1993).
†Ravi & Devaraj (1991).
‡Ding (1991).
¶Murthy & Varghese (1995, 1996a,b, 1998).
NA, data not available.

investigators. This is mostly a result of variations in experimental protocol. Based on available information, De Silva & Gunasekera (1991) estimated that the dietary protein level that results in maximum growth of major carps is 45% and the economically optimal dietary protein content is 31% (section 8.7.3 and Fig. 8.2).

The information available on the requirements of other nutrients is very scant and all the nutrient requirements for any one species of major carps are not known. As pointed out above, this probably reflects the mainly semi-intensive mode of culture of the Chinese and Indian major carps. As such, the need for development of complete diets does not arise. It therefore follows that there is no dire need to determine all the nutrient requirements for these species.

14.5 Culture

14.5.1 Larval rearing

Eggs of cultured cyprinids hatch out in 2–3 days at temperatures between 23°C and 27°C, and the yolk sac continues to provide nourishment for a further 3 days or so, at which time the larvae then require an exogenous food supply. It is desirable, if not essential, to expose the larvae to an external food source before yolk sac resorption is complete. This entails removing the larvae from the hatchery jars and introducing them into a fry-rearing facility. The young hatchlings, which mostly move vertically, tend to change to a horizontal movement, which indicates their readiness to ingest food particles.

Larval rearing in carp culture has two distinct phases:

- rearing postlarvae to the fry stage: usually carried out in nursery ponds, or in *hapas* (fine mesh enclosures) suspended in ponds or channels
- rearing fry to fingerling stage: most effectively done in well-fertilised rearing ponds.

It is, however, not uncommon to combine the above two stages in one pond.

Preparation of nursery and rearing ponds often involves sowing a short-term crop of a leguminous plant (e.g. beans, clover), and ploughing and levelling the pond once the crop has grown to 6–10 cm. This process is known as 'green manuring' and is believed to enhance pond productivity. In most instances, unwanted organisms in the ponds are eradicated using a biodegradable toxicant or quicklime. This procedure is carried out at least a fortnight before stocking. Commonly used toxicants are:

- Derris powder (4–20 mg/L)
- oilcake of the plant *Bassia latifola* (200–250 mg/L) (= mahua oilcake)
- tea-seed cake (525–674 kg/ha)
- quicklime (900–1050 kg/ha).

The next stage is to prepare the pond with a view to ensuring a good production of small zooplankters, such as rotifers, which provide a food source for growing larvae. Ponds are often treated with either organic or inorganic fertilisers. The quantity of manure to be used is related to the toxicant used earlier. For example, if mahua oilcake was used, a dose of dry cow manure at the rate of 5000 kg/ha 2 weeks before stocking and a similar dose 1 week after hatching are desirable. However, with toxicants that have no fertiliser value, doses of 10 000–15 000 kg/ha initially, and 5000 kg/ha later, are desirable. These manuring doses are sufficient for 1.5×10^6 larvae/ha.

Fertilisation with a mix of organic and inorganic fertilisers is undesirable, as more often than not it results in harmful planktonic blooms. Despite the early preparation of the ponds, undesirable predatory insects such as water spiders and water skaters may colonise the ponds. Therefore, the ponds have to be regularly treated for insect control, particularly before stocking. Jhingran & Pullin (1988) recommend any one of the four treatments given below:

- spraying an emulsion of 56 kg of mustard or coconut oil and 18 kg of washing soap per hectare
- spraying an emulsion of 56 kg of mustard oil and 560 mL of Teepol (detergent) per hectare
- A 0.01 ppm dose of pure gamma isomer of benzene hexachloride dissolved in ethyl alcohol
- application of 0.25–3.0 ppm organophosphate such as fumadol, sumithion or diptrex.

The prepared ponds are stocked when it is certain that a substantial zooplankton population (particularly small zooplankters such as protozoans and rotifers) is established. Abrupt changes in quality and temperature between hatchery water and nursery water are avoided when stocking. Stocking is best done in the evening, which hopefully gives the larvae sufficient time to acclimatise themselves before any possible predation.

The stocking rate depends on the proposed management practice and, if the following conditions are met, a stocking rate of 10×10^6/ha can be used:

- continued and repeated fertilisation to produce and maintain good plankton production
- supplemental feeding
- facilities to remedy oxygen deficiencies that may occur.

Postlarvae of carps are voracious grazers. Supplemental feeding and manuring, when carried out concurrently, result in better survival and growth. The commonly used supplemental feeds in carp culture are rice bran, and oilcakes of peanut (= groundnut), coconut and mustard in India. These are used in China, together with soybean milk and meal, and egg yolk paste. It is very rare for carp culture to be based on complete formulated feeds, except in the case of common carp culture in some countries. Feeds are often dispersed as a crude mix, in either dry (e.g. meals or pellets) or moist form. A summary of commonly used feeds and feeding schedules for fry of Chinese carps is shown in Table 14.6. However, it should be noted that in the case of common carp, which is cultured fairly intensively on a small scale in ponds in Israel and in cages in China, formulated diets (pellet feeds) are used.

The fish in rearing ponds are harvested with sieve nets when they reach 4–6 cm. Periodical harvesting may be carried out to avoid overcrowding. Often, rearing postlarvae to fingerling size is undertaken in larger, earthen ponds and polyculture is practised. Stocking densities of fingerlings in polyculture range from 100/m^2 to 2500/m^2 with a mean of about 800/m^2 (Table 14.7). Size at harvesting ranges from 7 to 20 cm, with grass and black carps tending to be the largest.

14.5.2 Grow-out

A number of distinctive general features are

Table 14.6 Some feeds and feeding rates used for Chinese carp fry and fingerling rearing

Country	Species	Pond area (m^2)	Depth (m)	Fish length (mm)	Age (days)	SD (per m^2)	Feed	Feeding rate
China	Big head, grass carp	–	0.5–1.0	Up to 20	Up to 30	100	Egg yolk paste or soybean milk + peanut cake after 10 days	1 egg/2500–7500 fry/day or milk from 300–500 g beans/50 000 fry/day
	All species	–	0.5–1.0	23.1	Up to 30	–	Soybean meal	45 kg/5000 fry/month
Hong Kong	All species	1000	0.8	8–30 (3 mg to 1 g)	Up to 25–30	150	Soybean milk and peanut cake meal	100 kg soybean milk or 200 kg peanut cake meal/month
	All species	1400	1.0	31 (1.5 g)	30–70	35	Peanut cake, rice bran or soybean cake	Start at 1.5 kg/day, build up to 5 kg/day

Reproduced from Jhingran & Pullin (1988) with permission from ICLARM.

Table 14.7 Examples of stocking rates and size at harvesting of carp fingerling in polyculture

Stocking density (× 100/m^2)						Size at harvesting (cm)					
gc	bc	sc	bhc	cc	wf	gc	bc	sc	bhc	cc	wf
4–6	–	20–25	–	–	–	13–15	–	8–10	–	–	–
10–25	–	4–5	–	–	–	8–13	–	11–13	–	–	–
2–4	–	–	8–12	–	–	16–20	–	–	11–13	–	–
4–6	–	–	15–20	–	–	13–15	–	–	8–10	–	–
10–25	–	–	4–5	–	–	8–13	–	–	11–13	–	–
–	5–6	4–5	–		–	–	13–15	13	–	–	–
–	5–6	–	4–5	–	–	–	13–15	–	11–13	–	–
1	–	–	4	5–6	–	13	–	–	13	8–10	–
1	–	–	0.08–0.1	–	15–20	16–20	–	–	0.25–0.5 kg	–	7

gc, grass carp; bc, black carp; sc, silver carp; bhc, bighead carp; cc, common carp; wf, Wuchang fish (*Parabramis pekinensis*).

recognisable in major carp culture practices. These practices:

- tend to be semi-intensive
- almost always use polyculture
- may be integrated with other forms of farming
- are carried out in earthen ponds or in pens, but rarely in cages and raceways.

However, the exception could be common carp. In certain instances they are intensively cultured in ponds and cages (Fig. 14.5).

Polyculture is thought to have originated in China, when various combinations of seven basic species with widely different food habits were cultured together, i.e. black carp (eat snails), grass carp (eat coarse vegetable matter), silver carp (eat phytoplankton), bighead carp (eat zooplankton and omnivorous) and mud carp (bottom scavenger). A typical species combination used in a polyculture practice with an approximate indication of the niches occupied by each of the species is shown in Chapter 2 (Fig. 2.7). The number of species used, and the ratio of each species, varies from region to region. Polyculture, apart from ensuring that most of the food resources in the system are efficiently utilised, offers other advantages, including higher yields, reduced susceptibility to disease and better growth rates of some species than in monoculture.

Polyculture maximises the synergistic fish–fish and fish–environment relationships and minimises antagonistic relationships. Milstein (1992) dealt in detail with synergistic interactions in polyculture systems and pointed out that synergism among fish species can be explained on the basis of two inter-related processes: increase in food resources and improvement of environmental conditions. However, antagonistic interactions occur between incompatible species and when stocking rates are unbalanced.

A concerted experimental effort occurred in India to develop suitable polyculture systems using both Chinese and Indian carps. This concept was termed composite fish culture. The basic species combination in Indian composite polyculture were catla, rohu, mrigal, silver carp, bighead carp and common carp. When stocked at a density of 5000/ha (120–250 kg/ha) the yield was nearly 9 mt/ha/year when fertilised and provided with simple supplemental feed, such as a mixture of rice bran and oilcake.

Despite various experimental findings in both China and India, farming activities tend to depend on the indigenous species of each country. This trend is primarily influenced by the preferred consumer acceptance of indigenous species.

In China and India, where carp culture is the predominant form of fish culture, two or three species of either Chinese or Indian carps are polycultured. In these polyculture systems, the dominant species in China is silver carp, and in India it is rohu. The actual culture practices vary from region to region and country to country. The primary variables are the size at stocking, stocking density, fertilisation regimes, and the nature and quantity of supplementary feeds.

Fig. 14.5 Intensive culture of common carp in cages in a reservoir in China. Each cage is equipped with an automatic feeder.

In Chinese systems, for example, fingerlings are generally stocked at a size of about 15–20 g (>10–12 cm total length). In Andhra Pradesh, India, rohu are stocked in grow-out ponds when they are more than 2 years old (between 80 and 100 g), as it is believed that this is when they approach their maximum growth rate.

Over the last two decades, the main carp culture countries have developed their own culture protocols. This is best exemplified in Andhra Pradesh, a coastal state in south-east India. In this state, only two species of Indian major carps, catla and rohu (the latter being the dominant species at 80%), are cultured. The ponds often exceed 1 ha and are stocked at a density of 5000 fish/ha with 6- to 12-month-old (100–150 g) juveniles. Ponds are generally fertilised with poultry manure and inorganic fertiliser, and are provided with supplementary feed, often consisting of simple mixtures of rice bran (de-oiled) and oilcake (mustard, peanut). The feed mixture is suspended in perforated polythene bags from bamboo poles at a number of locations in the ponds (20–25 poles/ha), from which the fish soon learn to feed. In this region, production averages about 8000 kg/ha with a range of 5300–14 620 kg/ha. Fish are harvested when they are over 1.5 kg.

In China, on the other hand, polyculture is practised with Chinese carps in conjunction with common carp. There are also significant differences in regional culture practices within China. The most important difference is the dominant species in polyculture systems. For example, grass carp is the main species used in southern China, whereas silver and bighead carp dominate in central China. Li (1987) pointed out that in the recent years there has been a trend towards increasing the proportion of grazing fish such as grass carp, black carp and blunt snout bream (*Megalobrama amblycephala*), and a corresponding decrease in filtering fish such as bighead and silver carp.

More importantly, carp pond culture in China and, more often than not, in other East Asian nations, is integrated with other forms of animal husbandry, such as swine, duck and poultry. Li (1987) considered the energy structure and efficiency of a typical Chinese fish farm integrated with animal husbandry, consisting of a comprehensively managed system of aquaculture and agriculture. The structure of such a system is schematically depicted in Fig. 14.6. Accordingly, the crop subsystem produced feed for the animal husbandry subsystem (e.g. a 2.0-ha dairy farm with 182 cattle and 2.0-ha brooder chicken farm) and ingredients for the preparation of pelleted fish feed. The aquaculture subsystem consisted of 18.3 ha of grow-out ponds, 5.5 ha of nursery ponds and 3.5 ha of bivalve (pearl) cultivation ponds. Manure from the dairy farm was used as fertiliser for the fish ponds, and chicken manure as a component of the pelleted feed. The net yield from this system was 142 070 kg (168 151–26 081 kg), equal to 7760 kg/ha/year (Table 14.8). More importantly, such systems are energy saving and sustainable over the long term.

The wide range of culture practices adopted in carp culture, within and between regions, makes it almost impossible to assess the potential yield from any one practice. For example, Chen *et al.* (1995), in their study of 1013 ponds in 101 farms in eight provinces in China, classified the farms into three productivity classes based on net yield. The mean yield in low-, medium- and high-productivity farms was 3321, 4981 and 7958 kg/ha/year respectively. A detailed study on carp farming systems in Andhra Pradesh, India, was conducted by Veerina *et al.* (1993), who observed two operational systems that utilised two species (rohu and catla) or three species (rohu, catla and mrigal). There was a very wide range in the yields among farms, and the gross yield ranged from 1730 to 14 830 kg/ha/year. However, there was no evidence to indicate that two- or three-species systems performed better than the other (Table 14.9).

14.5.3 Food and feeding

Natural food availability

In view of the fact that the great bulk of carp culture is semi-intensive, increasing the availability of natural food types in the culture systems plays a crucial role in enhancing yields. As pointed out earlier, the commonest method used for increasing natural food supply in carp ponds is through the application of common fertilisers and/or organic manures. The commonly used organic manures include cow dung, poultry litter and pig dung, and the inorganic fertilisers are superphosphate and ammonium sulphate.

Table 14.8 Initial number and weight of fingerlings stocked, and gross yield, from 38 ponds at Nanhui Fish Farm, China

Species	Number	Weight (kg)	Gross yield (kg)
Silver carp	116 870	8766	77 858
Bighead carp	20 620	1684	15 631
Crucian carp	68 570	1719	14 551
Tilapia	25 600	26	227
Grass carp	15 950	4167	16 101
Common carp	55 560	4167	21 355
Bream	56 570	3337	9540
Others		250	12 888
Total		26 081	168 151

Reproduced from Li (1987) with permission from Elsevier Science.

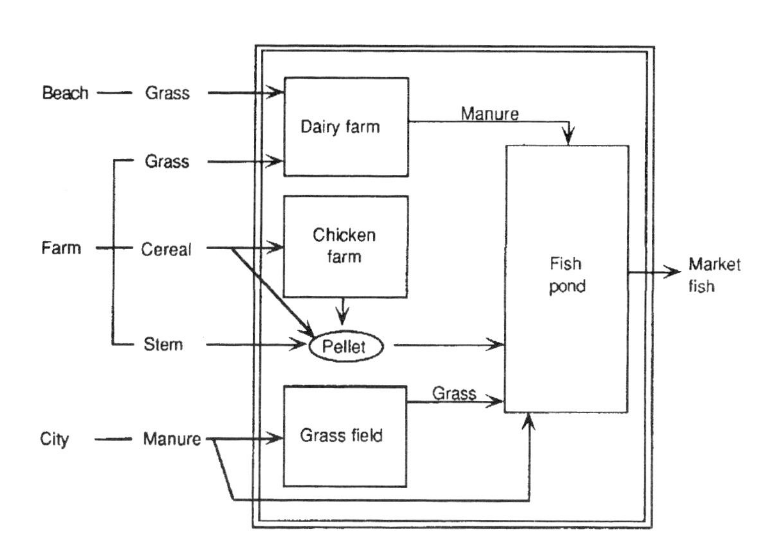

Fig. 14.6 A schematic representation of the interactions in an integrated carp pond culture and terrestrial animal rearing system. Reprinted from Li (1987) with permission from Elsevier Science.

There have been many studies conducted on the effects of fertilisation and manuring in carp polyculture practices. However, apart from the fact that such practices result in increased algal production, it is impossible and impractical to make a set of general conclusions from the findings. This is for a number of reasons, the foremost among these being:

- differences in the fertilisation and/or manuring regimes used in the different studies
- different stocking densities and species combinations used
- climatic differences
- management differences utilised (such as the time of fertilisation and/or manuring, length of trials, etc.).

Hepher (1988) presented the most comprehensive review of the influences of manuring and fertilisation of fish ponds on the production of herbivorous and omnivorous fish. Further studies have specifically considered manuring and fertilisation of ponds with polycultured carp. For example, the efficacy of fertilisers, such as rockphosphate, has been considered in polyculture systems with common carp and Indian major carps. There was also manipulation of the ratio of surface to bottom grazing species (Sahu & Jana, 1996). A ratio of 1:3 between surface feeders (catla, silver carp and rohu) and bottom grazers (mrigal, common carp and *Puntius sarana*) yielded the best results in ponds that received 100 kg/ha of rockphosphate, twice monthly. Sahu & Jana (1996)

Table 14.9 Mean size (kg) at harvest and production (kg/ha/year) of carp farming systems in Andhra Pradesh, India. The ranges in harvesting sizes and production are given in parentheses

	Two-species systems		Three-species systems	
	Size (kg)	Yield (kg/ha/year)	Size (kg)	Yield (kg/ha/year)
Gross yield	–	5900 (1730–11 112)	–	5857 (1730–14 830)
Rohu	1.8 (0.8–3.2)	4109 (1580–7710)	1.7 (1.0–2.8)	3690 (1850–9270)
Catla	2.7 (0.9–5.4)	1794 (440–4690)	2.4 (1.2–5.1)	1612 (400–3710)
Mrigal	–	–	1.7 (1.0–3.0)	550 (60–1660)
Net yield		5378 (1360–10 380)	–	5296 (1360–14 620)
Rohu	–	3835 (1380–7120)	–	3352 (1580–9120)
Catla	–	1544 (440–4690)	–	1520 (370–3420)
Mrigal	–	–	–	468 (60–1410)

Modified from Veerina *et al.* (1993).

suggested that the bioturbation activity due to dominance of bottom grazers induced a greater release of phosphorus from bottom sediment, which resulted in higher phytoplankton production. This was utilised by surface feeders and resulted in better growth and production of the latter.

Carp culture may also benefit from the provision of suitable substrates for periphyton growth in rearing ponds.

Supplementary feeds

The supplementary feeds used in carp culture are diverse. Most supplementary feeds are simple mixes of agricultural by-products, which are readily available at a relatively low cost. The most common of these are brans of rice and wheat, often mixed with cakes or meals of various oilseeds such as mustard, canola and soybean (Table 14.10). Most farmers tend to use some sort of supplementary feed, which could be either a single ingredient or a mixture of two or three, at most. The quantity of feed as well as the amount of individual ingredients used in the feed mixes could vary greatly (Table 14.10). Obviously, this is an area that needs further research, which in the long run could reduce feed use and thereby increase profitability. In addition, it could also lead to improved water quality in the ponds and cleaner pond effluent. This trend is indicative of a potential constraint to expansion of culture activities due to increasing competing demands for the same food ingredients from other animal husbandry activities and from other users (Veerina *et al.*, 1999). In Chinese polyculture, a wide range of ingredient mixes is also used as feed, the type and quantity often being dictated by availability and price. Soybean meal, sesame cake, silkworm pupae powder and canola meal are more commonly used in major carp farming systems in China.

The 'nutritional dynamics' of semi-intensive systems are complex and incompletely understood. For example, there is very little information on the extent to which the supplementary feeds provide direct nutrition to the target species. A schematic representation of the qualitative changes that are expected to occur in a semi-intensive culture pond and the potential utility of supplementary feeding strategies is provided in Fig. 14.7.

On the other hand, because the great bulk of carp farming depends on supplementary feeds, when compared with other cultured fish species, there have been relatively few attempts to develop pellet feeds, except for common carp. Barlow (2000) predicted that, in 2010, 50% of carp culture (predominantly common carp) will be based on commercial feeds and will require 675 000 mt of fish meal. That is, the fish meal requirement for carp culture will amount to approximately 24% of the total fish meal requirement for the aquaculture industry in 2010.

Despite the estimated fish meal requirement for carp culture, and that required for culture of carnivorous fish species, research into the potential of alternative protein sources in feeds for fish is relatively limited (section 8.7.3). Substituting fish meal

Table 14.10 Supplementary feed combinations used in Andhra Pradesh (India) carp farming systems

Ingredient	% farmers	Input (kg/ha/year)	
		Range	Mean (SD)
DOB	8	5000–33 000	2030 (7091)
DOB + CSM	4	10 000 40 000	25 280 (8187)
DOB + DOC	3	10 000–30 000	24 830 (5548)
DOB + PNC	75	5000–50 000	27 650 (6849)
DOB + RB	1	22 000– 22 260	–
DOB + SM	1	34 000–59 000	46 800 (2293)
RB + DOB	8	20 000–39 000	29 790 (7148)
No feed	1	0	0

DOB, de-oiled rice bran; CSM, cotton seed meal; PNC, peanut cake; RB, rice bran; SM, soybean meal.
The data are based on a survey of 189 farms, and modified after Veerina *et al.* (1993).

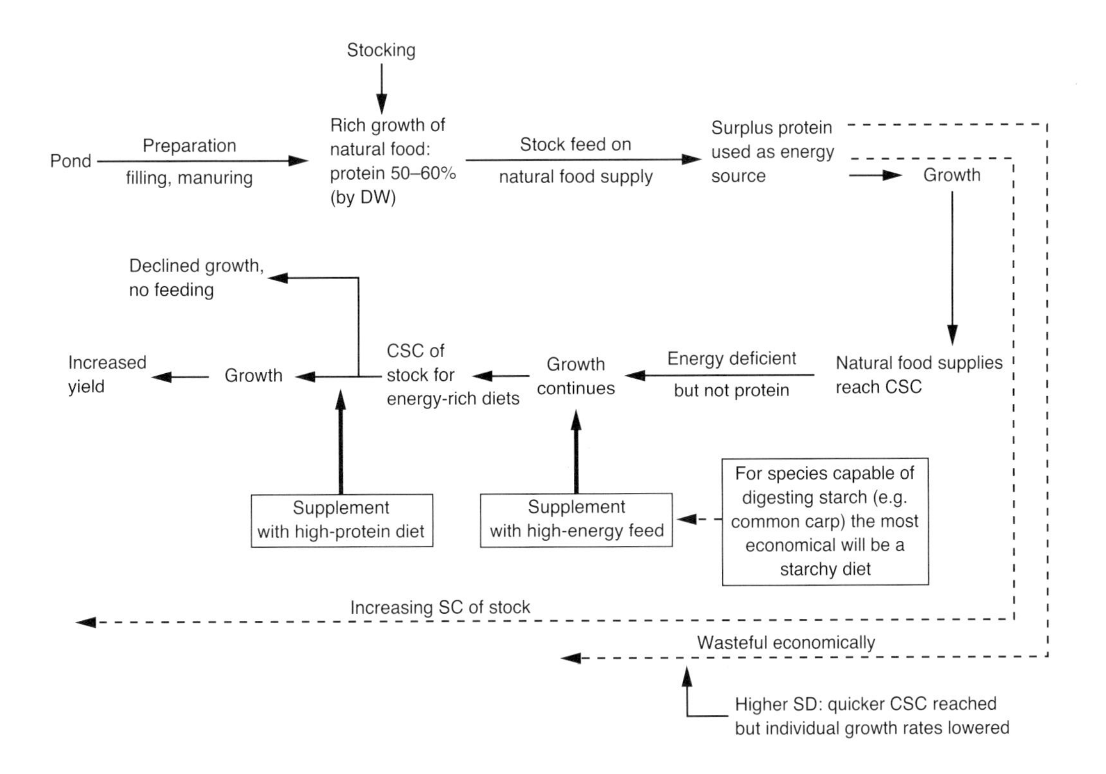

Fig. 14.7 A schematic representation of the qualitative changes that are expected to occur in a semi-intensive culture pond, and the potential utility of supplementary feed strategies. CSC, critical standing crop; SC, standing crop; SD, stocking density. Reprinted from De Silva (1995) with permission from the FAO.

protein in feeds with plant protein has so far been unsuccessful.

Feeding

In general, feed management is relatively poor in carp culture. The main reason for this is that most practices depend on supplementary feeds, which are simple mixes of agricultural by-products. Perhaps the only exception is the feeding practice adopted in common carp farming in eastern Europe and in some parts of Asia. A wide range of feeding practices are used by carp farmers, from simple hand broadcasting to tying perforated bags containing food to sticks in the pond and allowing the stock to obtain feed through the perforations. It is almost impossible to assess the food lost in the latter practice, and indeed it is possible that the food has a greater effect as a fertiliser than providing direct nutrition to the fish.

Mixed feeding schedules, i.e. use of different feeds at different feeding times, have been found to have beneficial effects on growth and food cost reduction. However, there is little information on its use in farm practices.

14.5.4 Harvesting

Pond-cultured carps are harvested when individual fish reach a weight of about 1–1.5 kg unless they are destined for a specialised market. Harvesting in most instances is done by seining, and is labour intensive. Major carp culture ponds are rarely flow-through and are rarely completely drainable either. These factors together with the large size of ponds make harvesting by seining almost an imperative. Major carps are generally marketed fresh. It is not uncommon to retain portions of the catch, live, in temporary net pens, to minimise market saturation within a short period of time (Fig. 14.8). This practice is followed in most rural areas when the distances to population centres are high and the total production in an area does not justify transportation to such centres.

Fig. 14.8 Temporary net pens used for retaining the harvest for a few days to minimise an oversupply of fish in rural China.

14.6 Diseases

It has to be conceded that, in the past decade or so, disease-related research in aquaculture has focused mainly on shrimp culture, because of its economic importance (Chapters 10 and 19). However, carps, like most fish, are susceptible to infectious (viral, bacterial, fungal and parasitic) as well as non-infectious diseases such as carcinomas, injuries, swim-bladder disorders, etc. (section 10.5.3). There are also environmentally induced disorders caused by excessively high or low pH, nitrogenous wastes, heavy metal pollution, etc. A very comprehensive monograph on carp diseases, diagnostic procedures and treatments was provided by Hoole *et al.* (2001), and readers are recommended to refer to this rather extensive work.

14.7 Genetic improvement

As described in Chapter 7, apart from the genetic improvement of salmonid stocks and, more recently, the Nile tilapia (*Oreochromis niloticus*) and channel catfish, in general genetic improvement of cultured fish and shellfish species has lagged far behind that of farmed terrestrial animals.

Some of the cyprinid species, in particular common carp and crucian carp (*Carassius auratus*), have been domesticated for centuries. This domestication and consequent selection, have resulted in the development of a number of strains, generally selected for aesthetic purposes rather than to augment food fish production. Examples of this are discussed in section 7.2.1 in relation to selection for scale type in common carp.

The lack of genetic selection of cultured stocks for favourable traits, and inbreeding of hatchery stocks, have come to light since the early 1990s. Eknath & Doyle (1990), in their study on 18 hatcheries in the south-west state of Karnataka, India, showed that the rate of inbreeding of the Indian major carps ranged from 2% to 17% per year. They suggested ways of reducing further accumulation of inbreeding in cultured broodstock of these species. The genetic resources of Indian major carps were reviewed recently (Reddy, 1999). This synthesis included biochemical and molecular genetic studies, hybridisation among Indian major carp and common carp, chromosomal manipulations and selective breeding. A concerted effort is currently being made on rohu and catla from founder populations in different river systems. A cooperative research programme for selection in five cultured carp species in Asia is ongoing, and a synopsis of the traits to be selected and the methods of selection adopted is given in Table 14.11 (Anon, 2000).

14.8 Economic viability

As carp culture is the largest and most widespread practice of animal aquaculture in the world, and since carps contribute most of the world's inland aquaculture production, it is implicit that carp culture is economically viable. If this was not the case, it would not have developed and continued to develop to the extent it has, despite the relatively low market price of the product. On the other hand, there have been very few studies on the economics of carp culture practices. In a broad study encompassing 101 farms in eight provinces of China, Chen *et al.* (1995) recognised that the stocking model utilised was related to the economic status in the region. Generally, farms in poorer areas tended to stock predominantly filter feeders, as opposed to 'feeding fish', such as grass carp and black carp (*Mylophryngodon piceus*) in the

Table 14.11 A summary of the breeding programmes that are ongoing for cultured carp species in Asia, under the auspices of the International Network of Genetics in Aquaculture

Country	Species	Traits	Selection method
Bangladesh	Silver barb (*Barbodes gonionotus*)	Size at harvest	Mass
	Catla	Size at harvest	To be decided
China	Common carp	Size at harvest	Mass
	Blunt snout bream	Size 1 (3 cm)	Mass
	(*Megalobrama amblycephala*)	Size 2 (10–15 cm)	Two separate lines
		Size at harvest (> 300 g)	
India	Rohu	Size at harvest	Combined mass
	Common carp	Size at harvest/late maturity	To be decided
Indonesia	Common carp	Size at harvest	To be decided
Thailand	Silver barb	Size at harvest	Mass
	Common carp		Mass
Vietnam	Common carp	Size at harvest (early survival)	Family
	Silver barb	Size at harvest	Mass

Reproduced from Anon. (2000) with permission from ICLARM.

Table 14.12 Balance sheet of the integrated carp farming systems in China. All weights and monetary values are in kg/ha/year and yuan/ha/year. SDs are given in parentheses. Low, medium and high refer to the classification based on level of productivity (from Chen *et al.*, 1995)

	Low	Medium	High	Overall
Fish yields				
Grass carp	999	1468	2444	1865
Filter feeders	2125	3032	3716	3168
Black carp	3	53	513	282
Omnivores	638	1119	2766	1865
Gross yield	3765	5672	9439	7180
Stocking weight	444	690	1481	1045
Net yield	3321 (1498)	4982 (2006)	7958 (3113)	6135 (3250)
Revenues				
Grass carp	3514	4245	8721	6416
Filter feeders	4500	4761	6626	5677
Black carp	6	134	2808	1508
Omnivores	2057	2926	7901	5325
Animal production	101	591	30	175
Other income	16	221	121	118
Total income	10 194 (5852)	12 878 (5142)	26 207 (13 117)	19 219 (12 638)
Costs				
Fish stocking	1597	1903	4768	3332
Animal stocking	27	179	11	53
Feed	1889	2112	6388	4303
Fertiliser	359	520	298	364
Fuel	218	195	516	369
Miscellaneous	664	1468	2154	1630
Total (non-labour)	4754 (4397)	6377 (2881)	14135 (7903)	10051 (7432)
Net income	5440 (4397)	6501 (3769)	12 072 (7586)	9168 (6905)
Labour	1252 (697)	3851 (3021)	4004 (3887)	3289 (3390)
Profit	4188 (4482)	2650 (3871)	8068 (7323)	5879 (6488)

more well-to-do areas. A balance sheet of the production systems (Table 14.12) indicates that the mean profitability in low-, medium- and high-production systems was 4188, 2650 and 8068 yuan/ha/year respectively. Interestingly, the medium-production systems were less profitable than the low-production systems, and this possibly reflects relatively high labour inputs in the former.

Veerina *et al.* (1999) conducted an analysis of production factors of the carp farming systems in Andhra Pradesh, India, with a view to assessing the inter-relationships among farm inputs and fish yield. The analysis showed a strong relationship of fish yield to stocking density, buffalo manure and poultry manure inputs, and the amount of peanut oilcake (a supplementary feed) used. In addition, other supplementary feeds and additives, such as salt and mineral mix, were found to have a positive influence on fish yield. Management inputs, such as the rate of water exchange, frequency of medication and the mean weight at stocking, also had a positive influence on fish yield. Older ponds were more productive, whereas sociological variables such as the age and literacy level of the farmer had less influence on fish productivity.

14.9 Culture-based fisheries

Culture-based fisheries fall within the realm of aquaculture, with regard to the intervention in the life cycle and the defined ownership, either singly or collectively, of the stock. Culture-based fisheries are considered to have very high potential, particularly in the light of increasing demand for primary resources such as land and water. They are often recognised as an important avenue for increasing inland fish production, particularly in developing countries (De Silva, 2001). Culture-based fisheries use existing water resources, which may be natural oxbow lakes or water resources created for other purposes, such as reservoirs and farm dams. As such, they compete minimally, if at all, with other uses. Moreover, culture-based fisheries often do not involve external inputs, such as feed, and therefore are more environmentally 'friendly' than traditional aquaculture practices.

Apart from all of the above, culture-based fisheries have significant relevance for carp culture, because the majority of such fisheries are based on Chinese and Indian major carps, occasionally augmented with tilapia and other minor species.

The culture-based fisheries in China are the most developed in the world. This fishery is confined to small- and medium-sized reservoirs throughout the country. It is estimated that the fishery yielded 815 100 mt in 1995 from 1 515 600 ha of reservoirs (Ceng & Zhang, 1998), approximating 538 kg/ha/year. Song (1999), who considered the culture-based fishery in reservoirs as a separate entity, estimated the total production from this activity in 1997 as 1 165 075 mt (from a total area of 1 567 971 ha), approximating a production of 743 kg/ha/year. It was also pointed out by Song (1999) that the culture-based fishery recorded an annual growth rate of 52% between 1979 and 1997.

The culture-based fishery practices in China are based primarily on the major Chinese carps, grass carp, bighead carp, silver carp and common carp. In addition, species such as the Wuchang fish (*Megalobrama amblycephala*), black bream (*Megalobrama terminalis*) and mud carp (*Cirrhina molitorella*) may be used. In southern Asia, culture-based fisheries are primarily based on a combination of Chinese and Indian carps, the latter being predominant. De Silva (2001) recognised a number of features that are responsible for the immense success of the culture-based fishery practices in China. Foremost of these are:

- consideration (at the planning stage of reservoir construction) of those factors that enhance fishery production
- relatively large and uniform size of fish at stocking
- minimising the number of escapees
- a staggered but complete harvesting of the stock
- adoption of marketing strategies that minimise an oversupply of fish within a narrow time-frame.

14.10 Conclusions

This chapter has highlighted the importance of carp culture, some of the key features of carp culture practices and the potential of carp culture as a food source. Living standards are generally increasing throughout the world, and it is often suggested that the demand for carp species will gradually decline as a result. However, production trends do not support such a contention. On the other hand, in the light of increasing environmental concerns related to the culture of carnivorous fish species, it may be that forms of carp culture could become even more important in the coming years. One of the major constraints to further intensification of carp culture is increasing competition for supplementary food sources, which are primarily agricultural by-products. As such, a concerted effort may be required to develop suitable feeds and to develop more prudent strategies of feed management. Apart from yield increases through intensification and better pond culture practices, popularisation and development of culture-based fisheries using carp species appears to have the greatest potential to augment inland fish production. This is particularly important to developing countries, thereby making available a good-quality source of animal protein, at an affordable price, to the poorer sectors of the community.

References

Anon (2000). Progress of the carp genetic project. *Naga*, **23** (3), 35–7.

Barlow, S. (2000). Fishmeal and fish oil: sustainable feed ingredients for aquafeeds. *The Advocate*, **3**, 85–9.

Bruzuska, E. (1999). Artificial spawning of herbivorous fish: use of an LHRH-a to induce ovulation in grass carp *Ctenophyryngodon idella* (Valenciennes) and silver carp *Hypophthalamichthys molitrix* (Valenciennes). *Aquaculture Research*, **30**, 849–56.

Ceng, D. & Zhang, D. (1998). Development and status of the aquaculture industry in the Peoples' Republic of China. *World Aquaculture*, **29**, 52–6.

Chang, W. Y. B., Diana, J. S. & Chuapoehuyz, W. (1983). Workshop Report on Agency for International Development, 19–29 April, 1983, p. 30. *Strengthening of South-east Asian Aquaculture*, Manila, Philippines.

Chen, H., Hu, B. & Charles A. T. (1995). Chinese integrated farming: a comparative bioeconomic analysis. *Aquaculture Research*, **26**, 81–94.

De Silva, S. S. (1995). Supplementary feeding in semi-intensive aquaculture. In: *Farm-made Aquafeeds* (Ed. by M. B. New, A. G. J. Tacon & I. Csavas), pp. 24–60. FAO-RAPA/AADCP, Bangkok, Thailand.

De Silva, S. S. (2001). Reservoir fisheries: broad strategies for enhancing yields. In: *Reservoir and Culture-based Fisheries: Biology and Management* (Ed. by S. S. De Silva), pp. 7–15. ACIAR Proceedings No. 98, ACIAR, Canberra, Australia.

De Silva, S. S. & Gunasekera, R. M. (1991). An evaluation of the growth of Indian and Chinese major carps in relation to dietary protein content. *Aquaculture*, **92**, 237–41.

Ding, L. (1991). Grass carp, *Ctenopharyngodon idella*. In: *Handbook of Nutrient Requirements of Fish* (Ed. by R. P. Wilson), pp. 89–96. CRC Press, FL.

Eknath, A. E. & Doyle, R. W. (1990). Effective population size and rate of interbreeding in aquaculture in Indian major carps. *Aquaculture*, **85**, 293–305.

FAO (1992). Aquaculture *Production Statistics. 1984–1990.* FAO Fisheries Circular No. 815 Revision 4. FAO, Rome.

FAO (1999). *Aquaculture Production Statistics. 1988–1997.* FAO Fisheries Circular No. 815 Revision 11. FAO, Rome.

FAO (2000). *FAO Yearbook. Fishery Statistics. Aquaculture Production. 1998. 86/2.* FAO, Rome.

Hepher, B. (1988). *Nutrition of Pond Fishes.* Cambridge University Press, Cambridge, UK.

Hoole, D., Bucke, D., Burgess, P. & Wellby, I. (2001). *Diseases of Carp and Other Cyprinid Fishes*, p. 264. Fishing News Books, Oxford.

Jhingran, V. G. & Khan, H. A. (1979). *Synopsis of Biological Data on Catla, Catla catla (Hamilton 1822).* FAO Fisheries Synopsis 32, Revision 7. FAO, Rome.

Jhingran, V. G. & Pullin, R. S. V. (1985). *A Hatchery Manual for the Common, Chinese and Indian Major Carps.* ICLARM Studies and Reviews. No. 11. Asian Development Bank & International Centre for Living Aquatic Resources Management, Manila.

Khan, H. A. & Jhingran, V. G. (1975). *Synopsis of Biological Data on Rohu, Labeo rohita (Hamilton 1822).* FAO Fisheries Synopsis, 11. FAO, Rome.

Li, S. (1987). Energy structure and efficiency of a typical Chinese integrated fish farm. *Aquaculture*, **65**, 105–18.

Lin, H. R. & Peter, R. E. (1991). Aquaculture. In: *Cyprinid Fishes. Systematics, Biology and Exploitation* (Ed. by I. J. Wingfield & J.S. Nelson), pp. 590–622. Chapman & Hall, London.

Milstein, A. (1992). Ecological aspects of fish species interactions in polyculture ponds. *Hydrobiologia*, **231**, 177–86.

Murthy, H. S. & Varghese, T. J. (1995). Arginine and histidine requirements of the Indian major carp, *Labeo rohita* (Hamilton). *Aquaculture Nutrition*, **1**, 235–39.

Murthy, H. S. & Varghese, T. J. (1996a). Quantitative dietary requirement of threonine for the growth of the Indian major carp, *Labeo rohita* (Hamilton). *Journal of Aquaculture in the Tropics*, **11**, 1–7.

Murthy, H. S. & Varghese, T. J. (1996b). Dietary requirement of the Indian major carp, *Labeo rohita* (Hamilton), for total aromatic amino acids. *Bamidgeh*, **48**, 78–83.

Murthy, H. S. & Varghese, T. J. (1998). Total sulphur amino acid requirement of the Indian major carp, *Labeo rohita* (Hamilton). *Aquaculture Nutrition*, **4**, 61–5.

NRC (1993). *Nutrient Requirements of Fish.* National Research Council. National Academy Press, Washington DC.

Peter, R. E., Chang, J. P., Nahamiak, C. S., Omelijaniuk, R. J., Solkolwska, M., Shih, S. H. & Billard, R. (1986). Interactions of catecholamines and GnRH in regulation of gonadotropin secretion in teleost fish. *Recent Progress in Hormone Research*, **42**, 513–48.

Ravi, J. & Devaraj, K. V. (1991). Quantitative essential amino acid requirements for growth of catla, *Catla catla* (Hamilton). *Aquaculture*, **96**, 281–91.

Reddy, P. V. G. K. (1999) Genetic resources of Indian major carp. *FAO Fisheries Technical Paper 387.* FAO, Rome.

Sahu, S.N. & Jana, B. B. (1996). Manipulation of stocking ratios between surface- and bottom-grazing fishes as a strategy to increase the fertiliser value of rockphosphate in a carp polyculture system. *Aquaculture Research*, **27**, 931–36.

Satoh, S. (1991). Common carp, *Cyprinus carpio.* In: *Handbook of Nutrient Requirements of Fish* (Ed. by R. P. Wilson), pp. 55–67. CRC Press, Boston.

Song, Z. (1999). *Rural Aquaculture in China.* RAPA Publication 1999/22. RAPA, FAO, Bangkok, Thailand.

Takeuchi, T. & Watanabe, T. (1977). Requirement of carp for essential fatty acids. *Bulletin of the Japanese Society for Scientific Fisheries*, **43**, 541–51.

Veerina, S. S., Nandeesha, M. C., De Silva, S. S. & Ahmed, M. (1999). An analysis of production factors in carp farming in Andhra Pradesh, India. *Aquaculture Research*, **30**, 805–14.

15
Salmonids

John Purser and Nigel Forteath*

15.1 Introduction

> The salmons having spent their appointed time and done their natural duty in the fresh waters, they then haste to the sea before winter, both the melter and spawner …
>
> Izaak Walton, *The Compleat Angler*

Worldwide, there are numerous species and strains of salmon, trout and charr within the family Salmonidae. Many of the important commercial species belong to the genera *Salmo*, *Oncorhynchus* and *Salvelinus*, examples of which are listed in Table 15.1. They are all either anadromous (ascending rivers from the sea to breed) or undertake their entire life cycle in freshwater. No species is solely marine. Perhaps the most important species are Atlantic salmon, rainbow trout, chinook salmon, coho salmon and brown trout. More recently, there has been a significant research focus on Arctic charr. The Pacific salmon, within the group *Oncorhynchus*, constitute most of the worldwide salmon production, but this is largely through fisheries activities rather than aquaculture (Table 15.2). Production figures for various salmonid species in 1999 are shown in Tables 15.3 and 15.4 (FAO, 2001a).

Freshwater facilities usually rear salmonids to the fingerling or smolt stage for transfer to marine farms (seafarms), although some specialise in plate size (300–400 g) or larger fish reared solely in freshwater. In contrast, seafarms grow out small fish to market size (2+ kg). Cage culture has proved the most successful strategy for marine production. The alternative pump-ashore facilities, whereby water is pumped into land-based culture facilities, are largely uneconomical for the production of market-size fish, but they are used to support a few broodstock programmes. In recent years, large recirculating shore-based marine systems have been proposed as a way of controlling the culture environment and avoiding threats such as predators, algal blooms and pollution experienced in cage culture. Similarly, cages based on solid bag net configurations offer protection from threats experienced by conventional mesh nets, but they rely on close monitoring and low-head pumping to circulate water. The solid bag excludes algae, predators and jellyfish, and allows control of lighting using covers and temperature by altering the depth of seawater intake.

In this chapter, general husbandry practices and techniques used for rearing salmonids are outlined. It is acknowledged that many variations of these

*The authors dedicate this chapter to Dr Lindsay Laird (1949–2001), who worked closely with the salmon industry in Europe and made a significant contribution to salmonid research, teaching and training.

Table 15.1 Salmonid species that are important to fisheries and aquaculture

Common name	Scientific name
Atlantic salmon*	*Salmo salar*
Brown trout* (sea trout)	*Salmo trutta*
Arctic charr	*Salvelinus alpinus*
Brook trout/charr	*Salvelinus fontinalis*
Rainbow trout* (ocean, steelhead trout)	*Oncorhynchus mykiss*
Pink salmon	*Oncorhynchus gorbuscha*
Chum salmon (dog, sake)	*Oncorhynchus keta*
Sockeye salmon (red, beni-masu)	*Oncorhynchus nerka*
Coho salmon* (silver, gin-maru)	*Oncorhynchus kisutch*
Chinook salmon*(king, quinnat, spring, masunosuka)	*Oncorhynchus tshawytscha*

*Species that are commercially cultured in marine, brackish or freshwater environments.

Table 15.2 Global salmon production from aquaculture and fisheries in 1999 showing percentages of totals in brackets (FAO, 2001a,b)

Species	Aquaculture	Wild harvest	Total (mt)
Atlantic salmon	797 560 (99.5%)	4287 (0.5%)	801 847
Pacific salmon	104 283 (11.3%)	822 460 (88.7%)	926 743

Table 15.3 Worldwide production figures from all sources of various salmonid species in 1999 (FAO, 2001a,b)

Species	Production (mt)
Atlantic salmon	797 560
Sea trout (*S. trutta*)	5892
Rainbow trout	418 654
Trout: general	50 432
Chinook salmon	14 708
Coho salmon	89 575
Brook trout	683
Arctic charr	990
Charr: general	869
Total	1 391 615

exist and that strategies differ between farms, countries and species cultured. The basic life cycle and production techniques for salmonids as a group are, however, very similar. The life history characteristics of three commercial species are used to illustrate these similarities and differences. This section is followed by a discussion of freshwater and marine husbandry techniques for salmonids using Atlantic salmon production in Tasmania, Australia, as the prime example.

Approximately 60 countries are involved in culturing salmonids. The major producing countries of Atlantic salmon and rainbow trout, the two species that together constitute almost 90% of salmonid production, are shown in Table 15.4. International salmonid production and general culture techniques are covered comprehensively in the excellent texts of Piper *et al.* (1982), Laird & Needham (1988), Heen *et al.* (1993), Sedgwick (1995), Pennell & Barton (1996) and Willoughby (1999).

15.2 Biology

15.2.1 Distribution of species

The natural range of the Salmonidae covers freshwater lakes, rivers and marine coastal regions of the northern hemisphere temperate zone. However, their range has been greatly extended as a result of a demand from recreational fisheries and aquaculture. The major species and their native regions are:

Table 15.4 Countries producing more than 20 000 mt/year of Atlantic salmon and rainbow trout in 1999 (FAO, 2001a)

	Atlantic salmon (mt)	Rainbow trout (mt)
Canada	61 990	
Chile	103 242	50 414
Denmark		39 729
Faeroe Isles	37 473	
France		44 498
Germany		25 027
Italy		44 000
Norway	418 758	45 276
Spain		30 000
UK	126 686	
USA		27 344
Total world production	797 560	418 654

- rainbow trout (*Oncorhynchus mykiss*) – Pacific coast of North America
- Atlantic salmon (*Salmo salar*) – Atlantic Ocean from the latitude of northern Spain to Iceland
- Pacific salmon (*Oncorhynchus* species) – Pacific basin between 35°N and 65°N
- brown trout (*Salmo trutta*) – Eurasia and North Africa
- brook trout (*Salvelinus fontinalis*) – north-eastern North America
- Arctic charr (*Salvelinus alpinus*) – highlands of central Europe, northern Britain, Scandinavia and the Arctic basin.

15.2.2 Life cycles of three farmed species

A generalised Atlantic salmon production cycle is illustrated in Fig. 15.1, which shows an outline of the life history and terminology used to describe the various stages.

Atlantic salmon (Salmo salar)

The life cycle of Atlantic salmon is typically anadromous, although some landlocked strains exist. Mature fish return to freshwater during the 12 months before spawning, which usually takes place in late autumn and early winter.

(1) The female uses body movement to make a 'redd' (a hollow) in silt-free gravel. The gravel varies between 2.5 and 15 cm in diameter and the redd is positioned to receive a relatively rapid water flow of at least 0.5 m/s. Male fish (Fig. 15.2) fertilise the eggs as they are released by the female into the redd, by releasing milt into the water column. Ova are buried to a depth of 30–40 cm. Female salmon produce about 1500 eggs per kilogram body weight.

(2) On hatching, the alevins remain within the redd, utilising their substantial yolk reserves and showing a negative phototaxis. Their movement within the redd is probably negligible, and little, if any, energy is spent maintaining their position. This permits maximum growth on yolk reserves, an important factor with respect to subsequent survival.

(3) Emergence of fry from the gravel occurs in the spring about 800 degree-days post fertilisation (e.g. 80 days at 10°C or 100 days at 8°C). The fry are carnivorous and territorial. As they grow, fry develop 'thumb-print' lateral markings that are characteristic of the parr stage.

(4) The parr stage may last for up to 8 years in cold environments, such as northern Norway, or as little as 1 year towards the southern part of the natural range of the species. During the parr stage, some males may mature early as '0+ fish' ('precocious' male parr) and may play an important role in fertilising ova from adult females.

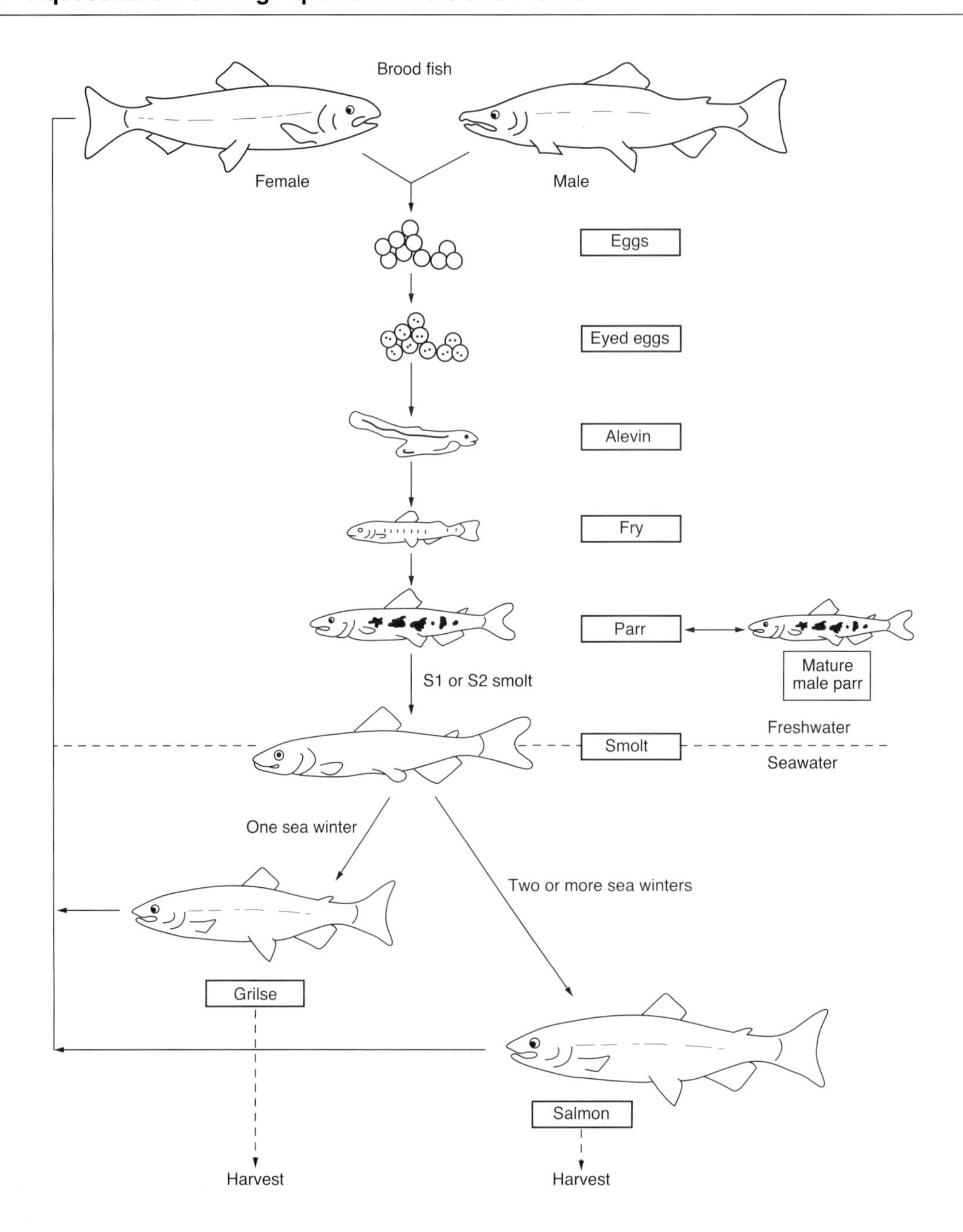

Fig. 15.1 Production cycle of Atlantic salmon, *Salmo salar*, showing specific life stages. Most anadromous salmonid species display this generalised cycle although the duration of stages is species and temperature specific (not to scale). Reproduced from Forteath *et al.* (1996) with permission from the Fishing Industry Board of Tasmania.

(5) The parr metamorphose into the smolt stage during smoltification or the parr–smolt transformation, a process that is initiated several months before the seaward migration, usually at an age of 1–3 years. The smolts, which may weigh up to 60 g or more, enter the sea in spring and summer, depending on latitude. The smolts enter the sea in spring at the more southerly distributions

Fig. 15.2 Male Atlantic salmon brood fish in spawning condition.

in the northern hemisphere. Once at sea, the fish migrate to rich feeding grounds, although feeding occurs throughout the migration. A number of these feeding grounds have been discovered off western Greenland and the Faeroe Isles.

(6) Some fish will return to their natal streams and rivers after 18 months. These fish are known as 'grilse' (one sea winter fish) and weigh between 1 and 7 kg. The remaining fish may spend two to five winters at sea, returning as 'salmon' of 4–30 kg in weight.

(7) Once spawned, the salmon is known as a 'kelt'. Females may spawn two or three times, but few males live to spawn more than once. Survival after spawning is generally low, but river conditions may play an important role in subsequent survival.

Rainbow trout (Oncorhynchus mykiss)

(1) Rainbow trout spawn between autumn and spring, with females producing about 2000 mature ova per kilogram body weight. These hatch some 4–7 weeks post fertilisation, depending on water temperature, e.g. eggs will hatch within 30 days at 10°C. Considerable egg mortality may occur at higher temperatures around 15°C. Eggs are deposited in gravel (redds), usually in riffles.

(2) Optimum temperatures for growth after hatching are between 12°C and 18°C. Males mature as 2-year-olds, although precocious males mature at 1 year of age, and the majority of females mature as 3-year-olds. Age at maturity depends on growth rates. Two-year-old mature females are quite common in some strains and environments.

(3) Spawning survival is low, with few individuals spawning more than once or twice in the wild. Some strains are anadromous ('steelheads') and undertake extensive oceanic migrations, whereas others remain in freshwater throughout their lives. It is not known if anadromy is a truly genetic adaptation or simply opportunistic behaviour that is based on a genetic predisposition (Gall & Crandell, 1992).

Chinook salmon (Oncorhynchus tshawytscha)

This species has several common names, i.e. chinook, king, spring, tyee, masunosuka, chavycha and quinnat. The species is the largest of the Pacific salmon and the least abundant. The name 'spring' salmon arises from the early season run of the fish entering the Columbia River in Canada. In fact, adult fish enter rivers over most of the spring, summer and autumn periods but distinct runs are often noted.

(1) Adult fish are inhibited from spawning until the temperature falls below 12°C. An average female weighs 10 kg, although 40- to 50-kg specimens have been recorded. They lay between 3000 and 12 000 ova, depending on size of the female. Eggs are deposited in a redd.

(2) Chinook salmon are able to migrate to sea as small fingerlings, 4 months after hatching, but some stocks remain in freshwater for 1 or 2 years as parr. Also, there are precocious males that spawn and then migrate to sea.

(3) The chinook salmon is a coastal species and remains at sea for 1–7 years; the duration of the sea-going life stage is determined by geographical distribution. Female fish of the northernmost rivers remain at sea for the longest period. However, many males return prematurely as precocious adults after only one winter at sea. These are called 'jacks' (grilse). They are rarely more than 30 cm in length and weigh less than 1.5 kg. All adults die after spawning.

15.3 Freshwater farming

15.3.1 Establishing a freshwater salmonid farm

Establishing a salmonid farm requires a host of factors to be taken into account. Some of the principal considerations are:

- water flow and volume availability
- water quality
- existence of other farms upstream
- land use upstream
- access and location of utilities
- topography of site
- soil permeability
- slope of land
- legislation concerning water rights.

Temperature is of great importance as a controlling factor in aquaculture, and it is essential that accurate water temperature profiles be obtained throughout the year. If possible, data over several years should be available before a site is decided upon. The same must be said of water quality and flow rates. Water can be pumped or gravity fed from a bore, river, spring or lake. Water quality criteria for salmonid fish have received considerable attention, and typical data are given in Table 15.5.

Trout farmers may concentrate specifically on the production and sale of ova, fry or table-fish, but many design farms for all three. The types of buildings, culture system design and feeding equipment are dependent on the objective. The majority of salmon farms produce smolt for growing-out on seafarms, although some facilities produce large grow-out fish, eyed eggs or caviar.

15.3.2 The hatchery

General considerations

General considerations regarding hatchery design and operation are described in Chapter 2 (section 2.7). Water quality control in the hatchery is of paramount importance. Heavy metals, suspended solids and poor levels of dissolved oxygen (DO) must be avoided at all times. Cadmium, copper, lead and zinc are all toxic to salmonid eggs, although the toxicity of these ions can be reduced by increasing the hardness of the water. In Europe and Scandinavia, adding lime to the hatchery water supply has been used to counteract pollutants such as acid rain and associated release of heavy metals (Needham, 1988). Waters in contact with mine tailings, etc., are avoided at all costs.

Suspended solids accumulate rapidly on incubating salmonid eggs, effectively smothering them, and some degree of filtration or settlement is usually necessary for the influent water. There is an increasing demand for oxygen as eggs develop. The influent water must be at saturation and effluent water must contain at least 6 mg/L DO. It is not unusual to experience a significant increase in biological oxygen demand (BOD) during hatch, resulting from the release of vitelline fluids and discarded egg shells. The hatching troughs generally become difficult to clean during hatch and while maintaining alevins. However, every effort must be made to avoid disturbing newly hatched fish. It is not uncommon to warm hatchery water to enhance the developmental rate of fertilised ova, alevins and first feeding fry.

An increase in the level of dissolved gaseous nitrogen to above saturation is dangerous as it causes gas bubble trauma. This is characterised in its later stages by the presence of gas emboli associated particularly with the head and gills. The problem often is caused by:

- air leaks in intake pipes
- air leaks in water pumps, i.e. submersible pumps sucking up air when placed in a shallow sump
- rapid heating of the water (solubility of N_2 decreases as temperature increases).

Water in these circumstances, as well as groundwater (which may be high in dissolved carbon dioxide or nitrogen gas), must be de-gassed before entering the hatchery. Elevated levels of carbon dioxide occur in recirculation systems and must be maintained below 10 ppm through de-gassing processes. Pure oxygen may also be added to the water before delivery to the fish, especially where high stocking densities or water recycle/recirculating conditions are used or where DO levels fall below 80% saturation, thus influencing growth.

Table 15.5 Water composition for salmonid rearing in freshwater

Parameter	Suitable range
Temperature	–0.5°C to 24°C (range), 4–18°C (optimum)
Dissolved oxygen (DO)	Minimum at outflow 6 mg/L, > 70% saturation
Carbon dioxide (CO_2)	Maximum 10 mg/L
Ammonia (NH_3)	< 0.0125 mg/L
Nitrate (NO_3^-)	< 0.1 mg/L (soft water) < 0.2 mg/L (hard water)
NO_2-N	0.03 mg/L (soft water) 0.05 mg/L (hard water)
Nitrogen (gas pressure)	< 110% saturation
Hydrogen sulphide (H_2S)	< 0.002 mg/L
Suspended solids	< 50 mg/L
Chlorine	< 0.03 mg/L
pH	5.5–8.5
Minerals	
Aluminium	0.1 mg/L (low pH or low calcium)
Cadmium	< 0.004 mg/L (< 100 mg/L alkalinity) < 0.003 mg/L (> 100 mg/L alkalinity)
Copper	< 0.006 mg/L (soft water) < 0.3 mg/L (hard water)
Iron (total)	< 0.15 mg/L
Iron (ferrous)	0
Iron (ferric)	< 0.5 mg/L
Zinc	< 0.04 mg/L

Reproduced from Laird & Needham (1988) with permission from Horwood Publishing Ltd.

15.3.3 Broodstock and spawning

Broodstock

Broodstock should ideally be cultured as part of a selective breeding programme based on family lines involving desirable traits such as:

- body shape
- skin and flesh coloration
- disease resistance
- consistency in growth.

More commonly, fish are simply selected from the current production run and conditioned before spawning. Where sea-grown fish are used, they are normally conditioned in freshwater for several months before stripping. Mature fish remaining in seawater will usually die, although some will survive and recondition. Male and female broodstock develop asynchronously under some environmental conditions, and hormone preparations such as 'Ovaprim' are used to synchronise gamete production. During the conditioning period, males and females may be kept separately or together in culture tanks or raceways (some of which are designed as continuous raceways) at low densities and in high water quality. Chemicals released by fish into the water may assist spawning synchrony and so some systems use water recycling through fish tanks rather than flow-through. Elevated holding temperatures in salmonid broodstock tanks can be detrimental to the quality of the developing ova, producing atresia pre-ovulation or poor fertilisation rates of ovulated eggs (Pankhurst *et al.*, 1996).

After stripping, many species can be reconditioned for the next spawning season. The exceptions are old fish or species of Pacific salmon, which may be euthanased before egg removal. Broodstock are valuable fish and must be handled carefully when checking for egg release.

Broodstock trout and salmon in Tasmania are conditioned at both freshwater and marine sites, although all fish undergo final conditioning in freshwater from about March (autumn) each year. Although a selective breeding programme would be desirable and has been proposed, brood selection is currently limited because of the small size of the industry and the relatively narrow genetic range of the stock.

Spawning

Spawning of salmon in Tasmania occurs in May (autumn), whereas trout are spawned from the end of May to July (autumn–winter). Most brood fish are used over one to two seasons at 3–5 years of age. Although hand stripping after anaesthesia (Fig. 15.3) is the most common method, ova may also be expressed by injecting air into the body cavity using a needle (air stripping) and through dissection after euthanasia. About 1200–1500 eggs per kilogram fish at a diameter of 4–7 mm are removed from the female stock of 4–14 kg. Fish are towelled before stripping to minimise water contamination of eggs and premature activation of milt. Sperm cells are inactive in expressed milt, and it is important that activation through contact with water is avoided before egg fertilisation as sperm motility, once activated, occurs over a short duration of only 1–2 min.

Under standard fertilisation techniques, ova and ovarian fluid are stripped into a dry bowl and milt is expressed onto the eggs and mixed with the ova. Variable fertilisation rates can occur using this technique, and some farms remove the ovarian fluid from the ova, wash with bicarbonate solution to minimise the effects of broken eggs and replace the solution with artificial ovarian fluid to maintain consistency between batches before the addition of the milt.

Fig. 15.3 Hand stripping ova into a dry bowl after anaesthesia.

Broken eggs may impact negatively on fertilisation rates; ~3% broken eggs in a batch may reduce the fertilisation rate by as much as 80–90%. Milt and ova are mixed gently by hand and allowed to stand undisturbed for 2–3 min. After fertilisation, eggs are washed gently with water and left undisturbed for 45–60 min to water harden. During this stage, the eggs increase in size and become hard to the touch. While taking up water they become slightly 'sticky' for a short period of time. The water-hardened eggs are laid down in the egg incubators.

15.3.4 Incubator systems

Factors in choice

Choice of incubator depends on the *raison d'être* of the hatchery. Upwellers (vertical cylinders) are preferred when large numbers of ova are being produced for sale in the 'eyed' state. Upwellers require minimal floor space and can hold up to 100 L of eggs. Water is forced up through the eggs (but without moving the eggs) to overflow over the lip at the top into a drain. Eggs may also be hatched in trays and trough systems. Trays such as the Heath system (Fig. 15.4) can be stacked in banks of up to 20 per unit and are space efficient. Water enters at the top of the stack, where it is first forced up through a perforated base supporting about 1 L of eggs spread two layers in thickness. It is then collected and gravity fed to the tray below and so on. Oxygen levels are maintained by water turbulence at the point of entry to each tray.

Troughs containing eggs baskets are often the preferred method of hatching, but these take up a considerable amount of space. Furthermore, sluggish water flows will cause detritus to accumulate on the eggs, and this may reduce oxygen availability. The Californian trough system contains up to seven baskets each holding about 1 L of eggs, i.e. 6000–8000 Atlantic salmon or 10 000–20 000 rainbow trout eggs. The eggs lie on a perforated base and water is forced up through them by a flange on each basket that is flush with the base of the trough.

Fig. 15.4 Vertical tray egg incubation system. Water flows from the top to the bottom of the rack via each tray.

Saprolegnia species fungus is a constant problem in hatcheries: it spreads from dead eggs to viable eggs. It is controlled using zinc-free malachite green at a concentration of 2 mg/L for 1 h. Biweekly treatment is necessary; however, this drug is carcinogenic and must be used with caution. It has been banned for use in some countries, being replaced by alternative treatments such as formalin, H_2O_2 or ozonised water.

Egg picking is possible in Californian troughs, reducing the use of chemicals, but care must be taken to create as little disturbance as possible before the eyed stage, as eggs are susceptible to handling and bright light during this time. Once eyed, picking is quite safe. Mechanical pickers are used to sort healthy eggs from non-viable ones. Eyed eggs may be subjected to physical shock (e.g. trays of eggs are 'banged' on the side of the troughs) in conjunction with sorting, which aids the hatching process, avoiding a prolonged hatch period.

Water flow

Water flow rates through the incubator systems require constant monitoring. Upwellers require considerably less water than trays or troughs. Indeed, only 25 L/min water may be sufficient in an upweller holding up to 100 L of eggs, but this will only be adequate if water quality is particularly good. Most hatcheries use 3–4 L/min water flow per 4–5 L of eggs. Water demand in a seven-basket trough system or a 16-tray incubator is approximately 15–20 L/min.

Temperature control

It is becoming increasingly popular to manipulate water temperature in hatcheries to modify spawning time or accelerate or delay embryonic development in fertilised eggs. Bromage & Cumaranatunga (1988) described a protocol for delaying development of rainbow trout eggs, but temperature manipulation must be avoided for the first 10–12 days. Subjecting the eggs to water temperatures of 2°C or 3°C between days 13 and 70 results in a hatching period of about 80 days. By comparison, eggs will hatch in 30 days at a constant 10°C. Photoperiod or day length manipulations of spawning time have the advantage of reducing the incidence of poor egg quality that is often associated with water temperature fluctuations (Bromage *et al*., 1992). High summer temperatures will cause problems with egg viability in out-of-season spawners, and it will be necessary to cool the water for brood fish. This, however, may be uneconomical.

Egg development in Atlantic salmon can be accelerated by increasing water temperature. A typical regime is 8°C for eggs to eyed stage, and 10°C for eyed eggs to hatch. Hatching will then take place in about 45 days. In contrast, at 3°C, hatching will take place in about 145 days. Great care must be taken to avoid accelerating development too rapidly as this may result in increased egg loss or poor-quality alevins.

Light

As salmonid eggs are buried in gravel in the wild, care is taken to ensure that lighting in the hatchery is limited to a few dim sources or to red lights. UV light is harmful and daylight bulbs are avoided. Consequently, egg trays and troughs are covered and exposed to low light levels only during periods of inspection.

15.3.5 Alevins and first feeding fry

Under natural conditions, salmonid eggs are deposited in gravel and fry do not emerge from the redd until the yolk sac is all but depleted. Alevins are stressed by bright light and frantically seek sanctuary beneath each other. As with eggs, very dim or red lights are used. It is important to ensure that maximum growth is achieved during the alevin stage. First feeding fry are much easier to culture if they are about 0.2 g or more. This weight is best achieved by providing a suitable substrate, permitting alevins to lie quietly for long periods. Various materials have been trialled as substrates, including corrugated sheeting, meshes, artificial turf, plastic grid, gravel and plastic biomedia.

Significant differences in weight are achieved between fish supported in this manner and those forced to lie on a smooth substrate. In addition, artificial substrates may protect the alevins from suffocation, caused by overcrowding (Fig. 15.5), and reduce the incidence of yolk sac constrictions.

Unlike Atlantic salmon, rainbow trout and Pacific salmon fry swim to the water surface to inflate their swim bladder ('swim-up') when ready to commence feeding. Water depth at this time must be shallow and not exceed 10 cm. As a rule of thumb, small quantities of feed should be delivered at frequent intervals once 50% of the fry have 'swum up'. A light with a 25-W red bulb suspended over the trough is recommended. Once the fry are feeding, starter feed must be spread over the water surface in numerous meals at a rate of ~8% body weight/day. Furthermore, feeding over a 20- to 22-h period with 2–4 h of darkness is beneficial as it maximises the time available for food intake and hence growth. The penalty of such an intensive feeding regime may be the deterioration of water quality. Low water velocities do not permit waste to be deposited in the drain at this early feeding stage and gentle brushing of the tank bottom is necessary to remove waste materials. The fry can be damaged very easily; therefore, violent brushing or sudden increases in water flow are avoided. Extreme care is required to prevent disease outbreaks and hypoxia. Cleaning is necessary twice per day to prevent rapid increases in BOD.

Atlantic salmon need to be weaned onto artificial diets before their yolk sacs are absorbed. It is quite difficult to determine the best time to start artificial feeding, but useful indicators are the onset of normal dorsal–ventral swimming attitude (alevins rest on their sides) and reduction of the yolk-sac to about 20% of its original size. Needham (1988) recommended regular sampling of the alevins to determine the developmental stages and to observe food particles in the gut. Moribund fry close to the water surface are often a sign of starvation. Starving fish are obvious, having large heads, emaciated bodies and black body colour. At a water temperature of 12°C, healthy fish will start stacking in the water column by day 10 of feeding.

First feeding may be established in troughs or fry tanks. Fry tanks are favoured for large Atlantic salmon hatcheries. These tanks are usually circular, but rounded-corner square (Rathbun) tanks have been used extensively. Fry on the point of taking

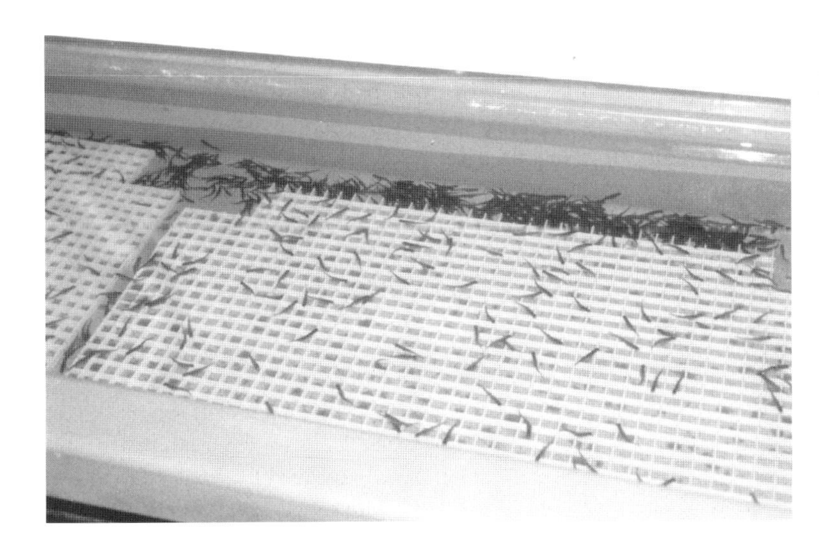

Fig. 15.5 Artificial substrate supports and spaces alevins across the trough floor compared with fish crowding induced by a smooth floor evident to the rear of this trough.

artificial food are carefully immersed in the tank and allowed to swim out of the shallow net or tray used for transfer. Recent designs have incorporated hatching trays and alevin incubation substrate within the alevin/fry tanks to reduce the frequency of handling and transfers.

Water temperatures between 12°C and 15°C are ideal for growth, but difficult to achieve unless artificial heating is available. Groundwater at 12°C may be available in some areas, precluding heating. However, this water may need to be aerated to remove high levels of nitrogen, carbon dioxide or dissolved iron and to improve oxygen levels. The farmer must strive to achieve maximum growth from a very early stage in order to maximise economic production in the grow-out phase.

Water temperature should ideally remain constant during the first 2–3 weeks of feeding. Light levels remain critical and must be < 50 lux at the water surface. A 25-W red bulb works well and ensures no shadows or bright spots. Lights not only serve to permit feeding, but also ensure a relatively even distribution of the fish over the tank bottom. In complete darkness, fry tend to form dense congregations, which can result in damage to fins and opercula.

Stocking levels used in Rathbun tanks are between 15 000 and 20 000 fry per tank. First feeding fry benefit from moderate crowding, which appears to encourage feeding in these early stages. The water flow rate initially is about 25 L/min. Stocks of rainbow trout and Atlantic salmon fry may need to be graded 4–5 weeks after first feeding.

At this point there is a major divergence in farming aims and objectives for the grow-out phase of the two species. The trout farmer will aim to grow fry to market size according to a plan designed to satisfy weekly production demands or to produce fish of 300–800 g for grow-out in seacages. The salmon farmer will endeavour to ensure that as many fish as possible will become smolts. In both cases, husbandry of stocks in the first 2–3 months after feeding is of great importance if subsequent goals are to be achieved.

15.3.6 Culture systems for juveniles

Vessels

Many 'low-cost' trout farms keep their stock in pond systems, which are usually based on the Danish earthen pond system. Construction is relatively cheap, but little attention is sometimes paid to design efficiency. Current flow can be irregular, uneaten food and faeces accumulate on the bottom, and the sidewalls are susceptible to erosion. The recommended design dimensions for ponds are 30 m × 10 m × 1 m sloping to 1.5–2 m depth at the outflow. A Herrguth-monk is commonly used at the outlet, although wide-bore pipes are sometimes installed. The latter increase the area of dead spots in the pond. Stocking rates in earthen ponds are less than in raceways or tanks.

A typical raceway design (Fig. 2.3) is in the order of 30 × 2.5 m × 0.7 m and is used for on-growing trout parr, Pacific salmon and, to a lesser extent, Atlantic salmon. One raceway often empties into another, making water quality control difficult. Fish densities must reflect the changing water quality towards the distal end of each raceway and any subsequent raceway. Fish tend to congregate in the upper portion of the raceway, and fin damage may occur due to nipping at feeding time or rubbing on the concrete walls. Fish must be graded more frequently in raceway systems than in ponds, as the aggregations of fish tend to hinder even food distribution, which results in growth divergence. Water exchange is more easily controlled than in ponds, and build-up of organic material can be avoided by increasing flow velocities using baffles. Hygiene in general is better in raceways than in earthen ponds. Although concrete is commonly used for raceway construction, earthen races are also used.

Tank farms, which can be based on 'modular' units, are expensive to purchase but upkeep is relatively less costly than other systems. Furthermore, well-designed tanks contain water that is better mixed and shows a more homogeneous water quality and flow profile than raceways, enabling the fish to use the entire water volume, and permitting efficiencies in water use and food distribution. A further advantage is better disease control than in either pond or raceway systems. Although raceways and ponds are used predominantly to rear trout, tanks are the preferred system for salmon (Fig. 15.6).

Water exchange in all three systems is an important consideration. In tanks and raceways it is usually sufficient to ensure 100–200% turnover per hour. The farmer must strive to ensure that the impoundment is 'self-cleaning' and this will require a flow of at least 2 cm/s in a raceway. Tanks normally require a lower flow to ensure self-cleaning. Only carefully designed ponds will avoid the need for regular mechanical removal of organic mud. Indeed, it is good management practice to leave some ponds fallow for several weeks each year to permit cleaning and facilitate disease control.

Grading

An experienced farmer knows when to grade stock for optimum growth efficiency and, most importantly, to provide marketable fish throughout the year. Fry are usually size graded before grow-out in ponds, raceways or tanks. Subsequent grading will depend on the market plan. However, sorting stresses fish and increases oxygen demand and is generally avoided during hot weather or low water flows, immediately before transport or while treating disease. Grading may also impair growth for a period through handling stress and the need to starve fish for 1–2 days before handling. Grading equipment has been refined over the years and the need to hand-net fish has been reduced by the use of fish pumps and elevators. Compact, manoeuvrable graders are available and these handle fish from 1 g upwards. They usually rely on the variation in fish width (girth) to separate fish, using tapered or set spaces between bars or belts.

Feeding

No farmer needs reminding of the high cost of food in intensive salmonid culture. Surprisingly, wastage still occurs as a result of overzealous feeding, inefficient grading leading to the inability of small fish to consume the pellets, or poorly designed feeders and feeding regimes. Underfeeding may also occur, resulting in poor FCR, low growth and possible hierarchy formation, which in turn can lead to more variable fish size distribution. It is essential that a sampling programme be established on the farm to determine the FCR (section 8.12). Samples can be taken fortnightly during the colder months, but in high summer once a month is prudent. In practice, though, sampling may be more infrequent. FCR figures vary between species, with trout usually producing less efficient ratios than salmon. The use of poor-quality food increases the level of solid and soluble waste and the FCR. Quality (and high-energy) diets are available and are very favourable from growth, environmental and economic standpoints (section 4.2.1).

Most feeding strategies are based around satiation feeding, with feeding tables acting as feeding guides (section 9.11). Although automatic feeding systems

Fig. 15.6 Tank farm for Atlantic salmon, Silfurlax, western Iceland (photograph by Dr Valdimar Gunnarsson).

(auger, spinning disc, rotating plate, pulsed air and belt feeders) can distribute feed at pre-set intervals throughout the day, they are supplemented by hand-feeding to assess feeding response, with subsequent amounts being adjusted accordingly. Pellet sizes used in the hatchery range from 400 μm as starter feed to about 3–4 mm for juveniles.

15.3.7 Production of Atlantic salmon smolt

Smoltification is a process in which fish in freshwater undergo a number of morphological, physiological and behavioural changes that enable them to migrate to, and then survive and grow in, marine conditions (Langdon, 1985). This process is complex and occurs over a number of months before the obvious morphological and behavioural changes in the fish. These changes include:

- obscured lateral parr marks
- darkened fin margins
- silvering and easy removal of scales
- decline in condition index
- a preference to swim with rather than against the current
- tolerance of high salinities.

All of these changes occur in freshwater.

Physiologically, in the final stages of smoltification, there are increases in cortisol, growth hormone and Na^+,K^+-ATPase, which aid osmoregulatory and ionic regulatory processes in seawater, and a decrease in prolactin, which is involved in freshwater osmoregulation.

Smoltification is not evident in some landlocked strains of salmon or trout. For example, rainbow trout stocks used for marine farming in Tasmania to produce ocean trout are a non-smolting strain and are transferred at a large size of 300+ g. They are grown in brackish waters or are acclimated to marine conditions.

Under a natural thermal regime, Atlantic salmon parr must attain a threshold size by the start of their first winter to successfully undergo smoltification in spring (Thorpe, 1977). Hatcheries often observe a bimodality in length/frequency distribution in parr populations, the upper mode completing smolting during the first spring (S1) and the lower mode becoming smolt 1 year later (S2). In general, S1 smolts at the commencement of winter are approximately 10–15 cm in length, whereas S2 fish are small in the first spring, but much larger the following spring, at the time of transfer. In contrast, Pacific salmon have a less defined smolt window(s), are usually much smaller in size at transfer, cannot always tolerate direct transfer to seawater and may require an acclimation period in brackish water.

The aim of the majority of Atlantic salmon

hatcheries is to produce yearling smolts (S1). Unfortunately, S1 and S2 are similar in size until summer, when S2 fish become anorexic (Huntingford & Metcalfe, 1993). However, S1 smolt production can be promoted by manipulating light and particularly temperature (and feeding) to optimise growth during the early stages of development. These two environmental variables play an important role in the smolting process. Seasonally changing temperature regulates metabolism, which affects appetite, consumption of food and growth rate (Saunders & Duston, 1992), whereas photoperiod signals seasonal cycles. If the light and temperature are held constant in the early stages to optimise growth, they must be resynchronised with natural photoperiod several months before the smolt window (i.e. in time to enter the sea when salinity tolerance is at a peak). Both incorrect photoperiod and high water temperatures will adversely affect salinity tolerance in potential smolts.

One strategy to allow year-round harvests involves the use of out-of-season smolts. Although spring smolts are produced largely under the influence of natural light, out-of-season smolts are produced through photoperiod manipulation, effectively advancing the autumn–winter–spring light signal artificially. Transfer to marine conditions of out-of-season smolt in March–May, followed by winter smolt and pre-smolt in June–August, before the spring smolt transfer in September–November, spreads the smolt intake period during the year, thus allowing a corresponding spread of production on the marine farms.

In Tasmania, the most popular smolt types produced are:

- all-female diploid and all-female triploid out-of-season smolts
- all-female diploid and mixed-sex diploid marine pre-smolts
- mixed-sex diploid and all-female diploid pre-smolts
- mixed-sex and all-female triploid spring smolts.

Indoor recirculation systems using controls over water quality and temperature and altered photoperiod regimes are being used to accelerate growth and manipulate egg development and smolting.

15.3.8 Grow-out of rainbow trout

Planning for grow-out of rainbow trout must be initiated before the spawning season. The farmer needs to draw up a production schedule ensuring a continuous supply of marketable fish. Pivotal in preparing the production strategy is knowledge of:

- consumer demand
- peak demand periods
- seasonal price fluctuations
- biological strengths and weaknesses in the stock
- knowledge of market competitors.

European farmers are able to extend egg production over at least a 9-month period. Photoperiod manipulation of broodfish permits about 6 months' supply of eggs. Temperature control of the hatchery environment may add an additional 2 or 3 weeks, and the importation of ova from Australia or other southern hemisphere countries adds a further 2 or 3 months' supply. It is quite feasible to operate a hatchery for 12 months in Europe if desired.

Australian trout producers have fewer techniques at their disposal to ensure continuous market supply than their European or North American counterparts. Australian hatcheries are prohibited by law from importing ova, and few hatcheries employ photoperiod manipulation of broodstock. Delaying or advancing egg development by temperature control can extend the hatching period by a few weeks. However, Australian farmers have fish that spawn naturally between late May and mid–late August. Ingram (1988) pointed out that careful selection of the offspring of early and late spawners as future broodstock will ensure a natural spread (genetically determined) of stock. At least 30 fish in each group should be used to reduce the possibility of inbreeding.

Although some flexibility exists to manipulate availability of rainbow trout fry to ensure an extended supply of marketable fish, the Australian farmer relies predominantly on two strategies: first, compensatory growth, whereby growth is slowed then increased to coincide with market demand; and, second, taking advantage of variable growth between fish stocks. Although the latter may be due to genetic factors, there are several parameters, acting alone or in combination, which also control growth including:

- water temperature
- oxygen levels
- tank or pond design
- flow rates
- feed
- grading (efficiency and interval)
- health maintenance.

There is an inverse relationship between water temperature and DO. As water temperature increases, less oxygen is available to the fish to metabolise its food (section 3.2.3). Overfeeding during periods of low DO may result in death or chronic stress. The latter often results in non-infectious disease or poor growth. The optimum temperature for growing rainbow trout is 15°C; they will, however, survive at 25°C for several weeks as long as there is sufficient DO. The farmer must have facilities available to supply additional oxygen when water temperatures are above 17°C. A plethora of commercial oxygenation systems and aerators are available to supplement low DO levels, which not only result from increasing temperatures but occur in conjunction with factors such as high stocking densities and reduced water flows (section 3.5.2).

Determining the required level of DO is very important for meeting an annual production schedule. Strategies for supplying oxygen range from gravity aerators (e.g. spray bars, water falls) between raceways to mechanical aerators (e.g. surface agitators, aspirators) and oxygen injection systems (e.g. low-head oxygenators, packed columns). At 15°C rainbow trout will survive at 3 mg/L DO for a while, but die if this level persists. This so-called 'level of oxygen resistance' (i.e. 3 mg/L) rises with temperature, and therefore 5 mg/L may prove fatal at 22°C. An oxygen meter is essential to determine DO levels throughout the warm summer months.

The most commonly used data to estimate mean daily oxygen requirements are those derived by Liao (1971). It has been calculated that 200–300 g of oxygen must be supplied for each kilogram of food. Because of the low solubility of oxygen in water, this amount of oxygen requires 100 000 L of water if the fish are extracting 3 mg/L DO through respiration without supplemental aeration. This clearly demonstrates the desirability of, or essential need for, aeration. However, knowledge of mean oxygen requirements is not sufficient when quantifying the amount of aeration to be provided (Petit, 1990). This is because instantaneous oxygen requirements of the fish may increase dramatically over periods of 20 min to several hours. Above 18°C, the amount of food must be reduced to avoid hypoxia and stress, as oxygen requirements of the fish rise sharply after feeding.

Timmons & Youngs (1991) gave an array of stocking densities for trout ranging from 60 to 234 kg/m^3. Stevenson (1980) correctly pointed out that density depends largely on oxygen content of the water, although experience and attention to the fish are important. At 15°C, 25 kg/m^3 is sensible, increasing to 35 kg/m^3 with aeration. In reality, it is the environment of the tank or pond that determines the stocking density that in turn permits maximum growth. Flow rate and water exchange are tools used to ameliorate the environment, but tank and pond design will determine the ease with which this is achieved.

15.4 Marine farming

Over recent years, great advances have been made in the technology and husbandry techniques required for farming salmonids in cages at sea. Cage structures have been developed to withstand more exposed sites, and they have generally increased in size and carrying capacity. Feeding, grading and pumping systems have improved fish performance and farm efficiencies. Reduced losses due to disease and parasite infestation and improved growth rates have resulted from:

- adoption of site rotation and fallowing practices
- species and age separation of leases
- reduction of stocking densities
- development of vaccines and alternative treatments.

The following section will discuss the general issues involved in salmonid farming in seacages, with a central focus on the Tasmanian industry.

15.4.1 Site characteristics

The majority of salmon sea farms are located inshore in high-salinity waters (32–35‰), although leases used as nursery areas or to inhibit amoebic gill disease (section 10.5.3) are located in estuaries influenced by freshwater. In contrast, trout are grown most successfully in estuarine areas. The location of low-cost cages in sheltered coastal waters (in bays, fjords and lochs) is coming under increasing pressure from recreational users and groups concerned by organic and inorganic enrichment from the farms (section 4.2.1).

Alternative offshore farming technologies are being developed to overcome these concerns, but they are expensive and they create additional problems in the areas of security, navigational hazard, predator attack and servicing.

In coastal areas, different sites may be used for specific age classes to match the requirements of the fish (e.g. nursery site as smolt) or to minimise the impacts of algal blooms, predator attack or the spread of disease between established grow-out fish and newly introduced smolt.

The site characteristics may also govern the layout of the farm, cage size, net depth and management strategies used. For example, the current speed, water exchange, site depth and tidal range may determine the mooring configuration, cage linkage format and net configuration. These, in turn, determine the management strategy (e.g. cage servicing from boats or directly from shore), the fish density used, the type of feeding system and net changing regime.

In general, desirable site characteristics include:

- adequate depth (minimum 5 m beneath the net)
- minimum average current flow of about 3 cm/s
- a high oxygen concentration (> 80% saturation)
- temperature < 18°C
- relatively clear water to observe fish
- a good water exchange to flush nutrients from the area and supply DO
- a sandy substrate
- sufficient distance from predator colonies (e.g. seals).

In addition to the physical features of the site, socio-economic factors are important. Many sea sites are located in regions remote from large populations. Important site features that affect the efficiency and economic viability of the operation include:

- access to the site to deliver equipment, fingerlings or smolt and to remove harvested product
- availability of workers and their accommodation
- availability of utilities
- proximity of processing facilities and airports
- availability of servicing and maintenance facilities.

Although these requirements are related to the site requirements for the marine farming phase of salmonid culture, they include a number of the features for site selection that are discussed in Chapter 2.

15.4.2 Site rotation and fallowing

Prolonged use of sites, accompanied by poor husbandry practices (e.g. overfeeding) and suboptimal environmental conditions (e.g. poor water movement), may cause 'souring', a condition that reduces the productivity of the site and interferes with the performance of the fish. Consequently, growth and survival may decrease and the incidence of disease may increase. It is of utmost importance to the grower that sites are carefully chosen and cages are carefully managed as it is the farm that will bear the brunt of any negative impacts. Solid waste from uneaten food and faecal material may accumulate on the seafloor under the cages. Initially this may increase the invertebrate productivity in the substrate but, over time, the oxygen availability will decline, with a resultant decline in the diversity of invertebrates. In extreme cases, the accumulation of organic matter leads to the production of gases (hydrogen sulphide, methane and ammonia), which may rise through the water column as bubbles.

Cages may periodically be moved around the lease area or rotated between different leases to allow for periodic recovery of the substrate, even when careful management is used. The time required for recovery ranges from 6 months to several years and is dependent upon site characteristics. Several sites may be used and each site may be used for specific age classes of fish. This production strategy enables sites to be fallowed between production cycles. Site rotation and fallowing are strategies that can be used in conjunction with, not instead of, sound husbandry practices to manage sites. Good farming sites are uncommon and growers know that these must be farmed sustainably.

Independent monitoring by government authorities of sediment condition under cages is now standard practice in a number of countries. Visual assessment using underwater video, chemical assessment (e.g. redox potential) and biological assessment (benthic in-fauna) are components of such a monitoring programme. In addition, growers closely monitor water quality parameters such as:

- DO
- current speed and direction
- temperature
- salinity
- turbidity
- microalgal concentration and type.

Measurements are taken both manually and telemetrically.

15.4.3 Transfer to sea farm from freshwater

Smolt or fingerlings are transferred to the seawater facilities from freshwater at a particular size and at suitable times of the year. Smolting species or strains are transferred during their 'smoltification window', which is usually in spring for Atlantic salmon, but variable for Pacific salmon. In the non-smolting groups (e.g. many rainbow trout stocks), the success of transfer is size related: larger fish are more adaptable to higher salinities. Some growers do not culture rainbow trout in full salinities, preferring more brackish conditions. The timing and length of the 'smolt window' varies slightly from year to year in relation to environmental conditions such as temperature.

Smolts are introduced directly into seawater from freshwater in most cases, although acclimation regimes or slightly brackish water sites are used when pre-smolt or large parr are introduced outside the smolting window. The inclusion of salt (5–10%) in the diet of smolting and non-smolting fish a few weeks before transfer has been used to aid osmoregulatory conditioning (Salman & Eddy, 1990; King, 1992), but this has been discontinued.

The fish are normally graded to produce a uniform stock before transfer to seafarms. A variety of methods are used to assess the actual timing for transfer. Although the fish may display a wide array of smolt-like characteristics, the best indicators are the 24-h (up to 72 h), 35‰ salinity adaptability test and the 40‰ challenge test. The first method assesses the ability of the fish to osmoregulate successfully after direct transfer from freshwater to seawater. This is undertaken through osmolality or sodium readings of the blood, plasma or serum and the 24-h readings compared with baseline data. The second test assesses survival of fish after a 24-h exposure. The salinity is increased from seawater to 40‰ as large parr may survive exposure to 35‰ over a short period. The 40‰ challenge test has been discontinued on ethical grounds.

Fish are moved to sea sites in transport tankers on trucks and are released via pipes into cages after pens are towed inshore. Alternatively, the tankers may be barged out to the pens, or the fish are released into small tankers and airlifted from shore to pen by helicopter. Choice of operation depends on cost, site characteristics and site access. A number of tanker characteristics are important when fish are transported over lengthy periods. These are:

- stocking density
- oxygen concentration and distribution
- tank ventilation
- carbon dioxide de-gassing capability
- temperature fluctuations.

Stocking densities up to ~90 kg/m^3 are supported by the use of oxygen diffused into the water, the concentration being monitored by in-tank probes in combination with read-outs and alarms in the truck cabs. The water may be de-gassed of carbon dioxide using pumps and spray-bars, aeration and tank ventilation. Trips to seafarms typically take 3–6 h.

Fish are normally 'starved' for a few days before transport to reduce oxygen demand and excretion of carbon dioxide, ammonia and faecal solids. Fish are counted or weighed during the loading process, with grading taking place some time beforehand. It is important that, in relation to transport biomass and seafarm itinerary, an accurate count is achieved. To avoid additional handling, the fish are not usually recounted at the seafarm. Average sizes and the range of sizes are also of significance as feed sizes and net mesh sizes need to be selected for any particular batch of fish. These parameters are normally determined in relation to average size, and a large size range in the batch can lead to fish being unable to accept pellets or escaping through net meshes.

15.4.4 Cage systems

The majority of cages are similar in basic structure: a bag net suspended inside a floating collar, which is moored to the sea floor (section 2.2.6). There are many cage shapes, construction materials, linkage configurations and sizes, their form determining the management strategies and servicing practices used (Beveridge, 1996). Some industries use individually moored cages because of size or to better utilise low water flow conditions. Many areas link cages together into platforms with service walkways between pens. Some are moored offshore using boat access (Fig. 15.7) and some extend from the shore with direct access by personnel and vehicles. Individual cage sizes vary enormously from small 1- to 2-mt units (e.g. 1–2 m^3 cage volume) to large 100- to 150+-mt units. Advances in cage technology show trends to larger cages or cage systems, moored further offshore and possessing a number of features that enable the cages to withstand offshore conditions (e.g. flexibility, being semi-submersible).

In Tasmania, the three cage styles that are used are:

- polyethylene rings (polar circle type, 60, 80 and 120 m circumference)
- square steel platforms
- solid bag nets.

Net bags are generally 5–20 m in depth, the shape being maintained by the use of metal rings, weights (~50–200 kg each), lead-core ropes or positive displacement of water (in the case of solid bag net systems). Maintaining the shape of open mesh nets is particularly important in preventing 'bagging' or loss of shape and volume, which occur through the resistance of biofouled nets against water currents. Net meshes on seacages range in size from about 10 mm to 35 mm (bar length), depending on fish size. They are generally constructed of a woven, knotless (e.g. raschel) and synthetic material such as nylon or polyethylene. The material is treated chemically for resistance against UV light. In Tasmania, these nets are stocked with fish to a level of about 5–15 kg/m^3, producing an overall biomass at harvest of 20–100 mt fish/cage, depending on cage design.

15.4.5 Biofouling and net changing

Biofouling occurs on nets, ropes and cages, and creates problems such as increased drag, increased weight, reduced water flow (exchange) and increased BOD. An electron microscopic study by Hodson & Burke (1994) showed that groups such as bacteria, protists and diatoms are early colonisers, with the form and orientation of the net mesh playing major roles in this colonisation process. These microscopic groups are usually followed through successive colonisation by the macroscopic algae, hydroids and other invertebrates such as amphipods.

Fouled nets are removed and exchanged for clean nets on cages more frequently in summer than in winter owing to the more rapid biofouling processes. For example, nets are exchanged every 7–25 days in summer and every 50–60 days in winter in Tasmania. The biofouling is removed from the net in a rotating washing machine. The nets are then checked for holes, repaired and folded ready for further use. Lifting equipment positioned on barges has become essential when servicing nets from large cages.

The net exchange process can be labour intensive. Consequently, alternative prevention and servicing methods have been developed to combat the biofouling problem. Rather than removing biofouled nets from cages, some companies use a dual net system on each cage: one net in and one net out of the water. Periodically, the nets are alternated so that the biofouled net is air-dried on the cage. In other areas, nets emptied of fish are raised out of the water onto the handrails and are cleaned with high-pressure jets of water, or rotating brushes are used to dislodge fouling on nets *in situ*.

Biocidal and non-biocidal antifoulant paints applied to nets by dipping have been trialled as a means of inhibiting growth on netting. Biocidal paints containing tributyltin have been banned because of the deleterious effects on larval crustaceans and molluscs in the surrounding waters. Copper-based paints are relatively effective in slowing fouling and, like some of the non-biocidal paints, stiffen the net after application. Stiffened netting is helpful in deterring seals, but it makes the nets more difficult to handle when they are changed after a few months.

As an alternative, non-toxic wax- and

Fig. 15.7 Circular polyethylene cages are popular within the salmonid industry worldwide. There are many variations of the design involving configurations of flotation rings and stanchions (photograph by Professor J. Lucas).

emulsion-based paints, and paints based on some newly developed natural products, can be used on nets. As some of these measures are not necessarily successful or cost-effective, many farms still favour simple net changing.

15.4.6 Predation

The type and number of predators on seafarms vary considerably and are site and country specific. In Tasmania, seals can be a major threat, removing or damaging fish through the nets or creating holes in the mesh, which allow stock to escape. Many farms use physical deterrents such as anti-predator nets of 10- to 15-cm bar mesh around individual pens or lease areas, whereas others use traps (the seals are relocated) or sophisticated electronic acoustic deterrents. Farmers who have found double nets to be labour intensive and awkward have developed stiffened heavy-ply fish nets that are weighted taut to prevent seals from entering the nets. Seals that are relocated to other regions after trapping will sometimes return to the farms within a few days, and can be trapped and relocated repeatedly. Seals have also developed behavioural strategies to overcome the deterrents:

- They swim with their heads above the water to avoid the acoustic deterrents.
- They jump over handrails to enter the cages.
- They gain entry to the space between the predator and fish nets.
- They establish territories on cage walkways and sometimes attack workers and divers.

Even when deterrents are successful in preventing losses of fish, the presence of the seals can depress the feed intake and subsequent growth potential of the fish.

Seal nets around individual cages also assist in excluding diving birds, such as cormorants, which cause damage by attacking fish through the nets. Bird nets are also used over cages, particularly smolt pens, to prevent bird attack. Gulls are normally present around farms in large numbers, scavenging any dead fish or feed. Droppings from birds may be a means of parasite and disease transfer; however, this is area specific. Droppings also cause work surfaces to become slippery, a potential occupational health and safety issue.

Some farms are troubled by scavenging fish, such as dogfish, sharks and eels, which are attracted to any dead fish in the base of the net. Such attacks can result in net damage and fish escape. Poaching and vandalism appear to be universal problems, being tackled by growers through the use of dogs, night patrols, radar or electronic surveillance techniques. Perceived problems from dolphins and large whales do not appear to be a justified risk, although their presence, like seals, can stress the fish, in turn reducing feed intake and growth.

15.4.7 *Fish transfer on and between sea farms*

Some companies use fishing boats to tow cages of fish between leases for grow-out and for harvesting. This procedure is slow, the speed dependent upon the swimming ability of the fish, currents, tides and bagging of the net. In other situations, fish are transferred from small to larger cages during grow-out and from one cage to another because of structural damage or when splitting batches of fish. These 'swim-throughs' normally involve the nets being connected and the fish being forced to swim between cages as the nets are raised.

Although small farms may still utilise hand-operated dip nets to move fish, they are impractical on larger operations or when transferring large numbers of fish. Instead, large canvas-lined brail nets, which are lifted by cranes, or fish pumps are used to move fish during grading, harvesting or freshwater bathing for amoebic gill disease (section 10.5.3).

Pressure-vacuum Venturi-driven or air-lift pumps are the most commonly used fish pumping methods; in most cases they are capable of handling fish up to ~5 kg in size and have a pumping rate of about 20–30 mt fish/h. Problems encountered with eye, skin and scale damage have been overcome by system automation, operator experience or new designs that do not contain closing valves in contact with the fish. Many growers also favour continuous-flow pumps over ones on a stop–start cycle.

15.5 Feeds

15.5.1 *Characteristics*

As feed costs constitute ~40–50% of the production costs, a significant amount of attention has been paid to feed composition, feed distribution, feeding behaviour, waste levels and feed ingestion.

'Moist' and 'wet' feeds (e.g. trash fish) were initially used during the early development of salmonid seafarming. With the development of formulated 'dry' diets (pelleted feeds), the problems of water quality deterioration, nutritional deficiencies and feed distribution associated with trash fish diets have largely been resolved. Further refinement of dry diets through the extrusion process (section 9.10.2) has provided benefits over steam-pressed pellets:

- lower dust content
- consistent pellet size
- high lipid content
- higher digestibility
- prolonged pellet integrity in water
- variable sinking rate.

Much of the research being undertaken by feed companies concentrates on the nutritional content of the pellet, particularly the level of lipid and protein, and its digestibility. Proximate composition of salmonid diets varies in relation to life cycle phase and pellet form, but a typical breakdown of energy sources is 40–55% protein, 15–28% lipid and 8–12% carbohydrate. The lipid levels used in Tasmania tend to be lower than those used in Norway, which may be as high as 40%. Trout diets tend to have lower energy contents. Digestibility of carbohydrates, optimisation of lipid level and replacement of fish meal with plant proteins are some fields of current research.

Feed pellet diameter sizes range from 400 μm to 12 mm, corresponding to increasing fish size from first feed to grow-out. Feed rates per day depend on fish stock, water temperature, location, feed strategy, usable energy content and feed type, but they typically range from ~7% for 1-g fry to < 1% for grow-out (3–4 kg).

With the development of more digestible feeds and technologies to maximise feed distribution to fish, FCRs have steadily improved to figures such as 0.8:1 to 0.9:1 for initial smolt growth and 1.2:1 to 1.4:1 for overall grow-out in seawater.

Specific feed additives may also be used:

- attractants/stimulants to improve feed intake and digestibility
- astaxanthin for flesh pigmentation in grow-out fish and egg requirements in broodstock.

Pigment is included in the diets at 64–80 mg/kg to produce flesh pigmentation levels of 7–10 mg/kg for salmon and 12–15 mg/kg for trout (Pivot Aquaculture, 2000). Although laboratory-based techniques may be used to quantify flesh pigment levels, they are usually assessed visually with the aid of colour cards. The deposition of pigment in the flesh is affected by fish size, whereas its loss is related to maturation status.

Although the use of antibiotics has been reduced drastically in many countries in recent years, these drugs, when required, are incorporated into the pellet during manufacture or added later using oil or gelatin-based coating. Many Tasmanian seafarms are antibiotic free, using management techniques to prevent or minimize the impact of diseases.

15.5.2 Feed distribution

Farms have developed various feeding strategies and these predominantly govern the means of distributing feed to fish. Some farms feed once or twice a day, others three to six times and yet others on a 'continuous' basis. Generally, smaller fish require more meals per day than larger fish. Pelleted feeds are distributed mechanically or by hand (section 9.11.2). Hand feeding, with the assistance of water cannons or air-blowers, enables personnel to assess the feeding response of the fish and other aspects of husbandry, such as dead or diseased fish, general behaviour and net or cage damage. It is, however, costly because it is labour intensive and limits the number of meals per day or simultaneous distribution of meals to all cages. Fish are normally fed two to four times per day on a rotational basis using this method, although smolt may be fed more frequently.

Mechanical feeders consist of storage hoppers with associated spinning discs (Fig. 15.8), vibrating plates, demand feeder pendulums or augers, or air blown through pipes from land-based hoppers to cages. They tend to deliver pellets using timers or computerised controls. Many of these delivery systems operate intermittently throughout the day, irrespective of fish requirements. Some farms combine hand feeding with mechanical feeders so that the advantages of both are used: observation by staff and numerous feeds throughout the day.

Inherent in all of these methods is the potential to overfeed or underfeed; however, new technologies are being developed to overcome these problems:

- adaptive feeding systems (Blyth *et al.*, 1993) (Fig. 15.8)
- acoustic detection systems (Bjordal *et al.*, 1993; Dunn & Dalland, 1993)
- air-lift
- underwater camera systems.

Fig. 15.8 The adaptive feeding system is one of a number of feed-monitoring devices being used to minimise pellet wastage. This system has been connected to a feed hopper with a spinning disc pellet distributor; it is able to self-regulate feed distribution and record fish feeding patterns.

These are now being used to optimise feed distribution to the fish by monitoring pellet wastage, fish distribution in the water column or consumption of pellets.

More specifically, adaptive feeders are self-regulating and record the amount and time of feed delivery through the use of computerised controls. They also are being used to determine, in more detail, the diurnal and seasonal feeding cycles of the fish and feeding patterns in relation to changes in environmental conditions. This improved understanding of the feeding requirements of the fish growing through their cycle and being exposed to different environmental conditions can be used in turn to refine automatic feeder and hand-feeding delivery strategies. All of these approaches aim to minimise feed wastage and optimise FCR, thereby reducing costs and the impact of the farm on the environment.

15.6 Grading and stocking densities

15.6.1 Grading

Grading is undertaken in both fresh and marine facilities. It aims to reduce size variability and feeding hierarchies in cohorts of fish or to separate maturing from non-maturing fish. Low size variability also directly aids other management tasks such as food pellet size and net mesh size selection.

Usually, the decision to grade fish is made after the size variability of a sample of fish is determined or at a particular time of the year when maturing fish can be identified. Mechanical graders, in conjunction with fish pumps (Fig. 15.9), are used during size grading, whereas hand grading combined with visual assessment is used to separate the maturing grilse or jacks from the immature fish. It is widely accepted that these mechanical operations inhibit fish growth for days or weeks and, instead, some operators either avoid grading or use swim-through bar graders to separate size groups.

As the task of weight and length sampling can inhibit growth, and add to the risk of stress and damage, techniques have been developed for measuring fish in their cages. This equipment uses the processes of acoustics, incorporating a measurement of the swim bladder (Dunn & Dalland, 1993) or infra-red fish imaging to construct, via computer technology, size–frequency histograms. Thus, the decision to grade the fish can then be made without handling the fish. The technology can be used frequently with little effect on the fish and is useful for large cage systems; however, the equipment needs to be positioned and calibrated correctly to be accurate.

Fig. 15.9 Adjustable bar graders are used in conjunction with fish moving devices, such as fish pumps, and are used to obtain more even size distribution of fish in pens.

15.6.2 Stocking densities

Stocking densities are governed by the species of salmonid reared (e.g. Arctic charr preferentially grow at much higher densities than other species), site characteristics and environmental conditions. The trend in recent years appears to favour lower densities in seacages with a reduction from around 30–40 kg/m^3 to 10–20 kg/m^3 maximum. The advantages of this strategy include:

- better growth
- lower incidence of disease
- better utilisation of feed
- lower concentrations of waste
- better survival.

Initial stocking of smolt into cages is much lower at around 0.5–1 kg/m^3, although very low densities must be avoided as these cause poor feeding responses and growth.

15.7 Maturation, sex reversal and triploidy

15.7.1 Maturation

Salmonids begin to mature during seawater grow-out from spring/summer and they develop gonads ready for spawning in autumn/winter. Energy from tissue reserves and from feed intake is directed toward gonad production to the detriment of somatic or flesh growth. Survival of maturing fish in seawater becomes increasingly more difficult approaching spawning and many fish die. In addition, development of undesirable characteristics, such as the following, reduce the fish's marketability:

- hooked jaw (kype) development
- milt and egg development
- change of body shape
- mucus production
- darkening of the skin
- loss of flesh colour
- softening of flesh.

The harvest period, therefore, is determined by growth to an adequate market size and the onset of maturation. Normally, all fish are removed from the water and sold while in good condition. The age of maturity varies between species and stocks.

In some regions, usually associated with higher water temperatures, fish may mature after one sea-winter. This is referred to as 'grilsing' in Atlantic salmon. The strategies to deal with these fish vary between countries. Where there is a significant proportion of the fish maturing, these fish may be graded from the cages and sold before deterioration of the quality. In other areas (e.g. Australia) fast growth to an average of 3–5 kg is achieved, but is also associated with almost complete maturation after one sea-winter. In this situation, all fish are harvested. At the other extreme, almost no grilsing occurs in colder regions of the northern hemisphere and the fish remain on the farms for up to 3 years.

Restricted feeding at specific times of the year and photo-manipulation using underwater lights are being trialled experimentally in some areas in an attempt to reduce maturation.

The production of all-female stock and sterile triploid fish (particularly all-female) are strategies that can be used to extend the availability of market fish to year-round, thus overcoming the limitations caused by maturing fish (Chapter 7). Triploidy rates of 95–100% are currently being achieved, with all-female stock being used to avoid pseudo-maturation characters seen in triploid males. Triploidy has declined in popularity in many areas, but it is still used in some industries. Improved culture strategies, such as out-of-season smolt, have given the growers some of the advantages gained previously from triploidy.

15.7.2 Sex reversal

Sex reversal of fish is discussed in detail in section 7.5.3. In salmonid culture, removing male fish that potentially mature earlier is a technique used to produce all-female stock. It is a multistep procedure that initially produces all-male fish through the addition of 17α-methyltestosterone in the feed for 70 days (at 10°C) at the first-feeding fry stage. Male fish (XY) develop normally, but the female fish (XX) develop to also produce sperm and are termed 'neo-males'. These fish must be dissected at maturity (usually after 2–3 years) to remove the testes, as the neo-males do not possess sperm ducts. The resultant XX milt is used to fertilise eggs, resulting in all-female (XX) fish which are grown to market size. It is essential that the hormone used in this procedure is not offered to market fish, only to the broodstock. It is for this reason, and the need to use the fish for this specific task, that neo-male stock must be identifiable and maintained separately during grow-out.

15.7.3 Triploidy

The triploidy-inducing process occurs at the fertilised egg stage and is outlined in section 7.6.1. It may be used on eggs in conjunction with sex reversal techniques to produce all-female triploids or separately to produce mixed-sex triploids. As it induces sterility it can be used to overcome problems normally associated with maturation – male precocity, grilsing and early maturation in seacaged fish – which temporally restricts harvesting. In Tasmania, triploidy rates are now close to 100%, whereas a few years ago triploid batches contained a significant number of diploids that ultimately would mature.

Eggs at approximately 30 min post fertilisation (at 10°C) are subjected to heat or hydrostatic pressure shock for a specific period. Details are related specifically to individual species, sizes of eggs and method of shock treatment, but as an example thermal shock occurs at 28.5°C for 10 min and pressure shock at ~9000 psi for 5 min, followed by re-exposure to 10°C water. After this treatment, normal rearing conditions are used. Although sterility is a characteristic of triploid fish, some male fish display pseudo-maturation features. This may be overcome by producing all-female triploids; the eggs are fertilised by sperm from neo-males and then subjected to the triploidy process.

Tripolid fish sometimes suffer from deformities (e.g. jaw deformities or 'drop-jaw') and may possess different physical characteristics (e.g. flesh characteristics and body form). Alternative strategies are being developed to combat the maturation problem and the associated restricted harvests. These strategies include the use of out-of-season smolt, controlled feeding strategies, use of lights to modify photoperiod and selective breeding programmes.

15.8 Fish health

A number of disease pathogens are prevalent in the culture of salmonids worldwide; many are widespread, whereas some are specific to certain regions. An overview of the major diseases affecting salmonids is given in section 10.5.3. Relatively isolated regions, such as Tasmania, are largely 'disease free' in terms of the incidence of serious bacterial and viral conditions. In general, the incidence of disease is decreasing, the use of preventative management strategies, many of which minimise fish stress, being favoured over antibiotic treatments. The use of embayment or coastal management strategies with restricted stock movement between sites and reduced on-site stocking densities, together with site fallowing strategies and the development of vaccines, have all improved fish survival and condition. Some of the more serious diseases that have been encountered worldwide include:

- sea-lice (*Lepeopthierus* and *Caligus* species)
- furunculosis (*Aeromonas salmonicida*)
- *Vibrio* species
- infectious salmonid anaemia (ISA)
- bacterial kidney disease (BKD)
- infectious pancreatic necrosis (IPN)
- amoebic gill disease (AGD)
- piscirikettsiosis (salmonid rikettsial septicaemia)
- enteric redmouth.

In Tasmania, AGD, caused by *Neoparamoeba pemaquidensis*, has a major impact on the industry. This disorder contributes up to 20% of production costs. This is primarily due to high labour costs associated with frequent bathing of the fish in freshwater that is used to reduce the incidence of the disease and its resultant mortalities. Freshwater bathing uses tarpaulin liners positioned in new cages and filled with freshwater. Fish are pumped from the seawater cages through a de-watering apparatus to the freshwater bath in the tarpaulins, where they remain for a few hours before the tarpaulins are removed. Originally experienced only in summer, it is now prevalent throughout the year, mainly at full seawater sites rather than brackish water localities.

Salmon in Tasmania also suffer in seawater from a wasting condition called 'pinheading' (Fig. 15.10), evident a few weeks or months after transfer. The cause is unclear, but is probably a combination of osmoregulatory stress and reduced appetite or poor feed intake. The level of pinheading varies between batches of fish transferred to the sea and appears to be different to the runting condition seen in many industries.

The pathogenic bacterium *Vibrio anguillarum* has been controlled by the development of a vaccine that is administered to the fish as a bath at the freshwater stage. *Yersinia ruckeri* (a strain of enteric redmouth) outbreaks occur occasionally in freshwater and after transfer to seawater, but these are not serious. They are currently controlled by management techniques and limited antibiotic treatments, although the development of a vaccine is being investigated. Other health conditions occur occasionally as a consequence of algal blooms, periodic influxes of jellyfish and nutritional problems such as those associated with vitamin deficiency and food spoilage.

15.9 Harvesting and products

The average size of salmonids at harvest is determined largely by market demand balanced with the cost of production. As examples, plate-sized freshwater trout may be harvested at 200–400 g, whereas sea-grown trout may occupy a different market niche at 2–4 kg. Atlantic salmon may be harvested as one-winter fish or 'grilse', which may vary considerably in quality and size, depending on country, and range from 1.5 kg to

Fig. 15.10 Representative salmon from a seacage cohort showing normal growth (top), runting (middle) and pinheading (bottom).

6 kg. Fish that are two winters or older or 'salmon' are usually 3 kg and above.

Before the harvest, samples may be taken to determine the quality of the fish in terms of fat content, flesh colour, average size, size range, shape and occurrence of marks. During the harvest process, fish are removed from cages using fish pumps or brail nets, then stunned, bled and chilled (Fig. 11.1), lowering the core temperature to preserve carcass quality. The ice slurry/carbon dioxide combination commonly used in industry to stun the fish may produce elevated lactic acid and lowered pH levels in the flesh, leading to issues of gaping and softness. Consequently, carbon dioxide is being replaced by anaesthetic baths using AQUI-S or by stun guns. Once harvested, the fish are placed in ice slurry to cool their core temperature until they are processed.

Salmonid products that are sold around the world include:

- fresh fish
- whole head-on, gilled and gutted fish (HOGG)
- fresh fillets and cutlets
- frozen HOGG fish
- smoked sides or portions
- gravadlax (lightly pickled flesh)
- pâta
- caviar.

This is an excellent range of products from an aquaculture species and it explains some of the attraction of salmonids for aquaculturists over many decades.

References

Beveridge, M. C. M. (1996). *Cage Aquaculture*, 2nd edn. Fishing News Books, Oxford.

Bjordal, A., Lindem, T., Juell, J. E. & Ferno, A. (1993). Hydroacoustic monitoring and feeding control in cage rearing of Atlantic salmon (*Salmo salar* L.). In: *Proceedings of the First International Conference on Fish Farming Technology, Trondheim* (Ed. by H. Reinertsen, L. A. Dahle, L. Jorgensen & K. Tvinnereim), pp. 203–8. Balkema, The Netherlands.

Blyth, P. J., Purser, G. J. & Russell, J. F. (1993). Detection of feeding rhythms in seacaged Atlantic salmon using new feeder technology. In: *Proceedings of the First International Conference on Fish Farming Technology, Trondheim* (Ed. by H. Reinertsen, L. A. Dahle, L. Jorgensen & K. Tvinnereim), pp. 209–16. Balkema, The Netherlands.

Bromage, N. R. & Cumaranatunga, P. R. C. (1988). Egg production in the rainbow trout. In: *Recent Advances in Aquaculture*, Vol. 3 (Ed. by R. J. Roberts & J. F. Muir), pp. 63–138. Croom Helm, London.

Bromage N., Jones J., Randall C., Thrush M., Davies B., Springate J., Duston J., & Barker G. (1992). Broodstock management, fecundity, egg quality and the timing of egg production in the rainbow trout (*Oncorhynchus mykiss*). In: *The Rainbow Trout* (Ed. by G. A. E. Gall), pp. 141–66. Elsevier, Amsterdam.

Dunn, M. & Dalland, D. (1993). Observing behaviour and growth using the Simrad FCM 160 fish cage monitoring system. In: *Proceedings of the First International Conference on Fish Farming Technology, Trondheim* (Ed. by H. Reinertsen, L. A. Dahle, L. Jorgensen & K. Tvinnereim), pp. 269–74. Balkema, The Netherlands.

FAO (2001a). *1999 Fisheries Statistics: Aquaculture Production*, Vol. 88/2. FAO, Rome.

FAO (2001b). *1999 Fisheries Statistics: Capture Production*, Vol. 88/1. FAO, Rome.

Forteath, N., Purser, J. & Daintith, M. (1996). Fin fish husbandry. In: *Fin Fish Farm Attendant's Manual* (Ed. by M. Daintith & M. George). Fishing Industry Training Board of Tasmania, Hobart.

Gall, G. A. E. & Crandell, P. A. (1992). The rainbow trout. In: *The Rainbow Trout* (Ed. by G. A. E. Gall), pp. 1–10. Elsevier, Amsterdam.

Heen K., Monahan R. R. & Utter F. (1993). *Salmon Aquaculture*. Fishing News Books, Oxford.

Hodson, S. & Burke, C. (1994). Microfouling of salmon-cage netting: a preliminary investigation. *Biofouling*, **8**, 93–105.

Huntingford, F. & Metcalfe , N. (1993). How to cash in on life-history variations. *Fish Farmer*, **16**, 50–51.

Ingram, M. (1988). Farming rainbow trout in freshwater tanks and ponds. In: *Salmon and Trout Farming* (Ed. by L. Laird & T. Needham), pp. 155–89. Ellis Horwood, Chichester.

King, H. R. (1992). The value of high salt diets in smolt

transfers. In: *Proceedings of the Saltas Research Review Seminar*, pp. 7–16, April 1992, Hobart.
Laird, L. & Needham, T. (1988). *Salmon and Trout Farming*. Ellis Horwood, Chichester.
Langdon, J. (1985). Smoltification physiology in the culture of salmonids. In: *Recent Advances in Aquaculture*, Vol. 2 (Ed. by J. F. Muir & R. J. Roberts), pp. 79–118. Croom Helm, London.
Liao, P. B. (1971). Water requirements of salmonids. *Progressive Fish Culturist*, **33**, 210–5.
Needham, T. (1988). Salmon smolt production. In: *Salmon and Trout Farming* (Ed. by L. Laird & T. Needham). Ellis Horwood, Chichester.
Pankhurst, N., Purser, G.J., Van Der Kraak, G., Thomas, P. and Forteath, N. (1996). Effect of holding temperature on ovulation, egg fertility, plasma levels of reproductive hormones and *in vitro* ovarian steroidogenesis in the rainbow trout *Oncorhynchus mykiss*. *Aquaculture*, **146**, 277–90.
Pennell, W. & Barton, B. A. (eds) (1996). *Principles of Salmonid Culture*. Developments in Aquaculture and Fisheries Science Vol. 29. Elsevier, Amsterdam.
Petit, J. (1990). Water supply, treatment and recycling in aquaculture. In: *Aquaculture* Vol. I (Ed. by G. Barnabe). Ellis Horwood, New York.
Piper, R. G., McElwain, I. B., Orme, L. E., McCraren, J. P., Fowler, L. G. & Leonard, J. R. (1982). *Fish Hatchery Management*. US Department of the Interior, Fish and Wildlife Service, Washington, DC.
Pivot Aquaculture (2000). *Feed Management Guidelines for Salmonids*. Pivot Aquaculture. Cambridge, Tasmania.
Salman, N. A. & Eddy, F. B. (1990). Increased sea-water adaptability of non-smolting rainbow trout by salt feeding. *Aquaculture*, **86**, 259–70.
Saunders, R. L. & Duston, J. (1992). Increasing production of Atlantic salmon smolts by manipulating photoperiod and temperature. *World Aquaculture*, **23**, 43–6.
Sedgwick, S. D. (1995). *Trout Farming Handbook*. Fishing News Books, Oxford.
Stevenson, J. P. (1980). *Trout Farming Manual*. Fishing News Books, Farnham.
Thorpe, J. E. (1977). Bimodal distribution of length of juvenile Atlantic salmon under artificial rearing conditions. *Journal of Fish Biology*, **11**, 175–84.
Timmons, M. B. & Youngs, W. D. (1991). Considerations on the design of raceways. Aquaculture Systems Engineers. *Proceedings of the World Aquaculture Society*. American Society of American Engineers Publications, St. Joseph, MI.
Willoughby, S. (1999). *Manual of Salmonid Farming*. Fishing News Books, Oxford.

16
Tilapias

Victor Suresh

16.1 Introduction

> Tilapias: the 'aquatic chickens' that are living up to their promise.

Fish resembling tilapias were featured in paintings in the Egyptian tombs of 2500 BC. Some believe that it shows that tilapias were cultivated in ancient Egypt. It is also claimed that the fish caught by Jesus's disciples in the Sea of Galilee and fed to the multitudes were tilapias. Such claims may be disputed because of the lack of hard evidence, but there can be no disputing that tilapias are one of the most successfully cultured fish of modern times. Current global production of tilapias exceeds 1 million mt/year (Fig. 16.1). China alone contributes 51% of the global production (Table 16.1). In countries such as China, Egypt and the Philippines, most tilapias produced through aquaculture are used for local consumption. There is also a substantial international trade in chilled and frozen tilapia products. The largest importer of tilapias is the USA, where consumption of tilapias has increased almost 10 times in the last 10 years (Fig. 16.2) and has exceeded that of trout since 1995. Central American countries, particularly Costa Rica, produce relatively small volumes of tilapias, but most of it is exported as high-value, fresh-chilled fillets to the USA.

The remarkable success of tilapias as a farmed fish can be attributed to two main factors.

(1) Their desirable qualities as a food fish: white flesh, neutral taste and firm texture. As a result, tilapias have gained acceptance in a wide variety of human cultures with differing tastes and food preferences.
(2) They are an easily farmed fish. Tilapias are easy to hold and breed in a captive environment. They tolerate crowding, relatively poor water quality and other stress factors and are less susceptible to disease. They can be grown in a wide variety of aquaculture systems (Table 16.2). They eat algae and detritus naturally produced in culture systems as well as manufactured feeds containing ingredients derived from plants. They reach typical market size (500–600 g) in about 6–8 months under optimum temperature conditions for growth (30–35°C).

Based on these characteristics, Pullin (1984) termed tilapias 'aquatic chicken'– an animal that can be farmed as easily and economically, and with the same broad market appeal, as chickens.

The most important problem in growing tilapias to an acceptable market size (> 300 g) is their early and uncontrolled reproduction in culture systems, especially earthen ponds. Their early maturation and frequent reproduction directs a significant amount of energy towards reproductive development and activities, and thereby reduces the energy available for growth. Furthermore, uncontrolled recruitment of

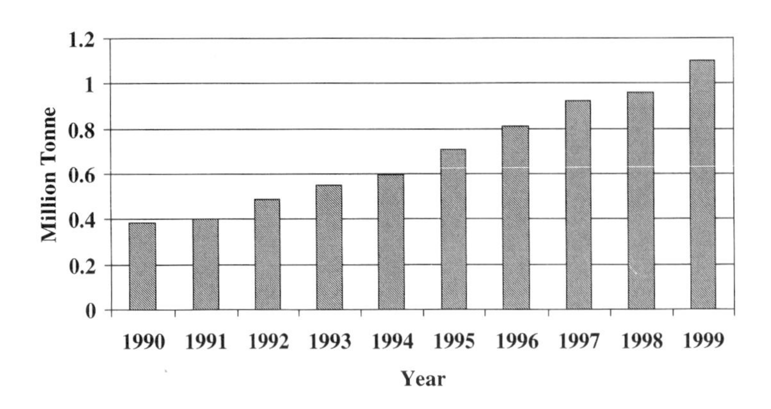

Fig. 16.1 Global production of tilapias in aquaculture, 1990–99 (FAO, 2001).

Table 16.1 Tilapia production in 1999 in the six major countries that produce tilapias (FAO, 2001)

Country	Volume (mt)	Share of global production (%)
China	561 794	51.1
Egypt	103 988	9.5
Thailand	98 250	8.9
Philippines	75 437	6.9
Indonesia	68 740	6.3
Colombia	57 183	5.2

offspring in the culture system increases demand for food and other resources, resulting in stunted growth. Thus, a number of solutions to this problem have been developed by researchers and tilapia farmers. Most solutions rely on the production and stocking of all-male tilapias. Male tilapias grow faster than females (Fig. 16.3) and, therefore, all-male tilapia populations provide an added growth factor, besides controlling unwanted recruitment.

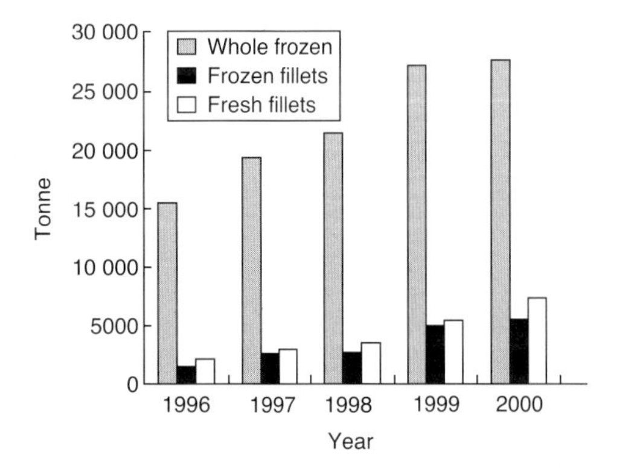

Fig. 16.2 Tilapia imports into the USA, 1996–2000 based on data provided by the Customs, USA.

Subsequent sections of this chapter provide a more detailed account of tilapia biology and culture. Tilapias are better understood than most aquaculture species because they have been a focus for researchers for over 50 years. In addition, a considerable amount of fundamental research has been done on tilapias because of the ease with which they can be maintained in the laboratory. Much of tilapia culture technology has resulted from understanding specific biological characteristics of tilapias. These are summarised in detail in a number of recent publications (Costa-Pierce & Rakocy, 1997, 2000; Beveridge & McAndrew, 2000).

16.2 Family, species and genetic variation

16.2.1 Family

Tilapias belong to the family of Cichlidae, a large family of tropical freshwater fish, which have a

Table 16.2 Characteristics of tilapia culture systems

	Extensive	Semi-intensive	Intensive
System type	Ditches, rice fields, backyard ponds, community ponds, reservoirs and tanks for irrigation	Ponds built specifically for fish farming	Small ponds, tanks, cages, raceways
Stocking density (no. of animals/m^2 or m^3)	< 1	2–3	> 5
Source of seed (fingerlings) for stocking	Wild fish, by-product of culture	Commercial hatcheries	Own hatchery
Reproductive control	None	All-male stock may be used	All-male stock
Fertilisation	None except incidental, run-off fertilisation	Manure and inorganic fertilisers applied	None
Feeds	None except occasional farm by-products and household wastes	Farm by-products such as rice bran, oilseed cakes or supplementary compound feeds	Complete, compounded feeds
Aeration/water exchange	None	Limited, occasional water exchange	Aeration and water exchange to manage water quality
Culture duration	Seasonal	6–9 months	4–6 months
Yield (mt/ha/year)	< 1	1–5	> 5
Market	Producers' own consumption and local, rural markets	Local and national markets	Urban, high-value, export markets

Fig. 16.3 Sexual dimorphism in growth: male (top) and female (bottom) of the same age (photograph by Dr C. Kwei Lin).

bilaterally compressed body and exhibit parental care. Cichlid fish have a wide natural distribution throughout the tropics, but the tribe, Tilapini, to which tilapias belong, occurred only in Africa and Palestine until translocations began. There are about 10 genera and over a hundred species within Tilapini. The common term 'tilapias' refers to pure species as well as hybrids belonging to the genera *Tilapia*, *Sarotherodon* and *Oreochromis*, especially the larger species that are commercially exploited. The three major genera are differentiated by the way they brood their eggs and larvae:

(1) Tilapias of the genus *Tilapia* lay their eggs on a substrate, which may be a depression on

the pond bottom or tree roots or submerged vegetation. Both parents care for the eggs until hatch. Females fan and clean the eggs with their fins while males guard the territory.

(2) Tilapias of the genus *Oreochromis* lay their eggs in a pit or nest prepared by the males. After the eggs are laid, the female parent incubates the eggs in her mouth. Parental care continues for several days even after the eggs hatch and the fry are free-swimming.

(3) Tilapias of the genus *Sarotherodon* are quite similar to *Oreochromis* in their reproductive behaviour, but both parents, or just the male parent, are responsible for mouth brooding.

Most farmed species of tilapias belong to the genus *Oreochromis*. Further details of reproduction in *Oreochromis* species are provided elsewhere in this chapter.

16.2.2 Species

Among the tilapias, members of the genus *Oreochromis,* such as *O. niloticus* and *O. aureus*, are favoured in aquaculture because of their performance under culture conditions (Table 16.3). Among the pure species, *O. niloticus* (common name, Nile tilapia) is the primary choice. Globally, 888 368 mt of *O. niloticus* was produced in 1999 (FAO, 2001). This accounted for nearly 81% of all tilapia production. The exceptional growth of this species in tropical freshwater conditions is the reason for its success. However, *O. niloticus* has poor cold tolerance and does not perform well in high-saline waters. So, in subtropical waters, the more cold-tolerant *O. aureus* (common name, blue tilapia) is the species of choice. Species such as *O. spilurus*, which tolerate and grow in high salinity, are considered for farming in seawater, particularly in the Middle East, where freshwater resources are limited. *O. mossambicus* (common name, Mozambique tilapia) was the first tilapia species to be introduced into various parts of Asia and beyond. It has, however, gradually lost favour for use in aquaculture as it developed a reputation as a fish that stunts early in pond culture and rarely reaches its genetic potential for growth and ultimate size.

16.2.3 Strains

The wide geographic presence of species such as *O. niloticus* in Africa has meant that there has been significant genetic divergence, resulting in distinct subspecies and strains. Their subsequent transfers outside Africa have also provided opportunities for further development of distinct strains. For example, about 200 individuals of *O. niloticus* were sent from Cairo to Japan in 1962. Out of these, 120 individuals survived and formed the basis of the Japanese stock that exists today. About 50 individuals of this Japanese stock were sent to Thailand in 1965 and placed in the royal pond. They formed the basis of the Chitralada strain, which has performed exceptionally well under Thai farming conditions. Similarly, a number of other strains of *O. niloticus* exist in Asia and elsewhere. These strains are known by their origin in Africa (such as Ugandan strain or Ghanaian strain) or the intermediate locations from which they were transferred (such as Israeli strain or Taiwanese strain) or the destination in which they were further domesticated (such as the Philippines strain or Chitralada strain). Eknath *et al.* (1993) conducted strain evaluations in the Philippines, comparing growth performance of farmed Asian strains of *O. niloticus* and wild *O. niloticus* from across the natural range of the species in Africa. They showed that wild strains, in general, grew better than domesticated strains and that wild strains from East Africa (Egypt and Kenya) were better than those from West Africa (Ghana and Senegal). Among the domesticated strains, the Chitralada strain, further domesticated in the Philippines, fared best. The worst performance was shown by the most common strain in the Philippines, originally imported from Israel.

16.2.4 Hybrids

There are several naturally occurring tilapia hybrids arising from two or more species sharing a geographic location. Some of these natural hybrids are good candidates for aquaculture. Intentional hybridisation for aquaculture purposes started with the accidental discovery that crossing female *O. mossambicus* with male *Oreochromis urolepis hornorum* resulted in all-male offspring (Hickling, 1960). Subsequent research showed that several other crosses also

Table 16.3 Characteristics of *Oreochromis* species suitable for aquaculture

Species	Growth	Critical environmental tolerance factors	Suitability
O. niloticus	Fastest growing species in many countries. Maximum size ~3 kg	Lower lethal temperature = 12°C; does not tolerate high salinity	Highly suitable for farming in tropical, freshwater and brackish water systems
O. aureus	Fast growing species. Maximum size ~ 3 kg	Tolerates cold temperatures relatively better than most species (lower lethal temperature = 8°C)	Best candidate for farming in subtropical freshwater and brackish water systems
O. mossambicus	Fast growth and large maximum size (~3 kg) observed in wild, but stunting common in culture	Lower lethal temperature = 10°C. Grows well and reproduces under salinities as high as 35‰	Suitability as a pure species is questionable. A good candidate for hybridisation if salinity tolerance is desired in the offspring generation
O. spilurus	Grows fast when young but slows down in adulthood. Maximum size ~1 kg	Low tolerance of cold temperatures like most *Oreochromis* species, but handles high salinity extremely well	A good candidate for marine systems

produce all-male or predominantly male offspring (Table 16.4). This was attributed to the differences in sex determination mechanisms between different tilapia species (section 7.3.3). However, commercial use of hybridisation as a method to produce all-male stocks has been limited to date. Successful production of all-male populations by this method requires a high level of vigilance to ensure that the parental species and strains are maintained.

No growth rate improvements have been observed from hybridisation (Bentsen *et al.*, 1998). Despite this, hybrids that exhibit commercially advantageous traits relative to their parental species have been developed. For example, Israel has developed an *O. niloticus* × *O. aureus* hybrid that combines the fast growth of *O. niloticus* and the cold tolerance of *O. aureus* (Wohlfarth, 1994). It is also believed that the primary candidate for tilapia culture in China is an *O. niloticus* × *O. aureus* hybrid.

16.2.5 Red tilapias

One of the significant advances in tilapia farming was the development of 'red tilapias' in the 1980s. Most tilapias, particularly *O. mossambicus*, which was then a widely cultivated species in Asia, have a dark grey-black skin colour. Their peritoneal cavity is also black in colour. This coloration was deemed unattractive in several markets, resulting in poor acceptance of the fish. Mutants that possessed red skin were observed in *O. mossambicus*, first in Taiwan, and later in the USA and Israel (Lovshin, 2000). These mutants were developed into red *O. mossambicus* strains (Table 7.6), which evoked strong commercial interest in the culture of tilapias. Red tilapias not only lacked the stigma associated with the coloration of the wild-type tilapias, but also resembled premium marine species such as sea bream and red snapper. They had the potential to achieve wider consumer acceptance and higher prices in many markets. Because pure *O. mossambicus* red strains had poor growth characteristics, they were hybridised with faster-growing tilapias, such as *O. niloticus*, *O. aureus* or their hybrids. As a result, a large number of red tilapia strains are available for fish culturists. There are differences between the strains in:

- growth rate
- tolerance of low temperature
- tolerance of high salinity
- other performance characteristics.

Table 16.4 Tilapia crosses that result in all-male or predominantly male offspring

Female	Male	Males in offspring generation (%)
O. mossambicus	*O. urolepis hornorum*	100 in pure strains
O. niloticus	*O. urolepis hornorum*	100 in pure strains
O. niloticus	*O. nyasalapia macrochir*	100 in selected strains
O. niloticus	*O. variabilis*	100 in pure strains
O. niloticus	*O. aureus*	100 in selected strains and 70–80 in mass spawning
O. spilurus niger	*O. nyasalapia macrochir*	100 in pure strains
T. zilli	*O. andersonii*	100 in pure strains

Because many red tilapia strains were derived from *O. mossambicus*, they perform well in saline environments but may have low tolerance of cold temperatures.

16.2.6 Genetically improved tilapias

Despite the impressive growth rates of some strains of tilapias, such as the Chitralada strain of *O. niloticus*, many fish geneticists believe, with some evidence, that the existing genetic base of tilapias in many countries is rather poor. Genetic improvement by means of selection or crossbreeding or both has been applied in developing strains of tilapias targeting a specific trait (mainly growth rate). A well-documented effort in tilapia genetic improvement is the development of the genetically improved farmed tilapias (GIFT) line of tilapias in Asia (Eknath *et al.*, 1993; Bentsen *et al.*, 1998; Dey & Gupta, 2000). The line was developed by crossing eight strains of *O. niloticus* followed by combined family and mass selection for body weight. Field trials in five Asian countries demonstrated that the GIFT line yielded 18–58% larger fish compared with local strains of *O. niloticus*.

At least two lines of true-breeding transgenic tilapias have been established (Mair, 2001). One is an *O. urolepis hornorum*-based hybrid that expresses a transgenic tilapia growth hormone gene and grows 55% faster than its non-transgenic counterpart. This was developed in Cuba and has been approved for use in commercial production in that country. The other line is an *O. niloticus* that is transgenic for a salmon growth hormone and grows 3–4 times faster than its non-transgenic counterpart. This line was developed in the UK and is yet to be approved for commercial use.

16.3 Ecology and distribution

16.3.1 General

Africa, excluding Madagascar but including parts of the Middle East, is the home of all tilapias. In general, *Oreochromis* species are endemic to the central and eastern parts of Africa, whereas *Tilapia* and *Sarotherodon* species are more common in the western parts. However, species such as *Tilapia zillii*, *Sarotherodon galilaeus* and *O. niloticus* have a much larger native range. Another well-known species, *O. aureus*, is native to the Nile delta and the Middle East.

Tilapias live in a wide variety of ecosystems:

- slow-moving parts of rivers
- floodplain pools
- swamps
- lakes
- coastal lagoons.

They are considered to have evolved as riverine fish that eventually colonised lakes. They are strictly warmwater species. They stop growing at temperatures below 16°C and do not survive below 10°C. They survive and grow in brackish waters and many species can tolerate, and grow in, seawater. Adult tilapias primarily eat plant materials (phytoplankton, benthic algae, macrophytes, etc.) and detritus derived from plant materials. They are also highly opportunistic feeders that are capable of changing their choice of food items or their feeding habits.

16.3.2 Translocations

The ability of tilapias to adapt to a wide variety of environmental conditions has led to the successful translocation of many tilapias within and outside Africa. The first translocation outside Africa occurred in the early 1930s, when *O. mossambicus* was introduced into Java (part of present-day Indonesia). From there it was introduced into much of tropical Asia and eventually into the Americas. Most of the introductions were purposeful, mainly for controlling aquatic weeds and insect pests, farming or enhancing fisheries. But a number of accidental and undesirable introductions have also occurred. For example, *O. mossambicus* was introduced into Australia (Queensland) in the 1970s as an aquarium fish, which then escaped into natural waters and became established as a feral population. Currently, it is regarded as a pest species that threatens the natural ecosystem it occupies. Other tilapias that have been deliberately established over a wide geographic range include *O. niloticus, O. aureus, O. urolepis hornorum, Tilapia rendalli* and *T. zillii*.

16.4 Sex determination and reproduction

16.4.1 Sex determination

Sex determination in tilapias is highly complex, with gender being determined by genetic, environmental and hormonal factors. Genetic determination is primarily through sex chromosomes and two different mechanisms for sex determination through sex chromosomes have been proposed.

(1) Species such as *O. mossambicus* and *O. niloticus* possess a system similar to humans in which females are homogametic (XX) and males are heterogametic (XY).
(2) Species such as *O. aureus* and *O. urolepis hornorum* possess a system in which females are heterogametic (WZ) and males are homogametic (ZZ). A cross of the species with different systems results in all-male or predominantly male progenies as shown in Table 16.5.

Apparently, the male-determining gene (Z) in the WZ system dominates the female-determining gene (X) in the XY system, whereas the male-determining gene (Y) in the latter dominates the female-determining gene (W) in the former.

A number of studies have shown that sex chromosomes are not alone in determining sex in tilapias.

(1) Genes in autosomes are suspected of playing a role in determining sex in *O. niloticus* and *O. aureus*. This may explain the inconsistent results obtained in the production of all-male *O. niloticus* × *O. aureus* progeny.
(2) Temperature plays a significant role in tilapia sex determination. Exposure of *O. niloticus* fry to high temperature (34–35°C) results in significant sex reversal in both directions, whereas that of *O. aureus* results in masculinisation. Another case is *O. niloticus* × *O. aureus*, a commercially produced hybrid in Israel. Because of its cold tolerance and, therefore, culture at relatively low temperatures, significant declines in the proportion of males have been observed in successive generations (Wohlfarth, 1994).
(3) Steroid hormones have perhaps the most significant effect on sex determination in tilapias as well as in many other fish. Apparently there is a period in the development of these species during which the sex is determined by the level of androgenic and oestrogenic hormones circulating in the body. Exogenous administration of a specific hormone or its analogue during this labile period will therefore result in the production of monosex stocks (section 7.5.2).

16.4.2 Reproductive biology

All tilapia species mature early (4–6 months) and reproduce year-round under suitable environmental conditions. Tilapias are perhaps the only major group of fish in aquaculture that breed in captivity without any special inducement or modification to their environment (Little & Hulata, 2000). The most common tilapia species, *O. niloticus*, reaches sexual maturity at a size of 30–40 g. When mature, tilapias

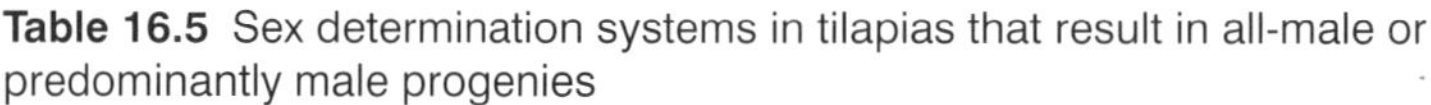

Table 16.5 Sex determination systems in tilapias that result in all-male or predominantly male progenies

O. mossambicus, female (XX) × *O. urolepis hornorum*, male (ZZ)

↓

XZ (all-male hybrid)

O. urolepis hornorum, female (WZ) × *O. mossambicus*, male (XY)

↓

WX (25% female hybrids)

WY, XZ, YZ (75% male hybrids)

can spawn year-round if the water temperature stays above 24°C. Typical breeding behaviour of *Oreochromis* species is shown in Fig. 16.4.

When a mature *Oreochromis* female is ready to spawn, she visits the breeding arena or 'lek'. The breeding arena consists of several males that form well-defended, individual nests. After brief courtship, the female lays her eggs while the male simultaneously fertilises the eggs. The female then picks up the fertilised eggs in her mouth for brooding and leaves the arena. Intensive parental care continues until the fry are large enough to be on their own. The female stands guard over the free-swimming fry. If the fry are threatened, they return to their mother's mouth until the threat passes. Mouth brooding lasts for 3 weeks, during which time the females eat little. Finally, the hatched fry are released in shallow waters. The female then resumes active feeding, which allows maturation of her ovaries. After a further 2–4 weeks, she is ready to spawn again.

Female tilapias lay their eggs in multiple batches. Typically, a female lays 8–12 batches per year under favourable temperature conditions. In the case of *Oreochromis* species, each batch contains about 2000 eggs. The eggs are large (3–5 mm) and contain sufficient yolk to sustain the newly hatched fry for 3–5 days after hatch.

To boost broodstock productivity, the time between two successive spawnings may be considerably shortened by removing the fertilised eggs from a female's mouth. Once the clutch of eggs is removed, the female assumes that the eggs are lost and quickly returns to reproductive mode. This is similar to removing eggs from a chicken to encourage it to lay more eggs. The eggs removed from the females are incubated and hatched in upwelling jars or trays with flowing water (Fig. 16.5).

16.5 Control of reproduction

16.5.1 Stunting

As outlined in section 16.1, early maturation and prolific spawning of tilapias in grow-out systems present two major problems for tilapia farmers:

(1) Energy resources are directed to the processes of sexual maturation and reproduction and become unavailable for somatic growth.
(2) Continuous recruitment of young tilapias into the grow-out systems means increasing competition for resources such as space and food (Fig. 16.6).

In the past few years, a number of methods have been developed to control tilapia reproduction and recruitment. Most of these methods are geared to producing all-male stocks of tilapias because males grow faster than females in almost all tilapia species (Fig. 16.3).

16.5.2 Monosex populations

Hand-sexing

Tilapias display distinct sexual dimorphism as they become juveniles. The males and females are differentiated by means of their genital morphology:

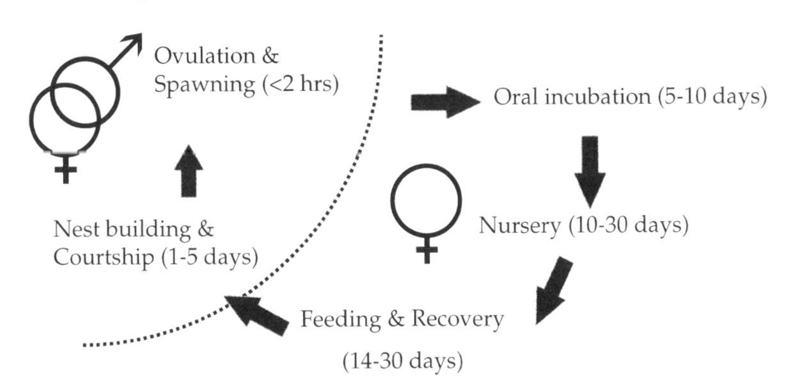

Fig. 16.4 Reproductive cycle of *Oreochromis* species.

Fig. 16.5 Clutch removal: removing fertilised eggs from a female *O. niloticus* mouth in a hapa-based hatchery (photograph by Dr C. Kwei Lin).

Fig. 16.6 Presence of male and female tilapias in culture systems results in reproduction before market size (photograph by Dr C. Kwei Lin).

- Males have a single opening.
- Females have two openings in the urinogenital papillae.

This differentiation becomes more obvious when the fish are 10 g and larger. The typical practice is to sex individual tilapias before stocking for grow-out and select only the males for stocking. It is a very labour-intensive method and wastes nearly 50% (females) of the seed. It is rarely practised these days.

Hybridisation

The use of hybridisation to produce all-male populations was considered in section 16.4.1.

Hormonal sex reversal

General techniques for hormonal sex reversal in fish were outlined in Chapter 7 (section 7.5). Most commercial all-male tilapia stocks are produced by treating tilapia fry with synthetic androgens, particularly 17α-methyltestosterone (MT) (Phelps & Popma, 2000). The most commonly used protocol is to incorporate MT into the fry diet at 40–60 ppm and feed it to the fry from the first feeding stage for 21–30 days. Seed stock containing more than 99% males can be obtained if the protocol is strictly adhered to. Synthetic androgens such as fluoxymesterone and trenbolone acetate are also highly effective in producing all-male tilapia stocks. Instead of adding to the diet, the hormones could also be applied as a bath. This method, however, produces highly variable results and therefore is not widely used.

Tilapias can be feminised using oestrogens. Non-steroidal oestrogens such as ethynyloestradiol and diethylstilboestrol are commonly used in feminising tilapias. The purpose of feminisation is to use feminised genetic males in breeding programmes to produce all-male stock (section 7.5.3).

However, use of hormones in sex reversal evokes environmental and food safety concerns because of the potential for hormones to enter water bodies and the human food chain. On this basis, genetic manipulation methods that do not require hormone use for masculinisation have been developed.

Genetic manipulation

Genetic manipulation of tilapias to produce either monosex or sterile stocks has nearly a 25-year history. Triploidy, gynogenesis and androgenesis are commonly used techniques that have been applied in the commercial production of tilapias. These techniques and the benefits and uses of triploids, gynogenesis and androgenesis are described in detail in section 7.6. Chromosome manipulation techniques are not appropriate for routine production of sterile or monosex stocks because it is difficult to induce 100% triploidy, gynogenesis or androgenesis at a commercial scale. Furthermore, growth and survival of triploid, gynogenic and androgenic fish are generally inferior to that of normal diploids during early life stages.

Production of genetically male tilapias

Production of genetically male tilapias eliminates the need for using hormones to mass produce tilapia seed. The method, however, requires the use of hormonal sex reversal at the initial stages of broodstock development (section 7.5.3). It is a laborious and time-consuming method because of the extensive progeny testing involved. But, once a broodstock population is founded, it can be used in any hatchery system by replacing normal males with YY super-males (section 7.5.3, Fig. 7.7). Hatcheries should then exercise sound broodstock management practices to prevent genetic contamination of the YY stock. Commercial stocks of genetically male tilapias are now available, including the pioneering stock produced by the University of Wales, Swansea. This stock has been shown to outperform equivalent mixed-sex tilapias and hormonally sex-reversed, all-male tilapias (Beardmore *et al.*, 2001).

16.5.3 Controlling recruitment

Although there is no recruitment in monosex populations, there is a method used for controlling recruitment in mixed-sex populations. The predator control method attempts to prevent overpopulation by using predators that eat newly recruited tilapia fry and fingerlings. A number of predator species have been tested:

- Asian catfish
- snakehead fish
- Nile perch
- some South American cichlid species.

To be used reliably, this method requires optimisation of predator–prey ratios, timing of predator release, and size of the predator at release. It has not achieved much success in practical tilapia culture.

16.6 Seed production

16.6.1 Harvesting from mixed-sex stock

Systems to produce seed stock for tilapia culture vary from location to location based on:

- local demand for seed
- geographic conditions
- environmental conditions
- economic factors.

The simplest form is to use fingerlings that result as the by-product of tilapia grow-out in ponds or tanks. Any system that uses both sexes of a tilapia species in grow-out will inevitably produce tilapia seed as a result of natural reproduction (Fig. 16.6). Although reliable technologies are available to control this reproduction through the production of all-male stock, many small farmers still use mixed-sex stocks in tilapia grow-out. For them, the fingerlings produced in their ponds are inexpensive stocking material for further grow-out. This system, however, suffers from many drawbacks, mainly inefficiency and unreliable supply of uniform-size seed. Hormonal sex reversal, the most common technique used in the production of all-male tilapias, cannot be applied in this system.

The result of this inefficient system is the development of dedicated seed production systems. These systems may use open ponds, tanks or net cages (hapas) in ponds or large water bodies (Table 16.6).

16.6.2 Pond systems

Earthen ponds are widely used in the production of tilapia fry or fingerlings. The ponds are typically small (0.01–0.1 ha) and well managed by means of fertilisation, water control, etc. Brood fish are stocked at low densities (0.5–1 tilapias/m^2) with a male–female ratio of 1:2–3, and they are fed. Fry and fingerlings are netted out on a periodic basis (daily, weekly or biweekly). This type of a system may yield fry at a rate of 0.1–3/m^2/day. The following factors influence fry yield in pond systems (Little & Hulata, 2000):

(1) Pond size. The smaller the pond, the better it can be managed and harvested, and therefore this increases its productivity. Ponds of < 1000 m^2 are preferable.
(2) Harvest interval. More frequent harvests result in removal of the larger seed, thereby reducing cannibalism on younger fry.
(3) Stocking density. Lower stocking density typically improves broodstock efficiency and produces larger seed.

However frequently and efficiently fry are harvested from breeding ponds, it is practically impossible to remove all fry from a pond. The pond is soon overpopulated with recruits from early spawns. This results in increased competition for food and space, which diminishes seed output. To overcome this problem, the breeding ponds have to be periodically drained to remove all fish. An alternative to this practice is to seine the pond and remove all broodstock, which are then transferred to another pond. The fry left in the pond are further nursed to a size suitable for stocking. Another alternative, practised in Israel, is to construct breeding ponds with two compartments: a spawning compartment and a fry-collecting compartment that is built at a lower level relative to the spawning compartment (Little & Hulata, 2000). The two compartments are separated by a sluice gate that retains the broodstock and allows collection of fry with minimal handling.

16.6.3 Tank systems

Concrete, plastic or fibreglass tanks may be used in tilapia seed production. A higher degree of

Table 16.6 Comparison of various representative hatchery systems to produce tilapia fry

	Pond	Tank	Hapa
Species	*O. niloticus*	Florida red tilapias	*O. niloticus*
Location	Philippines	Bahamas	Philippines
System size (m^2)	300–500	34	1
Stocking density (no./m^2)	1	7	2
Size (g)	Male and female 50–100	Male and female 143	Female 86, male 50–200
Sex ratio (F/M)	3:1	3:1	1:1
Harvest strategy	Harvest of fry, twice per day	Clutch removal every 15–16 days and artificial incubation	Clutch removal every 4 days and artificial incubation
Fry output (no./m^2/day)	0.7	91.7	20
Female productivity (fry/kg/female/month)	82.9	3021	6990
Source	Guerrero (1986)	Watanabe *et al.* (1992)	Mair *et al.* (1993)

control over the broodstock and seed, as well as the spawning environment, is a major advantage (section 16.5.1) in the use of tanks as seed production systems. Practices such as clutch removal and broodstock reconditioning, which improve hatchery efficiency, can be implemented in tank-based systems with relative ease. Tank systems are typically used when water or land resources are limited. This is because of their high seed output per unit of water and land area. The major disadvantage of tank systems is the high cost to build and operate them.

16.6.4 Hapa systems

Hapa-based hatchery systems provide some of the advantages of tank-based systems at a lower cost. A hapa is a cage made of netting that can be suspended in ponds or large water bodies such as lakes and reservoirs. Brood fish are stocked inside the hapa. Eggs and larvae are regularly collected by means of clutch removal. The hapa can also be designed to have two nested compartments. The inner compartment has a mesh that retains the broodstock but allows the fry to swim to the outer compartment, which is composed of fine-mesh netting. The inner cage, with brood fish, is removed every 10–15 days and the fry collected in the outer cage are allowed to grow to a size suitable for stocking. A major disadvantage of the hapa system is that the mesh openings of the hapas are easily fouled. This limits water exchange and supply of natural food organisms. So, hapas must be pulled out of the water periodically (every 15 days is recommended) and washed or sun-dried. This is a labour-intensive practice, but it also provides opportunities to rest and recondition the broodstock, thereby improving their productivity.

16.7 Nutrition, feeds and feeding

16.7.1 Diet and feeding habits

While tilapias in general are opportunistic omnivores, there is considerable specialisation in diet in some species. Young tilapias are carnivorous and prefer zooplankton. As they become juveniles, their diets shift to plant material or detritus of plant origin or both. Phytoplankton, benthic algae, macrophytes and periphyton are common foods of tilapias in nature. *Oreochromis* species feed primarily on microscopic plant materials, whereas *Tilapia* species prefer large plants. Tilapias use a wide variety of feeding methods:

- visual feeding
- suction feeding
- biting
- grazing.

The primary method of feeding in adult tilapias, particularly in *Oreochromis* species, is continuous suction feeding, in which food particles are entrapped by filtration as water is routinely passed over the gills. The food particles are crushed by the pharyngeal bones and then passed into the alimentary tract. Tilapias possess a stomach that can reach extremely acidic conditions (pH < 1) in proportion to the stomach fullness. This acidic condition lyses plant cells and prepares food material for further digestion in the intestine. Tilapias have a very long, coiled intestine. Intestinal pH is 6.8–8.8 and conducive to the action of digestive enzymes such as trypsin, chymotrypsin and amylase. Anaerobic fermentation may also occur in the hindgut. Crude protein digestibility and assimilation efficiency of plant matter (filamentous and planktonic green and blue-green algae) ranges from 50% to 80%.

16.7.2 Nutrient requirements

Protein and amino acids

Like most fish species, tilapias require a high concentration of proteins in their diets when they are young. Various assessments (30–56%) of the protein requirements of young tilapias have been reported. What is clear is that the protein level required for fry is greater than for adults. Field studies show that adult tilapias grow well when feeds containing 25–32% protein are used and that feeds containing higher protein levels do not add any substantial value in terms of growth. The recommended levels of protein in the diet range from 40–45% for fry to 25–30% for grow-out, with brood fish requiring a slightly higher level (25–35%) than other adults. There are limited and not entirely consistent data on the essential amino acid requirements of tilapias.

Energy

Although a feed that meets the protein requirements of tilapias is likely also to meet the energy requirements, balancing protein and energy is important in optimising feed cost relative to growth and other production parameters. Optimal protein–energy ratio in tilapias is 110–120 mg protein/kcal energy when the fish are young. This decreases to about 100 mg protein/kcal energy as they grow and reach adulthood.

Lipids and carbohydrates are cheaper sources of energy than protein. In general, warmwater omnivores such as tilapias utilise carbohydrates better than lipids. Whereas lipid levels above 12% cause reduced growth in tilapias, digestible carbohydrates as high as 40% are well utilised. Among carbohydrates, starch and dextrin are better utilised than glucose, whereas cellulose and other fibre components are not digestible. Inclusion of fibre above 5% causes depressed growth in tilapias.

Lipids and fatty acids

As noted above, tilapias do not appear to effectively use lipids as an energy source. Suggested maximum lipid levels in the diet for tilapias range from 5% to 12%. Excess lipids also result in substantial carcass and visceral deposition of fats. Vegetable oils such as corn oil and soybean oil are superior sources of lipids compared with animal fats, including fish oil. It appears that tilapias require low levels (≈1% or less) of linoleic acid (18:2*n*-6) and linolenic acid (18:3*n*-3) in their diet (section 8.9.7).

Minerals

In the typical hard water used in aquaculture, there is sufficient calcium to meet the calcium requirements of tilapias. In soft waters, dietary calcium is required. Similarly, a dietary supply of magnesium is important in waters that are low in this ion. Recommended dietary levels of minerals have been based on the requirements of other freshwater fish, especially channel catfish (section 17.6.1). Tilapias appear to have an unusually high requirement for chromium. Dietary chromium in the form of chromic oxide improves utilisation of dietary glucose in tilapias and a quantitative requirement of 140 ppm chromium has been established (Shiau & Shy, 1998).

Vitamins

The need for vitamin supplementation in semi-intensive pond culture of tilapias is questionable, because natural foods are rich in many vitamins. However, complete feeds for tilapias in intensive systems, particularly indoor tank systems, require comprehensive vitamin supplementation of the kind that is used in complete feeds for other fish in intensive culture (section 8.10).

16.7.3 Feeds and feeding

Feedstuffs

Tilapias are capable of utilising a wide variety of feedstuffs either as a single feed or as a part of compounded feed. Feedstuffs that are fed to tilapias in extensive grow-out systems include:

- rice bran
- various other grain by-products
- oil seed residues
- aquatic and terrestrial plants
- kitchen waste.

Numerous feedstuffs, including some of those above plus various animal by-products, tubers and fermentation by-products, are used in compounded feeds. Protein and energy digestibility for these feedstuffs ranges between 30% and 90%+ (Hanley, 2000) and, from this range, clearly some feed ingredients are to be preferred. Other factors that influence the choice of components used in feed formulations are:

- cost
- high fibre content
- amino acid deficiencies
- palatability
- toxins
- digestive enzyme inhibitors.

Table 16.7 shows examples of some feed formulations for tilapias. Once considered essential, fish meal has declined in its importance in tilapia feeds. Today, tilapia feeds can be formulated without any fish meal and still perform optimally in most culture systems. This is a major advantage of tilapias for aquaculture.

Feed forms and sizes

Tilapias accept feeds as dry meal, moist meal and pellets. Although dry meal is suitable for feeding fry, neither dry nor moist meals are appropriate for feeding growers in intensive culture systems. A considerable portion of meal-type feeds is wasted. Complete feeds that incorporate high-quality ingredients must be pelleted or extruded to minimise waste. Such processing also enables easy handling, storage and distribution of the feed by the farmer.

Particle size is an important consideration in selecting feeds for tilapias, which prefer smaller feed particles than many other cultured fish species. Unlike other species that swallow whole feeds, tilapias tend to chew large particles. These are repeatedly taken into the mouth and ejected until they are reduced to an appropriate size. This results in leaching of the nutrients and feed wastage. Table 16.8 presents the appropriate particle forms and sizes recommended for feeding tilapias of different body sizes.

Feed input

The extent to which natural foods are available to tilapias influences the amount and quality of feed input. In ponds, the amount of natural foods available to individual fish depends on a number of factors such as soil fertility, type and amount of fertilisers added, and the number and weight of fish stocked. Tilapias stocked at small size (< 10 g) and low density (1–2/m^2) typically grow very well on natural foods, and pond fertilisation may be sufficient to meet the nutrient requirements for growth. However, once the fish reach 40–80 g, satisfactory growth cannot occur on natural foods alone. At this stage, supplementary feeds are needed to attain good growth rates. As natural foods are high in protein, feeds rich in energy, such as rice bran, are ideal supplementary feeds. As the fish grow further to a larger size, e.g. > 300 g, natural foods and supplementary feeds are not sufficient to sustain growth, so complete feeds are required for further growth. As stocking densities increase, growth plateaux occur with smaller sizes of fish, so that complete feeds must be introduced earlier in the culture cycle. The potential of complete feeds to increase growth of tilapias in semi-intensive pond culture was demonstrated by Edwards *et al.* (2000) (Fig. 16.7). *O. niloticus* juveniles of ~25 g initial weight were stocked at 4/m^2 in 200-m^2 fertilised ponds for 4 months. The ponds were subjected to one of the following treatments:

- (F): fertilisation (4 kg N urea + 2 kg P triple superphosphate, TSP)

Table 16.7 Model tilapia feed formulations based on ingredient availability and costs in the USA

Ingredient	Semi-intensive ponds (26% protein)	Intensive ponds (32% protein)	Intensive tanks (36% protein)
Soybean meal	38.3	48.5	50.8
Wheat middlings	4	20	18
Fish meal	4	6	12
Corn	50.8	22.6	16.5
Dicalcium phosphate	1	1	0.8
Vegetable oil	1.5	1.5	1.5
Vitamin mix	0.2	0.2	0.2
Mineral mix	0.2	0.2	0.2

Modified from Lovell (1998) with permission from Kluwer Academic Publishers.

Table 16.8 Feed forms and particle sizes recommended for tilapias

Body size (g)	Particle size diameter (mm)	Recommended form
< 1	0.5–1	Meal
1–2	1–1.5	Crumbles
2–30	1–2	Crumbles
30–100	2.4	Pellets/extruded particles
100–250	3.2	Pellets/extruded particles
250 to market size	4.8	Pellets/extruded particles

Modified from Luquet (1991) with permission.

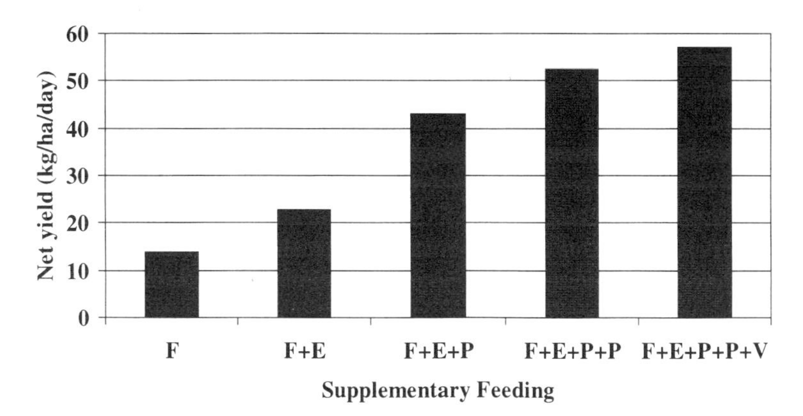

Fig. 16.7 Effect of supplementary nutrients on tilapia yield in semi-intensive ponds (see section 6.7.3 for more details). Modified from Edwards *et al.* (2000) with permission from Kluwer Academic Publishers.

- (F + E): fertilisation + energy (pelleted cassava starch + lipid)
- (F + E + P): fertilisation + energy + protein (fish meal and soybean meal)
- (F + E + P + P): fertilisation + energy + protein + phosphorus (as dicalcium phosphate)
- (F + E + P + P + V): fertilisation + energy + protein + phosphorus + vitamins.

Growth was significantly improved as each nutrient type was added into the feed, except for the vitamins. The addition of the energy source approximately doubled the net yield. Similarly, the addition of the protein meals approximately doubled the net yield. Addition of phosphate increased net yield by ~9 kg/ha/day. The increase in net yield with addition of vitamins, however, was not statistically significant, supporting the view that the natural feeds available to the animals in semi-intensive culture should provide sufficient vitamins.

Feeding allowance

There are two options in practical feeding of fish. One is to feed the fish to satiation and the other is to feed a restricted ration. Best growth is normally achieved by feeding to satiation. But satiation levels are not necessarily the most economic feeding levels, because food conversion at satiation levels is often poor. Also, it is difficult to determine satiation levels in fish because food consumption occurs in the water medium. This may lead to overfeeding, which is wasteful and deleterious to water quality. As a result, restricted rations are recommended for feeding fish. The choice between satiation and restricted feeding should be based on the protein and energy density of the feed. Although restricted feeding of high-protein, high-energy diets is beneficial, low-protein, low-energy diets have to be fed at satiation levels to meet the nutritional requirements of the species.

As discussed in the earlier section, natural availability of foods also determines feed allowance. Trials in Thailand showed that feeding at 50% of satiation rates in fertilised ponds gave similar growth and yield as full satiation feeding (Diana, 1997). Other factors that must be considered when determining feeding allowance include:

(1) Body size. As a fish increases in size, it is obvious that the absolute amount of feed it can consume increases. However, the amount of feed it can consume per unit of body weight decreases. Recommended feeding allowances for different body sizes are provided in Table 16.9.
(2) Temperature and other water quality factors. Feeding allowances presented in Table 16.8 are for tilapias grown at optimum temperature conditions (26–32°C). Feed consumption in tilapias decreases with decreasing temperatures and ceases at 16°C. Appropriate reductions in feeding allowance will be required under such conditions (Table 16.10). Reductions in feeding allowance are also required when temperature increases above optimum levels. At such conditions, dissolved oxygen levels decrease and the toxicity of ammonia increases in culture systems. So, feeding levels have to be lowered at high temperatures too.

Reduced feeding levels are recommended whenever water quality problems are anticipated or encountered. Adverse weather conditions such as high winds and heavy rainstorms cause changes in water quality in pond systems. Heavy algal blooms in pond systems, which may be due to high feeding rates, can cause severe oxygen depletion. In such events, feeding is suspended until conditions improve.

High stocking densities mean high feed input to the culture system. This leads to dissolved oxygen depletion and high ammonia levels. Stress due to poor water quality leads to poor feed consumption and conversion. So culturists using high stocking densities must use devices such as aerators and mixers or water exchange to maintain water quality at or above acceptable levels (Table 16.11). Pond trials in Honduras showed that maintenance of dissolved oxygen in excess of 10% saturation produces more rapid growth (Diana, 1997). In static ponds, feed application must not exceed 120–130 kg/ha/day.

Feeding frequency and time

In their natural habitat, tilapias are known to eat continuously through the day, so multiple feedings may be beneficial. Fry, which eat up to 16% of their body weight every day, need to be fed 8–12 times a day. This frequency is reduced as they grow (Table 16.9). At the grow-out stage, two or three meals per day would be sufficient for optimal growth. Daytime is the best time to feed tilapias. In their natural habitat, they eat during the day, with little or no feeding activity at night.

Feeding method

Hand feeding, although labour intensive, is considered to be the best method to feed tilapias because it allows the farmer to observe the animals and their feeding response. It is essential that the feed is distributed evenly over the water surface to allow all the fish to feed. Tilapias are socially aggressive and tend to develop social hierarchies in which one

Table 16.9 Feeding allowance and frequency for tilapias at different body sizes (modified from Lovell, 1998)

Body size (g)	Daily feeding allowance (% of body weight)	Feeding frequency (no. of meals/day)
<1	30–10	8 12
1–5	10–6	6
5–20	6–4	4
20–100	4–3	3–4
>100	3–2	2–3

Table 16.10 Feeding allowance as modified by water temperature

Temperature (°C)	Per cent of normal daily feeding allowance
> 32	80
24–32	100
22–24	70
22–20	50
20–18	30
18–16	20
<16	No feeding

Table 16.11 Desirable water quality for tilapia culture

Water quality parameter	Desirable level
Temperature	26–32°C
Dissolved oxygen	>3 ppm
Total ammonia	<1 ppm
pH	6.5–8.5
Alkalinity	>20 ppm
Hardness	>50 ppm
Salinity	0–20‰

individual dominates others and appropriates more feed. Uneven distribution of feed will result in dominant individuals occupying places that receive the most feed and eventually lead to large variation in fish size at harvest.

Automatic feeders and demand feeders may be used in cage culture, especially when access to cages is difficult or time-consuming. They may also be useful in large farms that would require extensive manpower to feed the fish. Automatic feeders must be able to distribute the feed evenly. Demand feeders, which are activated by the fish, often result in feed wastage as fish may activate the feeder without an intention to feed. The various method for food distribution in aquaculture systems are outlined in section 9.11.2.

16.8 Grow-out systems

16.8.1 Extensive systems

Extensive systems for tilapias include a broad range of culture units:

- backyard ponds
- roadside ditches
- irrigation tanks
- reservoirs
- rice fields
- wastewater treatment ponds (Table 16.2).

These systems are located in the tropics, operated by poor rural farmers for subsistence. They use basic aquaculture technology. Stocking is irregular and may consist of small tilapias harvested from the wild. No intentional fertilisation or feeding is carried out and yields are < 1 mt/ha/year. Although such systems are quite underproductive when compared with the yield potential of tilapias, they are important to their owners because they provide inexpensive animal protein and some cash income.

16.8.2 Semi-intensive pond systems

Semi-intensive pond systems represent a vast improvement over extensive systems of production.

The ponds are intentionally built for aquaculture. Stocking is planned. Seed is produced on the farm or bought from a hatchery. Methods to control reproduction in grow-out are applied. As in other semi-intensive pond culture, fertilisation of ponds is carried out. Feeds, prepared on-farm or at a feed mill, are used. Yields are typically 3 mt/ha/year, but some well-managed systems may yield up to 10 mt/ha/year. Most tilapia production in the world originates from semi-intensive pond systems.

Semi-intensive tilapia culture systems can be used effectively to satisfy the needs of subsistence farmers as well as the increasing desire of small-scale farmers to intensify production to generate cash income. Considerable research has been directed towards understanding the scientific basis of productivity in semi-intensive tilapia pond systems and has resulted in a vast improvement of our knowledge of optimising pond inputs such as fertilisers and feeds.

Fertilisation of ponds

The ideal levels of N and P in pond water to sustain production of phytoplankton are difficult to estimate. In general, an N/P ratio of 10:1 is recommended because this is the ratio between nitrogen and phosphorus in most phytoplankton. An N level of 1.3 ppm and P level of 0.15 ppm is recommended by some. Others favour a higher level of N (15–30 N to 1 P) to discourage the growth of nitrogen-fixing blue-green algae, which are low in nutritive value to fish, and to encourage the growth of green algae and diatoms, which have a high nutritive value. Loading rates of 4 kg N/ha/day and 1 kg P/ha/day are optimum for semi-intensive tilapia ponds. N loading rates in excess of 4 kg/ha/day, however, may be counterproductive as ammonia may increase to toxic levels. Lin *et al.* (1997) recommend 1.4–2 kg P/ha/day for new ponds because P is easily precipitated into insoluble forms by cations and is strongly adsorbed by pond soils. The high level of P fertilisation is also applicable to ponds in acid sulphate soils because of the soil's sequestering action.

The recommended levels of N and P may be applied in the form of organic fertilisers, such as livestock manures or crop residues, or as inorganic fertilisers. Livestock manures have traditionally been used in Asian aquaculture as fertilisers. In addition to supplying N, P and C, they may also contribute other trace elements. Furthermore, they may also play a role in the detrital food chain. However, the levels of N and P and the N/P ratio of most livestock manures are below optimum. Excessive application of manures to compensate for their low nutrient density results in severe oxygen depletion in the water and accumulation of organic matter in the pond bottom, which ultimately reduces yields. So, it is recommended that ponds are fertilised with livestock manures at lower rates and with inorganic fertilisers to compensate for the nutrient deficiency. One particular strategy that has proved effective is to apply chicken manure at a rate of 200–250 kg (dry-matter basis)/ha/week and to supplement it with urea and TSP at the rates of 28 kg N/ha/week and 7 kg P/ha/week respectively. At a stocking density of three tilapias/m^2, this fertilisation regime provided an extrapolated net yield of 8–11 mt/ha/year.

Supplementary feeding

The rationale and principles of supplementary feeding have been covered in detail above (see section 16.7.3). Supplementary feeding is important in semi-intensive pond culture of tilapias, and most nutrients required in complete feeds are also required in supplementary feeds (Edwards *et al.*, 2000). In practice, two options exist for supplementary feeding in semi-intensive pond culture of tilapias:

(1) Provision of feeds that complement the natural productivity of the ponds. In this approach, the animals are provided with feeds throughout the grow-out period.
(2) Commence feeding once fish growth plateaux on natural foods alone. Beginning feeding at a moderate size may be more economical than at smaller sizes (Diana *et al.*, 1996).

16.8.3 Polyculture

Polyculture of tilapias with other fish species is practised in many countries. Traditional polyculture is based on the premise that various species stocked together utilise different trophic niches that exist in a pond and therefore produce more biomass than if they were stocked alone in monoculture (section

2.3.4). Tilapias have been grown with carps in the Asian polyculture systems (extensive and semi-intensive ponds) for many decades (section 14.5.2). Similarly, tilapias have been grown along with common carp in semi-intensive pond systems in Israel (section 5.4). A recent trend in South and Central America is to grow tilapias with shrimp in brackish water ponds, as shrimp monoculture has been affected by a number of viral disease problems (sections 10.5.2 and 19.1.2).

The suitability of a given species in polyculture depends on its compatibility with other species, and tilapias have been shown to be compatible with some carp species and other fish. However, in polyculture, tilapias and other fish species are stocked at low densities: 5000–10 000 fish/ha. Compared with this, 30 000–40 000 fish/ha are stocked in semi-intensive monoculture of tilapias. Limited studies and field experience suggest that tilapias are not appropriate species for polyculture at high densities. Tilapias are typically more aggressive in their feeding and tolerate crowding and poor water quality conditions better than most tropical aquaculture species. When tilapias fetch an attractive price in the local or export markets, it is probably more profitable to grow them at high densities in monoculture than at low densities in polyculture. In many countries, such as Taiwan and Israel, tilapia farming began as polyculture but eventually developed into intensive monoculture. Tilapia polyculture is more desirable, however, when a farmer requires production of a variety of species for the market and where the other species used in polyculture also fetch an attractive price. Most tilapias produced in China are believed to be from carp polyculture systems, although there is also a trend of Chinese traditional carp polyculture systems shifting to tilapia monoculture. In rural SE Asia, polyculture of tilapias with Indian carps is practised because the latter also fetch an attractive price in the local markets.

16.8.4 Integrated farming

Tilapias play an important role in integrated terrestrial livestock/aquaculture systems in Asia. Integration of tilapia production systems with pig, chicken and duck production systems is practised mostly in SE Asia. Typically, the housings for the livestock are constructed adjacent to or above the tilapia pond (section 2.3.5 and Fig. 2.9). The pond, therefore, receives manure and uneaten feed from the livestock system on a regular basis. An extrapolated net fish yield of 10.1 mt/ha/year was reported in a tilapias–duck integrated system in Thailand, and much of the production was attributed to the inefficient feeding of ducks that resulted in feed wastage. In rural Nigeria, small farmers fertilise their tilapia ponds with excreta from their chicken farms. The ponds serve an important purpose in being the source of water for the farm crops and for poultry in the dry season (Njoku & Ejiogu, 1999). Ponds integrated at 1000 chickens/ha received an excreta load of 3600 kg/ha/month (dry-matter basis), which resulted in an extrapolated net yield of 18.25 mt/ha of an African catfish species and 14.9 mt/ha of *O. niloticus*. Integrated aquaculture practices are further discussed in Chapter 2 (section 2.3.6) and Chapter 4 (section 4.7.3).

16.8.5 Intensive pond systems

Semi-intensive pond systems in the tropics are typically stocked at a rate of two or three tilapias per m^2, and they yield an average of 3 mt/ha per crop, as previously indicated. In areas where land or water or both are limited, or climatic conditions restrict growing season to less than 1 year, it is desirable to achieve very high productivity by using high stocking densities. Experimental culture of tilapias in earthen ponds at stocking densities of 5–10/m^2 has demonstrated that intensive tilapia farming is feasible in earthen ponds. Such intensive pond systems, however, must be managed to maintain adequate water quality, otherwise significant retardation in growth and food conversion will occur. For management purposes, the ponds must be small (< 0.5 ha) and designed to drain and fill effectively within a short period. Aerators are used to maintain desirable dissolved oxygen levels, especially during the critical late night and early morning hours. Adequate circulation of the water must be provided to minimise accumulation of organic waste in the pond bottom. There may also be daily exchange of a part of the water to remove organic debris accumulating on the pond bottom. In Israel and Taiwan, where intensive pond systems were developed, the production systems are connected to a large reservoir that serves as a water treatment body (section 5.3.1). Wastes accumulating on the bottom of ponds are flushed to the reservoir periodically by means of water exchange. Tilapias and carps are stocked in the reservoir at low densities to harvest the algae and thereby reduce nutrient load in the water. Yields ranging from 10 to 30 mt/ha per crop are possible in intensive pond systems.

16.8.6 Cages

Cages provide the opportunity to grow fish in manageable units in large bodies of water such as lakes, reservoirs, and the open ocean, as well as running water bodies such as rivers and irrigation canals (Fig. 16.8). An additional advantage of using cages for tilapia grow-out is that it minimises unwanted recruitment as most eggs drop through the bottom of the cages (although the stock still engage in energy-wasting reproduction). There is commercial cage culture of tilapias in Colombia, Brazil, Indonesia, the Philippines and China. Reservoirs built for irrigation and hydroelectric power generation are used in many locations to grow tilapias in cages. The cages may be made of locally available bamboo, netting, etc., or constructed from commercial-grade materials such as PVC and steel. A wide range of cage sizes are used. Large cages (> 1000 m^3) are used, but are hard to manage. Such cages are typically stocked at 20–25 tilapias/m^3 and yield about 1 kg fish/m^3 per month. For intensive production, smaller cages (> 500 m^3) are preferred and stocking densities are usually more than 100 tilapias/m^3. The typical yield in such systems exceeds 2 kg fish/m^3 per month. Popma & Rodriquez (2000) reported that cages in two major reservoirs in Colombia produced ~2225 mt tilapias/year. Productivity ranged from 67 kg/m^3 per year to 116 kg/m^3 per year.

16.8.7 Raceways, tanks and water recycle systems

Intensive culture of tilapias in raceways is practised when there is an abundant supply of gravity-fed running water. There are a handful of commercial projects in Central America which grow tilapias in raceways. The source of water includes rivers, irrigation canals and reservoirs for hydroelectricity generation. The raceways are typical of this culture system, being generally long and narrow, concrete lined and having water exchange rates in the range of 300–2400% per day. High stocking density (> 50 tilapias/m^3) and biomass (> 20 kg/m^3) are used. Typical yields of more than 40 kg/m^3 per 6 months are possible in raceway systems.

Octagonal or circular tanks, lined with concrete or plastic, are also used in intensive tilapia farming. Stocking densities and yields are as high as those in raceway systems, but the tank systems are commonly used in conjunction with a water recirculation system to minimise water use. The recirculation system may involve exchange with an extensive pond for waste removal as described earlier in section 16.8.5 (Fig. 16.9) and in Chapter 5 (section 5.3.1). Alternatively, it may involve a sophisticated recirculating system with components for removal of solid waste, soluble nitrogenous compounds, addition of oxygen (Fig. 16.10), and disinfection of the water. These systems are generally used in temperate regions of the world, particularly the USA, for commercial production of tilapias. In many cases, the water for these systems is derived from either geothermal springs or power plant effluents. The systems are generally housed within a greenhouse or a similar enclosure to conserve heat (Fig. 16.10). The systems may also be coupled with hydroponic production of vegetables, such as lettuce or tomato, which takes advantage of the nutrient-rich effluents from the tilapia culture units (Chapter 5).

16.8.8 Growing tilapias in saline waters

The ability of tilapias to tolerate and grow in saltwater has resulted in the development of commercial tilapia grow-out systems that utilise brackish or seawater (Suresh & Lin, 1992). Although semi-intensive culture of tilapias in brackish water ponds is now widely practised in some South and Central American countries, there are only a handful of projects involving intensive tilapia culture in seawater in the Caribbean islands and the Middle East. Tilapias cultured in these systems have achieved growth rates and yields that are similar to tilapia cultured in equivalent freshwater systems. For example, Florida red tilapias grown in 0.2 ha brackish water (20–27‰

Fig. 16.8 Red tilapias reared in cages in Colombia (photograph by Dr George Chamberlain).

Fig. 16.9 Concrete tanks for growing tilapias intensively in Taiwan (photograph by Dr C. Kwei Lin).

Fig.16.10 Careful water quality management in an intensive tilapia culture system in southern California, USA (photograph by Dr C. Kwei Lin).

salinity) ponds at three fish per m^2 and fed a 25% protein diet reached an average weight of 452 g in 160 days and yielded a crop of 11.5 mt/ha in 220 days (Watanabe *et al.*, 1997).

16.9 Disease management

Tilapias are far more tolerant of adverse water quality conditions and other stress factors and are less prone to diseases than most other cultured fish. Wild tilapias growing at relatively low densities in their warm freshwater habitats rarely show serious disease. If they are stressed, however, due to crowding, low temperature or high salinity, they become susceptible to disease agents, but there are few infectious agents that are specific to tilapias (Plumb, 1997).

16.9.1 Common diseases

Bacterial diseases

Tilapias are most severely affected by bacterial diseases. Streptococcosis due to *Streptococcus* species, particularly *S. iniae*, *S. faecalis*, *S. facium*, *S. difficile* and *S. agalactia*, is a serious threat to intensive farming of tilapias in the Americas and elsewhere. Clinical signs include:

- lethargic swimming
- erratic swimming
- curvature of the body
- exophthalmic, opaque or haemorrhaged eyes
- accumulation of ascites in the abdominal cavity
- pale liver
- enlarged and black spleen.

Streptococcosis typically occurs at high salinity or cold temperatures. Mortality can be as high as 75%. Infections may be transmitted to humans who handle the tilapias.

Motile *Aeromonas* septicaemia due to *Aeromonas hydrophila* and related species is another common disease of tilapias. Clinical signs include:

- frayed fins
- haemorrhaged skin and fins
- inflamed skin and fins
- scale loss
- ulcerations on the body, head and mouth
- liver pale, with small red spots
- dark-red spleen.

Aeromonas infections occur in freshwater. Poor water quality, cold temperature and skin injury may facilitate infections. The mortality is typically chronic with low daily losses.

Other bacterial infections include vibriosis (due to *Vibrio* species), edwardsiellosis (due to *Edwardsiella tarda*) and columnaris (due to *Flavobacterium columnae*). Clinical signs of most bacterial infections in tilapias are quite similar, so it is necessary to isolate, culture and identify the pathogen for diagnosis and treatment purposes.

Viral diseases

Lymphocystis and infectious pancreatic necrosis-like virus have been known to occur in tilapias, but neither poses a major disease threat. There are a few isolated reports of diseases due to identified or unidentified viral pathogens, but, so far, no viral disease has been widespread in the tilapia industry.

Fungal diseases

Secondary infections due to the water mould *Saprolegnia parasitica* are common in many fish species, including tilapias. Infections occur after injury and particularly when water temperature drops below the optimum. It is a common problem in hatcheries, affecting eggs and reducing hatch rates.

Parasite problems

The protozoan parasite, *Ichthyophthirius multifiliis* (common name, Ich), can cause severe mortalities in tilapias and other freshwater fish (section 17.7.4). This parasite is most lethal at water temperatures between 20°C and 23°C, so it is not a problem when tilapias are farmed in their normal warmwater temperatures. Other ciliated protozoans that affect tilapias are *Trichodina, Trichodinella, Chilodonella, Apiosoma, Ambiphrya* and *Epistylis*. Common monogenetic trematodes such as *Dactylogyrus* species and *Gyrodactylus* species, and parasitic crustaceans such as *Lernaea*, *Ergasilus* and *Argulus* species, have been observed in tilapias. Tilapias grown in seawater are highly susceptible to an ectoparasitic marine monogenean flatworm, *Neobenedenia melleni* (Watanabe *et al.*, 1997). Although most parasites do not cause mortality, they predispose the fish to other serious infections such as streptococcosis, because they damage the scale, skin, fins or gills.

16.9.2 Control of diseases

Table 16.12 shows some of the chemotherapeutic treatments for common infectious diseases in tilapias. Although prophylactic therapies are recommended as part of a disease prevention programme, treatment therapies with antibiotics have been limited in success. Relapse of diseases after antibiotic therapy is common. Immunisation against streptococcosis may be possible in the future. Currently, however, the best option available for control of diseases in tilapias is to manage the environmental and stress factors that contribute to infections. Plumb (1999) recommended the following practices:

- Maintain the highest possible water quality.
- Maintain prudent stocking densities and standing crops.
- Disinfect the water supply and equipment used in the culture facility.
- Expedite removal of dead and moribund fish.
- Handle tilapias gently during stocking, sampling, etc.
- Use prophylaxis during and after handling to aid wound healing.

Diseases are also better managed by understanding the requirements of the disease agent. Marine flatworm may be controlled by the use of low-salinity water. Although this type of treatment is possible in a land-based system, it is impossible or very difficult for seacages, which must be towed into low-salinity water, if this is feasible. So, a novel method of using tropical cleaner fish such as the cleaning goby (*Gobisoma genie*) and the neon goby (*Gobisoma oceanops*) has been found effective in reducing the parasite load in such conditions.

16.10 Harvest, processing and marketing

Harvest size of tilapias varies from 150 g to > 600 g, depending upon the market requirements. Some markets, such as in the Philippines, prefer smaller size tilapias because the typical portion size is a whole fish per person. So, a family of four or five members prefers to buy four or five fish weighing a total of ~1 kg for a meal. Larger harvest sizes are commonly required for export markets. Tilapias have relatively low dressing yields: only 50–55% of the fish is

Table 16.12 Chemotherapeutic treatments for common infectious diseases of cultured tilapias

Drug/chemical	Concentration	Duration	Treatment purpose
Potassium permanganate	2–4 ppm	Indefinite immersion	Prophylaxis, external bacterial infections such as septicaemia and columnaris, and parasites
Formalin	20–25 ppb	Indefinite immersion	
Copper sulphate	0.5–3 ppm	Indefinite immersion	
Potassium permanganate	4–10 ppm	1-h immersion	
Formalin	167 ppb	1-h immersion	
Chloramine-T	10–20 ppm	1-h immersion for 3 days	External bacterial infections
Terramycin	50 mg/kg/day	In feed for 12–14 days; 21-day withdrawal period	Systemic bacterial infections such as vibriosis, edwardsiellosis and streptococcosis
Romet 30	50 mg/kg/day	In feed for 5 days; 42-day withdrawal period	
Erythromycin	50 mg/kg/day	In feed for 12 days	
Amoxycillin	50–80 mg/kg/day	In feed for 10 days	
Hydrogen peroxide	500 ppb	1-h immersion	Fungi

Modified from Plumb (1999) with permission from the World Aquaculture Society.

available as dressed carcass and 25–30% as fillets. So a tilapia weighing more than 500 g is required to produce a fillet weighing almost 100 g.

Seining is the most common method used to harvest tilapias from ponds. It is difficult, however, to harvest tilapias by seining alone because they tend to jump over the seine or escape underneath the seine. Partial or complete draining of the ponds is required for a complete harvest. Tilapias reared in more intensive systems are relatively easier to harvest because of the smaller size of the system and the high density of fish.

Tilapias reared in ponds often develop an off-flavour problem due to metabolites of bacteria and blue-green algae that thrive in nutrient-rich ponds. This problem is solved by stopping feeding and flushing water through ponds 3–7 days ahead of harvest. Alternatively, the fish could be harvested and stocked in tanks with a flow-through water supply.

Harvested tilapias may be sent directly to the market or to a processor for further processing and packing. When market conditions dictate that the tilapias are sold alive, extreme care must be exercised in harvesting and transporting the fish. Typically, the fish are transported in metal or plastic holding tanks with adequate aeration. Fish for further processing may also be transported alive or in ice. Tilapias are processed in a number of different ways depending upon market requirements. The whole fish may be cleaned and frozen whole or frozen as fillets. However, fresh-chilled products fetch higher prices than frozen products in most markets. For chilling, the fish may be gutted and deheaded before refrigerated packing, or converted into fillets. Fresh-chilled products have a shelf-life of 10–15 days post slaughter.

Market value of tilapias ranges widely. Although whole tilapias in a local market in a developing country may cost less than US$1/kg, the retail price of fresh tilapia fillets in a supermarket in a developed country may exceed US$10/kg. This enormous range of prices provides opportunities for producers as well as processors to maximise their returns by keeping their operations flexible, which, in turn, is beneficial from an economic perspective (section 12.5). Producers may serve the entire spectrum of the market or target a specific market niche and optimise their resources to meet the needs of the chosen niche.

References

Beardmore, J. A., Mair, G. C. & Lewis, R. I. (2001). Monosex male production in finfish as exemplified by tilapia: applications, problems, and prospects. *Aquaculture*, **197**, 283–301.

Bentsen, H. B., Eknath, A. E., Palada-de Vera, M. S., *et al.* (1998). Genetic improvement of farmed tilapias: growth performance in a complete diallele cross experiment with eight strains of *Oreochromis niloticus*. *Aquaculture*, **160**, 145–73.

Beveridge, M. C. M. & McAndrew, B. J. (2000). *Tilapias: Biology and Exploitation*. Kluwer Academic Publishers, London.

Costa-Pierce, B. A. & Rakocy, J. E. (1997). *Tilapia Aquaculture in the Americas*, Vol. 1. The World Aquaculture Society, Baton Rouge, LA.

Costa-Pierce , B. A. & Rakocy, J. E. (2000). *Tilapia Aquaculture in the Americas*, Vol. 2. The World Aquaculture Society, Baton Rouge, LA.

Dey, M. M. & Gupta, M.V. (2000). Socioeconomics of disseminating genetically improved Nile tilapia in Asia: an introduction. *Aquaculture Economics and Management*, **4**, 5–12.

Diana, J. S. (1997). Feeding strategies. In: *Dynamics of Pond Aquaculture* (Ed. by H. S. Egna & C. E. Boyd), pp. 245–62. CRC Press, Boca Raton, FL.

Diana, J. S., Lin, C. K. & Yang Yi (1996). Timing of supplemental feeding of tilapia in fertilised ponds. *Journal of the World Aquaculture Society*, **27**, 410–19.

Edwards, P., Lin, C. K. & Yakupitiyage, A. (2000). Semi-intensive pond aquaculture. In: *Tilapias: Biology and Exploitation* (Ed. by M. C. M. Beveridge & B. J. McAndrew), pp. 377–403. Kluwer Academic Publishers, London.

Eknath, A. E., Tayamen, M. M., Palada-de Vera, M. S., *et al.* (1993). Genetic improvement of farmed tilapias: the growth performance of eight strains of *Oreochromis niloticus* tested in different farm environments. *Aquaculture*, **111**, 171–88.

FAO (2001). *1999 Fisheries Statistics: Aquaculture Production*, Vol. 88/2. FAO, Rome.

Guerrero, R. D. (1986). Production of Nile tilapia fry and fingerlings in earthen ponds at Pila, Laguna, Philippines. In: *Proceedings of the First Asian Fisheries Forum* (Ed. by J. L. Maclean, L. B. Dizon & L. V. Hosillos), pp. 49–52. Asian Fisheries Society, Manila, Philippines.

Hanley, F. (2000). Digestibility coefficients of feed ingredients for tilapia. In: *Tilapia Aquaculture in the 21st Century* (Ed. by K. Fitzsimmons & J. Carvalho Filho), pp. 163–72. Proceedings from the Fifth International Symposium on Tilapia Aquaculture Panorama de Aquicultura, Rio de Janeiro, Brazil.

Hickling, C. F. (1960). The Malacca tilapia hybrid. *Journal of Genetics*, **57**, 1–10.

Lin, C. K., Teichert-Coddington, D. R., Green, B. W. & Veverica, K. (1997). Fertilisation regimes. In: *Dynamics of*

Pond Aquaculture (Ed. by H. S. Egna & C. E. Boyd), pp. 73–108. CRC Press, Boca Raton, FL.

Little, D. C. & Hulata, G. (2000). Strategies for tilapia seed production. In: *Tilapias: Biology and Exploitation* (Ed. by M. C. M. Beveridge & B. J. McAndrew), pp. 267–326. Kluwer Academic Publishers, London.

Lovell, T. (1998). Feeding tilapias. In: *Nutrition and Feeding of Fish* (Ed. by T. Lovell), pp. 215–25. Kluwer Academic Publishers, Boston, MA.

Lovshin, L. L. (2000). Criteria for selecting Nile tilapia and red tilapia for culture. In: *Tilapia Aquaculture in the 21st Century* (Ed. by K. Fitzsimmons & J. C. Filho), pp. 49–57. Proceedings from the Fifth International Symposium on Tilapia Aquaculture, 3–7 September 2001, Rio de Janeiro, Brazil.

Luquet, P. (1991). Tilapia, *Oreochromis* spp. In: *Handbook of Nutrient Requirements of Finfish* (Ed. by R.P. Wilson), pp. 169–79. CRC Press, Boca Raton, FL.

Mair, G. C. (2001). Tilapia genetics: applications and uptake. *Global Aquaculture Advocate*, **4**(6), 40–3.

Mair, G. C., Estabillo, C. C., Sevilleja, R. C. & Recometa, R. D. (1993). Small-scale fry production systems for Nile tilapia, *Oreochromis niloticus* (L.). *Aquaculture and Fisheries Management*, **24**, 229–35.

Njoku, D. C. & Ejiogu, C. O. (1999). On-farm trials of an integrated fish-cum-poultry farming system using indigenous chickens. *Aquaculture Research*, **30**, 399–408.

Phelps, R. P. & Popma, T. J. (2000). Sex reversal of tilapia. In: *Tilapia Aquaculture in the Americas,* Vol. 2 (Ed. by B. A. Costa-Pierce & J. E. Rakocy), pp. 34–59. The World Aquaculture Society, Baton Rouge, LA.

Plumb, J. A. (1997). Infectious diseases of tilapia. In: *Tilapia Aquaculture in the Americas,* Vol. 1 (Ed. by B. A. Costa-Pierce & J. E. Rakocy), pp. 212–28. The World Aquaculture Society, Baton Rouge, LA.

Plumb, J. A. (1999). Infectious diseases of tilapia and their management. In: *Central American Symposium on Aquaculture*, 18–20 August 1999, San Pedro Sula, Honduras (Ed. by B.W. Green, H.C. Clifford, M. McNamara & G.M. Montaña), pp. 125–113. Asociacion Nacional de Acicultores de Honduras, Latin American Chapter of the World Aquaculture Society, and Pond Dynamics/Aquaculture Collaborative Research Support Program, Choluteca, Honduras.

Popma, T. J. & Rodriquez, R. B. (2000). Tilapia aquaculture in Colombia. In: *Tilapia Aquaculture in the Americas,* Vol. 2 (Ed. by B. A. Costa-Pierce & J. E. Rakocy), pp. 141–50. The World Aquaculture Society, Baton Rouge, LA.

Pullin, R. S. V. (1984). Tilapia – potentially an international food commodity. *Infofish Marketing Digest*, **3**, 35–6.

Shiau, S.-Y. & Shy, S.-M. (1998). Dietary chromic oxide inclusion level required to maximize glucose utilisation in hybrid tilapia *Oreochromis niloticus* × *O. aureus*. *Aquaculture*, **161**, 355–62.

Suresh, A. V. & Lin, C. K. (1992). Tilapia culture in saline waters: a review. *Aquaculture*, **106**, 201–26.

Watanabe, W. O., Smith, S. J., Wicklund, R. I. & Olla, B. L. (1992). Hatchery production of Florida in brackish water tanks under natural-mouthbrooding and clutch-removal methods. *Aquaculture*, **102**, 77–88.

Watanabe, W. O., Olla, B. L., Wicklund, R. I. & Head, W. D. (1997). Saltwater culture of the Florida red tilapia and other saline-tolerant tilapias: a review. In: *Tilapia Aquaculture in the Americas,* Vol. 1 (Ed. by B. A. Costa-Pierce & J. E. Rakocy), pp. 55–141. The World Aquaculture Society, Baton Rouge, LA.

Wohlfarth, G. W. (1994). The unexploited potential of tilapia hybrids in aquaculture. *Aquaculture and Fisheries Management*, **25**, 781–8.

17
Channel Catfish

Craig Tucker

17.1 Introduction

> Sometimes we vary our diet with fish – wall-eyed pike, ugly, slimy catfish, and other uncouth finny things.
>
> Theodore Roosevelt, 26th President of the USA (1858–1919)

> Everything has beauty, but not everyone sees it.
>
> Confucius, Chinese sage (551–479 BC)

President Roosevelt's sentiments, although perhaps overstated in the quotation above, illustrate a major obstacle faced by farmers trying to market catfish in the USA. Quite simply, catfish had a public relations problem – until recently. Through an extraordinarily effective advertising and promotional campaign, farmers have managed to change the public's image of catfish from that of a strange, bottom-dwelling scavenger to that of a superior food fish. Against all odds, the farm-raised channel catfish (*Ictalurus punctatus*) is now the most important aquaculture species in the USA. In 2000, over 270 000 mt of channel catfish were processed, representing about half the total USA aquaculture production. Over 95% of channel catfish aquaculture occurs in four states (Mississippi, Alabama, Arkansas and Louisiana) located in the south-eastern USA. Mississippi is by far the leading channel catfish producing state, accounting for over 70% of the total production.

Channel catfish are in the family Ictaluridae (bullhead catfishes) within the order Siluriformes. They are native to central North America between the Rocky and Appalachian mountains, from the Gulf of Mexico north to the Hudson Bay drainage. Channel catfish are popular sport fish, and they have been widely introduced throughout North America.

The fish is slender and scaleless, with a gently sloping dorsal profile anterior of the dorsal fin. Normally pigmented fish are white to silvery on the undersides, shading to greyish blue or olivaceous to nearly black dorsally. Albino channel catfish are a beautiful peach colour (Fig. 17.1). Albinism is rare in wild fish but is not uncommon in certain domesticated strains. Irregular dark spots are present on the sides of young fish, but are absent in albino fish and are often lost in normally pigmented fish over ~0.5 kg. Eight barbels, four dorsal and four ventral, are located around a subterminal mouth. Channel catfish have soft-rayed fins; however, the pectoral and dorsal fins contain sharp spines. The anal fin, which is used to distinguish channel catfish from closely related species, is rounded and contains 24–30 rays. Another distinguishing characteristic is the deeply forked tail, although the depth of the fork may be much reduced in older fish, especially in large breeding males.

Channel catfish thrive in a variety of habitats, from clear, swiftly flowing streams to sluggish rivers, lakes and ponds. They are bottom dwellers

Fig. 17.1 Normally pigmented and albino channel catfish fingerlings weighing ~50 g.

that prefer a substrate of sand and gravel. In nature, channel catfish are opportunistic omnivores and use food items in proportion to availability. Young fish feed primarily on aquatic detritus, aquatic insects and zooplankton; adults feed primarily on aquatic insects, freshwater crayfish and small fish.

Channel catfish tolerate a wide range of environmental conditions (Tucker & Robinson, 1990; Hubert, 1999). The optimum water temperature for growth is 25–30°C, but fish can survive at temperatures from just above freezing to over 34°C. Growth is slow at temperatures less than 20°C and feeding activity essentially stops at temperatures below ~10°C. The maximum water temperature at which channel catfish can survive indefinitely is ~34°C, and fish survive only briefly at temperatures above 39°C. Adult channel catfish tolerate salinities from near 0 to 11‰, but growth is slowed at salinities above ~6‰. Eggs tolerate salinities as high as 16‰ but tolerance decreases to 8‰ at hatching.

Age at sexual maturity varies from 2 to 12 years, depending on the length of the growing season (fish generally mature faster in warmer climates). In nature, 2–4 years may be required to reach a weight of 0.5 kg, although growth rate depends on temperature and food availability. Channel catfish may live for over 20 years and attain weights in excess of 20 kg.

17.2 Commercial culture

Channel catfish have always been a popular food fish in the south-eastern USA, and traditional markets were supplied by harvest from local rivers and lakes. The regional popularity of the fish stimulated interest in pond culture in the 1950s and 1960s. By 1965, ~5000 ha of ponds were devoted to channel catfish culture, and production totalled ~7500 mt. Local markets remained the outlet for all production.

The industry began to expand at a rapid rate in about 1975. Expansion was stimulated, in part, by declining profits from traditional agriculture (mostly cotton and soybeans), and a desire to diversify agricultural production. Cooperation among farmers, particularly in the development of large feed mills and fish processing plants, and an effective national advertising campaign facilitated rapid growth of the industry. In 1980, ~15 000 ha of ponds produced 35 000 mt of channel catfish; by 2000, ~80 000 ha of ponds were in production and over 270 000 mt of fish were processed.

Taken as a whole, channel catfish possess a combination of biological and cultural attributes that make them excellent fish for commercial aquaculture. Channel catfish do not reproduce in culture ponds (owing to the absence of nesting sites), so the producer has control over pond populations. This is important because uncontrolled reproduction can lead to overabundance of fish, which may reduce the yield of marketable fish. Channel catfish are easy to spawn, and large numbers of fry are readily obtained using simple hatchery methods. The fish do not require special food at any life stage, are hardy, tolerate a wide range of temperatures and environmental conditions, and adapt well to all commonly used culture systems. The flesh of channel catfish grown in aquaculture is firm and white, with a mild flavour that is highly esteemed by American consumers.

17.3 Culture facilities

Over 98% of the channel catfish produced in the USA are grown in earthen ponds because production costs are generally lower for catfish grown in ponds than for those grown in any other culture system. Production of channel catfish in systems other than ponds is profitable only when some special circumstance exists, such as the opportunity to sell fish to a local market at an exceptional price or the availability of an unusual resource. For instance, a unique channel catfish aquaculture industry is present in the arid, inter-mountain region of the western USA. In the Snake River Canyon of Idaho, large artesian springs supply geothermal water, which allows rearing of catfish in flow-through raceways under nearly optimum temperature conditions throughout the year.

The remainder of this chapter focuses on facilities and production practices used in pond culture of channel catfish. Detailed information on these topics and reviews of other culture systems used to produce catfish are provided by Tucker (1985, 2000a), Stickney (1986) and Tucker & Robinson (1990).

Two hydrological types of ponds are used in catfish farming. The first type – called watershed ponds – are built in hilly areas by damming a small stream. In the long term, the major source of water is runoff from the drainage basin above the dam, although a source of pumped water is desirable to help offset evaporation and seepage during droughts. It may be possible to build a closely spaced series of several watershed ponds in the same catchment basin, with one pond discharging into the next pond downstream, but large groups of watershed ponds generally cannot be built in close proximity to one another. This is a serious disadvantage in large-scale operations because management of many ponds scattered across the countryside is difficult. Watershed ponds represent less than 10% of the total pond area devoted to channel catfish farming but are common in some regions, such as western Alabama.

Embankment ponds are the most common types of pond used in channel catfish farming. Embankment ponds are built on flat land by removing soil from the area that will be the pond bottom and using that soil to form levees or embankments around the pond perimeter (section 2.2.1). The ideal size and shape of embankment ponds are a compromise between construction cost, operating costs, management ease, and existing topography and property lines. Small ponds are easiest to manage, and there is evidence that fish production per unit area is greater in smaller ponds. Pond construction costs, however, increase dramatically as pond size decreases, and small ponds have less water area on a given amount of land because more area consists of levees and roads. As a compromise, commercial production ponds average between 4 and 8 ha of water surface, built on 4.5–9 ha of land. Most ponds are rectangular with a 4:1 to 6:1 ratio of length to width. Ponds are no more than ~200 m wide and between 1 m and 2 m deep to facilitate fish harvest by seining. Embankment ponds can be built in large contiguous tracts, which aids in pond management (Fig. 17.2).

Embankment ponds built on flat land have little watershed to supply water by runoff, so a source of pumped water is needed to fill ponds and maintain water levels during dry periods. The source of pumped water can be either surface water or, preferably, groundwater. Groundwaters are dependable throughout the year, free of wild fish, and are of consistent quality over time and less prone to pollution than surface waters. Shallow wells (less than 100 m) pumping from coarse aquifer materials are preferred because they yield abundant water at low cost. The most important quality criteria for the water supply are salinity (it should be less than ~5‰) and the absence of pesticides or other potentially harmful pollutants. It is also helpful if the water has moderately high total alkalinity and calcium hardness (both at least 20 mg/L as $CaCO_3$). Ample total alkalinity provides pH-buffering capacity to the pond water; calcium benefits fish osmoregulation and stress resistance.

The well should provide enough water to replace evaporation and seepage losses and to fill ponds in less than 2 weeks. If ponds fill slowly because the supply is inadequate, production time is lost and infestation of the pond with noxious aquatic weeds is more likely. The minimum acceptable water supply

Fig. 17.2 A large channel catfish farm in north-west Mississippi.

for ponds in the south-eastern USA is ~150 L/min per ha of pond. This is roughly twice the maximum daily rate of pond evaporation in the south-eastern USA, and it allows for maintenance of pond levels during periods of drought with some excess to meet moderate seepage losses.

17.4 Production practices

A typical production sequence for channel catfish farming begins with spawning of brood fish. Spawning begins in the spring when water temperatures increase to above 20°C. At that time, brood fish held in ponds randomly mate and the fertilised eggs are collected from spawning containers and moved to a hatchery. Eggs hatch after 5–8 days of incubation at 25–28°C and fry are reared in the hatchery for an additional 4–10 days. Fry are then transferred to a nursery pond, fed daily through the summer, and harvested in autumn or winter as fingerlings weighing 20–40 g each. Fingerlings are then stocked into foodfish grow-out ponds, fed daily, and harvested when they reach a size desired for processing (0.4–0.8 kg). In the catfish-producing areas of the south-eastern USA, where water temperatures are below 20°C for about 5 months out of the year, roughly 18–24 months are required to produce a food-size channel catfish from an egg.

The simple production sequence described above is complicated by a number of management decisions that must be made to optimise the production strategy for each farm. A few farmers specialise in producing fingerlings, which are then sold to farmers specialising in production of food-size fish. Many farmers combine all aspects of production. Those farms will have broodfish ponds, a hatchery, fry nursery ponds and foodfish grow-out ponds. Specific management practices also vary among farms. This is particularly evident in the variety of management schemes used in foodfish grow-out ponds.

17.4.1 Reproduction and breeding

An important consideration in choosing a fish species for aquaculture is the ease with which the reproductive phase can be controlled or manipulated. In that respect, channel catfish are ideal for aquaculture. Channel catfish broodstock are easy to maintain in pond culture, and spawning efficiency (percentage of female brood fish spawning in a year) is reasonably good without any special manipulation of environmental conditions or the need for hormone treatments to induce ovulation. However, the farmer has control over reproduction because sexually mature channel catfish will not reproduce in ponds unless they are provided with a nesting container in which to mate. Eggs and fry of channel catfish are relatively large and easy to handle. They develop rapidly at the proper temperature and they are hardy. Excellent egg hatchability and fry survival can be obtained in simple, inexpensive hatcheries. Fry accept relatively simple manufactured feeds at first feeding, which is important because it precludes the need for expensive, special feeds during the early life stages.

Until recently, most producers selected brood fish subjectively, with little systematic development of genetic potential. Brood fish were commonly obtained from foodfish grow-out ponds that contained large fish, or from existing broodstock that appeared to perform well on other farms. Fish were selected for general health, size and development of robust secondary sexual characteristics. Beginning in 1986, the United States Department of Agriculture has conducted research with the goal of improving the genetic potential of channel catfish. In 2001, the programme made the first general release of improved broodstock to the industry. That line of fish, called NWAC-103, has excellent growth compared with other strains used in the industry and marks the beginning of widespread availability of scientifically improved germplasm in channel catfish aquaculture.

Although channel catfish may mature sexually at 2 years, when weighing as little as 0.3 kg, they must be at least 3 years old and weigh at least 1.5 kg for reliable spawning. Fish of 4–6 years old, weighing between 2 kg and 4 kg, are considered prime spawners. Older fish produce fewer eggs per unit body weight, and larger fish may have difficulty entering the containers commonly used as nesting sites. During brood fish selection, sex of fish is determined so that females and males can be stocked into brood ponds in the desired ratio. Adjusting sex ratios is important because mostly males, which grow faster than females, may be selected if only the largest fish from a population are chosen as brood fish. The sex of mature channel catfish can be accurately determined by an experienced culturist by examination of external urogenital features and secondary sex characteristics, which include a wide muscular head in the male and a narrower head and a soft, full abdomen in the female.

Brood fish are maintained at relatively low standing crops (less than 2500 kg fish/ha) to provide good environmental conditions and minimise suppression of spawning by overcrowding. Brood fish are seined from ponds and inspected every year or two. Large fish, which may be poor spawners, are culled and replaced with smaller, younger brood fish. Broodstock replacement ensures a vigorous brood population and re-establishes proper fish standing crops.

Periodic inspection of brood fish also provides an opportunity for adjusting the sex ratios within brood populations. Female channel catfish spawn once a year, but males can spawn two or more times. Therefore, stocking more females than males make more efficient use of pond space. A good sex ratio in brood ponds is about two males for every three females.

Brood fish nutrition is important because a poor diet may result in poor egg quality or reduced spawning success. Limited ration can affect female brood fish in particular because males are more aggressive feeders and may out-compete females for scarce food. When water temperatures are consistently above 20°C, brood fish are fed a nutritionally complete manufactured feed of at least 28% crude protein daily at ~1–2% of body weight. At water temperatures of ~13–20°C, less feed is offered and feeding frequency is reduced to every other day. Channel catfish do not actively feed at water temperatures below 10–13°C. Some producers stock forage fish, such as fathead minnows (*Pimephales promelas*), into brood ponds to provide food in addition to manufactured feeds.

The reproductive cycle in channel catfish is controlled by seasonal changes in water temperature. Exposure to relatively cool water temperatures (less than ~15°C) for 1 month or more stimulates gametogenesis. A subsequent slow rise in average water temperature to 20–25°C usually initiates spawning in the spring. Water temperatures of around 25–27°C are considered optimum for spawning.

Channel catfish must be provided an enclosed nesting site for spawning. Various containers have been used successfully as artificial nesting sites (Fig. 17.3). Most containers have an internal volume of ~75 L and an opening of 15–25 cm across. Containers of that size are adequate for brood fish up to ~5 kg. Containers are placed in the brood pond shortly before water temperature is expected to rise into the range for spawning.

When water temperature reaches the spawning range, the male chooses a container and cleans the inside of debris and sediment. The female is attracted to the container by chemical signals and mating begins. Spawning occurs over a period of several hours as several layers of adhesive eggs are deposited. Females of between 2 and 5 kg typically lay between 6500 and 9000 eggs per kg body weight. Once spawning is complete, the male chases the female from the nest and guards the egg mass. The eggs are ~4.5 mm in diameter and initially are light yellow, becoming brownish yellow with age. Spawning success (percentage of females spawning) ranges from 40% to 80% each year, and depends mainly on the condition and age of the female brood fish and water temperatures during the spawning season.

Nesting containers are checked every 2 or 3 days for the presence of eggs. The container is gently brought to the surface and drained of water. If the male remains in the container, he is allowed to swim out before the egg mass is retrieved (Fig. 17.4). The egg mass usually sticks to the floor of the container

Fig. 17.3 Containers used as artificial nesting sites for channel catfish.

Fig. 17.4 A typical channel catfish egg mass retrieved from the spawning container.

and is gently removed by scraping it from the container. The eggs collected from the brood pond are placed in an insulated, aerated container and transported to the hatchery.

17.4.2 Hatchery practices

Hatcheries used to produce catfish fry are simple facilities that use single-pass, flow-through tanks for egg incubation and fry rearing. The most critical factor for a successful hatchery is a dependable supply of high-quality water. The desirable characteristics of channel catfish hatchery water are summarised in Table 17.1. Perhaps the most important characteristic is temperature. If water temperature is below 25°C, egg hatching and fry development are prolonged and fungi may invade the egg masses. If the water temperature is above 30°C, embryos may develop too fast and there will be a high incidence of malformed or non-viable fry. Also, bacterial diseases of

Table 17.1 Desired water quality parameters for channel catfish hatchery water supplies

Variable	Desirable level
Temperature	26–28°C
Total dissolved gases	< 105% total gas pressure
Dissolved oxygen	6 mg/L to saturation
Carbon dioxide	< 10 mg/L
Calcium hardness	> 20 mg/L as $CaCO_3$
Ammonia (unionised)	< 0.05 mg/L
Total iron	< 0.5 mg/L
Hydrogen sulphide	< 0.005 mg/L

eggs and fry and channel catfish virus disease of fry (section 17.7.1) are more common at higher water temperatures. Heating or cooling water is expensive, so it is desirable to use a source that supplies water with a stable temperature near optimum (27°C). Most hatcheries use deep wells (300–400 m deep) to obtain water that is warmed by the internal heat of the earth, making it unnecessary to heat or cool the water.

Shallow tanks holding ~400 L of water are used to incubate eggs and to rear fry. Water flows through the tank at 10–15 L/min. Egg hatching tanks are equipped with a series of paddles attached to a shaft running the length of the tank; the paddles are spaced along the length of the tank to allow wire-mesh baskets to fit between them. One or two egg masses are placed in each basket and the paddles gently rotate through the water to provide water circulation and aeration. The motion of the paddles simulates the action of the male parent, who, under natural conditions, circulates water over the egg mass by fanning the water with his tail.

The incubation time for channel catfish eggs varies from 5 to 8 days, depending upon water temperature. At hatching, the fry (called sac-fry at this point) fall or swim through the wire-mesh basket and school in tight groups in the corners of the tank. Fry are easily siphoned into a bucket and transferred to a fry-rearing tank. Fry rearing tanks are usually the same dimensions as egg-hatching tanks but are not equipped with paddles. Aeration in fry-rearing tanks is provided by surface agitators or by air bubbled through air-stones.

Sac-fry are initially golden in colour and are

not fed because they derive nourishment from the attached yolk sac. Over a 3- to 5-day period after hatching, they absorb the yolk sac and turn black. At that time, fry (now called swim-up fry) swim to the water surface seeking food. Swim-up fry must be fed 6–12 times per day for good survival and growth. Fry feeds are generally dry, finely ground, nutritionally complete feeds containing at least 50% crude protein. Most of the protein should be supplied as fish meal. Fry are fed in the hatchery for 2–10 days before they are transferred to a nursery pond.

17.4.3 Nursery pond management

After a brief stay in a hatchery, fry are moved to a nursery pond for further growth. Fry stocking density determines the average size of fingerlings harvested from nursery ponds after one season of growth (150–180 days). To produce a 20–40 g fingerling, fry are stocked at 200 000–300 000 fry/ha.

Nursery ponds should be fertilised to provide abundant natural foods because it is difficult to feed fry for the first several weeks after stocking. The best fertilisation programme uses a combination of high-phosphorus inorganic fertiliser and an organic fertiliser, such as cottonseed meal, rice bran or alfalfa pellets. Fertilisation should begin 2–3 weeks before fry are added and be continued until fish are actively accepting manufactured feed. Although fry may not be seen feeding on manufactured feed for several weeks after they are added to the nursery ponds, a finely ground, high-protein feed (40–50% crude protein) should be offered once or twice daily to train fish to accept the feed. As the fish grow, feed particle size is increased and the protein level of the feed is decreased. One month or so after stocking, the fish (now called fingerlings) are fed once or twice daily to satiation, using a small floating pellet with 32–35% crude protein.

Fingerlings of 5–9 months of age are harvested from nursery ponds in the autumn, winter and early spring for transfer to foodfish grow-out ponds. Survival of fry to the fingerling stage in excess of 75% is considered very good. Fingerlings from nursery ponds vary considerably in size and are usually graded to obtain a more uniform population for stocking. Grading is accomplished by using a seine with a mesh size that selectively retains fish of a certain minimum size.

After all fingerlings are harvested, the pond is drained and allowed to dry. Removing all fish from the pond is important to prevent cannibalism of fry stocked in the subsequent cycle of fingerling production. Allowing the pond bottom to dry helps reduce populations of predacious aquatic insects, such as dragonfly nymphs (order Odonata) and backswimmers (order Hemiptera, family Notonectidae), which prey on catfish fry.

17.4.4 Grow-out

In contrast with cultural practices used in hatcheries and nursery ponds, which are relatively standardised across the industry, management of foodfish grow-out ponds varies greatly from farm to farm. It is therefore impossible to describe a typical management scheme for production of food-sized channel catfish. Wide variation in cultural practices is due, in part, to differences in production goals between farms and to the lack of information on the economics associated with various production strategies. Individual farmers have developed and used various production schemes based on experience, personal preference, and perceived productivity and profitability.

In the early years of the catfish industry, most farmers used some variation of the 'clean-harvest' cropping system, wherein the goal was to have only one year class of fish in the pond at a given time. Fingerlings were stocked, grown to some desired harvest size (usually 0.4–0.8 kg) and then completely harvested before the pond was restocked with new fingerlings to initiate the next cropping cycle. Fish were either harvested at one time or harvested in two to four separate seinings, spaced over several months. After as much of the crop as possible had been harvested by seining, the pond was either drained and refilled, or restocked without draining, to conserve water and decrease lost production time between crops.

The clean-harvest cropping system is still used by some farmers and remains common in areas where watershed ponds are used. Some watershed ponds are too deep to allow harvest without draining the pond, making it necessary to raise single crops of fish interrupted by pond drawdown and draining. However, most watershed ponds are now constructed to avoid excessively deep areas near the dam and with smooth bottom profiles that allow fish harvest without pond drawdown.

In the mid-1970s, the catfish industry was expanding past the point at which sales were primarily to local markets, and a new cropping system was developed to provide a year-round supply of food-size fish to meet the demand of the new regional and national markets. This cropping system has become known as 'understocking', 'topping' or 'multiple-batch' (Tucker *et al.*, 1994). In the understocking system, several different year classes of fish are present after the first year of production. Initially, a single cohort of fingerlings is stocked. The faster-growing individuals are selectively harvested using a large-mesh seine and fingerlings are added ('understocked') to replace the fish that are removed plus any losses incurred during grow-out. The process of selective harvest and understocking continues for years without draining the pond. After a few cycles of harvest and understocking, the pond contains a continuum of fish sizes ranging from recently stocked fingerlings (20–40 g) to fish over 1 kg.

Oddly, the clean-harvest system has several advantages over the understocking system, even though the understocking system is far more common in the present-day catfish industry. Fish harvested from clean-harvested ponds tend to be more uniform in size than those from understocked ponds, and uniform fish size is highly desired by fish processors. It is also easier to maintain accurate inventory records when using the clean-harvest cropping system because fish populations are 'zeroed out' after each round of grow-out. Feed conversion efficiencies also tend to be better for fish in clean-harvested ponds, probably because there is little or no year-to-year carryover of big fish, which convert feed to flesh less efficiently than small fish.

On the other hand, when all ponds on a farm are managed with understocking, more ponds will contain fish of marketable size at any one time than if the clean-harvest system were used. This is important because pond-raised catfish are often temporarily unacceptable for processing because of algae-related off-flavours (section 17.5.3). So, if timely harvest of fish from a particular pond is constrained by the presence of off-flavours (or other factors, such as ongoing losses to an infectious disease), there is a greater probability of having acceptable fish to sell from another pond when ponds are managed with the understocking strategy. Economic analyses indicate that clean-harvest cropping systems generate greater net revenues than understocked ponds when fish can be harvested and sold without constraint. But, in the real world, where the presence of off-flavours and other market constraints often prevent timely fish harvest, the understocking system is the most desirable of the two cropping systems (Engle & Pounds, 1994). This is probably why the understocking system is the most common cropping system in use and it is likely to remain the cropping system of choice until solutions to off-flavour problems are developed.

The other important management decision that farmers must make, besides choice of cropping system, is the fish stocking density. There is no consensus on the best stocking density, and this is evident from the wide range of stocking densities used in the industry, which varies from less than 10 000 fish/ha to more than 35 000 fish/ha. The lack of consensus on optimal stocking densities is, in part, related to the counterintuitive fact that profit-maximising management practices are not necessarily the same as yield-maximising practices.

High stocking densities and corresponding high feeding rates produce the highest fish yields, although fish grow slowly and degraded pond environmental conditions may put fish at greater risk of loss to infectious diseases. In the absence of information to the contrary, most farmers tend to manage for maximum yield, believing that profits are simultaneously maximised. However, in two studies that examined the relationship between stocking density and economic return (Tucker *et al.*, 1994; Losinger *et al.*, 2000), profits were maximised at stocking densities well below yield-maximising stocking densities. For example, Losinger *et al.* (2000) used econometric modelling and farm survey data to identify profit- and yield-maximising stocking densities for catfish farming in the south-eastern USA. Maximum yield occurred at a stocking density of ~30 000 fish/ha, whereas profit-maximising stocking densities ranged from 17 000 to 21 000 fish/ha. This study did not address the effect of fingerling size at stocking, which is the third important management decision for foodfish grow-out (cropping system and stocking density being the first two). Obviously, much additional work needs to be conducted to determine the best grow-out procedures for food-sized catfish.

Despite considerable variation in fundamental production variables, daily management of foodfish grow-out ponds is similar from farm to farm. When water temperatures are above ~15°C, fish are offered an extruded floating feed of 26–32% crude protein. Feed is blown from mechanical feeders that are mounted on, or pulled by, vehicles. Feed is scattered over a wide area of the pond; usual practice is to feed fish once daily to near satiation based on visual assessment of fish feeding activity. Because fish densities are high in commercial ponds, relatively large amounts of feed are offered daily. Feed allowances in commercial grow-out ponds average between 75 and 125 kg feed/ha/day during the late spring to early summer period of maximum feeding activity. Feeding activity declines as water temperatures drop in late autumn and feeding rates average less than 30 kg feed/ha/day during midwinter, although feed allowances may be considerably higher during abnormally mild winters. Of course, feed allowances vary depending on fish biomass and fish health. In fact, sudden changes in feeding activity are an important indicator of the general health of fish and reduced feeding activity is often the first sign of an infectious disease outbreak.

Foodfish grow-out ponds managed with the understocking cropping system are drained only for purposes of pond renovation. Most commercial ponds remain in production for 3–20 years between pond renovations, with an average lifespan of 6.5 years. Renovation usually involves partial drying of the bottom and scraping to smooth the bottom profile and restore the drainage slope. Accumulated sediment on pond bottoms is derived from erosion of levee slopes by waves. Thus, material scraped from the bottom is used to rebuild the levee and restore the proper slope. Other than measures needed to restore proper pond morphology, no treatment of the bottom soils is undertaken during renovation.

17.5 Water quality management

High fish densities and large additions of manufactured feed are necessary for profitable culture of channel catfish in ponds. These practices result in large quantities of fish metabolic waste entering the water. Because channel catfish ponds are operated with no water exchange during the culture period, waste accumulation directly or indirectly causes deterioration of environmental conditions, which ultimately limits the production of fish. Availability of dissolved oxygen, accumulation of nitrogenous waste products (primarily ammonia) and development of environmentally derived 'off-flavours' in fish are the primary factors that limit the intensification of channel catfish culture in ponds. Reviews of water quality management in channel catfish ponds are provided by Tucker & Robinson (1990), Tucker (1996, 2000b) and Boyd & Tucker (1998).

17.5.1 Dissolved oxygen

Phytoplankton metabolism is the most important variable affecting dissolved oxygen budgets of channel catfish ponds (section 3.2.3). In ponds with dense phytoplankton blooms, dissolved oxygen concentrations often fall to critically low levels at night during the warm months of the year. The dissolved oxygen budget differs from pond to pond because ponds contain different standing crops of fish, different plankton densities and different densities of benthic organisms. The key to successful dissolved oxygen management in channel catfish ponds is early identification of those ponds that may require supplemental mechanical aeration to keep fish alive. Aeration is initiated (during darkness) when dissolved oxygen concentrations fall to a level considered critical by the individual farmer (usually around 3–4 mg/L) and continues until measurements indicate that dissolved oxygen concentrations are increasing as a result of photosynthesis during daylight.

Rates of photosynthesis and respiration decrease as water temperatures decrease; consequently, the amplitude of daily changes in dissolved oxygen decreases during the cool season. Problems with low dissolved oxygen are rare in channel catfish ponds when water temperatures fall below 15°C, and most producers discontinue dissolved oxygen monitoring when water temperatures are expected to remain below that value.

Paddlewheel aerators are the most common

type of aerator used in channel catfish ponds. Most paddlewheel aerators are powered by an electric motor (typically 7.5 kW) and are mounted on floats and anchored to the pond bank. Current practice is to provide aeration at ~2–4 kW/ha; for instance, two 7.5-kW aerators may be placed in a 6-ha pond.

17.5.2 Ammonia and nitrite

The use of mechanical aeration allows farmers to use high fish stocking densities and feeding rates without fish kills attributable to dissolved oxygen depletion. Nevertheless, aeration does not allow unlimited fish stocking densities and feeding rates because accumulation of potentially toxic nitrogenous waste products affects fish health and decreases fish growth rates as the intensity of the culture increases.

Virtually all the combined inorganic nitrogen in channel catfish ponds originates from nitrogen in feed protein and is excreted by fish as ammonia. Ammonia excretion by fish is proportional to the amount of feed consumed and amounts to ~0.03 kg of ammonia nitrogen per kg of feed consumed. At a feeding rate of 75 kg feed/ha/day, input of ammonia nitrogen to a typical catfish culture pond amounts to ~0.25 mg/L/day. If ammonia was not lost or transformed to other nitrogen-containing compounds, unionised ammonia concentrations (the form of ammonia toxic to fish) would increase in a few days to lethal levels. The fact that culture is possible at that feeding rate (and even higher) indicates that transformations and losses of nitrogen act to reduce ammonia concentrations and allow continual input of relatively large amounts of feed to ponds (Boyd & Tucker, 1998; section 3.2.7).

The maximum nitrogen loading rate (in the form of fish feed) possible without unreasonable accumulation of ammonia in the water is governed in the long term by the rate at which natural physical and microbiological processes remove nitrogen from the pond. Practical experience shows that feeding rates up to ~75–125 kg of feed/ha/day are possible during the summer growing season without unreasonable accumulation of ammonia. This is generally accepted as the upper range for feeding rates in channel catfish culture ponds managed with no water exchange.

Nitrite occasionally accumulates in channel catfish pond waters and poses a threat to fish health (section 3.4.2). Nitrite from the water enters the bloodstream of channel catfish and oxidises haemoglobin to methaemoglobin. Methaemoglobin is a brown-coloured pigment incapable of binding oxygen. Fish with high levels of methaemoglobin in their blood may suffocate even when dissolved oxygen concentrations are high.

In south-eastern USA, episodes of elevated nitrite levels are most common in the spring and fall when environmental conditions are rapidly changing. Fortunately, nitrite toxicosis is easily prevented by assuring that adequate chloride is present in the water. Chloride competes with nitrite for uptake at the gills and maintaining a ratio of 10 mg/L of chloride for every 1 mg/L of nitrite nitrogen prevents nitrite toxicosis in channel catfish. Chloride is inexpensively added to channel catfish ponds as common salt, sodium chloride.

17.5.3 Off-flavours

Objectionable flavours ('off-flavours') in processed channel catfish may be caused by feed ingredients, post-mortem rancidity or odorous compounds absorbed from the environment (section 11.11). The most common off-flavours described from pond-cultured channel catfish are attributed to geosmin and 2-methylisoborneol, which are two earthy-musty smelling metabolites produced naturally by certain species of cyanobacteria (blue-green algae). Species of the cyanobacterial genera *Oscillatoria* (= *Planktothrix*) and *Anabaena* are most commonly implicated in producing odorous compounds in channel catfish ponds (Tucker, 2000b).

Off-flavours are common in pond-raised catfish and all fish are 'taste-tested' before scheduling pond harvest. Testing is conducted by trained personnel at the fish processing plants. Typically, fish samples are submitted a week or two before the desired harvest date, a day or two before harvest, and immediately before unloading live fish from transport trucks at the plant. The entire pond population of fish is rejected for processing if off-flavours are detected in any sample. The development of off-flavour is a severe economic burden for catfish farmers because it prevents the timely harvest of fish, increases production costs, disrupts cash flows and interrupts the orderly flow of fish from farm to processor. Furthermore, if off-flavoured fish are inadvertently marketed, the negative reaction of consumers may adversely affect market demand, with the overall effect of reducing profits for all segments of the industry.

The occurrence of earthy-musty off-flavours in channel catfish is episodic and coincides with the appearance and eventual disappearance of the cyanobacteria responsible for synthesis of the odorous compounds. At present it is not consistently possible to prevent off-flavours because the specific environmental conditions leading to the occurrence of odour-producing cyanobacteria are not known. Use of algicides provides some measure of relief but is not always successful because available algicides (such as copper sulphate) are not specifically algicidal or algistatic against nuisance species.

Management of off-flavours in fish relies upon natural elimination of the odorous compound from the flesh to improve flavour quality once the fish is no longer in the presence of the organism producing the compound. Fish with off-flavours often are simply left in the culture pond until the odour-producing cyanobacteria disappear from the plankton community. Elimination of geosmin or 2-methylisoborneol from fish flesh is rapid; however, an unpredictable period (weeks to months) may be required for disappearance of the odour-producing algae from the community. A somewhat more dependable approach is to transfer fish to an odour-free environment, such as a different pond. Flavour quality often improves within 4–7 days after moving the fish. However, moving fish from pond to pond to eliminate flavour problems is labour intensive and stresses the fish.

17.6 Nutrition, feeding and feed formulation

Feed cost is the largest cost of producing channel catfish, accounting for about half of the total cost input. Efficient and economic feeds, together with effective feeding practices, are therefore essential in catfish farming. Knowledge of fish nutrition is the most advanced area of channel catfish culture. The general nutrient requirements of channel catfish have been established, and practical diets can be formulated from relatively few ingredients to meet those requirements. Feeding practices are less well standardised than nutrient requirements or feed manufacture and some consider feeding channel catfish to be much more of an art than a science. Feeding practices for each stage of channel catfish production were summarised above. Reviews of nutrition and feeding of channel catfish are provided by Robinson (1989) and Robinson & Li (1996).

17.6.1 Nutritional requirements

Forty nutrients have been identified as essential for normal physiological function of channel catfish. The qualitative nutritional requirements of channel catfish are similar to those of other animals, but the absolute and relative amounts of those nutrients needed by catfish may differ from those needed by other animals. Also, nutrient requirements for channel catfish may vary with fish size, growth rate, stage of sexual maturity, water temperature, diet formulation, feeding rate and other factors. Many of these interrelationships are poorly understood. Nutrients recommended for channel catfish grow-out feeds are shown in Table 17.2.

The range of dietary protein levels in Table 17.2 has been found adequate for grow-out of fingerlings to food-size fish. Generally, the dietary protein level is higher in feeds used for smaller fish (Table 17.3). Also, the vitamin and mineral requirements for channel catfish were determined with small, rapidly growing fish and may not accurately reflect the requirement for older, slower-growing fish. Mineral requirements are particularly difficult to quantify because part or all of the requirement for some minerals may be met by absorption of dissolved minerals from the water.

17.6.2 Feed formulation and manufacture

Feeds used in intensive pond culture of channel catfish are formulated to provide all required nutrients in the proper proportions. No single foodstuff can supply the optimum mix of nutrients, so a combination of feed ingredients is used to meet nutritional needs. Fortunately, the number of different foodstuffs needed to formulate affective channel catfish feeds is small. Examples of formulations for practical channel catfish feeds are shown in Table 17.4.

Feeds for fry are prepared as meals or flours of small particle size (< 0.5 mm). They are formulated

Table 17.2 Recommended nutrient levels for extrusion-processed channel catfish grow-out feeds (adapted from Robinson & Li, 1996)

Nutrient	Recommended level
Protein	26–32% of diet
Lysine	5.1% of dietary protein
Methionine and cystine	2.3% of dietary protein
Digestible energy	8–10 kcal/g protein
Lipid	≤ 6% of diet
Carbohydrate	25–35% of diet
Vitamins	
Thiamine	5.5 mg/kg of diet
Riboflavin	13.2 mg/kg of diet
Pyridoxine	11 mg/kg of diet
Pantothenic acid	35 mg/kg of diet
Nicotinic acid	22 mg/kg of diet
Folic acid	2.2 mg/kg of diet
B_{12}	0.01 mg/kg of diet
Ascorbic acid	50–100 mg/kg of diet
A	2200 IU/kg of diet
D_3	1100 IU/kg of diet
E	66 IU/kg of diet
K	4.4 mg/kg of diet
Minerals	
Available phosphorus	0.3% of diet
Cobalt	0.05 mg/kg of diet
Zinc	200 mg/kg of diet
Selenium	0.1 mg/kg of diet
Manganese	25 mg/kg of diet
Iodine	2.4 mg/kg of diet
Iron	30 mg/kg of diet
Copper	5 mg/kg of diet

to contain high dietary protein levels and high levels of fish meal. Most feeds for fingerlings and foodfish grow-out are prepared by extrusion cooking and drying (section 9.10.2). That process produces a hard expanded pellet that floats in water. Extruded feeds must contain ~25% corn or other high-carbohydrate grains for proper gelatinisation and expansion of pellets during extrusion. Soybean meal provides most of the protein in fingerling and foodfish feeds. Small amounts of fish meal or other animal protein may be added to improve amino acid balance and enhance palatability. Commercial trace mineral and vitamin premixes are added to all channel catfish feeds to ensure nutritional adequacy.

Table 17.3 Recommended dietary protein levels for various sizes of channel catfish (adapted from Robinson & Li, 1996)

Fish weight (g)	Per cent of diet
0.02–0.25	52
0.25–1.5	48
1.5–5.0	44
5.0–20.0	40
20.0 g and above	26–36

17.7 Infectious diseases

The high fish densities and marginal environmental conditions in channel catfish culture ponds are conducive to the outbreak and rapid spread of infectious diseases. Mortality due to infectious diseases has not been quantified but may exceed 20% of annual production. Losses are greatest in the fingerling phase of production. Mortality underestimates the true economic cost to producers because infectious diseases also cause poor growth and feed conversion in affected fish populations.

Infectious diseases may be caused by viral, bacterial, fungal, or protozoan pathogens. Bacterial diseases account for most of the losses of pond-raised channel catfish. Infectious diseases of channel catfish are reviewed by Johnson (1993) and in various chapters in Stoskopf (1993).

17.7.1 Viral diseases

There are two known viral diseases of channel catfish: channel catfish virus and channel catfish reovirus. Channel catfish virus disease (CCVD) is an important infectious disease of young catfish and can cause large losses in hatcheries or fry nursery ponds. Channel catfish reovirus appears to be of low pathogenicity and is not considered economically important.

Channel catfish virus disease is caused by a highly virulent herpes virus and channel catfish is the only species known to sustain natural epizootics. Fish with CCVD usually have a distended abdomen and exophthalmia due to accumulation of fluid in the body cavity. Fish feed poorly and swim erratically, often in a spiralling fashion. The severity of the disease is strongly affected by fish size and water

Table 17.4 Examples of practical catfish feed formulations. Two examples of food fish grow-out feeds with different crude protein levels are shown (adapted from Robinson & Li, 1996)

	Feed type			
Ingredient	Fry	Fingerling	Food fish (32% protein)	Food fish (28% protein)
Fish meal	60	6	4	–
Meat/bone meal	15	6	4	4
Soybean meal	–	39	23	26
Wheat middlings	20	20	18	23
Corn	–	16	21	34
Cottonseed meal	–	10	27	10
Dicalcium phosphate	–	1	1	1
Oil or fat	5	2	1.5	1.5
Mineral/vitamin premix	Include	Include	Include	Include

temperature. Fish less than 1 month old are considered very susceptible to the disease. Older fish are more resistant and fish over 20 cm long (typically ~9 months old) are considered resistant to CCVD. Fish do not develop CCVD at water temperatures below 15°C; disease occurrence is irregular and mortalities are usually low at temperatures between 20°C and 25°C. Losses can be devastating at temperatures above 30°C.

There is no cure for CCVD, but losses can be reduced by minimising stress in susceptible fry or fingerling populations. Disease incidence can be reduced by using hygienic hatchery practices to reduce the opportunity for spread of the disease.

17.7.2 Bacterial diseases

Three bacterial diseases are significant in channel catfish aquaculture: enteric septicaemia of catfish (ESC), columnaris disease and motile aeromonad septicaemia. Enteric septicaemia of catfish and columnaris disease are especially important in the nursery phase, and can cause large losses of fingerlings.

Enteric septicaemia of catfish is caused by the Gram-negative bacterium *Edwardsiella ictaluri*. In practice, the bacterium is host specific for channel catfish, although experimental infections can be established in several other fish species. The disease may manifest as an acute gastrointestinal septicaemia with rapid mortality or as a chronic form characterised by an ulcerative lesion on the top of the head, which may proceed to septicaemia and death. Fish with ESC are listless and may swim in slow erratic spirals at the surface. Infected fish often have a rash of pinpoint haemorrhages around fins and on the ventral surface.

The occurrence of ESC is highly temperature dependent. In south-eastern USA, epizootics occur only in the late spring and early autumn when water temperatures are between 22°C and 28°C. Channel catfish of all sizes and ages are susceptible to ESC, but fingerlings account for most of the fish lost to the disease. Vaccine development has been an active area of research and shows some promise of reducing losses to ESC. However, the only practical method of vaccinating channel catfish is by mass immersion of fry in the vaccine just before they are transferred from the hatchery to a nursery pond. This is a major constraint to the use of vaccines in channel catfish because immunocompetency may not be fully developed in very young fish. Until effective vaccines and vaccine delivery systems are perfected, prompt diagnosis and antibiotic therapy with medicated feed will remain critical to successful treatment of ESC.

Columnaris is an acute to chronic disease of channel catfish and many other freshwater fish. The disease is caused by the Gram-negative bacterium *Flavobacterium columnare*. The bacterium is ubiquitous in aquatic environments, and outbreaks of the disease are usually associated with some predisposing stress factor, such as poor environmental conditions. Epizootics occur throughout the warmer months of the year, particularly in spring and autumn, when water temperatures are between 20°C and 25°C. Columnaris usually begins as an external infection of the gills or body, which progresses to an internal, systemic bacteraemia. Gross lesions on the gills are characterised by yellow-brown areas of necrosis at the distal end of the gill filaments. Skin lesions appear as areas of depigmentation, which progress to large necrotic ulcers. Infections of *Flavobacterium columnare* can be controlled by antibiotic therapy using medicated feed. Long-term successful treatment of columnaris is, however, difficult unless the predisposing stressor is identified and corrected.

Motile aeromonad septicaemia (MAS) is caused by the Gram-negative bacteria *Aeromonas hydrophila* and *Aeromonas sobria*. Epizootics of MAS may occur throughout the year but are most common in spring and autumn, when water temperatures are between 15°C and 25°C. Bacteria in the genus *Aeromonas* are among the most common bacteria found in freshwater environments and, as with columnaris disease, outbreaks of MAS are usually associated with predisposing stress. Symptoms of MAS vary greatly depending on the virulence of the strain of *Aeromonas* involved in the infection. Clinical signs range from external manifestations such as areas of external haemorrhage and ulcerative skin lesions to internal lesions characteristic of systemic bacteraemia. Medicated feed therapy is used to treat MAS, but control may only be temporary unless the predisposing stressor is corrected.

17.7.3 Fungal diseases

Water moulds of the genus *Saprolegnia* are important pathogens of channel catfish eggs in hatcheries and are also associated with a serious disease problem affecting adult fish in ponds when water temperatures are below ~15°C. Water moulds are common, saprophytic micro-organisms that usually cause problems only when environmental conditions are less than optimal.

Problems with fungal infections of eggs are most common in hatcheries using water of less than 25°C. Fungal growth begins on infertile or dead eggs and may spread to and kill healthy eggs. Infected eggs can be treated in a 15-min bath of 100 ppm formalin to kill the fungus. Proper water temperature for egg incubation is, however, the best solution to the problem.

External fungal infections of channel catfish are common in the winter months when water temperatures are below 15°C. Infections are easily diagnosed by the presence of brownish, cottony patches on the external surfaces. Areas of depigmentation and loss of the mucous layer may be present before masses of fungal mycelia are obvious. It appears that the disease results from the combination of two factors: suppression of the catfish immune system by low water temperatures and coincident high concentrations of *Saprolegnia* zoospores in the pond water. When conditions are conducive for the disease, mortality rates may be high. There is no cost-effective treatment for fungal infections of pond-raised channel catfish.

17.7.4 Myxozoan and protozoan parasites

Proliferative gill disease is a common and often severe disease of pond-raised catfish. The disease is caused by a myxozoan parasite that is also found in the gut of an oligochaete worm (*Dero digitata*) that is commonly found in the bottom muds of catfish ponds. The oligochaete is either the primary host or the intermediate host for the parasite in ponds. Spores of the myxozoan are released from the oligochaete, infect catfish and encyst in the gills. The gills may then become severely inflamed, with breakage of the filamental cartilage and varying degrees of necrosis. The gill lesions reduce respiratory efficiency and death is presumably caused by tissue hypoxia.

Outbreaks of proliferative gill disease occur most commonly in spring and autumn, when water temperatures are between 15 and 20°C, although parasites may be found in fish throughout the year. There is no treatment for the disease once it occurs, although losses can be reduced by maintaining high dissolved oxygen levels during an epizootic.

'Ich' (*Ichthyophthirius multifilis*) is a common protozoan parasite of freshwater fish. Epizootics are rare in channel catfish aquaculture, but losses may approach 100% of the affected population when conditions are optimal for spread of the disease. Ich is a ciliated protozoan with a biphasic life cycle. The adults (called trophozoites) reach ~1 mm in diameter, are covered with cilia and have a C-shaped nucleus. Trophozoites cause great tissue damage as they migrate through the epidermis, feeding on tissue fluids and cell debris. The trophozoite leaves the fish after maturing and settles to the pond bottom, where it encysts and divides to produce up to 2000 infective cells, called tomites. Ich is an obligate fish parasite, and the tomites will die if they do not locate a fish host within a few days.

Epizootics of Ich are most common when water temperatures are between 20°C and 25°C; fingerlings are particularly susceptible because they are held at high densities that enhance the spread of the disease. The most obvious sign of Ich is the presence of many raised, white, pinhead-sized spots on the skin.

Ich is difficult to control because the parasite resides beneath fish skin. Treatment consists of breaking the life cycle by killing trophozoites after leaving the fish or killing the tomites before they infect the fish. Five to seven treatments of copper sulphate applied to the water every other day usually controls the disease if treatment is initiated early in the epizootic.

Species of *Trichodina*, *Trichophrya*, *Ambiphrya* and *Ichthyobodo* are commonly found in low numbers on the external surfaces (particularly the gill filaments) of channel catfish and cause insignificant damage to the fish. Under certain conditions, however, the abundance of protozoans increases dramatically, possibly killing the host fish. Factors regulating fish parasitism by these protozoans are poorly understood, but epizootics are commonly associated with overcrowding and poor environmental conditions. Most of these infestations can be treated with chemicals, such as copper sulphate or potassium permanganate, applied to the water. Such treatments must be carefully considered, however, because the chemicals are toxic to fish at concentrations only slightly higher than the therapeutic dosage. Prevention of disease outbreaks by maintaining good environmental conditions is the best management practice for these protozoans.

17.7.5 Metazoan parasites

A wide variety of metazoan parasites may occasionally infest the gills, skin, gut or internal tissues of channel catfish. These include various species of monogenetic nematodes (gill and skin flukes), digenetic trematodes ('grubs'), nematodes (roundworms) and cestodes (tapeworms). Metazoan parasites usually cause few problems, although there is one important exception. In the late 1990s, infestations of channel catfish with the digenetic trematode, *Bolbophorus* species, caused significant mortality and morbidity on several catfish farms in extreme southern Louisiana. Losses associated with the parasite have subsequently been confirmed outside southern Louisiana, including north-west Mississippi – the region with the highest concentration of catfish farms in the USA.

The life cycle of *Bolbophorus* species involves three intermediate hosts. The adult trematode develops in the final host, the American white pelican (*Pelecanus erythrorhynchos*). Eggs contained in the faeces of infected pelicans may be released into ponds or lakes, where they hatch into free-swimming miracidia. Miracidia infect the first intermediate host, the ram's horn snail (*Planorbella* species), which is common in the littoral of freshwaters in the south-eastern USA. If ram's horn snails are not present in the body of water, further development of the trematode cannot take place and fish cannot become infected. If, however, ram's horn snails are present and they become infected, the miracidia eventually develop into free-swimming larvae called cercariae, which leave the snail and infect the second intermediate host – fish. When cercariae penetrate and migrate into tissues of fish, they cause mechanical damage, haemorrhage and inflammation. The extent of damage depends on the level of infection and the size of the fish. Cercariae eventually encyst in the tissues of fish. The encysted stage is called a metacercaria. If levels of infection are low, there is little, if any, effect on the host after encystment; however, if great numbers of metacerciae are present, considerable damage can be done to affected tissues.

Catfish infected with *Bolbophorus* species have small pinpoint cysts located anywhere in the body. Cysts are often subdermal, and small raised bumps may be particularly noticeable in the caudal region. Heavy infections may result in extensive damage to the liver and kidneys. The farm-level results of *Bolbophorus* species infections of fish can range from no effect to large losses of fish, particularly if fingerlings are involved. Fish larger than 0.25 kg may not die even if heavily infected, but they feed poorly and grow slowly.

There is no effective therapy for fish after they have been infected. Management of the disease therefore depends on breaking the trematode life cycle. The most effective approach involves a combination of activities, such as:

- discouraging feeding and 'loafing' by pelicans
- reducing stands of aquatic vegetation around the pond perimeter to reduce snail habitat
- treating the pond margins with lime or copper sulphate to kill snails
- using biological control measures, such as snail-eating fish.

In certain areas, it may be possible to reduce the number of snails by increasing the salinity of pond water to around 2.5‰ – a level that is well tolerated by channel catfish but not by the ram's horn snail.

17.8 Harvesting and processing

After fish reach the desired market size, they are harvested, transported to the processing plant and processed. These practices are relatively standardised across the channel catfish industry.

17.8.1 *Harvesting*

Harvesting fish from levee ponds is simple because most ponds have regular shapes (usually rectangular) and are shallow with smooth bottoms devoid of snags. Seines used for harvest are purchased from commercial sources. A typical seine is ~2.75 m deep and ~1.5 times as long as the maximum width of the pond. Seines are usually made of knotted polyethylene and have a float line at the top and a weighted lead line on the bottom. The lead line is fitted with rollers or covered with a 'mud line' to help the bottom line slide over the mud bottom.

Seines are built with a tapered tunnel ~50 m from one end. The tunnel is fabricated with a closure that can be connected to the metal frame of a 'live car'. The netting used to construct seines and live cars is available in a variety of mesh sizes. The mesh is sized to capture the smallest fish size desired. Seines for food fish commonly use a square mesh size of

Fig. 17.5 A tractor pulling a seine reel.

~4 cm, which retains fish greater than 0.35 kg and allows smaller fish to escape and remain in the pond for further growth.

Two tractors are used to pull the seine through the pond. One tractor pulls a hydraulically operated seine reel that serves as a seine storage unit (Fig. 17.5). Once the two tractors have moved the length of the pond, the ends of the seine are brought together and the seine is gathered onto the seine reel. As the seine is reeled in, the fish become increasingly crowded in the remaining area. At this point, the live car is attached to the seine. The live car is an open-topped bag made of netting. One end of the live car is equipped with a metal frame that is attached to the drawstring closure on the seine. As the seine is spooled further onto the seine reel, the crowded fish flow into the live car (Fig. 17.6). The live car is detached from the seine and closed after it is filled to the recommended capacity. Another live car may then be attached to receive additional fish if the harvest is large. Fish are allowed to grade in the detached live car for several hours.

After grading is complete, fish in the live car are loaded onto transport trucks. The live car is positioned near the bank and fish are scooped into a loading net attached to a hydraulic boom (Fig. 17.7). In-line scales record the harvest weight.

17.8.2 *Processing*

Fish are transported alive to processing plants. Transport trucks carry four to nine aerated tanks, each ~3 m^3 in volume. Tanks are filled with ~1500 L

Fig. 17.6 Catfish crowded in the seine flow into the live car, where they are held until loaded onto transport trucks.

Fig. 17.7 Catfish are scooped from the live car and loaded onto live-haul trucks for transport to the processing plant.

of water and each tank can hold ~1000 kg of fish. At the processing plant, a sample of fish from the transport truck is tested for flavour quality. If the fish are of acceptable quality, they are unloaded into large concrete tanks containing flowing, aerated water. When needed in the processing plant, fish are removed from the tank, weighed and stunned with alternating current electricity. Stunning renders the fish immobile and makes them easier to handle in the processing line.

Much of the processing is done by hand, but there is a trend towards increased automation to reduce processing costs. A variety of products are marketed, with common products being fillets, fillet strips, nuggets (the 'belly flap' section removed from the fillet), steaks and whole dressed fish. Overall, ~53% of a live catfish is converted into saleable products after processing. Processed catfish are available either as fresh, ice-packed product or as various frozen products. Frozen fillets account for the largest portion (~40%) of total sales.

17.9 The future of channel catfish farming

Channel catfish aquaculture expanded at a remarkable rate in the period from 1980 to 2000. Industry growth was facilitated by a stable US economy, low feed prices, aggressive marketing and a general trend towards more fish in the diet of American consumers. Barring unforeseen changes in the domestic and global economy, the future of the industry should be bright, although it is doubtful if growth can continue at the rate seen in the last quarter of the twentieth century.

Availability of land and water will probably not constrain industry growth, at least in the near future. More likely, future growth will be limited by the large capital investment needed to enter the industry, and increasingly strict government regulations with respect to environmental issues. Pond aquaculture can be a relatively benign form of agriculture, but regulations promulgated without due consideration of the actual affects of aquaculture on the environment could adversely affect industry growth and economic performance.

Annual yields of catfish average ~4000 kg/ha under commercial conditions, which is about half of that attainable under controlled, experimental conditions. Losses to infectious diseases and bird depredation, and postponement of harvests by frequent episodes of off-flavour, account for the difference between potential and actual farm productivity. Clearly, technological improvements in those areas will improve farm economic performance. Advances in other areas also offer considerable potential for improving farm economic performance. Better yields and more efficient production may soon be realised, as improved germplasm becomes widely available. In addition, improvements in feeding practices can dramatically impact profitability because feed represents the largest single cost of production.

References

Boyd, C. E. & Tucker, C. S. (1998). *Pond Aquaculture Water Quality Management*. Kluwer Academic Publishers, Boston.

Engle, C. R. & Pounds, G. L. (1994). Trade-offs between single- and multiple-batch production of channel catfish *Ictalurus punctatus*: an economics perspective. *Journal of Applied Aquaculture*, **3**, 311–32.

Hubert, W. A. (1999). Biology and management of channel catfish. In: *Catfish 2000: Proceedings of the International Ictalurid Symposium* (Ed. by E. R. Irwin, W. A. Hubert, C. F. Rabeni, H. L. Schramm, Jr & T. Coon), pp. 3–22. American Fisheries Society, Symposium 24, Bethesda, MD.

Johnson, M. R. (1993). The veterinary approach to channel catfish. In: *Aquaculture for Veterinarians: Fish Husbandry and Medicine* (Ed. by L. Brown), pp. 249–70. Pergamon Press, Oxford.

Losinger, W., Dasgupta, S., Engle, C. & Wagner, B. (2000). Economic interactions between feeding rates and stocking densities in intensive catfish *Ictalurus punctatus* production. *Journal of the World Aquaculture Society*, **31**, 491–502.

Robinson, E. H. (1989). Channel catfish nutrition. *Reviews in Aquatic Sciences*, **1**, 365–91.

Robinson, E. H. & Li, M. (1996). *A Practical Guide to Nutrition, Feeds, and Feeding of Catfish*. Mississippi Agricultural and Forestry Experiment Station Bulletin 1041, Mississippi State University, MO.

Stickney, R. R. (Ed.) (1986). *Culture of Nonsalmonid Freshwater Fishes*. CRC Press, Boca Raton, FL.

Stoskopf, M. K. (1993). *Fish Medicine*. W. B. Saunders, Philadelphia.

Tucker, C. S. (Ed.) (1985). *Channel Catfish Culture*. Elsevier Science Publishers, Amsterdam.

Tucker, C. S. (1996). The ecology of channel catfish culture ponds in northwest Mississippi. *Reviews in Fisheries Science*, **4**(1), 1–55.

Tucker, C. S. (2000a). Channel catfish culture. In: *The Encyclopedia of Aquaculture* (Ed. by R. R. Stickney), pp. 153–70. John Wiley and Sons, New York.

Tucker, C. S. (2000b). Off-flavor problems in aquaculture. *Reviews in Fisheries Science*, **8**(1), 45–88.

Tucker, C. S. & Robinson, E. H. (1990). *Channel Catfish Farming Handbook*. Van Nostrand Reinhold, New York.

Tucker, C. S., Steeby, J. A., Waldrop, J. E., & Garrard, A. B. (1994) Production characteristics and economic performance of four channel catfish, *Ictalurus punctatus*, pond stocking density-cropping system combinations. *Journal of Applied Aquaculture*, **3**, 333–51.

18
Barramundi

Michael Rimmer

18.1 Introduction

> The Barramundi: north Australia's finest food fish.
>
> Gilbert Whitley, 1959

18.1.1 Distribution and habitats

Barramundi [or Asian sea bass or giant sea perch (*Lates calcarifer*)], is a euryhaline member of the family Centropomidae that is widely distributed in the Indo-West Pacific region from the Arabian Gulf, through South-East Asia to Papua New Guinea and northern Australia. Barramundi inhabit freshwater, brackish and marine habitats including streams, lakes, waterholes, estuaries and coastal waters. They are frequently found around cover, such as obstacles or rocky areas in rivers, or around rocky headlands in coastal waters. Barramundi are opportunistic predators, and crustaceans and fish predominate in the diet of adults (Copland & Grey, 1987).

Although barramundi inhabit a range of salinities, they can only breed successfully in saltwater. Generally, larger fish are found in estuarine or coastal areas, and juvenile fish inhabit rivers, freshwater lagoons, creeks and upper estuaries. Barramundi spawn after the new and full moons during the late spring–early summer breeding season, the timing of which varies according to location.

18.1.2 Attributes for culture

Throughout its range, barramundi has traditionally supported important commercial fisheries and, in northern Australia, it is a popular and economically important species targeted by recreational fishers. More recently, it has become an important aquaculture species. Barramundi has many attributes that make it ideal for aquaculture:

(1) It is a relatively hardy species that tolerates crowding and has wide physiological tolerances. Barramundi can tolerate a wide range of temperatures and salinities, from freshwater to hypersaline conditions. It is highly tolerant of poor water quality conditions, including high ammonia levels and low dissolved oxygen. The physiological tolerances of barramundi are listed later in this chapter.
(2) Larviculture is relatively easy, compared with many other marine fish species. Current commercial larval rearing procedures generally yield about 50% survival to metamorphosis.
(3) Barramundi are highly fecund: a single female can produce several million eggs at a single spawning, providing plenty of material for hatchery production of seedstock.
(4) Barramundi readily feed on artificial diets and juveniles are easy to wean from live feeds to pellet feeds.

(5) Barramundi grow rapidly, reaching a harvestable size (350 g to 3 kg) in 6 months to 2 years.

Compared with these advantages, the only major disadvantage of the species is that it is highly cannibalistic during the late larval and juvenile stages. Cannibalism and methods to reduce this problem are discussed later in this chapter.

Techniques for the culture of barramundi were originally developed in Thailand in the 1970s (Ruangpanit, 1987). The considerable research and development efforts directed into barramundi culture over recent decades (see Copland & Grey, 1987) have resulted in the development of reliable culture techniques. Consequently, barramundi is now cultured throughout most of its range and has been introduced to several countries, including French Polynesia and Israel, for aquaculture purposes.

18.1.3 Production

World production of cultured barramundi has increased steadily over recent years from ~4000 mt in 1985 to a steady level of ~20 000 mt since 1993 (FAO, 2001). Taiwan and Thailand are the major producers of cultured barramundi; the other major producing countries are Indonesia and Malaysia (Table 18.1). The estimated value of world production of cultured barramundi is ~US$80–90 million/year (Table 18.1), and this species is an important contributor to the economies of many South-East Asian countries.

Compared with some other fish species used for aquaculture, e.g. channel catfish, *Ictalurus punctatus* (Chapter 17), and Atlantic salmon, *Salmo salar* (Chapter 15), barramundi culture is at a relatively early stage of technological development. One reason for this is that barramundi culture has traditionally been carried out in developing countries, where funds available for research are much less than in the USA or in Europe. Also, much of the barramundi production in South-East Asia is carried out on small family-run farms, so the economies of scale that have proven so successful for channel catfish culture are not available for barramundi. However, many of the technologies that have proven successful with other fish species (such as selective breeding, development of more cost-effective feeds, etc.) are now being applied to barramundi with considerable success.

In Australia, barramundi is regarded as a premium angling and eating species. In contrast, in Asia barramundi is a relatively low-value species that is less highly regarded than many other marine fish such as snappers (family Lutjanidae) and groupers (family Serranidae, subfamily Epinephelinae). The relatively low value of barramundi is one factor that has led to a relatively stable production over the last few years; most aquaculture expansion has been with more lucrative species.

18.2 Broodstock management and spawning

Early work on culture techniques for barramundi relied heavily on obtaining fertilised eggs by stripping running-ripe males and females caught on estuarine spawning grounds. This approach is expensive and unreliable, and has largely been replaced by the development of controlled breeding techniques for captive broodstock.

18.2.1 Broodstock management

Barramundi broodstock are usually captured using commercial fishing techniques. In Australia, most broodstock are collected using gill nets set in estuaries or around coastal headlands. Increasing interest in the selective breeding of barramundi has led to the retention of farm-raised fish for use as broodstock, and some farms in Indonesia and Australia are up to their third or fourth generation of selectively bred broodstock. Newly captured barramundi are generally quarantined for several weeks before being introduced to tanks with established captive broodstock. This prevents the introduction of pathogens to the established broodstock. Barramundi broodstock may not feed for several weeks after capture and even broodstock that have been conditioned to captivity may occasionally cease feeding, particularly if water temperatures drop below about 22°C.

Barramundi broodstock may be held in cages, ponds or tanks. Cages used for broodstock range in size from 4 m × 4 m × 3 m to 10 m × 10 m × 2 m, with 4- to 8-cm mesh size. Generally, broodstock held in seacages or ponds are transferred to on-shore

Table 18.1 World production (mt) of cultured *Lates calcarifer* between 1995 and 1999 (FAO, 2001)

Country	1995	1996	1997	1998	1999
Australia	258	596	487	683	895
Brunei Darussalam	74	72	69	74	41
French Polynesia	3	10	2	1	0
Hong Kong	207	144	72	71	34
Indonesia	1815	4401	2483	2039	5741
Malaysia	2206	2417	3487	5246	5210
Philippines	0	0	0	0	0
Singapore	285	266	243	235	241
Taiwan	10 136	6981	5673	5733	4979
Thailand	3884	4087	4126	6844	6086
Total production	18 868	18 974	16 642	20 964	23 227
Total value (US$ millions)	86.5	86.7	65.3	77.0	87.1

tanks before breeding. However, broodstock may be spawned in seacages fitted with a hapa net (a fine mesh, usually ~0.5- to 0.6-mm-width mesh), which is set inside the cage to prevent the loss of spawned eggs. In Taiwan, barramundi, like many other marine fish species, are allowed to spawn naturally in saline ponds and the eggs are collected using a fine mesh net set downstream from a paddlewheel aerator (Parazo *et al.*, 1998).

Tanks used for barramundi broodstock are usually made from fibreglass or epoxy-coated concrete and range in volume from 20 m^3 to 200 m^3 (Fig. 18.1). Broodstock are kept at densities of around 0.5 kg/m^3 of water. Broodstock tanks may operate on either flow-through or recirculating water supply systems. The latter tanks use biological filtration and may also incorporate physical filtration systems such as sand filters to remove particulate matter. Sex ratios (M/F) usually range from 1:1 to 2:1 (Parazo *et al.*, 1998).

Barramundi are protandrous hermaphrodites. Most fish mature as males at 3–4 years of age and change sex to female at 5–7 years. Sex change occurs immediately after the spawning season. In a culture situation, this aspect of the species' biology requires the constant replacement of male broodstock because a proportion of captive males change sex each year unless the broodstock are held in constant photothermal conditions (section 18.2.2).

Barramundi broodstock are fed once daily at a rate of 1–2% of body weight: overfeeding is reported to reduce spawning success

Fig. 18.1 Barramundi broodstock become domesticated in captivity and can be handled readily.

(Ruangpanit, 1987). The usual feed is trash fish or commercially available baitfish. Pellets are not presently available in sizes large enough for broodstock and the nutritional requirements of barramundi broodstock have not been fully established. Because baitfish may not be well handled or stored between capture and sale, their nutritional quality is often poor. To improve the nutritional composition of the broodstock diet, the types of baitfish used are varied as much as possible and a vitamin supplement can be added to the baitfish before they are fed to the broodstock. The composition of one such vitamin supplement that is widely used in Australia is shown in Table 18.2. Optimal broodstock nutrition is important to produce high-quality eggs and larvae. However, there is little specific information on broodstock diet requirements for barramundi. For a more detailed review of marine fish broodstock nutrition and related issues, see Izquierdo *et al.* (2001).

Clear acrylic windows in tanks allow for easy examination of the skin and fins of broodstock (for protozoan parasites such as 'white spot', and the detection of unusual behaviour that may indicate the presence of fish health problems. Infestations of marine 'white spot' (*Cryptocaryon irritans*) are particularly common in barramundi broodstock kept in recirculating seawater systems. (Details of diseases affecting barramundi and other cultured fish can be found in Chapter 10.) Outbreaks of white spot are treated by lowering the salinity of the water in the broodstock tanks by adding freshwater and then maintaining the broodstock in freshwater for a period of 10 days.

18.2.2 Spawning

Barramundi broodstock may be kept in either fresh or saline water but must be placed in seawater (28–35‰) before the breeding season to enable final gonadal maturation to occur. Barramundi show no obvious external signs of gonadal development and must be examined by cannulation to determine their gender and reproductive status (section 6.2.1). The cannula is a 40- to 50-cm length of clear flexible plastic tubing (3 mm OD, 1.2 mm ID), which is inserted into the urinogenital orifice of males and the oviduct of females. Fish to be cannulated are anaesthetised and a wet cloth or towel is placed over the eyes to assist in calming the fish. The cannula is guided into the fish for a distance of 2–3 cm (males) or 6–7 cm (females) and suction is applied to the other end of the cannula as it is withdrawn. After withdrawal, the sample within the cannula is expelled onto a Petri dish or (in the case of eggs) into a vial containing 1% neutral buffered formalin for later measurement of egg diameter (Garrett & Connell, 1991; Parazo *et al.*, 1998).

In Australia, female and male barramundi broodstock can be induced to breed on at least four occasions over a 4-month period, with each spawning episode producing around 5 million larvae (Garrett & Connell, 1991). In many parts of Asia, particularly close to the equator where water temperatures remain stable all year, barramundi will spawn throughout the year. The breeding season of barramundi can be extended indefinitely by the provision of summer water temperatures (≥ 28°C) and day length (13L: 11D). Fish subjected to this regime can be spawned at monthly intervals throughout the year, although they still require hormonal induction to spawn. Although wild-caught barramundi do not change sex when subjected to this environmental regime, there are reports that captive-bred fish will change sex under these conditions (Garrett & O'Brien, 1993).

In Thailand, barramundi have been induced to spawn by manipulation of environmental parameters that simulate their natural migration to the lower estuary and the tidal regime in the lower estuary. Salinity in the broodstock tanks is reduced to 20–35‰ and the broodstock are introduced to the tanks. About 50–60% of the tank volume is exchanged to increase the salinity to 30–32‰, which simulates the migration from upper to lower estuarine areas for spawning. At the beginning of the new or full moon, the water level in the tank is lowered to allow the water to heat up to 31–32°C. The fish usually spawn immediately after the temperature manipulation, but, if spawning does not occur, the process is repeated for an additional 2–3 days (Ruangpanit, 1987).

In Asia, barramundi commonly spawn without external influences such as environmental manipulation or hormonal induction. In contrast, spontaneous 'natural' spawning of Australian barramundi has only been reported on a few occasions, in two commercial hatcheries. Consequently, Australian hatcheries routinely use hormonal induction techniques to spawn barramundi.

Where captive barramundi do not spawn spontaneously, they must be induced to spawn using hormone preparations (as in many cultured fish). The

Table 18.2 Vitamin premix formulation for use in soluble form for barramundi broodstock

Vitamin	Amount/kg premix	Allowance/kg brood fish/day
A	2×10^6 IU	80 IU
D_3	0.8×10^6 IU	32 IU
E (DL-α-tocopherol)	40 g	1.6 mg
K_3	2 g	0.08 mg
Ascorbic acid	40 g	1.6 mg
Thiamin	4 g	0.16 mg
Riboflavin	4 g	0.16 mg
Pyridoxine	4 g	0.16 mg
Pantothenic acid	10 g	0.4 mg
Biotin	100 mg	4.0 μg
Nicotinic acid	30 g	1.2 mg
Folic acid	1 g	0.04 mg
B_{12}	4 mg	0.16 μg
Choline chloride	200 g	8.0 mg
Inositol	50 g	2.0 mg
PABA	20 g	0.8 mg
Ethoxyquin	30 g	1.2 mg
Dextrose	(to 1.0 kg)	–

Allowance is based on a mixing rate of 100 g premix in 1 L of water and injected at a rate of 1 mL/50 g baitfish, with broodstock fed baitfish at a rate of 2% body weight/day. Formulation developed by Queensland Department of Primary Industries.
IU, international unit.

use of exogenous hormones effectively overrides the fish's own endocrine system and stimulates the production of natural hormones that cause final gonad maturation, ovulation and spawning (section 6.2.1). Barramundi females with eggs 400 μm in diameter or larger are suitable for hormonally induced spawning. Males that are suitable for spawning induction will indicate milt (dense sperm) when cannulated or may produce a small 'bead' of milt when moderate external pressure is applied to the fish's abdomen (Garrett & Connell, 1991; Parazo *et al.*, 1998). Barramundi broodstock are usually suitable for spawning induction when water temperatures reach or exceed about 28°C.

Barramundi have been successfully spawned using a range of hormones at various doses, which were administered by techniques including injection, slow-release cholesterol and polymer pellets, and osmotic pumps. Spawning induction is now generally carried out using the luteinising hormone-releasing hormone analogues (LHRHa) (Des-Gly^{10})D-Ala^6,Pro^9-LH-RH ethylamide and (Des-Gly^{10})D-Trp^6,Pro^9-LH-RH ethylamide. For details on spawning induction techniques for fish and recent developments in this field, see Zohar & Mylonas (2001) and Chapter 6. Hormones are injected intramuscularly at the base of the pectoral fin. LHRHa dosages of 3–5 μg/kg body weight usually produce a single spawning, whereas dosages of 10–25 μg/kg usually produce two to four spawnings on consecutive nights (Garrett & Connell, 1991; Parazo *et al.*, 1998). Although Parazo *et al.* (1998) recommend injecting male broodstock, in Australia generally only female barramundi are injected; males will usually participate in spawning without hormonal induction.

Barramundi broodstock are injected with hormones in the morning to allow for natural spawning in the evening of the next day. Pre-spawning behaviour involves the male fish pairing with a female and rubbing its dorsal surface against the area of the female's genital papilla, erecting its fins and 'shivering'. In the absence of such displays, egg release may occur but the eggs are not fertilised. Spawning occurs 34–38 h after injection, usually around dusk, and may be accompanied by violent splashing (Garrett & Connell, 1991).

Barramundi usually spawn for at least two, and for up to five, consecutive nights. Where spawning occurs on three or more consecutive nights, egg production, fertilisation rate and hatching rate are normally higher on the first and second nights' spawning than on subsequent nights; spawnings from the later nights are frequently discarded because of low fertilisation and hatching rates (Garrett & Connell, 1991).

Fig.18.2 Airlift-operated egg collector for barramundi eggs. Airlifts in each corner of the PVC frame draw water and eggs from the tank below and discharge them into the mesh collecting area.

18.2.3 Harvesting eggs

At spawning, the sperm and eggs are released into the water column and fertilisation occurs externally. Barramundi eggs are 0.74–0.80 mm in diameter with a single oil droplet of 0.23–0.26 mm in diameter. The fertilised eggs of barramundi are positively or neutrally buoyant, whereas unfertilised eggs sink (Parazo *et al.*, 1998).

The eggs are concentrated in the spawning tanks using egg collectors, either inside or outside the tanks. Internal egg collectors consist of bags of 300-μm mesh material, approximately 0.5 m^3 in volume, which are suspended from a PVC frame (Fig. 18.2). Eggs are concentrated in the net using airlifts fitted to the PVC frame. External egg collectors are placed in externally mounted tanks through which the tank effluent passes.

Where barramundi are spawned in hapa nets in seacages, the adult fish are removed from the cage after spawning, and the hapa net is lifted to concentrate the eggs to one side, where they are gently removed and transported to the hatchery. For all these egg collection techniques, the eggs are removed early in the morning of the day following spawning, before the larvae begin hatching.

18.2.4 Incubation

Eggs are placed in larval rearing tanks at densities of 100–1200 eggs/L for incubation and hatching. Dead and unfertilised eggs are removed by briefly turning off the aeration in the hatching tank and siphoning out the dead eggs, which sink rapidly to the bottom of the tank. The number of eggs or larvae is estimated volumetrically: a known volume of water is sampled and the number of eggs or larvae in that sample is counted. This is repeated until a consistent estimate of density is obtained and the mean sample density is used to estimate the number of eggs or larvae in the tank (Parazo *et al.*, 1998).

Fertilised eggs undergo rapid development, and hatching occurs 12–17 h after fertilisation at 27–30°C (Fig. 18.3). Newly hatched barramundi larvae have a large yolk sac that is absorbed rapidly over the first 24 h and is largely exhausted by 50 h after hatching (Fig. 18.4). The oil globule is absorbed more slowly and persists for about 140 h after hatching. The mouth and gut develop the day after hatching (day 2) and larvae commence feeding 45–50 h after hatching (Kohno *et al.*, 1986; Parazo *et al.*, 1998).

18.3 Larval rearing

Larval rearing of barramundi is undertaken using either intensive or extensive techniques. Intensive larval rearing involves the high-density culture of larvae in a controlled environment, such as a hatchery, where the fish larvae are supplied with prey organisms that are also cultured under controlled conditions (section 9.5.1). In contrast, extensive

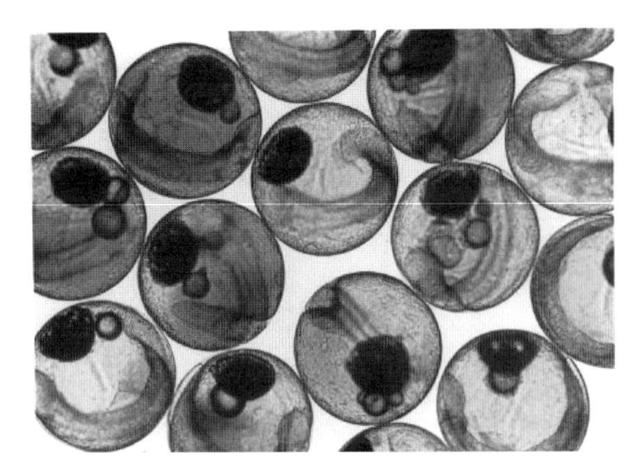

Fig. 18.3 Developing barramundi embryos before hatching (photograph by Dr Paul Southgate).

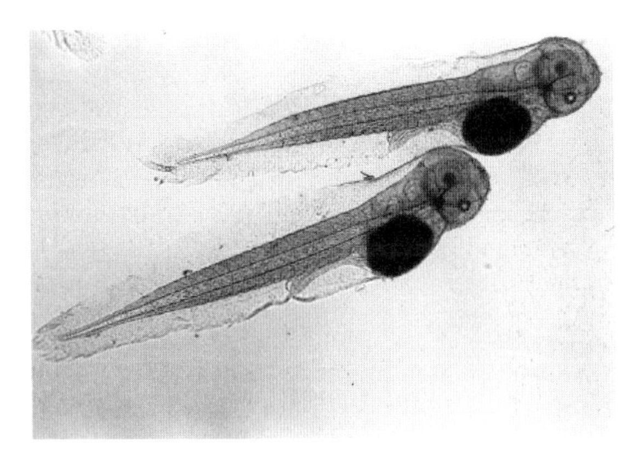

Fig. 18.4 Barramundi larvae at first feed (day 2) (photograph by Dr Paul Southgate).

larval rearing involves the culture of larvae in fertilised marine or brackish water ponds in which the culturist has little direct control over factors such as water quality and prey organism density.

18.3.1 Water quality

Despite the widespread culture of barramundi, the physicochemical tolerances of barramundi larvae are poorly documented. The recommended levels of these parameters for barramundi larvae are listed in Table 18.3. These are based on experience in the larval rearing of barramundi and on the few studies that have been carried out on the physicochemical tolerances of marine fish larvae. Newly hatched larvae have the lowest tolerances to various physicochemical parameters, and even moderate deviation from optimum conditions, although not directly lethal, may substantially reduce first feeding success, and hence survival. In addition, two or more factors that deviate slightly from optimal conditions may act synergistically to reduce survival; for example, increases in pH will increase the proportion of the toxic unionised form of ammonia (NH_3) (section 3.2.7).

18.3.2 Intensive larval rearing

Barramundi larvae are generally reared in circular or rectangular fibreglass or concrete tanks up to 10 m^3, although tanks up to 26 m^3 have been used (Fig. 18.5). Circular tanks with a conical base are preferred because water circulation and drainage is better than in rectangular tanks (Parazo *et al.*, 1998). Fibreglass larval rearing tanks are shaped with either a conical or hemispherical base so that debris and particulate matter tend to collect in the bottom centre area, and so can be readily removed from the tank by siphoning only a small proportion of the tank volume. Conical-based tanks are more suitable than hemispherical-based tanks in this regard. Larval rearing tanks are dark in colour (preferably grey or dark blue) to enable the larvae to more readily discriminate prey organisms against the background of the tank.

Rearing tanks are constructed with a central bottom outlet fitted with a removable screen to retain larvae (Fig. 18.6). Screens of various mesh sizes are used to retain larvae and to retain or flush prey organisms as necessary. The nominal mesh sizes used for larval rearing of barramundi are 62, 120 and 250 μm, with larger sizes necessary for juvenile fish.

Recommended stocking rates for barramundi larvae (up to about 10 mm total length, TL) are 10–40 fish/L. Some hatchery manuals recommend a gradual reduction in larval density as the larvae increase in size, but others recognise that normal mortality levels in the rearing tanks will achieve a similar reduction in density. Overall survival for intensively reared barramundi larvae from hatching to about 10 mm TL is usually between 15% and 40% (Parazo *et al.*, 1998).

Table 18.3 Recommended water quality criteria for larval rearing of barramundi based on various sources

	Optimum	Minimum	Maximum
Temperature (°C)	28–30	25	31
Salinity (‰)	28–31	20	35
pH	8.0	7.5	8.5
Dissolved oxygen (mg/L)	Saturation	2	–
Ammonia (NH_3) (mg/L)	0	–	0.1
Nitrite (mg/L)	0	–	0.2
Nitrate (mg/L)	0	–	1.0

Fig.18.5 Commercial barramundi hatchery in the Philippines.

18.3.3 Feeding and nutrition in intensive rearing

Barramundi larvae are reared using either 'clear water' or 'green water' techniques. Using 'clear water' techniques, optimal water quality is maintained by having high rates of water exchange to remove wastes, particularly ammonia. The water supply may be from flow-through or recirculating systems. In 'green water' culture, a microalgal culture (usually *Nannochloropsis oculata* or *Tetraselmis* species) is added to the rearing tanks at densities ranging from $8–10 \times 10^3$ cells/mL to $1–3 \times 10^5$ cells/mL. The microalgae aid in maintaining optimal water quality by utilising nitrogenous wastes and carbon dioxide, and producing oxygen. To maintain the desired density of microalgal cells, water changes are limited to 10–50% daily for the first 25 days of the rearing period and 50–75% daily thereafter. Rearing tanks are siphoned daily to remove particulate matter from the bottom. Microalgae cultures are added to maintain the required density of algal cells (Parazo *et al.*, 1998).

Intensively reared barramundi are fed on rotifers (*Brachionus plicatilis*) from day 2 (where day 1 is the day of hatching) until day 12 (or as late as day 15) and on brine shrimp (*Artemia* species) from day 8 onwards. Rotifers are fed two to four times daily, at a density of 15–20 individuals/mL. Brine shrimp are fed two or three times daily at 0.5–2 individuals/mL initially, gradually increasing to 5–10 individuals/mL. Prey consumption is closely monitored by routinely counting the density of prey organisms in the rearing tanks and by microscopically examining larvae to ensure that they are feeding. The precise demand for live prey organisms will depend on the design of the

Fig. 18.6 Tanks used for intensive rearing of barramundi larvae in a commercial hatchery in northern Australia. Tanks have a central bottom outlet and water drains through a removable screen (A), which fits onto the support in the centre of the tank (B). Screens can be covered with various sizes of mesh (photograph by Dr Paul Southgate).

rearing system, fish density and losses of live prey from the rearing tanks (Rimmer *et al.*, 1994; Parazo *et al.*, 1998).

An important requirement for intensively reared barramundi larvae is adequate nutrition. Barramundi larvae fed on diets deficient in highly unsaturated fatty acids (HUFAs), particularly 20:5*n*-3 (eicosapentaenoic acid or EPA), become pale and, when stressed, swim erratically and 'faint', after which they either recover (presumably temporarily) or die. Feeding procedures developed to overcome this deficiency provide prey organisms of suitable nutritional quality. Rotifers are cultured using microalgae high in HUFAs, such as *Nannochloropsis oculata* and *Isochrysis* species. Both rotifers and brine shrimp may be supplemented with a commercially available HUFA enrichment preparation in liquid, microcapsule, spray-dried or flaked microalgal form (Dhert *et al.*, 1990; Rimmer *et al.*, 1994) (section 9.4.7). Rainuzzo *et al.* (1997) provide further information on the nutritional requirements of marine fish larvae in general.

Freshly hatched brine shrimp are not used for feeding barramundi larvae because of their low levels of HUFAs. Instead, brine shrimp starved for 24 h to allow yolk absorption are fed to barramundi larvae until the larvae are large enough to ingest the larger supplemented brine shrimp (usually around day 13). Brine shrimp used for rearing barramundi larvae are also supplemented with commercial or home-made enrichment preparations after a period of starvation (~24 h) (Rimmer *et al.*, 1994).

The freshwater cladocerans *Daphnia* and *Moina* species have also been used to supplement, or replace, brine shrimp as prey for intensively reared barramundi larvae. *Moina* may be fed to barramundi larvae from day 15, although it is more often used as a substitute for brine shrimp from day 25 onwards. *Moina* are fed at least four times daily at densities of one individual per litre or greater and the salinity of the rearing water must be lowered to 10‰ or less to allow survival of the cladocerans (Parazo *et al.*, 1998).

During the later stages of larval rearing (from day 15 onward) barramundi may also be fed on minced fish flesh. The minced fish is screened to particle sizes < 2 mm and fed at a rate of about 10–15% fish body weight per day. Care must be taken not to overfeed the larval or juvenile barramundi using fish flesh, as decomposing fish tissue rapidly pollutes the rearing tanks, so 80–100% of water should be exchanged daily (Parazo *et al.*, 1998).

The use of artificial (microencapsulated or microbound) feeds (section 9.6) for rearing barramundi larvae has been the topic of some research but has not yet become commercially viable. Although barramundi larvae readily ingest artificial feed particles, they are apparently unable to digest these feeds in the absence of live prey, possibly because of the absence of exogenous enzymes that are found in live prey organisms and that assist with digestion. Although artificial feeds can be used to supplement live prey during the larval rearing process, the use of artificial feed as the sole feed type results in slow growth and high mortality of barramundi larvae (Walford *et al.*, 1991; Southgate & Lee, 1993).

18.3.4 Extensive larval rearing

Extensive larval rearing involves the production of juvenile fish from newly hatched larvae in marine or brackish water ponds. The ponds are managed to produce a 'food web' (section 9.9.2) that supports the continuous development of a zooplankton community, which in turn provides prey for the fish larvae.

Pond fertilisation

Ponds used for the extensive larval rearing of barramundi generally range from 0.05 to 1 ha in area and may be earthen or plastic lined. As with most pond culture (Chapter 2):

(1) The ponds are relatively shallow (< 2 m deep) to promote production of phytoplankton and to prevent stratification.
(2) The productivity of a pond is controlled by the addition of organic and inorganic fertilisers, which generate the blooms of phytoplankton, bacteria and protozoans that are food sources for zooplankton.

Although various inorganic and organic fertilisers can be used, the most commonly used fertilisers are diammonium phosphate (DAP) and lucerne (alfalfa) pellets. The recommended application schedule for these fertilisers is listed in Table 18.4. However, such schedules only provide a guide to fertiliser application (Boyd, 1990) because responses to various fertilisers vary according to:

- time of year
- salinity
- temperature
- the introduction of plankton into the pond
- the presence residual nutrients in the ponds from previous use.

Although phytoplankton is the major source of dissolved oxygen in larval rearing ponds, aeration may be used to moderate diel fluctuations in dissolved oxygen and to increase circulation within the pond. Fine diffusers (e.g. perforated water irrigation pipe) and centrifugal air-vanes or diaphragm pumps are used to aerate larval rearing ponds.

Table 18.4 Approximate schedule for pond fertilisation for extensive rearing of barramundi larvae

Day	Fertiliser
1–2	Fill pond; DAP (5.3 kg/10^6 L)
3	Lucerne (450 kg/ha)
6	DAP (5.3 kg/10^6 L)
8–9	Stock larvae
10	Lucerne (450 kg/ha)
12	DAP (5.3 kg/10^6 L)
14	Lucerne (450 kg/ha)
18	DAP (5.3 kg/10^6 L)
20	Lucerne (450 kg/ha)

DAP, diammonium phosphate.
Reproduced from Rutledge & Rimmer (1991) with permission.

Barramundi larvae are stocked into the pond at the time they are ready to begin feeding (usually day 2), to coincide with peak densities of the smaller zooplankton, i.e. rotifers and copepod nauplii, which are the initial prey of the larvae. Larvae are transported to the pond in plastic bags inflated with oxygen, at a maximum density of 5000 larvae/L. Upon arrival, the bags are placed in the pond to begin tempering: small amounts of pond water are added to each bag to slowly change the temperature and salinity to match that of the pond. The duration of the tempering procedure depends on the initial difference in the temperature and salinity of the bags and the pond; generally, 20–60 min is adequate. Barramundi larvae are stocked at densities of 400 000–900 000 fish/ha (Rutledge & Rimmer, 1991).

Zooplankton populations are monitored in each pond to ensure an adequate larval food supply. Small plankton nets (25-µm mesh) or vertical tube samplers are used to sample several sites within each pond. Subsamples are then counted using a microscope, to determine the density of each zooplankton type in the pond.

Water quality parameters, particularly dissolved oxygen, temperature, salinity and pH, are also monitored routinely. Larval and juvenile barramundi are sampled regularly to monitor growth and health and to provide a crude measure of fish abundance. Barramundi larvae can be readily sampled from aerated ponds using a zooplankton net with ~300-µm mesh. Juvenile barramundi may be sampled using square lift nets of 0.25–1.0 m^2, fitted with 1- to 2-mm mesh-size insect screen or similar material. The lift nets are left on the bottom of the pond and are lifted rapidly to trap any juvenile barramundi that are on or directly above the net.

Feeding and growth in rearing ponds

Barramundi in larval rearing ponds begin feeding on the smaller zooplankton, such as rotifers and copepod nauplii. As the larvae grow, they feed on larger organisms such as copepodites, adult copepods and, if present, cladocerans. As zooplankton densities decrease as a result of predation, the barramundi switch to benthic food sources, principally chironomid (midge) larvae ('blood worms') (Rutledge & Rimmer, 1991).

Extensively reared barramundi larvae grow faster than larvae reared intensively, possibly owing to better nutrition resulting from a more varied diet, and to greater prey availability throughout the day. Generally, extensively reared barramundi reach 20–30 mm TL after about 3 weeks in the pond, at which time they are harvested. In comparison, intensively reared barramundi reach about 10 mm TL after 3 weeks. Growth rates of up to 3.8 mm/day and specific growth rates (in length) of up to 28%/day have been recorded for extensively reared barramundi larvae. However, growth rates in ponds vary widely, and are particularly dependent on water temperature, salinity and food availability. Barramundi are harvested from the ponds when they reach 25 mm TL or greater (Fig. 18.7). The ponds are drained down so that the fish are concentrated in the concrete raceway. Water flow and aeration is maintained in the raceway during the harvesting process to ensure that the fish are in good-quality, well-oxygenated water. The fish are removed using hand nets and transferred to a transport vehicle, then transferred to nursery tanks (Rutledge & Rimmer,1991).

Survival of extensively reared barramundi now averages about 40%, but is highly variable, ranging from 0% to 90%. These figures correspond to production rates of up to 640 000 fish/ha. The lower costs of extensively reared fingerlings, estimated at 40–64% that of intensively reared fingerlings, and the lower infrastructure requirements, have resulted in the widespread use of extensive larval rearing techniques for barramundi in South-East Asia and northern Australia.

18.4 Nursery and grow-out

18.4.1 Nursery phase

Stocking and feeding

In South-East Asia, juvenile barramundi (1.0–2.5 cm TL) may be stocked in nursery cages in rivers, coastal areas or ponds, or directly into fresh or brackish water nursery ponds. Nursery cages are of floating or fixed design, and range in size from 1.8 m^3 (ca. 2 × 1 × 0.9 m) to 10 m^3 (ca. 5 × 2 × 1 m). Because of the small size of the net mesh used in nursery cages (1- to 2-mm mesh), these nets are easily damaged by strong currents, and foul rapidly. Barramundi juveniles are stocked in nursery cages at densities of 80–300 fish/m^3, and in nursery ponds at 20–50 fish/m^2. The barramundi are fed on minced trash fish (4–6 mm^3) at the rate of 100% of biomass twice daily for the first week, reducing to 60% then 40% of biomass for the second and third weeks respectively. Vitamin premix may be added to the minced fish at a rate of 2%. The fish are 'trained' to feed at the same site at the same time each day. This nursery phase lasts for 30–45 days and once the fingerlings have reached 5–10 cm TL they are transferred to grow-out ponds or cages (Kungvankij *et al.*, 1986).

In Australia, juvenile barramundi are transferred to nursery facilities after they have been harvested from the rearing ponds or from intensive culture

Fig. 18.7 Harvesting barramundi fingerlings from an extensive rearing pond.

tanks. Most nursery facilities use small cages (~1 m^3) made from insect screen mesh in concrete tanks or above-ground pools. Many barramundi farms use freshwater ponds for grow-out and thus operate freshwater nursery facilities. Juvenile barramundi (> 10 mm TL) can be transferred from saltwater to freshwater in as little as 6 h with no significant mortality.

Barramundi can be weaned to artificial diets when as small as 10 mm TL, although better survival and faster acceptance of artificial diets are obtained if weaning is delayed until the fish are at least 15–20 mm TL (Barlow *et al.*, 1996). These small barramundi may commence feeding on inert diets within a few hours of harvest and most fish commence feeding within a few days. High-quality weaning diets are available commercially and, although expensive, are preferable to the smaller grades of standard grower diets because they appear to be more attractive to the fish.

Most mortalities during the nursery phase occur during the weaning period. The cause of these mortalities varies, but most result from diseases caused by stress associated with:

- harvesting
- salinity conversion
- high stocking densities in the nursery tanks.

To reduce the incidence of pathogens (particularly protozoans) in freshwater nursery tanks, the fish are regularly bathed in a salt solution (10 g/L for 1 h). In seawater facilities, the fish may be treated by salinity reduction. Some fish do not learn to feed on inert diets and die of starvation, but these fish normally form only a small proportion of the population. A particular problem in nursery culture is outbreaks of columnaris disease, caused by bacteria of the *Flexibacter/Cytophaga* group.

Cannibalism

Cannibalism can be a major cause of mortalities during the nursery phase and during early grow-out. Barramundi will eat fish of up to 60–67% of their own length (Parazo *et al.*, 1998). Cannibalism may start during the later stages of larval rearing and is most pronounced in fish of less than about 150 mm TL. In larger fish it is responsible for relatively few losses. Cannibalism is reduced by grading the fish at regular intervals (usually at least every 7–10 days) to ensure that the fish in each cage are similar in size. In Asia, graders are generally constructed from plastic basins with holes drilled in the bottom or made from netting around a wooden frame. The latter design can be adapted to produce a nested set of graders that simplify the grading procedure. The size of the holes or the mesh ranges from 0.3 mm to 20 mm. Australian graders are usually basin-shaped, with a grid of parallel stainless-steel or Perspex rods in the base, and are designed to float in the nursery tanks. A range of grids is made up to grade different sizes of fish, with gap widths ranging from 2 to 10 mm. A useful ‘rule of thumb’ used by farmers is that the TL of barramundi averages about 10 times the gap width of the grader; thus barramundi retained on a 3.5-mm grader will be about 35 mm TL. Some Australian farms are successfully using mechanical graders originally developed for grading European sea bass (*Dicentrarchus labrax*) and sea bream (*Sparus aurata*).

18.4.2 Grow-out

Types

There are three main types of grow-out system used for barramundi culture:

(1) cages, either fixed or floating, in coastal waters
(2) fresh or brackish water ponds, usually with the fish in cages, but sometimes with the fish ‘free ranging’
(3) recirculating systems.

In Asia, most barramundi culture is undertaken using fixed or floating cages in coastal areas (Fig. 18.8). Such cages range in size from 3 m × 3 m up to 10 m × 10 m and are 2–3 m deep. Mesh sizes range from 2 cm to 8 cm (Cheong, 1990). Because most useable cage culture sites in the coastal zone are now taken, many countries are now using the larger circular cages originally developed for Atlantic salmon culture (Chapter 15). This enables the expansion of aquaculture activities further offshore.

In Australia, the cages commonly used in pond

culture are 8 m^3 in size (2 × 2 × 2 m), although larger cages (12–100 m^3) are also used (Fig. 18.9). The smaller cages used in ponds are usually constructed from a bag of knotless netting within which is placed a weighted square formed from PVC pipe and a floating square of the same material. In ponds, two rows of cages are floated either side of a central walkway, which allows access for feeding, sampling, harvesting and cage maintenance. Each cage is supplied with aeration to maintain a high dissolved oxygen concentration, and injector-type aerators are placed in the ponds to assist with water circulation and to increase dissolved oxygen levels. Water exchange rates in ponds vary considerably between different farms and with the biomass of the crop, but generally range from 5% to 20% of pond volume per day. Seacages used for barramundi culture in Australia are sited in estuarine areas where wind and wave action are greatly reduced, and are generally of the circular type originally developed for salmon culture (Barlow *et al.*, 1996). Biofouling of cages in ponds and estuaries causes blockage of the mesh openings, which reduces water movement through the cage and lowers water quality. Consequently, the mesh bags must be changed and cleaned regularly.

Barramundi are also farmed in earthen or lined ponds without cages, a technique known in Australia as 'free ranging'. This technique is reported to result in faster growth and better appearance and colour (silver rather than black) of the fish. The major disadvantage of this technique is the difficulty in harvesting the fish without draining the pond. Fish are captured using angling techniques, trapping or seine netting (Barlow *et al.*, 1996). In Asia, juvenile barramundi (20–100 g) are cultured in brackish water ponds at 0.25–2.0 fish/m^2.

In Australia, a number of barramundi farms have been established using recirculating fresh or brackish water systems with a combination of physical and biological filtration. Several of these farms are located in southern Australia where barramundi could not otherwise be farmed because of low temperatures. The major advantage of such culture systems is that they can be sited nearer to markets, thus reducing transport costs for the finished product.

Fish health and water quality are primarily monitored by observing fish behaviour, particularly during feeding. Lethargic feeding behaviour is a sign of poor health or water quality deterioration. Water quality requirements for grow-out of barramundi are listed in Table 18.5.

Stocking densities used for cage culture of barramundi generally range from 15 kg/m^3 to 40 kg/m^3, although densities may be as high as 60 kg/m^3. Generally, increased density results in decreased growth rates, but this effect is relatively minor at densities under about 25 kg/m^3. Barramundi farmed in recirculating production systems are stocked at a density of about 15 kg/m^3. Higher stocking densities require more monitoring of water quality and fish health, additional aeration and (in ponds) higher water exchange rates (Cheong, 1990).

In Asia, barramundi may be polycultured with tilapia in brackish water ponds. The ponds are stocked with tilapia broodstock at 0.2–1.0 fish/m^2 at a 1:1 or 1:3 (M/F) sex ratio about 2 months before stocking barramundi; the latter are stocked at 0.3–0.5 fish/m^2 (Cheong, 1990).

Feeding

Barramundi are fed on 'trash fish' or on commercially available pellets specifically formulated for this species. Trash fish is widely used in Asia, but only pellets are used in Australia. Barramundi fed on trash fish are fed twice daily at 8–10% body weight for fish up to 100 g, decreasing to 3–5% body weight for fish over 600 g. Vitamin premix may be added to the trash fish at a rate of 2%, or rice bran or broken rice may be added to increase the bulk of the feed at minimal cost. Food conversion ratios (FCRs) for barramundi fed on trash fish are high, generally ranging from 4:1 to 8:1 (Cheong, 1990).

Barramundi diets in Australia generally contain around 45% crude protein (CP) and 10% lipid. Semi-floating, extruded pellets are generally preferred because they are available to the fish (which feed only at the surface or in the water column) for longer than sinking pellets. The fish are fed to satiation twice each day in the warmer months and once each day during winter. Automatic feeders are not presently used, as farmers prefer to vary feeding rates by observing the feeding activity of the fish. Pellet-fed barramundi have achieved FCRs of 1.0:1 to 1.2:1 under experimental conditions, but in commercial farm conditions FCRs of 1.6 to 1.8:1 are usual. FCR

Fig.18.8 Floating cage farm in Indonesia.

Fig. 18.9 Grow-out cages in freshwater pond, Australia.

varies seasonally, often increasing to over 2:1 during winter (Barlow *et al.*, 1996).

There has been considerable research undertaken into feed development for barramundi. In Australia, a great deal of research has focused on replacing imported fish meal with terrestrial protein sources in barramundi diets. Barramundi diets containing little or no fish meal have been evaluated in both laboratory and on-farm conditions and shown to perform as well as, or better than, traditional diets with high inclusions of fish meal. Other research has demonstrated that high-density diets (≥ 55% CP, ≥ 20% lipid) produce better fish growth and higher farm profitability than the 'standard' diets. Economic modelling (section 18.4.2) has shown that the internal rate of return for a model barramundi farm increases from 8% to 23% owing to more rapid growth and lower FCRs resulting from the use of high-density diets (Williams *et al.*, 2000).

18.4.3 Diseases

Like other cultured fish species, barramundi are subject to a range of diseases. Larvae and juveniles may

Table 18.5 Summary of water quality parameters for grow-out of barramundi

	Optimum	Limit	Reference
Temperature (°C)	26–32	> 15	Cheong (1990)
Salinity (‰)	0–35		Cheong (1990)
pH	7.5–8.5	> 4	Cheong (1990)
Dissolved oxygen (mg/L)	4–9	> 1	Cheong (1990)
Ammonia (NH_3) (mg/L)	0	<0.46	Hassan (1992)
Nitrite (mg/L NO_3-N)			
Freshwater	< 1.5	< 14.5	
Saltwater	< 9	< 90	Woo & Chiu (1994)
H_2S (mg/L)	0	<0.3	Cheong (1990)
Turbidity (ppm)		<10	Cheong (1990)

suffer from viral nervous necrosis (VNN). This is more usually seen in intensive larval rearing, but has been reported in extensive rearing ponds, and some mortalities have occurred during the nursery phase. Affected larvae swim erratically in a corkscrewing motion before dying. Histological examination of the affected larvae shows extensive vacuolation of the retina, brain and central nervous system. The causative organism is a nodavirus, but it is probable that infections and mortalities are caused by predisposing environmental or physiological factors, such as stress or poor nutrition. Although improved hatchery hygiene can reduce the prevalence of VNN, there is currently little understanding of the routes of infection and the causes of disease associated with nodavirus.

Common bacterial pathogens that affect barramundi during the grow-out phase are *Vibrio*, *Aeromonas*, *Pasteurella* and *Streptococcus*. Columnaris disease (caused by bacteria of the *Flexibacter*/*Cytophaga* group) is a particularly important disease of barramundi in nursery facilities. Ciliated protozoans, including *Cryptocaryon irritans*, *Chilodonella* species, *Ichthyophthirius multifiliis*, *Trichodina* and *Amyloodinium* have been recorded from barramundi, and may cause significant losses in culture conditions, particularly among tank-reared fish. A range of other organisms infect barramundi, including:

- fungi (*Saprolegnia*)
- myxosporidians (*Henneguya*, *Kudoa*)
- monogenean (*Diplectanum*, *Dactylogyrus*, *Gyrodactylus*) and digenean (*Pseudometadena*) flatworms
- ectoparasitic and endoparasitic crustaceans (*Lernaea*, *Argulus*, *Ergasilus*, *Aega*, *Gnathia*)
- leeches (*Pontobdella*).

The viral disease lymphocystis is found in cultured barramundi, but, although disfiguring, is fatal only if the infection is severe and associated with very poor environmental conditions (Anderson & Norton, 1991). Further information on diseases affecting barramundi is given in Chapter 10 (section 10.5.3).

18.4.4 Growth

No significant differences in growth rate have been found in barramundi cultured in either fresh- or saltwater. Similarly, there appear to be no substantial differences in the growth rates of barramundi from different genetic stocks. Growth is highly variable, and depends on various factors including:

- water temperature
- feeding rate
- feed quality
- stocking density.

Generally, barramundi grow from fingerlings to 300–500 g in 6–12 months and to 3 kg in 2 years (Cheong, 1990; Barlow *et al.*, 1996).

Family-based selective breeding undertaken by farmers has been reported to result in significant improvement in growth rate, but no detailed results are available.

18.4.5 Harvesting and processing

Barramundi are marketed at a range of sizes, depending on demand. In Asia, barramundi are generally grown to at least 500 g and usually 800 g to 1 kg in weight, which takes 12–20 months. In Australia, there are two major market sizes: 'plate size' (i.e. 300–500 g) and fillet (2–3 kg). Fish are harvested from cages using dip nets, or from ponds by angling, seine netting or draining. In Australia, fish are killed and rapidly cooled by placing them in an ice and water slurry. They are then graded and packed with ice in plastic bags in polystyrene (Styrofoam) boxes. Most Australian farms do not process the fish before sale; processing is generally carried out by the wholesaler or retailer. In Asia and Australia, a proportion of farmed barramundi are sold live, but this varies between and within countries, depending on market requirements.

Most farmed barramundi are sold locally, but, owing to market saturation in many countries, there is an increasing trend to develop export markets. The largest producer of farmed barramundi, Taiwan, is already exporting substantial quantities (~250 mt/year) to Japan, the USA and Korea. Australian producers have begun exporting small quantities to Singapore and the USA.

18.4.6 Economics

Despite the importance of barramundi as an aquaculture commodity, relatively few economic evaluations of barramundi culture have been undertaken. In Australia, a sophisticated economic model ('Barraprofit') for barramundi farms has been developed using discounted cash-flow analysis to evaluate changes in a model farm (Johnston, 1998). Model inputs were obtained by surveying barramundi farmers throughout Queensland. 'Barraprofit' provides a starting point for potential barramundi farmers to gain a better understanding of what is involved in establishing and managing a viable aquaculture venture. It can also be used by established farms to evaluate changes in various parameters, such as:

- production parameters: farm biomass, FCRs, death rates, price, size of fish produced
- labour requirements: owner/operator, permanent and casual staff
- operating costs: feed licences, electricity, administration, etc.
- capital requirements: establishment and replacement costs.

18.5 Stock enhancement

A substantial number of barramundi juveniles are produced in Australia for stock enhancement purposes, primarily to enhance recreational fisheries. These fish are stocked into habitats where barramundi will not breed naturally, such as freshwater impoundments, or into coastal waterways where existing barramundi populations are believed to be in decline. Growth in some stocked barramundi populations is rapid: fish stocked in some northern Queensland impoundments grew to 10 kg in 3 years and some fish are now in excess of 35 kg in weight after 13 years. Rutledge *et al.* (1990) estimated that the overall economic benefit from stocking barramundi in an Australian freshwater impoundment could be 31 times the cost of stocking and raising the fish. Barramundi fingerlings are also sold to landholders for stocking farm dams that are primarily used for stock watering and irrigation. Barramundi stocked in such dams are used mainly for recreational fishing (Rimmer & Russell, 1998).

In northern Australia, stocking barramundi into coastal river systems remains a popular (among the general public) and contentious (among some fisheries managers and researchers) approach to enhancing barramundi populations. Barramundi as small as 30 mm TL survive when stocked into coastal rivers, but the type of habitat available at the stocking site is an important determinant of survival. Although about two-thirds of stocked barramundi initially remain within several kilometres of their release site, the remainder may move substantial distances within the catchment within the first few years after release.

Some fish move to adjacent catchments and foreshore areas. Stocked barramundi have been shown to survive, enter both the recreational and commercial fisheries, and reach sexual maturity so that they contribute to recruitment in the stocked population. In the Johnstone River in northern Queensland, stocked barramundi make up 20–30% of the barramundi population in a study area. Cost–benefit analysis indicates that less than 1% of stocked barramundi need to be recaptured to cover the costs of the stocking programme (Rimmer & Russell, 1998).

18.6 The future of barramundi farming

As noted earlier in this chapter, barramundi has a number of attributes that make it an ideal aquaculture species, and this is reflected in the widespread adoption of this species for aquaculture in tropical and subtropical areas of the Asia-Pacific region. Based on FAO data, world production of barramundi has stabilised at around 20 000 mt/year since 1993. It is likely that world production of barramundi will remain relatively stable, although production in some countries (e.g. Australia, Israel) is likely to increase. These increases are likely to be balanced by decreased production in Asia as many farms turn to more lucrative species such as groupers. Any expansion of barramundi culture depends on the development of new markets. *Infofish* predicts that the future markets for barramundi are the USA, Japan, Europe and Korea.

As with other products, the farm-gate price of barramundi has decreased as production levels have increased. The challenge for barramundi farmers is to reduce their production costs to maintain a viable price differential between production cost and sale price. Increased cost-effectiveness of production will be driven by improvements in growth rate through selective breeding programmes and by the development of more cost-effective diets. Barramundi farms in Asia, like other fish farms, will become larger as economies of scale force development away from smaller family-run farms to large corporate ventures.

References

Anderson, I.G. & Norton, J.H. (1991). Diseases of barramundi in aquaculture. *Austasia Aquaculture*, **5**, 21–4.

Barlow, C., Williams, K. & Rimmer, M. (1996). Sea bass culture in Australia. *Infofish International*, **2**(96), 26–33.

Boyd, C. E. (1990). *Water Quality in Ponds for Aquaculture*, p. 482. Auburn University, AL.

Cheong, L. (1990). Status of knowledge on farming of seabass (*Lates calcarifer*) in South East Asia. In: *Advances in Tropical Aquaculture Workshop*, held at Tahiti, French Polynesia, Feb. 20–Mar. 4 1989 (Ed. by J. Barret), *AQUACOP IFREMER Actes de Colloques*, **9**, 421–8.

Copland, J. W. & Grey, D. L. (Eds) (1987). *Management of Wild and Cultured Sea Bass/Barramundi* (*Lates calcarifer*). Proceedings of an international workshop held at Darwin, NT, Australia, 24–30 September 1986. ACIAR Proceedings No. 20. Australian Centre for International Agricultural Research, Canberra, Australia.

Dhert, P., Lavens, P., Duray, M. & Sorgeloos, P. (1990). Improved larval survival at metamorphosis of Asian seabass (*Lates calcarifer*) using ω3-HUFA-enriched live food. *Aquaculture*, **90**, 63–74.

FAO (2001). *1999 Fisheries Statistics: Aquaculture Production. Vol. 88/2*. FAO, Rome.

Garrett, R. N. & Connell, M. R. J. (1991). Induced breeding in barramundi. *Austasia Aquaculture*, **5**, 10–2.

Garrett, R. N. & O'Brien, J. J. (1993). All-year-around spawning of hatchery barramundi in Australia. *Austasia Aquaculture*, **8**, 40–2.

Hassan, R. (1992). *Acute Ammonia Toxicity of Red Tilapia and Seabass*. Fisheries Bulletin of the Department of Fisheries, Kuala Lumpur, Malaysia, No. 73.

Izquierdo, M. S., Fernández-Palacios, H. & Tacon, A. G. J. (2001). Effect of broodstock nutrition on reproductive performance of fish. *Aquaculture*, **197**, 25–42.

Johnston, W. L. (1998). *Commercial Barramundi Farming: Estimating Profitability*. Queensland Department of Primary Industries Information Series QI98042. QDPI, Brisbane, Australia.

Kohno, H., Hara, S. & Taki, Y. (1986). Early larval development of the seabass *Lates calcarifer* with emphasis on the transition of energy sources. *Bulletin of the Japanese Society of Scientific Fisheries*, **52**, 1719–25.

Kungvankij, P., Tiro, L. B., Pudadera, B. J. & Potestas, I. O. (1986). *Biology and Culture of Sea Bass* (*Lates calcarifer*). Network of Aquaculture Centres in Asia Training Manual Series No. 3. FAO, Rome, and Southeast Asian Fisheries Development Centre, Iloilo, Philippines.

Parazo, M. M., Garcia, L. Ma. B., Ayson, F. G., Fermin, A. C., *et al.* (1998). *Sea Bass Hatchery Operations*, 2nd edn. Aquaculture Extension Manual No. 18, Aquaculture

Department, Southeast Asian Fisheries Development Center, Iloilo, Philippines.

Rainuzzo, J. R., Reitan, K. I. & Olsen, Y. (1997). The significance of lipids at early stages of marine fish: a review. *Aquaculture*, **155**, 103–15.

Rimmer, M. A. & Russell, D. J. (1998). Survival of stocked barramundi, *Lates calcarifer* (Bloch), in a coastal river system in far northern Queensland, Australia. *Bulletin of Marine Science*, **62**, 325–35.

Rimmer, M. A., Reed, A. W., Levitt, M. S. & Lisle, A. T. (1994). Effects of nutritional enhancement of live food organisms on growth and survival of barramundi, *Lates calcarifer* (Bloch), larvae. *Aquaculture and Fisheries Management*, **25**, 143–56.

Ruangpanit, N. (1987). Developing hatchery techniques for sea bass (*Lates calcarifer*): a review. In: *Management of Wild and Cultured Sea Bass/Barramundi (Lates calcarifer)*, (Ed. by J. W. Copland & D. L. Grey), pp. 132–5. Proceedings of an international workshop held at Darwin, N.T., Australia, 24–30 September 1986. ACIAR Proceedings No. 20. Australian Centre for International Agricultural Research, Canberra, Australia.

Rutledge, W. P. & Rimmer, M. A. (1991). Culture of larval sea bass, *Lates calcarifer* (Bloch), in saltwater rearing ponds in Queensland, Australia. *Asian Fisheries Science*, **4**, 345–55.

Rutledge, W., Rimmer, M., Russell, J., Garrett, R. & Barlow, C. (1990). Cost benefit of hatchery-reared barramundi, *Lates calcarifer* (Bloch), in Queensland. *Aquaculture and Fisheries Management*, **21**, 443–8.

Southgate, P. C. & Lee, P. S. (1993). Notes on the use of microbound artificial diets for larval rearing of sea bass (*Lates calcarifer*). *Asian Fisheries Science*, **6**, 245–7.

Walford, J., Lim, T. M. & Lam, T. J. (1991). Replacing live foods with microencapsulated diets in the rearing of seabass (*Lates calcarifer*) larvae: do the larvae ingest and digest protein-membrane microcapsules? *Aquaculture*, **92**, 225–35.

Williams, K. C., Barlow, C. G., Rodgers, L., McMeniman, N. & Johnston, W. (2000). High performance grow-out pelleted diets for cage culture of barramundi (Asian sea bass) *Lates calcarifer*. In: *Cage Aquaculture in Asia: Proceedings of the First International Symposium on Cage Aquaculture in Asia* (Ed. by I. C. Liao & C. K. Lin), pp. 175–191. Asian Fisheries Society, Manila, and World Aquaculture Society–Southeast Asian Chapter, Bangkok.

Woo, N. Y. S. & Chiu, S. F. (1994). Effects of nitrite exposure on growth and survival of sea bass, *Lates calcarifer*, fingerlings in various salinities. *Journal of Applied Aquaculture*, **4**, 45–54.

Zohar, Y. and Mylonas, C. C. (2001). Endocrine manipulations of spawning in cultured fish: from hormones to genes. *Aquaculture*, **197**, 99–136.

19
Marine Shrimp

Darryl Jory and Tomás Cabrera

19.1 Introduction

Shrimp farmers represent the transition from sea hunters to farmers of the sea, a transition that has been and will be followed with many more commercially-important aquatic species.

19.1.1 History of shrimp farming

In the last three decades, farming of various marine shrimp species has developed tremendously. Shrimp farms now contribute a substantial proportion of the world's shrimp demand, rapidly replacing traditional fisheries as market demand suppliers. In addition to generating billions of dollars in trade, shrimp farming also provides employment for millions of people in developing nations, brings into production vast areas of previously unutilised coastal land unsuitable for other types of development, and produces a valuable export commodity that generates needed hard currency. The industry has also generated some environmental issues that are for the most part being acknowledged and addressed.

Marine shrimp have been grown in South-East Asia for centuries by farmers who raised them as incidental crops in tidal fish ponds. Modern shrimp farming began in the 1930s, when Motosaku Fujinaga, a graduate of Tokyo University, succeeded in spawning the Kuruma shrimp (*Marsupenaeus japonicus*). He cultured larvae through to market size in the laboratory and successfully mass produced them commercially. Dr Fujinaga generously shared his findings and published papers on his work during the next 40 years, and was honoured by Emperor Hirohito with the title 'Father of Inland Japonicus Farming'. In the early 1970s, researchers and entrepreneurs in various countries in Asia and Latin America became involved in promoting development of the industry, which grew steadily. Marine shrimp farming has come a long way in the last 25 years, and enormous progress has been made in developing technologies and methods to culture shrimp. The industry began a tremendous expansion in the early 1980s. Major references on global shrimp farming include Wyban (1992), Browdy & Jory (2001) and Rosenberry (2001).

19.1.2 Current status and production

Marine shrimp farming is currently practised worldwide in over 60 countries, but production is concentrated in about 15 nations in Asia and Latin America. Since 1992, 12 countries have contributed about 95% of farmed shrimp production. The top world producer of farmed shrimp has changed several times in the past, from Ecuador to Taiwan, Indonesia, China and Thailand. Thailand has been the world's leading producer of farmed shrimp in recent years, despite having to deal with serious shrimp disease problems.

Jory (1998) provided a detailed account of the industry in different nations (Table 19.1). Asia has much more land area and more farms than the Americas. The average farm size in Asia is 4.4 ha, compared with ~100 ha in the Americas. Mean annual production per hectare is slightly greater in the western hemisphere than in Asia, 1797 kg/ha vs. 1455 kg/ha. Farmed production could be increased by responsibly intensifying culture methods, without having to develop large areas. Most of the best locations for farms have already been developed, but the industry can still expand in several countries, including Brazil, Venezuela and several African nations. Shrimp culture is a minor activity in a few European countries (Spain, Italy), and Japan and the USA are relatively small producers owing to cool weather, high production costs and limited areas with suitable conditions. These and other developed countries are, however, significantly involved in the shrimp culture industry:

- as large consumers of shrimp
- as producers of supplies and materials used by the industry
- in providing substantial technical expertise on production and processing techniques.

Rosenberry (2001) reported, based on FAO data, that global production of farmed shrimp reached 1 130 000 mt in 1999, considerably higher than his estimated 660 200 mt of 2 years before (Table 19.1). The substantially greater FAO value for 1997 production, 1 000 000 mt (FAO, 2001), was published much later than Rosenberry (1997). The discrepancy between these 1997 data show how difficult it is to produce accurate and timely production data for cultured shrimp. The shrimp farming industry continued expanding in 2000 and 2001, especially in Vietnam, Taiwan and China, but also in Belize, Venezuela and Brazil. This is despite the white spot syndrome virus (WSSV) epidemic (section 19.8.4), which probably cost 300 000 mt of production in 2001, worth over one billion US dollars. Annual catch from the commercial shrimp fishery is ~2 000 000 mt, bringing total global production of shrimp to ~3 300 000 mt, with farmed shrimp contributing about 40% of this total. Shrimp farmers in the eastern hemisphere produce ~90% of the world's crop. Since the economic troubles and currency devaluations in Asia began in 1997, small-scale shrimp farmers throughout the region have adopted various management strategies to circumvent WSSV (Rosenberry, 2001).

Between 1975 and 1985, global production of farmed shrimp increased by 300%, and between 1985 and 1995 by 250%. Since 1995, however, industry growth has been slower due to viral and bacterial diseases. Costs have increased (also, market prices have dropped in the last 2 years), as the industry moves to comply with new international standards on product quality and the environment. If it were to increase by 200% in the 1995–2005 decade, production would be 2 100 000 mt in 2005. Production from the commercial shrimp fishery has averaged ~2 000 000 mt for the last 5 years, but has recently shown a declining trend. Global shrimp production by 2005 could be over 3 500 000 mt, with farmed shrimp contributing over half of this total (Rosenberry, 2001).

19.2 Cultured species

19.2.1 Current taxonomy

There is a new taxonomy for marine shrimp, prepared by Pérez Farfante & Kensley (1997). It recognises seven families and 56 genera of Penaeoidea and Sergestoidea shrimps. The main change for shrimp farmers in the western hemisphere is that *Penaeus vannamei* and *P. stylirostris* (Fig. 19.1) are now grouped in the new genus *Litopenaeus*. For shrimp farmers in the eastern hemisphere, *P. monodon* remains the same, whereas *P. japonicus* is now *Marsupenaeus japonicus*, and *P. indicus*, *P. penicillatus* and *P. chinensis* have been reclassified into a new genus, *Fenneropenaeus*. This new classification has its critics and is not totally accepted by the global shrimp farming community, but is used here.

There are about 2500 species of shrimp worldwide, but only 12 or so species are farmed to some degree. All belong to the family Penaeidae, characterised by a rostrum with ventral and dorsal teeth, and last thoracic segment with gills. Out of the

Table 19.1 Summary of global data for the shrimp farming industry in 1997

Country	Production (heads-on) (mt)	World total (%)	World rank	Grow-out area (ha)	Average production (kg/ha)	Estimated number of hatcheries	Estimated number of farms
Eastern hemisphere							
Thailand	150 000	22.7	1	70 000	2143	1000	25 000
China	80 000	12.1	3	160 000	500	1500	8000
Indonesia	80 000	12.1	3	350 000	229	400	60 000
India	40 000	6.1	4	100 000	400	200	100 000
Bangladesh	34 000	5.1	5	140 000	243	45	32 000
Vietnam	30 000	4.5	6	200 000	150	900	8000
Taiwan	14 000	2.1	8	4500	3111	200	2500
Philippines	10 000	1.5	10	20 000	500	90	2000
Malaysia	6000	0.9	12	2500	2400	60	800
Australia	1600	0.2	16	480	3333	12	35
Sri Lanka	1200	0.2	17	1000	1200	40	800
Japan	1200	0.2	17	300	4000	100	135
Other EH countries	14 000	2.1		20 000	700	30	2000
EH total	462 000			1 068 780		4577	241 270
Average					1455		
Global percentage		70		82		91	99
Western hemisphere							
Ecuador	130 000	19.7	2	180 000	722	350	1800
Mexico	16 000	2.4	7	20 000	800	23	220
Honduras	12 000	1.8	9	14 000	857	13	90
Colombia	10 000	1.5	10	2 800	3571	15	20
Panama	7500	1.1	11	5 500	1364	10	40
Peru	6000	0.9	12	3 200	1875	3	45
Brazil	4000	0.6	13	4 000	1000	18	100
Nicaragua	4000	0.6	13	5 000	800	4	25
Venezuela	3000	0.5	14	1 000	3000	5	8
Belize	2500	0.4	15	700	3571	1	7
United States	1200	0.2	17	400	3000	8	20
Other WH countries	2000	0.3		2 000	1000	5	15
WH total	198 200			238 600		455	2 390
Average					1797		
Global percentage		30		18		9	1
World totals	660 200			1 307 380	1626	5032	243 660

Jory (1998) adapted from Rosenberry (1997) with permission from *Shrimp News International*.

Fig. 19.1 The western or Pacific blue shrimp (*Litopenaeus stylirostris*).

commercially farmed species, two account for probably 90–95% of global production:

- the black tiger shrimp (*Penaeus monodon*) in Asia and Australia
- the Pacific white shrimp (*Litopenaeus vannamei*) in the Americas.

Penaeus monodon alone probably constituted 60–70% of the world's production of farmed shrimp in the last decade. Both species are popular in local and export markets, and are relatively easy to produce. Their seedstock can be produced in hatcheries using relatively simple technologies. Both can tolerate a wide range of salinity, from slightly greater than freshwater (1–2‰) to full-strength ocean water (35–40‰). Both species readily eat formulated commercially manufactured feeds, and require high-quality water for adequate health and growth.

Information on the major cultured species of shrimp, summarised from Rosenberry (2001) and others, is detailed below.

19.2.2 Black tiger shrimp (Penaeus monodon)

Penaeus monodon is the largest (363 mm maximum length) and fastest growing (up to 5.5 g/week) of the farmed shrimp. The species is native to the Indian Ocean and the south-western Pacific Ocean, from Japan to Australia, and dominates production everywhere in Asia except for Japan and China. This species can tolerate a wide range of salinities and is grown in inland low-salinity ponds in some countries. It is, however, very susceptible to two of the most lethal shrimp viruses, YHV (yellowhead virus) and WSSV (section 19.8.4). Breeding in captivity is difficult to induce and hatchery survival is low (20–30%). Seedstock production is mainly from wild broodstock, which are often in short supply in many areas. The protein requirement of *P. monodon* in formulated feeds is ~35%. They are widely exported from South-East Asia and marketed in Japan, the USA and Europe.

19.2.3 Western white shrimp (Litopenaeus vannamei)

This species, which is also known as the Pacific white or whiteleg shrimp, is the primary species farmed in the USA, Latin America and the Caribbean region. It ranges from the Gulf of California southward to northern Peru. Males reach a total length of 187 mm, and females 230 mm. It is well suited for farming, because it:

- breeds well in captivity
- can be stocked at small sizes
- grows fast and at uniform rates
- has comparatively low protein requirements (20–25%)
- adapts well to variable environmental conditions.

Litopenaeus vannamei is the leading farm-raised species in Ecuador and everywhere else in Latin America. It breeds in captivity better than *P. monodon*, but not as readily as *Fenneropenaeus chinensis* and *M. japonicus* (see following). In hatcheries, overall survival is relatively high at 50–60%. Some hatcheries in Latin America have captive stocks of *L. vannamei* broodstock, some pathogen free, some pathogen resistant, with some in captivity for 30 years. Production is mostly exported to the USA.

19.2.4 Western blue shrimp (Litopenaeus stylirostris)

Another important species in the western hemisphere is the western or Pacific blue shrimp, *Litopenaeus stylirostris*. It occurs from the Gulf of California south to Peru. Males reach 215 mm in total length, females 263 mm. They have a faster growth rate than *L. vannamei*, but their growth is not uniform, and this results in harvests with very broad size distributions. *L. stylirostris* tolerates lower water temperatures than *L. vannamei* but requires higher oxygen levels, turbidity, salinities, deeper ponds and formulated feeds with higher protein levels. It was often grown in combination with *L. vannamei* (at around 20% and 80% stocking respectively) and produced excellent results in many farms in several Latin American countries. It was a commonly farmed species until the late 1980s, when the IHHN virus appeared, to which *L. stylirostris* is highly susceptible. Captive stocks kept at several locations around the world, through selective breeding, developed resistance to the IHHN virus. During 1992–97, when *L. vannamei* stocks were being devastated by the Taura syndrome virus (TSV) everywhere in the western hemisphere, some shrimp producers determined that some of the captive stocks were resistant to both IHHN and TSV. Between 1997 and 2000, *L. stylirostris* made a comeback on farms throughout the western hemisphere, especially in Mexico. In the USA market, *L. stylirostris* and *L. vannamei* are often mixed together and sold as western white shrimp.

19.2.5 Chinese white shrimp (Fenneropenaeus chinensis)

Fenneropenaeus chinensis occurs off the coast of China and the western coast of the Korean peninsula. It tolerates muddy bottoms and very low salinities, and can withstand much lower water temperatures (down to 16°C) than *L. vannamei* and *P. monodon*. It readily matures in ponds, unlike most other commercially farmed species. However, it requires high protein levels (40–60%) in formulated feeds, reaches a relatively small size (183 mm maximum length) and has a lower meat yield (56%) than *P. monodon* (61%) and *L. vannamei* (63%). Between 1988 and 1993, during the boom years of Chinese shrimp farming, this species was marketed extensively around the world.

19.2.6 Japanese kuruma shrimp (Marsupenaeus japonicus)

Marsupenaeus japonicus is native to the Indian Ocean and the south-western Pacific Ocean from Japan to Australia, and is commercially cultured in Japan and Australia. Attempts to farm Kuruma shrimp elsewhere have met with limited success. They are relatively easy to ship live without water and packed in chilled sawdust and other materials (section 11.7), and they will mature and spawn in ponds. They need clean, sandy bottoms and high protein levels (55–60%) in formulated feeds. They can tolerate low water temperatures (down to 10°C) better than any other farmed species. Marketed live, they command extremely high prices in Japan, bringing up to US$200/kg.

19.2.7 Indian white shrimp (Fenneropenaeus indicus)

This species is native to the Indian Ocean from southern Africa to northern Australia and to all of South-East Asia. It is one of the major species in the commercial fisheries of these regions. *F. indicus* is primarily grown on extensive farms throughout South-East Asia, and it is widely cultured in India, the Middle East and eastern Africa. It tolerates low water quality better than *P. monodon*, grows well at high densities and salinities, reaches sexual maturity and spawns in ponds, and is widely available in the wild.

19.3 Grow-out systems

Shrimp grow-out operations span a wide continuum in terms of intensity, complexity and technology. Globally, four grow-out production systems are generally recognised, which share some characteristics, but differ in other aspects (Table 19.2).

Moving from the lowest towards the highest stocking systems, there is:

- progressive reduction in the area of ponds used
- progressive increase in capital, production costs and stocking densities
- a trend for increased use of intermediate nursery or multiple phases
- an overall intensification in production practices and technology.

The last involves a more thorough preparation of

Table 19.2 Various shrimp farming systems, based on stocking density, production inputs and operational parameters

	Intensity			
Parameter	Extensive (low density)	Semi-intensive (medium density)	Intensive (high density)	Intensive (very high density)
Stocking density (PL/m^2)	1–5	5–25	25–120	120–300
Construction cost (US$/ha)	> $5000	$5000–25 000	$25 000–200 000	> $200 000
Production cost (US$/kg)	$0.9–2.0	$2.5–5.0	$5.0–8.5	?
Pond/tank area (ha)	Ponds(5–100)	Ponds (1–25)	Ponds/tanks (0.1–5.0)	Tanks (0.1–1.0)
Seedstock source	Wild	Wild and laboratory	Laboratory	Laboratory
Water exchange (% daily)	Tidal (<5%)	Pumping (5–12%)	Pumping (up to 25%)	Pumping (25%+)
Management requirements	Minimal	Moderate	High	Very high
Feed	Natural productivity	Natural productivity and formulated feeds	Natural productivity and formulated feeds	Natural productivity and formulated feeds
Fertilisation	No	Generally yes	Sometimes	No
Mechanical aeration	No, minimal water exchange	Water exchange and some mechanical aeration	Mechanical aeration	Mechanical aeration
Energy use (hp/ha)	0–2	2–5	6–20	20–60
Production cycle (days)	100–140	100–140		
Annual production (kg/ha)	50–500	500–5000	5000–20 000	20 000–100 000 (estimated)

PL, postlarvae.
Modified from Jory (1993) and Fast & Menasveta (2000) with permission from CRC Press.

ponds, increased use of laboratory-produced seedstock, use of better formulated feeds, improved pond management practices and others. Up to the mid- to late 1990s, it also involved higher rates of water exchange, but for the last 3–5 years there has been a global trend towards reducing exchange rates and the volume of water required to produce shrimp (i.e. m^3 water/kg shrimp). This has even developed to the point of reaching zero exchange and recirculation to improve biosecurity and operational cost-efficiency. There are also trends towards increased use of technology, aeration, technical and professional labour, and a tendency to move away from the coastline to higher ground. Most farms produce at least two crops per year, although some farms can reach or surpass three or more annual crops, particularly those using nurseries or multi-phase production systems.

Early on, when the industry was young, farms were mostly extensive, and there are still many such farms in Thailand, Indonesia, India, Vietnam, Bangladesh, Central America and Ecuador, among other countries. Since around 1980, however, there has been a trend towards refitting farms to increase intensity of production. The more intensive farms are more typical in countries such as Taiwan, Japan and the USA (Jory *et al.*, 2001; Rosenberry, 2001).

19.3.1 Extensive systems (low stocking densities)

Shrimp farms with low stocking densities are typically located in tropical water impoundments ranging from ~2 ha to > 100 ha and located adjacent to estuaries, bays, and coastal lagoons and rivers. They

Fig. 19.2 A paddy in which shrimp are cultured extensively after rice harvest (Mekong, Vietnam) (photograph by Dr Michele Burford).

Fig. 19.3 Sieving wild-caught juvenile shrimp to sell for stocking extensive shrimp farms (Chalna, Bangladesh) (photograph by Professor Clem Tisdell).

frequently involve polyculture with herbivorous fishes such as mullet, milkfish and others. At some low stocking farms, liming materials are applied to ponds if soils are acidic, and they sometimes use animal manures or other organic materials to stimulate production of natural food for their shrimp (section 9.9). Ponds are filled by tides and any water exchange (typically < 5% per day) is also by tidal action. Ponds are stocked with wild shrimp postlarvae (PL), when naturally available, by opening pond gates to incoming tides. Fig. 19.2 shows a pond in the Mekong region, Vietnam, where shrimp are cultured extensively in ditches and paddies after the rice harvest. After the harvest, saline water from the Mekong River penetrates up into ditches and paddies containing rice stubble. The water brings shrimp PL for culture.

Alternatively, or additionally in various regions, juvenile shrimp may be netted from shallow coastal waters and stocked into ponds (Fig. 19.3).

In all cases, these shrimp feed on natural phyto- and zooplankton, small plants and animals living in or on the pond substrate, and particulate organic matter suspended in the water or lying on the bottom. This natural production may be promoted with applications of organic or chemical fertiliser. It is extensive aquaculture, as defined in Chapter 2 (section 2.3.3), e.g. the cultured shrimp are essentially part of a natural ecosystem that provides their nutritive and other requirements. Construction and operating costs are typically low, with the latter ~US$1–3/kg live shrimp, and production rarely surpasses 400–500 kg/ha in production cycles that last 100–140 days. Almost no new extensive shrimp farms are being built any more, because it is now illegal in several countries to build new shrimp farms in tidal and mangrove areas (Boyd, 1998; Rosenberry, 2001).

19.3.2 Semi-intensive systems (medium stocking densities)

Shrimp farms that operate at medium stocking densities (Table 19.2) are built above the high-tide line and include a pumping station and water distribution canals and reservoirs, and use of formulated feeds. In general, farm layout is relatively symmetrical and ponds are harvested by draining through a net or by using a harvest pump. Pond preparation may be elaborate, with dry-out once or twice a year, tilling and liming with various liming materials, and fertilisation with N, P and Si compounds to promote natural production. Producers may also apply various extracellular enzyme preparations and bacterial inocula to improve water quality, but the benefits of these treatments remain to be conclusively established. Organic fertilisers (mostly manure and agricultural by-products) are sometimes used. Although the water exchange rates typically used are 5–15% of pond volume per day, in recent years many farmers have adopted exchange practices of 2–5%. These lower exchange rates reduce pumping costs, and minimise fertiliser needs and the possibility of pathogen introduction. Formulated and pelleted feeds with 20–40% crude protein are usually applied

1–3 times per day, typically by manual broadcasting over pond surfaces from boats and levees. Feed quantity applied is calculated and adjusted based on the shrimp biomass estimated from results of cast net sampling and feeding charts typically provided by feed manufacturers. Natural productivity in the ponds is important for juvenile shrimp growth during the early weeks at all intensities of shrimp culture. Subsequently, there is variation in the degree of dependence on formulated feeds. Dependence on formulated feeds is not as great at this moderate density as it is at higher culture densities; hence, in the continuum of culture practices, this is semi-intensive culture (section 2.3.4).

19.3.3 Intensive systems (high stocking densities)

Shrimp farms using intensive culture (section 2.3.2) and high stocking densities (Table 19.2) typically have ponds of 0.1–2 ha, although various designs of raceways and above-ground tanks are also used, sometimes in greenhouses or other enclosures. Preparation before stocking is more meticulous, and management is often more elaborate, with feed applied 6–8 times a day. Mechanical aeration is needed to support the large shrimp biomass and heavy feeding rates needed. Aerators are placed in ponds and operated throughout the cycle, usually with increasing number of units and longer hours of operation as the cycle progresses. Generally, 4–12 hp/ha is used, with the amount increasing as the biomass of shrimp increases (Table 19.2). In Asia, several chemicals, including calcium peroxide, burnt lime, zeolite, chlorine, iodine, formalin and bactericides, are applied to ponds to improve water quality and prevent water quality deterioration and disease. Various probiotics, such as bacterial inocula and enzyme preparations, are added as in semi-intensive systems.

19.3.4 Intensive systems (very high stocking densities)

At the upper end of the continuum, systems with very high stocking densities (Table 19.2) include the highest level of environmental control, to the point of some being located indoors in greenhouses and other structures. Annual production can reach 20–100 mt/ha and higher, but there are currently only a few of these farms, in Thailand, the USA, and possibly several other countries (Rosenberry, 2001). Examples of these advanced farms and technology include the pioneer Belize Aquaculture Ltd (BAL) in Belize and OceanBoy Farms in Florida, USA (Fig. 19.4). Pond management at these farms is based on zero water exchange, heavy aeration (up to 50 or more hp/ha) and the promotion of a bacteria-dominated and stable ecological system. This compares with the highly unstable and traditional phytoplankton-dominated system. The feeding regime used promotes the growth of heterotrophic bacteria, and essentially makes the pond into a large outdoor bioreactor, akin to a sewage oxidation pond. In these high-density, zero-water-exchange systems, the pond ecology shifts (at weeks 9 to 10 after stocking) during the production cycle from an autotrophic phytoplankton-based community to a heterotrophic bacteria-based community. This shift improves water quality through fast digestion (oxidation) of organic waste and without production of toxic metabolites. At BAL, feeding rates have exceeded 350 kg/ha/day without resulting in deterioration of water quality once this heterotrophic bacterial community has

Fig. 19.4 The Labelle facility owned by OceanBoy Farms, Florida, shows recirculation and reduced water exchange technologies on a commercial scale. The treatment raceway (long and narrow on the right) is where suspended solids are removed and water is treated before re-use in lined and highly aerated ponds (photograph by OceanBoy Farms).

been developed and established in ponds. The shift also recycles wastes into nutritious bacterial flocs, the basis for 'natural production' in this system (McIntosh, 2000).

19.4 Preparation of ponds

General details of ponds, site requirements, layout, design and water turnover are given in sections 2.2.1–2.2.4.

Adequate pond preparation provides young shrimp with an environment that is relatively free of predators and competitors, with an ample supply of adequate natural food organisms, and environmental conditions that minimise stress and promote growth and survival. Pond preparation involves several sequential procedures including:

- pond draining and drying
- pH mapping
- soil tilling
- disinfection and liming
- fertilisation
- weir gate preparation and maintenance.

Some excellent reviews of these procedures are available, including Clifford (1992), Chanratchakool *et al.* (1998) and Cook & Clifford (1998).

19.4.1 Pond draining and sludge disposal

Effective removal of sludge that has built up on the pond bottom over a prior crop cycle is a very important step. If not removed, it will become an anaerobic, reduced sediment (Fig. 3.4). This will generate toxic metabolites such as methane, hydrogen sulphide, ammonia, nitrite, ferrous iron, etc., affecting pond water quality and production during the following crop cycle. This can be particularly relevant with some species of shrimp, which burrow periodically over the day/night cycle and after moulting. After a production pond is harvested, all water inlet/outlet gates are opened to generate a strong water flow through the pond, to resuspend and flush out as much accumulated sludge as possible. It is important to do this before the sludge dries out and solidifies. One technique that works in ponds with heavy sludge accumulation is to use heavy chains and rakes dragged along the bottom. Several people may walk along the internal canals to assist in resuspending sludge material as the pond is flushed. Most farms are located in areas with clearly defined dry and wet seasons, and it is not possible to dry pond bottoms effectively during the wet season. In addition, many ponds will not drain adequately because of defective design, construction or both. Low areas with standing water can be pumped dry or drainage ditches can be dug to facilitate drainage. These methods are only appropriate for production areas with sedimentation ponds and discharge canals that meander through coastal land, allowing sediment and organic matter to settle out before entering the coastal waters.

19.4.2 Pond drying and pH mapping

Earthen ponds must be allowed to dry out for 2–4 weeks to:

- promote decomposition of organic matter by bacteria
- eliminate pathogens and eggs, larvae and adults of predator and competitors
- dry out undesirable filamentous algae.

The rate of organic matter decomposition is greatest at a soil pH of 7.5–8.5 and, for ponds built on acidic soils, farmers add various lime products to improve

Fig. 19.5 Drying of pond bottoms until they crack and subsequent application of liming materials are important steps in pond preparation.

soil pH and promote decomposition of organic matter (Fig. 19.5).

Optimal microbial action for decomposition of organic matter occurs at about 20% soil humidity. Ponds infested with burrowing callianassid shrimp (snapping shrimp) must not be allowed to dry out, as this will cause these animals to burrow even deeper. Pond dry-out and disinfection may be the most effective methods for controlling epidemics of various shrimp pathogens, as well as various predators and competitors. UV radiation (via sunlight) and temperatures above 55°C will destroy several pathogens. If diseases have been detected within the last production cycle, a longer dry-out (2–3 months) may be helpful. The pond surface must be cracked to a minimum of 5 cm, to oxidise the soil and eliminate anaerobic conditions (Clifford, 1992). In acidic soils with pH < 7, pond bottom pH must be mapped promptly by sampling soil pH at various stations within the pond while soil humidity is about 30%, to calculate lime requirements. Once the pond surface is hard enough to walk on, soil pH is measured using standardised procedures. This technique involves collecting representative pond bottom soil from different sites throughout the pond. Appropriate addition of lime is then calculated (section 19.4.4).

Ploughing or turning the bottom soil of ponds (top 10–20 cm) is another optional step, which will depend on soil condition. Tilling the bottom soil of ponds can significantly promote oxidation of the lower layers of anaerobic sediments. Alternating cycles of tilling and flushing is a common action to reduce levels of iron compounds in acid-sulphate soils. It is also common to incorporate a N source such as urea into the soil before tilling, to increase the rate of organic matter decomposition. Some farmers will add 50% of the calculated lime requirements before tilling.

19.4.3 Disinfection

Disinfection is important to eliminate eggs, larvae, juveniles and adults of species of fish, crustaceans, insects, and other predatory and competitor species. Inadequate disinfection of ponds can considerably affect the yield of a pond. For example, the gobiid mudfish known as 'chame' (*Dormitator* species) can be a serious problem in Latin America. These fish can complete their life cycle in ponds, and will bury in the bottom mud and survive extreme conditions, including near desiccation, until the pond is filled again. They compete for feed and space, and can significantly affect the carrying capacity of a shrimp pond. Mudfish harvests of several mt have occurred where ponds were not adequately disinfected during preparation.

Several commercial products have been used to disinfect ponds before stocking shrimp. The use of pesticides (particularly chlorinated hydrocarbons and organophosphates) is not recommended because of their slow biodegradation, potential to accumulate in sediments and bioaccumulate in the food web. Chlorination has been the disinfection method of choice for many farmers, and its use has become prominent in the fight against WSSV, although lately it has been significantly discontinued in Latin America.

About 7 days into the dry-out period, all cement structures (inlets and outlets) must be cleaned, and mud, barnacles, oysters and algae removed. It is important to remove bivalves because they can reproduce in ponds, their filter feeding activity can quickly deplete plankton and they can be intermediate hosts of some shrimp parasites.

Applying calcium oxide or calcium hydroxide at 5000 kg/ha will raise the pH to greater than 10 and will destroy pathogens. Only muddy areas must be treated like this, not the entire pond bottom, as this process will destroy desirable bacteria needed to promote the development of productive benthos (section 19.4.6). Piscicides such as rotenone and teaseed cake are routinely used to eliminate fish, including mudfish.

19.4.4 Liming

Liming pond bottoms is a critical step in preparing earthen ponds. When calculating liming requirements the type of lime product to use must be considered, as this will influence the amounts required (Table 19.3). One lime application is generally added 2 days after tilling with lime spreaders, applying lime more heavily to wet low areas than to higher dry areas. One week is allowed for the lime to react with the soil before applying fertiliser. The calculated

Table 19.3 Neutralising values of various compounds used in pond liming

Compound	Chemical name	Neutralising value (%)
Agricultural limestone	Calcium carbonate	100
Agricultural limestone	Dolomite	109
Burnt lime	Calcium oxide	179
Burnt lime	Calcium magnesium oxide	208
Hydrated lime	Calcium hydroxide	135
Hydrated lime	Calcium magnesium hydroxide	151
Soda ash	Sodium carbonate	94
Baking soda	Sodium bicarbonate	59
Other compounds	Calcium silicate	86
Other compounds	Calcium phosphate	65

Reproduced from Boyd (1995) with permission of the World Aquaculture Society.

neutralising values of various compounds used in pond liming range from 59% (sodium bicarbonate) to 208% (calcium magnesium oxide) (Table 19.3).

When liming requirements cannot be calculated, the following liming rates, using calcium carbonate, $CaCO_3$, can be used (Boyd, 1995):

- for pH 6.5–7.5 (500 kg/ha)
- for pH 6.0–6.5 (1000 kg/ha)
- for pH 5.5–6.0 (2000 kg/ha)
- for pH 5.5–5.0 (3000 kg/ha)
- for pH < 5.0 (4000 kg/ha).

19.4.5 Weir gate preparation and entrance screening

One of the most important aspects of pond preparation is the maintenance and preparation of weir gates (monks). Configuration and placement of weir gate restriction boards and screens to prevent escapes and entry of predators, while at the same time allowing water to enter and exit continuously, is vital for cost-effective production. Entrance and effluent structures are scraped clean of barnacles, bivalves and filamentous algae. Screens with mesh of varying sizes are used to keep unwanted organisms out. Water filtration has become an important tool in recent years to exclude WSSV carriers from ponds. Clifford (1999) indicated that most WSSV carriers in the water can be selectively removed by properly screening intake water. Although it may not be feasible to totally eliminate all WSSV carriers from the ponds by screening, the more their numbers are reduced, the lesser the probability of WSSV transmission. Various sizes of mesh screens are used depending on their positions. Screens are placed after the discharge of the pumps (2-cm mesh), in the lateral branches of the main supply canal (1- to 2-mm mesh), in the inlet structure(s) of each pond (500-μm mesh as a secondary screen, followed by 300 μm or smaller as a primary screen), and in drainage canals (2- to 3-cm mesh). To initially fill ponds, very fine (200–250 μm) screens are installed in the inlet structure. When pond filling is complete, these filters are replaced with 300- to 500-μm screens. Bag net filters are used in the inlet structures to augment effective filtering surface area, and reduce screen clogging. Bag net filters of 2–5 m in length are the best, but there is no limit to the length or size of the bag net filter and, in general, the longer the better (Fig. 19.6). Multiple bag configurations (where a filter bag is placed inside another) are a cost-effective means of decreasing the mesh aperture size without installing a finer screen. Seines or gill nets (2- to 3-cm mesh) are installed in the drainage canals to prevent infected shrimp from escaping (Clifford, 1999). Ponds are inspected for proper completion of necessary preparation steps before filling with water.

19.4.6 Natural productivity

Proper management of natural productivity in shrimp ponds is critical to promote and sustain plankton blooms and microbial and benthic community productivity. A vigorous phytoplankton bloom will support a healthy benthic community and will contribute

Fig. 19.6 A battery of bag nets used to filter the inflowing water at a shrimp farm in north-west Mexico.

significantly to stabilising and maintaining adequate water quality in shrimp ponds. This happens through several mechanisms (Cook & Clifford, 1998):

- by increasing oxygen production through photosynthesis during daylight hours
- by decreasing levels of various metabolites and toxic substances such as ammonia, nitrate, nitrite, hydrogen sulphide, various heavy metals and other undesirable compounds
- by regulating pond water and bottom pH (extremely important in ponds with acid-sulphate soils)
- by providing shade to prevent the establishment of filamentous bottom algae
- by increasing turbidity with a consequent reduction in predation by diving birds, among other mechanisms.

Natural productivity is also important because it supports the generation of detritus, the particulate organic material produced from the dead bodies, non-living fragments and excretions of living organisms. In nature, organic detritus is an important food source for many estuarine organisms, and in shrimp ponds it can have an important role. Shrimp feed on detritus, and derive nourishment by stripping the micro-organisms from the detrital material as it passes through their gut. In addition, their faecal pellets may be recolonised and the process repeated until all the organic material has been utilised. Maximising the recycling capability of organic detritus within the culture environment, nutritionists and pond managers have an opportunity to reduce feed and production costs, improve FCR, and reduce environmental impacts.

As a source of carbon for growth, natural productivity is much more important in shrimp ponds with moderate stocking densities than in high densities. In the latter, formulated feeds provide most of the feed consumed by shrimp from the start of the production cycle. In comparison, natural productivity in moderately stocked ponds can sustain shrimp for about 30 days (ca. 20–30% of the duration of a typical grow-out cycle), depending on the stocking density (10–20 shrimp/m^2). This is until a critical shrimp biomass (100–300 kg/ha) is reached and additional subsidies, in the form of added formulated feeds, are needed to sustain shrimp growth and production. There is extensive evidence that juvenile shrimp feed on plant material and, although generally via intermediate prey species, algae are probably a major source of carbon for these shrimp.

19.4.7 Initial fertilisation

Fertilisation of shrimp ponds is an effective means of stimulating natural food production that can help reduce feed costs (e.g. Chanratchakool *et al.*, 1998; section 9.9). Through fertilisation, managers can promote those pond ecosystem components that are beneficial to shrimp production and discourage those that are detrimental. Several organic and inorganic nutrient sources can be used to fertilise shrimp ponds. They vary in their effectiveness because of differing nutrient density, solubility in water, potential toxicity, C/N ratios and other factors. Appropriate fertilisation rates vary depending on fertiliser type and the ambient nutrient concentration in the water. Various dynamic processes in ponds can also affect fertilisers, producing dissolved nutrient concentrations after fertilisation that may substantially differ from calculated concentrations.

Fertilisation involves the application of fertilisers over the entire pond bottom before filling with water. The objective of inorganic fertilisation during preparation and during the grow-out cycle is to promote and maintain populations densities of diatoms and green algae of at least 80–100 000 cells/mL. These algae groups are reportedly the most nutritionally desirable for shrimp (Clifford, 1994). Desirable

Table 19.4 Desirable phyto- and zooplankton densities in shrimp ponds with medium culture densities

	Minimum number of organisms per mL	Maximum number of organisms per mL
Phytoplankton		
Diatoms: bacillariophytes and chrysophytes	20 000	–
Chlorophytes (green algae)	50 000	–
Cyanophytes (blue-green algae)	10 000	40 000
Dinophytes (dinoflagellates)	–	500
Total phytoplankton cells	80 000	300 000
Zooplankton (copepods, rotifers)	2	50
Protozoans	10	159

Adapted from Clifford (1994) with permission.

phyto- and zooplankton densities in semi-intensive shrimp ponds are shown in Table 19.4.

There is no universally optimal N/P ratio, and each farm must determine by trial and error the most suitable inorganic fertilisation regime for its prevailing conditions and needs. In areas that have markedly different dry and wet seasons, there are different optimal fertilisation ratios for each season, as well as different optimal stocking densities and other significant management differences. It is only through experience and monitoring that the best fertilisation regime is determined for the prevailing conditions.

When selecting inorganic fertilisers, the type (nutrient class, composition and solubility), the N/P ratio, the daily dosage rates and the frequency of application must be considered. The main nutrients needed in shrimp ponds to promote phytoplankton blooms are N, P and Si. Si compounds stimulate production of the very desirable diatoms. Commonly used inorganic fertilisers are urea and sodium nitrate as N sources, and monoammonium phosphate (MAP), diammonium phosphate (DAP) and triple superphosphate (TSP) as P sources. Urea is the most commonly used source of N because it is widely available, inexpensive and effective. Nitrate-based fertilisers are very effective, but more expensive and less available. The most common silicate fertiliser used is sodium metasilicate, which is expensive and not widely available. A N/P ratio of 15–20:1 promotes the development of diatom blooms; however, the N/P ratios used in many shrimp farming areas worldwide can vary from 1:1 up to 45:1.

Organic fertilisers promote production through direct consumption by the shrimp. Shrimp graze on the bacterial detritus that forms on the fertiliser particles and on other natural food organisms whose populations are promoted by fertilisation. Chicken manure was the main organic fertiliser used in shrimp ponds for many years, with various other manures (duck, cattle and swine) also used to some extent. In recent years, however, farmers have switched to finely ground vegetable meals (wheat, soybean, rice, sorghum and others). These agricultural by-products serve as substrate for bacteria, zooplankton and meiofauna, and they can be consumed directly by shrimp (Cook & Clifford, 1998).

Initial fertilisation enhances the natural productivity of a newly filled pond, reducing the need for costly high-protein start-up feed, and also supplying richer more digestible protein, lipids, phospholipids, fatty acids, cholesterol and trace nutrient sources through planktonic and benthic organisms. The juvenile shrimp in a well-fertilised pond that was stocked at 15 PL/m^2 can usually reach an average weight of 2 g in 30–35 days with minimal supplemental feeding.

19.5 Reproduction and maturation

19.5.1 Hatchery production and the life cycle

The typical life cycle (Fig. 19.7) of a marine shrimp in nature begins with adult animals migrating up to several kilometres offshore, maturing and spawning. The eggs initially sink, but after a few hours they

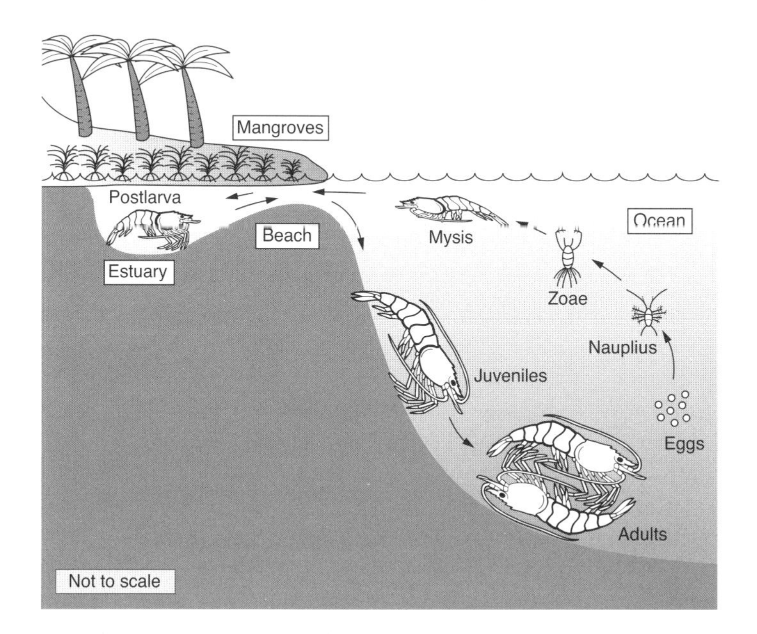

Fig. 19.7 A typical life cycle of a marine shrimp (from Rosenberry, 2001).

hatch and the nauplii (the first larval stage) float to the surface. Marine shrimp usually spawn in areas where favourable currents will eventually bring the developing larval stages inshore into nursery areas such as estuaries, large bays and coastal lagoons. These areas provide abundant natural food and adequate conditions for survival and growth. Developing shrimp remain in nursery areas for several months, and then begin maturing and moving offshore to spawn and complete their life cycle.

Hatcheries in the western hemisphere are generally large (Fig. 19.8), often belong to a vertically integrated operation that includes a grow-out farm and a processing plant, and they frequently produce excess nauplii, which are sold to smaller hatcheries. Eastern hemisphere backyard and medium-scale hatcheries produce most of the seedstock used by farmers (Rosenberry, 2001). Hatcheries produce seedstock ready for stocking into nursery or grow-out ponds and tanks. Shrimp are very fecund animals, producing 100 000–1 000 000 eggs per spawning in the wild, and 50 000–400 000 eggs per spawning in captivity (depending on species, size, water temperature, wild or captive origin, and the number of previous spawns). In the hatchery, gravid female shrimp (wild or matured in captivity) spawn at night, and eggs hatch the following morning.

Larval development from egg to PL is complex and involves three stages: nauplius, zoea and mysis.

Fig. 19.8 Outdoor tanks for larval culture in a hatchery in Brazil.

The first has five or six substages, and the next two generally have three substages each, with each substage lasting some to many hours. The larval development process takes ~15 days. As larvae develop and consume the yolk sac, their diet switches to phytoplankton and then to zooplankton. After the mysis stage, they consume a variety of organisms, including the brine shrimp (*Artemia* species). All these live organisms are produced in the hatchery, but major advances have been made recently to develop alternative inert diets (section 9.6). After they become PL, the animals look like small adult shrimp and are able to feed on zooplankton, detritus and commercial feeds. After several days as PL, they are ready to be stocked into nursery or grow-out systems. With the increasing importance of genetic improvement programmes to address biosecurity concerns for various shrimp diseases, hatcheries will play an increasingly important role in the support and expansion of the industry (Jory, 1996, 1997). For detailed information on hatchery procedures refer to Bray & Lawrence (1991) and Treece & Fox (1993).

It is commonly believed that PL from females matured in the wild are superior and are generally preferred over PL from eyestalk-ablated females (section 19.5.4), which are considered of inferior quality. This 'lower' quality of PL from ablated shrimp could be due to less than optimal treatment of broodstock at a time (after ablation) when there are increased energy requirements (proper nutrition) and disease susceptibility. Nauplii production from captive broodstock maturation varies considerably, but has several advantages, including known origin, species and age, increased biosecurity, and relatively predictable availability. These are all very important for production planning and forecasting, and even more relevant in genetic improvement programmes. The use of wild seedstock can also introduce pathogens, competitors and predators into grow-out systems (Jory, 1996, 1997). The dependence on wild broodstock is more serious in Asia than in the western hemisphere, because there has been less interest in captive reproduction of cultured broodstock in the former. Hatcheries in Asia have had little interest in using domesticated broodstock, as they have had few problems sourcing wild broodstock for their operations. The domestication of shrimp stocks in Asia generally trails that of the Americas, and the majority of *P. monodon* broodstock used by Asian hatcheries are from wild stocks. *P. monodon* has been reared in captivity many times since the 1970s, but a combination of cost, technical constraints and a lack of demand for domesticated stocks are the reasons for a lack of application on a commercially significant level. There also remains a widespread belief that domesticated shrimp are somehow not as hardy or as well suited for culture as their wild counterparts.

19.5.2 Broodstock maturation

The ability to produce seedstock on demand, consistently and in sufficient numbers to support the industry is extremely important. Maturation of shrimp broodstock is undertaken routinely in many countries, and at least 26 penaeid shrimp species have been matured and spawned in captivity to produce viable eggs. The life cycle has only been closed consistently and dependably for a few shrimp species, however, and most shrimp hatcheries around the world still depend significantly on wild broodstock. Some shrimp species appear more suited to captive maturation than others: *P. monodon* is comparatively difficult to mature in captivity, whereas *L. vannamei* and *L. stylirostris* are relatively easier. There are three approaches to maturation:

- Procure wild, sperm-bearing females that can spawn right away. This usually produces high quality eggs and larvae.
- Obtain wild adults, and mature and spawn them in captivity.
- Mature and spawn adult shrimp that have been grown in captivity.

The first two approaches are widely practised in most shrimp farming countries. The third strategy, producing succeeding generations in captivity, is the only one that will lead to successful stock domestication and selection for commercially important traits such as fast growth and disease resistance (Jory, 1996).

Large hatcheries typically have separate infrastructure to carry out broodstock maturation. The maturation section of a hatchery is normally isolated from other sections, to reduce noise levels and stress caused by human activity (section 2.7). Maturation tanks are typically round, about 3–5 m in diameter and 60–100 cm in height, and gently sloped towards a central drain to facilitate removal and siphoning of uneaten food and other undesirable debris. They are generally arranged in batteries or rows in a dedicated room. Tanks are generally black on the inside, and vinyl or PVC liners are commonly used to reduce construction costs, reduce injuries to animals caused by jumping into tank walls, and to facilitate cleaning and disinfection. A sand or pebble substrate is sometimes provided for burrowing species.

Environmental conditions in maturation facilities duplicate or even intensify conditions known to stimulate reproduction (section 6.3). Parameters such as water temperature, salinity, photoperiod, light intensity, lunar phase, maturation room and tank characteristics, and diet may be manipulated and adjusted. Each species has optimum ranges at which maturation will be facilitated. The value of these parameters and their rate of variation over time are critical to stimulate the reproductive process. The water supply system, closed or open, must continually provide maturation tanks with clear unpolluted water with oceanic characteristics and a daily exchange capacity of 200–300%. Optimum water temperature for maturation is typically around 28–29°C. An oceanic salinity of 30‰ is considered optimum, although maturation may occur between 28‰ and 36‰, and pH must be maintained between 8.0 and 8.2.

19.5.3 Open vs. closed thelycum species

Penaeid shrimp belong to two groups based on the structure of the female thelycum. This is a receptacle structure on the ventral thorax of females where the spermatophore is deposited by the male at mating. The open-thelycum species are indigenous to the western hemisphere and include *L. vannamei*, *L. stylirostris* and *L. occidentalis*. The closed-thelycum or 'brown' shrimp species include *P. monodon*, *M. japonicus*, *F. merguiensis*, *F. indicus* and several others.

- Open-thelycum species: moult → mature → mate → spawn.
- Closed-thelycum species: moult → mate → mature → spawn.

Consequently, the management of maturation procedures is different for open- and closed-thelycum species. For example, natural photoperiod is normally used in maturation facilities for most closed-thelycum species, whereas for open-thelycum species a reversed photoperiod regime is typically used (with artificial lights) so that animals will spawn during daylight or normal working hours.

19.5.4 Maturation procedures

The maturation process is relatively simple and includes selecting or sourcing good prospective broodstock (screened for absence of viruses) and holding them under stable, optimal environmental conditions, minimal stress and adequate nutrition using both natural and artificial diets. Exclusion and control of opportunistic shrimp pathogens, such as various bacteria, fungi and protozoea, is critical. It is accomplished by maintaining the best water quality possible and by periodic prophylactic treatments. Depending on animal size, maturation tanks are stocked at 3–8 animals/m^2 and at ratios from 1:1 to 1:4 males to females for closed-thelycum species, or 1:1 to 3:1 males to females for open-thelycum species. Eyestalk ablation is a relatively simple procedure and is normally done only on female shrimp that are hard shelled (section 6.2.2), not on animals that are about to moult or that have recently moulted. Prospective broodstock are normally acclimated for several weeks in maturation tanks before undergoing ablation. It is possible to mature, mate and spawn several shrimp species without resorting to eyestalk ablation, by subjecting animals to a controlled, manipulated photoperiod and water temperature regime (Bray & Lawrence, 1991; Treece & Fox, 1993; Jory, 1996).

Adequate nutrition is another factor critical for shrimp maturation, promoting sexual maturation and mating, and improving fertility and offspring quality. Maturation diets typically include combinations of commercial dry pelleted feed supplements and natural organisms such as various molluscs (clams and other bivalves, squid), crustaceans, fish and bloodworms. Bloodworms are marine polychaetes rich in the polyunsaturated long-chain fatty acids (PUFAs) that are essential for shrimp to mature.

There are five ovarian maturation stages in female shrimp: from stage I (immature) to stage IV (ripe or mature) and stage V (spent after spawning). Assessment of the progression in ovary development can be performed by holding animals against a light source, or by shining a flashlight from above over animals in the tanks to determine ovary size and colouration. In a stage I animal, the ovaries are very thin and look like almost invisible strands. As the ovaries develop through stages II to IV, they become progressively thicker and darker as seen through the animal's exoskeleton.

19.5.5 Mating and spawning

Mating in shrimp is characterised by particular courtship behaviour, and various pheromones are involved in sex attraction. Elaborate behaviour with extended and elaborate chasing, contortions and other activities has been described for different species. Mating in closed-thelycum shrimp normally happens at night. In closed-thelycum species, a hard-shell male mates with a soft-shell female that has just moulted, inserting its spermatophore into the female's thelycum near the female's gonopores. The female carries the sperm packet internally until it spawns or moults, and mating typically occurs again shortly. Mating in open-thelycum shrimp normally occurs late in the day between hard-shell males and hard-shell females. Males attach the spermatophore to the ventral surface of the female's body and close to the gonopores. As females of both types spawn and discharge their eggs through their gonopores, sperm exude from the spermatophore and fertilise the eggs. The species and size of the spawner determines spawn size, with larger species and larger animals producing more eggs per spawn. Ablated females will normally spawn multiple times over a period of several weeks. Nauplii can be collected directly from maturation tanks, but mated females are usually removed and placed in individual spawning tanks (100–500 L).

19.6 Hatchery design and larval culture

19.6.1 Hatchery design

Shrimp hatcheries can be set up in a warehouse structure or building shell, and have indoor and outdoor infrastructure and facilities. Hatchery design follows the typical pattern (section 2.7), with various areas for microalgae culture, laboratory, brine shrimp production, seawater treatment and holding, spawning, larval and PL rearing. Support infrastructure includes storage space, offices, and aeration and electrical power generating capability. There are small, medium and large shrimp hatcheries, and all have the same basic infrastructure.

(1) Small hatcheries are usually a family operation, with very low set-up and operating costs, using simple techniques and untreated water, low culture densities, and with larval culture tank capacity under 50 000 L. These generally produce either nauplii or PL, and often suffer disease outbreaks and water quality problems, but can readily shutdown and restart in a relatively short time. These hatcheries have been very successful in South-East Asia, but are not so common in Latin America.
(2) Small- and medium-scale hatcheries are usually based on the Taiwanese design (section 19.6.2), with large culture tanks between 5000 and 50 000 L.
(3) Large hatcheries use elaborate techniques to operate controlled culture environments. They target annual production of 100 million PL or more and are typically based on the 'Galveston' system (section 19.6.2). When large hatcheries have water quality and disease problems, they usually require several months to get back on line (Rosenberry, 2001).

19.6.2 Larval culture methods

Worldwide, various methods are used to produce shrimp larvae in hatcheries. These are all modifications of the two basic methods, the Taiwanese and the Galveston methods. The latter is a modification of the former and differs from it in that microalgae are cultured outside the larval tank and are added as required. Over the years, many researchers have helped modify and refine these techniques.

(1) The Galveston or 'clear water' method is based on the procedures developed at the Galveston Laboratory (National Marine Fisheries Service) in Texas in the 1970s, and has been successfully adapted to local conditions and implemented throughout the western hemisphere. It uses high densities of 100 or more larvae per litre, relatively high water exchange rates and elaborate water filtration and conditioning, and produces 6- to 12-day-old PL. Microalgae and brine shrimp nauplii are cultured separately and added to larval culture tanks as required.
(2) In the Taiwanese method, lower stocking densities and larger larval culture tanks are used. This method requires large broodstock numbers, and the microalgae blooms are promoted '*in situ*' by

directly fertilising larval culture tank water. The Taiwanese method performs well in temperate areas, where there is a short production time, but it can be comparatively harder to manage because of the large water volumes involved. This method is generally not used in tropical areas because of the management difficulties inherent to tropical conditions (Treece & Fox, 1993; Jory, 1997).

19.6.3 Larval nutrition

The larval feeding regime of shrimp hatcheries is typically based on microalgae and brine shrimp (section 9.5) or on live feeds combined with formulated/prepared diets (Table 19.5). The latter can be produced at the hatchery, although in recent years several commercial brands have become available. Most are 'dry diets' packaged under vacuum or in nitrogen-filled cans or pouches, and are produced in a range of particle sizes suited to the feeding habits of the larval stages for which they are intended. Although the sizes of particles may vary from one manufacturer to another, in general zoea diets are < 50 μm, mysis 50–100 μm, PL1–PL4 100–150 μm and PL4–PL8 150–250 μm. Crude protein levels typically range from 42% to 48%, whereas lipids are 12–16% and fibre is less than 5%. Artificial diets are not a complete replacement for natural food, but when used supplementary to natural food they can increase both survival rate and growth, when compared with the feeding of natural diets (algae and brine shrimp) alone. Determining which diet to use on a cost–benefit basis should be a high priority with the hatchery manager, and diets have been used successfully by several hatcheries feeding several dry formulations (Laramore, 2000). It is likely that in the near future these diets will totally replace live feeds in hatcheries (Jory, 1997).

Several species of microalgae are commonly produced in shrimp hatcheries, including species of the genera *Tetraselmis*, *Isochrysis*, *Chaetoceros*, *Skeletonema* and others. Table 19.6 shows some of the microalgae species typically cultured and used in shrimp hatcheries. *Isochrysis galbana* and *Chaetoceros gracilis* are among the best, because of their relatively small size and significant content of highly unsaturated fatty acids (HUFAs). It is common to enrich brine shrimp nauplii with HUFA before feeding these to larval shrimp (section 9.4.7).

19.6.4 Probiotics, vaccines and immunostimulants

The ability to control pathogenic organisms, particularly bacteria, has been very important to the success of commercial-scale hatcheries. Until recently, the general method to control pathogenic and undesirable bacteria was through various water filtration and disinfection techniques (e.g. UV, chlorination). Other common procedures include implementation of hygienic procedures (personnel and equipment disinfection) and use of non-infected cultures of microalgae and brine shrimp. Various biocides and chemotherapeutics are also used to indiscriminately eliminate or reduce the numbers of pathogenic bacteria, frequently as a routine procedure and not necessarily when needed. This practice can bring serious consequences, including development of resistance, chemical traces in shrimp tissues, and environment impacts. Probiotics are single or mixed cultures of some harmless or beneficial bacteria bacterial strains that are widely used to promote survival and growth of larval shrimp. Probiotics can be an effective hatchery tool by:

- competitively excluding pathogenic bacteria
- producing substances that inhibit growth in opportunistic pathogen species
- providing essential nutrients
- promoting digestion by supplying essential enzymes
- direct uptake of dissolved organic material mediated by bacteria.

There is much potential for improvement of probiotics, but more research is needed to identify suitable species and how to promote their population growth in larval culture tanks (Jory, 1997).

The immune system of crustaceans does not produce antibodies and is relatively primitive, with non-specific immune responses, which means that they cannot be 'vaccinated' in the traditional sense. In recent years, several commercial products have claimed the ability to stimulate disease resistance

Table 19.5 Feeding schedules of microalgae and brine shrimp nauplii used successfully by several hatcheries for rearing shrimp larvae and PL. In addition, the diet included a dry commercial diet. Z, zoea; M, mysis, PL, postlarvae

			Cells/mL*		
Day	Stage and substage	Estimated survival (%)	*Chaetoceros gracilis*	*Tetraselmis chuii*	Brine shrimp/mL†
1	Z1	90	40 000	–	–
2	Z1/Z2	85	50 000	–	–
3	Z2	82	60 000	–	–
4	Z2/Z3	80	70 000	–	–
5	Z3	78	80 000	–	–
6	Z3/M1	75	70 000	10 000	0.1
7	M1	70	60 000	15 000	0.2
8	M2	65	50 000	20 000	0.4
9	M3	62	40 000	25 000	0.7
10	M3/PL1	60	20 000	25 000	1
11	PL1–2	58		20 000	2
12	PL2–3	56		10 000	3
13	PL3–4	54			3
14	PL4/5	52			4
15	PL5/6	51			3
16	PL6/7	50			2
17	PL7/8	49			0
18	PL8/9	48			0
19	PL9/10	47			0

*Algae count expressed as desired number of cells.
†Use frozen or killed *Artemia* species to feed Z3/M1.
Temperature, survival and water quality may affect consumption and, therefore, feeding rates may be modified to prevent overfeeding and tank fouling.
Reproduced from Laramore (2000) with permission.

and promote shrimp health by preventing and minimising the impact of outbreaks of pathogenic bacteria both in hatcheries and during grow-out. In hatcheries, these products can be administered to late mysis and PL by immersion or by microencapsulation using brine shrimp. Results reported from field-testing indicate that some of these products can stimulate non-specific immune responses and improve survival and growth of treated animals, thus enhancing production (Jory, 1997).

19.7 Seedstock quality and stocking

19.7.1 Seedstock packing, transportation and reception

It is important that the delivery of seedstock (PL) is well coordinated with the farm, and good planning and communication are required to ensure a smooth transition from the protected conditions of a hatchery to the conditions of an outdoor pond, tank or raceway. Timing is critical to ensure that the PL are going into a grow-out environment that has been adequately prepared and has the proper conditions to promote acceptable survival and production. Of particular importance, in the case of outdoor ponds, is that there is sufficient natural productivity to provide adequate nutrition to the animals being stocked. There is a window of about 10 days, starting about 10 days after pond filling begins, when pond conditions should be adequate for stocking. In general, ponds should not be stocked beyond 20 days after filling has begun.

Seedstock of great quality may be severely stressed by inadequate water quality or high packing densities. Inadequate packing and transportation of

Table 19.6 Microalgae species typically used in penaeid shrimp hatcheries

Class	Species
Bacillariophyceae	*Skeletonema costatum* *Thalassiosira pseudomonas, T. fluviatilis* *Phaeodactylum tricornutum* *Chaetoceros calcitrans, C. curvisetus, C. neogracile, C. simplex* *Ditylum brightwelli* *Scenedesmus* sp.
Haptophyceae	*Isochrysis galbana*, *Isochrysis* species (Tahitian) *Dicrateria inornata* *Cricosphaera carterae* *Coccolithus huxley*
Chrysophyceae	*Monochrysis* species
Prasinophyceae	*Pyramimonas grossii* *Tetraselmis suecica, T. chuii* *Micromonas pusilla*
Chlorophyceae	*Dunaliella tertiolecta* *Chlorella autotrophica* *Chlorococcum* species *Nannochloris atomus* *Chlamydomonas coccoides* *Brachiomonas submarina*
Chryptophyceae	*Chroomonas* species
Cyanophyceae	*Spirulina* species

Reproduced from Treece & Fox (1993).

shrimp seedstock will significantly affect survival and health. Care must be taken to minimise stress and to provide the best handling and conditions possible. When possible, a farm representative should be present during harvesting of larval tanks and packing of PL for transport to the farm. They should also be involved during counting of PL, to calculate numbers as accurately as possible, and to observe any undesirable conditions and practices. The farm representative may also accompany the shipment, to ensure fast movement without unnecessary delays and to maintain adequate transportation parameters (mainly temperature and dissolved oxygen, DO) during transit.

Methods of packing and transportation of shrimp PL can vary significantly, depending on origin, species, distance to farm, resources available and other factors. PL are typically transported to farms from hatcheries or from collection and consolidation stations (wild PL):

- in plastic bags with water and added oxygen, within polystyrene foam or cardboard boxes and cooled to 18–22°C (down from the typical 28–29°C in hatchery larval culture tanks)
- in tankers ranging in water capacity from 2 to 20 mt and set up to provide water aeration/oxygenation.

Feed (e.g. frozen brine shrimp) is added to prevent cannibalism. Packing densities vary between 500 PL/L and 2000 PL/L, depending on species, age and estimated time in transit. When possible, PL are transported early in the morning or in the evening, to avoid high temperatures. Transport time should be as short as possible, ideally not exceeding 6–8 h. Reduced water temperature (18–22°C) and the addition of various compounds to shipping water (including ammonia suppressants, buffers and activated carbon) increase PL survival over extended shipping schedules.

Upon arrival, parameters such as water temperature, salinity, dissolved oxygen, pH and alkalinity must be determined to serve as the baseline for acclimation adjustments to match pond water characteristics. Upon arrival, PL may be transferred to counting or acclimation tanks, which can be located next to the pond(s) to be stocked. Care must be taken to maintain water parameters and avoid any sudden drastic changes.

19.7.2 Counting and quality control

There are various methods used to count shrimp PL, and controversy regarding the accuracy of each. According to Cook & Clifford (1998), volumetric subsampling is the most common technique. It involves concentrating all or most of the PL in a known volume of water, and drawing a fixed number of subsamples (around five) using a beaker or similar container of known volume. The subsamples are counted and their average is extrapolated to the larger volume.

Stocking only the best-quality shrimp PL is critical to the success of a shrimp farm. Several well-established criteria are used to assess PL quality, including:

- their origin and hatchery reputation
- visual evaluation
- stress tests
- various tests to detect the presence of pathogens.

Strict use of PL quality assessment criteria in the evaluation and selection of PL for stocking, and a careful acclimation procedure using the best-quality seedstock available, will have a significant effect on the production and profitability of a shrimp farm. Detailed information on PL quality assessment procedures are given in Clifford (1992).

The strength or 'hardiness' of PL from different hatcheries or batches can vary significantly and the acclimation schedule must be tailored to the PL 'fitness'. Stronger animals can be acclimated at a faster rate than weaker PL. Various stress tests are used to challenge PL and determine a suitable acclimation schedule. These tests typically involve subjecting a PL sample of 100–200 individuals to a thermal, osmotic or chemical shock for 1–4 h and counting the survivors. For example, Clifford (1992) proposed a standardised 'stress test' method whereby a sample of PL is placed in a container and the salinity and temperature are simultaneously brought down to 20‰ and 10°C, respectively, for 4 h (a test lasting under 4 h does not adequately account for lingering PL mortalities). Survival of 80–100% of the test animals indicates high-quality PL, but 60–79% survival is considered acceptable.

19.7.3 Acclimation and stocking

Most shrimp farmers spend substantial resources and effort during pond preparation to enable them to stock their PL into a grow-out environment with the best possible environmental conditions, as free of predators, competitors and stress as possible, and with ample supply of adequate food organisms. Still, the transition from relatively benign conditions in hatcheries to those prevailing in open grow-out systems, such as tanks and ponds, where water conditions continually or unpredictably change (day/night, dry/rainy seasons over the production cycle) can be a traumatic experience for PL unless the transition is gradual and stress is minimised.

Proper preparation of the acclimation station and equipment is a critical step in PL acclimation. The acclimation station, including all tanks and other water reservoirs and equipment (nets, siphons, buckets, tubing, others), is thoroughly cleaned and disinfected by scrubbing with chlorine or other disinfecting agent. Well-functioning and calibrated equipment to monitor water parameters (temperature, salinity, pH and dissolved oxygen) before the PL arrive at the acclimation station is critical.

The typical acclimation process involves holding the PL for a period in tanks and slowly adding water from the pond to be stocked to equalise various parameters (mainly salinity and temperature). General acclimation recommendations that have been used for many years include:

- Increase/decrease salinity by no more than 3‰/h.
- Avoid sudden temperature changes (> 3–4°C).
- Maintain DO levels at 6–7 ppm.

Acclimation densities should not exceed 300–500 PL/L depending on animal size and duration of acclimation.

Salinity is probably the most critical parameter to manipulate during PL acclimation. Table 19.7 suggests salinity acclimation rates for various scenarios, including when handling PL of different ages and condition (strong, weak). These recommended schedules may be used as the basis for developing in-house acclimation procedures. (Note that 'acclimation' is used here in the sense of rapidly developed tolerance of changed conditions, rather than long-term mechanisms of developed tolerance such as are described for temperature acclimation in section 3.3.1.)

Acclimated PL may be released into the pond using buckets or other containers at points at least 50 cm in depth, at 50-m intervals and on the upwind side of the pond to maximise PL distribution throughout the pond. Excessive turbulence at release must be avoided to prevent damage to animals.

Survival cages, buckets or net enclosures are commonly used to estimate PL survival 24–48 h after stocking. This information is very important because it is the basis for additional compensatory stocking if initial survival is not satisfactory. Chanratchakool *et al.* (1998) described the use of net enclosures (no less than 2 m^2 and 1 m deep) to monitor survival of stocked PL in Thailand. About 1000–2000 PL are placed in the enclosure and fed normally, and counted after 3–5 days. Clifford (1992) provides detailed descriptions of the use of survival cages in Latin America.

19.8 Production management and harvest

19.8.1 Water and sediment quality

Some water quality parameters must be routinely monitored to effectively manage shrimp production systems during grow-out. Controlling various parameters can be difficult, but, if not managed properly, pond carrying capacity can be rapidly exceeded and ponds can crash within a few hours or days. Shrimp mariculture requires high water quality to function efficiently and maintain farm productivity and profitability. Therefore, monitoring the quality and properties of intake water, pond water/soil conditions and organics in effluent are essential for good animal husbandry. It is also important for farms to be environmentally aware and maintain water management programmes to minimise any potential 'downstream' impacts (section 4.2). Table 19.8 shows the general range and monitoring frequency recommended by Clifford (1997) for various water quality parameters that are important in shrimp production systems. These standard parameters are described in section 3.2. They are measured in the field with probes and meters or by taking water samples that are analysed in a laboratory on site. Measurements and samples are taken at several places within a pond, including the water intake, the central region, drain, and surface and bottom layers. This will provide a more representative assessment of the parameter. Boyd (1990) provides detailed information on pond water parameters and their monitoring.

DO is one of the most critical physical parameters of a pond culture. Low DO is one of the most common causes of mortality and poor growth in high-density shrimp ponds. Lethal levels seem to vary from about 0.5 to 1.2 ppm, depending on the species and hardiness of a particular population. DO levels can be relatively unstable as a result of wide fluctuations in photosynthetic oxygen production and the bacterial population affecting BOD. DO levels in the water column fluctuates over a day/night cycle, from a low at dawn to a high in mid-afternoon, particularly due to phytoplankton photosynthesis during the day and then respiration at night. Hence it is appropriate to sample DO at dawn and mid-afternoon. DO readings in the water column immediately adjacent to the bottom are essential, because shrimp spend most of their time feeding and resting in or on the sediment.

Salinity and temperature are extremely important parameters that affect several biotic and abiotic processes in the production system environment, but there is usually relatively little that can be done to modify them. Water temperature, however, may be controlled in indoor systems of high stocking density. Farms with free access to both good-quality seawater and freshwater can mix these to desired salinities.

pH does not usually reach levels that affect the shrimp. Values seldom exceed the range of 6–9 in sediments or the water column (except in acid-sulphate soils). High pH (> 9) is a lesser risk than low pH (< 6). Low pH may affect the mineral deposition of the shrimp's exoskeleton after moulting,

Table 19.7 Suggested salinity acclimation rates for 'strong' and 'weak' PL. 'Strong' refers to animals older than 10 days (> PL10) or strong larvae. 'Weak' describes animals younger than 8 days (< PL8) or weak animals

Salinity change (‰)	Suggested acclimation rate (‰/h)	
	Strong PL	Weak PL
From 35 to 20	5	3
20–15	4	2
15–10	3	2
10–5	2	1
5–2	1	0.5
2–0	0.5	0.2
30–40	4	2
40–50	2	1

Reproduced from Clifford (1992) with permission of the World Aquaculture Society.

Table 19.8 General range and recommended monitoring frequency for various water quality parameters in shrimp ponds

Parameter	Minimum value	Maximum value	Monitoring frequency
Water temperature (°C)	24	30	2×/day
Salinity (‰) (very species specific)	15	45	1×/day
DO (ppm)	3	12	2×/day
pH	8.1	9.0	2×/day
Secchi disc (cm)	30	50	1×/day
Alkalinity (mequiv.)	100	200	1×/week
Total ammonia-N (ppm)	0.1	1.0	2×/week
Non-ionised ammonia-N (ppm)	–	0.2	2×/week
NO_3 (ppm)	0.6	1.2	2×/week
NO_2 (ppm)	–	0.5	1×/week
Total N (ppm)	0.6	2.5	2×/week
Phosphate (ppm)	0.2	0.5	2×/week
Silicate (ppm)	1.0	4.0	1×/week
H_2S (ppm)	–	0.1	1×/week

Adapted from Clifford (1997) with permission.

resulting in 'soft' shrimp, and can destabilise some phytoplankton species that prefer alkaline conditions. pH in the water column fluctuates with phytoplankton photosynthesis and respiration in proportion to dissolved CO_2, which is the reciprocal of DO. pH is lowest at dawn when dissolved CO_2 is highest and highest in mid-afternoon when dissolved CO_2 is lowest. Sediment pH varies less on a diel basis than water column pH, but sediment pH tends to decrease progressively towards the end of the production cycle, due to accumulation of organic acids and nitrification of ammonia.

Nitrogenous wastes resulting from protein digestion can accumulate and even reach dangerous concentrations, particularly at high stocking densities. Nitrite and unionised ammonia can accumulate to toxic levels periodically, particularly during massive die-offs of phytoplankton. Nitrate is usually not toxic at levels typically found in ponds. Total ammonia and nitrate are managed through water exchange.

Water transparency is an index of plankton biomass present and is measured with a Secchi disc, typically once per day at mid-morning (Fig. 3.1). Acceptable values are between 30 cm and 50 cm. Readings of < 30 cm indicate high phytoplankton biomass or suspended sediment or both in the water column. Pond water colour usually reflects the predominant phytoplankton species, i.e. golden brown = diatoms; green = green flagellates; blue-green = blue-green algae; and red = dinoflagellates.

Pond sediments tend to deteriorate with successive cycles and within production cycles due to accumulation of organic matter and sludge deposition. Sediment quality can be determined from hydrogen sulphide level and redox potential (E_h) (section 3.2.8), and by visual examination of sediment cores. Smelling a scoop of surface mud from various areas of the pond bottom is a quick check for hydrogen sulphide sediment. Redox potential of sediment is another quantitative indicator of sediment conditions. It is a measure of the proportion of oxidised to reduced substances, and is an indicator of the relative activity of aerobic and anaerobic bacteria in the sediment profile (Fig. 3.4).

19.8.2 Water management

Water exchange is an effective tool for pond management, i.e. to flush out wastes and, in some cases, to improve DO levels. At many large farms, water exchange often follows a fixed schedule rather than being a flexible alternative used when pond conditions require it. High exchange rates can be wasteful and have a negative effect on fertilisation and natural productivity by flushing out nutrients. As much as 35–40% of pond volume may be exchanged daily in some ponds stocked at high and very high densities. More typical exchange rates for medium and high stocking densities are, however, about 5–25%. With increasing incidence of various shrimp viruses and declining shrimp prices, there is a global trend to reduce water exchange rates to maximise biosecurity through exclusion of pathogens, to increase mechanical aeration and to promote improved assimilation of nutrients into shrimp biomass (section 19.10).

Fertilisation of ponds with inorganic and organic nutrients is an effective means of stimulating the production of natural foods and reducing feed use. In the first few weeks, it may be necessary to fertilise a pond every day, then every other day, and eventually biweekly. Routine fertilisation should be enough to maintain adequate water transparency (30–50 cm). Proper maintenance fertilisation can provide a continual natural food supplement and improve water quality. Once feed applications reach about 25–30 kg/ha/day, no additional N or P fertilisation is usually needed, because uneaten feed and shrimp faeces supply enough. Liming compounds are added to ponds in some farms (~20–100 kg/ha/week) during the cycle to increase alkalinity, particularly during the rainy season or in areas where low-salinity waters are prevalent. Liming compounds are also added under the presumption that water quality will be improved and pathogens in the water reduced.

19.8.3 Water aeration and circulation

Pond aeration is the primary life support system of many aquaculture ponds, including shrimp ponds. Management of DO aims at maintaining DO levels over 3.0 ppm throughout the pond, using the least equipment and with minimum cost. Efficient use of aeration and circulation equipment should take advantage of the oxygen supplied by photosynthesis during the day and by diffusion across the pond surface at night. Aerators are mechanical devices that act as the heart and lungs of an aquaculture pond, providing distinctive circulation and aeration effects, increasing the rate at which oxygen enters water. There are two basic techniques for aerating pond water: one is to splash water into the air and the other is for air bubbles to be released into the water, so there are 'splasher' and 'bubbler' aerators (Boyd, 1990). The splash created by paddlewheels and the bubble-jet from aspirators improve the exchange of dissolved gases between air and water, allowing oxygen to enter the pond and carbon dioxide and ammonia to escape. Splasher aerators include vertical pump, pump-sprayer and paddlewheel aerators. Pump-sprayer aerators use a centrifugal pump to spray water at high velocity through holes in a manifold and into the air. Vertical pump aerators comprise a motor and an impeller (propeller) attached to the shaft. The motor is suspended below a float with a centre opening and the impeller jets water into the air. Paddlewheel aerators splash water into the air as the paddlewheel rotates (Fig. 3.10). Bubbler aerators include diffused-air systems and propeller-aspirator pumps. In a diffused-air aeration system, an air blower or air compressor is used to deliver air through an airline, and the air is released through air diffusers located on the pond bottom or suspended in the water. Propeller–aspirator pump aerators have a high-velocity impeller at the end of a hollow shaft and housing. During operation, air moves down the shaft by the Venturi principle and flows into the water as fine bubbles. Most aerators are powered by electric motors.

Shrimp ponds require mechanical aeration when production biomass exceeds 2 mt/ha and additional aeration at a rate of 2 kW for each additional mt. The moderate circulation effect provided by one or two aerators is desired from the time of stocking. Surface circulation must be ~4 cm/s during feed applications. This speed will not disturb fresh pellets, but will keep finer particles and organics in suspension. Excessive water circulation may be managed by arranging aerators in uniform configurations or by other means. In a typical configuration, aerators generally create a central area of sediment deposition, and sometimes the sludge deposited may be removed by using suction hoses. There are central drains in some ponds with high stocking densities. Proper positioning of aerators and circulators is very important, because the operating efficiency of aerators is highly dependent on achieving adequate water circulation (Fig. 19.4). They must be positioned to generate a whole-basin flow pattern and produce an efficient water movement. This is achieved by aligning the current from each unit so that it supplements rather than conflicts with the overall flow pattern in the pond. Aeration equipment must be placed at the points of lowest DO within a flow pattern that encompasses the entire pond.

One of the needs for circulation and aeration is thermal stratification. Thermal stratification can occur in ponds when surface waters warm faster than deeper waters during the day. This may lead to DO depletion in the bottom water, because most of the DO in pond water originates from photosynthesis in the upper stratum of water or by diffusion from the air through the water surface. One kind of mechanical device often accomplishes aeration and circulation. Paddlewheel and propeller–aspirator pump aerators are especially efficient in circulating pond water. There are also circulators that supply little aeration but can prevent stratification. These are generally large, slowly revolving (50–150 rpm) propellers installed in ponds to quietly move large volumes of water.

19.8.4 Population sampling and health assessment

Periodic sampling of shrimp populations is an important management tool during the production cycle. Cast netting is about the only effective sampling tool currently available and it is widely used. The process of sampling has the objective of generating information about a large group of individuals, a shrimp population in a pond, by looking at a small number of individuals. Most shrimp farms routinely carry out sampling programmes to:

- monitor population size
- monitor individual and average size/weight of animals
- evaluate the animals' physical condition, appearance and product quality
- assess the animals' overall health, and to test for the possible presence of known pathogens or diseases.

In the last few years, and as their use becomes widespread, it has been realised that properly managed feed trays (section 19.9.8) or lift nets can provide adequate population estimates by combining daily feed consumption rates with percentage body weight curves.

One of the most important considerations is to eliminate or minimise sampling bias. In general, the biases are relatively consistent and, with experience, correction factors can be developed to compensate and produce good workable approximations of shrimp size and populations. Estimates of shrimp population size and survival rates can be remarkably accurate if properly carried out or very inaccurate owing to variable shrimp activity, e.g. moulting, lunar and tidal cycles of behaviour, size differences in habitat preference and catchability. In general, to improve the validity of a shrimp population sampling programme, sample collection must be carried out after lowering the pond water level by experienced personnel using a large and heavy cast net, and the number of sampling stations and frequency of sampling must be as large as possible. It is important for a farm to establish adequate in-house sampling methods that adequately reflect its needs and capabilities (Clifford, 1997).

Health assessment and management on shrimp farms has become an important issue in the last 12 years or so, because of the increased importance of various diseases, particularly of viral origin. There are ~20 distinct viruses (or groups of viruses) known to infect marine shrimp. Viral diseases have had a severe impact on the shrimp farming industry worldwide, causing very important production and economic losses. Viruses belonging to the WSSV, MBV (monodon baculovirus), BMN (baculoviral mid-gut gland necrosis), HPV (hepatopancreatic parvovirus), IHHNV (infectious hypodermal and haematopoietic necrosis virus) and yellowhead virus (YHV) groups have been important pathogens of cultured shrimp in Asia and the Indo-Pacific regions, whereas TSV (Taura syndrome virus), IHHNV and BP (baculovirus penaei) have been the most important in the Americas. Once thought to be limited to Asia, WSSV and YHV have been found in cultured and wild shrimp in the USA, and both viruses have been shown to be present in commodity shrimp imported from Asia and sold directly on the USA market. WSSV was first reported in Central America in 1999, and from there it expanded to most of the shrimp-farming countries in the region, causing severe economic losses.

Several strategies have been tried to control viral diseases in shrimp farming, ranging from improved husbandry practices to stocking 'specific pathogen-free' (SPF) or 'specific pathogen-resistant' (SPR) species or stocks. Further information on shrimp diseases is provided by Lightner (1996) and Alday de Graindorge & Flegel (1999) and in section 11.5.2.

A cost-effective health management and biosecurity programme requires reliable diagnostic tools that shrimp farmers can use to make adequate and timely decisions on management procedures to control or exclude pathogens. Virulent pathogens can produce catastrophic mortalities very rapidly, and shrimp farmers need this fast diagnostic capacity in order to respond effectively. Practical diagnostic methods that are accurate, sensitive, rapid and economical to conduct are already available, including polymerase chain reaction (PCR), dot-blot gene probes and various methods for rapid fixation and staining (Fegan & Clifford, 2001).

Early signs of many health problems can be promptly observed by examining shrimp during regular feed tray monitoring or from weekly sampling. Some of these signs include:

- loss of appetite (empty guts)
- changes in colour (blueish or reddish)
- persistent soft shells or shell fouling
- red discolouration (particularly in appendages and uropods)
- lethargic or disoriented, behaviour
- fouled or discoloured or black gills
- blackened lesions on shell
- opaque or white tail muscle
- various morphological deformities, such as cramped tail.

19.8.5 Harvest and transport to processing plant

Shrimp are highly perishable and delicate, and no amount of manipulation can restore product quality once it is lost. Therefore, proper preparation, harvesting and preservation are critical for the product to be the best quality and command the best price. This preparation can be quite elaborate, because a pond ready to be harvested may have 10–20 mt or more of shrimp, which must be properly collected, handled and packed (Fig. 19.9). The most important objectives at harvest are to:

- minimise the quantity of shrimp left on the pond bottom
- immediately chill shrimp to near freezing
- pack the shrimp in a manner that avoids physical damage.

The most important considerations of when to harvest a pond are shrimp size, price and maximising the economic return of the production cycle. Many shrimp farmers use the moon phase to programme their harvests, targeting periods of full and new moon. About 3–5 days before the harvest date, the

Fig. 19.9 Harvesting procedures are critical for the best-quality product that commands the best price.

texture (representative of the stage in the moulting cycle) of shrimp to be harvested is monitored by daily collection of a sample (100–300 animals/pond) to determine the percentage that are moulting. The particular requirements of the processing plant and the intended market determine what percentages of hard, soft and semi-hard (post-moulting) shrimp, as determined during pre-harvest sampling, are acceptable to proceed with a planned harvest. For example, for shrimp to be marketed whole, typically less than 5–8% of the animals sampled should be soft, and less than 15–20% should be semi-hard. Some farmers also stop feeding a few days before the harvest, but this is a decision that really depends on each pond, its biomass and its overall condition (Clifford, 1997).

Most shrimp are harvested by draining the production ponds, tanks and raceways. In preparing ponds for harvest, water levels are typically lowered to ~70% of their operational levels, beginning 24–48 h before the harvest. In larger ponds it is often necessary to reduce water levels to 50% or so before beginning the harvest, so that the harvest does not last excessively long, which would stress the shrimp and reduce their quality. Most pond harvests are undertaken during the night to avoid high water temperatures. A net is placed covering the out-flowing water and any shrimp remaining are picked up by hand or dip nets. The quality of these residual animals is generally reduced because of mud embedded in their gills and joints between tail segments, and delayed chilling. Harvested shrimp are immediately separated into size classes and layered with ice or immersed in an ice slurry for transport to the processing plant. Within minutes after death, shrimp begin to deteriorate at normal pond temperatures (25–30°C) that promote bacterial decomposition. Initial spoilage, however, is due to digestive enzymes from the hepatopancreas. These break down proteins and reduce product weight, and quickly break down tissue at the junction of the head and tail, making the head appear loose and sagging, which is unacceptable to the European market for head-on shrimp. Bacterial and enzymic activity can be stopped by immediately reducing temperature to near freezing. It is very important that shrimp are killed by thermal shock when immersed in chilled water (4°C is maintained from the time the animals are harvested to the time they reach the plant). Often additional processing takes place at the pond bank. This includes dipping the shrimp in chlorinated water (50–60 ppm) or a solution of sodium bisulphate to reduce blackening of the head for head-on shrimp product, removing crabs, fish and plant material, and selecting the largest animals to be used as potential broodstock (if needed). Shrimp are handled carefully during packaging to minimise damaging the product and reducing its acceptability in markets. Mechanical harvesting systems are being developed to harvest shrimp ponds and assure product quality.

Further details of post-harvest technology for shrimp, including safety and health, live transport, post-mortem processing and chilled storage life, are provided in sections 11.3, 11.7, 11.9 and 11.12 respectively.

Shrimp are marketed in a variety of forms: heads-on shrimp (including live), shell-on tails, peeled tails (included canned), breaded tails and other value-added products. These are transported to various markets in a variety of forms and packaging.

19.9 Nutrition, formulated diets and feed management

19.9.1 Nutritional requirements and formulated diets

The nutritional requirements of marine shrimp, their feed formulation and manufacture have been extensively discussed and documented, but much remains to be learned. With industry expansion there has been an intensification of production and an increased dependence on the use of manufactured dry feed, which often represents the highest production cost. Protein is typically the most expensive macronutrient in shrimp feeds, and dietary protein levels from 18% to 60% have been recommended for various species and sizes of marine shrimp species, possibly because of their wide range of natural feeding habits. PL shrimp require a higher dietary protein level than older shrimp. Formulated shrimp feeds are complex products and the main components typically are wheat flour (20–35%), soybean meal (15–25%) and fish meal (15–30%). These few ingredients contribute most or all the protein, amino acids and energy. The remaining ingredients include various lipids and micro-ingredients that provide essential fatty acids, vitamins, minerals, attractants, binders, preservatives, pigments and health additives (Table 19.9).

Chamberlain & Hunter (2001) recently reviewed additives used in formulated shrimp feeds. In total, ~110 additives are commonly used as ingredients in shrimp feeds today.

(1) Attractants include animal by-products (crustacean tissue meals, squid by-products, low molecular weight fish and meat extracts), and purified compounds (free amino acids, artificial flavours, betaine and nucleotides).
(2) Enzymes supplements improve digestibility of phytate, fibre, indigestible sugars and other components.
(3) Various additives, including coccidiostats, antibiotics and hormones, have been used as growth promoters in shrimp.
(4) Other additives, including immunostimulants, probiotics and vaccines used to stimulate the immune system of shrimp and improve resistance to disease. This has become especially important to reduce outbreaks of shrimp viral diseases. The immunostimulants used include beta-glucans, bacterial extracts, blood plasma, seaweed by-products and yeast.
(5) Wheat flour is the principal binder used in shrimp feeds but, by itself, may not provide proper water stability, especially if the manufacturing process does not include fine grinding, extended conditioning, and post-pellet conditioning. Therefore, additional binders are often needed to assure water stability.
(6) Additives are used to improve shrimp pigmentation, which is significantly determined by dietary

Table 19.9 Recommended nutritional composition of shrimp feeds. Only critical nutrients for feed formulations are listed on as-fed basis

Nutrient	Shrimp 0–3 g	Shrimp 3–15 g	Shrimp 15–40 g
Protein (minimum)	40.0	38.0	36.0
Lipid (minimum)	6.2	5.8	5.5
Lipid (maximum)	7.2	6.8	6.5
Fibre (maximum)	3.0	4.0	4.0
Ash (maximum)	15.0	15.0	15.0
Calcium (maximum)	2.3	2.3	2.3
Phosphorus available (minimum)	0.8	0.8	0.8
Potassium (minimum)	0.9	0.9	0.9
Lysine (minimum)	2.12	2.01	1.91
Arginine (minimum)	2.32	2.20	2.09
Threonine (minimum)	1.44	1.37	1.30
Methionine (minimum)	0.96	0.91	0.86
Methionine/cysteine (minimum)	1.44	1.37	1.30
Phospholipid (minimum)	1.0	1.0	1.0
Cholesterol (minimum)	0.35	0.3	0.25
20:5*n*-3 (minimum)	0.4	0.4	0.4
22:6*n*-3 (minimum)	0.4	0.4	0.4

Reproduced from Akiyama (1992) with permission of the World Aquaculture Society.

carotenoids. Some markets have a premium demand for shrimp with strong natural pigmentation, and ovarian pigmentation is important for maximum reproductive performance.

(7) Finally, because shrimp feeds are typically stored for several weeks in humid tropical environments, antioxidants (BHT, BHA and ethoxyquin) and preservatives are generally added to prevent oxidation of fats and mould infestation (Chamberlain & Hunter, 2001; section 9.10.4).

19.9.2 Feed management

Shrimp feed management is a critical aspect for cost-efficient, environmentally responsible shrimp production. Appropriate practices will produce maximum shrimp growth and survival concurrent with the lowest feed conversion, while reducing feed inputs and minimising impact of effluents. Efficient feed management is the summation of several sequential steps, including feed selection, storage and handling, application methods, and feeding regimes. Determining when to feed requires knowledge of diel activity patterns, feeding frequency and time (subject to change with geographical location, season, species, size, age, stocking density, unusual environmental conditions and other stimuli). Calculating feed rations involves estimating survival, population size and biomass, and size distribution. Monitoring and continuously adjusting the amount of feed applied, according to changes in consumption caused by various biotic (e.g. amount, quality and availability of natural food items) and abiotic (water quality and other environmental parameters) factors, is important for effective feed management. Evaluating and adjusting feed input involves regular population sampling and proper monitoring of various water quality parameters. Shrimp are bottom feeders, and it is difficult to estimate their feed consumption, unless feed trays (section 19.9.8) or lift nets are used. Inadequate feed management may promote the onset of various diseases and water quality-related problems, and may adversely affect production. Ineffective practices often include:

- applying feed during times convenient for employees (during daylight hours), but not necessarily at the best times for shrimp
- inadequate handling and storage practices during bulk feed storage and after feed distribution to pond side
- overfeeding.

The best shrimp feed in the world will yield poor results if it is not handled, stored and used properly.

19.9.3 Factors that affect feed consumption

Several factors can affect shrimp feeding behaviour, and it is important to understand these factors to make proper and timely management adjustments (Clifford, 1992, 1997; Jory *et al.*, 2001). The major factors affecting feeding behaviour are:

(1) Species, age and size. There are marked differences between species. Some species are much more active and aggressive while foraging for food, and this has to be incorporated into a feeding strategy. Feeding rate is a physiological function dependent on the growth stage of the animal; it decreases as the animal grows and approaches maturity. Growth immediately after pond stocking can be 12–15%/day, decreasing to 1–2%/day towards the end of the production cycle.

(2) Availability of natural food. When natural food is very available, the demand for formulated feeds is reduced. This is typical when the biomass of stocked shrimp is low during the first few weeks after stocking and until the natural carrying capacity of the pond is reached.

(3) Water quality. The most important parameters are temperature, DO, pH and salinity, but other parameters also influence the shrimp (Table 19.8). For each parameter, animals have a range of tolerance and a narrower optimum range that promotes optimum feeding, growth and overall well-being (section 3.3).

(4) Moulting. Shrimp moult periodically (days–weeks) throughout their lives, and this is a stressing period during which their appetite diminishes markedly. It can take 2–5 days for normal feeding to resume after moulting. Thus, it is important to recognise when there is a significant reduction in feed consumption (use of feed trays is a good method) indicating high incidence of moulting in a pond.

(5) Quality of commercial feeds. Shrimp eat to fulfil their nutritional needs, and if the feed they consume does not have enough energy or appropriate nutritional profile, their feeding activity will increase. Feed attractability and palatability are also important factors.

19.9.4 Feed handling and storage

Feed management at a shrimp farm begins upon arrival of a feed shipment. Poor storage and handling of feeds will result in product deterioration, reduced feed attractability and palatability, possible nutritional deficiencies and disease outbreaks, and reduced growth rates and overall production (sections 9.10.3 and 9.10.4). Upon reception of a feed batch, a few randomly selected bags are examined for physical integrity. In addition, twice a year feed samples are collected from newly received shipments (or when using a new feed) and analysed for proximate composition, mycotoxins and selected pesticides if pertinent (Jory, 1995; Jory *et al.*, 2001).

Feed is ideally used within the first 2–4 weeks after manufacture and must not be stored for more than 2–3 months. At farms that feed several times over 24 h, the total feed ration is often distributed from the farm warehouse to the ponds once, usually early in the morning. The feed bags must then be protected from sunlight and rain by storing them off the ground in simple pond-side sheds.

19.9.5 Application and distribution

Formulated feeds can be applied in a variety of ways to shrimp production systems (section 9.11.2). Feeds may be distributed manually in large ponds from boats or mechanically using blowers mounted on vehicles and boats (Fig. 19.10). Even crop-dusting aeroplanes have been used to distribute feed to very large ponds. In small ponds, tanks and raceways, automatic feeders with timing mechanisms can be used (section 9.11.2). At many farms in several countries, all feed is applied exclusively using feeding trays. Broadcasting feed from paddle- or motorboats is typical at many farms in Latin America. The use of feed blowers is not widespread yet, but could gain popularity in coming years because of their efficiency.

Fig. 19.10 A feed blower used to feed shrimp in ponds on a farm in Mexico.

It is important to distribute the feed evenly early in the grow-out cycle, but, as the cycle progresses, shrimp react to changing pond microhabitats. They avoid areas where anaerobic sediments accumulate and noxious compounds, such as H_2S, are produced, including internal drainage canals and areas close to the drainage structures. In addition, many shrimp will move to the deeper areas of ponds during the day to avoid light. Therefore, it may not be appropriate to provide feed to very shallow areas during daylight hours, because it is unlikely shrimp will consume it. It is also important to recognise that shrimp distribution in ponds is generally not uniform along the linear dimensions of the pond.

19.9.6 Frequency and timetables

All formulated feeds have an ideal feeding rate range that optimises growth and feed efficiency. This range varies with species, age and weight, stocking density, water quality, availability of natural foods, stress and other factors. At many farms, feeding is based on tables that do not properly consider natural feeding habits of shrimp or their physiological state. Increasing the frequency of feeding generally produces immediate benefits, including reduced nutrient and feed loss, and increased growth and feed utilisation efficiency. In Asia, it is common practice to feed up to six to seven times over a 24-h period. Because *L. vannamei* appears to be less nocturnal than *P. monodon*, most shrimp farms in Latin America feed one to three times/day, usually between the morning and late afternoon. Because of the higher degree of activity shown by *P. monodon* at night, many farms in Asia provide most of the daily ration during night hours.

How many times and when to feed is an important decision that each shrimp farm must determine, based on experience, production system used, season, environmental conditions, species, stocking densities and available resources. Feeding during the night typically becomes more important as the production cycle progresses and the availability of natural feed diminishes. In general, a minimum of two feed applications per day are needed at the beginning of the production cycle, increasing as the cycle progresses to at least three or four applications per day.

19.9.7 Feed rations

Feed rations can be calculated using a set schedule that accounts for animal weight and estimated biomass/survival in the pond. Feeding based on tables is still widely practised (section 9.11.1).

There are several problems with relying on feeding tables.

(1) There are problems in accurately estimating survival, particularly when dealing with small animals in large (< 5 ha) ponds.
(2) Various factors (section 19.9.3) affect feeding rates.

Feed consumption changes can usually be detected by proper monitoring of feed trays (section 19.9.8). Feeding rates are adjusted periodically (usually weekly), based on sampling estimates for individual average body weight, population size distribution and pond shrimp biomass. As shrimp grow the feed amounts used decrease as a percentage of the total shrimp biomass, but the absolute amount of feed increases together with increasing shrimp biomass (Jory *et al.*, 2001).

19.9.8 Use of feed trays

Feed trays are also called feed monitoring trays, observation or lift nets, feed inspection trays or umbrella nets. They are simple devices usually consisting of a frame and fine mesh that is lowered to the floor of the pond. There are several ways feed trays can be used, including:

(1) as indicators of feed consumption, population and health assessments and other observations
(2) to apply all feed (Peruvian system)
(3) a combination of these two.

In using the feed trays as indicators of feed consumption, about four to eight trays per hectare are used in ponds < 5 ha, whereas two to five trays per hectare are used in larger ponds (10–20 ha). A small percentage of the ration is placed in the trays and the ration is distributed throughout the pond. Trays are checked after 2–3 h and the data are used to adjust rations. In the Peruvian system, the entire ration is applied in trays, which requires many more trays per pond. Disadvantages of using feed trays to feed the entire pond include high costs and increased training and supervision in feeding practices, because two people are required daily for 12-h shifts for every 10 ha of pond. But additional construction, labour and equipment costs are reportedly covered by the resulting reduction of feed costs, precise feed consumption estimates and good FCRs. Finally, the Peruvian system may be used initially and then trays are just used as consumption indicators (Clifford, 1997).

The reliability of feed trays for estimating food consumption has been questioned by some, because shrimp may use trays as a refuge from reduced pond sediments. Crabs are attracted to feed trays, where they feed on shrimp pellets or prey on shrimp. Their presence in or near feed trays can keep shrimp away from trays, resulting in an underestimate of feed consumption and underfeeding. The main argument against using feeding tables is that it is very difficult to continuously estimate with accuracy the survival of animals in a pond. Therefore, it has become common for most farms, when feeding a pond for the first time, to initially follow their own or the feed manufacturer's guidelines and tables, and then begin adjusting daily rations using feed trays (Jory *et al.*, 2001).

In general, observation of feed consumption from a small number of trays is not an adequate measure of actual feed consumption, especially in large ponds. This is because there are considerable day-to-day variations in feed consumption as outlined in section 19.9.3.

19.10 Emerging production technologies and issues

19.10.1 Diseases and biosecurity

Viral diseases have had a considerable impact on commercial shrimp farming during the last decade (section 19.1.2), significantly affecting the operation, management and design of shrimp farms. Another resulting consequence is increased awareness of the need for better husbandry methods to reduce the risks of exposure to pathogens, and also of the need for improved management practices to enhance shrimp health. Shrimp farming is a relatively new industry and has lagged behind in the development of practices for standard health management. This in part results from the relatively poor level of understanding of shrimp physiology and their production systems. Also, there has been limited involvement by veterinarians and animal health specialists to develop health management practices like those in use in terrestrial animal husbandry.

The modern poultry industry, with its considerable advances in production and disease control, best exemplifies the advantages of biosecure animal production, and shrimp farming must adopt similar practices to become more competitive. Biosecurity in shrimp aquaculture involves those practices that will reduce the probability of introduction and dissemination of a pathogen. Shrimp producers often give only limited attention to routine biosecurity on their farms. This is because of the misconception that the potential costs of implementing biosecurity measures will outweigh the benefits or because they do not have appropriate knowledge. Effective implementation of biosecurity protocols requires awareness, discipline and a commitment by farm owners to implement them. We have improved our knowledge of shrimp viral diseases significantly during the past 15 years, mainly due to their negative impact on the industry, but biosecurity is still very new to shrimp farming (Fegan & Clifford, 2001).

A cost-effective, biosecure shrimp-farming protocol involves:

- aggressive methods for pathogen exclusion from production systems
- effective screening of seedstock
- appropriate environmental management
- effective health management, integrating genetic selection
- specific pathogen-free and pathogen-resistant stocks
- limited or zero water exchange
- stocking strategies
- feed management and use of immune stimulants to increase host defences
- strict and proactive health monitoring and farm management strategies.

Also important are farm location (site selection) and design. Currently, only a few shrimp farms have been specifically designed to prevent diseases, although it is more cost-effective to incorporate disease prevention and treatment aspects during the planning stage than to redesign or refit existing farms. Eliminating or reducing water exchange is an important aspect to prevent viral diseases, to exclude viral carriers or free virions from production systems, and to minimise stressful variations in water quality, which may trigger disease outbreaks. There are several examples of successful commercial farms in Asia and Latin America which use low or zero water exchange (section 19.10.5).

19.10.2 Probiotics and microbial management

As described, the use of probiotics is relatively common to limit pathogenic bacteria in the disease-prone intensive systems used to produce shrimp PL in commercial hatcheries (section 19.5.3). For some time now, the use of bacterial amendments has been recommended for use in aquaculture ponds to obtain a number of benefits (Boyd, 1995), including:

- reducing blue-green algae populations
- preventing off-flavour
- reducing N and P levels
- increasing DO
- promoting decomposition of organic matter.

There is ongoing research on the use of probiotics,

which could be important in promoting shrimp yield in intensive systems, in the implementation and optimisation of low- and zero-water exchange production systems, and in managing the quality of pond effluents.

The pond microbial community plays a major role in the natural food availability, mineral recycling rates and DO dynamics in shrimp ponds. Effectively managing the microbial community can help prevent or reduce the risk of a disease outbreak, but, if mismanaged, the microbial community can also promote disease by creating conditions that facilitate growth of pathogenic bacteria.

19.10.3 Nursery systems

Jory & Dugger (2000) recently reviewed the use and advantages of nursery systems for PL shrimp. These systems generally produce higher overall survival rates per production unit area and more efficient capital utilisation than direct stocking into grow-out systems. They also provide better management of environmental conditions, feeding, and exclusion of pathogens, predators and competitors. It is critical to know the survival, quantity, condition and quality of seedstock before stocking into ponds before investing several months of effort and resources into a production cycle. Nursery strategies involve holding seedstock at very high densities (5000–10 000 PL/m^2) in specially designed facilities for 20–40 days, with precise technical management, feeding and water quality monitoring (Fig. 19.11).

A two-stage grow-out system using a nursery phase as a quarantine area increases biosecurity. Indoor nursery systems increase turnover (number of grow-out production cycles) by reducing culture time to market size in grow-out ponds. Therefore, the grow-out pond is being used more efficiently as a biological system, and with greater capital and operating efficiency. A nursery system also provides improved accuracy to estimate the juvenile population before actual stocking in grow-out ponds. Thus, stocking juveniles allows for a more accurate estimate of the initial population and biomass, and improving feeding rate estimates when formulated feed becomes up to 60% of the production cost. Indoor, intensive nursery systems can further broaden the effective temperature, stocking windows for seasonal hatchery outputs, allowing greater efficiency for the hatchery and farm. Shrimp farms in areas of lower salinities can use the nursery as an acclimation system. Nursery head-start strategies may allow farms without hatcheries to purchase seedstock in advance of the peak demand periods, possibly at lower cost and with improved certainty of seedstock delivery.

Fig. 19.11 An intensive nursery system culturing shrimp seedstock before transfer to ponds (photograph by OceanBoy Farms).

Managing nursery systems in tanks and raceways is relatively more difficult than standard grow-out ponds stocked directly, but the many benefits derived from a two-phase grow-out strategy, using first a nursery system (indoor) followed by final grow-out to market size (outdoor pond) can significantly improve production and profitability.

19.10.4 Inland shrimp production

Marine shrimp farms have traditionally been built in tropical coastal areas, very close to the ocean or to an estuary or river. Recently, shrimp farms have been built in other environments such as inland areas that, in principle, do not appear suitable for this activity. Some of these farms are located in inland deserts with available undergroundwater having specific chemical characteristics. They could provide a new direction for the expansion of the industry, because deserts and other dry lands constitute over 40% of the global land area.

Establishing shrimp farms far from the ocean using low-salinity waters has been successfully implemented with *P. monodon* in South-East Asia. The effluent can be used to irrigate various crops, thus minimising effluent disposal efforts. These emerging technologies offer opportunities to establish shrimp culture operations on marginal arid land or agricultural sites, reducing demand for shrimp farming on limited, high-cost coastal land. Limited seawater use during a relatively short acclimation phase and complete re-use of effluent water for irrigation of agricultural crops can provide for environmentally friendly integrated systems. *L. vannamei* is being successfully raised in low-salinity groundwater with varying ionic compositions and salt concentrations in several regions of the world, including the USA and Ecuador (McMahon *et al.*, 2001).

Chapter 5 gives further examples of inland aquaculture in saline water and its integration with agriculture.

19.10.5 Recirculation and reduced water exchange systems

The shrimp farming industry can benefit significantly from improved water management regimes. In particular, these can help address viral disease problems and also issues raised by environmental groups that have targeted marine shrimp farming, pressuring the industry to adopt more sustainable production practices.

Large-scale application of zero-exchange and recirculation technologies on existing farms has already increased producer confidence in the potential for reducing or eliminating routine water exchange through most or all of the growing season. The Arroyo Aquaculture Association facility in southern Texas has been successful in recirculation and reduced water use so that the water required per kilogram of shrimp production has been reduced by 96%, from 37.6 m^3 to 1.5 m^3 (Hamper, 2000). It is interesting to consider that the ratio of water consumption to shrimp production was initially 37 600:1 in what was regarded as intensive culture at high stocking density. At a further stage, there are several successful examples of the implementation of zero-exchange production systems, including those in Belize, Panama and Florida (Fig. 19.4). In particular, McIntosh and co-workers at Belize Aquaculture Ltd have shown the tremendous potential these production systems have (section 19.3.4).

Browdy *et al.* (2001) recently reviewed various aspects of intensive closed technologies for shrimp production. Nutrient-rich effluents from intensive production systems can contribute to the eutrophication of receiving waters, potentially affecting both natural biota and local culture operations. Water exchange can be reduced or eliminated, and supplementary aeration can have an essential role in the successful operation of intensive closed systems. Paddlewheel aeration must be increased by 10% or more over levels traditionally applied in intensive culture to maintain appropriate DO levels. Better placement of mechanical aerators and use of backup aeration and alarm systems are also necessary. Formulated feeds are the main nutrient input into shrimp production systems and, as water exchange is reduced or eliminated, feed formulations and feed management fast become critical factors as stocking densities increase. The design and management of production facilities to re-use water, minimise exchange and eliminate discharge will improve the outlook for more profitable and sustainable production technologies.

19.10.6 Effluents

General impacts of effluents from coastal aquaculture farms on marine environments are considered in section 4.2.1. This is an issue that shrimp producers and processors need to address while they are discharging effluents.

Settling basins are especially efficient for treating shrimp farm effluents because the high concentrations of cations in seawater and brackishwater tend to neutralise the negative charges on suspended clay particles, which will flocculate and settle. Plankton cannot be removed efficiently by sedimentation, and products such as aluminium sulphate, lime and selected organic colloids, often used in wastewater treatment to promote sedimentation, are not needed in shrimp farm settling basins.

According to Burford *et al.* (2001), the shrimp farming industry will face increasing pressure to develop new practices and technologies that more efficiently convert the N in shrimp feeds to shrimp biomass, and to minimise or eliminate residual N waste before effluent discharge into receiving waters. Meeting these goals will require an understanding of the fate and transformations of dietary N in shrimp production ponds and effluent treatment systems. These authors discussed the development of new technologies to reduce N waste from shrimp ponds, including improvements in feed formulations and feed management, genetic selection for improved FCR, improved in-pond processing of N, and improved design and management of effluent treatment systems (section 23.3.5).

19.11 Responsible shrimp farming and the challenge of sustainability

19.11.1 Domestication and genetic improvement

During its first three decades, commercial shrimp farming has depended significantly on wild seedstock and broodstock. Wild seed supply has often been unreliable and limited, and PL shortages have severely afflicted the industry. Many factors may affect broodstock and wild larvae supply, from global weather phenomena, such as El Niño and annual monsoons, to localised pollution and environmental degradation, and overfishing or over-regulation of fisheries.

Marine shrimp are promising candidates for domestication and genetic improvement, because of their high fecundity, short generation interval and the presence of additive effect of genetic variance for growth rate (Jory, 1996) (Fig. 19.12). When compared with most livestock industries, however, shrimp farming is still at an incipient stage of domestication and selective breeding. International discussion and collaboration in both conventional selective breeding techniques and the application of the tools of modern molecular biology are needed to promote global progress and efficiency of genetic improvements in the industry (Preston & Clifford, 2002).

The global survey of shrimp farming practices recently conducted by the Global Aquaculture Alliance showed the significant differences in progress in the genetic improvement of farmed shrimp between geographic regions and among the dominant farmed species. In the western hemisphere, the domestication and selective breeding of the dominant species (*L. vannamei* and *L. stylirostris*) have become a widespread practice with clearly demonstrated benefits. The benefits include improved production efficiency and a more reliable supply of seedstock, the results of selection of genotypes with improved tolerance to disease and faster growth rates. Domesticated lines of *L. vannamei* have been reared in captivity for > 15–30 generations in various countries. In the eastern hemisphere, progress in the domestication and selective breeding of the dominant farmed species, *P. monodon*, has been very slow. Significant advances have been made, however, in the domestication and selective breeding of species that are less commonly farmed there, including native species such as *M. japonicus*, and *F. chinensis*, and introduced western species such as *L. stylirostris* and *L. vannamei*.

Achieving domestication of selected shrimp species, together with selection, genetic improvement and, possibly, hybridisation and ploidy manipulation, should be major research objectives. An industry with the global importance that shrimp farming has achieved cannot depend on nature to supply its seedstock reliably.

19.11.2 Nutritional requirements and formulated feeds

The development and use of compound feeds have been major factors in the expansion of the industry and will continue gaining importance. Growing demand for formulated feeds will increase competition for component resources, particularly fish meal. There is an enormous but still unrealised potential to reduce the production cost and improve the nutritional performance of compound feeds for shrimp. Extensive research must continue to improve our knowledge of the nutritional requirements of shrimp and to develop new diets that are species, area and even season specific. These diets could include innovative processing methods, lower-cost ingredients, health additives and growth promoters that improve survival, growth, yield, FCR and disease resistance

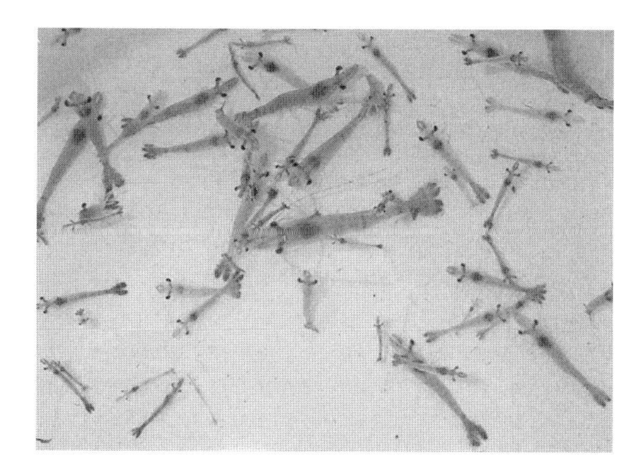

Fig. 19.12 Genetically based variation in size among sibling *Penaeus monodon* postlarvae (photograph by Dr Nigel Preston).

while reducing environmental concerns. Reducing the cost of feeds is an aspect critical to further expanding the industry and improving its competitiveness relative to other protein sources, such as beef, pork and poultry. Low-pollution or 'environmentally friendly' feeds must be developed to reduce the environmental impacts of shrimp feeds. Currently, emphasis is on 'least cost' feed formulations for minimising feed cost while optimising aquaculture production. In the future, emphasis will also be on 'least polluting' feed formulations for minimising environmental impacts with the greatest compatible aquaculture production.

19.11.3 Disease prevention, diagnosis and control

As described earlier, there have been spectacular collapses of shrimp farming industries in a number of countries, including the top producing countries, China, Thailand, Indonesia, Taiwan and Ecuador. Standard denominators among the shrimp farming industries of these countries were very fast, unregulated development and an increased incidence of diseases, particularly viral diseases. There are no known treatments for viral infections, and the best procedure for disease management is exclusion (section 19.8.4). There will undoubtedly be new pathogens that the industry will have to confront and manage. Accurate and prompt diagnosis of infectious agents has to be a research priority (Jory, 1996, 1997).

The application of effective pathogen detection and disease diagnostic methods, particularly those based on molecular biology and recently developed by the industry (section 19.8.4), are essential to better understand and prevent losses due to disease. Much progress is evident in the last few years: there were only a few shrimp disease diagnostic laboratories a decade ago in the Americas, but there are many more serving the industry today.

19.11.4 Best management practices

The resolution of environmental and social conflicts can be accomplished through regulations, technical assistance, education and voluntary measures. Some local and regional governments are already exercising increased pressure in regulating the shrimp farming industry. For example, there are regulations governing water quality parameters in effluent waters from shrimp aquaculture operations in a number of areas, and these will become increasingly strict. On the other hand, because shrimp farming is such a young industry, there are inadequate regulations and guidelines in most countries. Most countries will take years before they can formulate and enforce reasonable shrimp farming regulations, so the industry needs to be proactive and voluntarily adopt measures of self-regulation.

Best management practices (BMPs) are a practical way to approach environment management for shrimp farming. They are practices considered the most effective methods of reducing environmental impacts, while being compatible with resource management goals. Producers may adopt them voluntarily to show environmental stewardship and reduce the urgency for governmental regulations. BMPs may also be the backbone of environmental management in activities whose effects are diffuse and located over large areas. The shrimp farming industry, by voluntarily preparing and adopting BMPs, is demonstrating environmental responsibility to reduce the need for future regulations, and to provide a basis for the form of future regulations. A system of BMPs, however, is typically needed to prevent a particular type of farming from causing negative impacts. Shrimp farming is conducted over a wide range of coastal environments, with significant differences in resource patterns, and physical, chemical and biological conditions, so a single system of BMPs for use in all situations is not possible. Systems of BMPs must be developed by country or by region, and the national or regional system must be customised for site-specific conditions on each farm. Where voluntary adoption of BMPs is used to improve environmental management, it is convenient to establish it as a Code of Practice. This document usually asserts the commitment of adherents to the code to use environmentally and socially responsible management, states the purpose of the code and provides a list of BMPs to be used by the industry. Codes of Practice for shrimp farming already exist in Australia, Belize, Thailand and some other countries, and shrimp farming organisations in several more countries are intent on producing their own Codes of Practice (Boyd, 1998).

References

Akiyama, D. (1992). Future considerations for shrimp nutrition and the aquaculture feed industry. In: *Proceedings of the Special Session on Shrimp Farming* (Ed. by J. Wyban), pp. 198–205. World Aquaculture Society, Baton Rouge, LA.

Alday de Graindorge, V. & Flegel, T.W. (1999). *Diagnosis of Shrimp Diseases with Special Emphasis on the Black Tiger Shrimp (Penaeus monodon)*. CD-ROM. FAO, Rome, and Multimedia Asia Co. Ltd, Bangkok, Thailand.

Boyd, C. E. (1990). *Water Quality in Ponds for Aquaculture*. Alabama Agricultural Experiment Station, Auburn University, Auburn, AL.

Boyd, C. E. (1995). Chemistry and efficacy of amendments used to treat water and soil quality imbalances in shrimp ponds. In: *Swimming Through Troubled Water, Proceedings of the Special Session on Shrimp Farming* (Ed. by C. L Browdy & J.S. Hopkins), pp. 183–99. Aquaculture '95, World Aquaculture Society, Baton Rouge, LA.

Boyd, C. E. (1998). *Codes of Practice for Responsible Shrimp Farming*. Global Aquaculture, Alliance, St. Louis, MO.

Bray, W. A., Jr & Lawrence, A. L. (1991). Reproduction of *Penaeus* species in captivity. In: *Culture of Marine Shrimp: Principles and Practices* (Ed. by A. Fast & L. J. Lester), pp. 93–170. Elsevier Scientific Publications, Amsterdam.

Browdy, C. L. & Jory, D. E. (Eds) (2001). *The New Wave, Proceedings of the Special Session on Sustainable Shrimp Culture*. Aquaculture 2001, World Aquaculture Society, Baton Rouge, LA.

Browdy, C. L., Bratvold, D., Stokes, A. D. & McIntosh, R. P. (2001). Perspectives on the application of closed shrimp culture systems. In: *The New Wave, Proceedings of the Special Session on Sustainable Shrimp Culture* (Ed. by C. L. Browdy & D. E. Jory), pp. 20–34. Aquaculture 2001, World Aquaculture Society, Baton Rouge, LA.

Burford, M. A., Jackson, C. J. & Preston, N. P. (2001). Reducing nitrogen waste from shrimp farming: an integrated approach. In: *The New Wave, Proceedings of the Special Session on Sustainable Shrimp Culture* (Ed. by C. L. Browdy & D. E. Jory), pp. 35–43. Aquaculture 2001, World Aquaculture Society, Baton Rouge, LA.

Chamberlain, G. W. & Hunter, B. (2001). Global shrimp OP: 2001 – preliminary report on feed additives. *Global Aquaculture Advocate*, **4**(4), 61–65.

Chanratchakool, P., Turnbull, J. F., Funge-Smith, S. J., MacRae, I. H. & Limsuwan, C. (1998). *Health Management in Shrimp Ponds*. Aquatic Animal Health Research Institute, Department of Fisheries, Katsetsart University Campus, Jatujak, Bangkok, Thailand.

Clifford, H. C., III (1992). Marine shrimp pond management: a review. In: *Proceedings of the Special Session on Shrimp Farming* (Ed. by J. Wyban), pp. 110–37. World Aquaculture Society, Baton Rouge, LA.

Clifford, H. C., III (1994). El manejo de estanques camaroneros. In: *Proceedings of the International Shrimp Farming Seminar "Camarón '94"*, 10–12 February 1994 (Ed. by J. Zendejas-Hernández), pp. 24–43. Mazatlán, Sinaloa, Mexico.

Clifford, H. C., III (1997). *Manual de Operación para el Manejo de Super Shrimp en Estanques*. Super Shrimp, S.A. de C.V. División de Servicios Técnicos. Mazatlán, Mexico.

Clifford, H. C., III (1999). A review of diagnostic, biosecurity and management measures for the exclusion of White Spot Virus Disease from shrimp culture systems in the Americas. In: *Proceedings of the Third World Aquaculture Society/Latin American Congress* (Ed. by T. R. Cabrera, M. Silva & D. E. Jory), pp. 134–71. Puerto La Cruz, Venezuela, 17–20 November 1999. Latin American Chapter of the World Aquaculture Society, Baton Rouge, LA.

Cook, H. L. & Clifford, H. C., III (1998). Fertilization of shrimp ponds and nursery tanks. *Aquaculture Magazine*, **24**(3), 52–62.

FAO (2001) *1999 Fisheries Statistics: Aquaculture Production*, Vol. 88/2. Food and Agriculture Organization of the United Nations, Rome.

Fast, A. W. & Menasveta, P. (2000). Some recent issues and innovations in marine shrimp pond culture. *Reviews in Fisheries Science*, **8**, 151–233.

Fegan, D. & Clifford, H. C., III (2001). Health management for viral diseases in shrimp farms. In: *The New Wave, Proceedings of the Special Session on Sustainable Shrimp Culture* (Ed. by C. L. Browdy & D. E. Jory), pp. 168–98. Aquaculture 2001, World Aquaculture Society, Baton Rouge, LA.

Hamper, L. (2000). Reducing water use and water discharge at a south Texas shrimp farm. *Global Aquaculture Advocate*, **3**(3), 30–1.

Jory, D. E. (1993). An overview of marine shrimp culture in

Latin America and the Caribbean. In: *Proceedings, First Saudi Aquaculture Science & Technology Symposium*, pp. 176–95. Riyadh, Kingdom of Saudi Arabia, 14–16 April 1993.

Jory, D. E. (1996). Marine shrimp farming development and current status, perspectives and the challenge of sustainability. *Aquaculture Magazine Annual Buyer's Guide and Industry Directory 1996*, 35–44.

Jory, D. E. (1997). Current issues in marine shrimp farming. *Aquaculture Magazine Annual Buyer's Guide 1997*, 39–46.

Jory, D. E. (1998). Status of world shrimp culture. *Aquaculture Magazine Buyer's Guide 1998*, 32–41.

Jory, D. E. & Dugger, D. M. (2000). Intensive nursery uses strategy for improved shrimp health. *Aquaculture Magazine*, **26**(6), 67–72.

Jory, D. E., Cabrera, T. R., Dugger, D. M., Fegan, D., Lee, P. G., *et al.* (2001). A global overview of current shrimp feed management: status and perspectives. In: *The New Wave, Proceedings of the Special Session on Sustainable Shrimp Culture* (Ed. by C. L. Browdy & D. E. Jory), pp. 104–52. Aquaculture 2001, The World Aquaculture Society, Baton Rouge, LA.

Laramore, R. (2000). Shrimp larval nutrition: getting the most out of your formulated diets. *Global Aquaculture Advocate*, **3**(1), 70–2.

Lightner, D. V. (Ed.) (1996). *A Handbook of Shrimp Pathology and Diagnostic Procedures for Disease of Cultured Penaeid Shrimp*. World Aquaculture Society, Baton Rouge, LA.

McIntosh, R. P. (2000). Changing paradigms in shrimp farming. Part V. Establishment of heterotrophic bacterial communities. *Global Aquaculture Advocate* **3**(6), 52–4.

McMahon, D. Z., Baca, B. & Samocha, T. M. (2001). Florida's first inland, commercial-scale shrimp farm: developing protocols for inland culture of pacific white shrimp (*Litopenaeus vannamei*) with zero discharge in low-salinity ponds. *Global Aquaculture Advocate* **4**(5), 66–8.

Pérez Farfante, I. & Kensley, B. (1997). Penaeoid and Sergestoid shrimps and prawns of the world. Keys and diagnoses for the families and genera (Les crevettes peneides et sergestides du monde. Cles d'identification et diagnose des familles et genres). *Memoires du Museum National d'Histoire Naturelle*, **175**, 1–235.

Preston, N. P. & Clifford, H. C. (2002). Genetic improvement of farmed shrimp: summary and implications of a global survey. *Global Aquaculture Advocate* **5**(1), 48–50.

Rosenberry, B. (1997). *World Shrimp Farming 1997*. Annual Report by Shrimp News International, San Diego, CA.

Rosenberry, B. (2001). *World Shrimp Farming 2001*. Shrimp News International, San Diego, CA.

Treece, G. D. & Fox., J. M. (1993). *Design, Operation and Training Manual for an Intensive Culture Shrimp Hatchery (With Emphasis on Penaeus monodon and P. vannamei)*. Texas A&M University Sea Grant College Program, Publication TAMU-SG-93–505, Galveston, TX.

Wyban, J. (Ed.) (1992). *Proceedings of the Special Session on Shrimp Farming*. World Aquaculture Society, Baton Rouge, LA.

20 Freshwater Crustaceans

Freshwater crayfish: David 'Dos' O'Sullivan and Clive Jones

Freshwater prawns: Don Fielder and Clive Jones

20.1 General introduction

Their biology and culture have unique characteristics.

Freshwater aquaculture is basically concerned with fish species; crustaceans play only a minor role. Total production of freshwater crustaceans is very low compared with fish, and many fewer species are involved. Nevertheless, the species' biology and methods of production have unique characteristics that make them a particularly interesting group. Although national production of mitten crabs (*Eriochier sinensis*) is a substantial aquaculture industry in China (177 955 mt in 1999) (FAO, 2001b), international aquaculture of freshwater crustaceans comprises almost entirely prawns and crayfish (some species of which are known as crawfish in southern USA). This chapter will consider prawns and crayfish.

These freshwater crustaceans generally have a broader diet and simpler life cycle than brackishwater and marine shrimp and, consequently, may be reared with cheaper feeds, and culture methods that are less technically demanding. Freshwater prawns have free-living larval stages, which often require brackishwater, necessitating specialised hatchery facilities. Freshwater crayfish, however, have no free-living larval stages, and complete their entire life cycle in freshwater. Their aquaculture is relatively simple. Unfortunately, technical simplicity often leads to negligent practice and the benefits of simple biology of the species involved are lost due to a lack of commitment to basic aquaculture principles. This is not an issue if extensive aquaculture practices are used, but the future for freshwater crustacean aquaculture is in intensified culture techniques, and the potential for significant growth in production will be realised only through application of professional and industrial practices like those used for marine crustaceans.

The two groups are dealt with separately below.

Freshwater crayfish and prawns differ in morphology and life history, but they also have much in common. Consequently, the future development of culture techniques will no doubt involve some cross-fertilisation of ideas between the two groups.

20.2 Freshwater crayfish

20.2.1 Introduction

Freshwater crayfish (Fig. 20.1) are crustaceans collectively grouped as the Astacida within the order Decapoda. They are closely related to clawed marine lobsters, including the commercially significant *Homarus* and *Nephrops* genera. More than 500 species of freshwater crayfish are distributed around the world from polar regions to the tropics. They belong to three families (Hobbs, 1988):

(1) Family Astacidae, native to Europe and

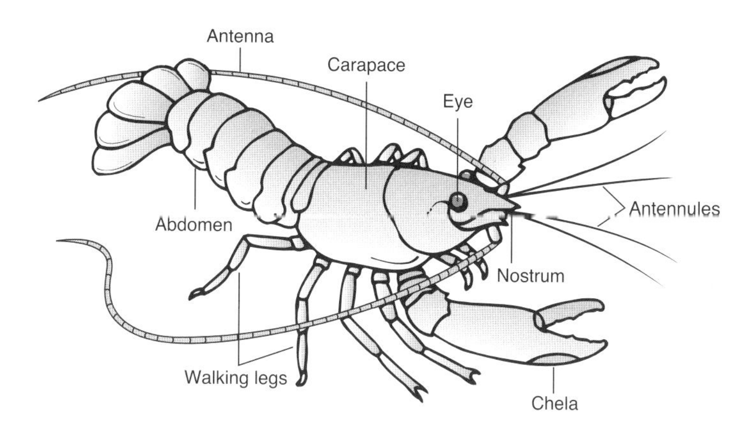

Fig. 20.1 External morphology of freshwater crayfish as illustrated by *Cherax quadricarinatus* (redclaw crayfish) (after Jones, 1990).

North America, is the least diversified family with only three recognised genera, *Astacus*, *Austropotamobius* and *Pacifastacus*, and 12 species.

(2) Family Cambaridae, native to eastern Asia and North and Central America, is the most diversified with 12 genera and at least 360 species, although representatives of only three genera, *Cambarus*, *Orconectes* and *Procambarus*, are of major economic importance. Some cambarid species have been translocated to Africa, Europe, Asia and South America.

(3) Family Parastacidae, native to the Australia, New Guinea and New Zealand region, southern South America and Madagascar, has the greatest number of crayfish genera (14), but contains only 145 species. Some of the *Cherax* species have been translocated to North, Central and South America, South-East Asia, various Pacific Islands and Africa.

Scholtz (2002) should be consulted for additional information on crayfish taxonomy and natural distribution. Introductions and translocations of several North American and Australian species have been documented in several studies (Huner, 1992a; Lee & Wickins, 1992; Holdich, 2002). Species of current or developing economic importance are listed in Table 20.1.

20.2.2 Aquaculture species

Crayfish are of significant aquaculture importance in the USA, western and northern Europe, and parts of China, and of relative importance in Australia. Out of the more than 500 species of crayfish, only 12 are cultured commercially. Aquaculture attributes of these species are reviewed in Morrissy *et al.* (1990), Lee & Wickins (1992), Holdich (1993, 2002) and Huner (1994).

20.2.3 Production status

World production

Most crayfish are used for human consumption and much smaller quantities are used as bait for recreational fishing. Huner (1992a) and Lee & Wickins (1992) reviewed the different crayfish products, which include:

- whole animals
  - live/dead
  - raw/cooked
  - purged/un-purged
  - fresh/frozen

and a range of value-added forms, such as:

- fresh or frozen tail meat
- soft shelled
- pâté.

The tail meat represents 10–40% of total body weight, depending on species, size and stage of maturity. The claw (chela) meat represents an additional 10% of body weight in some species, although it is generally difficult to remove this meat, despite its reputed superior taste.

Table 20.1 Freshwater crayfish of significant or developing aquaculture importance

Family/species	Common name	Regions*
Astacidae		
Astacus astacus	Noble crayfish	Europe
Astacus leptodactylus	Narrow-clawed crayfish	Europe
Pacifastacus leniusculus	Signal crayfish	Europe, USA
Cambaridae		
Orconectes limosus	Spiny-cheek crayfish	Europe
Orconectes rusticus	Rusty crayfish	Canada, USA
Procambarus zonangulus	White river crayfish	USA
Procambarus clarkii	Red swamp crayfish	Africa, Asia, Europe, USA
Parastacidae		
Cherax albidus/destructor	Yabbie	Australia
Cherax tenuimanus	Marron	Australia
Cherax quadricarinatus	Redclaw	Australia

*Regions do not necessarily reflect distributions due to widespread translocation.

Table 20.2 Production estimates for world freshwater crayfish in 2001

Region	Total production (mt)	Aquaculture production (mt)
USA	50 000	35 000
Canada	10 000	< 1000
Europe	5000	< 500
China	70 000	35 000
South-East Asia	1000	1000
Australia	1000	500
Central/South America	< 500	< 500
South Pacific Islands	< 500	< 500
Southern Africa	< 500	< 500
World total (approximate)	138 500	74 500

As wild fishery and aquaculture products are often sold within the same markets, and in similar product forms, production from both sectors may not be distinguished within broader production statistics. World production of freshwater crayfish, identifying wild-caught and cultured production, has been reviewed in detail by Avault (1992), Huner (1992a), Lee & Wickins (1992) and Ackefors (2000).

More contemporary, although less accurate, statistics are available in FAO publications. The difference is because aquaculture production of freshwater crayfish is an emerging sector in many parts of the world, and mechanisms for collection and reporting of accurate production statistics are not well established. Production figures appear to be under-reported. Estimates of current production are shown in Table 20.2.

Out of the estimated 138 500 mt annual production, ~54% comes from aquaculture. The USA (particularly Louisiana) and China account for the bulk of this production, which in both countries is based on the red swamp crayfish (*Procambarus clarkii*). This species is endemic to the south-eastern USA, but has been widely translocated throughout the world. Enterprising aquaculturists in China are now cultivating this native American crayfish and exporting it to the USA at a price with which the domestic USA industry has difficulty competing.

North America

Red swamp crayfish and, to a lesser extent, the white river crayfish (*P. zonangulus)* are the primary culture species in the USA. Aquaculture of *Orconectes* species in Canada and northern parts of the USA make up a small proportion of total North American production. Annual production sometimes exceeds 60 000 mt/year but fluctuates significantly from year to year, depending on rainfall and other environmental conditions. This production represents both wild caught and cultivated crayfish, the former of which is the more heavily influenced by climate. Because the cultivation of *Procambarus* species is based primarily on semi-intensive culture techniques and is vulnerable to climatic factors, it is also characterised by substantial annual fluctuations. Greater than 90% of the north American production consists of the red swamp and white river crayfish, and it comes from Louisiana and surrounding states bordering the Mississippi River (Huner, 2002).

About 80% of the total production is consumed within the state producing it. Much of the production area for these species is integrated aquaculture, with rice and soybean operations in large (> 1 ha), shallow ponds. The farms produce cereal crops in the warmer months and crayfish in the cooler months. The main production season is from December to April, although freezing of the product allows for year-round marketing. Crayfish production elsewhere in the USA and Canada is based upon *Orconectes* species cultured on a smaller scale, in smaller ponds and using more intensive techniques. Production of these species generally does not exceed 500 mt/year. They are primarily cultured for use as fishing bait.

North American crayfish achieve relatively small maximum size and are generally marketed in several grades less than 40 g. Local prices are quite low (US$1–1.50/kg); however, value-added products such as soft-shell crayfish achieve much higher prices (US$15–20/kg). About 50% of the Louisiana crop is sold alive for boiling, whereas the remainder is peeled for tail meat.

Europe

Consumption of native European crayfish is steeped in longstanding tradition, particularly in northern Europe, where the noble crayfish (*Astacus astacus*) is held in the highest culinary regard. Great demand during the traditional crayfish eating season and limited supply generate high prices, which sometimes exceed US$40/kg (farm-gate price). Noble crayfish are primarily sought in Sweden, Norway, Finland, Denmark and Germany. Significant demand also exists in southern Europe (France, Spain and Italy) for other species, both native and introduced.

At its height, in the late nineteenth century, natural production of the noble crayfish was around 3000–4000 mt/year. Introduction of disease (crayfish plague) (Ackefors, 2000) by translocations of American crayfish species wiped out many of the natural populations of noble crayfish (and other native European species), and they have remained at low levels ever since. Natural production is now less than 400 mt/year, despite substantial hatchery production of fingerlings and restocking programmes. Aquaculture of noble crayfish has generated some increased production, but the species biology, particularly its slow growth, is not amenable to large-scale commercial production. Much of the demand for noble crayfish is now met by aquaculture production of other species, primarily *Astacus leptodactylus*, *Pacifastacus leniusculus* and *Procambarus clarkii.*

Total European crayfish production is currently around 5000 mt/year; however, less than 500 mt of this is cultured. As recently as the early 1980s, Turkey was still producing (through wild catches) as much as 8000 mt/year of signal crayfish (*Pacifastacus leniusculus*), but this species was almost eliminated by the gradual spread of crayfish plague, and production is now insignificant.

Aquaculture production of noble crayfish in Europe (mostly Sweden, Finland and Germany) is currently 30 mt, with market prices up to US$30–40/kg. Production of the signal crayfish is around 50 mt/year. Prices are usually around US$20–30/kg. Red swamp crayfish are produced primarily in Spain, although only through extensive culture in natural waterways. Nevertheless, production exceeds 200 mt/year and is growing. Several hundred hatcheries produce crayfish for restocking throughout Europe (Holdich, 1993).

Three North American crayfish species are now widely distributed in Europe (Ackefors, 2000):

- red swamp crayfish (*P. clarkii*), responsible for the original introduction of crayfish plague to Europe
- signal crayfish (*P. leniusculus*)
- *Orconectes limosus*.

All have been developed to some extent as aquaculture candidates in parts of Europe because of their resistance to crayfish plague and their perceived commercial benefits. Not unexpectedly, they have each established wild populations, with consequent impacts on the native crayfish fauna. Bans exist in some European countries on the importation of live, non-endemic crayfish species, but this essentially symbolic gesture has come far too late for the protection of the native crayfish. An on-going polarisation of views exists on the development of aquaculture of exotic species that serve market demand, as opposed to removal of all non-endemic species with a long-term view to re-establishing viable populations of the native species.

Australia

The diverse freshwater crayfish fauna of Australia has attracted attention from aquaculturists around the world, primarily because of the large size that some of the species can attain. Several species within the genera *Cherax*, *Astacopsis* and *Euastacus* can attain weights exceeding 1 kg and up to 4 kg. However, owing to other less suitable characteristics, such as a low fecundity or slow growth, only the marron (*Cherax tenuimanus*), redclaw (*C. quadricarinatus*) and yabbie (*C. destructor*) are considered to have good aquaculture potential (Mills, 1989; Jones, 1990; Morrissy *et al.*, 1990; Semple *et al.*, 1995).

Natural fisheries for *Cherax* species are around 500 mt/year, but production from aquaculture is increasing rapidly. In 1988–89, aquaculture production was 37 mt, worth US$0.4 million, whereas in 1999–2000 more than 400 mt were produced, worth US$5 million (Jones, 2001).

Yabbies, which have a local market price of US$4–8/kg, constitute most of the above production. The second most important species is the redclaw, which sells for US$5–12/kg, whereas marron sells for US$5–15/kg. With increasing production, average domestic prices are expected to drop to US$4–8/kg.

Several production approaches have been applied to Australian crayfish aquaculture, ranging from extensive (harvesting of farm dams), through semi-intensive (managed, purpose-built earthen ponds), to intensive culture (battery systems). The semi-intensive approach appears to be the most economic, and most Australian crayfish are cultured by this method at present.

The Australian crayfish aquaculture industry (Edgerton, 1999) is subject to some health and disease issues, although most appear to be manageable. Crayfish plague is not present in Australia (Unestam, 1975). Although widespread translocations of Australian species to most parts of the world have occurred, no significant production of these species from outside Australia has yet been established. Nevertheless, given the excellent attributes of these species, such production is likely to occur in the future.

Asia

The red swamp crayfish was introduced into Japan in the 1930s to serve as a food for imported bullfrogs. It subsequently became popular as an aquarium species, and has now been spread throughout East and South-East Asia (Huner, 1992b). This species is now considered a pest in most regions owing to its destructive burrowing habit and the impact it has on native stream fauna. Red swamp crayfish were first taken into China over 50 years ago, but it is only in the past decade that commercial production and marketing of the species has become significant. Although substantial quantities (up to 40 000 mt/year) are harvested from rice fields, much of this is consumed locally. Managed aquaculture production has developed to the extent that in excess of 50 000 mt/year is now produced and exported to markets including the USA, where it competes directly with the domestic production (Xingyong, 1995; Ackefors, 2000).

Reports of production of redclaw and yabbies from China, Thailand, Malaysia and Singapore are now relatively common, although unsubstantiated. Clearly, production of other species in this region is likely in the near future.

Central and South America

Although South America has populations of endemic crayfish species, these have not been developed for aquaculture. Introductions of exotic species include the red swamp crayfish, which is now widespread in Central and South America. It has not developed as an aquaculture species. In the past decade, there has been considerable interest in the Australian *Cherax* species, and they have been introduced to several countries including Argentina, Chile, Ecuador, Costa Rica and Mexico. In Ecuador, in particular, significant promotion of redclaw as an aquaculture candidate has occurred, driven by large enterprises producing juveniles. Despite the intrinsic suitability of redclaw, the promotion was misdirected and only a small industry has thus far developed. Redclaw aquaculture has also now become established in Mexico, although production statistics are unavailable.

Africa

Huner (1992b) suggested that introduction of red swamp crayfish into Africa may have been for public health reasons, such as the control of parasites. It was thought that these crayfish might consume snails, which are vectors of serious human parasites. Harvesting of wild populations of red swamp crayfish in Kenya generates up to 500 mt/year, but no documented evidence exists of aquaculture operations for this species. As for other parts of Africa, the Australian *Cherax* species have been actively sought to investigate their potential for aquaculture. Translocations have occurred to Kenya, South Africa, Zambia and Zimbabwe, but production at this time is understood to be insignificant.

20.2.4 Biology

External anatomy

The body plan of the freshwater crayfish is typical of all decapod crustaceans such as shrimp, prawns and crabs. The exoskeleton is composed of chitin hardened with calcium carbonate. The body is clearly divided into the anterior cephalothorax and posterior abdomen (Fig. 20.1). The cephalothorax is largely enclosed in a carapace. Detailed morphology and functioning of body parts are well covered in general biological texts (Holdich, 2002). Fig. 20.1 illustrates the more prominent anatomical features of crayfish.

Posterior to the antennae and mouthparts, the appendages of the five thoracic segments consist of large chelipeds or claws and four pairs of walking legs. The chelipeds are used for defence and manipulation of food particles. Apart from functioning in locomotion, the first two pairs of walking legs have small chelae, used to assist with manipulation of food, and the last two have simple pointed ends, used for cleaning the body. Female reproductive openings (gonopores) are found at the base of the second pair of walking legs. Males have genital papillae, or penes, at the bases of the last (fourth) pair of walking legs. Males generally grow to a larger size than females.

Six articulated segments make up the abdomen, terminating with the telson, which has four appendages called uropods. The telson and uropods make up the tail fan. Sudden flexion of the well-developed abdominal muscle, the tail flick, allows the crayfish to swim backward, usually to escape a predator or other threat. A series of appendages, called pleopods, are found on the underside of abdominal segments two to five. These may be used for locomotion in small crayfish and are the site for attachment of eggs in adult females. The gentle waving of the pleopods also serves to aerate the eggs.

Vogt (2002) provided a full description of the internal body systems (circulatory, reproductive, excretory, respiratory, digestive, nervous and muscular systems).

Fig. 20.2 Berried female redclaw (*Cherax quadricarinatus*).

Reproduction

Some freshwater crayfish live for 30 or more years; however, most species live for < 10 years. They have similar life cycles, but differ in features such as age of sexual maturity and reproductive seasonality. The females of some species spawn in their first or second year, whereas in other species female maturity is not achieved until the third or fourth year. Some species mate in early winter, incubating the eggs through the winter for spring hatching, whereas others mate in spring, briefly incubate their eggs and release them in early summer. Some species spawn once each year, whereas others may breed three or four times within one season.

Environmental stimuli for breeding are primarily related to day length and water temperature (as for many other aquaculture animals; see Chapter 6). Females release pheromones to attract males and, after varying degrees of courtship, mating occurs. Behavioural aspects of the mating and spawning process differ between species, but share the following broad principles. The male actively grasps the female with his chelae and manipulates her into a suitable position for depositing a spermatophore. In all species, the spermatophore is deposited as a sticky white mass, containing non-motile sperm, on the sternum of the female, between the bases of her walking legs.

Eggs are released soon after mating and are fertilised by sperm released from the spermatophore by the female scratching its surface with the sharp tips of her last walking legs. A temporary brood chamber is formed by the tightly curled tail, into which the eggs and sperm are drawn with a current of water created by the beating of the pleopods. A sticky mucus known as the glair is released from glands on each of the abdominal segments and enables the fertilised eggs to become attached to the fine hairs (ovigerous setae) along the margins of the pleopods. The swirling nature of the current within the brood chamber causes the hairs to plait themselves into rope-like tethers to which the eggs become attached. At this stage the females are said to be 'in berry' or 'berried' (Fig. 20.2). The eggs are large and yolky. Development is lecithotrophic and without free larval stages. At hatching, the egg case splits open, but the hatchling crayfish remains attached to the female. After it breaks free, the hatchling grasps whatever surface it can beneath the female's tail using specialised leg hooks. Only after the first two moults does the hatchling begin to pursue its independence as a juvenile crayfish (Gherardi, 2002).

Moulting and growth

Moulting, or ecdysis, and growth of decapod crustaceans are described in Chapter 6 (section 6.4.2). Freshwater crayfish conform to this pattern of moult cycles and series of instars of increasing size. Their moult cycle is influenced by age, day length and water temperature. During the pre- to post-moult period, behaviour is modified to minimise interaction with other crayfish and to avoid exposure to predators. This is to avoid being attacked and eaten during the brief period immediately following moult, when the new shell is very soft and the crayfish is unable to defend itself. Often the risk of cannibalism by the same species is higher than that of predation by other species and, consequently, pre- and post-moult crayfish may move away from typical habitat or shelter where their con-specifics are located. Because moulting is such a demanding process, death of crayfish stressed or compromised by other causes often occurs at this time.

Food and feeding

Freshwater crayfish have a well-developed alimentary system and most are generalised omnivores. They are commonly opportunistic forag-

Fig. 20.3 Trapping boat with hopper for dispensing bait pellets, pyramid traps in foreground and broad-scale production pond in background. Red swamp crayfish aquaculture in Louisiana, USA.

ers, consuming whatever organic materials they encounter that can be manipulated to the mouth. Consequently, the diet may consist of:

- detritus and decomposing matter
- fresh plant material
- microscopic organisms associated with detritus and decaying organic matter
- benthic invertebrates, including con-specifics
- planktonic organisms
- fish.

Detection of potential food items is performed by the antennae, antennules and the large chelae, which prod the substrate. They are all equipped with tactile and chemosensory organs. Food items are then manipulated to the mouth with the chelate walking legs. A broad array of mouthparts is then used to shred, tear and partially filter materials before ingestion. From the mouth, food is moved through a short oesophagus to a large muscular foregut with an anterior 'cardiac' chamber that is equipped with calcified grinding teeth (gastric mill), which crush and macerate the food. Food particles then pass into the posterior or 'pyloric' chamber of the foregut, where further breakdown occurs, aided by the release of digestive enzymes from the adjacent hepatopancreas. Digestible components and partially digested material move into the midgut, where complex filters allow nutrient-rich fluids to flow into the tubules of the hepatopancreas for absorption. Indigestible material is moved along through a valve, into the hindgut and, once there, no further digestion takes place. A peritrophic membrane is laid down over the particles, forming them into faecal pellets, which are transported along a simple tube by peristalsis until they are released through the anus beneath the tail fan.

20.2.5 Culture intensities

Freshwater crayfish culture has most notably been exemplified by the red swamp crayfish industry of the southern USA, which was developed in the 1950s (Huner & Barr, 1991). The approach taken has been typically extensive or semi-intensive. It has proven to be extremely successful as evidenced by the volume of production from the USA and more recently from China using the same approach. In the past two decades, interest in culturing a range of other crayfish species around the world has developed. These involve a range of other approaches that suit the biology and economics of the particular species and location. These approaches are broadly summarised in Table 20.3.

Specific culture techniques vary significantly between species and localities in aspects such as:

- site selection
- pond specifications and farm design
- juvenile production
- stock management
- pond management
- harvesting
- post-harvest handling.

Lawrence *et al.* (1989), Lee & Wickins (1992), Huner (1994) and Holdich (2002) provide detailed information on specific techniques. The following discussion will be deliberately broad and will use the Australian crayfish species and their culture techniques for illustrative purposes.

Extensive culture

Extensive aquaculture of crayfish follows the general pattern of extensive culture outlined in Chapter 2 (section 2.3.3). The culture environment is nearest to that experienced by crayfish under natural conditions. Often the difference between extensive culture and a wild fishery is the difference between privately owned and publicly accessible stock, e.g. areas within the natural distribution of the red swamp crayfish (*Procambarus clarkii*) in Louisiana, USA. Extensive culture is based upon establishment and maintenance of natural populations. A nucleus stock may be introduced in cases where crayfish have never been present or where they have been overfished or completely removed. Growth of the population is left entirely to natural dynamics i.e. no effort is made to manipulate food, shelter, predation and water quality. Density of stock is generally low ($< 2/m^2$), reflecting the natural carrying capacity of the water body.

Various types of water bodies are used for this culture system, including sections of rivers, lakes,

Table 20.3 Characteristics and examples of the various approaches to freshwater crayfish aquaculture

Production approach	Extensive	Semi-intensive	Intensive
Culture environment	Natural/seminatural water bodies Dams and ponds*	Ponds	Tanks, raceways
Reason for activity	Opportunistic	Planned	Planned
Control over water availability and quality	Very limited Limited*	Significant	High
Control over production environment	Very limited Limited*	Significant	High
Harvesting	Trapping Primarily trapping*	Total harvest, trapping and draining	Manual
Annual productivity (kg/ha)	100–400 300–700*	700–4500	Unproven
Viability	Viable for a range of species around the world, usually on a small scale Very successful*	Proven viability when appropriately planned, sited and managed	Unproven
Commercial activity	*Only with red swamp crayfish in the USA, China	*Cherax* species in Australia; small scale in Europe	None

*Details for red swamp crayfish.

marshes and swamps, and man-made structures such as dams, gravel pits, reservoirs and drainage ditches. Cost of production is low and productivity is generally low (< 200 kg/ha), although in uncommon circumstances where the ecology of the system particularly suits the crayfish species, larger crops (> 500 kg/ha) can be produced. A primary difficulty associated with extensive culture is unpredictability of production due to the influence of the natural environment and its inherent variability.

Harvesting is achieved with baited traps deployed from trapping boats (Fig. 20.3) and it is often restricted to times when large crayfish are likely to be plentiful. The largest crayfish are removed, leaving smaller stock to replenish the population. Regular harvesting of the larger animals assists in increasing productivity by reducing stocking densities and reducing intraspecific competition.

Semi-intensive culture

As indicated in Chapter 2, a continuous spectrum of culture intensities exists, from extensive to intensive, with semi-intensive culture covering the middle ground. Semi-intensive cultures are conducted in the same water bodies as extensive cultures, with the addition of some measures of control or management to improve productivity. Such measures may include:

- stocking additional quantities of preferred stock
- supplemental feeding
- providing enhanced habitat or shelter
- water quality management measures
- predation control.

At this level of culture, it is difficult to balance

the necessary input costs against the unpredictable increase in productivity. Nevertheless, the bulk of the world's crayfish production from aquaculture is generated with some level of semi-intensive techniques. Red swamp crayfish production in the southern USA, China and Spain, *Orconectes rusticus* production in northern USA, and yabbie production from the farm dams of Western Australia are all examples of a level of semi-intensive culture. Yields vary substantially, but are typically in the range of 200–600 kg/ha, a range that overlaps with the best productivity from extensive aquaculture, reflecting the continuity of culture levels.

Although a further level of semi-intensive culture is considered to be the method of choice for a broad variety of crayfish species, it has yet to prove itself in terms of significant production from any one species. Although the full range of culture intensities are used for *Cherax* species in Australia, and in the various countries to which they have been translocated, a higher level of semi-intensive culture is the one that economic modelling has suggested will be the most viable. Similarly, crayfish culture throughout Europe is accomplished most successfully with a higher level of semi-intensive techniques.

This form of semi-intensive culture is typically conducted in earthen ponds, as in extensive and in other semi-intensive aquaculture, but is most clearly distinguished by the carefully planned control over all aspects of the production cycle and the culture environment provided. The physical facilities are clearly defined through aspects such as:

- selection of a suitable site
- design of the farm layout
- detailed pond specifications
- provision of purpose-built shelters
- the ability to provide aeration
- predator-proof fencing and netting (Fig. 20.4)
- the harvesting devices used.

Similarly, all aspects of stock and water are managed to optimise productivity. Juvenile production is often managed as a separate 'hatchery' process, and it may involve deliberate stock selection and controlled reproduction, possibly in tanks. Grow-out may involve stocking particular densities of graded, uniform-sized juveniles, sometimes sorted by gender for separate culture of males and females. Grading of harvested stock may involve distinguishing superior broodstock for selective breeding, marketable stock, under-sized stock for further grow-out and runt stock that might be culled and discarded. Harvesting involves techniques that include:

Fig. 20.4 Typical aquaculture pond for redclaw crayfish, approximately 0.1 ha, fully fenced and netted to exclude predators, in north Queensland, Australia.

- drain-harvesting
- baited traps
- flow traps, whereby water is run down an angled walkway to encourage the upstream behaviour of crayfish, which crawl out of the pond and up the walkway into a collecting tank (Jones & Grady, 2000).

Water quality is managed to promote desirable planktonic species (both as a food source and for shading the crayfish), to maintain optimal pH and dissolved oxygen levels and to minimise build-up of ammonia and nitrite. This is achieved by measured application of chemical and organic fertilisers, flushing of water and mechanical aeration devices.

Yields from this level of semi-intensive culture are typically in the range of 1000 kg/ha to over 5000 kg/ha. A farmgate price of at least US$5/kg must be achieved to cover the relatively high cost of production. The price is often considerably higher than this for large size-grades (> 90 g), given the increasing world demand for premium-grade crustaceans.

Intensive culture

Intensive crayfish culture is, at best, developmental and most typically still experimental. There has been considerable interest in the approach, but no substantial or sustained production for any species to date. This type of culture takes the level of control to its upper limits. It is a system approach, involving recirculation, where management of every significant variable is sought (section 2.3.2). Successful examples for fish are plentiful, particularly in Europe. For crayfish, intensive systems for hatchery production of juveniles have been developed, most notably in the USA for *Cherax* species and in Europe for *A. astacus* and *P. leniusculus*. Even the hatcheries that have operated successfully persisted for no more than a few years, probably as a result of economic factors and lack of commercial viability.

A further example of developing an intensive system for crayfish has been the strong interest in marron, in Western Australia, involving battery structures with individual housing. No successful model has been generated, despite some 20 years of experimentation (Fig. 20.5). There appear to be intrinsic biological constraints that render crayfish unsuitable for this approach, at least for grow-out.

An intensive approach was applied to the development of 'shedding systems' in the southern USA during the 1980s (Huner & Barr, 1984), to generate soft-shelled crayfish (*Procambarus clarkii*) for specialised niche markets both locally and abroad. They were biologically successful but uneconomic and, despite some short-lived prominence, have all but disappeared. Clearly, intensive systems for crayfish would have to be biologically successful and specifically target production of a niche product that has a greatly enhanced market value, to offset the very high costs of production.

Fig. 20.5 Experimental intensive system with individual compartments and recirculation (tested with marron, *Cherax tenuimanus*).

20.2.6 Culture methods

Production criteria depend on the species being cultured and the intensity of culture. For extensive culture, by definition, effectively no controls are applied. As the level of farming intensity increases there will be substantially more specifications to consider. The following production criteria are broadly generic to provide an introduction to the types of considerations that may be applied to crayfish aquaculture. Examples from particular species are given where appropriate, but for detailed specifications for particular species and the approach applied to them, other references detailed throughout this chapter should be sought.

Farm specifications

The two most fundamental criteria that must be satisfied for all approaches are availability of suitable soil and water. As in other pond aquaculture, the soil must have sufficient clay content and particle size distribution to hold water throughout the culture period. Water must be available in sufficient quantity and quality to sustain the crop and the management applied to it. Large volumes of high-quality water will be required for some semi-intensive approaches. It is necessary to calculate a water budget that accounts for:

- standing volumes
- evaporation
- flushing
- drain harvesting
- ancillary applications (tank systems, irrigation of surrounds, etc.)
- ensuring that this volume of water is available even when drought conditions apply.

In terms of farm design, a systematic layout is important to optimise the cost-effectiveness of the farm's operation. This applies particularly to use of gravity for filling and draining ponds.

Cost of establishing the operation needs to be considered in determining an appropriate scale of farm. For example, a 3- to 4-ha area of grow-out ponds is the minimum for commercial viability of redclaw aquaculture in Australia using semi-intensive techniques. In comparison, semi-intensive culture of red swamp crayfish in southern USA typically involves operations with total pond areas in excess of 50 ha.

Optimal pond specifications have been defined for a number of species, and they are typically defined in terms of surface area, depth, shape, and degree of upward slope of outer perimeters (batters). These in turn can be related to primary productivity of the water, standing stock of crayfish, and ease and speed of drainage. The areas of individual ponds in semi-intensive crayfish farming vary between < 0.2 ha and 5 ha according to the level of culture intensity (an inverse relationship).

Infrastructure

A number of infrastructure considerations can be critical to the success of crayfish aquaculture at more than an extensive level.

(1) Most crayfish species require some form of shelter and under aquaculture conditions this must be provided to ensure that the desired density can be maintained. Without such shelter, aggressive interactions between individuals will increase and survival of the stock will be greatly diminished. For semi-intensive culture, shelter may be provided by the presence of living or dead vegetation. Red swamp crayfish culture is typically undertaken as a rotational crop with a cereal plant such as rice or soybean. The bases of the plants are left standing after harvest and become an effective shelter (and food source) when the field is flooded, becoming a pond for the subsequent crayfish crop. *Cherax* species are typically cultured semi-intensively, using artificial materials to form shelters, which are positioned throughout the pond. These are fabricated from a range of materials including plastic pipes, plastic mesh, netting and car tyres (Fig. 20.6). Key attributes of effective shelters are:

Fig. 20.6 Car tyres provided as hides on a gravel substrate pond.

- They be free-standing and occupy at least some of the water column.
- They must provide separate dark spaces for the crayfish to isolate themselves.
- Each shelter must be surrounded by some open space for foraging.
- They must drain freely and completely of water when the pond is drained, so the crayfish will leave them to be harvested.
- They must be sufficiently abundant to support the standing stock (e.g. ~1 shelter/4 m^2 of pond bottom).

(2) A facility to maintain adequate dissolved oxygen levels in the culture water is fundamental to semi-intensive approaches. This can be achieved most simply by delivering intake water to a pond through an aeration tower in which water cascades vertically through a series of mesh baffles of decreasing mesh size, breaking the water into small droplets and allowing natural inward diffusion of oxygen. For more intensive approaches, more sophisticated aeration equipment is used. Airlift pumps, paddlewheels and aspirators are most commonly used and are essential where yields greater than 1000 kg/ha are targeted.

(3) The third crucial item of infrastructure for many crayfish aquaculture operations is predator proofing. Predation by birds and terrestrial animals (e.g. rats) can decimate the standing stock of the farm. This impact on productivity is often not seen until harvest, when a small proportion of what was expected is revealed. A small number

of predators over the entire culture period can wreak havoc. For semi-intensive operations requiring large pond areas, cost-effective options are limited. Scare devices are only partially effective and must be continually moved and modified to have maximum impact. Trapping or shooting the predators is now generally outlawed, as the animals involved are usually protected native species. Despite their high initial cost, fencing and netting enclosures are cost-effective for semi-intensive farms. Economic modelling for several species has demonstrated that the cost, which may range from US$10 000 to US$20 000/ha for a fully enclosed netting structure over the entire ponded area, may be recouped within one season where predators are prevalent.

Juvenile production

Some semi-intensive crayfish culture relies on natural reproduction and juvenile production. In the red swamp crayfish industry, adult crayfish left behind after harvesting breed and then burrow into the pond bottom. A rotational cereal crop may then be grown in the soil through summer and, when the pond is again flooded in autumn, juveniles emerge from the burrows to stock the pond.

Varying levels of managed juvenile production are used for other semi-intensive culture. Most commonly, mature stock or, in some cases, berried females are obtained from the harvest of a normal production pond and then stocked into a newly prepared pond. This juvenile production pond is maintained for 3 to 4 months and then harvested to obtain juveniles, which are then sorted and stocked elsewhere for grow-out. This process can benefit from increasing levels of management control. For example the following measures can increase productivity to the extent of 200 000 juveniles/ha or more:

- specific preparation of the pond, with provision of shelters that particularly suit juveniles (mesh bundles that simulate aquatic plants are very effective)
- fertilisation of the pond (lime, inorganic and organic fertilisers) to promote planktonic productivity
- use of selected broodstock (see below).

A managed juvenile production process is considered essential for semi-intensive culture of *Cherax* species in Australia. Some farmers simply utilise as 'juveniles' the small crayfish that are produced incidentally in their normal production ponds. In subsequent grow-out with these 'juveniles' they find that their yields progressively decline. They are effectively selecting genetically inferior runts.

In Europe, specialised hatcheries have been developed particularly for the production of the native species for restocking to natural waterways (Fig. 20.7). These generally consist of raceways with flow-through or recirculating water supply. Broodstock are placed in the tanks, and provided with artificial shelter and supplemental feeding. Breeding occurs naturally and juveniles are generated for stocking elsewhere. Because the species have a protracted incubation period, and relatively low fecundity, these hatcheries are generally uneconomic. They serve a community service in maintaining or bolstering natural crayfish populations.

Breeding

Managed breeding programmes for crayfish can only be justified for semi-intensive or intensive culture. Given the fundamental economics of all industrialised countries and costs of production, it is inevitable that crayfish aquaculture will need to apply principles of genetic selection to improve productivity, just as it is applicable to all other sectors of livestock and fish production. Indeed, crayfish lend

Fig. 20.7 Backyard hatchery pond for small-scale production of juveniles of the noble crayfish (*Astacus astacus*) in Bavaria, Germany.

themselves to selection programmes because they have a simple life cycle and do not have technically demanding free-living larval stages.

At its simplest level, the deliberate selection of superior mature stock as breeding stock, applied successively over several generations, will produce significant benefits. The application of sophisticated molecular tools to such breeding programmes, such as microsatellite marker assisted selection, may provide even further or more expeditious improvements. Genetic selection programmes, primarily based on superior growth rate, have been applied successfully to each of the three *Cherax* species cultured in Australia (Jones *et al.*, 2000).

It is anticipated that for other crayfish species besides red swamp crayfish, development of significant production will necessitate on-going breeding programmes and, therefore, dedicated hatcheries.

Grow-out

For some semi-intensive culture, grow-out represents the primary phase during which some management control is applied, juvenile production being left to natural processes. In more intensive culture, grow-out is managed as an entirely separate process from juvenile production and there may be good cause for several separate grow-out phases. Initially, grow-out involves the stocking of small juvenile crayfish, typically at a set density that the management is known to sustain. This may be in the order of 5–15 crayfish/m^2. Strong evidence suggests that stocking with uniform-sized crayfish will improve yield as it mitigates against development of size-related dominance hierarchies. In culture of *Cherax* species, short grow-out periods (3–6 months) are an advantage, as growth rates vary substantially between individuals and size variability can retard overall production. Furthermore, for redclaw and yabbies at least, sexual maturity will be achieved within 1 year, so that unplanned reproduction should be prevented by harvesting grow-out stock before this can occur. Secondary and tertiary phases of grow-out may then follow, with stock again graded by size and possibly by gender. Yields from monosex cultures may be higher than those from mixed-sex stocks, where reproduction begins to occur. When a large mean size is sought for grow-out (> 100 g), two or three successive grow-out phases are required, each managed separately and with successively reduced stocking density.

Feeding

Availability of food of sufficient quality throughout the production phases is critical to success. Type of food and its delivery method vary considerably within the range of crayfish aquaculture, but there is clear evidence that heavy reliance on the naturally available food in earthen ponds, rather than managed or supplemental food provision, will greatly diminish yields. As previously described, the remaining vegetation from rotational cereal crops is the primary food source in red swamp crayfish culture, along with naturally occurring aquatic plants. It is well understood that the lodging rate, i.e. the rate at which this vegetation decays and falls to the pond bottom, is sufficiently protracted to sustain the crayfish throughout grow-out (Huner & Barr, 1991).

Red swamp crayfish gain the bulk of their nutrition from the enriched microbial fauna and flora associated with the decay of the vegetation (Fig. 20.8). Cost–benefit analyses indicate that application of formulated foods does not increase yields.

Attempts have been made to extend the successful feeding models of red swamp crayfish culture to semi-intensive crayfish culture with little success. Although all cultured crayfish species may be sustained with a microbial food source, when densities

Fig. 20.8 A typical broad-scale pond of several hectares used for red swamp crayfish aquaculture (Louisiana, USA). It shows aquatic vegetation used as a food source for the crayfish.

are relatively high and fast growth rate is desired, provision of a supplemental food source is essential. In culture of *Cherax* species, use of formulated pellet diets has been resisted, owing to their cost, freight and storage considerations, and their unproven performance. Consequently, there is still great reliance on simple food items, including:

- whole or cracked grain
- legumes (e.g. sorghum, millet, lupins)
- hay
- reject vegetables (e.g. potatoes).

Although these can be effective, they do not provide a nutritionally complete diet and productivity based on them will not be high. Chicken pellets are used commonly as a food source, because of their low cost, wide availability and convenience. Again, they do not provide a nutritionally complete diet and because they are not stable in water they may have a significant negative impact owing to their consequent polluting effect.

High yields will only be attained from semi-intensive culture once pellet diets formulated specifically for crayfish are available. In addition to appropriate diet formulation, it is important that such a diet be reasonably fresh and that it is carefully stored to prevent spoiling (section 9.10.3). Frequency, rate and time of feeding are also important in optimising production and attaining the best food conversion ratio (FCR) possible. Generally, daily feeding at dusk at a rate based on a proportion of the standing stock of crayfish will produce optimal results.

Water management

Intensity of farming approach determines the degree to which the water quality needs to be managed. In semi-intensive farming, maintaining suitable dissolved oxygen is the primary issue of water quality management. At a minimum, discontinuous levee banks may be built into the pond to force water movement and prevent localised 'dead' areas where dissolved oxygen becomes limiting. Mechanical aerators are used to force aerated water around the pond in some sectors of the red swamp crayfish industry.

In higher levels of semi-intensive crayfish aquaculture, water management involves some regular monitoring, and adjustment of dissolved oxygen, pH, ammonia, hardness and plankton abundance. Each species has known optimal ranges for these parameters (Chapter 3). It is important that farmers not only have skills and knowledge of how to measure and understand water quality, but that they have contingencies for when a particular parameter diverges outside normal limits. Three rules of thumb that apply particularly to crayfish aquaculture water quality are:

(1) Measurements should be taken near the pond bottom, where the stock live.
(2) When aeration devices are used, they are deployed to maximise water circulation throughout the pond. Unlike fish, which can move quickly away from areas of compromised water quality, crayfish are benthic and slow moving, so uniformity of water quality is more important.
(3) Ensure drying and therefore some sterilisation of the pond bottom between crops. Again, this is particularly important for a benthic species that lives on the pond soil surface. Furthermore, it promotes better water quality for subsequent crops.

Harvesting

Methods of harvesting differ considerably among the various approaches and species of crayfish. As might be expected, the most labour-intensive are those associated with extensive culture. Manually set and retrieved baited traps are generally used over protracted periods. In semi-intensive farming, baited traps or nets may also be used for partial harvesting of ponds, but more commonly the entire pond is harvested at one time by draining it completely. If the pond has been designed appropriately, the crayfish will aggregate at the lowest point in the pond as the water is drained. This facilitates manual collection of the stock, which can then be crated away to a handling and sorting area. For redclaw in particular, flow-trapping is particularly effective, and it may be effective for other species. Jones & Ruscoe (2000) described the principles and specifications of flow-trapping.

Basically, it involves the deployment of a water-tight box with a ramp into the pond. Water, sourced from outside the pond being harvested, flows through the box and down the ramp, stimulating crayfish to move upstream. When flow trapping is combined with drainage of the pond, over 90% of the pond stock can be harvested. If the box is equipped with crates, the harvested stock can be transported quickly to other facilities for post-harvest handling. A major advantage of this method is that crayfish remain in clean, well-aerated water and are minimally stressed. In contrast, manual drain harvesting exposes crayfish to very dirty water for relatively long periods of time and results in stressful handling.

20.2.7 Post-harvest handling

Procedures for handling crayfish after harvesting are particularly important given that the bulk of product is marketed alive. To ensure maximum survival and quality of the live product, specific post-harvest procedures may need to be used to suit the biology of the species involved. Notwithstanding these considerations for live crayfish, post-harvest handling of crayfish for non-live product forms must also abide by quality assurance principles to maximise the product value.

Crayfish are normally moved to a processing facility immediately after harvesting. Whether the processing is as simple as brief size-grading before restocking to ponds, or is more involved, crayfish must be held in clean water at a cool temperature appropriate to the species. The first consideration is normally to facilitate removal of any dirty water from the gill chambers and to clean the crayfish of any mud or debris. More protracted holding (24–48 h) in clean, running and aerated water is sometimes used to allow voidance of the alimentary canal contents, a process known as purging. Further processing typically includes size and quality grading to suit restocking requirements on the farm or to meet particular market requirements. Bathing in saline is sometimes used to assist in removal of fouling organisms on the shell or to improve the flavour of the flesh. The concentration of saline and duration of the application depend on the desired outcome and the species tolerances.

Transporting live crayfish is generally quite easy, as most species tolerate exposure to air and can be held out of water for extended periods. This tolerance lends itself to transporting crayfish without water. Despite their robustness, air exposure is stressful. Every effort must be made to optimise the condition of stock before packing and to minimise the period for which the crayfish are held out of water. Maintaining a relatively low temperature within the transport container assists in lowering metabolic activity in the stock and minimising stress.

Although most crayfish are marketed alive, processed product is increasingly popular. Live product has some advantages in the market place, often achieving premium price. Costs of transport are, however, usually higher than for processed product and there is potential for mortality during transport if conditions are unfavourable. Processing is generally done to suit particular market requirements and ranges from simple cooking of whole animals through to production of value-added products such as pâté.

20.2.8 Pathogens and parasites

Crayfish plague caused by the fungus *Aphanomyces astaci* is perhaps the best known of crayfish pathogens, because of the devastating effect it has had on European crayfish populations. Although it remains a threat to crayfish aquaculture, and particularly the non-American species, which all appear to be highly susceptible, its impact has been of greatest significance to wild populations. Of equal or even more significance to crayfish aquaculture in Europe are:

- fungi, *Saprolegnia* and *Fusarium* species
- a microsporidian, *Thelohania* species
- a protistan, *Psorospermium haeckeli*.

Most of these are of moderate infectivity and are of greatest threat to crayfish already stressed by other factors.

Fungal infections are less common in the Australian species, which have a relatively higher incidence of viral and bacterial infections (Evans & Edgerton, 2002). Mass mortality associated particularly with parvo-like viruses has been recorded in farm stocks of *Cherax* species, although all have been isolated to particular farms and have not spread. Australian crayfish farmers are well informed of the threat of disease and are increasingly applying preventative measures. In addition to the viruses and bacteria which are truly pathogenic, several other parasitic organisms have been recorded that may not kill crayfish but may retard growth, suppress reproduction or reduce marketability. These are comprehensively reviewed by Evans & Edgerton (2002).

20.3 Freshwater prawns

20.3.1 Introduction

Freshwater prawns are decapod crustaceans within the infraorder Caridea. The Caridea contains 11 superfamilies, which include freshwater, brackish water and marine prawns/shrimp. Most of these have little aquaculture potential. Only species of the genus *Macrobrachium* belonging to the family Palaemonidae have been considered seriously for aquaculture.

More than 150 *Macrobrachium* species are distributed throughout the tropical and subtropical zones of the world. They occur in most inland freshwater areas, such as lakes, rivers and swamps, as well as in estuaries. Many species require brackish water for their early development, but others can complete their life cycle in freshwater.

20.3.2 Aquaculture species

By far the most widely cultured species is the giant Malaysian prawn or giant river prawn, *Macrobrachium rosenbergii* (Fig. 20.9), which is indigenous to the whole of south and South-East Asia, northern Oceania, and the islands of the western Pacific. Not only does this species form the basis of a significant industry over most of its natural distribution, but it has also been widely translocated to other locations and regions. These include Hawaii, South America and the Caribbean, where it also supports successful aquaculture industries.

Other *Macrobrachium* species have been cultured on a small scale or have shown promise as culture species on an experimental basis. These include *M. acanthurus*, *M. nipponense* and *M. australiense*; however, no significant farming operations have yet developed for species other than *M. rosenbergii*.

A primary reason for the development of *M. rosenbergii* as a successful aquaculture species is its ability to grow rapidly to a large size. Females may grow up to 120 g and males to 200 g. Local fisheries and, therefore, local markets have also existed for a very long time within the natural distribution of *M. rosenbergii*. However, their tail meat return is only approximately 35% of body weight compared with about 65% for marine shrimp.

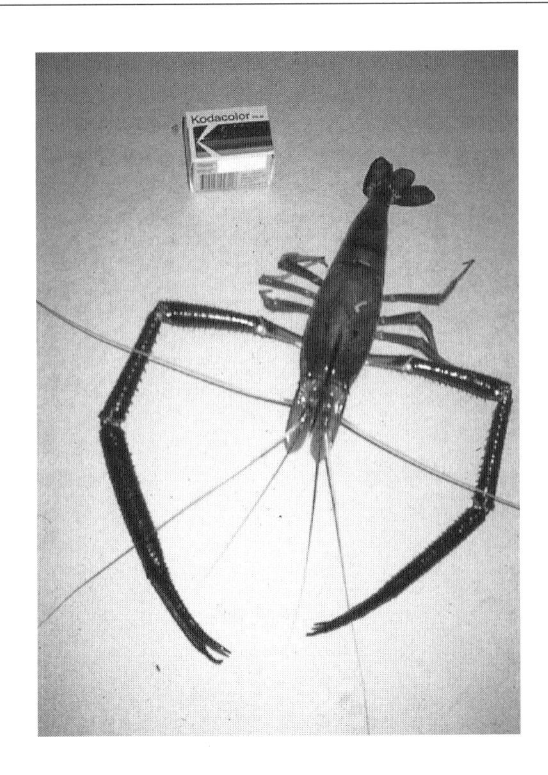

Fig. 20.9 A large male *Macrobrachium rosenbergii* from ponds at Walkamin, north Queensland, Australia.

M. rosenbergii has a long complex larval development that requires expensive hatchery culture with brackish water to provide postlarvae (PL) for grow-out. Water temperatures in excess of 22°C are required to maintain good growth rates. In grow-out ponds, a few fast-growing males ('bolters') soon dominate the rest of a population, leading to 'runting' and necessitating regular cull harvests of large prawns to allow smaller animals to increase in size. Other species, such as *M. australiense*, have less demanding environmental requirements and may prove to be good alternatives to *M. rosenbergii* in the future.

20.3.3 Production status

Total world aquaculture production of *M. rosenbergii* increased several-fold from 1990 to 1999 (Table 20.4). (Statistics are not available for China for 1990 and this may be inflating the difference in annual totals between these years.) The 102 000 mt production in 1999 was valued at US$400 million (FAO, 2001b). The great majority of this production was from China. Three Asian countries (Bangladesh, Taiwan and Thailand), together with China, contributed 97% of world production.

Annual production of *M. rosenbergii* exceeds that of freshwater crayfish (Tables 20.2 and 20.4), but total meat production from the former is substantially less than the latter. The proportion of meat return compared with body weight from crayfish is several times greater than that of the freshwater prawns.

Unlike freshwater crayfish, for which fisheries production contributes about 25% of world production, most of world production of *M. rosenbergii* is from aquaculture. World fisheries production of the species was 5500 mt in 1999, 5% of world production. The fisheries catch came exclusively from Indonesia (FAO, 2001a), where, surprisingly, *M. rosenbergii* is not cultured.

20.3.4 Biology

External anatomy

Although much of the anatomy of freshwater crayfish and freshwater prawns is similar, there are some differences. The exoskeleton of freshwater prawns is relatively much lighter and less pigmented than that of crayfish. The rostrum, which is quite small and dorso-ventrally flattened in crayfish, is accentuated in *Macrobrachium* species as a relatively long, laterally flattened and toothed extension of the carapace.

Characteristically, freshwater prawns possess a very long, slender, chelate second pair of legs, which are used in a similar manner to crayfish chelipeds for bulldozing gravel, in aggressive encounters, and for capture and manipulation of large food organisms. The first pair of legs are also chelate and are used primarily for picking small epiphytic food organisms from solid substrates. Males generally grow larger than females and are easily recognised by their much larger second legs (from which their generic name, 'long arms', is derived). The reproductive openings of males are located at the bases of the fifth pair of legs, whereas female reproductive openings are located at the bases of the third pair of legs.

Like crayfish, rapid contraction of the abdomen and tail allows rapid backward escape movements. However, *Macrobrachium* species are more mobile than crayfish in that their pleopods or swimmerets are large and allow effective swimming.

Life cycles

Macrobrachium rosenbergii grows to maturity in freshwater where, during the breeding season, hard-shelled mature males copulate with recently moulted, soft-shelled females. A spermatophore is deposited on the ventral thorax of the female between the walking legs. Eggs are extruded within a few hours of copulation and are fertilised as they pass posteriorly across the spermatophore. Eggs are then kept in a brood pouch under the female abdomen until they hatch. It appears that egg-bearing females move downstream into saline regions of estuaries where eggs hatch into planktonic zoea larvae.

Table 20.4 World aquaculture production of *Macrobrachium rosenbergii* in 1990 and 1999 and values for these years for the major producing countries (data from FAO, 2001b)

	1999 production (mt)	1990 production (mt)
China	–	79 055
Bangladesh	–	9008
Taiwan	11 607	7223
Thailand	6503	4341
World	20 842	102 124

M. rosenbergii may produce 80 000–100 000 eggs per brood when fully mature. Ovaries ripen again while eggs are still being carried, so that a series of broods is possible at approximately monthly intervals.

The zoeal larval stages (approximately 11) are active swimmers. They feed on zooplankton and metamorphose into PL within 3–7 weeks, depending on temperature, water quality and food supply. PL closely resemble adults. They become bottom crawlers and migrate upstream from saline conditions to complete their life cycle in freshwater. Sometimes mass migrations of PL are quite spectacular, particularly when an obstacle such as a weir slows their progress and aggregations occur (Fig. 20.10).

In contrast to *M. rosenbergii*, some species, such as *M. australiense*, do not require a period in brackish water to complete their life cycle. *M. australiense* has only three zoeal stages and are lecithotrophic until metamorphosis to PL, after about 1 week. As such, the larvae of *M. australiense* do not require feeding to complete larval development. Growth to adults can, therefore, take place near to the hatching site. In some rivers where winter dry seasons reduce rivers to a series of isolated water holes, maturing prawns migrate upstream after the first spring rains stimulate river flow. These mass migrations are equally as spectacular as the postlarval runs of *M. rosenbergii* and ensure that the mature prawns are well upstream before zoea larvae are released. This minimises the risk of being swept downstream into estuaries and being lost.

Fig. 20.10 Mass upstream migration of juveniles of *Macrobrachium australiense* at Glebe Weir, Queensland, Australia.

Mature males of *M. australiense* scrape a disc-shaped ‘nest’ in a gravel bed. A pre-moult mature female then enters the nest, to be actively defended against other males that take up a position around the nest perimeter. The female moults and copulation occurs with the nesting male within the nest while the female is still soft and relatively helpless. Sometimes one of the other waiting males may copulate with the female if the nesting male is distracted. Egg numbers depend on size of female but are never more than about 500. The three zoeal stages are completed rapidly, with low mortality. Providing water temperatures are maintained above 20°C, females produce serial broods at 3- to 4-week intervals during the warmer months of the year.

20.3.5 Culture systems

New & Valenti (2000) comprehensively reviewed farming techniques for *M. rosenbergii*.

Hatcheries

Because *M. rosenbergii* completes its larval life in brackish water, and PL and adults live in freshwater, a dedicated hatchery is required for its culture. Hatcheries must be located near sources of freshwater and seawater (although artificial seawater can be used). Larvae are obtained by allowing eggs on broodstock females to hatch. Larvae are then cultured in one of two types of hatchery systems.

Flow-through hatcheries. Flow-through hatchery systems were developed originally by Takuji Fujimura in Hawaii in the mid-1960s (New, 2000).

Tanks used are often rectangular (10 m^3 in volume) and are stocked at relatively low densities (30–50 larvae/L). 'Green water' is sometimes used in this approach by adding batch-cultured microalgae, such as *Chlorella*, *Nannochloropsis*, *Isochrysis* and *Tetraselmis* species, to larval cultures. The prawn larvae do not eat the algae, but the algae improve water quality by removing nitrogenous wastes excreted by the prawn larvae. Algae may also be eaten by zooplankton, which in turn are consumed by prawn larvae. Freshly hatched brine shrimp (*Artemia*) nauplii are added as supplemental food after day 2. Prawn larvae feed by chance encounters with the *Artemia* nauplii so the nauplii need to be present in sufficient density to allow such encounters to occur at regular intervals. For example, Roustaian *et al.* (2001) fed early-stage *M. rosenbergii* larvae at a rate of five to six nauplii/larva (0.15 nauplii/mL), whereas older larvae received 60 nauplii/larva (1.8 nauplii/mL). Uneaten food, faeces, etc. must be removed on a daily basis.

Recirculation hatcheries. Recirculation hatchery systems were developed subsequent to flow-through systems (Valenti & Daniels, 2000). They enabled hatcheries to be established inland, near grow-out facilities, using seawater transported to the site or artificial seawater. Recirculation systems incorporate a range of filtration technology to remove solids and nitrogenous wastes from the culture water, and to provide for disinfection (section 2.4.4). All tanks and associated equipment are normally sterilised between batches of larvae.

Nursery phase

Nurseries are designed to rear PL to the juvenile stage before stocking into grow-out ponds. The nursery phase is often omitted in tropical regions, but production from the grow-out phase is generally more predictable if juveniles are used to stock ponds. Tropical nurseries utilise earthen ponds stocked with 70–800 PL/m^2 and fed in similar fashion to grow-out animals. In temperate regions, nurseries are enclosed and tanks are stocked at much higher densities, in the range of 1000–1500 PL/m^2. Netting or palm fronds placed in nursery tanks provide added substrate for PL to crawl on.

Grow-out

Extensive culture. Monoculture of *M. rosenbergii* may be extensive, semi-intensive or intensive according to the standardised descriptions of Valenti & New (2000). Extensive culture of *M. rosenbergii* is usually conducted in earthen ponds, natural waterways and impounded areas built for other purposes. Ponds may be fertilised with organic materials before stocking to encourage an algal bloom. The increased turbidity produced by algal blooms reduces the possibility of bird predation and reduces growth of macroalgae, which can hinder harvest operations. Soil pH must be within the range 6.5–7.0, and rotenone is often applied to kill residual parasites and predators before ponds are filled.

Stocking rates are low (1–4/m^2). Water quality and stock are not managed and no supplemental feeding is applied. Controlled drainage is usually not possible and harvesting is difficult and inefficient. Productivity is generally less than 500 kg/ha. This system is most widely applied in Asia (Fig. 20.11).

Semi-intensive culture. Semi-intensive culture usually involves purpose-built ponds and a range of management protocols for water quality, stock, feeding and harvest. This is the most widely practised system for freshwater prawn culture throughout the world. Once stocked, grow-out ponds may be managed in either of two ways.

Fig. 20.11 A harvest of *Macrobrachium* species cultured extensively in rice fields during the dry season in the Khulna region of Bangladesh (photograph by Professor C. Tisdell).

- Ponds are harvested as batches.
- Ponds are harvested continuously.

Batch culture involves stocking with PL or juveniles and drain harvesting the whole pond after 6–9 months, when prawns have reached marketable size (20–60 g). Male prawns have variable growth rates and, as a consequence, batch culturing does not yield a crop of uniform-sized prawns. Production in batch culture systems ranges from 1000 kg/ha/year to 3000 kg/ha/year, depending on the length of grow-out period and initial stocking density.

In the tropics, where prawn growth continues throughout the year, ponds are often emptied only after several years. After initial stocking, cull harvesting begins after 4–9 months by seine netting, which captures only the largest prawns. This procedure, repeated once or twice per month, yields crops of large animals of uniform size. PL or juveniles may be added periodically to compensate for harvested prawns. This type of culturing may yield 2000 to 4000 kg/ha/year. Cull harvesting adds to the labour costs of harvesting, but it overcomes the 'bull/runt' growth syndrome associated with the male component of the population.

Males occur as three recognisable forms:

- Big clawed (BC) males or 'bulls'. These males are blue, sexually active, aggressive and collect harems of females.
- Small (SM) males. These males are capable of increased growth only when isolated from other types of males, particularly BC males. They are able to copulate with harem-held females if the dominant BC male is distracted.
- Orange-clawed (OC) males. These males are less aggressive and are not sexually active. They grow more rapidly than BC or SM males.

By removing the BC males at regular intervals, SM males have an opportunity to grow to market weight, intraspecific aggression is reduced, and less energy is channelled into fighting and reproduction. Consequently, greater productivity is possible.

Provision of supplemental food is necessary to produce good yields from grow-out ponds. Several prepared pelletised foods, containing about 25% protein, are available from feed-mills. Some farmers also use:

- poultry broiler feeds
- rice
- vegetable waste
- trash fish.

However, these alternative foods either produce lower growth than pelleted feeds developed specifically for crustaceans or they quickly reduce water quality. Feeding rates are usually 'rule of thumb', based loosely on the estimated prawn biomass in a pond (5–10% of body weight/week).

The culturing of monosex populations (all male or all female) rather than the traditional mixed-sex populations may have significant advantages. In the case of *M. rosenbergii*, an all-male culture should produce a greater biomass than a similar all-female population, as males grow to a larger size than females cultured for the same period. Growth should also be accelerated in all-male populations as no energy will be partitioned into reproductive activities. Support has been given to this proposal by at least one monosex production trial. All-male ponds yielded: larger animals, greater biomass, an estimated 24.5% revenue increase over mixed populations and an 85.7% revenue increase over all-female populations.

To make use of this type of management procedure, an easy and low-labour method of producing all-male populations needs to be developed. Hand-sexing is time-consuming, labour intensive and subject to errors. A promising method is currently being explored whereby the male-determining glands (androgenic glands) are surgically removed from juvenile males, resulting in the development of neo-females, which, when mated with normal males, produce all-male offspring. Treatment of juveniles with a male hormone such as testosterone may also be a fruitful pathway ultimately leading to culture of all-male populations as a normal farming protocol.

Intensive culture. Most intensive cultures of *M. rosenbergii* have been made under experimental conditions and involve stocking densities $> 20/m^2$. Establishment costs are high and a high level of technical skill is required, which so far has deterred investment. However, production levels of up to the equivalent of 10 000 kg/ha/year have been reported. As with many other aquaculture species, intensive culture may become more popular as technical experience with *M. rosenbergii* increases. Unfortunately, the biology of *M. rosenbergii* is not entirely compatible with intensive culture, particularly the heterogeneous growth and social behaviour aspects of its biology.

References

Ackefors, H. E. G. (2000). Freshwater crayfish farming technology in the 1990s: a European and global perspective. *Fish and Fisheries*, **1**, 337–59.

Avault, J. J. W. (1992). A review of world crustacean aquaculture, Part 2. *Aquaculture Magazine*, **18**, 83–92.

Edgerton, B. (1999). A review of freshwater crayfish viruses. *Freshwater Crayfish*, **12**, 261–78.

Evans, L. H. & Edgerton, B. F. (2002). Pathogens, parasites and commensals. In: *Biology of Freshwater Crayfish* (Ed. by D. M. Holdich), pp. 377–438. Blackwell Science, Oxford.

FAO (2001a). 1999 Fisheries Statistics: *Capture Production*. Vol. 88/1. FAO, Rome.

FAO (2001b). 1999 Fisheries Statistics: *Aquaculture Production*. Vol. 88/2. FAO, Rome.

Gherardi, F. (2002). Behaviour. In: *Biology of Freshwater Crayfish* (Ed. by D. M. Holdich), pp. 258–90. Blackwell Science, Oxford.

Hobbs, H. H. (1988). Crayfish distribution, adaptive radiation and evolution. In: *Freshwater Crayfish: Biology, Management and Exploitation* (Ed. by D. M. Holdich & R. S. Lowery), pp. 52–82. Croom Helm, London.

Holdich, D. M. (1993). A review of astaciculture: freshwater crayfish farming. *Aquatic Living Resources*, **6**, 307–17.

Holdich, D. M. (2002). *Biology of Freshwater Crayfish*. Blackwell Science, Oxford.

Huner, J. (1992a). Overview of international and domestic freshwater crawfish production. *Journal of Shellfish Research*, **8**, 259–65.

Huner, J. (1992b). Chinese crawfish and the Louisiana crawfish industry. *Aquaculture Magazine*, **18**, 6–13.

Huner, J. V. (1994). *Freshwater Crayfish Aquaculture in North America, Europe, and Australia: Families Astacidae, Cambaridae, and Parastacidae*. Food Products Press, New York.

Huner, J. V. (2002). *Procambarus*. In: *Biology of Freshwater Crayfish* (Ed. by D. M. Holdich), pp. 541–84. Blackwell Science, Oxford.

Huner, J. V. & Barr, J. E. (1984). *Red Swamp Crawfish: Biology and Exploitation*. Louisiana Sea Grant College Program, Louisiana State University, LA.

Huner, J. V. & Barr, J. E. (1991). *Red Swamp Crawfish: Biology and Exploitation*, Louisiana Sea Grant College Program, Baton Rouge, LA.

Jones, C. M. (1990). *The Biology and Aquaculture Potential of the Tropical Freshwater Crayfish, Cherax quadricarinatus*. Queensland Department of Primary Industries, Brisbane, Australia.

Jones, C. M. (2001). Australian freshwater crayfish aquaculture. In: *Queensland Crayfish Farmers Association Annual Conference* (Ed. by P. Long), Rockhampton.

Jones, C. M. & Grady, J. (2000). *Redclaw from Harvest to Market: a Manual of Handling Procedures*. Queensland Department of Primary Industries, Brisbane, Australia.

Jones, C. M. & Ruscoe, I. M. (2000). Assessment of stocking size and density in the production of redclaw crayfish, *Cherax quadricarinatus* (von Martens) (Decapoda: Parastacidae), cultured under earthen pond conditions. *Aquaculture*, **189**, 63–71.

Jones, C. M., McPhee, C. P. & Ruscoe, I. M. (2000). A review of genetic improvement in growth rate in redclaw crayfish *Cherax quadricarinatus* (von Martens) (Decapoda: Parastacidae). *Aquaculture Research*, **31**, 61–7.

Lawrence, W., de la Bretonne, J. L. & Romaire, R. P. (1989) Commercial crawfish cultivation practices: a review. *Journal of Shellfish Research*, **8**, 267–75.

Lee, D. O. C. & Wickins, J. F. (1992). *Crustacean Farming*. Blackwell Scientific Publications, Oxford.

Mills, B. J. (1989). *Australian Freshwater Crayfish. Handbook of Aquaculture*. Freshwater-Crayfish Aquaculture Research & Management, Lymington, Australia.

Morrissy, N. M., Evans, L. E. & Huner, J. V. (1990) Australian freshwater crayfish: aquaculture species. *World Aquaculture*, **21**, 113–22.

New M. B. (2000). History and global status of freshwater prawn farming. In: *Freshwater Prawn Culture. The Farming of Macrobrachium rosenbergii* (Ed. by M. B. New & W. C. Valenti), pp. 1–11. Blackwell Science, Oxford.

New, M. B. & Valenti, W. C. (2000). *Freshwater Prawn Culture. The Farming of Macrobrachium rosenbergii*, p. 443. Blackwell Science, Oxford.

Roustaian, P., Kamarudin, M. S., Omar, H. B, Saad, C. R. & Ahmed, M. H. (2001). Biochemical changes in freshwater prawn *Macrobrachium rosenbergii* during larval development. *Journal of the World Aquaculture Society*, **32**, 53–9.

Scholtz, G. (2002). Phylogeny and evolution. In: *Biology of Freshwater Crayfish* (Ed. by D. M. Holdich), pp. 30–52. Blackwell Science, Oxford.

Semple, G. P., Rouse, D. B. & McLain, K. R. (1995). *Cherax destructor, C. tenuimanus* and *C. quadricarinatus* (Decapoda: Parastacidae). A comparative view of biological traits relating to aquaculture potential. *Freshwater Crayfish*, **8**, 495–503.

Unestam, T. (1975). Defence reactions in and susceptibility of Australian and New Guinea freshwater crayfish to European-crayfish-plague fungus. *Australian Journal of Experimental Biological Medical Science*, **53**, 349–59.

Valenti, W. C. & Daniels W. H. (2000). Recirculation hatchery systems and management. In: *Freshwater Prawn Culture. The Farming of Macrobrachium rosenbergii* (Ed. by M. B. New & W. C. Valenti), pp. 69–90. Blackwell Science, Oxford.

Valenti, W. C. & New M. B. (2000). Grow-out systems – monoculture. In: *Freshwater Prawn Culture. The Farming of Macrobrachium rosenbergii* (Ed. by M. B. New & W. C. Valenti), pp. 157–76. Blackwell Science, Oxford.

Vogt, G. (2002). Functional anatomy. In: *Biology of Freshwater Crayfish* (Ed. by D. M. Holdich), pp. 53–151. Blackwell Science, Oxford.

Xingyong, W. (1995). Report on red crayfish (*Procambarus clarkii*) from Nanjing, Jiangsu Province People's Republic of China. *Freshwater Crayfish*, **8**, 145–7.

21
Bivalves

John Lucas

21.1 Introduction

Like the wounded oyster, he mends his shell with pearl.

Ralph W. Emerson (1841)

Culture of table oysters has a history going back to Roman times. Mussel culture in Europe began more than 700 years ago (Jeffs *et al.*, 1999). It was relatively easy to culture mussels with many shore-inhabiting species, such as table oysters and mussels, and to make the transition from harvesting wild stocks to farming them. Substrates, such as rocks and stakes, were added to promote larval settlement in areas of natural settlement. Then, where someone or some group had rights to harvest these stocks at consumable size, this constituted a simple form of aquaculture.

Bivalves are good aquaculture candidates because they can be reared using simple technology and because they are filter feeders. The simple technology and low labour inputs in the rearing process offset the low value of the product in many cases. Furthermore, as bivalves obtain their food by filter feeding phytoplankton and particulate organic matter (POM) from the water around them, there are no costs for feeds in the grow-out phase of culture. The different modes of life and range of commercial values of various bivalves, however, require different culture methods. The techniques for individually handling pearl oysters, a high-value group, are totally inappropriate for low-value species such as mussels, which must be cultured in huge numbers for an economically viable operation.

The production of molluscs was 10.1 million mt/year in 1999; valued at US$9 billion (FAO, 2001). This ranks them second in quantity of world aquaculture production and second in value (marginally greater than crustaceans) (Chapter 1 and Fig. 1.2). Their proportion of total world aquaculture production increased slightly from 21% to 24% over the decade 1990–99. Bivalves are by far the greatest source of mollusc production, with culture industries based on gastropods being largely in their infancy (Chapter 22).

21.2 Aspects of biology

21.2.1 Morphology

The name bivalve arises from these animals' most prominent and dominant feature: a shell consisting of two valves that are joined along a hinge and enclose the body. Each valve usually has a prominence near the hinge, the umbo, which is the oldest part of the valve. There are often concentric lines on the shell reflecting previous valve sizes. Each valve consists of three basic layers made up of calcareous and proteinaceous material. The most important layer for aquaculture is the inner nacreous layer, which is often lustrous and very hard (the 'pearly' layer of pearl oysters and pearl mussels).

Bivalves have no identifiable head. The body is laterally compressed within the valves and mainly consists of a soft visceral mass: gut, kidney, gonads, blood vessels, etc. There is a ventral foot, which may be insignificant in sedentary species or very well developed in burrowing species. Associated with the foot there may be a gland that secretes attachment threads (byssal threads). Some sedentary bivalves, such as mussels, are attached to the substrate by byssal threads.

Extending across between the valves are one or two large muscles, anterior and posterior adductor muscles, which close the valves. Lining the valves and enclosing the visceral mass are two thin lobes, the mantle lobes. These reach to the edges of the valves and secrete the material for shell growth and thickening. The cavity enclosed by the mantle lobes is called the mantle cavity and it contains the gills on each side of the body. The gills consist of fused parallel filaments.

This is a brief generalised description of bivalve morphology. Among the large group of bivalves there are impressive variations on this basic form (see, for example, Ruppert & Barnes, 1994).

21.2.2 Filter feeding

Bivalves obtain their nutrition by filter feeding on fine (2+ μm) particles, including phytoplankton and POM. They may also obtain some nutrition by direct uptake of dissolved organic matter (DOM) from seawater across the large permeable surfaces of their gills and mantle chamber. A water current is created through the mantle cavity by cilia on the gills. The gills' fine structure is basket-like and particles are trapped by cilia lining the microscopic apertures. These particles are bound in mucus and directed anteriorly to the mouth in food channels. Palps at the mouth sort the particles, rejecting large particles, inorganic particles and excessive numbers of particles as mucus-bound 'pseudofaeces'. Ingested food particles are digested within the gut and undigested material is compacted into faeces and expelled. The volume of water 'pumped' through the gills per unit of time may be impressive. Pearl oysters are among the highest, with values for large individuals being > 1000 L/day (Yukihira *et al.*, 1999). The ingestion rate is described by:

$$IR = [(FR \times PR) \times FC] - Ps$$

where IR is the ingestion rate of food, e.g. in mg/h, FR is the flow rate of water through the gills, e.g. in L/h, PR is the proportion of food in the water retained by the gills, e.g. 0.9, FC is the food concentration in the water, e.g. in mg/L, and Ps is the rate of pseudofaeces production, e.g. in mg/h.

There is often an optimum range of particle concentrations that maximises food intake while minimising food wastage. When feeding bivalves with unicellular algae cultures, e.g. in conditioning bivalve broodstock or rearing larvae and juveniles, it is not sufficient to load the water with as high a density of algae as possible to maximise food intake and growth. This is wasteful and may be detrimental to growth due to clogging of the feeding processes. The optimum cell concentration varies with algal species. This relates to the size, other physical characteristics, acceptability and age of the algal culture (section 9.3.3). The optimum cell concentration usually increases through the life cycle in bivalves. They may range from ~10 000 cells/mL for early larval stages, to ~50 000 cells/mL for late larval stages, to ~100 000 cells/mL for juveniles and 100 000+ cells/mL for adults.

21.2.3 Growth

Scope for growth (SFG) is used for physiological assessments of bivalves. It is a measurement of the surplus energy from feeding that is available for growth and reproduction. SFG results from the ingested energy (IR) that is assimilated into the body, less the energy costs of metabolism and excretion (section 6.4.3). Growth in bivalves is such that the size versus age relationship follows the pattern described in Chapter 6 (section 6.4.1).

Phytoplankton levels strongly influence growth rates and the final size (L_∞) to which bivalves grow. Low phytoplankton levels result in suboptimal food intake, slow growth, stunted size and reduced reproductive output.

21.2.4 Anaerobic metabolism

Anaerobic metabolism is a particular aspect of bivalve physiology. Bivalve morphology gives a level of protection against many adverse conditions: the valves may be held tightly closed for long periods as protection against adverse conditions such as desiccation, salinity variations, toxins and predators seeking entry into the mantle cavity. This behaviour, however, limits oxygen availability, and extended periods of valve closure are accompanied by reduced metabolic rate and anaerobic metabolism, which is well tolerated. It is best developed in intertidal and substrate-inhabiting species, such as oysters, mussels, clams and cockles. A practical consequence of this behaviour and biochemistry is that it is better to hold most bivalves in damp conditions out of water during transportation than to keep them crowded together in polluted conditions in water.

21.2.5 Reproduction

Some bivalve species are dioecious, some are hermaphrodites, and some change sex. Table oysters, pearl oysters, mussels and clams are usually dioecious, whereas scallops are hermaphrodites. Giant clams are an example of protandric hermaphroditism (initially male, then subsequently with ovary and testis tissue mixed together) (Lucas, 1994). Sexuality is not strongly fixed in some bivalves and individuals may change sex during a breeding season or between seasons, depending on prevailing conditions or as an age-related phenomenon. Pearl oysters are an example of variable sexuality.

Most temperate bivalve species and many tropical species have a distinct breeding season that is primarily related to water temperature. Typically, gametogenesis occurs mainly during the late winter to early summer months, with progressive accumulation of mature gametes in the gonad. Then spawning occurs, with numerous individuals spawning synchronously through being triggered by pheromones released with gametes by nearby individuals (section 6.3.2).

Although histological studies may be used in research on bivalve reproductive seasonality, such techniques are usually not available to the industry and simpler methods are used to assess sexual maturity. Condition index (CI) (Quayle & Newkirk, 1989) is a quantitative measure used in commercial aquaculture.

CI = [wet weight of meat (g)/volume of shell cavity (mL)] × 100

or

CI = [dry weight of meat (g)/volume of shell cavity (mL)] × 1000

Volume (mL) of shell cavity is determined from:

[whole wet weight* (g) – shell weight (g)]/1.025

*When there is no air in the mantle (shell) cavity.

CI relates the amount of body tissue to the volume of the shell cavity. For the dry weight CI, a value of 70 for a Pacific oyster indicates a 'thin' oyster in poor condition, whereas a value of 150 indicates a 'fat' oyster (Quayle & Newkirk, 1989). General CI levels can reflect the state of nutrition or health of the bivalve, but seasonal changes reflect the gametogenic cycle, because the mature gonad becomes a major component of the total visceral mass.

Even simpler than CI measurements, viewing the size and colour of the gonads through open valves (shell valves may have to be wedged open) may be sufficient to identify their degree of development, and sex, once one is familiar with the morphology.

21.2.6 Life cycles

Most bivalve life cycles consist of planktonic trochophore, veliger and pediveliger larval stages, and then metamorphosis to a juvenile, called a spat (Fig. 6.2). Some variations from this pattern occur in flat oysters and freshwater mussels, in which the early larval stages are retained within the mantle cavity of the female and released late in development. These species are exceptions to the normal pattern of eggs being shed into the water to be fertilised by concurrently released sperm.

Bivalve eggs are usually quite small, 40–100 μm in diameter, and very numerous. Fecundity ranges from about one million eggs per spawning, e.g. flat oysters, to tens of millions, e.g. cupped oysters, to hundreds of millions per spawning in the largest giant clam (*Tridacna gigas*). (In fact, this giant clam species may be the most fecund animal in existence with tens of billions of eggs released through its decades-long lifespan.) Bivalve sperm are also very numerous. They have a small head, a few microns in diameter, and long flagellum. The sperm swim vigorously when in good condition.

The mature sperm is haploid (n) but the mature egg is tetraploid ($4n$) until a sperm penetrates its outer membrane. The fertilised egg then undergoes two meiotic divisions. These are evident by the successive ejection from the egg of two small polar bodies, each containing the discarded half of the chromosomes from each meiotic division. The haploid egg nucleus is then ready to merge with the sperm nucleus to form a diploid embryo.

The embryo then commences mitotic divisions involving the whole cytoplasm and passing through two-cell, four-cell, eight-cell stages, etc. It develops into a ciliated trochophore larva, usually within 24 h at temperatures normal for the species. The trochophore rapidly develops into a bivalved veliger, which is also known as a D-stage because of its shape. The veliger swims and feeds on phytoplankton with a ciliated lobe, the velum. Towards the end of its planktonic life, which typically lasts 2–4 weeks, the veliger develops a prominent foot to become a pediveliger. The pediveliger settles on an appropriate substrate and metamorphoses into a spat. Metamorphosis involves absorption of the velum, development of gills and a gill-based feeding mechanism, and attachment to the substrate with byssal threads or other means. During this transformation from velum to gill feeding, they undergo a period when they are unable to feed. They must rely on the energy reserves built up in the later part of larval development.

There follows a period of growth without abrupt changes in morphology (Fig. 6.2). Sexual maturity is usually reached in about 2 years or more. Thereafter follows a long period of sexual maturity. Bivalves are quite long-lived and some are very long-lived, e.g. several temperate clam species have been shown to live for more than 100 years. Age can be measured from annual growth lines in cross-sections through a valve, like growth rings in a tree stump.

21.3 Cultured bivalves

There are five main 'groups' of cultured bivalves:

- table oysters
- marine mussels
- scallops
- clams and cockles
- pearl-producing oysters and freshwater mussels.

The annual production of some important and other bivalve species is shown in Table 21.1. These values, however, should be taken as indicative rather than definitive. They vary with different data sources, partly because there is difficulty in differentiating between aquaculture production of bivalves and fisheries. There are also no data on the value of cultured pearls, although this is a significant aspect of bivalve aquaculture in terms of commercial value.

It is notable that, after the Pacific oyster, which is clearly the bivalve of greatest importance and widest geographic consumption, the next most important species is the Manila clam (or Japanese carpet shell, *Ruditapes phippinarum*). This species is almost entirely cultured in China. In fact, China contributes at least three-quarters of world aquaculture production for each of the top three species, Pacific oysters, Manila clams and yesso scallops, and all of the fourth category, miscellaneous marine mussels, in Table 21.1. China's contribution to world bivalve productions, as for other groups, is massive. The third major bivalve, a scallop, *Pecten yessoensis*, is rather surprising. Scallops have not been regarded as easy material for aquaculture. The yesso scallop is, however, highly valued and produced in huge quantities in China and Japan. Another notable feature of Table 21.1 is the huge quantity of marine mussels produced in China for a small return. This is because much of the marine mussel crop is used for feed in shrimp culture. By contrast, far fewer Mediterranean mussels are produced for a greater return, and New Zealand mussels have been developed as a specialist product and exported to more than 50 countries (Jeffs *et al.*, 1999).

The relative importance of the five groups in quantity and value of aquaculture production is shown in Fig. 21.1. Oysters make up 40% of the quantity and value of bivalve production. They are the reference

Table 21.1 World aquaculture production of some bivalve molluscs in 1999 (FAO, 2001)

Species	Quantity (1000 mt)	Value (million US$)
Pacific oyster	3600	3313
Manila clam	1820	2195
Yesso scallop	929	1252
Marine mussels (China)	008	< 92
Blue mussel	498	272
Razor clams	479	383
Blood cockle	316	277
Mediterranean mussel	162	100
New Zealand mussel	71	21
Green mussel	69	8
American cupped oyster	58	39

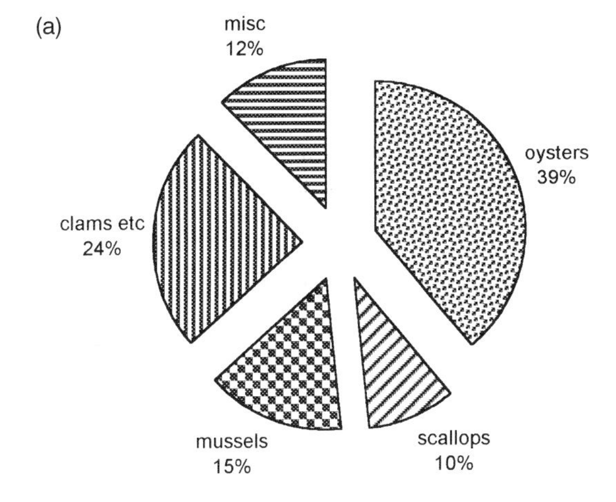

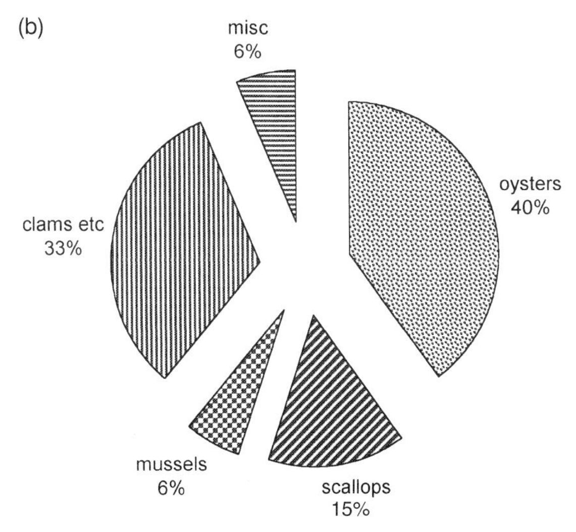

Fig. 21.1 Relative aquaculture production of the four major groups of bivalves in 1999. (a) Quantity. (b) Value (FAO, 2001).

in terms of value. On this basis, clams and scallops are more valuable, with about 50% greater value per unit quantity compared with oysters. On the other hand, mussels are far less valuable than oysters, being about 40% of their value.

21.3.1 Table oysters (family Ostreidae)

In oysters the left valve (called the lower valve) is cemented to a firm substrate. Oysters are divided into cupped oysters (e.g. *Crassostrea* and *Saccostrea* species) (also known as rock oysters) and flat oysters (*Ostrea* species). As well as shell shape, there are other differences, including mode of development. Cupped oysters are oviparous, spawning gametes into the water column, whereas flat oysters are larviparous, brooding embryos in their mantle cavity and releasing larvae.

21.3.2 Mussels (family Mytilidae)

There are two unrelated groups of mussels that are cultured. One group, family Mytilidae, is marine and cultured for human consumption. The other group, family Unionidae, is freshwater and cultured for pearl production. Both families have wide geographic distributions. As these groups share little in common, except that they both happen to be called 'mussels', the unionids will be treated with pearl oysters, for which they at least have a similar reason for culture.

Commercial mytilid species are mainly from two genera of large mussels:

- *Mytilus* species, which are temperate
- *Perna* species, which are mainly tropical.

Mussels have similar valves. They are not motile and attach to a firm substrate by strong byssal threads. Their foot becomes vestigial soon after settlement. As in oysters, the mouth is near the umbo and there is essentially one adductor muscle, the posterior one. Mytilid mussels have gonads located in their mantle tissue as well as in the visceral mass.

21.3.3 Scallops (family Pectinidae)

These include species of *Pecten* and other genera, such as *Chlamys* and *Aequipecten*. Most scallops are free-living on soft substrates. They respond to disturbance with swimming movements caused by rapid 'flapping' of their valves, expelling water jets. This is caused by vigorous contractions of a large single adductor muscle. Scallops tend to be circular in outline with a long prominent hinge, similar to pearl oysters.

21.3.4 Clams and cockles

The terms 'clam' and 'cockle' have no taxonomic significance, and bivalves from a range of families are included in this assemblage. They include:

- the blood cockles (*Anadara* species)
- the Manila clam or Japanese carpet shell (*Ruditapes philippinarum*)
- the North American hard clam or quahog (*Mercenaria mercenaria*) (of clam chowder fame).

Their common feature is that they inhabit particulate substrates, ranging from thick mud to sand. They have a well-developed foot for burrowing through the substrate. They have two adductor muscles. Some variation occurs in the incurrent and excurrent channels; but, because they are buried in the substrate, their mantle lobes are usually fused to create tubular incurrent and excurrent siphons that connect the mantle cavity to the surface seawater.

Giant clams (Tridacnidae) are an exception to the usual 'clams'. They are a small group of coral reef-inhabiting species that are exceptional in size and in obtaining nutrition by filter feeding and by photosynthesis of symbiotic algae (Lucas, 1994). Species of giant clam have been fished to extinction in many parts of their ranges and markets are being developed for cultured clams.

21.3.5 Pearl oysters and pearl mussels (families Pteriidae and Unionidae)

Pearl oysters and pearl mussels are unrelated and, apart from common bivalve morphology, they have little in common except for their use in cultured pearl production. Pearl oysters are marine species; they have a vestigial foot after settlement and, at least initially, attach to a firm substrate with byssal threads. They are mainly from the genus *Pinctada* (Gervis & Sims, 1992). The inner shell layer, the nacreous layer, is very lustrous and this is the basis of pearl production.

In contrast to pearl oysters, unionid mussels live in freshwater. They have no byssal threads and burrow in particulate substrates with a well-developed foot. They share with pearl oysters a nacreous layer that is very lustrous, and this again is the basis of pearl production. The life cycle of these mussels involves suppressed larval development, a feature in common with some other freshwater invertebrates (section 6.3.4). Embryonic development occurs in eggs retained in the gills. Subsequently, a well-developed and bivalved glochidium larva hatches from the egg and attaches to the gills, fins or surface of a fish host by an adhesive thread and possibly hooks on its valves. The glochidium larva feeds from the fish host as a parasite. The parasite gradually changes into a juvenile bivalve, detaches from the host, falls to the substrate and assumes the adult burrowing habit. Rearing these mussels through the glochidium stage without a fish host as a source of nutrients has proved to be very difficult.

21.4 Phases of bivalve aquaculture

The complete process of bivalve aquaculture involves the series of phases outlined in section 6.3 (Fig. 6.2). The period for this complete process, hatchery production to marketable size, is at least 18 months for most commercial bivalves. Much bivalve aquaculture, however, does not involve this complete process: the complete process is only used in culturing the higher-value species. For many cultured bivalves the process commences after natural spatfall, and the costly and technically demanding hatchery and nursery culture phases are omitted.

21.4.1 Culture from natural spatfall

As indicated in the introduction to this chapter, a very simple form of aquaculture for some bivalve species is to provide additional substrates in regions where natural spatfall occurs, i.e. in regions of the environment where larvae are completing development and seeking appropriate substrates on which to metamorphose. These added substrates may be natural surfaces such as rocks and wooden stakes that remain in place.

A further step in technical development is to provide spatfall substrates that can subsequently be moved to other regions or which will allow some thinning or sorting of the bivalves as they grow. These are known as spat collectors or, particularly in relation to oysters, as culch. They take a variety of forms depending on local industry practices and on the species' biology and its value (Figs 21.2 and 21.3). The amount of microalgae and fouling by larger organisms can have a significant effect on the amount of settlement. Settlement is also strongly related to the density of larvae in the water flow over the spat collectors. It is essential to deploy spat collectors at the appropriate time of late larval development and in areas where the highest concentrations of larvae are known to occur.

Oyster larvae tend to prefer a hard rough surface that is non-greasy and clear of silt and algae. Culch used for them includes oyster shells, cement tiles, lime- or tar-coated wooden sticks, mangrove tree branches and bamboo. Quayle & Newkirk (1989) emphasised that whatever material is used it should be low cost and locally available. There is also another feature of the culch material that is very important to the harvesting and market acceptability of the cupped oyster species. The oyster's lower valve assumes the shape of the surface on which it is growing. Thus, cupped oysters growing on a broad flat surface without near neighbours will be broad and flat. The desirable cupped form comes from growing on a narrow substrate and competing for surface space with neighbouring oysters.

Fig. 21.2 Groups of battens in Salamander Bay, New South Wales, to collect Sydney rock oyster spat (photograph by Dr John Holliday).

Fig. 21.3 Spat collectors consisting of mesh bags filled with fishing line or similar filamentous material. This kind of collector may be used for scallop and pearl oyster spat, which tend to prefer protected soft substrates (photograph by Mr John Mercer).

Culchless oysters, i.e. oysters that have been detached from or outgrown the original settlement substrate, also grow into the desirable cupped form. This is the main advantage, but they are also very simple to harvest. Producing culchless oysters in hatcheries is readily achieved (see p. 452). Methods to produce culchless spat from natural settlement include culch of plastic strips, which are flexed to dislodge the juvenile oysters.

Marine mussels are an example of bivalves that

are grown on fibrous substrates. Commonly used spat collectors are frayed ropes, which provide an attractive surface for the larvae, which attach to them in very high densities. The natural densities of spat on the settlement substrates are generally too high, which can jeopardise growth and survival. Furthermore, they may not be in the region for best subsequent growth. Therefore, when the juveniles grow to 20+ mm shell length, they are generally stripped from the settlement substrate, tearing their byssal thread attachments to this substrate. They are reattached to vertical ropes at controlled densities for further growth. This reattachment is achieved by passing the settlement ropes steadily into the water from the surface. Each rope is surrounded by a sleeve of fine thin mesh and mussels are fed in between the mesh and rope. The mesh and mussels are tied periodically into 'sausages' and pass over a roller into the sea. The mesh holds the mussels in place until they have attached to the rope by their byssal threads (a matter of hours rather than days) and the mesh subsequently disintegrates.

21.4.2 Culture from hatchery production

Spawning induction

Often, little more than stress is required to induce spawning of ripe bivalve broodstock. The stress may be exposure to elevated temperature for a tolerable period or to air for several hours. Addition of some chemicals to seawater containing the broodstock may induce spawning, e.g. hydrogen peroxide, potassium chloride and ammonium hydroxide. Ultraviolet irradiation of the seawater, which produces free oxygen radicals, may also stimulate spawning. The precise action of each of these chemical treatments is not clear.

Serotonin, a neurotransmitter substance, injected directly into the gonad (e.g. 0.2–0.5 mL of 10^{-3} molar solution) is effective in inducing spawning, both sperm and ova release, in a variety of bivalves (Citter, 1985; Crawford *et al.*, 1986). The action of serotonin is presumably to by-pass all other hormonal and central nervous system mechanisms associated with spawning and act directly on the gonad muscles, causing them to contract vigorously and expel gametes.

When one individual sheds eggs or sperm (usually sperm; males tend to be more easily triggered to spawn) (Fig. 21.4), the gametes are stirred into the vicinity of other individuals to stimulate them with the gamete pheromones.

Spawning induction may be by-passed completely with gonad stripping. When the broodstock are known to contain fully mature gametes, they may be killed, their gonads repeatedly lacerated with a sharp blade, and the gametes washed out into a beaker. This is only appropriate to dioecious bivalves as there would be self-fertilisation (and reduced genetic variability) from hermaphrodite gonads using this technique.

Fertilisation

Where the broodstock can be sexed, the sexes are often placed in separate containers for spawning induction. Alternatively, the animals are removed from a group as they commence to spawn and the type of gametes, eggs or sperm, is evident. The objective is to have dense suspensions of eggs in filtered seawater, up to 1000 or 2000 per mL, and very dense suspensions of sperm in separate containers. The eggs may be kept in suspension with gentle aeration and their numbers estimated from microscope counts of small subsamples in a Sedgwick-rafter

Fig. 21.4 A hard clam (*Mercenaria mercenaria*) releasing gametes in response to elevated water temperature. Spawning induction in bivalves is often conducted in containers with a black background to help identify spawning individuals (photograph by Dr Paul Southgate).

counting chamber. A few millilitres of sperm suspension, mixed from several males, is then added to the suspensions of eggs.

There are constraints to the amount of sperm to be added: too little results in low levels of fertilisation, too many sperm may result in embryonic abnormalities through polyspermy (section 6.3.3).

Triploidy

Triploidy is induced in bivalves by preventing the release of the second polar body resulting from the second meiosis (section 7.6.1). Triploidy has been particularly studied in relation to oysters, and methods for inducing triploidy are well known (Scarpa *et al.*, 1994). Triploid oysters differ from diploid oysters in having larger cell nuclei and cells to accommodate the additional set of chromosomes; but, more importantly, they show poor gonad development. For markets that prefer oysters without ripe gonads, e.g. the USA market, this is advantageous for product quality. Furthermore, with less energy diverted to gametogenesis, more energy should be available for other tissue growth. This has not been found in all industries using triploid oysters, but at least some triploid oysters show improved growth over diploids. Triploids of the commercial Sydney rock oyster (*Saccostrea* species) have at least three advantages over diploids (Kesarcodi-Watson *et al.*, 2001):

- They grow faster.
- They have higher total energy content at the same size.
- They have a higher soft tissue to shell energy ratio.

Triploid Sydney rock oysters have been reported to be 41% heavier than diploid siblings after 2–5 years of growth and to reach market size 6–18 months faster than diploids (Nell *et al.*, 1994). This study also noted that triploids maintained better meat condition and were just as resistant to disease as diploids.

Larval rearing

Modern bivalve hatcheries tend to use the following techniques:

(1) Seawater is generally filtered to around 1 μm to remove silt, organic debris, zooplankters, which may be predators or competitors, and phytoplankton that the larvae may not be able to feed on. This requires a comprehensive filtration system in the seawater intake.

(2) Water temperature is controlled. Temperature fluctuations reduce survival and temperature affects development rate.

(3) Larvae are kept at low densities, 1–5 larvae/mL, in large culture volumes, often 10–20 000 L (Fig. 21.5). The culture tanks are typically stocked with embryos at a level that is several times greater than the anticipated final density of larvae. Density is reduced during larval development by:
    - mortality
    - discarding the slower-growing individuals during the regular sieving programme.

    With a final density of about one larva per mL, a 20 000-L tank has an output of 20 million pediveligers.

(4) Mass-cultured unicellular algae are added as food for the larvae. Usually a mixture of species is supplied, because mixed-species diets give a better balance of nutrients (section 9.3.3). Small unicellular algae, about 5 μm, are used, such as *Isochrysis* species, *Pavlova* species and single-celled diatoms, e.g. *Chaetoceros* species. Long-chain, spiny diatoms, which are appropriate to the setal feeding structures of crustacean larvae, are quite unsuitable for the ciliary feeding mechanisms of bivalve larvae.

(5) Gentle aeration is used to maintain water movement, keeping the larvae in suspension.

(6) Culture water is changed regularly, usually every second day, to control the build-up of metabolites from larvae and algae, and especially the build-up of bacterial populations. The culture water and larvae are discharged through sieves of appropriate mesh size to collect the larvae (Fig. 21.5). The water may be discharged through several sieves of different mesh sizes, with coarsest mesh uppermost, to grade the larvae. Abnormally small larvae pass through the sieves and are lost in the effluent water. The larvae retained on different mesh sizes are transferred to new culture tanks with the objective of having uniform batches of larvae.

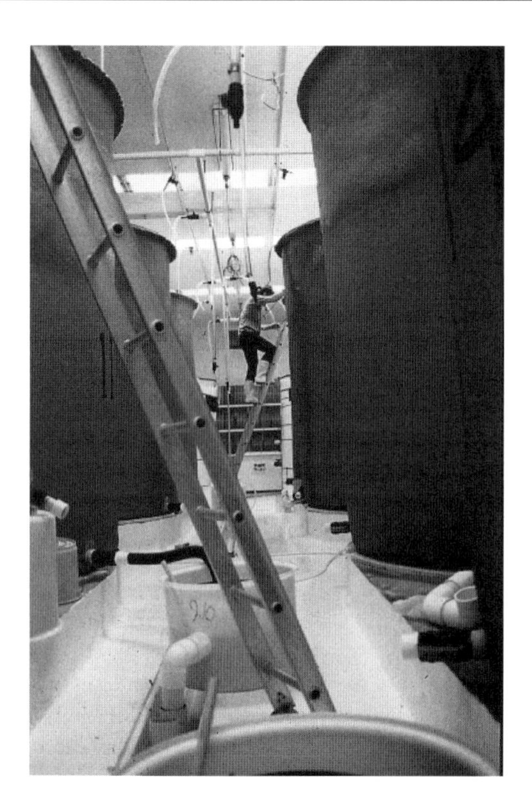

Fig. 21.5 Large tanks (20 000 L) for rearing bivalve larvae. The tank in the left foreground is being drained via a black pipe into sieves to collect bivalve larvae (Shellfish Culture Pty Ltd, Tasmania, Australia).

(7) Scrupulous attention is paid to cleanliness to avoid contamination and bacterial blooms. There is generally a separate larval culture or tank room with limited access. The inner surfaces of the culture tanks are scrubbed with hypochlorite solution to sterilise them, or with detergent, between uses. With a culture regime of water changes every second day, each tank is cleaned after it is emptied and then refilled with seawater to be back in use a day later. Sieves, hoses, buckets, etc., are all washed in hypochlorite solution after use.

Settlement

Late pediveliger larvae may be sieved from their culture water on an appropriate-sized mesh and transferred to settlement tanks with static water and settlement-inducing substrates. Alternatively, appropriate substrates may be added to the larval tanks. Oysters, scallops, clams and pearl oysters have their individual requirements for settlement-inducing substrates in the hatchery, as they have in the field. Chemicals may be added to the water in hatcheries to promote settlement. Adrenaline (epinephrine) at 100 μmol/L and GABA (gamma-aminobutyric acid) may be used.

Remote settling

In the final stages before settlement, oyster pediveligers are amazingly robust and can be collected into dense balls of millions of larvae, wrapped in damp material and stored in a refrigerator or freighted in small containers. In this way, pediveligers of the American cupped oyster (*Crassostrea virginica*) are sent from hatcheries to distant oyster farms in the USA ('remote settling'). The pediveligers are settled at the recipient farms.

Culchless spat

A technique used in settling cupped oyster larvae is to provide particles of shell, about 500 μm in diameter, as substrates. The pediveligers metamorphose on these but, within a few days, the spat shells have grown larger than the particles on which they settled. The shell particles with and without spat are then passed through an appropriate sieve that separates the spat from the plain particles. Thus, culchless spat are produced.

21.4.3 Nursery culture

Early bivalve spat are 200–500 μm in shell length and very vulnerable to predation and overgrowth by fouling organisms. They need good concentrations of phytoplankton to sustain their growth. Thus, best survival and growth may be obtained by maintaining them in on-shore facilities with cultured algae as food. This is not usually an option with juveniles resulting from natural spatfall.

The spat may be maintained in a flow-through system with high phytoplankton content or within a recirculating system with supplementary microalgae. In the latter case, the spat may be held on sieves at high densities with either vigorous up-welling or down-welling flows through the sieves. Up-welling has the advantage of lifting the spat off the mesh and stirring them, whereas down-welling packs them down onto the mesh. Stronger flows, however, can be used with down-welling systems. The mass-cultured microalgae added to the recirculation system do not need to be as 'clean' as those used for larval culture and may be cultured in open tanks (section 9.3.1).

When the spat are of sufficient size, they are transferred to appropriate sites for ocean-nursery culture. There is pressure to make this transition as it is costly to maintain the spat in an on-shore facility with feeding and maintenance.

21.4.4 Ocean nursery and grow-out

Various methods are used for culturing bivalves during their ocean phase. These reflect differences in biology and habitat of the cultured species; and also local factors in the farming process, such as labour and material costs, and market price. The basic difference is between culturing bivalves:

- within the substrate
- on or just above the bottom
- near the ocean surface.

Within particulate substrates

The substrate-inhabiting cockles, clams, etc., are reared within their natural muddy to sandy substrates with minimal intervention. In some cases, the substrate may be prepared for addition of seed bivalves by ponding and fertilisation or by harrowing it to loosen up the substrate and remove predators such as starfish. After seeding, the surface of the culture area may be covered with a mesh to deter predators and the area may be defined with a fence.

On or just above the bottom

Some bivalves are cultured on the surfaces of hard substrates, such as cement slabs and horizontal and vertical stakes, which require natural spatfall. Rows of horizontal or vertical stakes (usually wood or bamboo) are used for oyster and mussel culture. These bivalves tolerate intertidal exposure, and stake culture is particularly used for farming in the intertidal zone on sheltered muddy coasts with a good tidal range. There are considerable advantages, in that installation, maintenance and harvesting can be carried out at low tide.

Bivalves are also cultured on racks above the bottom in mesh boxes, mesh baskets, trays and horizontal wooden and asbestos–cement battens. Mesh boxes are used for seed bivalves, particularly culchless oysters, immediately after transfer from the land nursery. Later, when the oysters are less prone to predation, they can be transferred to open mesh baskets or trays. Horizontal battens are typically used where the battens were previously used to collect spat from natural spatfall (Fig. 21.2).

Scallops may be cultured on natural substrates in mesh enclosures. The enclosures may be complete boxes or vertical walls. Where only mesh walls are used they must be sufficiently high to exceed the vertical swimming ability of the scallops. Scallops are intolerant of intertidal conditions and are cultured on appropriate substrates subtidally, typically ca. 10 m in depth.

Bivalves are sometimes cultured attached to ropes that hang down from horizontal racks above the substrate. This is typically in the intertidal zone for mussels and oysters.

Surface or suspended culture

Culture methods at the surface include hanging the bivalves on ropes or in appropriate culture units from rafts, long-lines and floats.

(1) Rafts are rectangular metal, wooden or bamboo frames with buoyancy provided by large air-filled drums or floats. Ropes with bivalves attached hang down at intervals from cross-members on the raft (Fig. 2.5). In raft culture, the bivalves are typically mussels attached to vertical ropes in vast numbers (Fig. 21.6).
(2) Long-lines are 50+ m horizontal ropes, supported at the surface by floats at regular intervals (Fig. 4.8). Bivalves are cultured on ropes suspended vertically at regular intervals along the long-line. Attached to the vertical ropes are various culture structures depending on the species, such as:
   - 'ear-hung' pearl oysters (Fig. 21.7) and scallops

Fig. 21.6 Mussels (*Mytilus edulis*) cultured on ropes in western Scotland.

— scallops in lantern nets (elongate cylindrical nets that are suspended vertically and which have regular horizontal partitions on which the scallops rest)
— ropes packed with mussels
— rectangular frames with net pockets for pearl oysters
— vertical series of pyramid-shaped nets (Fig. 21.8).

Long-lines are held in place at their ends by terminal anchors or attachment to points of hard substrate below. Some long-lines have their supporting floats held below the surface by anchors, so that they may be invisible at the surface or only indicated by marker buoys. This system has the advantages of being resistant to storm damage, having less chance of damage from boating, and being less conspicuous to poachers and more aesthetically acceptable.

(3) Floats are single versions of the multi-float long-line system. The buoyancy floats are much larger and directly support the culture structure below. They are used in similar ways to long-lines in that they are often joined in lines.

21.5 Culture problems

21.5.1 Predators, parasites and diseases

Predators

Bivalves are prone to predation, especially during the early months of development while their shells are fragile. Major predators include gastropod molluscs, starfish, crabs and fish.

As an example of the range of predators, Gibbons & Blogoslawski (1989) listed 34 invertebrate species and 21+ species of vertebrate predators of the hard clam (*Mercenaria mercenaria*) (Table 21.2). It must be noted that the hard clam is a soft substrate inhabitant in the intertidal zone. A different suite of predators will prey on bivalves in other habitats and, of course, other regions.

(1) The gastropod predators feed using a toothed rasping radula, in an extensible proboscis. They either drill through one shell of the bivalve to

Fig. 21.7 'Ear-hung' blacklip pearl oysters, Takapoto Atoll, French Polynesia.

Fig. 21.8 Scallops (*Pecten fumatus*) in a pyramid net (photograph by Mr John Mercer).

Table 21.2 Predators of the hard clam (*Mercenaria mercenaria*)

Group of predators	Number of species
Decapod crustaceans (crabs, hermit crabs, snapping prawns, etc.)	19
Gastropod molluscs	12
Horseshoe crab	1
Starfish	2
Birds	8+
Rays	8+
Finfish	5

Reprinted in part from Gibbons & Blogoslawski (1989) with permission from Elsevier Science.

access the animal within or push their proboscis between the prey's valves. The major gastropod predators of bivalves include members of the families Naticidae, Ranellidae, Buccinidae and Muricidae.

(2) The crab predators include swimming crabs (family Portunidae), rock crabs (families Cancridae and Grapsidae) and stone crabs (family Xanthidae). Crabs are equipped with two chelate arms ('pincers'), the larger chela being used for crushing and the smaller for manipulating and picking. Other decapod crustaceans, such as hermit crabs, shrimp and snapping prawns, although not equipped with such powerful crushers, will readily consume early juvenile bivalves.

(3) Starfish predators of cultured bivalves are mainly a problem in temperate regions. They particularly feed on soft substrate-inhabiting bivalves, flat oysters, clams and scallops. The starfish attaches to both valves with its tube feet, pulling the valves apart with prolonged tension, and then inserting its extruded stomach between the valves.

(4) Octopuses have not been widely reported as predators of cultured bivalves, but they prey on bivalves through prolonged tension in a similar manner to starfish.

(5) A variety of rays and other fish have been reported to feed on cultured bivalves. The fish include flatfish and whiting (active on soft substrates), perch, bream, puffer fish, snapper, etc. These fish are bottom feeders that both crush and consume the whole prey or crop the siphons off burrowing bivalves. The latter activity is not necessarily lethal to the prey but must reduce its growth rate.

Measures against predators include excluding them with appropriate-sized meshes around the bivalves. There is, however, a problem in that meshes can be fouled by algae and epifauna, seriously reducing the flow of water to these filter feeders. Another measure is to culture the bivalves away from the predators' habitats (e.g. above the substrate), and regularly inspect for mortality and identification of its source (e.g. damaged mesh, concealed predatory gastropods).

The advantage of using racks and other methods for keeping the cultured bivalves off the substrate, where appropriate, is that it is more difficult for predatory gastropods, starfishes, crabs, etc., to access their prey. Then, when predators gain access there is less cover for them to hide. They can be located and removed during regular inspections when localised groups of dead bivalves are observed. The predatory fish that feed on benthic bivalves tend not to swim up off the benthos to structures above their habitat.

Ocean surface culture is a further step in removing cultured bivalves from benthic predators. There is no physical connection with the substrate by which predators can reach the bivalves. The problem here can be with predators, such as crabs and gastropods, which settle as larvae from the plankton and grow rapidly to a nuisance size.

Parasites and diseases

Chapter 10 describes some of the diseases of bivalves. Elston (1990) has detailed the major diseases of oysters and some other cultured bivalves. The parasitic and other disease-causing organisms of hard clams are summarised by Gibbons & Blogoslawski (1989). They include:

- herpes-like viruses and rickettsia
- eubacteria (e.g. *Vibrio* species)
- protozoans (e.g. *Perkinsus marinus*)
- fungi, sponges and polychaete worms
- flat, tape, nemertine and nematode worms
- pyramidellid gastropods (ectoparasites that suck body fluids)
- endoparasitic copepods
- pea crabs ('food-stealers' among the gills in the mantle cavity).

Some, such as copepods and pea crabs, may have minimal effect on the host, causing some loss of host condition due to loss of nutrients and local irritation. Others, such as some viral and protozoan diseases, can have devastating effects on bivalve aquaculture industries, through mass mortalities, resulting in the industries becoming uneconomic, being closed down or requiring alternative, more resistant, species. The traditional flat oyster (*Ostrea edulis*) industry in France is an example of the last case (Heral, 1990) (Fig. 10.2). The Pacific oyster was imported because of its resistance to two protozoan parasites that were causing havoc with the native flat oyster. Proportions of oyster species in the annual harvests varied but, in 1984, a particularly bad year for the native oyster, the annual harvest was 98% Pacific oysters.

21.5.2 Fouling

Like boat hulls and jetty piles, aquaculture structures become coated with fouling organisms, which may include communities of bivalves where these are not the target species. As described earlier, fouling is a particular problem for bivalve culture as it restricts water flow to these filter feeders. Furthermore, many of the fouling organisms are themselves filter feeders and compete for available phytoplankton with the culture species. Fouling organisms include bacteria, diatoms, macroalgae, sponges, hydroids, hard and soft corals, bryozoans, tube-inhabiting polychaetes, barnacles, mussels (feral species) and tunicates.

The problems of fouling may be most acute during the spat and seed stages of bivalve culture. Fouling of culch surfaces before settlement reduces the sites for pediveligers to settle and metamorphose; and thus spat collection on the culch is reduced. The initial phase of fouling by bacteria and diatoms may result in a slimy coating that is unattractive to the settling pediveligers of many species. Furthermore, spat and seed bivalves are particularly vulnerable to fouling overgrowth due to the relative size of macroscopic fouling organisms.

A further aspect of fouling of suspended culture systems, e.g. long-lines and rafts, is that it can add so much weight to the system that it threatens the system's buoyancy.

Many bivalves are subject to fouling of their shells. These fouling organisms include boring fungi, sponges and tube worms that may riddle the bivalves' shells. The borers may penetrate through the inner nacreous layer of the shell, necessitating the mantle to secrete new nacreous layers to keep the borers at bay. Borers weaken the shell, making the bivalve more prone to crushing predators, and their burrows detract from the shell's appearance. This is particularly a problem for pearl oyster culture as good-quality shells are valuable. (In the Australian cultured pearl industry, such riddled shells are known as 'chicken shell'.)

One of the simplest ways to control fouling is to expose the cultured bivalves and their fouled equipment to air and sunlight for a sufficient period to kill most of the fouling organisms without killing the bivalves. The period required, however, may be more than 1 day and this method is time-consuming and costly in requiring regular movements of large numbers of bivalves to hold them out of water. Culturing in the intertidal zone has the advantage that exposure to air occurs regularly without movements of stock and, generally, fouling is less of a problem in intertidal culture than in subtidal culture.

Other methods for controlling fouling are pressure hoses and mechanical scrubbers. As well as these physical methods of fouling control, chemical and biological methods may be used (Quayle & Newkirk, 1989). Most of these chemical methods, as with exposure to sunlight, rely on the bivalves' capacity to close valves and shut out deleterious environmental factors for some time. The fouled bivalves may be immersed for appropriate periods in freshwater or hypersaline seawater. Toxic solutions such as hypochlorite may also be used. Other antifouling chemicals may be used to treat the culture equipment. These chemicals include tar, some new synthetic organic coatings and, in the past, toxic paints containing copper, zinc or tin. Tributyl tin, in particular, was a very successful and widely used antifoulant. However, use of these toxic metals as antifoulants is now recognised as being environmentally unacceptable (section 4.4.3). Furthermore, they may cause abnormalities and deaths of cultured bivalves, and the metals can accumulate in bivalve tissues to levels unacceptable for human consumption.

Biological methods are appropriate when there are one or a few major fouling organisms to deal with. Biological methods require knowledge of the breeding season, patterns of larval settlement and life cycle of the fouling organism. For example, knowing the breeding period of the fouling organism, it may be appropriate to put culch out after this period, or it may be appropriate to use a chemical or exposure treatment while the fouling organisms are recently settled and more vulnerable. Local areas of heavy settlement of fouling organism larvae may be avoided during the period of settlement. If the fouling organism larvae tend to settle at particular depths, then these can be avoided in suspended culture.

21.5.3 Biotoxins and gut contents

Biotoxins

With appropriate hydrological conditions, some phytoplankton species bloom to cell densities of millions/mL, causing discolouration of inshore waters, a phenomenon often known as red tides. During red tides, the harvesting and consumption of bivalves may be suspended owing to accumulation of potent biotoxins in the bivalves from their filter feeding on toxic phytoplankters. More insidiously, at other localities there are blooms of toxic phytoplankters without obvious discolouration of the water.

Most sources of these biotoxins are dinoflagellates, and about 20 out of the total 1500 dinoflagellate species have been implicated. They include species of *Alexandrium*, *Dinophysis*, *Gymnodinium* and *Porocentrum*. Several species of the diatom *Nitzschia* are also toxic.

The biotoxins do not apparently poison the bivalves, although red tides may cause mass mortalities through effects on water quality. The toxins are assimilated and accumulate in the bivalves' digestive gland and other organs. Then the toxins may severely affect humans who subsequently consume the infected molluscs. Some of the sources and effects of biotoxins from bivalves are:

(1) Amnesic shellfish poisoning (ASP): gastrointestinal disorders such as nausea, abdominal cramps, diarrhoea. More severe symptoms are neurological, such as headaches, dizziness, disorientation and loss of short-term memory, caused by domoic acid, which is derived from several diatoms (*Nitzschia* species). Poisonings have primarily occurred on the north-eastern coast of North America but the diatoms have a wide distribution.
(2) Diarrhoeic shellfish poisoning (DSP): gastrointestinal disorders caused by okadaic acid, which is derived from species of *Dinophysis*. It has a wide occurrence but occurs especially in Europe and Japan.
(3) Neurotoxic shellfish poisoning (NSP): neurological symptoms that are milder than PSP. These are caused by a polyether toxin of the same class of compounds as ciguatoxins in fish flesh. The toxin is derived from a dinoflagellate (*Gymnodinium breve*) and primarily occurs in Florida and Texas.
(4) Paralytic shellfish poisoning (PSP) manifests as numbness of lips and tongue, fingers and toes. Extreme symptoms are muscular paralysis and breathing difficulties and, in severe cases, death. Variable complexes of toxins are involved and various dinoflagellates, including *Gymnodinium catenatum*.

In recent times, biotoxins in bivalves have been responsible for several thousand cases of severe poisoning and several hundred deaths throughout the world. Apart from these dire effects on seafood consumers, the biotoxins have severely affected bivalve culture industries through industry closures or loss of consumer confidence. For example, the Spanish mussel industry was devastated following loss of consumer confidence after PSP-infected mussels were exported throughout Europe in 1976, while the New Zealand shellfish industry, valued at NZ$100 million per year, was closed down after suspected biotoxin poisoning of over 170 people. A problem with biotoxins at one locality can result in loss of consumer confidence that extends to the whole industry.

Biotoxins in bivalves have been a long-standing problem for the bivalve culture industries of Europe, South-East Asia and North America, but the problem seems to have become more common and widespread (Hallegraeff, 1993). Suggested reasons for this are:

- increased scientific awareness of the problem
- increased utilisation of coastal waters for aquaculture
- increased algal blooms due to eutrophication of coastal waters
- unusual climatic conditions stimulating algal blooms
- transport of dinoflagellate cysts in ships' ballast water
- transport through transfers of bivalve stocks.

Discharge of ships' ballast is blamed for introducing the dinoflagellate *Gymnodinium catenatum* (a source of PSP) to Tasmania, where its first recorded bloom was in 1986, and where it has resulted in closures of Pacific oyster farms in some areas.

The safe level for PSP is 0.8 ppm shellfish tissue (80 μg/100 g). However, although toxins in infected bivalves can be magnitudes above safe levels, after the toxic algal bloom clears from their environment the animals lose their accumulated toxins over periods of weeks or months and become marketable.

Biotoxins are a major problem for worldwide bivalve industries and unfortunately there is no simple solution. The safest, but uneconomical, way to cope with biotoxins in bivalves is to bioassay all suspect product before it is marketed. Practical solutions are to have annual periods of closure in areas of long-standing bivalve culture where the occurrence of toxic algal blooms is regular and seasonal. For example, this applies to areas of the Atlantic coast of Canada and to parts of the Californian and Washington State coastlines (Quayle & Newkirk, 1989). Specific stations in these areas are monitored to adjust the timing of the closures. Regular samples of cultured bivalves from representative sites may be tested when the occurrence of biotoxins is suspected or anticipated. In other areas the occurrence of red tides or the particular conditions that lead to red tides in that locality are carefully monitored.

Gut contents

The other possible sources of health problems for bivalve consumers are live micro-organisms in bivalves' guts when the animals are consumed raw. (One does not wish to be accused of trying to turn away 'shellfish' consumers; this chapter is directed at those who are interested in bivalve aquaculture and it is important that consumers' needs for a safe product are identified.) Bivalve guts contain bacteria and viruses, which they have ingested through their filter feeding. Coastal waters may be subject to high levels of sewage input including inadequately treated sewage or there may be high levels of waste input from animal husbandry. Under these conditions there will be high levels of faecal bacteria and viruses in the water column and hence in the bivalves' guts. The bacteria include *Escherichia coli* and *Salmonella* species, which can cause gastroenteritis and typhoid fever respectively. Contaminated bivalves have also been responsible for transmitting viral hepatitis and cholera (Quayle & Newkirk, 1989).

These micro-organisms in the bivalves' guts can be dealt with much more readily than biotoxins accumulated in their tissues. It is a matter of exposing the animals to purified water for several days while they clear their gut contents – a process called depuration. It usually involves a large recirculating seawater system with a means of water treatment, such as ozone, chlorine or UV radiation, and a system of racks and trays to hold large numbers of bivalves in tanks within the seawater system. The animals are then marketed. In many areas, depuration is used as a routine practice between harvesting and marketing.

21.6 Introductions and other environmental issues

The Pacific oyster (*Crassostrea gigas*) has many desirable traits for culture and market acceptability, and it has been introduced over large distances from its original distribution, apparently without adverse environmental effects. It originated from the northern Pacific region, where it is cultured on both sides of the Pacific, e.g. in Japan, South Korea and North America. Subsequently, it has been translocated to Europe and to Australasia, where very successful industries have developed. This must, however, not be taken as support for introductions. The prevalence of diseases has major implications for introductions and reintroductions of cultured bivalves and other aquaculture organisms (section 4.4.2). There are temptations to introduce species that are being cultured successfully in other regions, and translocations are relatively simple because many bivalves tolerate extensive periods out of water if handled correctly. Introductions and reintroductions, however, cannot be undertaken without comprehensive quarantine preparations and assessment of potential ecological effects. This is in view of the suite of noxious organisms associated with bivalves, and their potential for devastating effects on bivalves at the recipient locations (Carriker, 1992). Translocations to regions where there are existing populations of a species also have genetic implications. It is possible to destroy regional genetic variability through this process (section 4.4.2). It is impossible to prevent spawning in translocated bivalves in the open aquatic environments where they must be cultured, and spawning will lead to vast numbers of 'foreign' gametes entering the environment.

There is a further aspect to deliberate or accidental introductions of bivalves that warrants caution. The introduced bivalve species can proliferate to the extent of becoming a pest in its new environment, where factors controlling it within its normal range are absent. The proliferation of introduced species can be at the expense of related local species. The zebra mussel in the Great Lakes system, North America, is a well-known example. Since it was introduced in the mid-1980s, probably from ballast water of a trans-Atlantic ship or ships, it has become a disastrous fouling organism in this environment (Kastner, 1996). Furthermore, it has the potential to cause the extinction of some endemic mussel species.

One potential environmental problem with many aquaculture operations, the input of large quantities of nitrogen-rich feeds, is not present in bivalve aquaculture. There may be artificial feeding of bivalve larvae and early juveniles in hatcheries, but this is only with relatively small biomasses of bivalves that are fed microalgae. By the time the cultured bivalves reach a substantial size, they are totally dependent on natural phytoplankton. There can, however, be deleterious environmental effects from the filter feeding. Large marine mussels may each filter 2–5 L/h. In very general approximates, a raft of mussels may filter 70 million L/day, ingest 180 mt of organic matter per year and produce 100 mt of pseudofaeces and faeces per year. The 'rain' of faeces and pseudofaeces from these rafts endangers the benthic environment below. It may transform the substrate into an anaerobic ooze, just as may occur below cages in fish culture with the 'rain' of uneaten feed and faeces. The solutions are to:

- locate the culture structures in regions of high current velocity to promote dispersal of faeces and pseudofaeces in the water column
- limit biomass/unit area of cultured bivalves to a level where dispersal processes are sufficient to minimise faecal and pseudofaecal accumulation on the substrate
- regularly move the rafts and long-lines to new sites before there is excessive accumulation of wastes on the substrate below.

Cleaning the fouling growths off the bivalves' shells and culture apparatus may also be a source of environmental problems where the fouling debris is discarded into the adjacent water.

A different kind of environmental problem arises because bivalves are usually cultured in coastal regions where there are other users of the environment. As with other aquaculture operations in these regions, farms consisting of rafts, racks, long-lines and floats may be perceived to be an environmental problem from an aesthetic, navigational or recreational viewpoint (section 4.4.5). On the other hand, there may be situations in which bivalves can be used with positive environmental effects, such as removing particles and nutrients from pond effluent (section 21.8).

21.7 Industry reviews

21.7.1 Table oysters

There are major industries culturing cupped oysters

on unpolluted coastlines of most developed nations in temperate regions. A number of these nations have commercial oyster hatcheries as a source of spat. This is because oysters have long been regarded as a seafood delicacy and culture methods are long established (reviewed by Matthiessen, 2001). On the other hand, although there are many species of cupped oysters in tropical and subtropical regions, few are commercially cultured and there are few hatcheries (Angell, 1986). This is because oysters are not highly regarded where animal protein production is more important than seafood delicacies, that is unless the oysters are exported.

By far the greatest proportion of world oyster production is from cupped oysters, with the Pacific oyster being pre-eminent (Table 21.1). Its growth rates, favourable morphology for marketing and relative resistance to disease compared with most other oyster species, and especially compared with flat oyster species, make it very attractive for culture.

Table oysters are cultured on the bottom, off the bottom and suspended from the surface. The spat may be obtained from natural settlement on culch, e.g. sticks, shells, tiles, etc., or they may be hatchery produced. The spat may be left on culch or they may be removed. The simplest level of culture is with natural spat in areas of firm substrate. These oyster farms are usually fenced off, predators are controlled and other management measures undertaken. Harvesting is often by dredge. Stick culch with seed oysters may be placed horizontally on racks above the bottom or driven vertically into the bottom. Culchless oysters and oysters on small pieces of culch may be kept in trays or mesh containers on racks above the bottom or suspended from the surface from long-lines or rafts.

Oyster fisheries and culture industries have been characterised by great fluctuations in annual production. Factors causing these fluctuations are:

- overexploitation to the extent of reducing the reproductive output and subsequent recruitment where the aquaculture industry relies on natural spatfall
- mass mortalities from various protozoan and viral diseases, apparently resulting from overstocking or introduced pathogens
- mass mortalities from unusual environmental conditions, including temperature extremes and red tides.

Heral (1990) provides an excellent example of wild fluctuations in annual oyster harvests from a culture industry. These data are for the French industry for over a century, 1865–1983. They result from the factors described above.

The Sydney rock oyster industry in eastern Australia is another example of fluctuating fortunes. The industry commenced mainly in southern Queensland during the late nineteenth century with a simple system of oyster banks. The flourishing industry, however, crashed through massive losses of oysters with QX disease (caused by a protozoan parasite, *Marteilia sydneyi*; section 10.5.1). QX continues to be a problem limiting the industry in southern Queensland and northern New South Wales. In southern New South Wales, the oysters suffer 'winter mortality' (caused by another protozoan parasite, *Mikrocytos roughleyi*). The Sydney rock oyster farms depend on natural spatfall and, as Pacific oysters spread north from Tasmania, the farms are now being 'contaminated' with unwelcome Pacific oysters.

21.7.2 Mussels

Mussels do not usually attract prices as high as those obtained for table oysters. In some cases, as in parts of coastal China, mussels are cultured to use as food for other more valuable aquaculture organisms, such as shrimp. Furthermore, the low economic return per unit mussel usually renders hatchery production of spat and seed uneconomic, and almost all mussel industries are based on natural spatfall.

The relatively low economic value of mussels is compensated by their ease of culture and high productivity. The highest productivity of bivalve culture per unit farm area has been reported for marine mussels. These mussel farms operate in the western Galicia region of Spain in a series of sunken river valleys (rias). Primary productivity in the rias was found to average 10.5 μg organic carbon/L over the year (Figueras, 1989) due to upwelling of nutrient-rich water. This provides copious phytoplankton for mussel culture. Thus, a 20 m × 20 m raft may produce 50+ mt of mussels in 12–18 months and there may be six rafts per hectare. Meat yield is 30–50% of whole weight, so this farming system potentially yields meat at ~100 mt/ha/year without feed or fertiliser inputs. There is no low-input farming on land, such as grazing, which can even vaguely compare with this production rate. Maximum meat production from unsupplemented grazing is in the order of 2 mt/ha/year.

Blue mussel (*Mytilus edulis*) production in Spain was ~260 000 mt/year in 1998 and 1999 (FAO, 2001). It is based on vertical ropes suspended from rafts. Seed mussels are attached to the ropes at ~10 mm shell length and are grown to their marketable size, 80–100 mm shell length (Fig. 21.6). This was often reached in less than a year in the past. Now it takes longer in the central regions of the rias, apparently due to the densities of mussels reducing phytoplankton levels.

As well as raft culture, mussels of both *Mytilus* and *Perna* species are grown in bottom culture, on vertical poles in the intertidal zone, and in long-line culture (Vakily, 1989; Dardignac-Corbeil, 1990) (Table 21.3).

Mussels are a traditional cultured seafood in Europe and some parts of Asia. Despite this, there have been major changes in mussel production over recent decades. As shown in Tables 21.1 and 21.3, China now produces > 600 000 mt of marine mussels of a number of species, being by far the largest producer. Production in Europe is generally steady, with marked fluctuations in individual industries, e.g. The Netherlands' mussel industry is strongly influenced by unpredictable storms that strip mussels from the farms. New Zealand has markedly expanded its mussel production, including specialised products such as freezer packs of mussel in the half-shell. On the other hand, Thailand's production of green mussels was more than 200 000 mt in 1971, but then declined precipitously, because of, among other factors, environmental deterioration and loss of farming areas to other purposes. The industry has somewhat recovered, but the history of mussel farming in Thailand reflects the kinds of problems faced by bivalve industries worldwide.

21.7.3 Scallops

The scallop's large adductor muscle is a highly valued seafood and scallop fisheries are often unpredictable and overexploited, so culture is an attractive alternative to fisheries (Hardy, 1991; Shumway, 1991; Hardy & Walford, 1994). Despite the interest in culture, world aquaculture production of scallops is almost exclusively from China and Japan. They produced 712 000 and 216 000 mt, respectively, of the yesso scallop (*Pecten yessoensis*) in 1999 (FAO, 2001). This is 97–98% of world scallop production based on this one species. Interestingly, at least in the past, the technology for culturing scallops contrasted between Japan and China in a way that one would not necessarily have predicted. The Japanese industry relied entirely on collecting bountiful natural spat, whereas China had a large number of hatcheries (Cropp, 1989).

The yesso scallop is a coldwater species and most production is from northern Japan. Although natural spatfall is used, it is not a casual operation. The reproductive state of scallop populations and progress of natural larval development are carefully monitored to accurately time the deployment of spat collectors in each region. The spat collectors (mesh bags with internal mesh, Fig. 21.3) are hung from subsurface long-lines. When the seed scallops are 10–15 mm, they are removed from the collectors and placed in pyramid-shaped mesh nets (pearl nets, Fig. 21.8), on long-lines. Juvenile scallops, 30 mm, may then either be released to prepared regions of the seafloor or placed into lantern nets. The former culture method involves high mortality from predators; no meshes are used and it is essentially enhancing natural populations. The latter method is more expensive owing to labour and equipment costs. Another less common culture method is 'ear hanging', whereby a small hole is drilled through the hinge 'ear' of the shells and the scallops are tied to supporting ropes (Fig. 21.7). Harvest size (> 100 mm shell length) is reached in 2–3 years (Cropp, 1989).

21.7.4 Clams and cockles

Some clams and cockles, like mussels, are the bases for large aquaculture industries that are often very simple and traditional in technology, e.g. the culture industries for blood cockles, *Anadara* species (Broom, 1985) (Table 21.1). They may consist of little more than collecting seed from areas of heavy natural settlement and then stocking the seed at appropriate densities, e.g. 10 million/ha, in defined

Table 21.3 Details of some mussel mariculture industries

Country	Mussel species	Main culture techniques
China	Blue, green, Chinese, etc.	Various
Spain	Blue	Raft
Italy	Mediterranean	Vertical poles
Thailand	Green	Vertical poles
France	Blue and Mediterranean	Vertical poles ('bouchets')
The Netherlands	Blue	Bottom
New Zealand	New Zealand	Long-lines
Germany	Blue	Bottom
Philippines	Green	Vertical poles

Blue mussel, *Mytilus edulis*; Chinese mussel, *Mytilus coruscus*; green mussel, *Perna viridis*; Mediterranean mussel, *Mytilus galloprovincialis*; New Zealand mussel, *Perna canaliculus* (FAO, 2000).

farm areas where there are favourable conditions for growth and predator control. Notwithstanding technical differences, there is a major difference between the two bivalve groups in that clam and cockle aquaculture results in products that are substantially more valuable than those from mussel aquaculture (Fig. 21.1).

Many clam industries are based on hatchery and nursery technology. For example, the hard clam or northern quahog (*Mercenaria mercenaria*) industry of eastern North America cultures the larvae and spat using similar hatchery methods and upweller systems to those used in oyster culture. Then the seed clams are transferred to the field at densities of 3000–4000/m^2 into either carefully prepared substrate areas, floating trays or trays on racks (Castagna & Kraeuter, 1981). After a year, when the juveniles reach about 25 mm in shell length, they are spread out in the grow-out area. The considerably smaller Manila clam (*Ruditapes philippinarum*) is being cultured commercially in the USA on an increasing scale and it is based on hatchery production (Nosho, 1998).

Aquaculture production of the hard clam in the USA amounted to 26 500 mt in 1999 (FAO, 2001). This is a substantial quantity, but it is only 1% of total world clam, cockle and arkshell production, which is derived from > 25 species of cultured bivalves in this category.

With a great variety of species, growth rates and commercial sizes within this group, there is obviously a wide range of culture periods. Clams and cockles are, however, generally slow growers compared with mussels, and periods of > 2 years to commercial size are not uncommon with the larger species. This adds to the necessity for low-cost technology in the grow-out phase.

Reference has been made to the most unusual group of clams, the giant clams. These have been the target of much culture-oriented research in recent decades. Although the industry is in the initial phase of development, the technology and 'how to' culture manuals are available, e.g. Heslinga *et al.* (1990), Braley (1992), Calumpong (1992).

21.7.5 Pearl oysters and pearl mussels

There is the common idea that pearls are formed in oysters, especially pearl oysters, when a sand particle gets into its mantle cavity and causes irritation. In fact, such particles frequently enter the mantle cavity of substrate-inhabiting bivalves and they are vigorously expelled. That pearls are formed when foreign particles lodge within the valves cannot be precluded, but most natural pearls probably arise from small parasites (such as endoparasitic copepods) invading bivalve tissues. These are encased in shell material, especially nacre, by the adjacent host tissue. Many bivalves produce natural 'pearls', e.g. it is a hazard of eating table oysters, and an enormous pearl was collected from a giant clam. However, only the mantle tissues of some bivalves secrete very lustrous nacre and some are used in cultured pearl production. Interestingly, some gastropod molluscs also secrete a very lustrous nacre and these are used commercially (Shirai, 1994; sections 22.2.1 and 22.4.1).

Pearl oysters

Before they declined or reached a level from which they could not recover, many pearl oyster stocks were intensely fished. The frequency of occurrence of natural pearls varies with the species and some stocks were fished for natural pearls. Other stocks were fished mainly for pearl shells, which were used for button manufacture and pearl inlay. Pearl culture methods for pearl oysters were developed and patented by several Japanese biologists early in the twentieth century. A spherical bead made from shell is used as the 'nucleus' for the cultured pearl. The bead, which must be of an appropriate size for the size of the oyster, is inserted into a pocket cut in the oyster's gonad (Fig. 21.9). Inserted with the bead is a small piece of outer mantle tissue removed from a donor oyster (Gervis & Sims, 1992). Cells of the transplanted mantle tissue spread over the bead surface to make a pearl sac and this secretes successive layers of nacre onto the bead to make the cultured pearl. Thus, cultured pearls are produced via host oysters supporting foreign cells that secrete the pearl.

The period between bead insertion and harvest varies between about 9 months and 3 years, with the thickness of nacre and quality of pearl being directly related to this culture period. Only a low percentage of the inserted beads result in high-value pearls. Losses occur from oysters that die from the insertion procedure, ejected beads and other processes that affect the shape and surface quality of the pearl (Norton *et al.*, 2000). After the harvest, the pearls must be graded before being presented for sale.

The cultured pearl industry has been dominated at all levels by the Japanese industry:

- technicians who inserted the beads
- hatchery operators
- managers of oyster farms
- company owners or joint owners
- pearl graders
- pearl marketing.

The largest component of world pearl production is in Japanese waters, from the Akoya pearl oyster (*Pinctada fucata* = *P. imbricata*). Japanese production of Akoya pearls peaked at about 70 mt/year in the late 1980s, but has declined in recent years, partly as a result of coastal pollution (Gervis & Sims, 1992). On the other hand, cultured pearl production has been increasing in Australia and Indonesia (large silver-white pearls from *P. maxima*), in Polynesia (large black pearls from *P. margaritifera*) and in other nations, including China.

Fig. 21.9 Operating to insert a bead and piece of mantle tissue for cultured pearl production in a blacklip pearl oyster, Manihiki Atoll, Cook Islands. The technician is the late Ian Turner.

Similar techniques to those used for scallops in suspended culture are used for pearl oysters, e.g. rafts or long-lines with suspended nets or mesh frames, or 'ear hanging' on ropes (Fig. 21.7). The techniques vary with location and species. Again, the sources of oysters for bead insertion vary with location and species: from direct collection of appropriate-sized wild oysters, from spat obtained from spat collectors and from hatchery rearing.

Pearl mussels

Production of cultured pearls from freshwater mussels is a very different process, which is largely conducted in China, but also in Japan. Pieces of donor tissue are used and the process requires a skilled technician, but there are three major differences from marine pearl culture:

- There is no bead and the resulting pearls have no foreign nucleus.
- The site of insertion is the recipient mussel's mantle, not the gonad (the mantle lobes are much thicker in these mussels than in pearl oysters).
- Up to 30 or 40 donor mantle pieces may be inserted into the two mantle lobes of the host mussel.

The resulting pearls, harvested after several years, are generally < 6 mm in diameter and irregular ('rice' or 'wrinkled' pearls). They are not very valuable individually, but their production is reliable and large numbers are produced per mussel. They have been used in multi-strand necklaces. Furthermore, after the pearls have been harvested, the mussels can be returned to culture and their pearl sacs will produce several further crops of these nucleus-free pearls. Recent technological improvements are resulting in larger and more spherical pearls, which are considerably more valuable.

In China, the people who engage in the culturing phase of pearl-producing mussels are often also engaged in agriculture and aquaculture of freshwater fish. The mussels with inserted tissue are cultured by suspension in freshwater tanks with flow-through or in freshwater environments. The mussels are suspended in groups: in netting baskets or netting bags or tied together via holes through their shells.

21.8 Conclusion

Aquaculture of bivalves originated in antiquity and has continued through the ages. With the general increases in world aquaculture production in recent decades, however, there have been large increases in bivalve production. The rate of increase, 11.6% per year on average over the decade to 1998, exceeded the general rate of aquaculture increase by almost 1%. The increases varied between bivalve industries, with the greatest growth, a 16.8% per year increase on average, being for clam and cockle production over the decade 1989–98 (FAO, 2001). Scallop and oyster production increased by 15.5% and 12.9% per year respectively over the same decade. Mussel production increased 3% per year, a more modest rate compared with the other three major groups. As with fish, crustaceans and seaweeds, China dominates bivalve production and it is undoubtedly China's lack of emphasis on mussel production that has comparatively limited its expansion.

There is every reason to predict that bivalves will continue to be a major component of world aquaculture production.

- Many species are regarded as premium seafoods and have high market demand.
- They have excellent nutritional properties.
- They filter feed and make no demands on artificial diets (except possibly on a modest scale in the hatchery phase) or on fish meals.
- They may be reared with relatively simple techniques.
- They feed very low in the food webs (phytoplankton → bivalve) and thus capture a very high proportion of the energy of primary production.
- They can be cultured at very high densities in appropriate areas of high phytoplankton production and water quality conditions, and, without supplementary feeding, achieve levels of production/unit area or volume that are not even nearly matched by any other farmed aquatic or terrestrial animals.

A picture of entirely vigorous growth of bivalve industries is not completely honest. As indicated in this chapter, disease, pollution and biotoxins are limiting bivalve industries in some areas and regions. These ongoing problems must be dealt with by management and research.

Finally, some further predictions for the future of bivalve aquaculture:

(1) Bivalve industries will increasingly develop hatchery production of spat, instead of the

traditional dependence on natural spatfall. Hatchery product will bring its immediate advantage of reliability of spat supplies and potential advantage of facilitating genetic selection.

(2) Genetics and stock selection will be researched and developed from the current limited database. There will be industries based on hatchery-produced seed stock reared from genetically selected broodstock to have particular traits that make them very viable commercially, e.g. stock that are disease resistant, and have high growth rates and highly acceptable taste. Through hatchery production and through genetic selection, many bivalve industries will no long use natural spatfall or wild broodstock as the basis for their industry.

(3) There will be technological developments in the grow-out culture phase to increase efficiency and make better use of available sites. Some of the traditional grow-out techniques, e.g. oysters in baskets on racks or sticks, mussels on poles, are labour intensive. There will be systems to achieve greater mechanisation without increasing production costs.

(4) Bivalves will have an important new role to play in the future of aquaculture: a role in treating aquaculture effluents and the recirculating water of large recirculating systems. The various components of effluent water from large pond farms, e.g. inorganic nutrients, suspended solids, bacteria, etc., must each be treated appropriately, before the water is discharged into the adjacent environment. Similarly, water being recirculated back into a large pond farm requires the same kind of treatment. A variety of ways of treating the various components of the effluent are being tested, e.g. seaweeds are being tested for removing inorganic nutrients (section 13.4.6) and bivalves are being tested for removing phytoplankton and other suspended organic particles. Jones & Preston (1999) found that filter feeding by Sydney rock oysters can substantially reduce the concentrations of total suspended solids, bacteria, phytoplankton, and particulate and dissolved N and P in shrimp pond effluents. Use of commercial organisms such as seaweeds and bivalves in this role gives double value in that they produce a commercial crop while performing their environmental function.

References

Angell, C. L. (1986). *The Biology and Culture of Tropical Oysters*. ICLARM Studies and Reviews 13. International Center for Living Aquatic Resources Development, Manila, The Philippines.

Braley, R. D. (Ed.) (1992). *The Giant Clam: Hatchery and Nursery Culture Manual*. Australian Centre for International Agricultural Research, Canberra, Australia.

Broom, M. J. (1985). *The Biology and Culture of Marine Bivalve Molluscs of the Genus Anadara*. ICLARM Studies and Reviews 12. International Center for Living Aquatic Resources Development, Manila, The Philippines.

Calumpong, H. P. (Ed.) (1992). *The Giant Clam: An Ocean Culture Manual*. Australian Centre for International Agricultural Research, Canberra, Australia.

Carriker, M. R. (1992). Introductions and transfers of molluscs: risk considerations and implications. *Journal of Shellfish Research*, **11**, 505–10.

Castagna, M. & Kraeuter, J. N. (1981). *Manual for Growing the Hard Clam Mercenaria*. Special Report in Applied Marine Science and Ocean Engineering No. 249. Virginia Institute of Marine Science, VA.

Citter, R. (1985). Serotonin induces spawning in many West Coast bivalve species. *Journal of Shellfish Research*, **5**, 55 (Abstract).

Crawford, C. M., Nash W. J. & Lucas, J. S. (1986). Spawning induction, and larval and juvenile rearing of the giant clam, *Tridacna gigas*. *Aquaculture*, **58**, 281–95.

Cropp, D. A. (1989). *Scallop Culture in the Pacific Region*. Technical Report 34. Department of Sea Fisheries, Tasmania.

Dardignac-Corbeil, M.-J. (1990). Traditional mussel culture. In: *Aquaculture*. Vol. 1 (Ed. by G. Barnabe), (Translation Ed. J. F. de L. B. Solbe), pp. 285–341. Ellis Horwood, West Sussex.

Elston, R. A. (1990). *Mollusc Diseases: Guide for the Shellfish Farmer*. Washington Sea Grant Program. University of Washington Press, Seattle, WA.

FAO (2001). *1999 Fisheries Statistics: Aquaculture Production*. Vol. 88/2. FAO, Rome.

Figueras, A. J. (1989). Mussel culture in Spain and France. *World Aquaculture*, **20**, 8–17. (Special issue: on mussel culture; Ed. G. Jamieson).

Gervis, M. H. & Sims, N. A. (1992). *The Biology and Culture of Pearl Oysters (Bivalvia: Pteriidae)*. ICLARM Studies and Reviews 21, International Center for Living Aquatic Resources Development, Manila, The Philippines.

Gibbons, M. C. & Blogoslawski, W. J. (1989). Predators, pests, parasites, and diseases. In: *Clam Mariculture in North America* (Ed. J. J. Manzi & M. Castagna), pp. 167–200. Elsevier, Amsterdam.

Hallegraeff, G. M. (1993). A review of harmful algal blooms and their apparent global increase. Phycological Reviews 13. *Phycologia*, **32**, 79–99.

Hardy, D. (1991). *Scallop Farming*. Fishing News Books, Oxford.

Hardy, D. & Walford, A. (1994). *The Biology of Scallop Farming*. Aquaculture Support, Kyle of Lochalsh.

Heral, M. (1990). Traditional oyster culture in France. In: *Aquaculture*. Vol. 1 (Ed. G. Barnabe) (Translation Ed. J. F. de L. B. Solbe), pp. 342–87. Ellis Horwood, West Sussex.

Heslinga, G. A., Watson, T. A. & Isamu, T. (1990). *Giant Clam Farming*. Pacific Fisheries Development Foundation (NMFS/NOAA), Honolulu, HI.

Jeffs, A. G., Holland, R. C., Hooker, S. H & Hayden, B. J. (1999). Overview and bibliography of research on the greenshell mussel, *Perna canaliculus*, from New Zealand. *Journal of Shellfish Research*, **18**, 347–60.

Jones, A. B. & Preston, N. P. (1999). Sydney rock oyster, *Saccostrea commercialis* (Iredale & Roughley), filtration of shrimp farm effluent: effects on water quality. *Aquaculture Research*, **30**, 1–7.

Kastner, R. J. (1991). What every fish farmer should know about the zebra mussel. *Aquaculture Magazine*, **27**(6), 36–48.

Kesarcodi-Watson, A., Klumpp, D. W. & Lucas, J. S. (2001). Physiological energetics of diploid and triploid Sydney rock oysters, *Saccostrea commercialis*. II. Influences of food concentration and tissue energy distribution. *Aquaculture*, **203**, 195–216.

Lucas, J. S. (1994). The biology, exploitation, and mariculture of giant clams (Tridacnidae). *Reviews in Fisheries Science*, **2**, 181–223.

Matthiessen, G. C. (2001). *Oyster Culture*. Fishing News Books/Blackwell Science, Oxford.

Nell, J. A., Cox, E., Smith, I. R. & Maguire, G. B. (1994). Studies on triploid oysters in Australia. I. The farming potential of triploid Sydney rock oysters *Saccostrea commercialis* (Iredale & Roughley). *Aquaculture*, **126**, 243–55.

Norton, J. H., Lucas, J. S., Turner, I., Mayer, R. J. & Newnham, R. (2000). Approaches to improve cultured pearl formation in *Pinctada margaritifera* through use of relaxation, antiseptic application and incision closure during bead insertion. *Aquaculture*, **184**, 1–17.

Nosho, T. (Ed.) (1998). *Clam and Oyster Farming Conference for Shellfish Growers*, Olympia, WA, 3–4 March 1997. Sea Grant Program, Washington University, Seattle, WA.

Quayle, D. B. & Newkirk, G. F. (1989). *Farming Bivalve Molluscs: Methods for Study and Development. Advances in World Aquaculture*, Vol. 1. World Aquaculture Society, Baton Rouge, LA.

Ruppert, E. E. & Barnes, R. D. (1994). *Invertebrate Zoology*. Saunders College Publishing, Fort Worth, TX.

Scarpa, J., Toro, J. E. & Wada, K. T. (1994). Direct comparison of six methods to induce triploidy in bivalves. *Aquaculture*, **119**, 119–33.

Shumway, S. E. (Ed.) (1991). *Scallops: Biology, Ecology and Aquaculture*. Development in Aquaculture and Fisheries Science 21. Elsevier, Amsterdam.

Vakily, J. M. (1989). *The Biology and Culture of Mussels of the Genus Perna*. ICLARM Studies and Reviews 17. International Center for Living Aquatic Resources Development, Manila, The Philippines.

Yukihira, H., Klumpp, D. W. & Lucas, J. S. (1999). Feeding adaptations of the pearl oysters, *Pinctada margaritifera* and *P. maxima* to variations in natural particulates. *Marine Ecology Progress Series*, **182**, 161–73.

22 Marine Gastropods

Laura Castell

22.1 Introduction

> Suppose we had tried selective breeding of frogs and found we were developing the head instead of the legs. But a snail is just a walking intestine [gastro-pod].
>
> Georges Auer (1980)

Many marine gastropod molluscs are valued for their shells ('seashells') and there is a large trade in shells for the collectors' market. Other marine gastropods are harvested for their meat or for both their shells and meat. The three most important gastropods in terms of large-scale fisheries are:

- abalone (*Haliotis* species)
- the trochus or top shell (*Trochus niloticus*)
- the queen conch (*Strombus gigas*).

Techniques for aquaculture have been developed for each of these. Furthermore, a number of other fished species are receiving increasing attention for their potential in aquaculture, including, to name a few:

- the green snail (*Turbo marmoratus*) and silver-mouth (*Turbo argyrostomus*), which are utilised on a small scale in the Pacific Islands and Japan (Yamaguchi, 1993)
- the red conch (*Rapana venosa*), mud snail (*Bullacta exarata*) and sea hare (*Notarchus leachii cirrosus*) in China (Guo *et al.*, 1999)
- the spotted babylon (*Babylonia areolata*) in Taiwan (Chaitanawisuti & Kritsanapuntu, 1998)
- the predatory muricids, *Chorus giganteus* in Chile and *Chicoreus ramosus* in Thailand.

This chapter will focus on the culture of abalone, trochus and the queen conch. The fisheries and culture of the three species have several common characteristics. These gastropods are particularly susceptible to overfishing because generally they live in relatively shallow waters accessible to divers using mask and snorkel; they move slowly and are easily targeted by an experienced diver. Thus, natural stocks of these three gastropods have been overfished in many places through commercial fisheries. This was the impetus for developing culture methods to supplement fisheries for these valuable products. Wild juveniles are generally very difficult to find and, therefore, collecting juvenile wild stock from their natural habitat to grow under controlled conditions until harvest size is impractical. As such, the culture of these species began in earnest when techniques for larval rearing were developed.

These gastropods take a relatively long period before reaching harvestable size in the field. For example, Californian red abalone require up to 15 years to grow to minimum legal size of 197 mm in shell length (SL) (Tegner *et al.*, 1992). Australian blacklip abalone take between 11 and 15 years to reach a size of 120 mm in Tasmanian waters, and this is still below the minimum legal size for harvest (Nash, 1992). The New Zealand abalone takes 3–5 years to reach 60–70 mm in size, but the legal size is 125 mm (Schiel, 1992). Trochus may reach 80 mm in shell diameter in ~3 years (see Nash, 1993), whereas queen conchs enter the fishery at a SL of 200 mm, when individuals are aged ~3 years (Weil & Laughlin, 1984). One of the advantages of culturing these gastropod species is that there are no size restrictions placed on the harvest of cultured individuals.

Until recent years, growing these gastropods to a large commercial size in on-shore facilities required large amounts of food, tank space and time, making the process labour intensive. Costly hatcheries for all three of these gastropods were initially established largely for research purposes and to culture juveniles for release into natural habitats (seeding). Intensive research into culturing techniques, however, has led to the rapid development of aquaculture for these species, especially abalone and conchs, which are now produced on a commercial scale. Most commercial farms of abalone were first established in the 1990s and are now well established in several countries, including Australia, Chile, China, Japan and USA. Aquaculture production of conchs and trochus has developed at a slower pace, focusing on research and seeding; but conch culture has recently reached commercial levels with at least one commercial farm for *S. gigas* in the Turks and Caicos Islands. There are currently no commercial farms producing trochus.

This chapter describes the culture techniques developed for abalone, trochus and conchs, focusing on the methods that are best described. Given the wide range of culture practices for some species, especially abalone, techniques are highly variable depending on species and location. Thus, more detail is given for those methods for which the literature is available.

22.2 Abalone

22.2.1 Commercial importance

There are over 90 species of abalone that occur along the west coast of North America, eastern and southern coasts of Asia, Pacific Islands, Australia, New Zealand, Africa and Europe. However, only some of these are commercially exploited (Table 22.1), and this is usually for the meat of their muscular foot. The beautiful nacre layer inside the shell of some species, such as the paua (*Haliotis iris*) from New Zealand, is also used in handicraft industries. The culture of pearls from abalone began development in the 1960s, but it was in the late 1980s and early 1990s that abalone pearls entered the market to become a very lucrative activity. Abalone pearl farms have been developed in several countries, including New Zealand, Korea and the USA.

Wild abalone stocks have consistently declined worldwide, reaching critically low levels in some areas, such as California, where commercial fisheries were closed for 10 years in 1997 (McBride 1998). Wild abalone stocks in regions previously thought to be sustainable, such as Australia, New Zealand and South Africa, are also showing significant declines in at least some locations (e.g. Cook, 1998; Shepherd *et al.*, 2000). Australia is regarded as having a well-managed industry, although stock depletion is apparent at some locations. Paralleling this worldwide decline has been a continuous increase in aquaculture production. It is estimated that between 1989 and 1999 the world abalone fisheries declined by 30% (from 14 830 mt in 1989 to 10 150 mt in 1999), whereas production of cultured abalone increased by 600% (from 1220 mt in 1989 to almost 8000 mt in 1999) (Gordon, 2000). Over 75% of cultured abalone production is from Asia, mainly from China and Taiwan (Fig. 22.1) and the main consumers of abalone (wild and cultured) are Hong Kong and Japan. Demand is still higher than supply, i.e. it is a demand-driven market (Chapter 12), so interest in the aquaculture industry remains very high.

There has been a steady increase in world abalone prices over the past 15 years. For example, prices for whole (in shell) wild abalone were over US$8/kg in California in the late 1980s and early 1990s (Tegner *et al.*, 1992). Prices for abalone 'in shell' are now more uniform worldwide, but still vary largely depending on how the product is sold. Cultured abalone are sold in a variety of sizes and styles (live, dried, frozen, canned), some more valued than others, depending on the country of destination. The live large abalone (60–85 mm) is one of the most highly prized in China, reaching values of up to US$55/kg for producers in Australia (Fleming, 2000). Although

Table 22.1 Commercially important species of abalone (*Haliotis* species)

Species	Common name	Locality
*H. asinina**	Donkey's ear	Philippines, Thailand, tropical Australia
H. cocinea canariensis		Canary Islands
H. corrugata	Yellow, pink	USA, Mexico
H. cracherodii	Black	USA
H. discus	Kuro awabi	Japan
H. discus hannai	Ezo awabi, wrinkled	China, Japan
H. diversicolor	Colourful	China
H. fulgens	Blue, green	USA, Mexico
H. iris	Paua, black foot	New Zealand
H. kamtschatkana	Northern, pinto	Canada
H. laevigata	Greenlip	Australia
H. mariae	Omani	Arabian Sea
H. midae	–	South Africa
H. roei	Roe's	Australia
H. rubra	blacklip	Australia
H. rufescens	Red	USA, Mexico, Chile†
H. tuberculata	Ormer	Europe

*One of the fastest growing species of abalone.
†*H. rufescens* was introduced to Chile in the early 1980s.

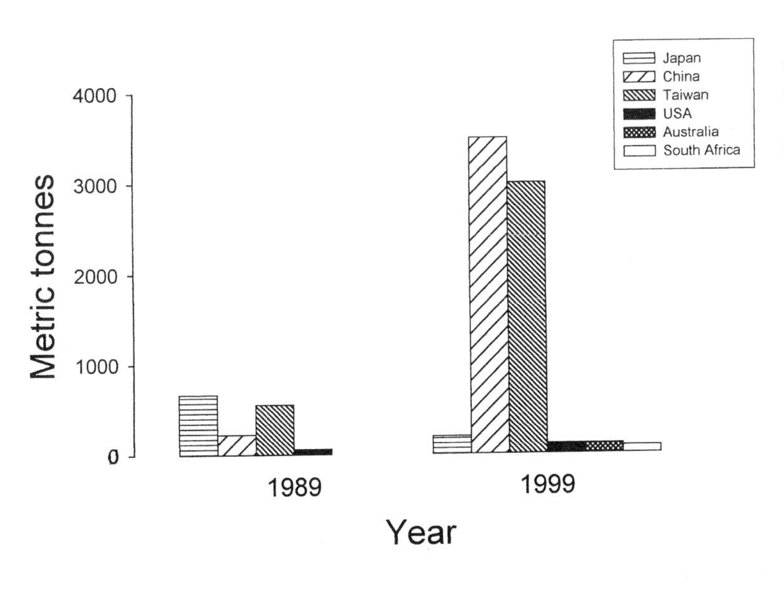

Fig. 22.1 Production of cultured abalone over a 10-year period in the countries with highest production. Modified from Gordon (2000) with permission.

abalone culture has been practised for more than 60 years, it was during the 1970s that a number of important aspects of abalone culture were successfully investigated. Japan was probably the first country to culture juvenile abalone on a large scale; however, the late 1980s and early 1990s witnessed the development of abalone farms in several other countries such as South Africa, Chile, Australia, USA, Mexico and China. There are now farms in practically all countries with a history of commercial fisheries.

22.2.2 Biology

Abalone are prosobranch gastropods that live attached to rocks and boulders on rocky shores exposed to moderate to strong wave action. Some species are restricted to shallow waters, but most species are found to a depth of several tens of metres. Sexes are separate and sex can be determined by the colour of the gonad, which can be viewed by turning an abalone upside-down and pushing the foot aside. The gonad develops on top of the digestive gland, showing a dark-brown colour when immature, and turning green in females and cream-white in males as they mature. A visual index to describe the degree of ripeness of the gonad in abalone has been developed, based on colour and size.

Abalones are broadcast spawners; the eggs and sperm are shed freely into the surrounding water, where fertilisation occurs (Fig. 22.2). Spawning occurs during the warmer months of the year in many species; however, there may be significant variation according to species, locality and food availability. For example, the red abalone is capable of spawning throughout the year at the lower latitudes of its range, whereas it spawns only in the warmest months at higher latitudes (Hooker & Morse, 1985).

22.2.3 Culture techniques

Broodstock

There are two options for obtaining mature broodstock.

(1) Collect mature abalone during the spawning season and keep them in tanks for a short period while spawning occurs. This is recommended for small-scale culture systems.
(2) Collect adults for conditioning in indoor tanks a few months before the spawning season.

As artificial diets for adults are developed, this second option is likely to become more popular. In areas where cultured abalone have been introduced, such as *Haliotis rufescens* in Chile, research is directed towards closing the life cycle under culture conditions, to minimise the risk of introducing diseases with further imported abalone (Godoy & Jerez, 1998).

Given adequate nutrition, water temperature seems to be the major environmental factor regulating the breeding cycle of abalone. Ambient temperature controls the rate of gonad development and therefore the timing of spawning. For example, daily increase in gonad index of *Haliotis discus* is greater as temperature increases (Fig. 22.3). Furthermore, there is a positive relationship between the effective accumulative temperature (EAT), which is a measure of the total amount of time at a given temperature, and the percentage of abalone spawning. For this species, the appropriate conditioning regime is 1000–2000 degree-days, achieved by keeping the abalone for 80 days at constant temperature of 20°C, or for longer at a lower temperature (Uki & Kikuchi, 1984). Thus, by keeping adults in good culturing conditions in terms of water quality and food availability, and at warmer than ambient temperatures, similar to those present during the spawning season, gonad development is accelerated. Up to 150 gravid animals can be kept for up to 18 months in a 1000-L tank provided with abundant food supply, adequate conditions of water flow and periodic cleaning (Tong & Moss, 1992). Adults are fed on seaweeds, such as *Macrocystis*, *Lessonia*, *Ulva* and *Gracilaria*, depending on availability and abalone species. Broodstock are successfully fed on locally manufactured artificial feed in Australia.

Spawning induction

Ripe abalones often spawn spontaneously when collected, probably in response to handling stresses. Spawning can also be induced using a variety of methods. The three most commonly used are the

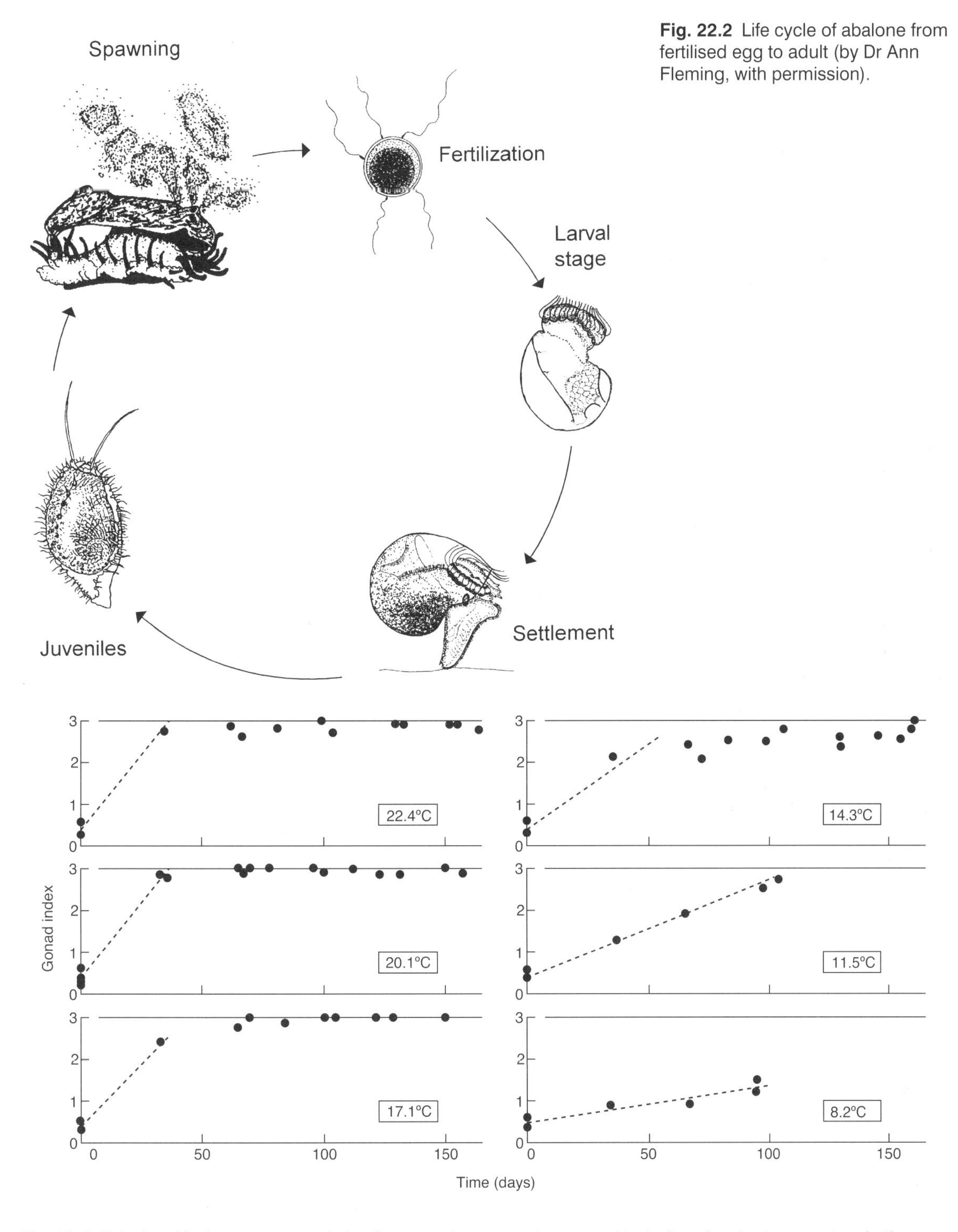

Fig. 22.2 Life cycle of abalone from fertilised egg to adult (by Dr Ann Fleming, with permission).

Fig. 22.3 Relationship between gonad development (expressed as gonad index) and water temperature in the abalone *Haliotis discus hannai* showing slower development at colder temperatures. Reprinted from Uki & Kikuchi (1984) with permission from Elsevier Science.

addition of hydrogen peroxide (H_2O_2) to the holding water, irradiation of holding water with ultraviolet (UV) light and increased water temperature.

(1) H_2O_2 added to alkaline seawater induces gravid individuals to spawn. The peroxide stimulates the synthesis of a prostaglandin that is thought to trigger the spawning response (see Hooker & Morse, 1985). Pairs of mature abalone of the same sex are placed in containers of ~12 L, filled with slightly alkaline seawater (pH 9.1), and H_2O_2 is added to a final concentration of 6% (v/v) and left for a maximum of 3 h (Tong & Moss, 1992). The water is then replaced with fresh clean seawater. Spawning generally occurs within 2 h and the time of response is positively related to temperature.
(2) Irradiation with UV light consists of passing water through a UV steriliser. The intensity of UV light required depends on the species, and the quality and temperature of the water. In Japan, an intensity of 2.42 W/L has been found to provide best results. Spawning response is generally observed within 3–8 h of exposure (Hahn, 1989; Fleming, 2000).
(3) Increasing the water temperature by a few degrees can induce spawning. A sudden change in water temperature (lower or higher) may also induce spawning in abalone; however, this 'thermal shock' method may result in variable spawning responses and immature gametes being spawned.

Desiccation or exposure to air is another induction method used in some farms. Generally, farms will combine two or three methods to induce spawning.

Released eggs are fertilised with a small volume of water containing sperm at a minimum density of 25 000/mL (Tong & Moss, 1992). After a maximum of 15 min, the eggs are repeatedly washed with filtered seawater and then transferred to clean hatching tanks.

Larval development

Larval development of abalone is illustrated in Fig. 22.2. The time between fertilisation and hatching varies between 10 h and 24 h, depending on species and water temperature. Generally, development is faster at higher temperatures, up to a level beyond which higher temperatures are detrimental to development, e.g. the optimum temperature for rate of development is about 24°C in *H. rufescens* (Leighton, 1974).

A few days after hatching, veliger larvae are collected and transferred to settlement tanks. Larval density varies between hatcheries and ranges from 4–8 larvae/mL to 20–30 larvae/mL. Under favourable conditions, veliger larvae reach metamorphic competence 3–8 days after fertilisation. Abalone larvae are lecithotrophic (section 6.3.4). As such, an exogenous food supply is not required; however, supplementing the water with dissolved organic matter (DOM) results in better growth. Larvae with sufficient yolk reserves and access to DOM in the water can extend their larval life; however, if they fail to settle after a certain period, they show low post-settlement survival (Kawamura *et al.*, 1998).

Settlement and metamorphosis

The highest rate of mortality occurs during this time, with survival typically ranging between 5% and 20%. This makes the period between settlement and the first few months one of the main milestones of abalone culture (Jarayabhan & Paphavasit, 1996; Cook, 1998; Fleming, 2000) (Table 22.2).

Various natural and chemical substances have been shown to induce settlement and metamorphosis of abalone larvae, but responses to the various stimuli differ between species. Encrusting coralline algae, microalgae (diatoms), bacterial films and abalone

Table 22.2 Example of survival rate at various stages of culture development of abalone

Period of development	Survival rate (%)
Eggs to fertilisation	80–85
Fertilisation to larvae	80
Larvae to settlement	50
Settlement to 150 days	5–20
150 days to first year	80–90
First year to second year	90–95
Second year to 3–4 years	90–95

The period between settlement and the first few months is the most critical.
Modified from Fleming (2000) with permission from the FRDC Abalone Aquaculture Subprogram.

mucus are among the natural inducers. The amino acid neurotransmitter, γ-aminobutyric acid (GABA), has also been found to induce settlement and metamorphosis in several species of abalone (Morse, 1992). Some hatcheries use conditioned settlement plates covered with diatoms and bacteria to induce metamorphosis, whereas others use GABA dissolved in culture water. Good results in terms of growth and survival have been demonstrated with the use of pregrazed settlement plates. These are prepared by first allowing small juvenile abalone to graze on plates conditioned with diatoms before placing them in the settlement tanks (Slattery, 1992). This technique is now widely used in Japan (Takami *et al.*, 1997). The bacteria present in the mucus left by juvenile abalone of the same species, combined with the mucus of the diatoms and the diatoms themselves, probably provide a good food source for the newly settled larvae as well as the older postlarvae (PL).

Settling tanks are conditioned for several days to allow build-up of bacteria and diatoms on the tank walls. In some cases, tanks are inoculated with a diatom slurry (e.g. *Navicula* or *Nitzschia*) ~24 h before introducing the larvae. Conditioned settlement plates are also generally placed inside the tanks as they increase the settlement surface.

The veligers are introduced into settling tanks at densities of 1000–2000/m^2 and left with minimal or no water flow for 1–4 days, until they have settled onto the plates or tank walls. There are different designs of settlement tanks used worldwide. For example:

- 900-L V-shaped tanks stocked with vertical conditioned plates are used in New Zealand (Tong & Moss, 1992)
- 300-L circular tanks arranged in tiers are commonly used in the USA and Mexico (Ebert, 1992)
- rectangular tanks (approximately 40 cm deep × 1.5 cm wide and up to 3–5 m long) with racks of conditioned plates suspended within are used in Japan, China and Australia.

Larvae are kept in settlement tanks (or raceways) with conditioned plates for ~2 weeks in Japan or until reaching 3 mm SL in China. They are then moved to other culture systems (Hahn, 1989; Guo *et al.*, 1999).

Juveniles

There is great variation in culture methods used for juvenile abalone. Very young abalone (2–5 mm SL), also referred to as PL, show highly variable patterns of survival and growth, which are greatly influenced by the quality and quantity of their food. It is now clear that not all species of diatoms are beneficial and the specific diet of diatoms available to postlarval abalone is very important for controlling their growth and survival (Roberts *et al.*, 1999). The different abilities of PL to digest various diatom species are largely related to characteristics such as the diatom cell size, thickness and composition of the cell wall, and how hard they are attached to the substratum. The mechanism by which young abalone select their food remains an area of much needed research.

Juvenile abalone are generally left in the settlement tanks until they reach a certain size and are then transferred to grow-out tanks, where they are left to grow until ready to sell. The size at which juveniles are transferred for grow-out culture varies. In China, 3-mm SL juveniles are moved from settlement plates to large punctured plastic plates (the holes improve water circulation). The plastic plates are supported in net-pans and placed in raceways until the juveniles reach up to 20 mm SL (Guo *et al.*, 1999). In New Zealand, juveniles are left in the settlement tanks for 6–9 months, when they reach 10–15 mm SL (Tong & Moss, 1992), whereas in the USA juveniles are harvested at 10 mm (Ebert, 1992). In Australia, juveniles are held in the settlement tanks until they reach 5–10 mm SL and then they are transferred to grow-out tanks (Fleming, 2000). In South Africa, juveniles are transferred at about 3–5 mm SL to an intermediate stage between settlement and grow-out. This stage consists of a small inverted cone sitting in a shallow tank. Animals hide under the cone and are fed algae, which is held down by the rim of the cone.

As mentioned before, small juveniles feed on diatoms and bacteria until they reach a size of around 5–10 mm SL, when they start feeding on macroalgae (seaweeds). Diatom culture is relatively easy, but in Japan and China gel-based artificial feeds are also used for the small abalone during the nursery phase (Hahn, 1989; Guo *et al.*, 1999).

The supply of natural seaweeds depends on access to reliable sources throughout the year, and this can sometimes be a problem, especially as juvenile abalone can consume between 10% and 30% of their body weight per day (Hahn, 1989; Tong & Moss, 1992). Numerous species of macroalgae have been used to feed juvenile abalone in Japan, California and Mexico. Species of the genera *Egregia*, *Macrocystis*, *Laminaria*, *Eisenia*, *Ecklonia*, *Undaria*, *Gracilaria*, *Pterocladia* and *Lessonia* are commonly used in Japan. Artificial feeds may be supplied as the only food source or as a supplement to a natural diet. As Australian abalone species prefer and grow best on red algae, the industry has not been able to access sufficient quantities of suitable algae all year round. The mass culture of red algal species has proved to be uneconomical and impractical. Consequently, the Australian industry has moved towards feeding their stock with manufactured feeds that are specifically suited to the nutritional requirements of abalone.

Grow-out culture

Juveniles are grown to market size in farms using a variety of methods including:

- long-line rafts
- seacages (or barrels)
- land-based tanks
- stock enhancement.

Factors such as the availability of land, the quality of surrounding coastal waters, and their degree of exposure to big swells and wave exposure determine the preferred culture method. Barrels and cages suspended in long-line rafts are the most common methods used in Mexico (barrels) (Aviles & Shepherd, 1996) and China (cages), although raceways stocked with plastic plates are also used in China (Guo *et al.*, 1999). In Japan, a complex raft system holding tanks and seacages is used to grow juveniles up to 30 mm. Each raft can hold 400 000 juveniles between 15 mm and 30 mm SL (Hahn, 1989). A multiple-tier basket system has recently been developed in Taiwan. Abalone are placed inside baskets and the baskets are stacked in several (up to 14) tiers in indoor cement ponds (Fig. 22.4) (Chen & Lee, 1999). Land-based tanks and a system of barrels suspended in the ocean are commonly used in California, whereas

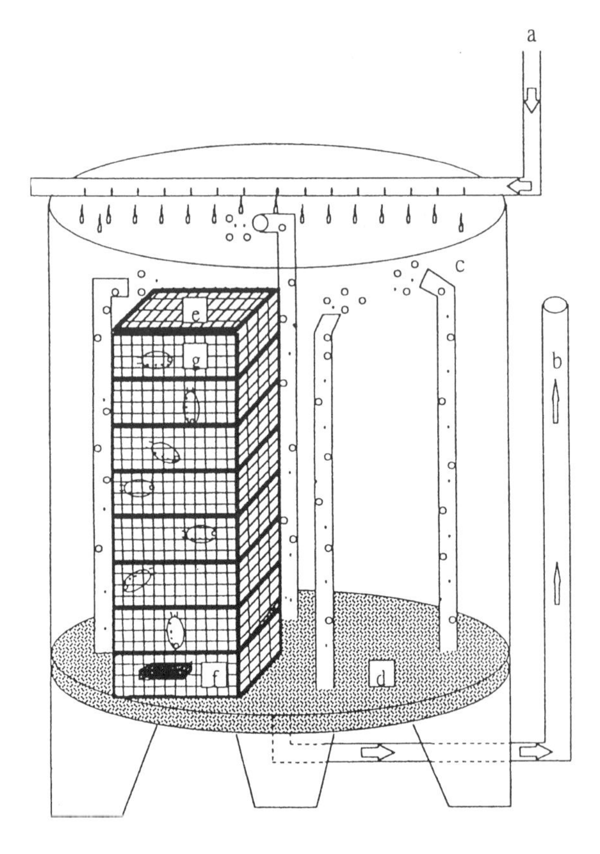

Fig. 22.4 Diagram of a multiple-tier basket system used to culture juvenile abalone (~ 26 mm SL) in Taiwan. (a) Water inlet; (b) water outlet; (c) airlift; (d) gravel; (e) basket; (f) cement block; (g) abalone. Reproduced from Chen & Lee (1999) with permission.

land-based tanks are the preferred culture method in Australia and South Africa, where coastal waters are generally exposed to strong wave action. The tier systems (using baskets or raceways) represent strategies to use space more efficiently in places where land is limited.

There has been intensive research into the design of grow-out tanks to maximise survival and growth rates of abalone, while at the same time minimising the need for labour. Early culture of abalone in Australia involved using deep rectangular tanks, but the design quickly evolved to round tanks, long raceways and a maze tank system (Fig. 22.5). In the maze system, the water flows as a unit through the tank, as in a raceway, minimising dilution and mixing, and therefore improving water quality. Each step in

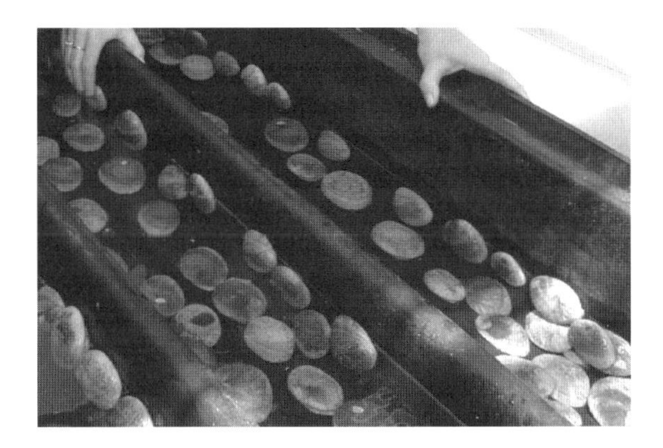

Fig. 22.5 Greenlip abalone, *Haliotis leavigata*, in a maze tank (photograph by Dr Sabine Daume, Department of Fisheries, Western Australia).

system design represented a significant improvement in water quality, maintenance requirements and ultimately growth rates (Fleming, 2000). Although it has been a common practice in many countries to import culturing techniques already developed in pioneering countries, it is now recognised that abalone species can differ in their biological requirements and the techniques most appropriate for one species may not necessarily be the same for another. Culture methods are quickly evolving to suit the specific needs of each farm in terms of location and cultured species.

Growth rates of cultured abalone vary with species and culturing conditions (e.g. size stocking density, temperature and salinity, water quality, food supply, and culturing method). Maximising growth rates is one of the main driving forces in the development of culturing techniques. Growth rates for most species, typically temperate species, are generally up to a maximum of 30 mm/year in SL. Some examples are *Haliotis laevigata* (20–30 mm/year), *Haliotis rubra* (15–25 mm/year) (Fleming, 2000) and *Haliotis iris* (~20 mm/year) (Tong & Moss, 1992). The fastest growth rates occur in the tropical abalone, *Haliotis asinina*, which can reach a size of 60 SL mm in 1 year (Capinpin *et al.*, 1999). Although the culture of this species is at a relatively early stage, its fast growth rate, small commercial size and potential to spawn throughout most of the year give it great culturing potential.

Natural seaweeds are still mainly used in abalone culture and alternatives to collection from the wild remain an important area of research. Culturing methods have been developed, in some cases for many years, for some species of seaweed that are also used for other purposes (Chapter 13). For example, the alga *Gracilaria* species has been commercially cultured in Taiwan since the early 1960s and now most of the produce is used to feed abalone (*Haliotis diversicolor supertexta*) (Chen & Lee, 1999). Seaweeds (*Ulva lactuca* or *Gracilaria conferta*) have been cultured in an innovative integrated system that also includes the culture of fish and abalone (Neori *et al.*, 2000) (section 13.2.4, Fig. 13.6).

Methods of feeding during grow-out vary considerably. In many regions, abalone are typically fed one main species of seaweed, generally the most abundant in their habitat or in the area surrounding the farm, with a few other species and/or gel-based artificial feeds used as needed if main supplies are low. For example:

- Fresh kelp (*Laminaria* species) is one of the primary foods used in China and Japan (Hahn, 1989; Guo *et al.*, 1999).
- Fresh kelp (*Macrocystis pyrifera*) meets most of the feeding needs of abalone in the Californias (McBride, 1998).
- Local kelp supplies are usually sufficient to meet demands in South Africa (Cook, 1998).

However, in many cases, other species of macroalgae and artificial diets are used, especially as local supplies of macroalgae may vary throughout the year. Artificial diets are the main method of feeding the larger abalone in Australia (Fleming, 2000).

Abalone can feed on a variety of seaweed species and they may grow better when feeding on a mixture of diets given on a rotation of one species at a time (Simpson & Cook, 1998). There is also evidence that a combination of natural seaweed and artificial diets can result in faster growth than a single-seaweed diet (e.g. Chen & Lee, 1999). Even though many farms have established methods for feeding abalone with reasonable results, there is awareness that much more can be done in terms of increasing the productivity of abalone through the choice of diets.

Artificial diets

As abalone culture increased towards commercial levels, research into the development of artificial diets intensified. Japan was probably the first country to develop and use artificial diets for abalone, but now most farms use artificial diets at some stage of their production. Despite their widespread use, commercial production of artificial diets is mainly restricted to Japan and China, with some smaller-scale production in South Africa, New Zealand and Australia.

The development of artificial diets remains an area of intensive research, as there is great pressure to reduce manufacturing costs and improve diet performance. Besides the importance of the nutritional content of diets, aspects such as making them more attractive to the abalone and reducing the rate of decomposition and leaching of nutrients are also critical to the success of feeds. These are therefore receiving increased attention. There is also the additional factor that abalone change their diet as they grow, and consequently different feeds need to be developed for small (< 10 mm SL) and larger juveniles.

Fleming *et al.* (1996) compared the composition of the many types of artificial feeds produced around the world. They found that feeds were relatively similar in their compositions, with characteristically high protein and carbohydrate content, and low levels of lipid and fibre (Table 22.3). They highlighted the fact that the choice of ingredients in the preparation of feeds has not often been based on sound scientific research, which leaves great scope for improvement in this area.

Diseases and pests

As the scale of abalone culture increased, problems caused by deterioration of environmental parameters such as water quality and contamination, together with diseases and pests, have also intensified. The advent of diseases and pests that seriously affect abalone culture occurred in the mid-1990s. Two main types of infectious diseases and two pests have been identified.

Table 22.3 Approximate mean proportions of the main nutritional components of artificial feeds for abalone (information collated from Fleming *et al.*, 1996)

Nutrient	Average (% dry matter)	Range of values (% dry matter)
Protein	30	20–50
Lipid	4	1.5–5.3
Carbohydrate	47	30–60

(1) Viral infections, one of which, the withering syndrome, is a progressive debilitating disease that usually leads to the death of the abalone. It was initially found to affect only the black abalone (*Haliotis cracherodii*) in Californian farms, but has recently been reported to also affect the red abalone (*H. rufescens*) (Moore *et al.*, 2000). The positive relationship between the intensity in the signs of the disease and the degree of infection by Rickettsiales-like prokaryote (RLP) indicates that RLP is responsible for this disease. Abalone exposed to high temperatures are at a greater risk of infection (Moore *et al.*, 2000).

(2) A bacterium (*Vibrio fluviales*) causes the pustule disease, which affects the Asian abalone (*H. discus hannai*), resulting in high mortality of abalone in hatcheries (Taiwu *et al.*, 1998).

(3) Infestation by boring polychaetes, a sabellid worm (*Terebrasabella heterouncinata*), was first discovered in the mid-1990s in Californian abalone farms as a result of its damaging effect on the abalone industry (Oakes & Field, 1996). The sabellid is endemic to South Africa and has since also been found to cause infestations in South Africa (Ruck & Cook, 1998). These pests bore tiny holes into abalone shells. Infected individuals show markedly reduced growth rates and meat yield, and they are generally weaker and more likely to die.

(4) Another boring polychaete, the mudworm (*Polydora* species), has affected mainly sea-based farms in Tasmania, Australia, leading to mortalities of more than 60% of the stock. Removal of infected stock in combination with fast growing conditions and good water quality are successful measures to combat this pest (Fleming, 2000).

22.2.4 Stock enhancement

Cultured juveniles (seed) have been released into their natural habitat in many regions, and in some regions for many decades, but the success of this stock enhancement (seeding) practice is highly variable and still questionable (Chapter 1). Many factors seem to affect the success of these releases, such as juvenile size, habitat characteristics, predator types and their abundances, and handling techniques. In most cases, seeding appears to offer little as a means to maintain wild populations, although there have been some examples with promising results. Despite the variability in results, mass stock enhancement with juvenile abalone is commonly practised in Japan, as it is with other shellfish and fish.

22.3 Conchs

22.3.1 Commercial importance

Although there are numerous *Strombus* species, the main commercially exploited species is the queen conch (*Strombus gigas*) (Fig. 22.6). Other harvested species are the white conch (*Strombus costatus*) in the Caribbean, *Strombus luhuanus* in the Indo-West Pacific and, more recently, the Florida fighting conch (*Strombus alatus*), which has very fast growth and is currently grown for the aquarium market. Conchs are primarily used for the meat of their muscular foot, but their large shell also has ornamental value. The queen conch is one of the most prized fishery resources in the Caribbean. The estimated landing of conchs between 1988 and 1990 was slightly over 4000 mt, with Cuba and Jamaica being the major producers. Fishery landings have increased significantly with recent estimates at 6000 mt and a corresponding potential value of US$60 million (Chakalall & Cochrane, 1997). Wholesale prices for wild harvested conchs for 1997 ranged between US$5.50 and US$9.50/kg. Wholesale prices for cultured queen conchs are considerably higher than those for wild conchs, with values for 1999 ranging between US$18 and US$48/kg of meat, depending on product type (e.g. shell size, live or shucked) (Davis, 2000a). Cultured conchs can be sold live and, perhaps because they are of much younger age than wild animals, they are described as sweeter and more tender.

The potential for conch mariculture was well recognised in the mid-1970s, but development of techniques to culture and manage the resource properly did not begin until the 1980s. Aquaculture of conchs has until recently focused on queen conchs.

22.3.2 Biology

Queen conchs are found throughout the Caribbean, from Bermuda and the Bahamas in the north, to the Brazilian coast of South America. The natural habitat of the adults is shallow (< 8 m) seagrass beds that are protected from wave exposure and which contain sand patches and coral rubble covered with turf algae. The natural habitat of juveniles is not well known. Juveniles are first observed when they are 35–40 mm SL, in sandy shoals and shallow seagrass meadows. The large variability in natural densities

Fig. 22.6 Size progression of the queen conch *Strombus gigas* showing changes in shell form and development of the flared lip in adults (photograph by Ms Katherine Orr).

of juveniles and survival of experimental releases of wild and cultured juveniles suggest that several factors determine the suitability of a particular habitat. These include intermediate density of seagrass, high algal productivity, high oceanic water quality, low predation intensity and abundant larval supply (Stoner *et al.*, 1995).

Queen conchs are dioecious and fertilisation is internal as the male copulates with the female. The female then lays egg masses within several weeks of copulation. Reproduction generally occurs during the warmer months of the year and lasts 6–8 months, depending on the locality. In Venezuela, the spawning season extends from late April to late November (Weil & Laughlin, 1984), whereas in the Turks and Caicos Islands it extends from late March to late October (Davis *et al.*, 1984). This variability is related to differences in environmental factors, mainly water temperature and photoperiod.

22.3.3 Culture

Dr Megan Davis-Hodgkins, Harbor Branch Oceanographic Institution, is an authority on conchs and has published a number of papers on their culture. Much of this information is contained in Davis (2000a) and this is the main source of the following sections on culture, unless otherwise indicated.

Selection of breeding site

Culture of conchs has relied on the collection of egg masses from the wild during the spawning season, as adults have not been bred in captivity until very recently. To collect eggs from the wild, it is necessary to identify good breeding habitats with natural resident broodstock or to move the broodstock to enclosed breeding sites (commonly known as 'egg farms').

The desired criteria for a breeding site are:

- areas naturally inhabited by conchs (presently or in the past)
- areas at least partially enclosed by natural boundaries such as coral outcrops, mounds or a steep slope
- substrate formed of calcareous sand and coral rubble with low organic content
- slow water movement (< 0.5 m/s)
- depths of between 4 m and 8 m
- within 3 km of the hatchery to facilitate continuous monitoring of the area.

The size of a breeding enclosure will depend on the required production capacity of the hatchery. Recommended stocking density of broodstock is one conch per 10 m^2, based on natural densities of undisturbed populations. An area of 1500 m^2 holding a broodstock of 150 would supply sufficient eggs for a commercial hatchery.

Spontaneous spawning of *S. gigas* and another three conch species in culture tanks was reported for the first time early in 2001 at the Harbor Branch Oceanographic Institute. This means that culture activities will very likely become independent of eggs from the wild and the need for egg farms will eventually disappear.

Broodstock

Queen conchs become sexually mature at 3–4 years of age. With the onset of sexual maturation, growth in shell length ceases and the pink flared-lip is formed by progressive thickening (Fig. 22.6). The thickness of the lip has been used as an indication of sexual maturity, with mature conch generally having a lip thicker than 4 mm. The size at which individuals become sexually mature can show great variation. For example, in Venezuela, size at maturation varies between 180 mm and 260 mm SL (distance from the tip of the siphonal canal to apex of the spire) (Weil & Laughlin, 1984).

Sexually mature conchs are collected and moved to the breeding enclosure 1 month before the beginning of the spawning season. Sex cannot be determined from characteristics of the shell, but is determined by turning the shell on its side with the lip pointing up, and waiting for the animal to right itself. When doing this, the reproductive structures (the verge or penis in males and the egg groove in females) become visible. The position during mating can also be used to determine sex, as females are positioned in front and males behind. Although sexual differentiation occurs at the juvenile stage, the reproductive structures only begin development with lip formation on the shell.

Production and collection of egg masses

Egg mass production shows seasonal variation and is positively related to water temperature and weather conditions. The crescent-shaped egg masses are generally deposited in sandy substrates with some coral rubble covered with algae or seagrass. It takes between 24 h and 36 h for a female to complete laying an egg mass (Randall, 1964) and several egg masses may be produced from a single copulation. The average number of egg masses per female is in the order of 1–4 per month or 8–25 egg masses per female per reproductive season. An egg mass of ~100 g wet weight contains 400 000 eggs (Weil & Laughlin, 1984).

The 'egg farm' is visited once or twice a week when females are first observed laying egg masses. Fresh egg masses are collected and transported to the hatchery in plastic bags filled with seawater. In the hatchery, egg masses are cleaned and disinfected with 0.5% sodium hypochlorite for 45 s. After thorough rinsing, they are left to hatch in filtered water with continuous water flow, or in static water tanks with periodic water changes. More detailed information on the treatment of egg masses is provided by Davis (2000a).

Larval rearing and settlement

Larvae commence hatching 3–5 days after the egg mass is produced. The egg mass is removed after hatching is completed and veliger larvae are siphoned out and moved to larval rearing tanks. Clean filtered seawater, periodic aeration and an adequate food supply are necessary for high larval survival. A detailed description on the handling of egg masses during hatching is given by Davis (2000a).

Unlike abalone and trochus, which have brief periods of lecithotrophic larval development, conchs have a long period of planktotrophic larval development. This means that, compared with the first two species, much more time and effort must be invested in rearing conch larvae, including the need for large-scale cultures of microalgae. Development of conch larvae to metamorphosis may take up to 40 days, depending on culture conditions. Under different culturing conditions, larval growth rates (SL increments) vary between 20 μm and 50 μm/day. Water temperature, water quality, salinity, food availability and quality, and culture density influence larval development rate. Davis (2000b) reported that larvae were able to grow and complete development between 16 days and 24 days, over a relatively large range of temperatures (24–32°C) and salinities (30–40‰). These are typical of the reproductive season in the natural environment. Optimal growth rates, however, were at 32°C and 40‰ salinity.

Larval culture is one of the most difficult aspects of conch culture. Three critical developmental stages have been identified in veligers and these are related to nutrition:

(1) at 3–5 days after hatching, when food becomes increasingly important
(2) around 6–11 days when yolk reserves are exhausted
(3) at the time of metamorphic competence.

Conch veligers are fed microalgae. The flagellate *Isochrysis* species and the diatom *Chaetoceros gracilis* are among the most commonly used species. In the commercial farm in the Turks and Caicos Islands, the alga *Isochrysis galbana* or Caicos *Isochrysis* is the only food supplied during the first 10 days. To feed 10- to 21-day-old larvae, a diatom (*C. gracilis* or *Chaetoceros muelleri*) (Davis & Shawl, 2002) is also included in the diet at a ratio of 6:1 (*Isochrysis* to *Chaetoceros*). The dense microalgae stock may be added by continuous dripping in a flow-through larval culture system or as a batch feed after the daily water change in a static larval culture. An example of some recommended stocking densities and feeding regimes used to rear larval and juvenile conch is summarised in Table 22.4.

Water changes are essential throughout larval development and they begin the day after hatching and continue every 48 h for veligers less than 16 days old and every 72 h for older veligers. These changes must be performed with extreme care as veligers are large and easily stressed during the process (Davis, 2000a; Davis & Shawl, 2002).

The larvae are competent for metamorphosis at ~1.2 mm SL and when they show characteristic features that reflect their subsequent mode of life (Davis *et al.*, 1990). Competence is generally reached 21

Table 22.4 An example of stocking densities and feeding procedures used to rear juvenile queen conchs in a static system (information collated from Davis & Shawl, 2002)

Stage	Density of individuals	Food concentration	Period
Veligers	Initial: 150–200/L Final: 30–60/L	*Isochrysis* (5000 cells/mL) for newly hatched larvae 25 000 cells/mL for metamorphically competent larvae + *Chaetoceros* (3300 cells/mL) from day 15	3 weeks
Postlarvae 3–4 mm*	3200/m^2	10 mg gel food/conch/day	2 weeks
Juveniles 4–12 mm	1600/m^2	Gel food†	2–4 weeks
Juveniles 12–30 mm	480/m^2	Gel food	4–7 weeks
Juveniles 30–40 mm	275/m^2	120 mg gel food/conch/day	

*Shell length.
†The daily amount of gel food provided to conchs is determined as a percentage of their soft-tissue wet weight. It is between 2% and 15%, with smaller juveniles getting a higher ration.

days after hatching. Metamorphosis can be induced using:

- water-soluble extracts of a natural inducer
- the red alga, *Laurencia* species
- neuroactive compounds such as KCl and H_2O_2.

Laurencia species extracts have been used for many years as inducers of metamorphosis of conch veligers, but the collection and processing of plants, as well as the natural variability in the extracts, can be reflected in their ability to induce metamorphosis (Davis *et al.*, 1990). This led to research into the use of cheaper and more readily available alternatives, and H_2O_2 is now most commonly used as the inducer.

Studies to improve larval feeding efficiency, growth and survival are continuing, and they are considered a priority area in conch aquaculture research.

Juveniles

Newly metamorphosed conchs or PL (3–4 mm SL) are placed in sand trays at densities of 3200 conchs/m^2. These trays have a window screen sewn to the bottom to allow water circulation, and they are filled with a layer of 1-cm-deep clean sand. The trays are placed in fibreglass tanks and cleaned periodically. Food is always provided after cleaning. Juveniles are generally active between sunset and sunrise, and activity during the day is a sign of insufficient food (Davis, 2000a; Davis & Shawl, 2002).

Conch PL graze on benthic algae. However, fast-growing planktonic algae, such as the diatom *C. gracilis*, can be used as a food source for conch of 1–4 mm SL after flocculation with chitosan (a macromolecule from the exoskeleton of crustaceans). Davis (2000a) provides a detailed description of this process. The size of individuals must be taken into account to determine how much food to supply (Table 22.4).

After reaching a more manageable size (12–15 mm SL), juveniles are cultured using a variety of methods, including shaded ponds and raceways. These may be made of concrete or with wooden walls around a sandy bottom. Densities need to be reduced as juveniles grow, from 275 conchs/m^2 for 12–15 mm SL to 30–50 conch/m^2 for conchs of 60 mm SL.

Artificial feeds for juveniles larger than 5 mm SL became available in the late 1990s. Two types of artificial foods are used to feed juvenile conchs, a gel-based diet and a pelleted conch chow (Davis & Shawl, 2002). They are now used in some farms and support growth at least comparable to that observed using natural diets. However, there is scope for much improvement as little research has been done on the comparison of growth rates and food conversion ratios of various artificial and conventional natural diets.

Grow-out

Sea enclosures can be used to grow the larger juveniles, from 9 cm to 16 cm in SL over 1 year. Protection from predators using appropriate fencing material and traps is needed. Densities in these enclosures are kept at natural levels (one to two conchs per m^2) and groups must be rotated around several enclosures to avoid overgrazing.

22.3.4 Stock enhancement

As with abalone and trochus, the decline in natural stocks and the rapid development of culturing techniques led to intensive research into the release of culture juveniles to enhance wild stocks. After numerous studies and experimental releases, it seems that prospects for successful enhancement are poor as a result of low growth rates and high mortality of released individuals (Stoner, 1997). Many factors beside juvenile size (arguably the most critical) have been identified as important in affecting the survival of released individuals. Some of these factors given by Stoner (1997) are:

- predators
- conch density
- habitat characteristics
- season
- even behavioural and morphological differences between wild and cultured conch juveniles.

This means that it is difficult to predict the outcome of releases. However, the Conch Heritage Network (CHN) comprises a group of scientists from various organisations working towards using a combination of good management, captive breeding, habitat conservation and community education to ensure the future of the conch resource (Davis & Shawl, 2002)

22.4 Trochus

22.4.1 Economic importance

Trochus (*Trochus niloticus*) (Fig. 22.7) is a prosobranch gastropod and the largest member of the family Trochidae (top shells), reaching a maximum shell diameter (SD) of 160 mm. Although there are several species in the genus *Trochus*, the name 'trochus' is used here only with respect to *T. niloticus*.

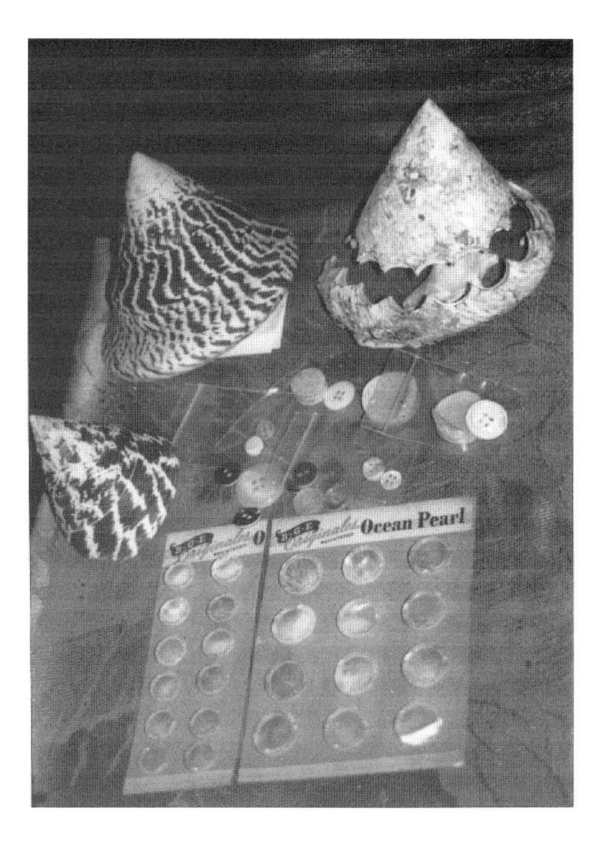

Fig. 22.7 Shells of *Trochus niloticus* showing the characteristic zig-zag pattern and their use in manufacturing buttons (photograph by Professor John Lucas).

Trochus are of major economic importance in the tropical Indo-West Pacific. There are commercial trochus fisheries in Australia, Fiji, French Polynesia, Indonesia, Japan, New Caledonia, Papua New Guinea, the Philippines, Vanuatu, and in the Cook, Caroline, Mariana, Marshall and Solomon Islands. In some Pacific islands, trochus have been, for many years, a major source of family income and a major export product. The pearly nacre inside the shell is used to make 'pearl' buttons and handicrafts, whereas the meat of the muscular foot is eaten fresh or dried.

Since the 1920s, trochus have been introduced to new areas, including the Cook Islands, Guam, Hawaii, Kosrae, Tahiti, Tuvalu, Yap and other Pacific islands. Not all introductions have been successful, but in the Cook Islands, 24 years after the release of

300 adults, the first year of harvest yielded 200 mt of shells.

Although trochus have been commercially fished for many decades, the fishery experienced a boom during the 1970s as a result of increasing demand for shells for button manufacture. World production of trochus shell is currently estimated to be ~3900 mt/year, 59% of which is produced by the Pacific islands, where shell is processed to produce 'blanks' mainly for Korea and Japan (ICECON, 1997). The second and third producers of trochus are Australia (500 mt) and Indonesia (475 mt) (ICECON, 1997). Demand, mainly from Japan and Europe, has been higher than supply, in the order of 7600 mt/year (FAO, 1992). The Pacific islands export prices for unprocessed shell vary between years and localities, as well as with trochus quality and shipping charges. Prices for unprocessed shell in 1995 were on average US$3.41/kg (range US$2.50–5.30/kg) (ICECON, 1997). An increase in the use of trochus pearly nacre is expected in the fashion industry in the near future with a return to 'natural' buttons.

Overexploitation of wild stocks has invariably been observed in areas with commercial fisheries. Interest in culturing trochus as an alternative to fisheries resulted in the first successful broodstock spawning and culture of larvae at the Micronesian Mariculture Demonstration Center, Palau (Heslinga & Hillmann, 1981). Since then, techniques have been developed for large-scale culture of juveniles. A few thousand juveniles up to 20 mm in SD can easily be produced using simple and relatively inexpensive techniques. The relatively simple requirements for trochus culture make it ideal for the development of small regional hatcheries. Trochus are or have been cultured in many regions, including the Cook Islands, Guam, Indonesia, Kosrae, Japan, New Caledonia, Palau, Solomon Islands, Vanuatu and Australia (Lee & Amos, 1997). Aquaculture production of trochus has focused on research into culturing techniques to produce juveniles largely for population enhancement. There are no commercial farms of large trochus at present, but these may develop in the future if techniques to feed and keep large individuals (> 50 mm SW) are developed. In 1996, a private commercial hatchery was opened at Seram, eastern Indonesia, to produce juveniles for reseeding onto reefs (Lee & Amos, 1997).

22.4.2 Biology

Trochus is a herbivore that grazes on the algal turfs covering coral rock. It inhabits the intertidal and subtidal zones of coral reefs exposed to wave action and is generally found at depths shallower than 10 m. The most favourable habitat for trochus is thought to be on reefs with an extensive flat of coralline rubble and boulders, exposed at low tide and sloping progressively offshore. Juveniles are most frequently found on the intertidal reef flat. In surveys along the north and central Great Barrier Reef, in Guam, and in Okinawa, juveniles of 2–70 mm in SD occurred intertidally, whereas larger individuals were subtidal (Smith, 1987; Nash, 1993; Castell, 1997). It is difficult to find juveniles smaller than 20 mm in SD in the wild. They are very cryptic and careful searching is required to find them.

Trochus are dioecious. To determine sex, the shell has to be broken to observe the colour of the gonad, which is green in females and pale brown to cream in males. Clearly, this method cannot be used when collecting broodstock for culture; however, when the animal is pushed back into the shell, extrusion of some gametes may occur from gravid individuals.

22.4.3 Culture methods

Broodstock and spawning

The reproduction of trochus has been investigated in recent years, but some aspects are still not completely established. These aspects include spawning periodicity and the relationship between spawning and the lunar cycle.

Although trochus may have periods of more intense spawning in the wild, it may be possible to obtain a successful spawning throughout most of the year (Nash, 1993). This is possible, with the exception of one or two months that are likely to coincide with the periods of lowest mean gonad index (gonad size/body size) (e.g. Pradina *et al.*, 1997). The general pattern is for spawning to occur over the warmer months, but variability in this pattern suggests that factors other than temperature may also play an important role in the reproductive activity of trochus in the wild.

Males held in tanks may spawn on successive nights and in successive months. Females, however, have not been observed to spawn more than once within a month. Furthermore, the proportion of males spawning in tanks is generally much larger than the proportion of females.

For spawning, trochus broodstock are best collected a few days before the new moon. Sexual maturity is reached in the range of 50–90 mm in SD and most trochus larger than 80 mm in SD are reproductively mature. Between 20 and 50 individuals are needed for small-scale production and, at a larger scale, 100 adults may be used. Spawning methods vary with locations and, although induction methods are becoming more widespread, small hatcheries can generally rely on spontaneous spawning or simple induction methods such as aeration and warming the water. The newly collected broodstock are kept in tanks with static and strongly aerated seawater. After 24 h, the animals are moved to a clean tank of filtered seawater where they are spread on the bottom. Spontaneous spawning is generally observed any time during the afternoon or evening. Periodic checks are necessary and surplus spawning males removed to avoid large volumes of sperm being shed into the water.

Spawning induction is practised for larger scale production of juveniles. In Okinawa, UV-irradiated water is used to stimulate spawning, and water temperature may also be raised by 5°C. Spawning usually occurs 1–3 h after induction. Success is not always guaranteed, with the percentage of trochus spawning varying from 0 to 83% (Murakoshi, 1991). In Australia, temperature control has been used successfully (Lee, 1997). The spawning tank is cleaned and drained shortly before sunset and the tank is then filled with heated water (2–3°C higher). If no spawning is observed, the process is repeated after 2 h. Observations on trochus spawned without induction in tanks in Australia (18–19°S) suggest that a minimum temperature of 25.5°C is required for spawning. Above 32°C, individuals are stressed and remain retracted inside their shells.

Males generally spawn first, releasing white aggregates of sperm. The quantities of sperm released can be very large and may rapidly cloud the water, impairing visibility and increasing the possibility of polyspermy (section 6.3.3). Spawning females may sometimes be identified because the gametes are expelled in a series of spasmodic contractions. At other times, spawning is only discovered after observing eggs in the water. Females release up to one million eggs per spawning event.

The eggs, ~180 μm in diameter, are siphoned from the spawning tank onto a submerged 100-μm screen. Sperm present in the spawning tank are usually sufficient to fertilise the eggs and no deliberate fertilisation technique is used. Eggs are then cleaned of excess sperm and mucus by repeated washing with filtered seawater. This step is important in minimising sperm concentration and bacterial contamination. Finally, the eggs are transferred into containers filled with filtered (1–10 μm) seawater, preferably at densities lower than 1/mL or approximately one single layer of eggs uniformly spread on the bottom of the container. The eggs can be reared in:

(1) Static water with light aeration.
(2) Small containers lined with < 100-μm mesh and partially immersed in a larger container or tank supplied with gently flowing filtered water. Water flows out through the mesh and the gentle water movement maintains the eggs in suspension.

Larvae

Larval development of trochus is outlined in Table 22.5 and Fig. 22.8. Trochus larvae hatch as trochophores, 10–24 h after fertilisation, depending on the temperature. At 27–30°C, trochophores hatch between 11 and 13 h after fertilisation. Hatched larvae swim towards the water surface, where they can be collected for transferring into larger tanks. If

Table 22.5 Timetable of larval development of *Trochus niloticus* for rearing temperatures between 25°C and 30°C

Time since spawning	Developmental stage
0	Egg (180–186 μm)
30–65 min	First cleavage
12–13 h	Hatching trochophore
20–22 h	Veliger with complete shell
60 h	Settlement and creeping
70–84 h	Metamorphosis

Reprinted from Heslinga & Hillmann (1981) with permission from Elsevier Science.

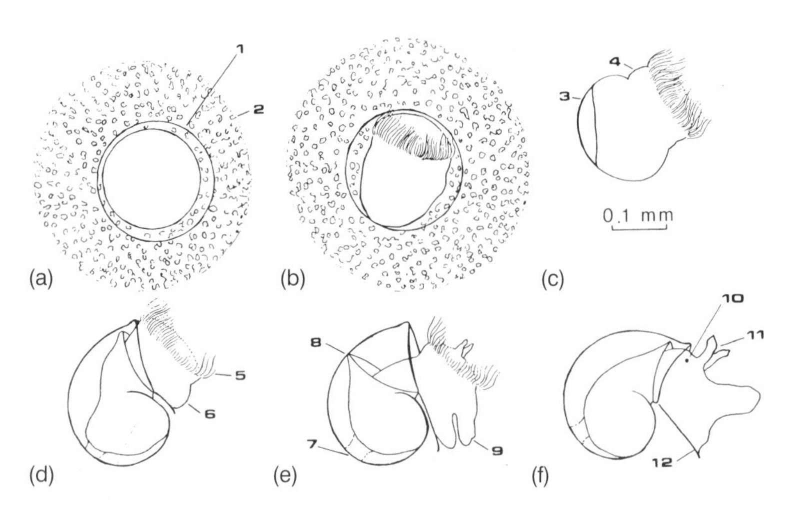

Fig. 22.8 Larval development of *Trochus niloticus*. (a) Fertilised egg; time (t) = 3 min after fertilisation. (b) Trochophore just before hatching; t = 11.5 h. (c) Pretorsional trochophore with embryonic shell growth; t = 14 h. (d) Veliger with complete larval shell, after 90° torsion; t = 20 h. (e) Veliger with reduced velum, approaching metamorphic competence; t = 65 h. (f) Metamorphosed juvenile; t = 73 h. (1) Vitelline membrane; (2) jelly layer; (3) embryonic shell; (4) foot rudiment; (5) velum; (6) foot; (7) attachment site of larval retractor muscle; (8) mantle margin; (9) propodium; (10) eye; (11) cephalic tentacles; (12) operculum (from Heslinga, 1981).

larval rearing is conducted using the continuous flow method, larvae can be left in the container until they are competent to metamorphose. Feeding during larval development is unnecessary, as trochus larvae, like abalone larvae, are lecithotrophic.

Settlement and metamorphosis

Veligers are ready to settle after ~3 days at 27–30°C. Well-developed veligers have dark pigment spots or 'eyes', a large propodium and a reduced velum (Fig. 22.8). They spend increasingly longer periods on the tank bottom with frequent but short bursts of swimming activity. At this stage the larvae are transferred to the juvenile rearing tanks (generally made of fibreglass), which have been conditioned for 2 or 3 weeks to allow a film of algae and bacteria to develop on the walls of the tank. More commonly, conditioned plastic plates or dried fragments of coral are also introduced into the tank. Water is kept static with gentle aeration for a few days, until no swimming larvae can be seen. The algal surface serves to induce metamorphosis and is utilised as a food source for newly settled individuals.

Juveniles

The nursery tanks where larvae settled are commonly used to grow juveniles, with increased water flow and continuous replacement of conditioned plates, minimising the need for handling small juveniles. After 1–2 months, the small white shells of the juveniles become visible. They graze on the algae-covered surfaces within the rearing tank. As juveniles grow, their food intake increases and care is required to ensure that food supply is abundant. Dwiono *et al.* (1997) found that 20 000 individuals of 4–7 mm in SD would graze out the algae growing on the tank surface (6.5 m^2) in just 1 week. In Okinawa, survival rates to 3.5 months of age of between 29% and 35% have been obtained rearing 100 000–200 000 larvae in 2.75-m^3 tanks (Murakoshi, 1991).

Juveniles rapidly deplete the algae within the tank and continuous replacement of algal plates is necessary. Plates are conditioned by suspending them in the sea for 1 or 2 weeks. In Okinawa, tanks with corrugated plates are inoculated with cultures of the diatom *Navicula ramosissima* (Murakoshi, 1991). There has been no development of artificial diets for trochus.

Growth rate of young juveniles varies with culturing conditions, but they grow at 1–2 mm/month and 30 mm SD can be reached in ~1 year (see Nash, 1993). The grazing substratum has an important effect on growth, e.g. conditioned bivalve shells and mixed dried coral pieces resulted in faster growth and better survival of 2-mm-SD juveniles compared with fibreglass plastic sheets (Gimin & Lee, 1997). Other factors such as culture density and husbandry techniques also affect juvenile growth rate (Clarke *et al.*, 2003). The fastest growth rate of cultured trochus was reported by Heslinga (1981), who found that juveniles reached an average 60 mm in SD in 1 year when growing at low densities in concrete raceways at 27–30°C. Such variable growth rates suggest that research into the factors that affect juvenile growth is likely to result in significant improvement of juvenile growth rates.

Grow-out

Although juveniles up to 30 mm in SD can be reared in tanks, labour requirements and running costs may be limiting when rearing on a large scale. Grow-out culture was first tried in Okinawa using floating cages made of plastic mesh (Murakoshi, 1991). Juveniles of 5 mm SD grew up to 22 mm in ~7 months. However, the main problems identified were:

- rapid clogging of the mesh
- the possibility of predators entering the cages as eggs or larvae
- poor resistance of cages to rough weather conditions.

Culturing systems in Japan have probably improved since then but such information is not available in the main published literature. Cages fixed to the reef flat have been tried in eastern Indonesia. These were made of 2-mm fishing net and concrete brick walls, and were stocked with 8- to 15-mm-SD juveniles at a density of 50 individuals/m^2. Juveniles reached an average of 44 mm SD in 37 weeks, showing faster growth than tank-reared juveniles (Dwiono *et al.*, 1997).

22.4.4 Stock enhancement

Juveniles have been released experimentally in numerous areas around the Indo-Pacific region; however, the feasibility of such practice is still under investigation. Survival rates of released juveniles have been very variable, ranging from practically no survival after only a few days to 28% after 13 months (Crowe *et al.*, 1997). There is increasing evidence that the larger the individuals are at the moment of release, the greater their chances of survival. However, large juveniles are more costly to produce. Stock enhancement with such large juveniles will only be feasible if the expenses associated with culturing techniques are minimised.

References

Aviles, J. G. G. & Shepherd, S. A. (1996). Growth and survival of the blue abalone *Haliotis fulgens* in barrels at Cedros Island, Baja California, with a review of abalone barrel culture. *Aquaculture*, **140**, 169–76.

Capinpin, E. C., Jr., Toledo, J. D., Encena, V. C., II & Doi, M. (1999). Density dependent growth of the tropical abalone *Haliotis asinina* in cage culture. *Aquaculture*, **171**, 227–35.

Castell, L. L. (1997). Population studies of juvenile *Trochus niloticus* on a reef flat on the north-eastern Queensland coast, Australia. *Marine and Freshwater Research*, **48**, 211–7.

Chaitanawisuti, N. & Kritsanapuntu, A. (1998). Growth and survival of hatchery-reared juvenile spotted babylon, *Babylonia areolata* Link 1807 (Neogastropods: Buccinidae), in four nursery culture conditions. *Journal of Shellfish Research*, **17**, 85–8.

Chakalall, R. & Cochrane, K. (1997). The queen conch fishery in the Caribbean: an approach to responsible fisheries management. *Proceedings of the Gulf and Caribbean Fisheries Institute*, **49**, 531–54.

Chen, J.-C. & Lee, W.-C. (1999). Growth of Taiwan abalone *Haliotis diversicolor supertexta* fed on *Gracilaria tenuistipitata* and artificial diet in a multiple-tier basket system. *Journal of Shellfish Research*, **18**, 627–35.

Clarke, P. J., Komatsu, T., Bell, J. D., Lasi, F., Oengpepa, C. P. & Leqata, J. (2003). Combined culture of *Trochus niloticus* and giant clams (Tridacnidae): benefits for restocking and farming. *Aquaculture*, **215**, 123–44.

Cook, P. (1998). The current status of abalone farming in South Africa. *Journal of Shellfish Research*, **17**, 601–2.

Crowe, T. P., Amos, M. J. & Lee, C. L. (1997). The potential of reseeding with juveniles as a tool for the management of Trochus fisheries. In: *Trochus: Status, Hatchery Practice and Nutrition* (Ed. by C. L. Lee & P. W. Lynch), pp. 170–7. ACIAR Proceedings No. 79, Australian Centre for International Agricultural Research, Canberra, Australia.

Davis, M. (2000a). Queen Conch (*Strombus gigas*) culture techniques for research, stock enhancement and grow-out markets. In: *Recent Advances in Marine Biotechnology, Seaweeds and Invertebrates* (Ed. by M. Fingerman & R. Nagabhushanam), pp. 127–59. Science Publishers, Enfield, NH.

Davis, M. (2000b). The combined effects of temperature and salinity on growth, development, and survival for tropical gastropod veligers of *Strombus gigas*. *Journal of Shellfish Research*, **19**, 883–9.

Davis, M. & Shawl, A. L. (2002). A guide for culturing queen conch, *Strombus gigas*. In: *Manual of Fish Culture*, Vol. 3. American Fisheries Society, Bethesda, MD.

Davis, M., Mitchell, B. A. & Brown, J. L. (1984). Breeding behavior of the queen conch *Strombus gigas* Linne in a natural enclosed habitat. *Journal of Shellfish Research*, **4**, 17–21.

Davis, M., Heyman W. D., Harvey W. & Withstandley C. A. (1990). A comparison of two inducers, KCl and *Laurencia* extracts, and techniques for the commercial scale induction of metamorphosis in queen conch *Strombus gigas* Linnaeus, 1758, larvae. *Journal of Shellfish Research*, **9**, 67–73.

Dwiono, S. A. P., Makaputi, P. C. & Pradina. (1997). A hatchery for the topshell (*T. niloticus*) in Eastern Indonesia. In: *Trochus: Status, Hatchery Practice and Nutrition* (Ed. by C. L. Lee & P. W. Lynch), pp. 33–7. ACIAR Proceedings No. 79. Australian Centre for International Agricultural Research, Canberra, Australia.

Ebert, E. E. (1992). Abalone aquaculture: a North America regional review. In: *Abalone of the World: Biology, Fisheries and Culture* (Ed. by S. A. Shepherd, M. J. Tegner & S. A. Guzman del Proo), pp. 570–82. Blackwell Science, Oxford.

Fao, B. (1992). Information on trochus fisheries in the South Pacific. *SPC Trochus Information Bulletin*, **1**, 12–5.

Fleming, A. E. (2000). The current status of the abalone aquaculture industry in Australia. *Proceedings of the 7th Annual Abalone Aquaculture Workshop, August 2000, Dunedin, New Zealand.* Fisheries Research and Development Corporation, Australia.

Fleming, A. E., Van Barneveld, R. J. & Hone, P. W. (1996). The development of artificial diets for abalone: a review and future directions. *Aquaculture*, **140**, 5–53.

Gimin, R. & Lee, C. L. (1997). Effects of different substrata on the growth rate of early juvenile *Trochus niloticus* (Mollusca: Gastropoda). In: *Trochus: Status, Hatchery Practice and Nutrition* (Ed. by C. L. Lee & P. W. Lynch), pp. 71–80. ACIAR Proceedings No. 79. Australian Centre for International Agricultural Research, Canberra, Australia.

Godoy, C. & Jerez, G. (1998). The introduction of abalone in Chile: ten years later. *Journal of Shellfish Research*, **17**, 603–5.

Gordon, H. R. (2000). World abalone supply, markets and pricing: historical, current and future prospectives. *4th International Abalone Symposium*, Cape Town, South Africa.

Guo, X., Ford, S. E. & Zhang, F. (1999). Molluscan aquaculture in China. *Journal of Shellfish Research*, **18**, 19–31.

Hahn, K. O. (1989). *Handbook of culture of Abalone and other Gastropods*. CRC Press, Boca Raton, FL.

Heslinga, G. A. (1981). Larval development, settlement and metamorphosis of the tropical gastropod *Trochus niloticus*. *Malacologia*, **20**, 349–57.

Heslinga, G. A. & Hillmann, A. (1981). Hatchery culture of the commercial top snail *Trochus niloticus* in Palau, Caroline Islands. *Aquaculture*, **22**, 35–43.

Hooker, N. & Morse, D. E. (1985). Abalone: The emerging development of commercial cultivation in the United States. In: *Crustacean and Mollusk Cultivation in the United States* (Ed. by J. V. Huner & E. E. Brown), pp. 365–413. AVI Publishing, Westport, CT.

ICECON (1997). Aspects of the industry, trade, and marketing of Pacific Island trochus. *SPC Trochus Information Bulletin*, **5**, 2–14.

Jarayabhand, P. & Paphavasit, N. (1996) A review of the culture of tropical abalone with special reference to Thailand. *Aquaculture*, **140**, 159–68.

Kawamura, T., Roberts, R. D. & Takami, H. (1998). A review of the feeding and growth of postlarval abalone. *Journal of Shellfish Research*, **17**, 615–25.

Lee, C. L. (1997). ACIAR Trochus Reef Reseeding Research: A simplified method of induced spawning in trochus. *SPC Trochus Information Bulletin*, **5**, 37–9.

Lee, C. L. & Amos, M. (1997) Current status of topshell *Trochus niloticus* hatcheries in Australia, Indonesia and the Pacific – a review. In: *Trochus: Status, Hatchery Practice and Nutrition* (Ed. by C. L. Lee & P. W. Lynch), pp. 38–42. ACIAR Proceedings No. 79. Australian Centre for International Agricultural Research, Canberra, Australia.

Leighton, D. L. (1974). The influence of temperature on larval and juvenile growth in three species of southern California abalones. *Fisheries Bulletin*, **72**, 1137–45.

McBride, S. C. (1998). Current status of abalone aquaculture in the Californias. *Journal of Shellfish Research*, **17**, 593–600.

Moore, J. D., Robbins, T. T. & Friedman, C. S. (2000). Withering syndrome in farmed red abalone *Haliotis rufescens*: thermal induction and association with a gastrointestinal Rickettsiales-like prokaryote. *Journal of Aquatic Animal Health*, **12**, 26 34.

Morse, D. E. (1992). Molecular mechanisms controlling metamorphosis and recruitment in abalone larvae. In: *Abalone of the World: Biology, Fisheries and Culture* (Ed. by S. A. Shepherd, M. J. Tegner & S. A. Guzman del Proo), pp. 107–19. Blackwell Science, Oxford.

Murakoshi, M. (1991). *Development of Mass Production Techniques of Trochus Seeds in Okinawa*. Okinawa Prefecture Sea Farming Center Report. [In Japanese: English translation by J. Isa] FAO/South Pacific Aquaculture Development Project, Suva, Fiji.

Nash, W. J. (1992). An evaluation of egg-per-recruit analysis as a means of assessing size limits for blacklip abalone (*Haliotis rubra*) in Tasmania. In: *Abalone of the World: Biology, Fisheries and Culture* (Ed. by S. A. Shepherd, M. J. Tegner & S. A. Guzman del Proo), pp. 318–38. Blackwell Science, Oxford.

Nash, W. J. (1993). Trochus. In: *Inshore Marine Resources of the South Pacific; Information for Fishery Development and Management* (Ed. by A. Wright & L. Hill), pp. 451–95. University of the South Pacific, Suva, Fiji.

Neori, A., Shpigel, M. & Ben-Ezra, D. (2000). A sustainable integrated system for culture of fish, seaweed and abalone. *Aquaculture*, **186**, 279–91.

Oakes, F. R. & Field, R. C. (1996). Infestation of *Haliotis rufescens* shells by a sabellid polychaete. *Aquaculture*, **140**, 139–43.

Pradina, Dwiono, S. A. P., Makatipu P. E. & Arafin, Z.

(1997). Reproductive biology of *Trochus niloticus* L. from Maluku, Eastern Indonesia. In: *Trochus: Status, Hatchery Practice and Nutrition* (Ed. by C. L. Lee & P. W. Lynch), pp. 47–51. ACIAR Proceedings No. 79. Australian Centre for International Agricultural Research, Canberra, Australia.

Randall, J. E. (1964). Contributions to the biology of the 'queen conch' *Strombus gigas*. *Bulletin of Marine Science of the Gulf and Caribbean*, **14**, 146–95.

Roberts, R. D., Kawamura, T. & Nicholson, C. M. (1999). Growth and survival of postlarval abalone (*Haliotis iris*) in relation to development and diatom diet. *Journal of Shellfish Research*, **18**, 243–50.

Ruck, K. R. & Cook, P. A. (1998). Sabellid infestations in the shells of South African molluscs: implications for abalone mariculture. *Journal of Shellfish Research*, **17**, 693–9.

Schiel, D. R. (1992). The paua (abalone) fishery of New Zealand. In: *Abalone of the World: Biology, Fisheries and Culture* (Ed. by S. A. Shepherd, M. J. Tegner & S. A. Guzman del Proo), pp. 427–37. Blackwell Science, Oxford.

Shepherd, S. A., Preece, P. A. & White, R. W. G. (2000). Tired nature's sweet restorer? Ecology of abalone (*Haliotis* spp.) stock enhancement in Australia. In: *Workshop on Rebuilding Abalone Stocks in British Columbia* (Ed. by A. Campbell). *Canadian Special Publication in Fisheries and Aquatic Science*, **130**, 84–97.

Simpson, B. J. A. & Cook, P. A. (1998). Rotation diets: a method of improving growth of cultured abalone using natural algal diets. *Journal of Shellfish Research*, **17**, 635–40.

Slattery, M. (1992). Larval settlement and juvenile survival in the red abalone (*Haliotis rufescens*): an examination of inductive cues and substrate selection. *Aquaculture*, **102**, 143–53.

Smith, B. D. (1987). Growth rate, distribution and abundance of the introduced topshell *Trochus niloticus*. *Bulletin of Marine Science*, **41,** 466–74.

Stoner, A. W. (1997) The status of Queen Conch, *Strombus gigas*, research in the Caribbean. *Marine Fisheries Reviews*, **59**, 14–22.

Stoner, A. W., Lin, J. & Hanisak, M. D. (1995). Relationships between seagrass bed characteristics and juvenile queen conch (*Strombus gigas* Linne) abundance in the Bahamas. *Journal of Shellfish Research*, **14**, 315–23.

Taiwu, L., Mingjin, D, Jian, Z., Jianhai, X. & Ruiyu, L. (1998). Studies on the pustule disease of abalone (*Haliotis discus hannai* Ino) on the Dalian Coast. *Journal of Shellfish Research*, **17**, 707–11.

Takami, H., Kawamura, T. & Yamashita, Y. (1997). Survival and growth rates of post-larval abalone *Haliotis discus hannai* fed conspecific trail mucus and/or benthic diatom *Cocconeis scutellum* var. *parva*. *Aquaculture*, **152**, 129–38.

Tegner, M. J., DeMartin, J. D. & Karpov, K. A. (1992). The California red abalone fishery: a case study in complexity. In: *Abalone of the World: Biology, Fisheries and Culture* (Ed. by S. A. Shepherd, M. J. Tegner & S. A. Guzman del Proo), pp. 370–83. Blackwell Science, Oxford.

Tong, L. J. & Moss, G. A. (1992). The New Zealand culture system for abalone. In: *Abalone of the World: Biology, Fisheries and Culture* (Ed. by S. A. Shepherd, M. J. Tegner & S. A. Guzman del Proo), pp. 583–91. Blackwell Science, Oxford.

Uki N. & Kikuchi, S. (1984). Regulation of maturation and spawning of an abalone, *Haliotis* (Gastropoda) by external environmental factors. *Aquaculture*, **39**, 247–61.

Weil, M. E. & R. G. Laughlin. (1984). Biology, population dynamics and reproduction of the queen conch *Strombus gigas* Linne in the Archipielago de Los Roques National Park. *Journal of Shellfish Research*, **4**, 45–62.

Yamaguchi, M. (1993) Green snail. In: *Inshore Marine Resources of the South Pacific; Information for Fishery Development and Management* (Ed. by A. Wright & L. Hill), pp. 497–51. University of the South Pacific, Suva, Fiji.

23
The Future

John Lucas

23.1 Introduction

> The growth industry for the next 30 years is going to be fish farming.
>
> Peter F. Drucker (1999)

When senior aquaculturists make predictions about the way that the aquaculture industry will greatly expand in the future decades (e.g. New, 1997, 1999), it is possible to treat them cautiously. (Although the statistics exist to support their predictions.) When, however, a guru like Dr Peter Drucker, author of 31 books and 'father of modern management', joins the predictions, it causes widespread attention. Dr Drucker is quoted in *The New York Times* as saying that 'his choice for the growth industry of the next 30 years is not e-commerce. "It's going to be fish farming" '(Pollack, 1999). As a simple illustration supporting his prediction, Drucker (1999) pointed out that 25 years ago salmon was a delicacy, and the choice of meats at gatherings such as conventions was beef or chicken. Now salmon is the other choice, much of it from salmon farms.

23.2 Future growth

23.2.1 Growth rate predictions

The rapid growth in aquaculture production over the past two decades, compared with the static level of production from capture fisheries, is shown in Chapter 1 (Table 1.1 and Fig. 1.1). Aquaculture's growth rate in the 1980s and 1990s was much more rapid than was ever anticipated or predicted. For example, Hempel (1993) predicted that aquaculture production of fish and shellfish would exceed 19.6 million mt by the year 2000. This value was exceeded the following year! Ratafia (1995) considered that, after aquaculture had 'skyrocketed' during the 1980s, it would show more moderate growth during the 1990s. The rate of growth, however, increased from the 1980s to the 1990s. New (1997) gave forecasts by six authors for annual aquaculture production by the year 2000. All of these forecasted values were reached within the period 1991–95.

It is highly questionable whether it is possible to make reliable predictions of the annual production of aquaculture for the next decade or so. Apart from other factors, aquaculture production will grow at different rates in different countries (New, 1999). However, the processes that are in place and that are driving the current growth of aquaculture should continue, and it is likely that high rates of aquaculture growth will be sustained in the immediate future. As described in Chapter 1, growth in aquaculture is largely from developments in the low-income, food deficit countries (LIFDC) (developing countries), particularly China. China has undertaken a deliberate massive expansion of aquaculture to increase its component of world aquaculture production from ~45% to ~70% in the course of a decade (Table 1.3).

The FAO has determined the world need for animal seafood in 2010 as 105–110 million mt/year for human consumption (Currie, 2000). Animal seafood production for human consumption from fisheries is ~60 million mt/year [30 million mt/year out of the ~90 million mt/year harvest (Table 1.1) is for fish meal] and equivalent aquaculture production was ~30 million mt/year in 1998–99. Thus, 1998–99 fisheries and aquaculture production was 15–20 million mt/year less than the FAO estimate for world seafood needs in 2010. As it is very unlikely that there will be any substantial increase in production from fisheries, the 2010 deficit will have to be made up by aquaculture, if it is made up. Considering the recent mean rate of growth of aquaculture, ~10% per year, and the fact that aquaculture production would need to continue to grow only at a more modest rate of 4–5% per year until 2010 to achieve the FAO-determined world need for animal seafood in that year, it seems likely that aquaculture will reach this level. It may even exceed it.

23.2.2 Unsustainable growth

As described in Chapter 1, some thousands of years ago on the various continents there were transformations from hunting–gathering societies to societies practising agriculture. This led to increased production per unit land area and higher densities of human population could be supported (Fig. 23.1). It was the first (1) in a series of stages that may be identified in the progressive development of agriculture. The adoption of farming (2) was followed by thousands of years over which the farmed plants and animals were progressively domesticated (3), and their productivity enhanced, through natural and deliberate stock selection processes. The Industrial Revolution (4) further increased production from the land with the introduction of mechanisation and, for example, tractors, threshing machines, etc., replacing horses and labourers. Finally, in the twentieth century, the Green Revolution (5) introduced higher levels of production through genetically improved stocks, heavy fertilisation, heavy pesticide use, irrigation and advanced feed formulations.

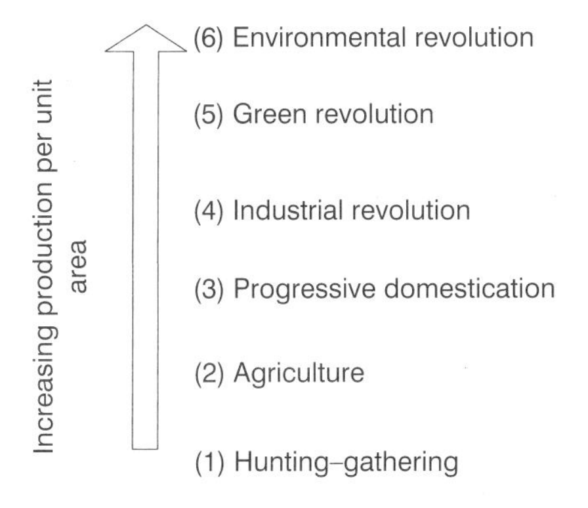

Fig. 23.1 Diagram of hypothetical stages in human development leading to increased production per unit area of land.

This evolution from hunting–gathering to Green Revolution has driven, and has been driven by, a huge expansion in the world's population. The world's population 2000 years ago was 300 million (UN, 1998), representing some thousands of years of population expansion from the hunting–gathering densities. The current world population is 6 billion, a further 20 times the AD 1 value, with most of this increase occurring in the last two decades. The United Nations estimates that the world's population will be about 9 billion by 2050 (UN, 1998). This is a 50% increase on the current population, but the UN estimates that three times the amount of basic food will be required. There must be enough to feed everyone at a level that eliminates countries from the food deficit category of LIFDC; hence, it is not sufficient to increase basic food production by 1.5 times. The problem with increasing food production is that there is a down-side. Increasing productivity per unit landmass through history (Fig. 23.1) has had increasing adverse impacts on the environment. Even early grazing and cropping led to widespread deforestation and loss of topsoil. If production of food is going to increase threefold in the next 50 years, it is going to have to do so without a threefold burden on the environment. Perhaps an 'Environmental Revolution' (6) is needed for the next half-century. This would involve increased production per unit area while environmental impacts are decreased: a reversal of the previous historical trend.

Aquaculture has not followed the pattern of development shown by agriculture:

- Capture fishing, a hunting–gathering activity, still remains the major source of world seafood production.
- The common carp is the only aquaculture species

that was progressively domesticated in the manner of plants and livestock.

None the less, the conclusion regarding an 'Environmental Revolution' for agriculture is applicable to aquaculture. Not all, but much, of the increased aquaculture production in LIFDCs has been achieved by dedicating aquatic environments to aquaculture. Over the next years, it is likely that major annual increases in world aquaculture production will continue to come from dedication of more environment to aquaculture in LIFDCs, especially China. This method of increasing production is not sustainable. The environment available for aquaculture is finite. There will have to be overall improved technology leading to greatly increased production/unit area or volume, i.e. a general shift of intensity as appropriate to each industry. Environmental impacts must be controlled, where these have been a problem, and environmentally sustainable aquaculture established on a long-term basis.

23.3 Factors in future growth

Technologies leading to increased aquaculture production and reduced environmental impacts have already been considered in various sections of this textbook, but some will be briefly reviewed here.

23.3.1 Genetics

Chapter 7 considered the techniques and status of genetics and stock improvement in aquaculture. As the quotation at the beginning of Chapter 7 indicates, far too much of the aquaculture industry is based on broodstock organisms from the wild, which are totally unselected.

Table 23.1 shows growth improvements in meat chicken production associated with breeding programmes and nutrition. The study compared the growth, food conversion ration (FCR) and cost/unit meat produced from 1957- and 1991-strain chickens, reared with 1957 and 1991 formula feeds. The major difference in this table is between the genetic strains of chickens, regardless of feeds. The 1991-strain chickens grew more than three times as fast as the 1957-strain chickens on both 1957- and 1991-formula feeds. Over a 30-year period, from 1957 to 1991, it was possible to increase their growth rate threefold by using genetics and stock selection. On the 'modern' diet, the 'modern' chickens were reaching more than 2 kg in 6 weeks. (It is sobering to consider that the chickens in the freezers at supermarkets probably had a lifespan of only about 6 weeks.) Yet these animals are probably distant descendants of Asian wild fowls (*Gallus gallus*) that weighed tens of grams at 6 weeks of age. Over the long centuries of domestication, their growth rate was increased by selection, and over recent decades it was possible to increase growth rate a further threefold.

These results:

- magnitudes greater growth rates than wild stock
- marked increases in growth rate of recently selected stock

indicate the outstanding improvements that may be obtained. These results from genetics and stock selection are not just for improvements in growth rate, they apply to FCR (compare the higher FCRs of the 1957 strains with the 1991 strains on the same diets in Table 23.1), environmental tolerances, marketability features, etc.

Fish have been shown to be capable of marked increases in growth rate in response to genetic manipulation. Research with transgenic fish has shown that 10-fold increases in growth rate can be achieved (Table 7.13). These responding fish include Atlantic salmon, which have already been subject to genetic and stock improvement programmes. However, as with other commercial transgenic plants and animals, there will be barriers of non-acceptance to overcome.

Focus on genetics and stock improvement must be of the highest priority in increasing the productivity of aquaculture.

23.3.2 Culturing new species

A substantial component of the continuing expansion of aquaculture must include the development of industries based on new species. There are undoubtedly many fish, invertebrate and algae species that are not cultured currently but which have features that make them very good candidates for

Table 23.1 Weights of 6-week-old chickens of 1957 and 1991 genetic strains fed 1957 and 1991 formula feeds. Also presented are changes in food conversion ratio (FCR) and cost of feed/kg body weight gain (US$)

	1957 genetic strain	1991 genetic strain
1957 formula feed		
Size at 6 weeks (kg)	0.508	1.775
FCR	3.00	2.29
Cost of feed/kg body weight (cents)	86	65
1991 formula feed		
Size at 6 weeks (kg)	0.627	2.134
FCR	2.52	2.04
Cost of feed/kg body weight (cents)	78	63

Data from Havenstein *et al.* (1994).

commercial aquaculture with appropriate technology. This reflects a major difference between aquaculture and agriculture. Because aquaculture is at a much earlier stage of development than its terrestrial counterparts, there has been less identification and testing of potential candidates for commercial developments in the former. Furthermore, the diversity of aquatic organisms of potential economic value through culture is large, and it will take considerable time to work through a major proportion of them. Their potential economic importance may not just be in terms of human consumption, e.g. some may be sources of pharmaceuticals and there is potential for non-traditional aquarium products such as corals.

Reflecting that the range of species in culture is not static:

(1) In Europe there are more than 15 species of fish and shellfish that are at an introductory phase of being developed for commercial aquaculture (Table 12.3). This is in countries with long histories of aquaculture.
(2) In China, where aquaculture extends back into antiquity, 20 native species have recently been domesticated from the wild for commercial aquaculture (Cen & Zhang, 1998).

The introduction of exotic species is completely different from developing new species within their geographical range. These introductions have the potential for seriously adverse effects on the environment, and several countries have legislated against such introductions.

23.3.3 Nutrition

The improvements in production of meat chickens were not entirely from genetics and stock improvement (Table 23.1). The 1957 strain increased from 0.51 kg at 6 weeks when fed on the 1957 diet to 0.63 kg when fed on the 1991 formula feed. At the same time, FCR decreased from 3.0 to 2.5 and cost of feed per kilogram body weight decreased from 86 US cents to 78 US cents. The 1991 strain, although it does well on the 1957 formula feed through its genetic capacity, needed the 1991 formula feed to achieve a FCR of 2.0 at a cost of US$0.63 feed/kg body weight.

Although this example shows the great importance of genetics, none the less it clearly demonstrates the importance of developing feed formulations, and this is an area of major importance in the future development of aquaculture.

Many feeds used in fish culture are formulated on the basis of extrapolations from the known nutritional requirements of a few groups, such as salmonids, tilapias and carps, for which there are comprehensive data. Thus, there is need for much broader and specific knowledge of the nutritional requirements of those fish groups and species for which 'borrowed' data are used for determining their nutritional requirements.

Many of the feed formulations used in aquaculture contain high levels of fish meal. These may lead to the release of large amounts of N and P into the adjacent environment. Furthermore, the increasing use of fish meal in animal feeds, in both animal husbandry and aquaculture, is unsustainable, as the supply of fish meal is limited. Probably the greatest continuing imperative in nutrition research for aquaculture is to find a substitute for fish meal protein (from terrestrial animal, plant or bacterial sources) or to reduce dependence on fish meal.

23.3.4 Disease control

As stated in the conclusion of Chapter 10, the nature of aquaculture and the best anti-disease technologies are the direct antitheses of each other. Much of aquaculture is about increasing culture density to increase profit, whereas anti-disease methodologies ideally include keeping culture densities as low as possible to break the host-to-host cycles. Instead, other management procedures must be followed to exclude pathogens or limit their deleterious effects. A variety of procedures, including chemotherapeutic treatments, probiotics, vaccines and diagnostics, have been developed, but disease, especially viral disease, remains a major problem for some industries. Much research is still to be done in this field.

23.3.5 Environmentally sustainable technology

There are three developments taking place in onshore and coastal aquaculture that are leading to environmental sustainability.

(1) The development of high-intensity culture (> 100 kg fish/1000 L of water) in recirculating systems using the various components of recirculating systems for water quality control (section 2.4.4). These systems are indoor systems with minimal effluent. They are highly conservative of water (section 5.5).
(2) There are movements back from the modern trend in aquaculture of monoculture, towards the polyculture origins of aquaculture. The additional productivity of polyculture is gained while recirculated or effluent water quality is enhanced. Some of the technologies developed or being developed include:
   - hydroponic plants included in recirculating systems with fish tanks (Fig. 5.4)
   - use of aquaculture effluent for irrigation of crops
   - bivalves and seaweed culture within the recirculation or effluent channels of shrimp farms (spatially separated polyculture and bioremediation)
   - seaweed culture adjacent to seacages (see following).
(3) Traditional aquaculture in Asia is integrated with agriculture and animal husbandry. The productivity of the ponds is supported with crop, animal and human wastes. An equivalent modern use of wastewater and domestic sewage on a large scale in aquaculture production of fish in some regions of Asia and Europe is a positive environmental role for aquaculture (section 4.7).

Seacage culture is another major aquaculture technology, contributing many hundreds of thousands of metric tonnes of production each year. Seacage farms produce effluent that disperses into the adjacent environment (section 4.3.1) and cannot be treated in a concentrated form in the manner of effluent from tanks and ponds. The effluent from seacages is best absorbed by the environment when the cages are well spaced and far above the substrate. Seaweed farming in the vicinity of seacages, however, is being developed as a means of absorbing N and P wastes, and producing an additional crop (i.e. polyculture of seaweeds and seacaged fish). Culture of sponges is also being considered for the vicinity

Fig. 23.2 A large partially submerged seacage being deployed in the Gulf of Mexico. The cage will be submerged until only the very uppermost portion is above the ocean surface (photograph by Dr Tim Reid, with permission for use by the Gulf of Mexico Offshore Aquaculture Consortium).

of seacages. Another way of reducing the impact of effluent is to move the cages offshore to open ocean positions. The cages must be engineered to withstand the extreme physical conditions in open seas. This involves greatly enlarging and reinforcing them, and designing them to float with only a minimal portion of the totally enclosed cage above the surface (Fig. 23.2). (Wave effects are less severe below the surface.) The economics of maintaining seacages a considerable distance offshore may be an issue, but it is conceivable that in the future there may be seacage farms sited tens or hundreds of kilometres offshore in the open ocean.

Several other alternatives to increase production from coastal waters culture, as an alternative to seacages and seapens, were reviewed by Currie (2000):

- sea fisheries stock enhancement
- fertilising the ocean.

The former was discussed in section 1.5 and, although it is widely practised in Japan and some other countries, there is currently no strong support for its effectiveness. There is need for more research into the factors affecting success. Furthermore, the ultimate primary productivity of the ocean is not increased by adding fishery recruits. The suggestion of fertilising the ocean would involve sealing off a large volume of the ocean, adding nutrients to promote primary production and, in turn, fish production. It runs contrary to all the processes whereby aquaculture is seeking not to add nutrients to the environment.

23.3.6 Public image

As already described, environmentally sustainable technology is the only responsible technology, and it is in the long-term interests of the aquaculture industry. However, in some regions, especially within developed countries, aquaculture has a poor public image because of its apparently inadequate concern for the environment (New, 1999). This leads to antagonism towards further developments. There are many current examples of heated controversies within communities about new aquaculture developments and their impact.

In the terrestrial environment, many traditional agricultural industries have caused from substantial to major environmental damage. Some impact is almost ineviatble because agriculture is not part of the natural environment. These impacts are generally accepted by the community because these are long-established, primary industries, most of their effects on the environment often having occurred centuries ago, and they are our main sources of food.

Modern aquaculture is rapidly developing as a latecomer for food production within a world that is very different from that in which much of agriculture development occurred. It is a world where in many countries there is great concern for the environment. As in other animal and plant production, aquaculture developments that are not completely integrated with or completely isolated from their environment will inevitably have some impact, however minor. These impacts must be deliberately minimised through, for instance, comprehensive environmenatal impact assessments for new developments and codes of practice for established industries (section 4.5).

To improve its image, the aquaculture industry must convince governments and communities by its actions and words that it is a responsible and environmentally sound industry. A poor public image may hinder the development of aquaculture in many countries as the public becomes increasingly environmentally aware.

23.4 Conclusion

The rate of growth of aquaculture production in the past two decades was unanticipated and exceeded all predictions. It came from the LIFDCs, especially China, and it is likely that the processes driving this growth will continue in the immediate future. This expansion in aquaculture production, however, is not sustainable in the long term. Continued expansion of aquaculture production will ultimately depend on:

- higher productivity/unit area or volume achieved through technological advances
- responsible and environmentally sustainable technology.

Genetics and stock improvement, development of new species, improved feed formulations and disease control are among areas of research for higher productivity. Technologies that increase productivity with improved environmental sustainability include high-intensity recirculating systems, linkage of recirculating systems to hydroponics, water-conserving culture systems and new forms of polyculture in large recirculating systems, or in the effluent of pond systems, or in the surrounds of seacage farms. Technologies for offshore culture are being developed. All of this involves much research and development, and this is what makes aquaculture such an exciting and dynamic field at this time. Aquaculture is not only expanding rapidly in terms of production, but is also developing rapidly from a technological perspective to meet new environmental, economic, genetic and other challenges.

23.5 References

Cen, F. & Zhang, D. (1998). Development and status of the aquaculture industry in the People's Republic of China. *World Aquaculture*, **29**(2), 52–6.

Currie, D. J. (2000). Aquaculture: opportunity to benefit mankind. *World Aquaculture*, **31**(1), 44–9.

Drucker, P.F. (1999). Beyond the information revolution. *The Atlantic Monthly*, **284**, 47–57.

Havenstein, G. B., Ferket, P. R., Scheideler, S. E. & Larson, B. T. (1994). Growth, livability, and feed conversion of 1957 vs 1991 broilers when fed "typical" 1957 and 1991 broiler diets. *Poultry Science*, **73,** 1785–94.

Hempel, E. (1993). Constraints and possibilities for developing aquaculture. *Aquaculture International*, **1**, 2–19.

New, M. B. (1997). Aquaculture and the capture fisheries: balancing the scales. *World Aquaculture*, **28**(2), 11–30.

New, M. B. (1999). Global aquaculture: current trends and challenges for the 21st century. *World Aquaculture*, **30**(1), 8–13, 63–79.

Pollack, A. (1999). Private sector; seeing the corporation's demise. *The New York Times*, 14 November 1999.

Ratafia, M. (1995). Aquaculture today: a worldwide status report. *World Aquaculture*, **26**(2), 18–24.

UN (1998). *Revision of the World Population Estimates and Projections*. Department of Economics and Social Affairs, United Nations, New York.

Index